64.4
N3-11
Locked
Cage

REGIONAL GROUND-WATER QUALITY

REGIONAL GROUND-WATER QUALITY

Edited by

William M. Alley
U.S. Geological Survey

VAN NOSTRAND REINHOLD
New York

Copyright © 1993 by Van Nostrand Reinhold
Library of Congress Catalog Card Number 92-36483
ISBN 0-442-00937-2

All rights reserved. No part of this work covered by the
copyright hereon may be reproduced or used in any form
or by any means—graphic, electronic, or mechanical, including
photocopying, recording, taping, or informational storage and
retrieval systems—without written permission of the publisher.

Chapter 20, "Implementation of a Statewide Survey for Agricultural
Chemicals in Rural, Private Water-Supply Wells in Illinois," is the
product of a tax-funded research program written by Illinois state
employees and no claim to copyright is made on it.

 Van Nostrand Reinhold is a division of International Thomson
Publishing. ITP logo is a trademark under license.

Printed in the United States of America

Van Nostrand Reinhold
115 Fifth Avenue
New York, NY 10003

International Thomson Publishing
Berkshire House
168-173 High Holborn
London WC1V 7AA, England

Thomas Nelson Australia
102 Dodds Street
South Melbourne 3205
Victoria, Australia

Nelson Canada
1120 Birchmount Road
Scarborough, Ontario
M1K 5G4, Canada

16 15 14 13 12 11 10 9 8 7 6 5 4 3 2

Library of Congress Cataloging-in-Publication Data
Regional ground-water quality/edited by William M. Alley.
   p. cm
  Includes bibliographical references and index.
  ISBN 0-442-00937-2
  1. Water, Underground—Quality.   2. Water, Underground—Quality—
Case studies.   I. Alley, William M.
TD403.R45   1993
628.1′61—dc20
                                               92-36483

To Rosemarie, Amie, and Erik

Contents

Preface, xiii
Contributors, xvii

PART I: INTRODUCTION

1 General Design Considerations 3
William M. Alley
Introduction, 3
Target and Candidate Populations, 4
Issues of Scale, 6
Use of Existing Wells Compared with Specially
   Constructed Wells, 14
Documentation of Wells and Well Sites, 17
Quality Assurance, 18
Concluding Remarks, 19
References, 19

2 Establishing a Conceptual Framework 23
William M. Alley
Introduction, 23
The Vadose Zone, 23
Ground-Water Flow Systems, 27
Uses of Geochemical Data in Regional Conceptualizations, 42
Concluding Remarks, 55
References, 56

PART II: STATISTICAL CONCEPTS

3 Ground-Water-Quality Surveys 63
William M. Alley
Introduction, 63
Some Preliminary Concepts, 64
Special Considerations for Ground-Water-Quality Surveys, 66
Probability Sampling Approaches, 69
Grid-based Search Sampling, 81
Concluding Remarks, 84
References, 85

4 Geostatistical Models 87
William M. Alley
Introduction, 87
Some Preliminary Concepts, 88
Regionalized Variables and the Intrinsic Hypothesis, 89
Variograms, 90
Ordinary Kriging, 94
Kriging Variants, 99
Concluding Remarks, 106
References, 106

PART III: WATER-QUALITY CONCEPTS

5 Scales in Chemical Hydrogeology: A Historical Perspective — 111
William Back, Mary Jo Baedecker, and Warren W. Wood
Introduction, 111
Historic Evolution of Scale in Chemical Hydrogeology, 113
Chemical Hydrogeology on a Regional Scale, 116
Chemical Hydrogeology on a Local Scale, 121
Chemical Hydrogeology at the Site Scale, 123
Concluding Remarks, 126
References, 128

6 Inorganic Chemical Processes and Reactions — 131
Donald D. Runnells
Introduction, 131
Physical and Chemical Properties, 133
Chemical Elements of Significance, 141
Types of Chemical Reactions, 142
Kinetic versus Equilibrium Considerations, 148
Concluding Remarks, 149
References, 149

7 Organic Chemical Concepts — 155
Paul D. Capel
Introduction, 155
Organic Chemicals and Their Nomenclature, 155
Physical/Chemical Properties of Organic Chemicals, 162
Environmental Processes Affecting Organic Chemicals in
    Ground Water, 166
Concluding Remarks, 176
References, 176

8 Subsurface Microbiology — 181
Francis H. Chapelle, Paul M. Bradley, and Peter B. McMahon
Introduction, 181
Microbial Ecology of Ground-Water Systems, 183
Distribution of Terminal Electron-Accepting Processes (TEAPs)
    in Ground-Water Systems, 186
Biodegradation of Atrazine in Alluvial-Aquifer Sediments, 194
Concluding Remarks, 196
References, 196

9 Geochemical Models — 199
D. L. Parkhurst and L. N. Plummer
Introduction, 199
Inverse Modeling, 199
Forward Modeling, 214
Concluding Remarks, 222
References, 223

10 Uses of Environmental Isotopes — 227
Tyler B. Coplen
Basic Concepts, 227

Review of Selected Isotopes, 231
Applications, 241
Concluding Remarks, 249
References, 249

11 Environmental Tracers for Age Dating Young Ground Water 255
L. N. Plummer, R. L. Michel, E. M. Thurman, and P. D. Glynn

Introduction, 255
Tritium, 256
Tritium–Helium-3, 265
Krypton-85, 266
Chlorofluorocarbons, 268
Comparison among Environmental Tracers, 277
Selected Nuclear Event Markers, 279
Organic Compounds as Event Markers: Selected Examples, 280
Concluding Remarks, 287
References, 288

PART IV: SELECTED WATER-QUALITY ISSUES

12 Nitrate 297
G. R. Hallberg and D. R. Keeney

Introduction, 297
Nitrogen Cycling, 297
Sources of Nitrate to Ground Water, 300
Nitrate Distribution and Variability, 307
Concluding Remarks, 316
References, 317

13 Organic Contaminants 323
Douglas M. Mackay and Lynda A. Smith

Introduction, 323
Hydrogeology, 324
Types of Organic Pollutants, 325
Pollutant Sources, 325
Approaches to Monitoring of the Subsurface, 328
National Surveys of VOC Contamination, 332
State Surveys of Ground-Water Contamination, 335
Limitations of Existing Ground-Water Quality Data, 339
Concluding Remarks, 340
References, 341

14 Pesticides 345
P. S. C. Rao and William M. Alley

Introduction, 345
Factors Influencing Contamination Potential, 345
Regional Assessment of Ground-Water Vulnerability, 360
Selected Studies of Pesticides in Ground Water, 366
Implications for Regional-Scale Studies, 373
References, 377

15 Pathogens
Marylynn V. Yates and Scott R. Yates — 383

Introduction, 383
Factors Affecting Microbial Fate and Transport, 388
Modeling Microbial Transport, 395
Concluding Remarks, 400
References, 400

16 Acid Precipitation
Gunnar Jacks — 405

Introduction, 405
Soil Acidification, 406
Weathering, 408
Plant Uptake of Nutrients, 410
Relation between Ground-Water and Surface-Water Acidification, 410
Observed Trends in Ground-Water Composition as Influenced by Acidic Deposition, 411
Health Effects of Acidic Ground Water, 413
Modeling of Soil and Ground-Water Acidification, 414
Critical Loads of Sulfur and Nitrogen for Ground Water, 416
Amendments for Acidic Ground Water, 417
Concluding Remarks, 417
References, 418

17 Natural Radionuclides
Richard B. Wanty and D. Kirk Nordstrom — 423

Introduction, 423
Radiochemistry, 425
Geochemistry of U, Th, Ra, and Rn, 430
Case Studies of Natural Radionuclides, 432
Concluding Remarks, 436
References, 436

18 Analysis of Ground-Water Systems in Freshwater-Saltwater Environments
Thomas E. Reilly — 443

Introduction, 443
System Definition, 443
Field Measurements Required for Analysis, 453
The Use and Misconceptions Regarding Freshwater-Equivalent Heads, 455
Summaries of Selected Studies Using the Different System Conceptualizations, 457
Monitoring Strategies in Freshwater-Saltwater Systems, 465
Concluding Remarks, 466
References, 466

19 Analysis of Karst Aquifers
William B. White — 471

Introduction, 471
Permeability in Karst Aquifers, 471
Karstic Ground-Water Basins, 472
Dynamical Response of Karst Aquifers, 480
Sediment Transport in Karst Aquifers, 484
Evolution of Karst Aquifers, 486

Water-Quality Problems in Karst, 487
Concluding Remarks, 488
References, 488

PART V: CASE STUDIES

20 Implementation of a Statewide Survey for Agricultural Chemicals in Rural, Private Water-Supply Wells in Illinois — 493
Dennis P. McKenna, Thomas J. Bicki, and Warren D. Goetsch

Introduction, 493
Project Organization and Responsibilities, 495
Study Design, 498
Implementation, 501
Results, 510
Concluding Remarks, 511
References, 511

21 Ground-Water-Quality Monitoring in The Netherlands — 515
Willem van Duijvenbooden

Introduction, 515
Environmental Monitoring, 515
Hydrogeologic Situation, 516
The Dutch National Ground-Water Quality
    Monitoring Network, 517
Provincial Ground-Water-Quality Monitoring Networks, 522
Soil-Quality and Shallow-Ground-Water Monitoring, 526
Monitoring of Drinking-Water Quality, 531
Interpretation Tools and Presentation Techniques, 533
Integrated Monitoring, 533
Concluding Remarks, 534
References, 534

22 Multiscale Approach to Regional Ground-Water-Quality Assessment: Selenium in the San Joaquin Valley, California — 537
Neil M. Dubrovsky, Steven J. Deverel, and Robert J. Gilliom

Introduction, 537
Valleywide Sampling, 539
Areal Distribution of Selenium in Shallow Ground Water, 543
Depth Distribution of Selenium in Ground Water, 548
Site-specific Study of Selenium Mobility in Chemically
    Reduced Sierra Nevada Sediments, 556
Concluding Remarks, 560
References, 561

23 Multiscale Approach to Regional Ground-Water-Quality Assessment of the Delmarva Peninsula — 563
Robert J. Shedlock, Pixie A. Hamilton, Judith M. Denver, and Patrick J. Phillips

Introduction, 563
Hydrogeomorphic Regions in the Surficial Aquifer, 564
Design of Study, 568
Ground-Water-Quality Patterns, 572
Concluding Remarks, 585
References, 586

24 Ground-Water Quality in the Oklahoma City Urban Area — **589**
Scott C. Christenson and Alan Rea

Introduction, 589
Design of the Sampling Program, 591
Methods Used for Characterizing Urban Land Use, 595
Ground-Water Quality, 598
Factors Related to Ground-Water Quality, 601
Possible Sources of Contaminants, 606
Concluding Remarks, 609
References, 610

25 Uses and Limitations of Existing Ground-Water-Quality Data — **613**
Pixie A. Hamilton, Alan H. Welch, Scott C. Christenson, and William M. Alley

Introduction, 613
Sources of Ground-Water-Quality Data, 616
Data Screening, 616
Suitability of Data for Regional Water-Quality Assessment, 616
Applications, 619
Concluding Remarks, 621
References, 622

Index, 623

Preface

Ground water is an important source of water supply in many parts of the world. As ground-water use has increased, issues associated with the quality of ground-water resources have likewise grown in importance. For many years, attention has been directed at contamination from point sources, such as landfills and hazardous-waste-disposal sites. More recently, concerns have increased about nonpoint sources of contamination and about the overall quality of ground-water resources. This has led to considerable interest in the design of investigative studies and monitoring programs to describe ground-water quality over regions that may range from tens of square kilometers to an entire country in area. These regional ground-water-quality investigations differ considerably from those designed to understand a specific point-source contaminant plume. The water-quality contaminants commonly are spatially dispersed and may be widespread and frequent in occurrence.

Proceedings of scientific meetings devoted to many specialized topics associated with regional ground-water quality have been published in recent years, but no comprehensive text exists on ground-water-quality investigations from a regional context. This book is intended to fill that gap.

This is not a cookbook or how-to manual. Given tremendous diversity in natural systems and in the human impacts on these systems, a meaningful cookbook-like approach is not possible. Instead, the book is intended to serve as a sourcebook of ideas on how to approach the study of regional ground-water quality in an efficient, creative, and meaningful way. The book should serve as background reading when first approaching a regional-scale study and as a reference in designing specific regional-scale studies.

The text brings together widely dispersed knowledge of practical use to those involved in studying regional ground-water quality. It should be of interest to a wide range of individuals in the disciplines of hydrology, water resources, agricultural sciences, and environmental sciences. The book will be of interest to individuals involved in the design and interpretation of regional ground-water-quality studies, as well as to those who must evaluate the utility and reliability of reported results from these studies. Because of its comprehensive nature, the book could serve as supplemental reading for senior or graduate-level courses in ground-water quality or environmental engineering.

Most readers of any chapter are not expected to be experts in that chapter's specialty. Hence, the authors were asked to write their chapters in a format useful for the wide range of interests represented in the intended audience. The authors also were selected to achieve a diversity of backgrounds, interests, and perspectives.

A few comments on the scope of the book are in order. This volume focuses on the scientific and technical elements of regional ground-water quality; it does not address management and policy elements except in a few places. General comments on appropriate field methods are made throughout the text; however, the text does not cover details of field methods for sampling, well installation, surface and borehole geophysics, and so forth.

Elements of these field methods often are required for the complete ground-water-quality investigation, but these topics are covered in detail elsewhere. Finally, the book does not treat in any depth ground-water-flow and solute-transport modeling. Again, this topic is covered in many other places.

The unique contribution of the text is its focus on regional water-quality studies. Inevitably, however, the findings from many different scales of study are discussed. Indeed, the utility of studying regional ground-water quality at multiple spatial and temporal scales is emphasized as a major theme of the book. Furthermore, principles developed over the years from detailed studies of ground-water quality must not be ignored but rather integrated into the design of regional-scale studies. Future progress in understanding regional ground-water quality will depend to a large extent on our ability to transfer the knowledge gained from finer scale studies to broader areas.

The volume has been organized into five parts containing 25 chapters. A brief synopsis follows.

Part I: Introduction   The book begins with two chapters on basic concepts for regional ground-water-quality studies. Chapter 1 discusses important differences in design and objectives among surveys of water at the point of use, well-water surveys, and ground-water resource assessments. In addition, issues of spatial and temporal scale, the choice between existing and specially constructed wells, documentation of sampling sites, and quality assurance are treated. Chapter 2 discusses the importance of establishing a conceptual framework for a regional study and some factors to consider in developing such a framework. A hopeful outcome is an appreciation of the multidisciplinary nature of regional ground-water-quality investigations.

Part II: Statistical Concepts   Chapter 3 deals with the design of ground-water-quality surveys. The importance of drawing upon both hydrogeologic knowledge and some basic statistical principles is reinforced throughout the chapter. Chapter 4 discusses geostatistical analysis. In addition to possible ground-water-quality applications, geostatistical models may be useful with spatial data, such as ground-water levels or soil characteristics, used in interpreting water-quality data.

Part III: Water-Quality Concepts   Chapter 5 returns to the important issue of scale and provides an overview of how scale has been treated in problems of chemical hydrogeology as the discipline has evolved. Chapters 6 and 7 then discuss key concepts associated with inorganic and organic chemicals, respectively. Chapter 8 focuses on the importance and methods of understanding subsurface microbiological processes at the regional scale. Chapter 9 reviews selected geochemical models and discusses their role in understanding regional ground-water quality. Chapter 10 provides a review of environmental isotopes and their uses in ground-water studies. Finally, Chapter 11 focuses on chemical indicators of young ground water. These indicators are particularly relevant to concerns about the presence of pesticides and other modern organic compounds in ground water.

Part IV: Selected Water-Quality Issues   Chapters 12 to 16 deal with problems that typically result from human sources of contaminants, including nitrates, organic contaminants, pesticides, pathogens, and acid precipitation. Chapters 17, 18, and 19 focus on natural radionuclides, analysis of ground-water systems in freshwater-saltwater environments, and karst environments.

Part V: Case Studies The six case studies selected to illustrate specific concepts are only a small subset of the regional-scale studies completed or presently underway. Many other studies are used as examples in the earlier chapters. Chapters 20 to 24 include a statewide well-water survey in Illinois, a national monitoring network in The Netherlands, and three resource assessments. The Illinois case study highlights many practical considerations and complexities that arise in the design of a large-scale ground-water-quality survey. The Netherlands case study illustrates the benefits of integrating examination of other environmental media, such as soil-water quality, with the study of ground-water quality. The three resource assessments focus, respectively, on a natural trace element (selenium), agricultural chemicals (mainly nitrate and inorganic chemical indicators of agricultural contamination), and contaminants in an urban area. A major emphasis of these three case studies is to illustrate how investigations at different scales can complement one another in providing a balanced and meaningful picture of regional ground-water quality, and how the scales of focus should be adjusted to the unique characteristics of the areas under study. Chapter 25 looks at the uses and limitations of existing data, based on evaluations of available ground-water-quality data in three study areas across the United States.

ACKNOWLEDGMENTS

This volume is immeasurably a benefactor of the ideas of numerous people. The initial proposal for this text was reviewed by David Freyberg, David Krabbenhoft and Joseph Devinny, all of whom provided insightful suggestions for the volume content and its organization. Mary Anderson reviewed the final draft and made a number of significant contributions for improvement.

Each chapter was reviewed by two to four persons. Usually, two of these reviewers were authors of other related chapters in the book to enhance the integration of the material in the various chapters. Other persons who substantially contributed during the review process are M. A. Ayers, J. E. Barbash, J. K. Böhlke, K. J. Breen, E. Busenberg, G. C. Buzicky, H. H. Cheng, J. A. Cherry, B. Ekwurzel, J. O. Englund, O. L. Franke, S. Gerould, D. A. Goolsby, D. R. Helsel, C. J. Hurst, A. G. Journel, C. Kendall, M. T. Koterba, W. W. Lapham, R. W. Lee, W. Low, G. E. Mallard, R. S. Mansell, D. J. Mulla, G. A. O'Connor, B. J. Ryan, and W. E. Sanford. The book also benefitted from early trials of some chapters at the Third and Fourth Summer School on Hydrogeological Hazard Studies held in Perugia, Italy, and sponsored by the National Research Council of Italy and the Water Resources Research and Documentation Center (WARREDOC) of the Italian University for Foreigners. Leslie J. Robinson contributed immensely to many of the figures in the book.

Contributors

William M. Alley, U.S. Geological Survey, 411 National Center, Reston, Virginia 22092

William Back, U.S. Geological Survey, 431 National Center, Reston, Virginia 22092

Mary Jo Baedecker, U.S. Geological Survey, 431 National Center, Reston, Virginia 22092

Thomas J. Bicki, Ocean Spray Cranberries, 1 Ocean Spray Drive, Middleboro, Massachusetts 02349

Paul M. Bradley, U.S. Geological Survey, 720 Gracern Road, Suite 129, Columbia, South Carolina 29210

Paul D. Capel, University of Minnesota, Gray Freshwater Institute, P.O. Box 100, Navarre, Minnesota 55392

Francis H. Chapelle, U.S. Geological Survey, 720 Gracern Road, Suite 129, Columbia, South Carolina 29210

Scott C. Christenson, U.S. Geological Survey, 202 NW 66th, Building 7, Oklahoma City, Oklahoma 73116

Tyler B. Coplen, U.S. Geological Survey, 431 National Center, Reston, Virginia 22092

Judith M. Denver, U.S. Geological Survey, 300 South New Street, Dover, Delaware 19901

Steven J. Deverel, U.S. Geological Survey, Room W-2234, Federal Building, Cottage Way, Sacramento, California 95825

Neil M. Dubrovsky, U.S. Geological Survey, Room W-2234, Federal Building, Cottage Way, Sacramento, California 95825

Robert J. Gilliom, U.S. Geological Survey, Room W-2234, Federal Building, Cottage Way, Sacramento, California 95825

Pierre D. Glynn, U.S. Geological Survey, 432 National Center, Reston, Virginia 22092

Warren D. Goetsch, Bureau of Environmental Programs, Illinois Department of Agriculture, State Fairgrounds, Springfield, Illinois 62794

George R. Hallberg, Iowa Department of Natural Resources, 123 North Capital, Iowa City, Iowa 52242

Pixie A. Hamilton, U.S. Geological Survey, 3600 West Broad Street, Richmond, Virginia 23230

Gunnar Jacks, Department of Land and Water Resources, Royal Institute of Technology, S100 44 Stockholm, Sweden

Dennis R. Keeney, Leopold Center for Sustainable Agriculture, Iowa State University, Ames, Iowa 50011

Douglas M. Mackay, Centre for Ground Water Research, University of Waterloo, Waterloo, Ontario, Canada N2L 3G1

Dennis P. McKenna, Montana Bureau of Mines and Geology, Montana College of Mineral Science and Technology, Butte, Montana 59701

Peter B. McMahon, U.S. Geological Survey, Box 25046, MS 415, Denver Federal Center, Lakewood, Colorado 80225

Robert M. Michel, U.S. Geological Survey, 431 National Center, Reston, Virginia 22092

D. Kirk Nordstrom, U.S. Geological Survey, Box 25046, MS 458, Denver Federal Center, Lakewood, Colorado 80225

David L. Parkhurst, U.S. Geological Survey, Box 25046, MS 418, Denver Federal Center, Lakewood, Colorado 80225

Patrick J. Phillips, U.S. Geological Survey, James T. Foley U.S. Courthouse, P.O. Box 1669, Albany, New York 12201

L. Niel Plummer, U.S. Geological Survey, 432 National Center, Reston, Virginia 22092

P. S. C. Rao, Soil and Water Science Department, 2171 McCarty Hall, University of Florida, Gainesville, Florida 32611

Alan Rea, U.S. Geological Survey, 202 NW 66th, Building 7, Oklahoma City, Oklahoma 73116

Thomas E. Reilly, U.S. Geological Survey, 431 National Center, Reston, Virginia 22092

Donald D. Runnells, Department of Geological Sciences, University of Colorado, Boulder, Colorado 80309-0250

Robert J. Shedlock, U.S. Geological Survey, 208 Carroll Building, 8600 La Salle Road, Towson, Maryland 21286

Lynda Smith, Metropolitan Water District, Water Quality Division, P.O. Box 54153, Los Angeles, California 90054

E. Michael Thurman, U.S. Geological Survey, Research Project Office, 4821 Quail Crest Place, Lawrence, Kansas 66049

Willem van Duijvenbooden, RIVM, National Institute of Public Health and Environmental Protection, P.O. Box 1, 3720 BA Bilthoven, The Netherlands

Richard B. Wanty, U.S. Geological Survey, Box 25046, MS 916, Denver Federal Center, Lakewood, Colorado 80225

Alan H. Welch, U.S. Geological Survey, 705 N. Plaza Street, Room 224, Federal Building, Carson City, Nevada 89701

William B. White, Department of Geosciences, The Pennsylvania State University, University Park, Pennsylvania 16802

Warren W. Wood, U.S. Geological Survey, 431 National Center, Reston, Virginia 22092

Marylynn V. Yates, Department of Soil and Environmental Sciences, University of California, Riverside, California 92521

Scott R. Yates, USDA/ARS Pesticides and Water Quality Research, Department of Soil and Environmental Sciences, University of California, Riverside, California 92521

REGIONAL GROUND-WATER QUALITY

I

Introduction

I

General Design Considerations

William M. Alley

INTRODUCTION

The protection and enhancement of the quality of ground-water resources is a high-priority environmental concern. Deterioration of ground-water quality may be virtually irreversible, and treatment of contaminated ground water can be expensive. Detection of ground-water contamination is complicated by the "out-of-sight" nature of ground water. Commonly, neither the sources nor the effects of contamination are easily observed or measured. Many contaminants are colorless, tasteless, and odorless. The degree of threat posed by ground-water contamination depends on many factors, including the concentrations of the contaminants, their toxicity (individually or in combination), the volume of ground water affected, the uses made of water from the aquifer, the population affected by these uses, and the availability of an alternative water supply (Patrick, Ford, and Quarles 1987).

The sources of ground-water contamination are numerous and diverse. Contaminants can enter ground water through many different routes, including

1. Downward percolation from a land-surface source
2. Downward percolation from sources in the shallow subsurface
3. Direct entry of contaminants from the land surface through wells
4. Cross contamination between aquifers in wells open to more than one aquifer
5. Flow of contaminated or saline water into freshwater aquifers as a result of pumping
6. Interactions of ground water with surface-water bodies
7. Percolation of acidic or contaminant-bearing precipitation
8. Interactions with geologic formations that contain natural contaminants, such as radon

Anthropogenic sources of ground-water contamination can be classified in a general way as either point sources (localized or individual sources) or nonpoint sources (activities or processes that introduce contaminants to ground water areally and that can consist of multiple point sources). In this text, regional ground-water contamination will broadly refer to the types of ground-water contamination that can be observed at the regional scale, recognizing that some of the observed contaminants originate from point sources and others from nonpoint sources. It is commonly difficult to differentiate point-source from nonpoint-source contamination. Nonetheless, the issue is an important topic of concern, and we will return to it in later chapters.

Some approximations of the extent of ground-water contamination in the United States have been made on the basis of estimates of the number and average spatial extent of different types of contamination. For example, Lehr (1982) estimated the volume of contaminated ground water resulting from waste-disposal sites of all types and compared this volume to an estimate of the volume of "available" ground water in the United States. He estimated that less than 1% of the ground-water resource has been contaminated by these sources. Lehr acknowledged that any of the large number of assumptions that were required to develop this estimate can be readily challenged. His key point was that a relatively small percentage of the ground-water resource has been contaminated by point sources and that use of appropriate preventive measures in the future could keep this overall percentage low. Lehr's results do not indicate the extent of nonpoint-source contamination. Furthermore, the calculated averages do not provide a perspective on the magnitude of individual incidences of contamination and their relation to water use; neither do these averages account for uneven geographic distribution, as evidenced by some areas where the degree of contamination greatly exceeds that of the country on the whole.

Investigations of regional ground-water quality are used to answer a number of questions, some of which are quite fundamental. As shown in Figure 1-1, general types of information that might be provided include statistical descriptions of ground-water-quality conditions and trends, descriptions of the spatial distribution of contaminants, and explanations of observed ground-water-quality conditions and trends. Different designs stemming from a wide range of possible objectives will result in different levels of each of these three basic types of information.

Programs designed to investigate regional ground-water quality must be guided by a carefully crafted set of objectives. In this chapter, some of the critical choices associated with establishing objectives for regional ground-water-quality studies are discussed. Concepts emphasized are determination of target and candidate populations, issues of spatial and temporal scales, and the choice between use of existing wells and specially constructed wells. The chapter concludes with brief discussions on the importance of careful documentation of sampling sites and on quality assurance.

TARGET AND CANDIDATE POPULATIONS

Closely associated with the objectives of a study are the concepts of the target and candidate populations. The *target population* is the aggregate about which inferences are to be made, whereas the *candidate population* is the set of locations available for sampling. Although these terms are traditionally used in a strict statistical sense, the basic ideas are relevant to the overall design of regional ground-water-quality studies. Ideally, the candidate and target populations are identical in a sampling program. Unfortunately, a number of characteristics of ground-water systems and of the limited points of access for ground-water sampling can result in substantial differences between candidate and target populations.

The definition of the target population depends on the type of investigation being designed, especially with regard to the location of the sampling point and the method of sampling. Types of investigations include (1) studies of water at the point of use, (2) well-water surveys, and (3) ground-water-resource assessments. A brief synopsis of the characteristics of each of these three types of investigations is given. This text will emphasize the latter two types.

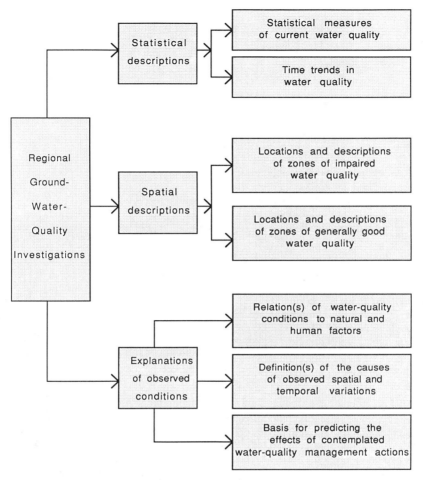

FIGURE 1-1. General types of information provided by regional ground-water-quality investigations.

Studies of Water at the Point of Use

Studies of the quality of water at the point of use can be designed to develop estimates of human exposure or risk assessments for selected contaminants. An advantage of sampling water at its point of use is that the target population can be defined simply as water at a particular point (i.e., at the tap). To relate the observed results to conditions in the aquifer(s) can be misleading, however, because many changes can occur in water quality between the ground and the point of use. Such changes are a particular limitation in studies of public water supplies, which commonly combine water from multiple wells or combine surface water and ground water.

Well-Water Surveys

Well-water surveys define the target population as a particular population of wells. Commonly, these studies have the dual purpose of providing descriptive statistics and identifying statistical relations between ground-water quality and selected human and natural factors. The target population typically consists of wells associated with a specific water use, such as domestic wells or public water-supply wells. An advantage of this designation is that it implicitly recognizes that the type of well will affect the results of the study, because well construction and well siting tend to differ among well types. Unlike studies of water at the point of use, water-quality sampling for well-water surveys is done as close to the wellhead as possible, before water reaches holding tanks, distribution systems, or treatment systems. Definition of the target population for well-water surveys can require substantial attention to detail, as noted in Chapter 20 for a well-water survey in Illinois.

Ground-Water-Resource Assessments

Ground-water-resource assessments attempt to describe ground-water quality in situ. Although the assessments focus on the ground-water resource, the results are unavoidably weighted toward representing the quality of water in and near wells. On the other hand, well-water surveys focus on wells but tend to relate the results to the quality of ground water and not to the water standing in a well. Thus, the differences between the two study designs are not as distinct as might first appear. The key difference is that well-water surveys focus exclusively on a particular target population of wells. Resource assessments tend to strive for greater explanation of observed conditions than do well-water surveys, but at the cost of a less clear relation between the target and candidate populations.

Ground-water-resouce assessment lies between the domains of process research and monitoring applied at a broad scale (Alley and Cohen 1991). As in process research, these assessments should be dynamic and inquiring and should involve continuous data evaluation and interpretation. On the other hand, if a long-term assessment is envisioned, some of the stability and consistency of monitoring must be incorporated to achieve sufficient repetition of sampling over the life of the study.

ISSUES OF SCALE

Ground-water-quality data are collected to address problems at a wide range of spatial and temporal scales. The scales of study control the issues that can be addressed and the level of understanding that can be gained. The following discussion centers on the value of conducting studies at multiple spatial and temporal scales.

Spatial Scales

It is useful to consider several spatial scales of ground-water-quality studies. Here, we will define "regional-scale" studies in a broad sense as those that attempt to characterize water-quality conditions within areas of tens to tens of thousands of square kilometers. Within the context of regional-scale studies, investigations can be done at several scales (see Figure 1-2). For example, "local-scale" studies can be designed to focus on ground-water quality within areas that might range from a few thousand square meters to a few square kilometers. These local-scale studies can be critical to the explanation of observed conditions at the regional scale. At the other end of the scales spectrum are "national-scale" studies, such as the national monitoring program of the Netherlands (see Chapter 21). National-scale studies of some small countries can actually cover less area than regional-scale studies in a large country, such as the United States.

Survey and Targeted Sampling

The coarsest spatial scale that might be considered is low-density *survey sampling* over a broad area. Survey sampling can provide (1) overall descriptive statistics on the occurrence and concentrations of a set of water-quality constituents, (2) the possibility of identifying statistical relations between water quality and selected human and natural factors, and (3) an initial, broad basis for describing the geographic distribution of water-quality conditions within the study area.

Issues of Scale 7

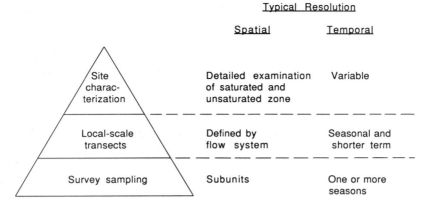

FIGURE 1-2. Typical spatial and temporal resolutions of regional ground-water-quality investigations.

The information available from survey sampling is useful for overall reference, but it does not address many basic questions about the spatial distribution of water-quality problems or the causes of this distribution. This is particularly true for constituents that are present at environmentally significant levels only sporadically. Furthermore, if prior knowledge of a constituent indicates that it is concentrated in a small part of the study area, then survey sampling is unlikely to contribute greatly to knowledge about the spatial distribution of that constituent. Survey sampling is considered further in Chapter 3 and in the case study of Chapter 20.

Many regional ground-water-quality investigations employ *targeted sampling* to make fuller use of knowledge about the likely controlling factors on water quality in particular areas. Targeted sampling can be done at a range of scales finer than survey sampling. Examples of targeted sampling include designs to test hypotheses about the distribution of contaminants as a function of specific land uses or other factors, and local-scale transects of wells along inferred flow paths in representative settings. Detailed site characterizations, including vadose-zone sampling and sediment and pore-water sampling of aquifers and confining units (cf. Pucci, Ehlke, and Owens 1992), can be done at selected localities.

Targeted sampling can be useful in identifying water-quality problems that would be missed or underrated by survey sampling and can provide enhanced detail on spatial distributions and variability. For example, a mix of survey and targeted sampling designs are embodied in the different types of pesticide monitoring studies discussed by Cohen, Eiden, and Lorber (1986), namely, large-scale retrospective, small-scale retrospective, and small-scale prospective studies. Examples of multiscale regional water-quality assessments are presented in Chapters 22 and 23.

In carrying out targeted sampling and interpreting its results, one must recognize that the relative importance of hydrologic processes varies with scale. A common scale for local-scale studies is that of local ground-water-flow systems evaluated by use of wells located along shallow ground-water-flow paths. Typically, vertical profiles are obtained at each site by means of two to four monitoring wells, each screened at a different depth.

One advantage of local-scale studies is that more detailed ancillary data bases can be developed for the local areas than is feasible for the entire study area. This can lead to identification of key factors that might not otherwise be considered. Many times, these factors can then be tested for correlations with water quality at the regional scale.

An example of how local-scale studies can yield useful insights about regional groundwater quality is provided by Robertson, Cherry, and Sudicky (1991), who studied the hydrologic character and geochemistry of plumes from two septic systems above shallow unconfined sand aquifers in Ontario, Canada. At both sites, the plumes had sharp lateral and vertical boundaries. The weakly dispersive nature of the aquifers was consistent with the results of recently reported natural-gradient tracer tests at Borden and Twin Lakes, Ontario, and Cape Cod, Massachusetts. The authors concluded that, in unconfined sand aquifers, long "pencil-like" plumes of mobile contaminants such as nitrate could extend far beyond the distance-to-well regulations typical of North America. An additional finding was that the only significant attenuation of nitrate in the plumes was the almost complete attenuation observed within the last 2 m before discharge to a river. This attenuation was attributed to denitrification within organic-matter-enriched riverbed sediments.

Further examples of the use of data from targeted sampling to draw inferences about regional ground-water quality are given throughout this book. From a scientific viewpoint, use of targeted sampling for regional assessment remains an area of relative infancy. Further research and practical experience are needed to advance methodologies for selection of targeted areas, design of studies within them, and regionalization of results.

Subunits
One approach commonly used to help guide the design of regional ground-water-quality studies is to define a set of subunits (or subregions) on the basis of similarities in hydrogeologic, geomorphic, and land-use considerations. The subunits can comprise areal and depth subdivisions. The set of subunits can be used to (1) help ensure that important areas are not underrepresented or that other areas are not overrepresented by the sampling program, (2) form the basis for survey sampling, and (3) facilitate the selection of representative areas for targeted sampling. Because of the limited number of water samples generally collected, the number of subunits selected in most regional studies is small, commonly less than 10.

One of the challenges in dividing a region into subunits is to group a number of different factors likely to affect ground-water quality into a small number of carefully chosen subdivisions. The unattractive alternative to such grouping is an attempt to design a ground-water-quality network on the basis of a multitude of different factors that could easily approach or exceed the number of sampling sites.

Within a given study area, one can usually identify a few major hydrogeologic settings from an analysis of general hydrogeologic information. These settings can constitute a first-level set of subunits. Indeed, the determination of differences in ground-water quality among a set of hydrogeologic settings might be one of the major objectives of the study. Without explicit consideration of the settings in the sampling design, the number of samples collected from each would be left to chance; conceivably, a very small number of samples—perhaps none at all—would be obtained from a particular setting of interest.

A second level of subdivision can be developed for shallow aquifer systems on the basis of natural and human-influenced characteristics of the landscape. These characteristics include soil characteristics, landform, land use, and agricultural practices. Thus, a hierarchy can be established in the definition of subunits: a first-level subdivision by major hydrogeologic settings and a second-level subdivision of shallow hydrogeologic settings related to landscape characteristics. Because landscape characteristics are generally interrelated (for example, land use can be highly correlated with soil types, landforms, and other factors), the spatial distributions of many landscape characteristics overlap. One guide to the selection of landscape subdivisions is to identify combinations of several key environmental characteristics for which significant overlap exists. Concepts developed about spatial patterns in the field of landscape ecology (Turner 1987) might have utility in the design of regional ground-water-quality studies. An example of landscape-based subunits are the hydrogeomorphic subregions described in Chapter 23 for the surficial aquifer of the Delmarva Peninsula.

Commonly, subunit boundaries are not discrete; rather, they represent zones of transition where the characteristics of one area blend with those of another. In general, subunits that can be precisely represented on a map are preferred. Subunit descriptions also are limited in their precision by the spatial variability of the subunit characteristics. Some subunits will consist of a distinct, homogeneous set of characteristics. Others will consist of a complex mosaic of features. Descriptions of a subunit represent characteristics typifying each subunit, but it is inappropriate to assume that all areas within a subunit possess all typifying characteristics (Gallant et al. 1989).

Sampling Density

It is not uncommon to describe a particular ground-water-quality network in terms of its sampling density, computed as the number of wells sampled divided by the size of the area. This numeric can have limited meaning, however. Obvious limitations are that in many areas the sampled sites are not distributed evenly, and that areal sampling density does not account for differences in water quality with depth. Furthermore, many, if not all, of the areas of contaminated ground water in a region are likely to be much smaller than the average area indicated by the sampling density. Thus, sampling density, per se, does not provide much information on the insights to be gained about the geographic distribution of contamination or its causes. In addition, if the purpose of the water-quality sampling network is to provide estimates of the statistical distribution of contaminant concentrations, then, as will be seen in Chapter 3, the number of sites sampled is a much more relevant measure of precision than is the density of sampling.

Temporal Scales

A sampling program can result in a one-time evaluation of ground-water quality, or it might be part of a continuing, perhaps long-term, evaluation. In either case, possible temporal variations in ground-water quality need to be considered. Decisions related to temporal variability include well-purging requirements, timing and frequency of sampling, and length of time during which sampling at different sites can be considered comparable despite differences in sampling dates. In general, shallow ground water exhibits the most pronounced temporal variability in water quality. Short-term variability (time scales of minutes to days), seasonal variations (time scales of months), and long-term trends (time scales of years) are considered in the following sections.

Short-Term Variability

In an effort to obtain water samples representive of water in the aquifer, wells are typically purged prior to sampling. The purpose of well purging is to remove water stored inside the well casing and water in the formation immediately adjacent to the well. Substantial variations in water quality typically occur during this process.

Simple well-purging criteria commonly are used to judge when chemical stability has been achieved. The most common criteria are to sample after (1) a specific number of well volumes have been withdrawn, (2) measurements of selected characteristics, such as specific conductance, pH, temperature, and turbidity, attain a specified stability in the discharge water, or (3) some combination of criteria (1) and (2) are met. These simple criteria continue in common use, but a number of investigators have noted their limitations. For instance, Keely and Boateng (1987) and Gibs and Imbrigiotta (1990) have questioned the use of such criteria as indicators of the stability of organic compounds. Robbins and Martin-Hayden (1991) used mass-continuity models of monitoring well purging to show that the number of casing volumes cannot be set without a priori information on the degree to which chemical stratification occurs in the aquifer.

An alternative approach is to examine the variation of concentrations with elapsed time at each well and to establish individual protocols (Rivett, Lerner, and Lloyd 1990). Of course, it could be difficult to devise purging criteria optimal for the diverse set of constituents sampled, and short-term variability at a site at one time may not be indicative of the variability at a later date. Furthermore, control over pumping before sample collection commonly is not practical at many water-supply wells. Nonetheless, some investigations of short-term variability might be useful in establishing purging criteria, particularly for wells to be sampled as part of a long-term sampling network. For monitoring wells, some authors (cf. Kearl, Korte, and Cronk 1992) suggest installation of dedicated sampling devices and limited purging of the well prior to sampling.

Seasonal Variations

Seasonal variations in water quality generally are less important to ground-water assessment than to surface-water assessment. Nevertheless, in some settings they may be more significant than commonly assumed. Wells chosen for seasonal sampling should be selected carefully because substantial costs can be involved. Furthermore, the high spatial variability of ground-water quality should be considered in making trade-offs between temporal sampling and enhanced spatial sampling (cf. Bjerg and Christensen 1992).

Some seasonal variations in ground-water quality are related to human activities, others to natural phenomena. Seasonal variability can be caused by (1) variations in the quantity and quality of recharge, (2) changes in ground-water-flow patterns, and (3) complex interrelations between the spatial distribution of water quality near the well and factors affecting the sources of water withdrawn from the aquifer.

Variations in shallow-ground-water quality due to temporal variations in recharge are most likely to be significant in near-surface water-table aquifers. Fluctuations in the quality of shallow ground water can be greatly affected by the flushing of constituents (or, conversely, by dilution) during individual recharge events and seasonal periods of recharge. Water-table fluctuations that result in the up-and-down movement of the saturated zone into different depth horizons also may have a significant effect on water quality in shallow aquifers.

As an example of the effects of recharge, Pettyjohn (1976, 1982) examined temporal variations in the quality of shallow ground water pumped from an alluvial aquifer in Ohio. The study site had been contaminated by oil-field brines several years earlier. Pettyjohn attributed continuing cyclic fluctuations in ground-water quality to the long-term storage in the unsaturated zone of a considerable amount of water-soluble substances from the brines; these substances were being flushed into shallow ground water during periodic recharge. An important contributing factor was hypothesized to be the rapid transport of contaminants through fractures and macropores.

The natural water quality of deep or confined aquifers tends to be less variable with time and, at any particular site, reflects the geochemical reactions that occurred as water migrated through confining layers and aquifers from its recharge area to the point of interception by the well. The quality of ground water can change in response to stresses on the aquifer system, however. Changes in hydrostatic head brought about by pumping, for example, can cause leakage of more highly mineralized water from adjacent units into the producing zone. In addition, continued pumping of a confined aquifer can cause inflow of water from nearby unconfined areas that differs considerably in quality from water originally in the confined aquifer (Schmidt 1977).

Temporal changes in ground-water quality due to pumping are particularly likely if vertical stratification of ground-water quality is significant and the well intercepts more than one aquifer. Temporal changes in water quality during pumping are affected by variations in the quality of ground water near the pumped well and by spatial variations in the permeability of individual strata that control temporal variations in the sources of water to wells during pumping. The contributions of water from different strata can be expected to vary as a function of the recent pumping history. Major differences in water quality can be expected, for example, from samples collected during the pumping and nonpumping seasons of a seasonally pumped well, simply because different parts of the aquifer system are likely to be represented by the water-quality samples collected during the two periods.

Examples of how the pumping history of a well can greatly affect observed concentrations over time are shown in Figures 1-3 and 1-4 (Rivett, Lerner, and Lloyd 1990). Figure 1-3 shows changes in concentrations of three organic compounds in response to a 10-week nonpumping period at a well that otherwise was pumped at $460 \, \text{m}^3/\text{d}$. Trichloroethylene (TCE) concentrations declined from 4,800 to 440 µg/L (micrograms per liter). When pumping restarted at a constant rate, TCE and trichloroethane (TCA) concentrations gradually rose, and tetrachloroethylene (PCE) concentrations declined. The reverse behavior of PCE suggests that its distribution in ground water may be

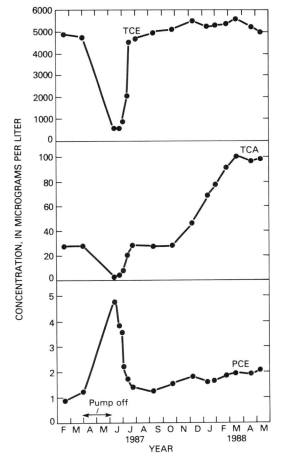

FIGURE 1-3. Example of effects of change in well pumping on three trace organic compounds. (*After Rivett, Lerner, and Lloyd 1990.*)

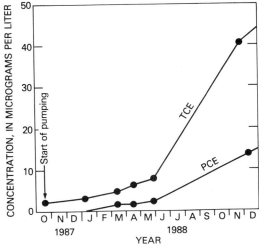

FIGURE 1-4. Example of increase in solvent concentrations during a 15-month period due to start of pumping at a well. (*After Rivett, Lerner, and Lloyd 1990.*)

different from that of TCE and TCA. Figure 1-4 shows the concentrations of TCE and PCE after the start of pumping in a well that had not been pumped for more than 10 years. TCE and PCE had been used on the site, and PCE was still being used in large quantities within 40 m of the well. In spite of the proximity of potential sources, concentrations remained low for many months after pumping started, and then began to rise and were still rising 15 months after pumping began.

Long-Term Trends

Long-term trends in ground-water quality reflect variations in the rate and quality of recharge, the time scales of subsurface transport of contaminants, and changes in water levels and in the direction of ground-water flow. Some results that might be expected are shown in Figure 1-5, a graph of the relation between nitrate and well depth from a set of wells in Nebraska sampled in 1974 and 1984. The data show an increase in nitrate concentration over time at all well depths, the largest increases occurring in shallow ground water.

The trends shown in Figure 1-5 are not always as evident in other studies. Complications in evaluating long-term trends include changes in field and laboratory procedures, changes in well construction or maintenance, climatic variability, and variations in long-term withdrawals. The first of these complications indicates the necessity of consistent application of field and laboratory procedures over time and of careful documentation of data-collection and analytical procedures. If procedures do change, then some period of overlap between the new and old procedures should be allowed.

Apparent trends in ground-water quality can be artifacts of changes in well construction or well maintenance such as those that result from partial filling, cleaning, or deepening of a well or, conversely, from well-screen encrustation or collapse. Thus, records of well maintenance and changes in construction are vital to valid interpretation of long-term changes in ground-water quality.

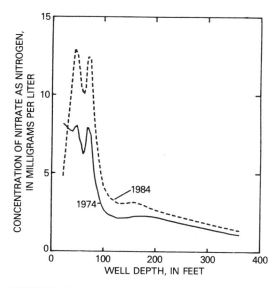

FIGURE 1-5. Change in nitrate-nitrogen concentrations in Nebraska ground water between 1974 and 1984. (*Compiled by D. R. Helsel, U.S. Geological Survey, from data in Exner and Spalding [1976] and Exner [1985].*)

Seasonal fluctuations can easily mask long-term trends. Thus, seasonal samples can be useful in determining the degree to which seasonal variability may confound identification of long-term trends and in determining the appropriate time(s) of year to sample over the long term. For example, Anderson (1989) found that the seasonal fluctuation of nitrate concentration in water from nine wells in surficial sand-plain aquifers in Minnesota was as great or greater than an observed long-term trend in the mean concentration of nitrate nitrogen (from 4.7 to 11.0 mg/L) observed by comparing nitrate concentrations in 1982 with concentrations at the same wells from 1965 to 1978. As a second example, data on annual precipitation, nitrate concentrations in shallow aquifers, wheat yields, and inorganic fertilizer applications in Czechoslovakia over a 20-year period are shown in Figure 1-6. The large nitrate concentration in 1981 coincided with a period of heavy precipitation.

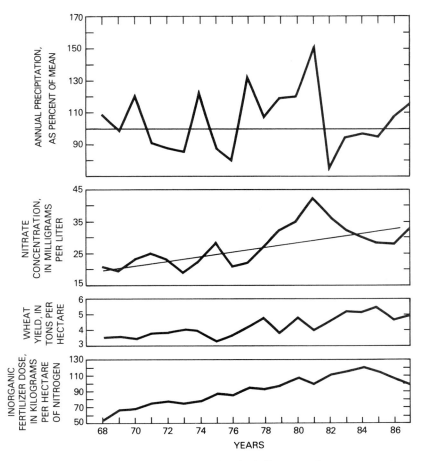

FIGURE 1-6. Relation of nitrate concentrations in shallow ground water to annual precipitation, wheat yield, and inorganic fertilizer doses in Czechoslovakia. (*After Vrba and Pěkný 1991.*)

Low precipitation and climatic conditions favorable for crop growth from 1982 to 1985 led to a marked decline in nitrate concentrations. These climate-induced variations are superimposed on a long-term trend of increasing nitrate concentrations.

Because of the sources of temporal variability mentioned above and the relatively slow rates of ground-water flow, trends in ground-water quality can take many years to be detected with any degree of confidence. Apparent trends over a few years time should be interpreted with great caution because of the seasonal variability and serial correlation typical of ground-water levels and ground-water quality (Montgomery, Loftis, and Harris 1987). Although trends in ground-water quality can be slow to evolve, periodic data interpretation is essential to avoid a common pitfall of long-term monitoring—the collection of data rendered useless from causes that could have been diagnosed early.

Any evaluation of trends in ground-water quality should be done to the extent possible in the context of the ground-water-flow system. The analysis of trends from wells producing water of significantly different ages can confound trend detection and lead to misleading conclusions. Phillips et al. (1992) showed how careful ground-water-age dating by techniques such as chlorofluorocarbons (see Chapter 11) can be used to help evaluate trends in ground-water quality.

Long-term sampling can involve local-scale and regional networks. Local-scale networks can provide enhanced understanding of the causes of the trends. Regional networks are necessary to determine regionwide trends and to verify that observed trends are not of a local character only.

In some circumstances, trends in streamwater quality during baseflow can provide evidence about trends in ground-water quality. For example, Katz, Lindner, and Ragone (1980) suggested that significantly lower ammonium and nitrate concentrations in baseflows in sewered areas of Long Island than in unsewered areas were evidence for improvements in ground-water quality in the sewered areas. Use of trends in baseflow quality to infer trends in ground-water quality should be approached with caution, however, because of physical, chemical, and biological transformations that occur in the stream and near-stream environment (see Chapter 2).

USE OF EXISTING WELLS COMPARED WITH SPECIALLY CONSTRUCTED WELLS

In practice, data collected from a sampling point are used to infer ground-water-quality conditions for some spatial domain about the sampling point. The extent to which "point" values represent the spatial distribution of a contaminant is a key factor in defining the regional implications of the data. A major problem in using wells to sample ground water is knowing the degree to which the sample from the well is representative of the in situ ground-water quality at the chosen site. The problem is analogous to the Heisenberg uncertainty principle in physics (Bueche 1969); that is, an exact measure of in situ ground-water quality is impossible because the measurement process, including the installation of a test hole or well, will always disturb the local system (although some recent techniques such as the use of fiber optics can minimize the disturbances). These considerations lead to the often difficult choice between use of existing wells or construction of new wells specifically designed for water-quality monitoring.

A number of factors favor the use of specially constructed wells:

1. Well-construction materials and design are known, can be tailored to the sampling-program objectives, and can be standardized to reduce the variability in sampling results due to differences in well construction.
2. Well-purging criteria can be more readily standardized, and water-quality samples can be properly collected by use of pumps suitable for the constituents of interest.
3. Information can be obtained on the characteristics of the vadose zone and aquifer materials during well construction.
4. Special considerations related to network design and depth of screen openings can be taken into account.
5. The wells can be sited to avoid disadvantages of location common to many existing wells (for instance, wells in rural settings usually are drilled near landowners' houses, and large areas of the water resource that underlie other land uses and land covers may not be adequately represented by these wells).
6. Wells can be designed to sample from near the water table in unconfined aquifers, thereby increasing the possibility of relating ground-water quality to land-use activities.
7. If wells are sampled repeatedly over time, the cost of installing the wells will become a decreasing proportion of the total investment in the study.

On the other hand, drilling requires specialized equipment and expertise and can greatly reduce the financial resources available for sampling. Installation costs for new wells, particularly to tap deep aquifers, may be prohibitive within the budget of a project. Limitations of access and permission to sample may greatly restrict where new wells can be installed.

It is often constructive to carry out a preliminary sampling of existing wells before committing costly resources to the installation of new wells. Initial sampling of existing wells can provide a preliminary characterization of the ground-water quality in an area and can help frame hypotheses that require further refinement and testing through use of specially constructed wells. As a general rule, existing and newly constructed wells should be considered as sampling different subpopulations.

Further considerations about the use of existing and specially constructed wells involve considerations of the role of wells as mechanisms for contaminant transport and the effects of the well open intervals.

The Role of Wells in Contaminant Transport

Wells can serve as major pathways for the movement of contaminants between aquifers by short-circuiting natural flow paths. This short circuiting can occur either in the wellbore or outside the well casing in an unsealed annulus. Contaminants can move through direct entry at the land surface or through cross contamination between aquifers by way of screens or perforations open to more than one aquifer. Special care is needed during construction of new wells to avoid cross contamination as part of the drilling process (Keely and Boateng 1987).

Causes of the direct movement of contaminants down wells include spills on the surface, back-siphoning during chemigation, entry of surface water due to faulty well construction, and introduction by way of drainage wells. These point sources of ground-water contaminants can be widespread. For example, in an inventory of wells in which pesticides were detected in the western United States, DeMartinis and Royce (1990) found that 47% of the wells had poor surface seals (or none at all) or had visibly cracked or corroded casings. Hallberg (1986) noted the prevalence of ground-water contamination of water-supply wells near farm-chemical dealerships. Point versus nonpoint sources of pesticides are discussed further in Chapters 11 and 14.

Exchange of water between different strata in the wellbore can take place as a result of vertical hydraulic gradients or shallow water cascading down the well casing during nonpumping periods. Cross contamination among aquifers can be a significant transport mechanism and can greatly complicate the interpretation of ground-water quality data. This exchange of water can occur in wells that are currently in use or in abandoned wells or test holes that have been improperly sealed. Abandoned wells screened in more than one aquifer, and perhaps covered over, can easily escape detection.

Effects of the Open Interval

Wells that are screened or open to more than one aquifer should be avoided because of uncertainties concerning the source of the sampled water. The length and vertical position of the well-screen interval must still be considered carefully when samples are taken from a single aquifer.

Monitoring wells sometimes are designed to have long well screens with the idea that samples from such wells will, at least, indicate the presence (or absence) and approximate average concentrations of possible contaminants. Data from such wells can be difficult to interpret, however, particularly in the presence of multiple water-bearing zones.

Reilly, Franke, and Bennett (1989) have shown that substantial wellbore flow can occur in monitoring wells with long screens, even if the wells are completed in homogeneous aquifers having very small vertical-head differences (less than 3 mm between the top and bottom of the screen, a difference much less than is commonly detectable in the field). As examples, these investigators describe two situations that can lead to misrepresentation of aquifer contamination in the presence of a downward gradient in a well. The first situation may occur if contamination intersects the top part of the well screen. Contaminated ground water may circulate through the wellbore and discharge into the aquifer in all directions from the bottom part of the screen. When the well is pumped for water-quality sampling, this contaminated water in the aquifer surrounding the bottom of the screen may result in a different proportion of contaminated water in well samples than would be present in the absence of wellbore flow. The second situation may occur if contamination intersects the bottom part of the well screen. Because of the downward flux of water in the well, the contaminant plume may not pass through the lower part of the borehole but may flow around it. This could lead to underestimation of the average contamination over the screen length, depending on the extent to which the well is purged before a water sample is collected and at what depth the pump intake is placed in the well. Many other complexities associated with long well screens are possible.

Interest in sampling near the water table can lead to the installation and sampling of wells in which screened intervals are partly above and partly below the saturated zone. The position of the water table with respect to the screened interval in these wells will vary as the water table rises and falls throughout the year. Because solute concentrations may decrease or increase with depth, the measured contaminant concentrations in water samples will depend on how much of the water column is intersected by the screen. If constituent concentrations decrease with depth below the water table, for instance, then the concentrations in sampled well water may be largest when the water table is lowest simply because the most concentrated part of the ground water is sampled. It may be difficult to differentiate these variations in constituent concentrations from seasonal variations in recharge water quality and long-term trends in shallow-ground-water quality. The problem can be exacerbated by water-level drawdowns during well purging. Possible sampling biases caused by well-screen placement have been evaluated mathematically by Robbins (1989) and Robbins et al. (1989), who concluded that these biases may be comparable to, or in excess of, concentration biases related to well-construction and sampling procedures.

In addition to the screened or open interval of a well, the gravel pack can have a major influence on the hydrogeologic units from which the well is producing and on the potential for cross contamination among aquifers. Many wells that have short screens or short open intervals are gravel-packed from the bottom of the well to the base of the seal, close to the land surface. Wells with short gravel-packed intervals are preferred to avoid possible avenues for vertical flow. Further considerations for well-screen placement associated with contaminants that are immiscible or whose density is significantly different from water are discussed in Chapters 13 and 18 on organic contaminants and freshwater–saltwater environments, respectively.

Cost considerations and the characteristics of existing wells in an area may limit a regional study to sampling existing wells with open intervals that draw water from multiple water-bearing zones or aquifers. In these situations, one might consider use of tools capable of direct measurements of the contribution of water to a well from individual aquifer layers. For example, flowmeter velocity logs coupled with depth-dependent water-quality data collected using a downhole sampling device may be useful in developing quantitative estimates of the water quality in individual water-bearing zones within multiple aquifer systems. Techniques, such as flowmeter velocity logs, have not found widespread use in regional ground-water-quality studies, but may hold future promise.

In some hydrogeologic situations (e.g., unconsolidated sand aquifers and clay/silt), low-cost techniques that require minimal or zero drill rig costs can be used to collect samples from small-diameter holes (Smolley and Kappmeyer 1991; Cherry et al. 1992). For example, Cherry et al. (1992) note that in parts of North America depths of 30 to 50 meters below ground surface are obtainable using drive-point devices. The reduced costs of such techniques over conventional monitoring techniques may make it possible to substantially increase the number of samples collected in some settings. Other advantages include minimal disturbance of geochemical conditions during installation and minimal purging requirements.

DOCUMENTATION OF WELLS AND WELL SITES

Thorough documentation of each sampled well is an essential part of any ground-water-quality study. Unfortunately, it is all too easily neglected. Documentation of wells may be needed for data interpretation and for site selection. In establishing criteria for the suitability of existing wells for inclusion in a sampling program, one should consider, among other factors, the well-construction features, the condition of the well, the existing pumping equipment, and accessibility for sampling and water-level measurement.

Well and well-site information can be derived from a number of sources. In general, wells selected as potential sampling sites should be visited before sampling, and existing information for these wells should be field-checked and verified to the extent possible. Inaccurate information about the characteristics of a site can be much worse than no information at all in the interpretation of water-quality data. Cartwright and Shafer (1987) note that private wells commonly are constructed with less adherence to quality-control procedures and are less well documented than are municipal and industrial wells.

Some of the most basic information needs are accurate latitude and longitude for the well, depth of the well, depth to the top and bottom of each open interval, hydrogeologic unit(s) to which the well is open, pump type, and static water levels. In establishing information needs for site characterization, a review of data used in other studies can be helpful. The USEPA has developed a

TABLE 1-1. Example Site Characteristics Recorded for Sampled Wells

Site ID (station number)
Local well number
Agency code
Project number
State code
County code
Latitude
Longitude
Latitude-longitude accuracy code
Altitude of land surface
Method used to determine altitude
Accuracy of altitude
Topographic setting
Primary use of site
Primary use of water
Depth of well
Source of depth data
Type of lift
Date of well construction
Method of construction
Type of finish
Type of surface seal
Casing material
Rated capacity of pump
Primary aquifer
Aquifer-type code
Depth to top of open interval (for each open interval)
Depth to bottom of open interval (for each open interval)
Depth to top of geohydrologic unit (for each unit)
Depth to bottom of geohydrologic unit (for each unit)
Lithologic unit identifier (for each unit)
Well owner
Water level
Date of water-level measurement
Status of well at time of water-level measurement
Method used to measure water level
Accuracy of water-level measurement
Beginning date for use of water-level measuring point
Ending date for use of water-level measuring point
Height of water-level measuring point above land surface
Description of water-level measuring point

Adapted from Hardy, Leahy, and Alley (1989).

"minimum set of data elements" for ground-water quality studies (U.S. Environmental Protection Agency 1992) to form a core set upon which individual studies can add additional elements to meet specific needs. An expanded set of key characteristics is listed in Table 1-1. Many investigators attempt to compile additional information from well owners, local experts, and land managers about agricultural practices, waste-disposal practices, and pesticide and chemical storage and handling practices in the vicinity of sampled wells.

Information pertaining to the accessibility of a potential sampling site should be carefully documented. Examples of pertinent information include the site owner, contacts necessary for permission to sample, special constraints that can affect when the well can be sampled, and procedures needed to gain access to the well.

Finally, if data from the sampling program are to be useful for future studies, the purpose for which each sampling site was selected must be documented. For example, Hardy, Leahy, and Alley (1989) established data elements to record the purpose of sampling each time a well is sampled in the USGS National Water-Quality Assessment Program. One element records if the well was selected randomly. A second element records the type of investigation (survey sampling, geochemical investigation, local-scale network, and so forth).

QUALITY ASSURANCE

A quality-assurance (QA) program needs to be an integral part of any ground-water-quality study to ensure that technically sound procedures are used for data collection, analysis, and interpretation and that these procedures are documented and capable of being verified. Quality-assurance programs are becoming increasingly important in light of the low analytical reporting limits for trace constituents. Quality assurance also needs to be an essential part of the entire process of accurately recording and documenting the potentially vast amount of data collected by a study. A review of the quality assurance of chemical measurements is given by Taylor (1987). A few general statements on laboratory and field QA are made here.

The laboratory QA program should have internal as well as external (interlaboratory) quality-control (QC) components. The internal QC ensures the reproducibility of data within a laboratory and, if suitable certified reference materials are available, yields information on the accuracy of analytical results. The interlaboratory QC is best conducted by an independent agency that evaluates the overall performance of different laboratories. This evaluation can be based on analyses of "blind" samples submitted to participating laboratories.

The field QA should help ensure that the actual sampling locations chosen are adequate for the purposes of the program, the samples collected are representative of that location, sample contamination and decomposition are not occurring, and all necessary field data and information are properly recorded. A key element of field QA is the submittal of replicate, spiked, and blank samples (see Chapter 24 for further discussion of these and other types of field QA samples).

Field QA should be a dynamic process. Initial guidelines on the number of field QA samples generally are intended only as a first-level screening process. Additional QA directed at particular locations or constituents may be necessary to uncover problems detected by the initial screening or to further validate key or unexpected findings. Additional recommendations for field QA are to (1) select the wells before sampling and distribute QA samples among a range of networks, hydrochemical conditions, and field-sampling conditions, as well as over time; (2) intensify QA when significant changes are made in sample-collection procedures or in the equipment used; and (3) conduct the different types of QA for a particular class of water-quality constituents at the same sampling sites to help in the interpretation of the results. For instance, submittal of duplicate spiked samples together with duplicate unspiked samples, one of which is later spiked in the laboratory, can be useful in simultaneously investigating analytical precision, interferences from the sample matrix, and the effects of sample storage and shipping on pesticide samples.

CONCLUDING REMARKS

Each regional ground-water system has its own set of hydrologic characteristics, and each responds differently to natural and human-induced stress. The challenge of regional ground-water-quality studies is to develop inferences about the primary effects of human activities, climate, geology, and soils on the water quality of individual regional systems. In this chapter, several key concepts necessary for developing a study of regional ground-water quality have been reviewed. Particular attention has been given to the target population, spatial and temporal scales of study, trade-offs in the use of existing wells as opposed to specially constructed wells, importance of careful documentation of wells and well sites, and need for quality assurance.

A regional ground-water-quality study with a specific purpose requires a survey or targeted sampling design tailored to the objective of the study; however, for many studies, considerable advantages can result from integrating survey and targeted sampling. In the absence of survey sampling, targeted sampling results may be largely anecdotal and lack a larger framework for interpretation. In the absence of targeted sampling, survey sampling may support little more than crude statistical inferences about the patterns and causes of water-quality variations.

Ideally, a ground-water-quality assessment would provide a detailed three-dimensional characterization of water quality throughout all aquifers in a study area. Because such a characterization is not feasible, trade-offs must be made in the proportion of sampling effort allocated to different parts of the ground-water system. Some of the strategies that can be used are to

1. Limit the sampling of all aquifers to broad survey sampling and use the remaining resources for targeted sampling of type settings.
2. Emphasize areas of greatest ground-water use.
3. Emphasize areas of greatest perceived vulnerability to ground-water contamination.
4. Focus initial sampling on shallow parts of the most intensively used aquifers or on aquifers that are important sources of recharge to intensively used aquifers.
5. Use isotopes and hydrogeologic knowledge to define zones of young ground water.
6. Combine information on the distribution of elevated concentrations of key constituents in shallow ground water with information on ground-water-flow paths to help guide sampling of deeper ground water.

These strategies are discussed in subsequent chapters. A prerequisite to their proper application is the development of a regional framework to serve as a paradigm for the design and interpretation of the studies. This is the topic of the next chapter.

References

Alley, W. M., and Philip Cohen. 1991. A scientifically based nationwide assessment of groundwater quality in the United States. *Environmental Geology and Water Sciences* 17(1):17–22.

Anderson, H. W. 1989. *Effects of Agriculture on Quality of Water in Surficial Sand-Plain Aquifers in Douglas, Kandiyohi, Pope, and Stearns Counties, Minnesota.* St. Paul, Minnesota: U.S. Geological Survey Water-Resources Investigations Report 87-4040.

Bjerg, P. L., and T. H. Christensen, 1992. Spatial and temporal small-scale variation in groundwater quality of a shallow sandy aquifer. *Journal of Hydrology* 131:133–49.

Bueche, Frederick. 1969. *Introduction to Physics for Scientists and Engineers.* New York: McGraw-Hill.

Cartwright, Keros, and J. M. Shafer. 1987. Selected technical considerations for data collection and interpretation—Ground water. In *National Water Quality Monitoring and Assessment: Report on a Colloquium Sponsored by the Water Science and Technology Board,* pp. 33–56. Washington, D.C.: National Academy Press.

Cherry, J. A., R. A. Ingleton, D. K. Solomon, and N. D. Farrow. 1992. Low technology approaches for drive point profiling of contaminant distributions. In *National Groundwater Sampling Symposium Proceedings,* pp. 109–11. Clovis, California: Grundfos Pumps Corporation.

Cohen, S. Z., C. Eiden, and M. N. Lorber. 1986. Monitoring ground water for pesticides. In *Evaluation of Pesticides in Ground Water,* eds. W. Y. Garner, R. C. Honeycutt, and H. N. Nigg, pp. 170–96. Washington, D.C.: American Chemical Society Symposium Series 315.

DeMartinis, J. M., and K. L. Royce. 1990. Identification of direct-entry pathways by which chemicals enter ground water. In *Ground Water Management, Proceedings of the 1990 Cluster of Conferences,* pp. 51–65. Kansas City, Mo.: National Water Well Association.

Exner, M. E. 1985. *Concentration of Nitrate-Nitrogen in Groundwater, Central Platte Region, Nebraska 1984.* Lincoln, Nebraska: Conservation and Survey Division, Institute of Agriculture and Natural Resources, University of Nebraska (map).

Exner, M. E., and R. F. Spalding. 1976. *Groundwater Quality of the Central Platte Region, 1974.* Lincoln, Nebraska: Conservation and Survey Division, Institute of Agriculture and Natural Resources, University of Nebraska, Resource Atlas No. 2.

Gallant, A. L., T. R. Whittier, D. P. Larsen, J. M. Omernik, and R. M. Hughes. 1989. *Regionalization as a Tool for Managing Environmental Resources.* Corvallis, Oregon: U.S. Environmental Protection Agency EPA/600/3-89/060.

Gibs, Jacob, and T. E. Imbrigiotta. 1990. Well-purging criteria for sampling purgeable organic compounds. *Ground Water* 28(1): 68–78.

Hallberg, G. R. 1986. From hoes to herbicides: Agriculture and groundwater quality. *Journal of Soil and Water Conservation* 41(6):357–64.

Hardy, M. A., P. P. Leahy, and W. M. Alley. 1989. *Well Installation and Documentation, and Ground-Water Sampling Protocols for the Pilot National Water-Quality Assessment Program.* Reston, Va.: U.S. Geological Survey Open-File Report 89–396.

Katz, B. G., J. B. Lindner, and S. E. Ragone. 1980. A comparison of nitrogen in shallow ground water from sewered and unsewered areas, Nassau County, New York, from 1952 through 1976. *Ground Water* 18(6):607–16.

Kearl, P. M., N. E. Korte, and T. A. Cronk. 1992. Suggested modifications to ground water sampling procedures based on observations from the colloidal borescope. *Ground Water Monitoring Review* 12(2):155–61.

Keely, J. F., and Kwasi Boateng. 1987. Monitoring well installation, purging, and sampling techniques—Part 1: Conceptualizations. *Ground Water* 25(3):300–13.

Lehr, J. H. 1982. How much ground water have we really polluted? *Ground Water Monitoring Review* 2(1):4–5.

Montgomery, R. H., J. C. Loftis, and Jane Harris. 1987. Statistical characteristics of ground-water quality variables. *Ground Water* 25(2):176–93.

Patrick, Ruth, Emily Ford, and John Quarles. 1987. *Groundwater Contamination in the United States.* 2nd ed. Philadelphia: University of Pennsylvania Press.

Pettyjohn, W. A. 1976. Monitoring cyclic fluctuations in ground water quality. *Ground Water* 14(6):472–80.

Pettyjohn, W. A. 1982. Cause and effect of cyclic changes in ground-water quality. *Ground Water Monitoring Review* 2(1):43–49.

Phillips. P. J., L. N. Plummer, E. Busenberg, and S. A. Dunkle. 1992. Use of chlorofluorocarbon dating to determine age of shallow ground water in a sandy surficial aquifer. *Transactions, American Geophysical Union* 73(14):130.

Pucci, A. A., T. A. Ehlke, and J. P. Owens. 1992. Confining unit effects on water quality in the New Jersey Coastal Plain. *Ground*

Water 30(3):415–27.

Reilly, T. E., O. L. Franke, and G. D. Bennett. 1989. Bias in groundwater samples caused by wellbore flow. *Journal of Hydraulic Engineering* 115(2):270–76.

Rivett, M. O., D. N. Lerner, and J. W. Lloyd. 1990. Temporal variations of chlorinated solvents in abstraction wells. *Ground Water Monitoring Review* 10(4):127–33.

Robbins, G. A. 1989. Influence of using purged and partially penetrating monitoring wells on contaminant detection, mapping, and modeling. *Ground Water* 27(2):155–62.

Robbins, G. A., R. D. Bristol, J. M. Hayden, and J. D. Stuart. 1989. Mass continuity and distribution implications for collection of representative ground water samples from monitoring wells. In *Proceedings of the Conference on Petroleum Hydrocarbons and Organic Chemicals in Ground Water: Prevention, Detection, and Restoration*, pp. 125–40. Dublin, Ohio: National Water Well Association.

Robbins, G. A., and J. M. Martin-Hayden. 1991. Mass balance evaluation of monitoring well purging. Part I. Theoretical models and implications for representative sampling. *Journal of Contaminant Hydrology* 8:203–24.

Robertson, W. D., J. A. Cherry, and E. A. Sudicky. 1991. Ground-water contamination from two small septic systems on sand aquifers. *Ground Water* 29(1):82–92.

Schmidt, K. D. 1977. Water quality variations for pumping wells. *Ground Water* 15(2):130–37.

Smolley, M., and J. C. Kappmeyer. 1991. Cone penetrometer tests and HydroPunch sampling: A screening technique for plume definition. *Ground Water Monitoring Review* 11:101–6.

Taylor, J. K. 1987. *Quality Assurance of Chemical Measurements*. New York: Lewis.

Turner, M. G., ed. 1987. *Landscape Heterogeneity and Disturbance*. New York: Springer-Verlag.

U.S. Environmental Protection Agency. 1992. *Definitions for the Minimum Set of Data Elements for Ground Water Quality*. Washington, D.C.: U.S. Environmental Protection Agency EPA 813/B-92-002.

Vrba, Jaroslav, and Vladimir Pěkný. 1991. Groundwater-quality monitoring—Effective method of hydrogeological system pollution prevention. *Environmental Geology and Water Sciences* 17(1):9–16.

2

Establishing a Conceptual Framework

William M. Alley

INTRODUCTION

Two general statements can be made about regional ground-water-quality studies (Romijn and Foster 1987): (1) the establishment and maintenance of a ground-water-quality monitoring network is expensive, and (2) data are always lacking of the desired type at the desired time and place.

Typically, a few tens to several hundred sites are sampled during an investigation of regional ground-water quality. Even when supplementing data from these sites with existing data, one is faced with a sparse network relative to the spatial variability of ground-water quality. Thus, some unifying concepts are needed to interpret regional ground-water quality data in a meaningful manner.

This chapter reviews some key considerations in establishing a conceptual framework for regional ground-water-quality studies. It is divided into three parts dealing with the vadose zone, ground-water-flow systems, and uses of geochemical data for regional conceptualization. An objective of the chapter is to illustrate how each of these three must often be considered in order to provide meaningful explanations of regional ground-water quality. The chapter is interspersed with examples selected to show the relevance of the different factors discussed in a variety of situations. Additional examples are given throughout the book.

THE VADOSE ZONE

The vadose zone is the terrestrial part of the Earth's crust between the land surface and the water table. This zone normally is unsaturated, but it also includes the capillary fringe, and possibly, perched water. The upper part of the vadose zone—the root zone—is defined as that part occupied by plant roots. At the regional scale, the part of the vadose zone that lies between the root zone and the water table is defined as the intermediate vadose zone. More precise distinctions within the vadose zone become important as more detailed studies are undertaken. For instance, the capillary zone can be significant in local processes affecting contaminant transport.

Relation of the Vadose Zone to Ground-Water Quality

The vadose zone not only occupies a strategic position between the land surface and the water table, it is also generally a favorable environment for attenuation of contaminants. Examples of active roles played by the vadose zone are (1) attenuation of trace elements and other inorganic chemicals through precipitation, sorption, cation exchange; (2) sorption and biodegradation of organic compounds; and (3) interception, sorption, and elimination of pathogenic bacteria and viruses. Many of the processes causing contaminant attenuation or degradation within the vadose zone proceed at their highest rates in the soil solum (the upper part of the soil profile above the C horizon), which is the most biologically active part of the vadose zone and commonly has the highest organic-matter content in the subsurface and a high clay mineral content.

The vadose zone also plays a key role in nutrient cycling. A surplus of available nutrients in the vadose zone can result in deterioration of ground-water quality as unused substances are leached downward (see Chapter 12).

The "filtering" function of the vadose zone from processes such as sorption and microbial decay is often far from complete and is highly dependent on soil properties. In the case of persistent, mobile contaminants, limited attenuation may occur, with the vadose zone merely introducing a lag time before contaminant arrival at the water table. In addition, rapid downward percolation of infiltrating water can occur through preferential flow paths.

Regional Characterization of the Vadose Zone

A challenge of regional ground-water-quality studies is to relate the highly variable and usually poorly known characteristics of the vadose zone to observations of ground-water quality. The high variability of the physical, morphologic, and chemical properties of the vadose zone (Warrick and Nielsen 1980) hinders precise characterization of small field plots, let alone large, complex regions.

Most of the regional-scale information that is available on properties of the vadose zone is referenced to soils; thus, the following discussion centers on soils data.

Soils

Despite the variability in soil characteristics, where the major soil-forming factors (climate, parent material, living organisms, time, and relief) are alike, similar soils are expected. This predictability of the relation of soil type to the landscape is the basis of soil mapping.

Zoeteman (1987) describes soil characteristic maps available at the scale of 1:400,000 for the Netherlands that include thickness, organic matter content, clay content, $CaCO_3$ content, and cation exchange capacity (CEC). In the United States, soil surveys by the National Cooperative Soil Survey Program administered by Soil Conservation Service (SCS) are a major source of information about soil properties. Soil taxonomic classification in these soil surveys is a hierarchical system that establishes several successive categories of soils (Soil Survey Staff 1975). Taxonomic levels in the soil taxonomy of the United States include, from higher (greater aggregation) to lower levels: order, suborder, great group, subgroup, family, and series. Classification criteria include a wide range of physical, morphologic, mineralogic, and chemical characteristics.

The SCS has established three soil geographic data bases related to the soil taxonomy (Reybold and TeSelle 1989): the Soil Survey Geographic Data Base (SSURGO), the State Soil Geographic Data Base (STATSGO), and the National Soil Geographic Data Base (NATSGO). Each of these geographic data bases is linked to a soils interpretations data base that contains information on the soil components in each map unit, the proportionate extent of each soil component in each map unit, and the properties of each major layer of each soil component. Soil components are typically phases of soil series (i.e., locally defined soil series).

The three soil geographic data bases

have been compiled at different scales. The SSURGO data base contains the soil mapping units that have been delineated in soil-survey reports at scales ranging from 1:15,840 to 1:31,680. The STATSGO data base (Bliss and Reybold 1989) is derived by generalizing the soil-survey maps; where soil-survey maps are not available, soils are classified using geology, topography, vegetation, and climate information together with satellite imagery. The data are compiled at 1:250,000 scale by State. The NATSGO data base provides a general description of soils on the basis of sampling by the National Resources Inventory (U.S. Department of Agriculture 1981). The NATSGO map has been digitized at a scale of 1:7,500,000.

In several recent studies, investigators have attempted to relate observations of regional ground-water quality to soils data. For example, Meeks and Dean (1990) applied a simple leaching index for the pesticide DBCP (1,2-dibromochloropropane) to 381 sections 1 mi^2 each (2.59 km^2) in an agricultural area of central California. A strong correlation was found between detections of DBCP in ground water and sections with high values of the leaching index. The positive detections of DBCP appeared to be more closely related to the leaching index than to pesticide-use records.

Teso et al. (1988) used discriminant analysis with soils data to develop statistical functions to predict the occurrence of wells contaminated by DBCP in part of California. The dominant soil mapping units were identified in 835 sections of 1 mi^2, which contained wells that had been sampled for DBCP. The presence or absence of DBCP in wells in each section was related to the soil mapping units and to associated higher taxa of the soil taxonomy. The resulting predictive functions for the series and family taxa levels were used to predict areas of contamination in nearby Merced County, where DBCP also had been measured in ground water. The overall level of success in predicting the presence or absence of DBCP contamination in Merced County with the series- and family-level soil-survey data was 57% and 61% respectively. Although this is only a moderate improvement over the 50% success rate that would presumably occur by chance, the results indicate that soils information that is related to ground-water quality in one county could increase the probability of identifying likely areas of contamination in another county.

Users of soil-survey information for studies of regional ground-water quality should keep in mind the following factors:

1. Soil mapping units are not part of the soil taxonomy. They are working representations of soil series that are identified on soil-survey maps. Constraints limit the number of delineations that can be made on a map; thus, a soil mapping unit may represent a dominant soil series and lesser amounts of other series having different characteristics.

2. The dominant soil series may not properly reflect the potential for ground-water contamination within a soil mapping unit. For example, even though the dominant soil series in a soil mapping unit may be effective at retarding pesticide movement, leaching of pesticides to ground water may be common in some areas of the soil mapping unit because of inclusion of a soil series with low organic-carbon content. Thus, the most vulnerable soil component within a soil mapping unit may be of greatest interest in relating soils to ground-water quality.

3. The variability in soil properties within a map unit will increase in going from the SSURGO to NATSGO data. For example, a soil association (the map unit for STATSGO) can consist of as many as 21 soil types with widely differing characteristics (Lytle and Mausbach 1991).

4. The usefulness of soil surveys and soil taxonomy for ground-water-quality assessments is limited by the extent to which the criteria used to classify soils and to delineate soil mapping units are relevant to the processes of chemical transport and transformation. In general, those soil characteristics that are used to differentiate among the units in mapping and in classifying soils will be less variable than those that are indirectly used, or not considered, in making soil surveys. As an example, soil organic-carbon content is a principal soil property influencing the sorption and mobility of many organic chemicals in soils. Soil organic carbon is used as a differentiating criterion at several, but not all, taxonomic categories, and soils are not specifically mapped on the sole basis of this criterion.

5. Limited information on soil macropores exists with the soil-survey data. Yet, field studies have demonstrated that the presence of macropores in the vadose zone results in rapid, downward movement of contaminants that can greatly exceed the rates explained by considering standard transport and sorption properties of the vadose-zone materials (cf. Germann 1988). Bouma (1991) provides a review of the influence of soil macropores on contaminant transport in the vadose zone and suggests procedures at different levels of sophistication for describing macropores and predicting their influence on contaminant transport. We will return to complexities introduced by macropores in Chapters 12 and 14 on nitrate and pesticides, respectively.

6. Spatial variability in soils is related to the nature of the parent material from which the soils are formed. Variability would be expected to increase, for example, in going from soils of loess parent material to those of glacial parent material.

7. Soil mapping units are an expression of hypotheses about soil genesis. Landform characteristics relating to water movement are considered less in defining these units. For example, the same mapping unit is often delineated on convex, concave, and linear slopes, resulting in inclusion of areas having large differences in moisture behavior within a given mapping unit (Hall and Olson 1991). Thus, landform position may need to be considered jointly with the soil mapping units.

8. Much of the basis for describing the properties of a particular soil type is based on pedons, which are corings or excavations at selected locations. Pedons tend to be in undisturbed areas; thus, they may not reflect important human-induced modifications to soil properties, resulting, for example, from a plowed layer or hardpan in farmed areas.

9. Most knowledge about the factors that influence contaminant pathways is from examination of the top meter or two of the soil. Likewise, most of the data available for estimating degradation rates of pesticides and other chemicals are from measurements with surface soils (see Chapter 14). These data will likely not be representative of deeper parts of the soil profile or of the intermediate vadose zone.

Intermediate Vadose Zone

The intermediate vadose zone can possess unique characteristics that exert major controls on contaminant transport. For example, the intermediate vadose zone can contain extensive joint systems, which serve as major pathways for water and contaminant movement in fine-textured, low-permeability materials such as glacial till, loess, alluvium, and lacustrine deposits. Pedologists and Quaternary geologists have attempted to develop more unified approaches to characterizing both the root zone and intermediate vadose zone in some areas (Hallberg, Fenton, and Miller 1978; Tandarich et al. 1990); however, few data still exist to characterize the intermediate vadose zone at a regional scale. Because the intermediate vadose zone is less biologically active than the root zone, chemical degradation pathways may become relatively more important. Also, the importance of clay mineral surfaces relative to organic matter for contaminant sorption may increase in going from the root zone to the intermediate vadose zone.

Overlying Low-Permeability Units

It is becoming increasingly obvious that overlying low-permeability units, which are relatively impermeable in a hydraulic context, can be transmissive for contaminant migration of both dissolved-phase (Harrison, Sudicky, and Cherry 1992) and immiscible-phase (Kueper and McWhorter 1991) contaminants. Vertical fractures as small as 10 to 20 microns (on the order of the diameter of a human hair) have been found to be important for contaminant transport to aquifers (Cherry 1991; Harrison, Sudicky, and Cherry 1992). Characterization of the potential for contaminant movement through overlying low-permeability units remains problematic, however, particularly at the regional scale.

Ruland, Cherry, and Feenstra (1991) describe several approaches to study the depth of water flow in fractures in clayey till deposits in an area of southwestern Ontario. These approaches included visual evidence from excavated test pits, seasonal variations in hydraulic head profiles in piezometer nests, and tritium sampling. Cherry (1991) and Ruland, Cherry, and Feenstra (1991) discuss the use of angled boreholes to investigate areas with widely spaced vertical fractures.

Overlying low-permeability units can also greatly influence natural water quality in an aquifer. For example, Fortin, van der Kamp, and Cherry (1991) demonstrated the use of detailed data from an overlying till aquitard in investigating the hydrology and hydrogeochemistry of a glacial drift aquifer in the western glaciated plains region of Canada. Hydrologic and chemical processes within the overlying till were found to control the recharge rate, major-ion chemistry, and isotopic composition of water in the aquifer. The aquifer acted largely as a chemically inert lateral transmission and mixing zone for reaction products transported from the overlying till by advection and diffusion.

Summary Remarks on Characterizing the Vadose Zone

Overall, much remains to be learned about how to relate vadose-zone properties to ground-water quality at the regional scale. Part of the difficulty lies in the limited data available to characterize the vadose zone relative to the large spatial variability of vadose-zone properties. Inadequate knowledge of vadose-zone processes is also a significant limitation. Even at the site scale, currently used methods are inadequate to describe processes such as preferential flow, nonlinear and nonequilibrium sorption, and transport of contaminants on colloidal organic carbon.

The relations between the vadose zone and ground-water quality are likely to be easiest to observe in the upper part of water-table aquifers. As part of such studies, it may be advantageous to make a special effort to characterize the vadose zone near sampled wells to supplement the information available from existing spatial data bases. A generalized surficial geology map also might be beneficial in characterizing vadose-zone properties.

GROUND-WATER-FLOW SYSTEMS

Ground-water-flow systems consist of recharge and discharge areas separated by transition zones, with the spatial distribution of the recharge and discharge areas commonly controlled largely by topography and geologic heterogeneity. Some aspects of ground-water-flow systems particularly relevant to regional ground-water-quality studies are discussed here as background for the examples cited in later chapters.

Scales of Flow Systems

Ground-water velocities generally are low. Shallow ground water typically moves less than a few centimeters to as many as several meters per day, although water can move much faster in karst or fractured-rock terraines. Deep circulating ground water may move very slowly, sometimes as little as a few meters or less per century.

Figure 2-1 is a schematic illustrating concepts of flow-system scale and the relative time required for ground water to move along different flow paths from recharge to discharge areas. Clearly, in situations such as illustrated in Figure 2-1, the position

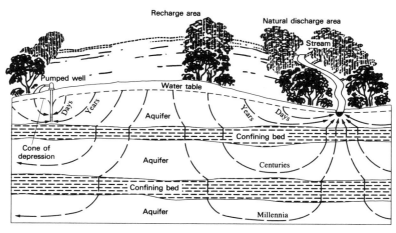

FIGURE 2-1.   Rates of ground-water flow. (*After Heath 1989.*)

within the flow system that is sampled will have a strong bearing on the types of contaminants observed, their concentrations, and seasonal variability. Depending on their entry point to the ground-water system, contaminants may move only a short distance before being discharged, or they may become part of a larger flow system and, with time, be detected at great depths and over a wide area.

Toth (1963) observed that, in many cases, ground-water-flow systems can be qualitatively subdivided into paths of local, intermediate, and regional flows. He defined a local flow path as a relatively shallow path extending from a recharge area to an adjacent lower discharge area, sometimes hundreds of meters or less in length. Intermediate flow paths include at least one local flow system between their respective points of recharge and discharge and are somewhat deeper and longer than local flow paths, perhaps on the order of one or a few kilometers. Regional flow paths are the longest and deepest paths. They begin at the major ground-water divide and traverse the entire region to the major drain.

The development of different scales of flow systems can be significantly affected by the areal patterns of ground-water recharge as illustrated in Figure 2-2. This simple example consists of a two-layered ground-water system with relative permeabilities of 1 and 10 for the top and bottom layers, respectively, and with impermeable boundaries on the bottom and sides. The top configuration of the water table reflects different recharge conditions in the two cross sections. With a linear water table and recharge mainly at the right end of the system (Figure 2-2a), a relatively smooth regional flow system develops. With a more undulating water table (Figure 2-2b), a much more complex pattern of small, local flow systems has developed.

Geologic Controls on Flow Patterns

Ground-water-flow patterns and velocities are highly sensitive to the hydraulic conductivity of the medium. Some knowledge of this property is important in developing a regional sampling program and in considering the potential areal extent of contamination from different types of sources. Unfortunately, hydraulic conductivity can vary by more than 12 orders of magnitude (Heath 1989), complicating its spatial characterization. The heterogeneity in hydraulic conductivity can be viewed as occurring in a hierarchy of scales, with the types of heterogeneity being as varied and complex as the

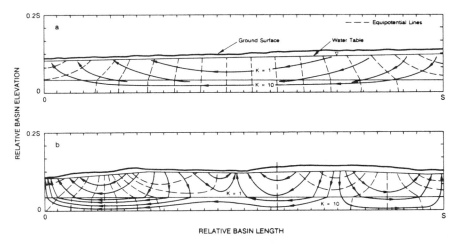

FIGURE 2-2. Effects of the recharge rate, as reflected by the configuration of the water table, on regional ground-water-flow patterns. (*After Freeze 1969.*)

processes by which the geologic materials were deposited (Gillham et al. 1983).

At the regional scale, stratigraphic variations that represent significant depositional events or structural alterations of the original deposits can exert major controls over ground-water flow and quality. A simple example of the effects of broad-scale heterogeneity on regional ground-water-flow systems is illustrated in Figure 2-3. The only difference between cross sections 2-3a and 2-3b is the pattern of geologic layering, defined in terms of the relative hydraulic conductivities shown. With simple layering (Figure 2-3a), a regional flow system develops that discharges at one end of the section. With a high-conductivity basal aquifer of limited extent (Figure 2-3b), recharge is directed toward the higher-conductivity layer but then diverges at the downgradient end, resulting in a ground-water discharge zone about midway along the cross section. (Note: This latter effect is accentuated by the specified constant head boundary used for the water table in the simulation.)

It may be useful to develop conceptual models to characterize spatial trends in hydraulic conductivity from limited field data. An example of such an approach is given by Anderson (1989), who applied hydrogeologic facies models to describe the geometry and large-scale spatial trends in hydraulic conductivity in glacial and glaciofluvial sediments. Although such conceptual models commonly are limited in their ability to make predictions at a specific site without substantial fine-turning to local data, they can be helpful to identify predictable sequences of deposits at the regional scale, type settings that might be studied, and the extent to which different settings might have greater or lesser variability in water quality. These conceptual models can be refined iteratively as data are collected and new information obtained. The utility of conceptual models of ground-water-flow systems for ground-water modeling is discussed by Anderson and Woessner (1992).

In addition to stratigraphic features, geologic structures can exert major controls on ground-water-flow systems and their water quality. An example are the effects of high-angle faults on salinity in the Edwards aquifer of central Texas (Maclay and Small 1984). The faults, which in some places act as barriers to ground-water flow, have prevented flushing of saline water from parts of the aquifer downdip of the faults, causing a marked change in water quality, referred to locally as the "bad-water" line. Updip of the

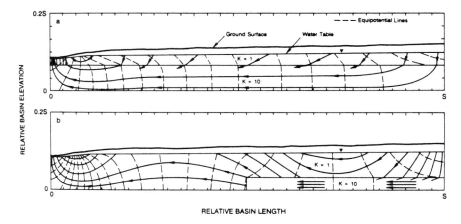

FIGURE 2-3. Effects of broad-scale heterogeneity on regional ground-water-flow patterns. (*After Freeze 1969.*)

bad-water line, the total dissolved solids is less than 1,000 mg/L (milligrams per liter); downdip of the line, total dissolved solids increase rapidly to about 9,000 mg/L.

The previous examples illustrate the effects of broad- or intermediate-scale heterogeneity. At the other end of the scale spectrum are the small-scale variations in hydraulic conductivity that exist at scales of millimeters to meters. Although generally not a major consideration in regional flow, small-scale heterogeneity is typically considered the major cause of dispersion at the field scale and results in contaminants spreading over a greater volume of the medium than would be predicted on the basis of ground-water-flow lines in an assumed homogeneous medium.

Effects of Transient Flow

At a regional scale and over several months or years, a given quantity of infiltrating water can be viewed as a somewhat uniform movement of water through the vadose zone to the water table. Much of the apparent uniformity, however, is due to the scale of observation which can smooth out local differences in recharge rates and to the lateral redistribution of water in the saturated zone with time.

At the local scale, the areal and temporal distribution of recharge are not uniform. In fact, highly dynamic local flow conditions can be superimposed on a larger regional ground-water-flow system (e.g., Figure 2-2b). Because the altitude of the water table is constantly changing in response to recharge, it is the most dynamic boundary of the ground-water-flow system in both time and space.

Topographically high areas commonly are recharge zones for underlying ground-water-flow systems, and topographically low areas commonly are ground-water-discharge zones. In contrast, recharge in areas of low-to-moderate relief may be focused where the thickness of the vadose zone is least—directly adjacent to surface-water bodies or at depressions in the landscape. The volume of recharge at these depressions will vary areally, depending on the size of the depression, the permeability of the soil, and the amount and type of vegetation. The volume of recharge at a depression also will be highly variable in time, depending on precipitation, antecedent soil-moisture conditions, and evapotranspiration demands. Further, in areas where deep soil frost forms in winter, most of the overland runoff during the spring can concentrate in land-surface depressions before the frost melts (Lissey

1971). This concentration of water can result in relatively little ground-water recharge beneath topographic highs.

Focused recharge causes the formation of transient, local flow systems that can result in complex movement of water in the upper-most parts of the ground-water system. Water in these flow systems may move first in one direction as a result of localized recharge, and then move in the opposite direction as the water-table mound associated with that recharge dissipates. These local flow systems can last from a few hours to several months in more permeable settings, and for years in some less permeable settings. Examples of complex variations in flow direction caused by focused recharge near surface-water bodies are presented by Winter (1983).

Understanding the effects of local-scale flow systems takes on increased significance as attention is directed toward shallow-ground-water quality. The possibility of small local flow systems forming and dissipating within the ground-water system introduces a number of complexities in interpreting ground-water-quality data. Because of possible flow reversals, water sampled from nearby wells with similar completion depths can have significantly different residence times. Likewise, water samples from a given well might be largely from a small, local flow system at one time and from a regional flow system at other times.

Although there are significant limitations to the extent to which local-scale flow systems can be defined by regional-scale water-quality studies, the possible effects of these systems should be considered in selection of sites, timing of sampling surveys, and design of local-scale networks for shallow-ground-water quality. For example, topographic setting can be a key factor when selecting a regionally distributed set of wells to be sampled. Likewise, for local-scale networks, extensive synoptic and seasonal ground-water-level measurements may be needed to evaluate the possible effects of flow-path dynamics and gradients on water quality.

Surface-Water–Ground-Water Relations

Seepage to and from streams commonly is a major component of a basin's internal hydrologic budget. Ground-water discharge can make up most of the streamflow during dry months. On the other hand, a substantial percentage of total recharge to ground water in some arid areas can be derived from stream leakage. Many perennial streams lose water to the subsurface over some portion of their total reach, while gaining ground-water discharge elsewhere. Gaining and losing reaches can be influenced by human activities such as irrigation or ground-water pumpage. Also, the locations and extent of recharge and discharge can have significant seasonal variations. For instance, a stream can be a losing stream at high-runoff stages but a gaining stream at low flow. The discharge of ground water to a gaining-stream reach occurs as saturated seepage and the stream and aquifer are hydraulically connected (Figure 2-4A). On the other hand, a losing-stream reach can be hydraulically connected (Figure 2-4B) or disconnected (Figure 2-4C) from the aquifer.

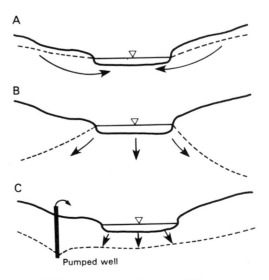

FIGURE 2-4. Schematic diagram of (A) gaining stream, (B) hydraulically connected losing stream, and (C) hydraulically disconnected losing stream. (*After Spalding and Khaleel 1991.*)

The following discussion centers on interactions of ground water with streams to illustrate some basic concepts. However, in addition to streams, the quantity and quality of other surface-water bodies, including lakes (cf. Anderson and Bowser 1986) and wetlands (cf. Winter 1988) can be greatly affected by ground water, and vice versa.

In a very simplified conceptualization, streamflow has two components: (1) baseflow (or delayed flow), which is derived primarily from seepage of ground water to the stream, and (2) direct runoff (or quick flow), which can result from several hydrologic processes during and soon after a storm event. Because ground water generally contains higher concentrations of natural solutes than surface runoff, an inverse relation commonly exists between discharge and solute concentrations in streams. Many complications affect this simple model, however. For example, minimum solute concentrations tend to occur after stream discharge has peaked. Moreover, the increase in solute concentration during the falling stage may be relatively slow. Thus, solute concentrations during the falling stage can be substantially lower than those at the same discharge during the rising stage. These effects occur in part because of dynamic interactions between the river flow and water in the banks and streambed. Some of the more dilute runoff water during the rising stages of streamflow can infiltrate into the banks and streambed to return later to the stream as the stream stage declines. This process can prevent the stream from returning to pre-event baseflow conditions for some time. Another complication, as noted in the previous section, is that recharge commonly is quickly focused adjacent to surface water after the beginning of a precipitation event, resulting in time-varying near-stream saturated conditions and the rapid movement of ground water and displaced soil water from different source locations to the surface-water system.

Stream and ground-water interactions affect the distribution of contaminants introduced by humans in several ways. First, ground water can be a principal medium through which contaminants are transported from the land surface to streams. The importance of this function is indicated by the frequent detection of water-soluble pesticides and high nitrate concentrations in streams during baseflow conditions. Second, bank storage of surface water during high-flow conditions, and its subsequent slow release to the stream, can be a source of contaminants during low-flow periods. Third, large ground-water withdrawals can result in the movement of contaminated water from streams into ground-water systems for later release downgradient in the stream or at a well.

An example of the third effect is illustrated by ground-water quality in an alluvial aquifer along Fountain Creek between Colorado Springs and Pueblo, Colorado (Cain, Helsel, and Ragone 1989). Figure 2-5 shows the concentration of nitrite plus nitrate in the aquifer as a function of miles downgradient from Colorado Springs. The line shown in Figure 2-5 represents locally weighted, scatterplot smoothing (Cleveland 1979) of the concentrations. Urban land is located mostly in the northern (upgradient) one third of the alluvial aquifer area; agricultural land is located mostly in the southern (downgradient) two thirds of the aquifer area. Thus, based on land use, the opposite relation of that shown in Figure 2-5 might be expected. The large concentrations of nitrate in water from wells in the upgradient end of the study area appear to have resulted from extensive pumping from the alluvial aquifer. The pumping induced aquifer recharge from Fountain Creek that contained large concentrations of nitrogen from upstream wastewater discharges (annual induced recharge is equivalent to about 10–15% of the volume of water in storage in the aquifer). Similar downgradient decreases were observed for detergents. Thus, the land-use effects on ground-water quality in this narrow alluvial aquifer appear to be dominated by surface-water–ground-water interactions as op-

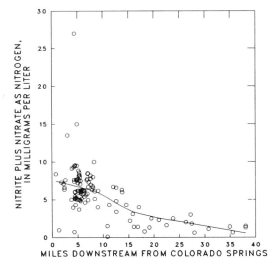

FIGURE 2-5. Decrease in concentration of nitrite plus nitrate in a downgradient direction in water from the Fountain Creek alluvial aquifer, Colorado. (*From Cain, Helsel, and Ragone 1989.*)

posed to leaching through the vadose zone.

Hydrologists have developed a number of methods for analyzing stream-aquifer interactions. Useful methods include minipiezometers and seepage meters (Lee and Cherry 1978; Winter, LaBaugh, and Rosenberry 1988), seepage runs (measurements of baseflow discharge at several locations along a stream during a short time interval), field tracer experiments, and analysis of recession hydrographs. Chemical indicators of groundwater contributions to surface water include concentrations of individual dissolved inorganic constituents and total dissolved solids (Freeze and Cherry 1979, pp. 221–24), radon (Lee and Hollyday 1987), volatile organic compounds (Vroblesky and Lorah 1991), and various isotopes (see Chapters 10 and 11). Careful stream-temperature measurements also can be used to indicate whether a stream is gaining or losing water. The temperature of shallow ground water tends to reflect the mean annual temperature of the region and to fluctuate less than the temperature of surface water. During the warm season, surface water tends to be warmer than shallow ground water, and the reverse is true during the cold season.

The effects of stream-aquifer interactions can be analyzed at the local scale using transects of wells installed along flow paths to streams. Considerable care is needed in selecting sampling sites along such transects. It may be advisable to operate a water-level monitoring network for several seasons before selecting sites for chemical analysis. Although commonly visualized as perpendicular to stream channels, flow paths typically do not intersect the stream channel at right angles and are three-dimensional in nature. As an example, a schematic flow path near a gaining stream is shown in Figure 2-6. The position and shape of the flow path shown would be modified by heterogeneity and anisotropy of the earth materials comprising the aquifer in actual systems. Also, ground water can flow under, and hence bypass, streams.

A recent local-scale study of stream-aquifer interactions is presented by Gburek and Urban (1990), who investigated the effect of the shallow weathered fracture layer on ground-water movement to streams in east-central Pennsylvania, using transects of wells across two valleys. Under wet conditions, the stream at each section served as a point of ground-water discharge, while under dry conditions the more upland section showed no ground-water discharge to the land surface. Instead, subsurface flow appeared to occur beneath and parallel to the channel controlled by a discharge point at some downstream location. An implication of this is that, in the upper part of the watershed, the contributions of ground water to stream-water quality during dry periods may not reflect the immediately adjacent land use, but instead reflect the land use upgradient of a point further upslope in the stream. That is, the contributing land-use areas to a point in the intermittent part of a stream vary seasonally as the stream-aquifer connections change. Further complications occur for perennial reaches of a

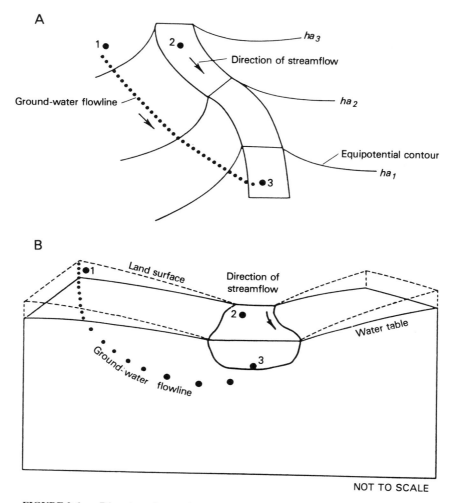

FIGURE 2-6. Direction of ground-water movement near a gaining stream (A) in plan view and (B) in three dimensions. The equipotential contours (ha_1, ha_2, and ha_3) shown in A are omitted from B because in three dimensions the contours would be complex surfaces. (*After Harbaugh and Getzen 1977.*)

stream that alternate between gaining and losing conditions.

Particular attention should be directed to sampling ground-water flow near and within the streambed in carrying out local-scale studies of stream-aquifer interactions. One reason is that much of the ground-water flow to a stream may be largely vertical. A second reason is that large changes in water quality can take place in the last few meters of ground-water discharge to a stream. This latter concept is expanded on here.

The water-saturated region immediately beneath and adjacent to a stream in which active exchange takes place between surface and subsurface waters has been referred to as the hyporheic zone. The hyporheic zone is capable of supporting a very rich biota, and it can be characterized by large concentration gradients in oxygen and dissolved organic carbon (Triska et al. 1989).

The near-stream environment, or hyporheic zone, can be a significant source or sink of nutrients. For example, denitrification in the near-stream environment can greatly reduce nutrient loads of ground water to streams or lakes (cf. Robertson, Cherry, and Sudicky 1991). Potential factors affecting the importance of the denitrification process include the presence or absence of peat layers or organic matter imbedded in the streambed materials. On the other hand, Triska, Duff, and Avanzino (1990) found that the mixing of stream water containing dissolved oxygen with ground water containing ammonium supported hyporheic nitrification in a small, gravelcobble bed stream in Redwoods National Park, California. The near-stream environment also can affect the ground-water contribution of organic contaminants to surface water. For example, Fusillo et al. (1991) found evidence of near-stream degradation of TCE to DCE near Picatinny Arsenal in New Jersey.

The spatial and temporal variability of areas where ground water discharges to streams and where streams discharge to ground water are great and may be difficult to predict. For example, in studies in three States of stream reaches with net discharge of ground water to surface water, Harvey (1991) found localized zones of seepage of surface water to ground water. Surface water entered subsurface sediments at these locations, mixed with subsurface water, and discharged at rates exceeding the areal net ground-water discharge rate.

Overall, some key features of the near-stream environment are (1) ground-water quality measured even within a few meters of a stream may be quite different from the quality of subsurface water actually contributing to the stream as a result of significant chemical and biologic changes occurring in the near-stream environment, (2) the first interaction between surface water and ground water may take place in the subsurface, and (3) zones of surface-water seepage to ground water may be interspersed with zones of ground-water seepage to surface water in a complex pattern. The hyporheic zone can also have ecological significance and support a host of insects and microorganisms.

Contributing Areas to Pumping Wells

Much of the current interest in regional ground-water quality centers on the effects of human activities at the land surface. Thus, the *contributing area* of a well is commonly of interest, where this term generally refers to the land area that has the same horizontal extent as that part of the ground-water system from which flow is diverted to the well. Recharge that enters an aquifer through the contributing area of a well will eventually be discharged by the pumping well. Often, the contributing area is defined on the basis of a time-of-travel criterion, such as the recharge area contributing water to a well within "X years."

It is important to distinguish the contributing area of a pumping well from the *area of influence*, which is the land area that directly overlies and has the same horizontal extent as the part of the water table (or other potentiometric surface) that is perceptibly lowered by the withdrawal of water. The distinction between these terms is illustrated in Figure 2-7. The zone of diversion (or *capture zone*) shown in Figure 2-7 is simply the volumetric extension of the contributing area (i.e., that portion of the ground-water system through which ground water flows on its path to a pumping well). Where the water table has a gradient, as in most situations, the contributing area will extend a greater distance on the upgradient side than on the downgradient side of the well. The contributing area may not include the well itself, particularly for wells with open intervals near the bottom of an aquifer.

The hydrogeologic factors that affect the flow field around a pumping well also affect the contributing area to the well, depending on specific conditions at the site. These include (1) rate and duration of pumping, (2) aquifer properties (transmissivity, storage coefficient, specific yield), (3) proximity of the pumping well to aquifer boundaries, (4) spatial and temporal variations in recharge, (5) partial penetration of the pumping well,

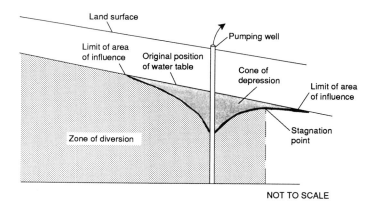

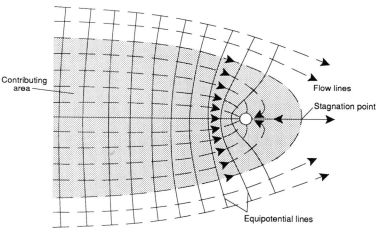

FIGURE 2-7. Sketch illustrating contributing area, area of influence, and zone of diversion for a pumping well. (*After Morrissey 1989.*)

and (6) presence of extensive confining layers (Morrissey 1989).

Analytical and numerical models have been widely applied in recent years in the analysis of contributing areas (capture zones), particularly to define wellhead protection areas (zones) around water-supply wells for the purpose of aquifer management. The methods vary greatly in their level of sophistication, simplifying assumptions, and data needs. Most methods involve hydraulic analysis and neglect solute-transport phenomena, such as sorption and dispersion.

Analytical flow models use a well-hydraulics equation that is deemed appropriate for the particular hydrogeologic setting to compute the distribution of drawdown surrounding the pumping well. Definition of the contributing area is then based on subtraction of the simulated drawdown distribution from a measured or assumed regional hydraulic-head distribution. The analytical models commonly require many simplifying assumptions regarding the nature of ground-water flow. The Theis equation (Theis 1935) is an example. The effects of multiple wells and simple aquifer geometry and boundary conditions can be handled using image-well theory and superposition.

Numerical flow models may be two- or three-dimensional and allow incorporation of complex aquifer geometry and boundary conditions, heterogeneous and anisotropic conditions, multiple partially penetrating wells, and, possibly, three-dimensional flow in the aquifer and confining layers. Particle-tracking algorithms use the simulated head distribution from the numerical flow model to trace out fluid-particle pathlines, by tracking the movement of imaginary particles placed in the flow field. Recent papers discussing the use of particle-tracking models include Morrissey (1989), Buxton et al. (1991), and Bair and Roadcap (1992) and a chapter in the text by Anderson and Woessner (1992). Example applications of particle-tracking models to define relations between land use and ground-water quality are given later in this chapter and in Chapter 24. Buxton and Modica (1992) provide a useful illustration of how the distribution of ground-water travel times calculated from a 2-D finite element model can be used to explain the distribution of nitrate concentrations observed in samples from existing wells along a cross section through Long Island, New York.

Effects of Large-Scale Ground-Water Development

Ground-water development can have a pronounced effect on ground-water-flow systems, and on the quality of water in them. This is perhaps best illustrated through examples, adapted from Johnston (1988), for two of the major aquifer systems in the United States.

Central Valley Aquifer System

The Central Valley of California is an area of long-term, intensive development of ground water. The aquifer system of the Central Valley is composed primarily of alluvial deposits with minor amounts of volcanic deposits. A comparison of regional hydrologic budgets for predevelopment and development (1961–77) conditions is shown in Figure 2-8A.

Before development, much of the recharge to aquifers in the Central Valley came from infiltration of streamflow at the heads of alluvial fans in the foothills that surround the valley. The ground water then moved toward the center of the valley and discharged as evapotranspiration and seepage to streams. This situation changed as ground-water pumpage increased steadily throughout the 1940s, 1950s, and 1960s, primarily in the San Joaquin Valley (see Figure 22-1 for location of San Joaquin Valley). Ground-water development during this period caused water-level declines of tens to hundreds of meters. Importation of surface water, beginning in the late 1960s, led to further increases in the application of irrigation water, a reduction in ground-water pumpage, and rising ground-water levels.

During the period of large declines in ground-water levels in the Central Valley, particularly in the San Joaquin Valley, subsurface clays were compacted, resulting in a loss of ground-water storage and a lowering of the land surface (Ireland, Poland, and Riley 1984). Although the vertical permeability of the clays has probably been decreased by compaction, the vertical hydraulic connection across the aquifer system has actually increased as a result of the construction of about 100,000 wells with long sections of perforated casing (Williamson, Prudic, and Swain 1987).

The extensive ground-water development greatly altered the ground-water-flow system. As a result, the previous discharge area of the aquifer in the San Joaquin Valley is now largely a recharge area. Most recharge is now supplied by irrigation return flow (of imported surface water) rather than infiltration of streamflow in upland areas; most ground-water discharge is now pumpage.

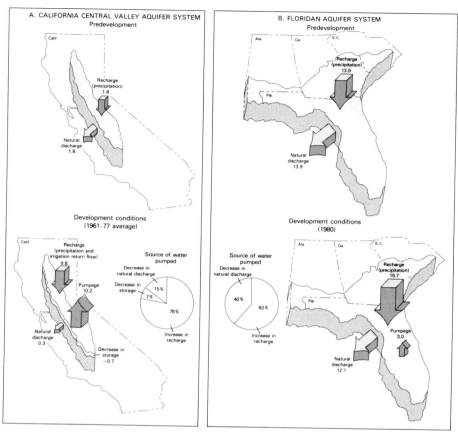

FIGURE 2-8. Comparison of aquifer response before and after development in (A) the Central Valley aquifer system of California, and (B) the Floridan aquifer system of the southeastern United States. Values are in billion gallons per day. (*After Williamson, Prudic, and Swain 1987; Bush and Johnston 1988; Johnston 1988.*)

The estimated recharge rate in the Central Valley during 1961–77 was more than five times the predevelopment recharge rate (see Figure 2-8A). Some of the effects of these changes on water quality are in the following list.

- The changes in the ground-water-flow system can redirect poor-quality water in some areas toward pumping centers. This redirection includes the vertical movement of saline water found beneath the freshwater in the Central Valley and poor-quality water found above, or within, the fresh ground water.
- In parts of the San Joaquin Valley, selenium is being leached into ground water from seleniferous soil by applied irrigation water (see Chapter 22).
- Because of the increase in the recharge area and rate of recharge in the Central Valley, chemicals applied at the land surface have a greater potential to contaminate ground water.
- The downward hydraulic gradient resulting from pumping, and the increased hydraulic connection among aquifer layers provided by multilayer-screened wells, creates an opportunity for poor-quality water to move from shallow aquifers into deep aquifers.

Floridan Aquifer System

Compared to the Central Valley aquifer system, distinctly different effects of ground-water development are illustrated by the Floridan aquifer system in the south-eastern United States. The Floridan aquifer system is the principal source of public, industrial, and agricultural water supply in the area shown in Figure 2-8B, except in south Florida where it contains saline water. The Floridan contains thick beds of highly permeable limestone, and transmissivity generally is very large. Large amounts of rainfall, with little surface runoff, provide abundant recharge.

Ground-water development has not altered the overall flow system of the Floridan to the extent that it has in the Central Valley of California. To some degree, the differences are a reflection of differences between the humid southeastern and the arid western United States.

Overall, pumpage from wells is balanced by increased recharge from and decreased discharge to surface-water bodies, and the change in ground-water storage has been negligible (see Figure 2-8B). However, the pumpage is distributed unevenly throughout the Floridan aquifer system, and large withdrawals have caused long-term water-level declines in some areas (Bush and Johnston 1988).

Figure 2-9 shows a generalized hydrogeologic section that extends from the outcrop area of central Georgia to the coast and passes through Brunswick, Georgia, which is a center of heavy pumpage. Before development (Figure 2-9A), the flow system was comparatively simple; recharge occurred directly in the outcrop areas or by downward leakage further downgradient, and discharge occurred as upward leakage near the coast. Postdevelopment changes to the flow system (Figure 2-9B) include formation of a shallow cone of depression at Brunswick with the water level in the upper Floridan currently below sea level at the center of the cone. The vertical hydraulic gradient is downward, making the area of the cone a potential recharge area; however, recharge by vertical leakage is impeded by a thick layer of clay overlying the Floridan (Krause and Randolph 1989). The thick clay beds, combined with the great distance to the outcrop recharge area, help provide natural protection in this area against the infiltration of contaminants introduced at the land surface. However, ground-water quality in the upper Floridan aquifer in Georgia has been degraded by upward migration of saline water induced by pumping (Wait and Gregg 1973).

Changes in water quality resulting from changes in the regional flow system are less dramatic for the Floridan aquifer system than the Central Valley aquifer system. Major factors affecting degradation of water quality in the Floridan aquifer are the land-use practices, the proximity of the aquifer to the land surface, and the presence or absence of confining units or breaching by sinkholes and other karst features. For example, the pesticide EDB has been found in numerous samples of ground water in central and northwest Florida.

Fractured-Rock and Karst Systems

In recent years, the hydrogeologic character of "multiple-porosity" systems, such as fractured-rock and karst aquifers, has received increasing attention. In some settings (e.g., shallow fractured bedrock and well-developed karst), these aquifers can be quite vulnerable to contamination.

Fractured-Rock Systems

Many rock masses consist of rock blocks bounded by discrete fracture planes. The blocks may be porous, permeable rock, such as sandstone, or may have low porosity and very low permeability, such as granite. The ratio of the permeability of the rock blocks to the permeability of the fratures determines the significance of fractures in a given flow system (Gale 1982).

Ground-water studies in more permeable rocks have generally assumed that the rock block–fracture system is a porous medium,

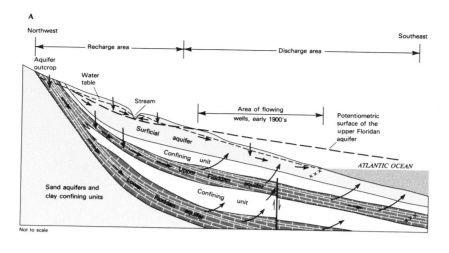

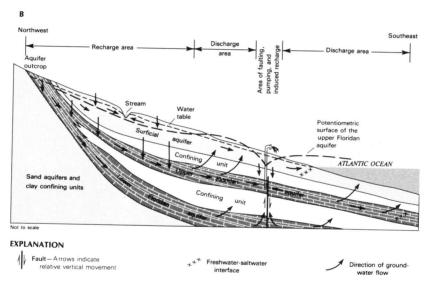

FIGURE 2-9. Comparison of ground-water-flow conditions (A) before development (pre-1900) and (B) after development (early 1980s) in the Floridan aquifer system, southeastern Georgia. (*After Krause and Randolph 1989; Johnston 1988.*)

and little attempt has been made to distinguish the contribution to flow of the fracture system from that of the rock blocks. In metamorphic and granitic rocks, the primary hydraulic conductivity of rock blocks generally is less than 10^{-8} cm/s (Freeze and Cherry 1979), and significant flow can occur only through the fracture system. Hence, interest has focused on the effect of fractures on flow in these systems. Likewise, as discussed previously in the section "Overlying Low-Permeability Units," fracture flow in shales and other low-permeability argillacious rocks is of considerable contemporary interest.

In the most general case, fracturing adds secondary porosity to some original porosity. Pores within the rock blocks have length and width of similar dimension in a highly tortuous pattern of interconnection, while the fractures provide more continuous openings with lengths far in excess of their widths (Shapiro 1987). The discontinuities in the fractured media can be conceptualized as joints, fracture zones, and shear zones (Gale 1982). *Joints* are individual fractures that are usually discontinuous in their own plane. When several closely spaced families (sets) of joints are present, they can form a highly interconnected three-dimensional network for flow. Within a given rock or sediment, a relatively large number of joint sets can be present, each with its own unique orientation in space. *Fracture zones* are zones of closely spaced and highly-interconnected discrete fractures that are generally not filled with clay or other material. *Shear zones* provide a large-scale discontinuity with a permeability that can be either higher or lower than the rock mass, depending on filling materials, age, and stress.

Prediction of fluid movement and chemical transport in fractured rock is a complex task because of the difficulty in characterizing the spatial variability of hydraulic properties over the various length dimensions. This problem is common to all subsurface flow regimes, but it is particularly severe in fractured rock because of extreme spatial variability and abrupt spatial changes in hydraulic properties.

The presence of fractures in geologic units adds a significant complexity to understanding fluid flow and contamination of ground water. The fractures can represent avenues of high permeability within an otherwise low-permeability matrix. Thus, fractures have the capacity to transport fluid and contaminants rapidly over large distances, and the majority of fluid and chemical constituents can be conducted through a small volume of the rock. Even in cases of highly fractured rock, the formation is not necessarily analogous to a porous media, because not all fractures are capable of conducting fluid.

The relatively large number of geologic processes that can cause fracturing (e.g., tectonism, weathering, glacial stresses, thermal stresses, etc.), coupled with a tendency for older fractures to be propagated into unfractured units, means that fractures are a dominant element controlling fluid migration in all kinds of geologic settings (National Research Council 1990). Thus, many aquifer systems are to some extent controlled by fracturing, whether or not this fact is explicitly recognized.

Theoretical interpretations of fluid movement and chemical transport have progressed significantly in the relatively short history of fractured-rock hydrogeology (cf. Shapiro 1987; National Research Council 1990). It is difficult, however, to quantitatively transfer the information at this time to regional studies of ground-water quality.

Karst Aquifers

The picture of fractures presented so far is one end member in a hierarchy of multiple porosity systems. In the case of soluble bedrock, such as limestone, dolostone, or evaporites, conduit flow can develop as original fracture systems are enlarged by solution. In many situations, networks analogous to a river system develop in which smaller tributaries supply water to a succession of larger and larger conduits. Greater difficulty in understanding this conduit network than a stream network results because it is hidden from view underground.

Water moves at highly variable rates through karst aquifers, rapidly through solution-enlarged fractures and conduits, and more slowly through fine fractures and pores. The interaction and relative flow rates of water in conduits and pores can lead to complex ground-water-flow behavior.

In many regions with well-developed karst features, large springs act as outlets or drains for regional ground-water movement. As a result, the rate and chemical composition of spring discharge reflect basin-scale water and solute transport in the aquifer system. The water discharging from the spring outlet is a combination of water draining from the pores and relatively dilute storm-derived water associated with recent rainfall events.

Tracing methods and water chemistry are widely used for studying karst aquifers. Extensive tracer studies combined with field work to locate points of recharge and discharge have been used to estimate the recharge areas of springs, rates of ground-water movement, and the water balance of aquifers. Variations in parameters such as temperature, hardness, Ca–Mg ratios, Ca and Mg saturation indices, and carbon-14 have been used to describe sources and rates of ground-water movement, differentiate rapid and slow karst ground-water components, and compare spring-flow characteristics in different regions.

Rapid transport of contaminants within karst aquifers and to springs has been observed in many locations. Because of the rapid movement of water in many karst aquifers, water-quality problems that may be localized in other aquifer systems can become regional in nature in karst systems. Unique aspects of karst aquifers and special considerations for ground-water monitoring are reviewed in Chapter 19.

USES OF GEOCHEMICAL DATA IN REGIONAL CONCEPTUALIZATIONS

The kinds of ions in solution and their concentrations result from chemical processes relating to the lithology and hydrologic flow pattern of a particular hydrologic system. Thus, an analysis of geochemical data, both existing and newly collected data, is useful in developing a regional hydrogeologic framework. A few commonly used approaches for the display and analysis of geochemical data will be briefly reviewed with examples of their applications to regional interpretation of ground-water quality. The related topics of geochemical models and environmental isotopes are covered separately in Chapters 9 to 11. Descriptions of ground-water quality associated with different lithologies are given by Freeze and Cherry (1979), Matthess (1982), and Hem (1985).

Map and Graphical Displays

Simple Map Displays

One of the simplest means of displaying geochemical data is a dot map of the quality of ground water prepared by entering numbers or symbols at well and spring locations to represent concentrations of constituents. Through shading, color coding, or by their size, dots displayed at the sampled locations can be used to indicate different concentration ranges. In general, dot maps have limited ability to illustrate spatial patterns in water quality. If many analyses exist, and the ground-water body exhibits relatively distinct patterns in water quality, then maps showing isopleths of chemical characteristics within certain formations can be constructed. Care must be exercised in all types of map presentation to consider possible depth variations in ground-water quality.

Stiff Patterns

A common method of presentation of geochemical data is the Stiff (1951) pattern. A polygonal shape is created from the plotting of chemical values along horizontal axes which are separated from each other by

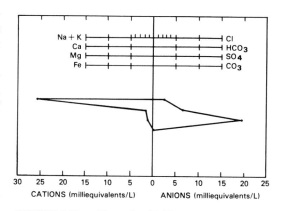

FIGURE 2-10. Example of Stiff pattern.

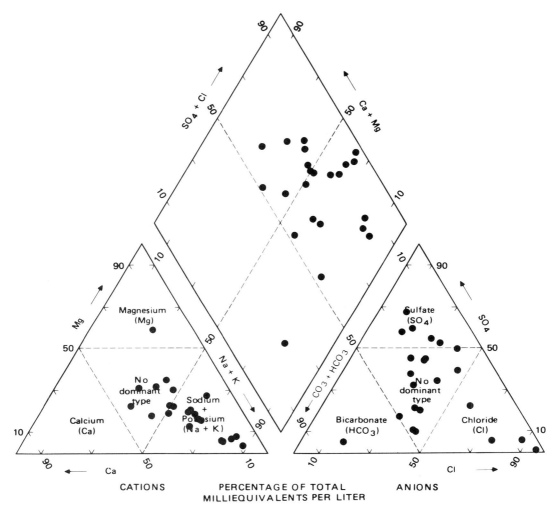

FIGURE 2-11. Example of a trilinear diagram showing chemical composition of ground water in confined zone in northwestern part of San Joaquin Valley, California. (*After Dubrovsky et al. 1991.*)

equal distances and are divided by a vertical center line. Cations are plotted in milliequivalents per liter on one side of the vertical axis and anions on the opposite side. An example Stiff pattern is shown in Figure 2-10. The use of the lower horizontal bar with iron and carbonate is optional, because the concentrations of these two constituents can be close to zero. Sometimes other constituents, such as nitrate, are shown on the lower horizontal bar. Stiff patterns facilitate rapid comparisons among water from different sources as a result of their distinctive graphic shapes. They are useful for illustrating chemical composition in hydrogeologic cross sections and as a symbol on maps. They are not, however, well suited for graphical presentation of large numbers of analyses.

Trilinear Diagrams

A commonly used method of graphical presentation is the trilinear diagram (Hill 1940; Piper 1944). Several variants of the trilinear diagram exist; a frequently used version is shown in Figure 2-11. Two triangles are displayed: one for major cations and the other for major anions. The composition of the water with respect to cations is indicated by a point plotted in the cation triangle, and the composition with respect to anions by a point plotted in the anion triangle. The values are presented as percentages of total milliequivalents per liter of the cations or anions depicted. The points in the two triangles corresponding to a single sample are projected to the diamond-shaped field. Sometimes only the diamond-shaped field is shown without the triangles, and is referred to as a quadrilinear diagram.

Trilinear diagrams permit the cation and anion composition of many samples to be represented on a single graph in which major groupings or trends in the data can be discerned visually. Because the concentrations are represented as composition percentages, waters with very different total concentrations can have identical representations on the diagram. Trilinear diagrams are useful to illustrate how the ionic composition of water varies among different hydrogeologic settings, and they can be used to define hydrogeochemical facies (Chapter 5) on the basis of the dominant ions. Furthermore, trilinear diagrams are sometimes convenient for showing the effects of the mixing of two waters from different sources; the mixture of two different waters will plot on the straight line joining the two points, if the ions are not added or removed from solution by processes such as ion exchange, precipitation, or solution of salts.

Durov Diagrams

The Durov diagram (Chilingar 1956; Zaporozec 1972) is a method of presentation similar to the trilinear diagram. An example is shown in Figure 2-12 for ground water in two parts of the Carson River basin in eastern Nevada and western California.

The Durov diagram consists of five fields, two triangular and three rectangular. Each chemical analysis is plotted as five points on the diagram. The relative percentages of major cations and anions are shown on the left and upper triangles, respectively. These triangles are similar to those of the trilinear diagram. Two other properties are plotted in the two outside rectangles. These properties are selected from possibilities such as total dissolved solids, specific conductance, pH, hardness, and total dissolved inorganic carbon. Total dissolved solids and pH are presented in Figure 2-12. The central square serves primarily as a transitional area to connect the four outside triangular and rectangular plots. The primary advantage of the Durov diagram is that it provides on a single illustration a visual characterization of eight major ions and two properties for the ground water in an area.

The overprinting of data on diagrams such as Figure 2-12 presents difficulties in interpretation when many data are plotted. One approach to resolving this problem is to plot polar-smoothed curves of the data (Helsel and Hirsch 1992), as is done in Figure 2-13 for the data set plotted in Figure 2-12.

Boxplots

A useful way of presenting data on individual constituents is through the use of boxplots (Tukey 1977; Chambers et al. 1983), which provide visual summaries of the statistical distribution and key statistical characteristics of a data set. Boxplots commonly are placed side-by-side to compare and contrast groups of data. In this way, they are useful for comparing individual water-quality constituents among different settings.

Boxplots shown in Figure 2-14 show statistical distributions of nitrate and boron in shallow ground water underlying five land-use types on Long Island. The box part of the boxplot is drawn from the 25th percentile (lower quartile) to the 75th percentile (upper quartile) of the data, and a line is drawn across the box at the median value. The box illustrates several key

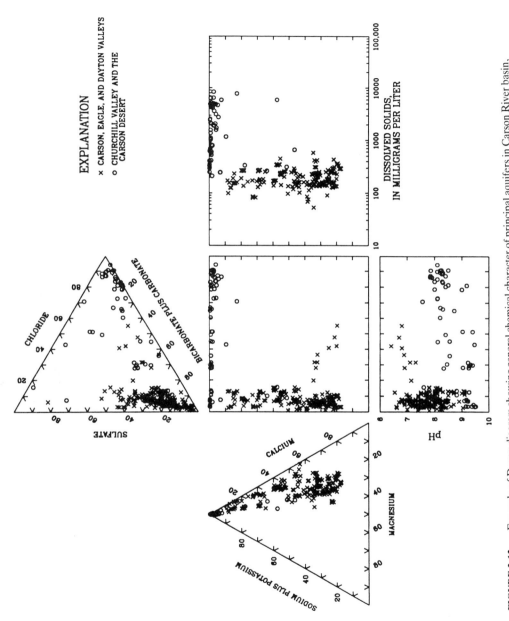

FIGURE 2-12. Example of Durov diagram showing general chemical character of principal aquifers in Carson River basin, Nevada and California. (*From A. H. Welch, U.S. Geological Survey, written commun., 1991.*)

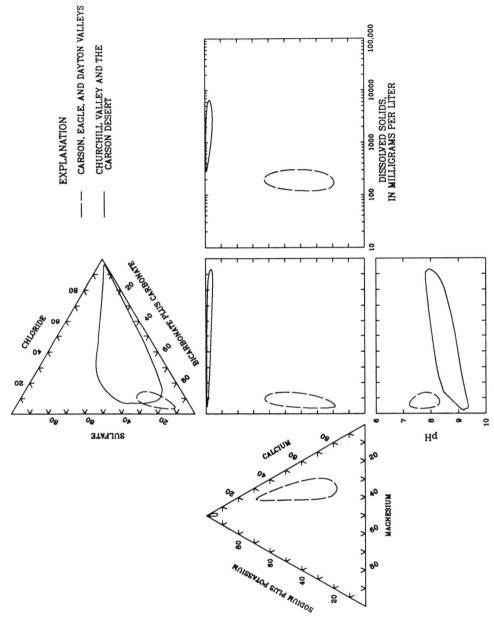

FIGURE 2-13. Example of Durov diagram with 50% polar smoothing showing general chemical character of principal aquifers in Carson River basin, Nevada and California. (*From A. H. Welch, U.S. Geological Survey, written commun., 1991.*)

features of the distribution of the data. First, the sample median is a robust measure of the central tendency of the data that is not influenced by extreme values. Second, the difference between the top and bottom of the box, known as the interquartile range (IR), is a robust measure of the spread of the data. The box measures the range of the central 50% of the data and is not influenced by the 25% on either end. Third, the distance from the top of the box to the median compared with the distance from the median to the bottom of the box is a measure of the skewness of the data. The two distances should be about equal for data derived from distributions that have zero skew such as the normal distribution.

"Whiskers" are drawn from the top and bottom of the box to "extreme" values. Several variants of boxplots have been used, depending on how the extreme values are chosen. The schematic, or standard, boxplot shown in Figure 2-14 is perhaps the most commonly used version. The attempt here is to distinguish unusual values from the rest of the plot; the whiskers are shortened to extend only to the last observation within 1.5 times the IR beyond either end of the box. Observations farther than this are plotted individually. For data from a normal distribution, these values will occur less than 1% of the time. Observations farther than 3 times the IR from either end of the box are additionally distinguished by a different symbol (e.g., a square is used in Figure 2-14). For data from a normal distribution, values more extreme than 3 times the interquartile range beyond the box limits will occur fewer than once in 300,000 times.

Other versions of the boxplot are the "simple" boxplot in which the whiskers are simply drawn from the ends of the box to the maximum and minimum data values, and a version in which the whiskers are drawn only to the 10th and 90th percentiles of the data set; the largest 10% and smallest 10% of the data are not shown. This latter type should be used only when the extreme 20% of the data are not of interest. Features sometimes added to boxplots include confidence limits about the sample median and key limits such as water-quality standards.

Statistical tests of differences among the data groups commonly are reported with the boxplots. Usually, the Mann-Whitney and Kruskal-Wallis tests (Helsel and Hirsch 1992) are used in cases of two or more boxplots, respectively. These tests yield information on whether the groups are different overall—that is, whether at least one group is significantly different from the others. If overall differences are detected, the next step is to ascertain which groups differ from one another using multiple-comparison tests (Helsel and Hirsch 1992). This procedure is analogous to the parametric approach of analysis of variance followed by pairwise-contrast tests (Neter and Wasserman 1974). The letters shown above the boxplots of nitrate concentrations in Figure 2-14 indicate that nitrate concentrations are not significantly different among the three urban land uses and agricultural lands, but that nitrate concentrations in ground water underlying undeveloped lands do differ significantly from these four groups. A slightly more complicated pattern of differences among the five land-use types is shown by the letters for the multiple comparison tests of boron concentrations. The results indicate, for example, that boron concentrations are significantly higher in the recently sewered area (perhaps from laundry detergents) and the agricultural area (perhaps from fertilizers) than in the unsewered area, but that the longterm sewered area cannot be distinguished from any of these three.

Multivariate Statistical Analyses

A wide range of multivariate statistical techniques can be used to explore the joint behavior and spatial patterns of water-quality constituents. In some circumstances, these techniques can be useful in developing and testing regional concepts for the study area. All of the methods should be ap-

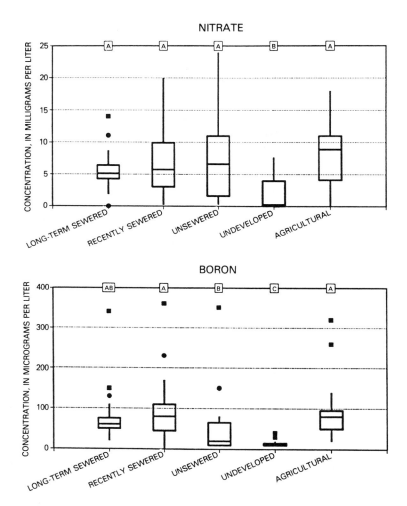

FIGURE 2-14. Boxplots of nitrate and boron in ground water underlying five land-use types on Long Island, New York. Letters above boxplots indicate results of multiple comparison tests (groups of data with common letters do not differ statistically). Sample sizes are 20 for first three groups and 15 for last two groups. (*After Eckhardt, Siwiec, and Cauller 1989.*)

proached as exploratory in nature rather than leading to hard-and-fast rules on regionalization. Typically, the techniques require standardization (conversion of the data to mean zero and standard deviation of one), if the variables being examined are of varying units or orders of magnitude.

Two major types of multivariate statistical methods are considered: (1) principal components and factor analysis, and (2) cluster analysis. A brief discussion of each is followed by an example application. First, however, a few general statements are made about the use of statistical techniques for analyzing ground-water-quality data.

Because ground-water-quality data typically are highly skewed and contain infrequent high values as outliers, they frequently violate the normality assumptions required by many statistical methods. For this reason, nonparametric procedures which do not assume the data follow any specific distributional shape commonly are used. Alternatively, logarithmic or other transformations are used in attempts to meet normality assumptions. Some caution in using these transformations is advised. A number of statistical tests exist for detecting departures from normality of the original or transformed data; however, for small sample sizes (say less than 30), all tests for normality have quite low power to reject the null hypothesis of normality when the data are truly nonnormal. In addition, biased estimates of statistical moments such as the mean and standard deviation result from retransforming logarithmic or other transformed data to original units (although methods exist to reduce these biases; e.g., Miller 1984).

A particular limitation of parametric procedures for analysis of water-quality data arises when some data are at concentrations less than the analytical detection or reporting limit (less-than values or "censored data"). These data are becoming increasingly important as trace elements and organic compounds are investigated (Helsel 1990). To use parametric procedures, the less-than values either have to be deleted or values fabricated for them before statistical analysis. The fabricated values might be zero, one half the detection limit, or the detection limit. Nonparametric procedures typically use the ranks of the data in place of their actual values and assume only that all censored data are equivalent and are lower in value than the lowest detected observation. They thus avoid arbitrary deletion of values or assignment of fabricated values.

Principal Components and Factor Analysis

Principal components and factor analysis are similar data transformation techniques used to search for structure in multivariate data sets. The methods are treated in detail in many texts (cf. Joreskog, Klovan, and Reyment 1976; Davis 1986; Howard 1991). Many variants of principal components analysis and factor analysis exist.

Principal components analysis (PCA) operates by transforming a set of interrelated variables into a new coordinate system in which the axes are linear combinations of the original variables and are mutually orthogonal or uncorrelated. The new transformed variables account for the same amount of variability in the data, but in such a way that the first axis accounts for as much of the total variance as possible; the second axis accounts for as much of the remaining variance as possible while being uncorrelated with the first axis, and so forth. The axes are weighted in proportion to the amount of the total variance that they describe. Typically, the first few axes account for most of the variance.

Factor analysis uses a similar mathematical model to that of PCA, but usually operates within a statistical framework that includes analysis of a residual term. Factor analysis can be much more complex than PCA. Either technique can involve rotation of axes.

Different modes of PCA and factor analysis operate on different covariance (or correlation) matrices. For spatial analysis of geochemical data, two modes, R-mode and

Q-mode, are of particular interest. Differences exist in the definition of the "loading" and "score" matrices in these two modes. The discussion focuses on R-mode PCA and Q-mode factor analysis (Although called Q-mode factor analysis, the application described is actually closer to Q-mode PCA).

The basic idea in applying R-mode PCA to ground-water-quality data is that much of the variation in a large number of water-quality constituents can be defined by a smaller number of principal components. These components can be useful in describing regional water-quality variations and in defining hydrogeochemical facies or subregions. The usefulness of R-mode PCA depends on the extent to which the loadings (correlation coefficients between the water-quality constituents and each principal component) and the scores (the numerical value of each component at each sampling site) can be interpreted in a geochemically meaningful way. Commonly, the loadings are used to interpret the geochemical significance of the principal components by observing which group of constituents are most correlated with each component. Inspection of the loadings may reveal constituent groupings that were not foreseen. The scores for a particular component typically are plotted on a map or other figure to graphically depict spatial patterns. Inspection of the scores may reveal spatial patterns that can be related to flow paths, hydrogeochemical facies, and geochemical processes.

Q-mode factor analysis is useful for studying the interrelations among sampling sites with respect to a set of end-member compositions identified by the analysis. In this instance, Q-mode factor analysis is applied to compositional data (data that sum to a constant for each sample; cf. Miesch 1976). A goal is to derive a relation of the form

$$X_{ij} = \sum_{m=1}^{r} a_{im} F_{mj},$$

$i = 1, \ldots, I$ (sampling sites);

$j = 1, \ldots, J$ (constituents)

where X_{ij} is an estimate of the concentration of the jth constituent in the ith sample, the F's are the hypothetical water compositions (scores) of the end members, and the a's are the loadings giving the proportions for combining the hypothetical end members to approximate the ground-water composition at the ith sampling site. The number of sampling sites and the number of water-quality constituents are equal to I and J, respectively, and r is the number of factors. Typically, r is about 3 or 4.

As an example, Kimball (1992) used Q-mode factor analysis and geochemical reasoning to determine the source of saline water to aquifers in part of Utah. Three hydrogeochemical facies were identified by Kimball in the factor analysis: recharge water, "diagenetic" water that has evolved from the recharge water through chemical reactions, and saline water resulting from mixing between diagenetic water and a brine. Major changes in the relative amounts of these facies were evident along hypothesized flow paths. The compositions of the three end members, expressed as normative-salt compositions (Bodine and Jones 1986), are shown as pie diagrams in Figure 2-15 and correspond to the F's in the preceding Eq.; the percentages of each of these end members are plotted on the triangular diagram for each site based on the composition loadings for that site (the a's in the above Eq.).

Cluster Analysis

Cluster analysis is a collection of statistical methods whose purpose is to divide a data set into groups of similar variables, hopefully reflecting underlying structure present in the data. The goal is to have group members differ as little as possible and to have each group be distinct from the other groups.

Cluster analysis involves a number of subjective decisions regarding the choice of the variables to cluster, whether the

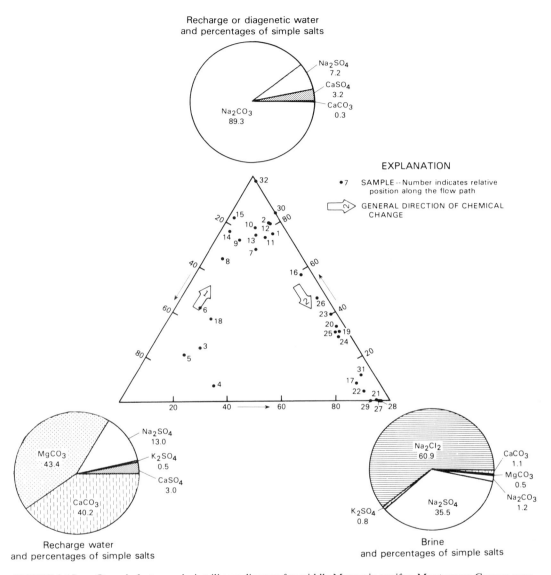

FIGURE 2-15. Q-mode factor analysis trilinear diagram for middle Mesozoic aquifer, Montezuma Canyon area, Utah. Site numbers indicate position along flow path. (*After Kimball 1992.*)

clustering is to be hierarchical (clusters, once merged, cannot be separated) or non-hierarchical (iterative), and at which step to terminate the clustering process (i.e., how many clusters to define). A method of determining the similarity between individuals (sites) and a method of determining the similarity between clusters also must be chosen. Discussions of cluster analysis are available in many texts, including those of Anderberg (1973), Davis (1986), and Kaufman and Rousseeuw (1990).

In using cluster analysis to search for structure in multivariate data, any given set of data can admit several different but meaningful partitions. Each partition may pertain to a different aspect of the data. A set of clusters largely reflects the degree to which the data set conforms to the structural forms embedded in the clustering algorithm (Anderberg 1973). Many methods of cluster analysis fail to identify particular types of clusters that are obvious to the eye. Thus, a partition obtained from a cluster analysis procedure has no inherent validity. The value and explanatory structure of the partition must be justified by its consistency with known facts.

Pedroli (1990) applied cluster analysis to classify shallow-ground-water types in a sandy, lowland area of the Netherlands. The water samples were initially separated into two groups: those with pH < 5 and those with pH ≥ 5. Pedroli (1990) explored both the original water-quality constituents and the first six principal components as variables in the cluster analysis. Solutions of the clustering algorithm with eight clusters were chosen for each pH group. The resultant 16 water types were interpreted with respect to hydrologic and landscape settings, as mineralized local seepage water, nutrient-poor seepage water, and similar categories.

Cluster analysis does not make use of previous subjective information about constituent groups. Alternately, a related set of techniques, known as discrimination methods, can be used to test the significance of differences among a priori defined groups. The most commonly used discrimination method, discriminant function analysis, requires rather restrictive assumptions, including that the data within each category are multivariate normal and often that the clusters share a common covariance structure (Klecka 1980). If the data are divided into only two categories, then logistic regression (Hosmer and Lemeshow 1989) may be useful to avoid such strict distributional assumptions. Helsel (1987) illustrated the advantages of using the ranks of the data in a discriminant function analysis of stream-water-quality data. An application of discriminant analysis by Teso et al. (1988) to predict the occurrence of the pesticide DBCP in wells was discussed previously in the section on the vadose zone.

Relating Ground-Water Quality to Land Use

A major objective of many regional ground-water quality studies is to examine relations between ground-water quality and land use. Examples of how shallow-ground-water quality has been quantitatively related to land use are provided throughout the text. The purpose of this section is to briefly review some considerations and complexities in relating ground-water quality to land use.

Land use usually is not homogeneous throughout large areas, and more than one land use can exist near a well. In evaluating the effects of land use on ground-water quality, the most reliable approach is to select wells that are located in recharge areas and directly downgradient from a single land-use setting. This approach will help avoid the influence of other land-use activities and complications from upward movement of water that originated in distant areas. Depending on the locations of cells and the degree of homogeneity of land use, these rules will be easier to follow in some areas than in others. Furthermore, relating ground-water quality to land use may be difficult for all but the uppermost part of an aquifer system.

In many cases, methods will have to be developed to assign a measure of land use to each well. Commonly, the land-use type assigned is the predominant land use within some specified area or estimated contributing area around the well. In some cases, the presence or absence of a particular land use can be more relevant than the predominant land use; for example, commercial development may not be the principal land use, but it may have the dominant effect on certain aspects of the ground-water quality in an area that is largely rural.

Land use can be defined simply by different categories or by measures of the intensity of land use. The boxplots in Figure

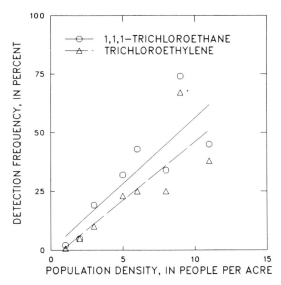

FIGURE 2-16. Relations between population density and frequency of detection of 1,1,1-trichloroethane and trichloroethylene in water from sampled wells on Long Island. (*From Cain, Helsel, and Ragone 1989.*)

2-14 illustrated an example of how shallow-ground-water quality can be related to different categories of land use. The relations shown in Figure 2-16 show how a different set of water-quality constituents in the same region can be related to a land-use intensity factor, in this case population density. The compounds shown were used as industrial solvents and degreasers such as dry-cleaning agents and septic tank cleaners. Some of these uses were more common in areas of high population density and result in an increased likelihood of their detection in ground water.

A critical element in relating land use to ground-water quality is the method of characterizing the land use near a well. The method can range from simply plotting the well location on a land-use map to the use of a particle-tracking model to delineate the contributing area to the well. The utility of different approaches will depend on the heterogeneity of land use, the level and

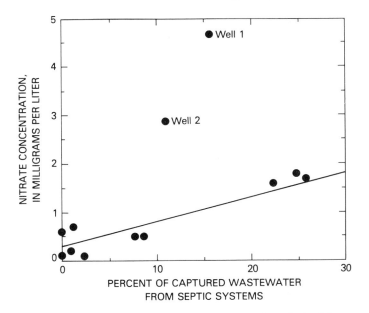

FIGURE 2-17. Plot of nitrate concentrations in 12 Cape Cod public water-supply wells and estimated percentage of well pumpage composed of captured wastewater from septic systems. The estimates of the percent captured wastewater were determined using a three-dimensional ground-water flow model and particle-tracking algorithm. (*After Barlow in press.*)

detail of the geohydrologic data available, and the project resources. Several methods are reviewed in Chapter 24.

An example of the possible utility of particle tracking to define relations of ground-water quality to land use is shown in Figure 2-17 from a study by Barlow (in press) of various techniques to delineate contributing areas to public-water-supply wells in Cape Cod, Massachusetts. Figure 2-17 shows the observed nitrate concentrations measured in 1987 at 12 public water-supply wells plotted against estimates of the percent of the well discharge composed of captured wastewater from septic systems. Estimates of recharge from septic systems were based on water-use estimates and the density of residential housing and commercial facilities; the percent captured wastewater for each well was predicted using a particle-tracking model (Pollock 1989). Wells 1 and 2, shown as outliers in Figure 2-17, are downgradient from a wastewater treatment facility which is expected to be a large potential source of nitrogen. The particle-tracking model estimated that wells 1 and 2 derive about a quarter of their water from this wastewater treatment facility. The line of correlation shown in Figure 2-17 was determined using the 10 wells that capture wastewater from septic systems alone (i.e., wells 1 and 2 were omitted). A second example of the use of particle-tracking models is given in Chapter 24.

In attempting to statistically relate ground-water quality to land use, several caveats should be kept in mind. First, if enough land-use variables are tested, some statistical relations may be found, but these relations may be spurious. Second, if the land-use variables included in a regression analysis sum to 100% this creates problems in interpreting the results, because the presence of one land use is associated with the absence of another (Barringer et al. 1990). For instance, suppose all land uses fall into one of three categories: A, B, or C. Positive correlation between a constituent X and say percent of area in land-use A simply may reflect a negative correlation between constituent X and land-use B or C. Additional problems can occur if sampled wells are so close that spatial correlation exists in the data. Barringer et al. (1990) suggest that one corrective strategy is to reduce the spatial correlation by subsampling the data set to include only a set of wells with sufficient interwell distances. For new studies, a preferable option is to avoid these difficulties by using some of the grid-based site-selection approaches discussed in Chapter 3.

Usually, land-use information is available for a particular time. The age of the land-use information may or may not correspond to the age of the ground-water sampled; usually it will not. Thus, some review of the stability of the land use over time should be made.

Ground-water quality may be difficult to relate to land use because the resolution of the land-use data is coarser than the contributing area of the sampled well. The results of a simple experiment serve to illustrate this point: Nitrate concentrations in shallow wells (well depths less than 10 m) were compared among three land-use categories on the Delmarva Peninsula using four different sources of land-use data (Hamilton and Shedlock 1990). The four sources of land-use data were (1) land-use maps at a scale of 1:250,000, (2) topographic maps (with woodlands and urban areas shown by shading on the map) at a scale of 1:24,000, (3) field observations of the predominant land use within a quater-mile radius of the well, and (4) field observations of the predominant land use upgradient of the well. The results are shown in Table 2-1. When using the 1:250,000 land-use data, the median nitrate concentration associated with woodlands is larger than the value associated with agricultural lands. A much different result is shown for the other three sets of land-use data, which indicate that nitrate concentrations are much greater in ground water underlying agricultural lands. The relatively coarse scale of the land-use data at the 1:250,000 scale probably leads to the in-

TABLE 2-1. Comparisons of Median Nitrate Concentrations Among Land Uses Using Different Sources and Scales of Land-Use Data, Delmarva Peninsula, Delaware, Maryland, and Virginia

Type of Land-Use Data	Sample Size	Median Nitrate Concentration (mg/L)		
		Urban Areas	Agricultural Areas	Woodlands
1:250,000 land-use maps	62	0.9	5.8	7.0
1:24,000 topographic maps	62	—	6.7	0.6
Predominant land use within quarter-mile radius	32	—	4.2	0.1
Predominant upgradient land use	32	—	3.7	0.1

Data from Hamilton and Shedlock (1990).

consistent result. At the 1:250,000 scale, the minimum differentiated size for agricultural and woodland areas is about 16 ha; thus, a well located in the middle of a 10-ha farm surrounded by a forest might easily be categorized as woodland. Note that values for the urban land use are not shown in Table 2-1 for the latter three scales, because only two samples in this category were available in each case.

One approach to improve the information available on local land use in the vicinity of shallow sampled wells is to document the land use in the field at each well. For example, Hardy, Leahy, and Alley (1989) present a special land-use–land-cover form that is filled out for each well and updated each time a well is resampled. The form includes a checklist of land-use types and local features that might affect ground-water quality. In addition, they recommend that a series of photographs be taken of each well and its surrounding area to serve as a record of the local land use and to help locate the well and the water-level measuring point in the future.

CONCLUDING REMARKS

A conceptual regional framework should be a central feature in the design and interpretation of any regional ground-water-quality study. It should be developed at the outset of a study with continuing review and refinement. The conceptual framework can take many forms; initially, it may be largely qualitative. It is advantageous to begin along several lines of inquiry, including first-hand viewing of the study area; analysis of available soil, hydrogeologic, and water-quality information; review of existing hydrologic models for the area; and sketching of maps, depth profiles, and block diagrams that depict current conceptions of the regional ground-water-flow systems. Although a universal approach to regional conceptualization would be ideal, such an approach does not exist. Each regional system is inherently different, and the approaches used must be adjusted to each.

The foregoing discussion has considered how understanding properties of the vadose zone, the ground-water-flow system, and natural and human-influenced geochemical patterns can contribute to an evolving picture of ground-water quality in a region. The key to success in understanding regional ground-water quality lies in linking observable spatial patterns in these attributes with water-quality characteristics and processes observed in the field. Because the sources of ground-water contamination are spatially diffuse and can be characterized only indirectly, the apparent causes of a particular problem may not turn out to be the primary ones.

References

Anderberg, M. R. 1973. *Cluster Analysis for Applications*. New York: Academic Press.

Anderson, M. P. 1989. Hydrogeologic facies models to delineate large-scale spatial trends in glacial and glaciofluvial sediments. *Geological Society of America Bulletin* 101:501–11.

Anderson, M. P., and C. J. Bowser. 1986. The role of groundwater in delaying lake acidification. *Water Resources Research* 22(7):1101–8.

Anderson, M. P., and W. W. Woessner. 1992. *Applied Groundwater Modeling*. New York: Academic Press.

Bair, E. S., and G. S. Roadcap. 1992. Comparison of flow models used to delineate capture zones of wells: 1. Leaky-confined fractured-carbonate aquifer. *Ground Water* 30(2):199–211.

Barlow, P. M. in press. *Particle-Tracking Analysis of Contributing Areas to Public Supply Wells in Simple and Complex Flow Systems, Cape Cod, Massachusetts*. Marlborough, Massachusetts: U.S. Geological Survey Open-File Report.

Barringer, Thomas, Dennis Dunn, William Battaglin, and Eric Vowinkel. 1990. Problems and methods involved in relating land use to ground-water quality. *Water Resources Bulletin* 26(1):1–9.

Bliss, N. B., and W. U. Reybold. 1989. Small-scale digital soil maps for interpreting natural resources. *Journal of Soil and Water Conservation* 44:30–34.

Bodine, M. W., Jr., and B. F. Jones. 1986. *THE SALT NORM: A Quantitative Chemical-Mineralogical Characterization of Natural Waters*. Denver, Colorado: U.S. Geological Survey Water-Resources Investigations Report 86-4086.

Bouma, J. 1991. Influence of soil macroporosity on environmental quality. In *Advances in Agronomy*, ed. D. L. Sparks, pp. 1–37. New York: Academic Press.

Bush, P. W., and R. H. Johnston. 1988. *Ground-Water Hydraulics, Regional Flow, and Ground-Water Development of the Floridan Aquifer System in Florida and in Parts of Georgia, South Carolina, and Alabama*. Denver, Colorado: U.S. Geological Survey Professional Paper 1403-C.

Buxton, H. T., and E. Modica. 1992. Patterns and rates of ground-water flow on Long Island, New York. *Ground Water* 30(6):857–66.

Buxton, H. T., T. E. Reilly, D. W. Pollock, and D. A. Smolensky. 1991. Particle tracking analysis of recharge areas on Long Island, New York. *Ground Water* 29(1):63–71.

Cain, Doug, D. R. Helsel, and S. E. Ragone. 1989. Preliminary evaluations of regional ground-water quality in relation to land use. *Ground Water* 27(2):230–44.

Chambers, J. M., W. S. Cleveland, B. Kleiner, and P. A. Tukey. 1983. *Graphical Methods for Data Analysis*. Boston: Duxbury Press.

Cherry, J. A. 1991. Groundwater monitoring: some deficiencies and opportunities. In *Hazardous Waste Site Investigations; Toward Better Decisions, Proceedings of the 10th ORNL Life Sciences Symposium, Gatlinburg, Tennessee*, eds. B. A. Berven and R. B. Gammage. Chelsea, Michigan: Lewis.

Chilingar, G. V. 1956. Durov's classification of natural waters and chemical composition of atmospheric precipitation in USSR: A review. *Transactions, American Geophysical Union* 37:193–96.

Cleveland, W. S. 1979. Robust locally weighted regression and smoothing scatterplots. *Journal of the American Statistical Society* 74:829–36.

Davis, J. C. 1986. *Statistics and Data Analysis in Geology*. New York: Wiley.

Dubrovsky, N. M., J. M., Neil, M. C. Welker, and K. D. Evenson. 1991. *Geochemical Relations and Distribution of Selected Trace Elements in Ground Water of the Northern Part of the Western San Joaquin Valley, California*. Denver, Colorado: U.S. Geological Survey Water-Supply Paper 2380.

Eckhardt, D. A., S. F. Siwiec, and S. J. Cauller. 1989. Regional appraisal of ground-water quality in five different land-use areas, Long Island, New York. In *U.S. Geological Survey Toxic Substances Hydrology Program—Proceedings of the Technical Meeting, Phoenix, Arizona*, pp. 397–403. Reston, Virginia: U.S. Geological Survey Water-Resources Investigations Report 88-4220.

Fortin, G., G. van der Kamp, and J. A. Cherry. 1991. Hydrogeology and hydrochemistry of an aquifer-aquitard system within glacial deposits, Saskatchewan, Canada. *Journal of Hydrology* 126:265–92.

Freeze, R. A., and J. A. Cherry. 1979. *Groundwater*. Englewood Cliffs, New Jesey: Prentice-Hall.

Freeze, R. A. 1969. *Theoretical Analysis of Regional Groundwater Flow*. Inland Waters Branch, Department of Energy, Mines, and Resources, Canada, Scientific Series No. 3.

Fusillo, T. V., B. P., Sargent, R. L. Walker, T. E. Imbrigiotta, and W. H. Ellis, Jr. 1991. Investigation of the discharge of ground water containing volatile organic compounds into a stream at Picatinny Arsenal, New Jersey. *Transactions, American Geophysical Union* 72:185–6.

Gale, J. E. 1982. Assessing the permeability characteristics of fractured rock. In *Recent Trends in Hydrogeology*, ed. T. N. Narasimhan, pp. 163–81. Geological Society of America, Special Paper 189.

Gburek, W. J., and J. B. Urban. 1990. The shallow weathered fracture layer in the near-stream zone. *Ground Water* 28(6):875–83.

Germann, P. F., ed. 1988. Rapid and far-reaching hydrologic processes in the vadose zone (15 papers). *Journal of Contaminant Hydrology* 3:115–380.

Gillham, R. W., M. J. L. Robin, J. F. Barker, and J. A. Cherry. 1983. *Groundwater Monitoring and Sample Bias*. Washington, D.C.: American Petroleum Institute Publication 4367.

Hall, G. F., and C. G. Olson. 1991. Predicting variability of soils from landscape models. In *Spatial Variabilities of Soils and Landforms*, eds. M. J. Mausbach and L. P. Wilding, pp. 231–42. Madison, Wisconsin: Soil Science Society of America, SSSA Special Publication Number 28.

Hallberg, G. R., T. E. Fenton, and G. A. Miller. 1978. Part 5. Standard weathering zone terminology for the description of Quaternary sediments in Iowa. In *Standard Procedures for Evaluation of Quaternary Materials in Iowa*, ed. G. R. Hallberg, pp. 75–109. Iowa City, Iowa: Iowa Geological Survey Technical Information Series 8.

Hamilton, P. A., and R. J. Shedlock. 1990. Relations between land use and nitrate concentrations in shallow ground water, Delmarva Peninsula. In *U.S. Geological Survey Yearbook Fiscal Year 1989*, pp. 38–41. Reston, Virginia: U.S. Geological Survey.

Harbaugh, A. W., and R. T. Getzen. 1977. *Stream Simulation in an Amalog Model of the Ground-Water System on Long Island, New York*. Denver, Colorado: U.S. Geological Survey Water-Resources Investigations Report 77-58.

Hardy, M. A., P. P. Leahy, and W. M. Alley. 1989. *Well Installation and Documentation, and Ground-Water Sampling Protocols for the Pilot National Water-Quality Assessment Program*. Reston, Virginia: U.S. Geological Survey Open-File Report 89-396.

Harrison, B., E. A. Sudicky, and J. A. Cherry. 1992. Numerical analysis of solute migration through fractured clayey deposits into underlying aquifers. *Water Resources Research* 28(2):515–26.

Harvey, J. W. 1991. Localized recharge in net discharge environments: importance to solute transport in wetlands and streams. *Transactions, American Geophysical Union* 72:168.

Heath, R. C. 1989. *Basic Ground-Water Hydrology*. Denver, Colorado: U.S. Geological Survey Water-Supply Paper 2220.

Helsel, D. R. 1987. Advantages of nonparametric procedures for analysis of water quality data. *Hydrological Sciences Journal* 32:179–90.

Helsel, D. R. 1990. Less than obvious: statistical treatment of data below the detection limit. *Environmental Science and Technology* 24(12):1767–74.

Helsel, D. R., and R. M. Hirsch. 1992. *Statistical Methods in Water Resources*. New York: Elsevier.

Hem, J. D. 1985. *Study and Interpretation of the Chemical Characteristics of Natural Water*, 3rd ed. Denver, Colorado: U.S. Geological Survey Water-Supply Paper 2254.

Hill, R. A. 1940. Geochemical patterns in Coachella Valley, California. *Transactions, American Geophysical Union* 21:46–53.

Hosmer, D., and S. Lemeshow. 1989. *Applied Logistic Regression*. New York: Wiley.

Howard, P. J. A. 1991. *An Introduction to Environmental Pattern Analysis*. Park Ridge, New Jersey: Parthenon.

Ireland, R. L., J. F. Poland, and F. S. Riley. 1984. *Land Subsidence in the San Joaquin Valley, California, as of 1980*. Denver, Colorado: U.S. Geological Survey Professional Paper 437-I.

Johnston, R. H. 1988. Factors affecting ground-water quality. In *National Water Summary 1986—Hydrologic Events and Ground-Water Quality*, pp. 71–86. Reston, Virginia: U.S. Geological Survey Water-Supply Paper 2325.

Joreskog, K. G., J. E. Klovan, and R. A. Reyment. 1976. *Geological Factor Analysis*. New York: Elsevier.

Kaufman, L., and P. J. Rousseeuw. 1990. *Finding Groups in Data: An Introduction to Cluster Analysis*. New York: Wiley.

Kimball, B. A. 1992. Geochemical indicators used to determine the source of saline water in Mesozoic aquifers, Montezuma Canyon Area, Utah. In Selected Papers in the Hydrologic Sciences, ed. Seymour Subitzky, pp. 89–106. Denver, Colorado: U.S. Geological Survey Water-Supply Paper 2340.

Klecka, W. R. 1980. *Discriminant Analysis*. Newbury Park, California: Sage University Paper Series on Quantitative Applications in the Social Sciences 07-019.

Krause, R. E., and R. B. Randolph. 1989. *Hydrology of the Floridan Aquifer System in Southeast Georgia and Adjacent Parts of Florida and South Carolina*. Denver, Colorado: U.S. Geological Survey Professional Paper 1403-D.

Kueper, B. H., and D. B. McWhorter. 1991. The behavior of dense, nonaqueous phase liquids in fractured clay and rock. *Ground Water* 29(5):716–28.

Lee, D. R., and J. A. Cherry, 1978. A field exercise on groundwater flow using seepage meters and mini-piezometers. *Journal of Geological Education* 27:6–10.

Lee, R. W., and E. F. Hollyday. 1987. Radon measurement in streams to determine location and magnitude of ground-water seepage. In *Radon, Radium, and Other Radioactivity in Ground Water*, ed. B. Graves, pp. 241–49. Chelsea, Michigan: Lewis.

Lissey, A. 1971. Depression-focused transient groundwater flow patterns in Manitoba. *Geological Association of Canada Special Paper* 9:333–41.

Lytle, D. J., and M. J. Mausbach. 1991. Interpreting soil geographic databases. In *Proceedings: Resource Technology 90, Second International Symposium on Advanced Technology in Natural Resource Management. Georgetown University Conference Center, November 12–15, 1990*, pp. 469–76.

Bethesda, Maryland: American Society for Photogrammetry and Remote Sensing.

Maclay, R. W., and T. A. Small, 1984. *Carbonate geology and hydrology of the Edwards Aquifer in the San Antonio Area, Texas*, Austin, Texas: U.S. Geological Survey Open-File Report 83–537.

Matthess, Georg. 1982. *The Properties of Groundwater*. New York: Wiley.

Meeks, Y. J., and J. D. Dean. 1990. Evaluating ground-water vulnerability to pesticides. *Journal of Water Resources Planning and Management* 116(5):693–707.

Miesch, A. T. 1976. *Q-Mode Factor Analysis of Geochemical and Petrologic Data Matrices With Constant Row-Sums*. Denver, Colorado: U.S. Geological Survey Professional Paper 574-G.

Miller, D. M. 1984. Reducing transformation bias in curve fitting. *The American Statistician* 38(2):124–26.

Morrissey, D. J. 1989. *Estimation of the Recharge Area Contributing Water to a Pumped Well in a Glacial-Drift, River-Valley Aquifer*. Denver, Colorado: U.S. Geological Survey Water-Supply Paper 2338.

National Research Council. 1990. *Ground Water Models: Scientific and Regulatory Applications*. Washington, D.C.: National Academy Press.

Neter, John, and William Wasserman. 1974. *Applied Linear Statistical Models*. Homewood, Illinois: Irwin.

Pedroli, Bas. 1990. Classification of shallow groundwater types in a Dutch coversand landscape. *Journal of Hydrology* 115:361–75.

Piper, A. M. 1944. A graphic procedure in the geochemical interpretation of water analyses. *Transactions, American Geophysical Union* 25:914–23.

Pollock, D. W. 1989. *Documentation of Computer Programs to Compute and Display Pathlines Using Results from the U.S. Geological Survey Modular Three-Dimensional Finite-Difference Ground-Water Flow Model*. Reston, Virginia: U.S. Geological Survey Open-File Report 89-381.

Reybold, W. U., and G. W. TeSelle. 1989. Soil geographic data bases. *Journal of Soil and Water Conservation* 44:28–9.

Robertson, W. D., J. A. Cherry, and E. A. Sudicky. 1991. Ground-water contamination from two small septic systems on sand aquifers. *Ground Water* 29(1):82–92.

Romijn, E., and S. S. D. Foster. 1987. Conclusions on topic 2: Monitoring strategies for the quality of soil and groundwater. In *Vulnerability of Soil and Groundwater to Pollutants*, eds. W. van Duijvenbooden and H. G. van Waegeningh, pp. 43–44. The Hague, The Netherlands: TNO Committee on Hydrological Research Proceedings and Information No. 38.

Ruland, W. W., J. A. Cherry, and Stan Feenstra. 1991. The depth of fractures and active ground-water flow in a clayey till plain in southwestern Ontario. *Ground Water* 29(3):405–17.

Shapiro, A. M. 1987. Transport equations for fractured porous media. In *Advances in Transport Phenomena in Porous Media*. eds. J. Bear and M. Y. Corapcioglu, pp. 407–71. NATO Advanced Study Institutes Series. Dordrecht, Netherlands: Martinus Nijhoff.

Soil Survey Staff. 1975. *Soil Taxonomy*. Washington, D.C.: Soil Conservation Service, U.S. Department of Agriculture Handbook No. 436.

Spalding, C. P., and Raziuddin Khaleel. 1991. An evaluation of analytical solutions to estimate drawdowns and stream depletions by wells. *Water Resources Research* 27(4):597–609.

Stiff, H. A., Jr. 1951. The interpretation of chemical water analysis by means of patterns. *Journal of Petroleum Technology* 3:15–17.

Tandarich, J. P., T. J. Bicki, D. P. McKenna, and R. G. Darmody. 1990. The pedo-weathering profile and its implications for ground water protection. In *Ground Water Management, Proceedings of the 1990 Cluster of Conferences*, pp. 893–900. Kansas City, Missouri: National Water Well Association.

Teso, R. R., T. Younglove, M. R. Peterson, D. L. Sheeks III, and R. E. Gallavan. 1988. Soil taxonomy and surveys: classification of areal sensitivity to pesticide contamination of groundwater. *Journal of Soil and Water Conservation* 43(4):348–52.

Theis, C. V. 1935. The relation between the lowering of the piezometric surface and the rate and discharge of a well using ground water storage. *Transactions, American Geophysical Union* 2:519–24.

Toth, J. 1963. A theoretical analysis of groundwater flow in small drainage basins. *Journal of Geophysical Research* 68(16):4795–4812.

Triska, F. J., J. H. Duff, and R. J. Avanzino. 1990. Influence of exchange flow between the channel and hyporheic zone on nitrate production in a small mountain stream. *Canadian Journal of Fisheries and Aquatic Sciences* 47(11):2099–111.

Triska, F. J., V. C. Kennedy, R. J. Avanzino, G. W. Zellweger, and K. E. Bencala. 1989. Retention and transport of nutrients in a third-order stream in northwestern California: hyporheic processes. *Ecology* 70(6):1893–1905.

Tukey, J. W. 1977. *Exploratory Data Analysis*. Reading, Massachusetts: Addison-Wesley.

U.S. Department of Agriculture. 1981. *Land Resouce Regions and Major Land Resource Areas of the United States*. Washington, D.C.: Agricultural Handbook 296.

Vroblesky, D. A. and M. M. Lorah. 1991. Prospecting for zones of contaminated ground-water discharge in streams using bottom-sediment gas bubbles. *Ground Water* 29(3):330–40.

Wait, R. L., and D. O. Gregg. 1973. *Hydrology and Chloride Contamination of the Principal Artesian Aquifer in Glynn County, Georgia*. Georgia Geologic Survey Hydrologic Report No. 1.

Warrick, A. W., and D. R. Nielsen. 1980. Spatial variability of soil physical properties in the field. In *Applications of Soil Physics*, ed. Daniel Hillel, pp. 319–44. New York: Academic Press.

Williamson, A. K., D. E. Prudic, and L. A. Swain. 1987. *Ground-Water Flow in the Central Valley, California*. Denver, Colorado: U.S. Geological Survey Professional Paper 1401-D.

Winter, T. C. 1983. The interaction of lakes with variably saturated porous media. *Water Resources Research* 19(5):1203–18.

Winter, T. C. 1988. A conceptual framework for assessing cumulative impacts on the hydrology of nontidal wetlands. *Environmental Management* 12(5):605–20.

Winter, T. C., J. W. LaBaugh, and D. O. Rosenberry. 1988. The design and use of a hydraulic potentiomanometer for the direct measurement of differences in hydraulic head between ground water and surface water. *Limnology and Oceanography* 33(5): 1209–14.

Zaporozec, Alexander. 1972. Graphical interpretation of water-quality data. *Ground Water* 10(2):32–43.

Zoeteman, B. C. J. 1987. General introduction. In *Vulnerability of Soil and Groundwater to Pollutants*, eds. W. van Duijvenbooden and H. G. van Waegeningh, pp. 17–27. The Hague, Netherlands: TNO Committee on Hydrological Research Proceedings and Information No. 38.

II
Statistical Concepts

3

Ground-Water-Quality Surveys

William M. Alley

INTRODUCTION

Regional ground-water-quality surveys may be undertaken for many different purposes and with varying degrees of complexity. The simplest survey is a reconnaissance with wells selected for sampling based on professional judgment about where contamination is most likely. In this case, no attempt is made to portray the results as representative of ground-water quality; the results simply provide a perspective on whether selected contaminants are present in ground water at the times and places where they are most expected. Such preliminary reconnaissance surveys may be useful before committing costly resources to more detailed studies. Their design is highly dependent on the situation of interest. Examples are described by Walker and Porter (1990) for a survey of pesticides in upstate New York, and by Roux, Balu, and Bennett (1991) for a large-scale retrospective ground-water-monitoring study for metolachlor.

If more than a reconnaissance survey is planned and one goal is to make statistical inferences about a population, then certain general principles of probability sampling must be considered in the survey design. At the same time, application of probability sampling theory to surveys of ground-water quality is subject to many complications. These include bias resulting from the restricted points of access to ground water, biases resulting from alterations of the water chemistry by well installations and by the equipment and methods used to obtain the water-quality sample, difficulties identifying the population of existing wells, and the effects of temporal variability. This chapter reviews these complications and methods of dealing with them. Emphasis is placed on achieving a balance between statistical and hydrogeological considerations.

The chapter serves, in part, a tutorial role for describing probability concepts relevant to ground-water-quality surveys. Space limitations, however, preclude a thorough treatment of probability sampling, and only basic concepts are highlighted. Considerable reference is made to texts on probability sampling by Cochran (1977) and Gilbert (1987) for further details. Some reviews of related statistical concepts have been presented previously in ground-water journals (e.g., Sgambat and Stedinger 1981; Nelson and Ward 1981; Harris, Loftis, and Montgomery 1987).

The discussion is divided into four parts. First, some key underlying statistical precepts are briefly reviewed. Second, special considerations for ground-water-quality surveys are discussed. Third, selected probability sampling approaches are presented in the context of regional ground-water-quality surveys. Finally, grid-based search sampling is briefly reviewed.

SOME PRELIMINARY CONCEPTS

A Note on Terminology

The words *sampling* and *sample* are used in both water-quality and statistical senses. Where confusion may occur, a modifier such as "water-quality" is attached to these terms. In addition, the term *random selection* is substituted for the term *random sampling*.

Target Populations

It is not uncommon for a study to report statistics at great length, yet devote little attention to carefully describing the population to which these statistics apply. As noted in Chapter 1, the target population for ground-water-quality surveys may be ground water at the point of use, a population of wells, or the ground-water resource. Each type of target population requires a different set of design considerations. This chapter focuses on well-water surveys and survey sampling conducted as part of ground-water-resource assessments.

The Role of Random Selection

To the ground-water hydrologist well-versed in deterministic hydrology, a study design wherein wells are randomly selected may appear to be an abdication of responsibility. The response to such a reaction must consider several aspects, including an appreciation of (1) how random selection and nonrandom selection can complement one another in ground-water-quality investigations, (2) the underlying purpose and attributes of random selection, and (3) how hydrogeologic knowledge can be integrated into some of the choices involved in setting up a random selection design.

Establishing a random process for selecting sampling sites can minimize many subtle biases that can enter during the process of selecting sites, including hidden biases that result from "convenience sampling." For instance, there is a natural tendency to sample from wells that are logistically convenient, or, perhaps, have been sampled before. Yet, some of these wells may have been sampled originally because of suspected problems, or they may be situated in environments unrepresentative of the full target population. For instance, if previous well sampling had been concentrated in valleys, a nonrandom selection of wells may inadvertently focus on areas of a particular soil type having sorptive properties much different than soils in the remaining area of interest.

The focus of random selection is on the selection process; it does not guarantee that the actual selection achieved will be representative, but instead attempts to ensure that the selection process is unbiased. Thus, any property of bias by which one may label the sample is really a property of the process that generated it. The virtue of a random selection process is "less in its achievement (... the actual sample it produces on a particular occasion) than in its promise (to be impartial between samples)" (Stuart 1976, p. 13).

A primary motivation of random selection is that it is a prerequisite to making assumptions about the sampling distribution of statistical parameters. For example, much of the interest in the normal distribution for survey sampling rests on the central limit theorem result that the sampling distribution of the sample mean tends to approach the normal distribution as sample size increases. Although this result holds for all distributions with finite variance (i.e., virtually all distributions of interest in hydrogeology), the more irregular the population distribution, the larger the sample size needed for the normal approximation of the sample mean to apply. For many populations, a sample size of about 30 is considered large enough for the sampling distribution of the sample mean to be approximated by the normal distribution (Stuart 1976), but there are certainly no guarantees.

Types and Sources of Errors

In natural systems, the distribution of a random variable is not known and a set of observations $\{x_1, \ldots x_n\}$ is used to estimate the population parameters of interest. For example, the following two estimators commonly are used to estimate the mean and variance:

$$\bar{x} = \frac{1}{n} \sum_{i=1}^{n} x_i \qquad (3\text{-}1)$$

$$s^2 = \frac{1}{n-1} \sum_{i=1}^{n} (x_i - \bar{x})^2 \qquad (3\text{-}2)$$

The values of $\bar{x}$ and s^2 are referred to as the sample mean and sample variance, respectively. Their deviation from the population values may be large, particularly for small sample sizes n.

The accuracy of sample estimates of a parameter is a function of both random errors and bias (see Table 3-1). Bias is the deviation from the true parameter value that would occur on average if one could repeatedly draw sets of n units from the population. Random errors, on the other hand, affect the degree of closeness or agreement that would occur among the individual sample estimates.

Random errors include both random sampling error and random measurement error. *Random sampling error* is the variation attributed to the random selection process. It arises from the variability inherent in ground-water quality. Unless the entire resource could be sampled, different estimates

TABLE 3-1. Errors in Estimating Statistical Properties of Regional Ground-Water Quality

Type of Error	Sources of Error	Methods Used to Reduce Errors
	Random Errors	
Random sampling errors	Inherent variability in ground-water quality	Increase number of wells sampled and frequency of sampling
Random measurement errors	Variability in measurement techniques and well designs	Use consistent techniques for water-quality sampling and well construction
	Bias	
Selection bias	Wells selected for sampling do not represent target population	Randomized well selection
Measurement bias	Alterations of the chemistry of the water by well installations and by the equipment and methods used to obtain the water-quality sample	Careful adherence to established sample collection, handling, and analysis techniques
		Existing wells—exclude wells that do not meet specified criteria for obtaining a representative sample
		Newly constructed wells—follow well-construction guidelines appropriate for problem under study

of statistical parameters will be derived, depending on the set of sites chosen for the measurements. One of the goals of probability sampling schemes is to maintain random sampling error within acceptable limits. Temporal variability also may contribute to random sampling error as discussed later.

Random measurement errors occur from the inherent random variability that occurs in taking measurements of any phenomenon. These can be reduced by maintaining consistency in ground-water-sampling procedures from location to location and over time. The basic idea is to keep the variability caused by measurement techniques small in relation to the natural variability among the population units. Measurement errors include both field-sampling errors and laboratory errors.

Bias also can be grouped into two major categories. The first, *selection bias*, concerns the degree to which the selected sampling sites in aggregate are representative of the region of interest. Selection bias is a potential major contributor to bias and can arise in any number of ways, some of which can be quite subtle, as noted previously. The primary objective of probability sampling schemes is to minimize selection bias by using randomization.

The second category of bias, *measurement bias*, concerns the degree to which the measured water quality is representative of the ground-water quality at the chosen site. Measurement bias can result from alteration of the in situ geochemical environment by the well installation, alteration of the water as it moves from the aquifer into the well, characteristics of the well-construction materials that alter the composition of the water, changes caused by the equipment and methods used to remove the sample from the well, chemical and biological changes that occur during sample transport and storage, and biases in analytical procedures. Gillham et al. (1983) note than an obvious way to reduce measurement bias caused by human factors is to provide adequate training and supervision of field personnel. In addition, sampling procedures that are relatively simple are expected to have the lowest potential for human-induced variability in the data, even though the absolute accuracy may not be as great. Because of the difficulty of ensuring that no unsuspected bias enters into estimates, one usually speaks of the precision (a function of random errors only) of a statistical estimate instead of its accuracy.

In addition to the above sources of error, one must also consider gross errors and mistakes that can occur during sample collection and transport, laboratory analysis, data entry, and so forth.

Effects of Spatial Correlation

Survey sampling methods usually assume that the data are uncorrelated over space. This assumption applies only if the distance between sampling sites is sufficiently large. In general, if the errors have a positive spatial correlation, estimates of uncertainty such as the standard error of the mean will tend to underestimate the true uncertainty, making the estimate of the mean (or other parameter) seem more precise than it really is. Methods exist for dealing with spatial correlation; however, these commonly require quite restrictive assumptions and estimation of parameters that may be highly uncertain. It is best to minimize spatial correlation from the outset in the design of a survey sampling network. A common approach is to use one of the grid-based sampling schemes discussed later. Geostatistical methods for spatially correlated data are treated in Chapter 4.

SPECIAL CONSIDERATIONS FOR GROUND-WATER-QUALITY SURVEYS

Ground-water-quality surveys differ in several respects from other types of statistical surveys. Some special considerations include the suitability of wells for collecting water-quality samples, difficulties in identifying candidate sites, temporal variability, and the possibility of future network modifications.

Suitability of Wells for Collecting Water-Quality Samples

Survey sampling may use existing wells or newly constructed wells. If newly constructed wells are used, then these can be designed to enhance their suitability for obtaining representative ground-water samples. On the other hand, when using existing wells, a major decision involves determining whether to exclude wells that do not meet specified criteria for obtaining a representative sample of ground-water quality.

If the study goal is to make estimates (inferences) regarding the quality of the ground-water resource, the preferred policy is commonly to sample only from those existing wells for which information on well construction and local hydrogeology is sufficient (within a reasonable degree of confidence) to determine the hydrogeologic unit from which the well is producing and to ensure that the well is suitable for sampling the constituents of concern. The assumption in this case is that reductions in measurement bias from carefully choosing wells suitable for water-quality sampling will more than offset any increase in selection bias that occurs as a result of rejecting wells.

Some surveys, however, have chosen not to exclude wells. For instance, the U.S. Environmental Protection Agency (USEPA) National Pesticide Survey (U.S. Environmental Protection Agency 1990), a water-well survey of pesticides and nitrate in domestic and community drinking-water wells throughout the United States, included all wells from the target population as candidates for sampling, regardless of the well characteristics. This choice was made because of the concern that the exclusion of wells for any reason would bias the overall statistics on the quality of drinking water *in* wells. On the one hand, this choice leads to a set of statistics that may more closely represent the overall quality of water in wells; on the other hand, it makes it more difficult to relate the observed results to conditions in the ground-water resource and to explain the factors causing the observed conditions.

The Sampling Frame

Before selecting the sample, the population must be divided into a set of candidate sampling units known as the *sampling frame*. Different sampling frames are constructed for well-water surveys and resource assessments.

Ideally, the sampling frame for a well-water survey consists of a list of the target population of wells. Such a list may not be readily available, particularly for domestic supply and other nonpubic supply wells. Thus, methods must be devised so that each well in the population or statistical stratum has about an equal (or known) chance of being selected.

The sampling frame for a resource assessment typically is represented on a map. Generally, the goal is to sample the ground-water resource of interest uniformly over space; thus, a grid-based approach commonly is used. Because the locations of existing wells may not be known, a set of randomly selected points may be chosen first. Then, either suitable wells nearest to the chosen points are selected, or sampling sites are randomly selected from wells located within a certain radius of the point.

A grid-based approach also can be employed for a resource assessment that uses newly constructed wells, but decision rules must be established to translate a point on the map to an actual site in the field. Several problems are likely to occur, not the least of which is obtaining permission to install the well at the desired location. Prior to visiting the field, one should develop a specific set of instructions about how to select the sites from the points identified on the map. The instructions should include a set of required distances from various types of physical structures, such as houses, ditches, and roads. In essence, these instructions are part of the definition of the target population.

For any of the above types of ground-water-quality investigations, the possibility exists that a suitable well cannot be found or installed at or near the location selected. This may occur for a number of reasons, including a lack of suitable wells, problems with access, the site in the field turns out to be located in a different setting than it is supposed to represent, and so forth. Also, when constructing new wells some locations may have to be abandoned because an insufficient water yield is obtained or problems occur in developing the well satisfactorily. For these reasons, locations for alternative sites commonly should be selected as part of the random selection process. For example, when using grid-based sampling, one might select an alternative site in each grid cell. Records should be maintained of reasons for the nonselection of primary sites in order to identify and evaluate possible bias in the final selection of sites.

Temporal Variability

The possible influences of temporal variability in ground-water quality should be considered in any ground-water-quality survey and should be explicitly recognized in sampling designs. For logistical reasons, several months or longer may be required to carry out surveys of ground-water quality. Constraints may include the (1) long distances between wells, (2) lengthy times required to gain access and do the water-quality sampling, (3) limited number of qualified people and equipment available for water-quality sampling, and (4) limited laboratory capacity available to process the samples within the allowable sample holding times.

The long time periods required to complete a sampling survey may result in confounding influences due to temporal variability. For instance, observed differences in concentrations between two hydrogeologic settings can result from intrinsic differences in water quality between the two settings or from differences in the time when samples were collected. These types of uncertainty are of particular concern for studies of shallow-ground-water quality.

Several approaches may be used to deal with temporal variability (Liddle et al. 1990). The appropriate approach depends on whether the intent is to represent average conditions over the year or to represent a particular time of year or hydrologic condition.

If average conditions over the year are to be represented and seasonal variations in water quality are likely, then the sampling can be spread out over the year or among seasons. One approach would be to divide the wells to be sampled into 12 groups and then randomly select a group of wells for water-quality sampling during each month of the year. This design complicates making comparisons among subsets of wells that were sampled during different seasons, because seasonal differences cannot be separated from between-well differences.

If ground-water-quality conditions during a particular time of year or hydrologic condition are to be represented, the sampling survey should be carried out within a specific index period. The index period should be defined in light of general knowledge about recharge periods, management practices such as agricultural chemical applications, and other factors likely to result in measurable seasonal differences in ground-water quality. Tests for differences in water quality between seasons may be done through stratification into two or more index periods with water-quality samples collected from the same wells during each of the periods. Some surveys have chosen a subset of sampled wells for seasonal sampling (cf. Hallberg et al. 1990).

One possible approach to reduce the time required to collect the water-quality samples is to use several teams that operate simultaneously. If this option is taken, extensive effort should be made to ensure consistency among the different teams. In general, consistency is more likely if a single team collects all samples.

Temporal variations in water quality during the index period contribute to random sampling errors. These errors will be of unknown magnitude if only a single sample is collected from each site during the index period. Thus, it may be useful to collect replicate samples from a few wells at random times during the index period to evaluate the possible magnitude of these errors. It may be most efficient to explore these and other temporal variations in conjunction with targeted sampling where the data are likely to be most interpretable.

Future Network Modifications

A characteristic of virtually all water-quality sampling programs is that knowledge is attained about a more efficient design after sampling is completed and the results are analyzed. For long-term studies, the anticipation that modifications may be made to the network at a future date favors the utilization of fairly simple designs at the outset. These designs can be more easily modified in the future while continuing to comply with the essential rudiments of probability sampling theory. Because of the logistical complexity involved in ground-water-quality surveys, it is often useful to carry out a pilot run of the process first.

PROBABILITY SAMPLING APPROACHES

Figure 3-1 shows several probability sampling plans for spatially distributed variables (Gilbert 1987). These are reviewed individually here. The sampling designs considered are rather simple. Studies with very specialized objectives, such as risk assessment, may require more complicated designs. In the ensuing discussion, uppercase letters are used to refer to characteristics of the population, and lowercase letters to those of the sample.

Survey sampling may be undertaken to estimate any of several possible characteristics of the population. Invariably, interest in ground-water-quality surveys extends well beyond simply estimating mean values. Furthermore, as noted in Chapter 2, nonparametric procedures are preferred for many statistical analyses of ground-water-quality data, because the data commonly are highly skewed and may contain high outliers and "nondetects." The following discussion treats the estimation of sample means, medians, and proportions. Discussion of sample means is used largely to introduce basic concepts.

Simple Random Selection

The principles of simple random selection underlay the other probability sampling schemes discussed herein. The basic idea is to select n units out of a population of size N such that every unit has an equal and known chance of being drawn.

In its simplest form, the units of a population are numbered from 1 to N. A series of random numbers between 1 and N is then drawn by means of a computer or table of random numbers. The sample of n units is drawn unit by unit, generally *without replacement*. That is, once a unit has been selected it is no longer a candidate for future draws.

Random sampling *with replacement* also is possible. In this case, each of the N members of the population has an equal chance of being selected at each draw no matter how many times it has already been drawn. Thus, some sites may be measured more than once. The formulas for the estimated variances of estimates from the sample are often simpler for sampling with replacement than without replacement. For this reason, sampling with replacement is sometimes used in more complex sampling plans.

Estimation of Means

For a random sample of size n from an infinite population, the variance of the mean is simply

$$\text{Var}(\bar{x}) = \frac{\sigma^2}{n} \qquad (3\text{-}3)$$

where σ^2 is the population variance of the random variable X. When the population is finite, a *finite population correction* (fpc) factor is needed to modify this result. The

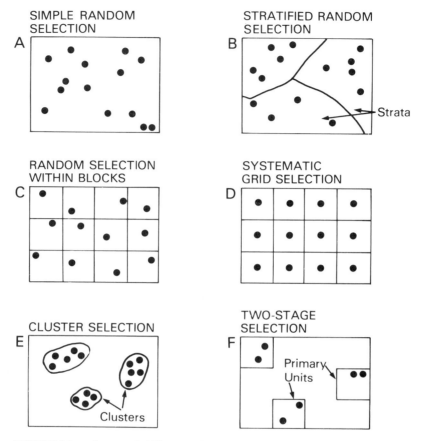

FIGURE 3-1. Some probability sampling designs for spatial sampling. (*After Gilbert 1987.*)

fpc is equal to $(N - n)/N$ or, equivalently, to $(1 - f)$, where f is the sampling fraction n/N. Thus,

$$\text{Var}(\bar{x}) = \frac{\sigma^2}{n}\left(\frac{N - n}{N}\right) = \frac{\sigma^2}{n}(1 - f) \quad (3\text{-}4)$$

The finite population correction factor results in a decrease in the estimated variance for larger values of the proportion of the population sampled. In the extreme case where all members of the population are sampled (i.e., $f = 1$), the estimated variance of the mean is equal to zero, as expected. For ground-water-quality studies, use of the fpc is only appropriate for water-well surveys where the target population consists of a known number of wells. For resource assessment, the sampling fraction, f, is assumed equal to zero.

Computation of the fpc requires an estimate of the total number of wells in the target population. As an example, McKenna et al. (1989) estimated the number of wells using a computer data base of the number of private water wells by zip code compiled by the National Water Well Association (1986) from U.S. Census Bureau data.

Equation 3-4 is appropriate only when sampling without replacement from a finite population of N units. Sampling with replacement is analogous to sampling from an infinite population; thus, the sampling fraction f is set to zero, and Eq. 3-4 reduces to Eq. 3-3.

An unbiased estimator for Var($\bar{x}$) is obtained by replacing σ^2 in Eqs. 3-3 and 3-4 by s^2 from Eq. 3-2. Thus, for example, Eq. 3-4 becomes

$$\text{var}(\bar{x}) = \frac{s^2}{n}(1 - f) \qquad (3\text{-}5)$$

The estimate for the standard error of the mean (se) is then

$$\text{se}(\bar{x}) = \frac{s}{\sqrt{n}}\sqrt{1 - f}$$

Note the use of the lowercase "var" and "se" to denote that these are sample estimates.

If either the population is infinite or the sampling fraction is small, the size of the population has minimal, if any, direct effect on the standard error of the sample mean. For instance, if σ is the same in two populations, a sample of size 100 from a population of 1 million gives almost as precise an estimate of the population mean as a sample of size 100 from a population of 10,000. The correction factor $\sqrt{1 - f}$ is .9999 and 9950, respectively, for these two cases.

This property is important when sampling from large populations. It is not uncommon to hear sampling surveys criticized because they only sampled a very small percentage of the population. As demonstrated, the size of the sample, not the proportion of the population it contains, generally determines the precision of estimation.

Figure 3-2 shows the standard error of the mean for samples of varying sizes drawn using simple random selection from populations with variances of .25, .5, and 1. It illustrates that standard errors rapidly increase as sample sizes decrease below about 20.

If an estimate is available for the variance of X, then one can estimate the sample size required to achieve some prespecified precision in the sample mean. For example, suppose that the true mean μ_x will be estimated by $\bar{x}$ and that Var($\bar{x}$) should be no larger than a prespecified value V. Setting $V = \text{Var}(\bar{x})$ in Eq. 3-4 and solving for n yields

$$n = \frac{\sigma^2}{V + \sigma^2/N} \qquad (3\text{-}6)$$

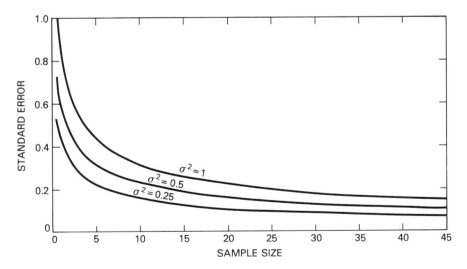

FIGURE 3-2. Standard error of the mean for random samples of varying size drawn from populations with variances (σ^2) of 0.25, 0.5, and 1. (*From Webster 1977, by permission of Oxford University Press.*)

If N is very large relative to σ^2, or for infinite populations or small sampling fractions f, Eq. 3-6 reduces to

$$n = \frac{\sigma^2}{V} \qquad (3\text{-}7)$$

Other equations can be used to determine n based on a prespecified absolute or relative error that can be tolerated when estimating the mean with a specified probability of exceeding that error (Cochran 1977; Gilbert 1987).

Sample size equations, such as Eqs. 3-6 and 3-7, should be used with care since they only account for random sampling error. In addition, popular sample size formulas may tend to underestimate required sample sizes, because the final reported precision also depends on the sample variance of the data values collected (Kupper and Hafner 1989; Blackwood 1991). Also, an a priori estimate of the variance of X may be difficult to obtain. Furthermore, data usually are being collected for more than one water-quality constituent, and these constituents may exhibit a wide range in variability. Calculations of sample size can lead to a set of conflicting values among constituents, and some method must be found for reconciling these values. Because of these complexities and others, estimates of sample sizes required to meet specified levels of precision should be developed in consultation with someone well versed in probability sampling theory. Perhaps one of the primary benefits of making preliminary sample size calculations is that it leads to a healthy degree of skepticism about the likely precision of results attainable within the resources available to a project.

Estimation of Medians

For skewed data with considerable potential for outliers, the median is a more robust measure of the central tendency of a distribution than is the mean. For data that may be approximated by a lognormal distribution, sample sizes required for estimating the median within a desired degree of precision can be estimated using an expression derived by Hale (1972) and reported in Gilbert (1987, p. 174). Conover (1980, pp. 111–12) and Iman and Conover (1983, pp. 199–201) present a method for obtaining the confidence interval of the median and other quantiles from sample data. An example application to ground-water-quality data is given by Spruill and Candela (1990).

Estimation of Proportions

Sometimes the goal of a sampling program is to estimate the proportion, or percentage, of units in a population that possess some characteristic or fall into some defined class. For instance, one may be interested in an estimate of the proportion of ground water that contains detectable concentrations of a selected suite of pesticides or in the percent exceedance of a water-quality standard.

Consider the case in which every unit in the population falls into one of two classes, C and C'. For any unit in the sample or population, define x_i as 1 if the unit is in C and as 0 if it is in C'. Denote P as the proportion of units in the population that fall into class C. An unbiased estimator, p, is then simply the average value of x_i. That is,

$$p = \frac{1}{n} \sum_{i=1}^{n} x_i \qquad (3\text{-}8)$$

An unbiased estimate of the variance of p derived from the sample is

$$\text{var}(p) = \frac{p(1-p)}{n-1}(1-f) \qquad (3\text{-}9)$$

If n is small relative to N, so that f can be ignored, this reduces to

$$\text{var}(p) = \frac{p(1-p)}{n-1} \qquad (3\text{-}10)$$

Confidence limits for P are sometimes calculated as

$$p \pm \left\{ z\sqrt{\frac{p(1-p)}{n-1}} \sqrt{1-f} + \frac{1}{2n} \right\} \quad (3\text{-}11)$$

where z is the normal variate for the chosen probability (e.g., $z = 1.96$ for a 95% confidence interval). Equation 3-11 assumes either an infinite population or a small sample size relative to the number of population units (i.e., the sampling fraction f is small). It also requires a normal approximation to the binomial distribution. Cochran (1977, p. 58) gives a table of "working rules" for deciding when the normal approximation may be used. The term $1/2n$ in Eq. 3-11 represents a correction for continuity that is needed because the normal distribution is continuous, whereas the binomial distribution is not.

Equations 3-9 and 3-10 show that, for a given sample size, the variance of the estimated proportion is simply a function of its magnitude. For illustration purposes, estimates of the standard error of the estimate p (the square root of Eq. 3-10) expressed as an absolute magnitude and as a percent of the value p are listed as follows for a sample size of 100 from an infinite population.

p	Standard Error	Standard Error as a Percentage
0.02	0.014	70
0.10	0.030	30
0.50	0.050	10
0.90	0.030	3
0.98	0.014	1

Although, in an absolute sense, the greatest error is likely to occur when the proportion equals .50, relative error increases as p decreases and can become quite large for small values of p (i.e., rare events). This implies that large sample sizes may be required to estimate reliable occurrence percentages for constituents that are detected infrequently.

Estimates of sample sizes required to achieve certain levels of precision in the estimate p can be made just as in the case of the mean (cf. Chapter 20). An advantage is that Eqs. 3-10 and 3-11 are simply a function of p itself and n. Again, care is required in understanding the limitations behind any estimates made of required sample sizes.

Stratified Random Selection

Stratified random selection (see Figure 3-1B) is a useful extension of simple random selection. The approach makes use of prior supplementary information to subdivide the target population into subpopulations that are likely to be more homogeneous than the population as a whole. Each subpopulation (*stratum*) is sampled individually by simple random selection. The results can be used to estimate statistical properties for each stratum, as well as for the full target population.

The supplementary information used to subdivide the target population into strata may be somewhat subjective. However, a reasonable stratification based on sound judgment will almost invariably lead to more precise estimation of population statistics than will simple random selection (Stuart 1976). Stratification tends to produce large gains in precision of the sample estimates for the population when the strata have been defined so that (1) the average values of the characteristic of interest are as different as possible among the strata, and (2) the within-stratum variances of the characteristic are as small as possible.

It is important to distinguish between the concept of subunits introduced in Chapter 1 and statistical strata. Considerable latitude exists in the choice of subunits, and they may or may not completely encompass a groundwater system. Statistical strata, on the other hand, must be defined so that the entire population of interest is completely specified by a set of nonoverlapping strata. Although estimation of the characteristics of each statistical stratum may be important, a key purpose of the stratification is to improve estimation of characteristics for the overall population that has been stratified.

Let N be the total number of units in the target population, and L the number of non-overlapping strata into which the population is subdivided. The strata are denoted by the subscript h. The number of population units in the hth stratum is thus denoted by N_h, and the number of units measured in the hth stratum by n_h. The values of N_h and n_h for the L strata sum to N and n, respectively.

Unbiased estimates of the mean and variance of x within the hth stratum are obtained from simple random selection and are computed as

$$\bar{x}_h = \frac{1}{n_h} \sum_{i=1}^{n_h} x_{hi}$$

$$s_h^2 = \frac{1}{n_h - 1} \sum_{i=1}^{n_h} (x_{hi} - \bar{x}_h)^2$$

where x_{hi} denotes the ith unit within the hth stratum.

Estimation of Means

An unbiased estimator of the population mean is obtained by weighting the individual stratum means. That is,

$$\bar{x}_{st} = \sum_{h=1}^{L} W_h \bar{x}_h$$

where the weights $W_h = N_h/N$ represent the relative sizes of the strata and are assumed to be known.

If the samples are drawn independently from the different strata, then the variance of the sample mean for the target population is the sum of the variances of the sample mean for the strata multiplied by their squared weights. That is,

$$\text{Var}(\bar{x}_{st}) = \sum_{h=1}^{L} W_h^2 \text{Var}(\bar{x}_h)$$

$$= \sum_{h=1}^{L} W_h^2 \left(\frac{\sigma_h^2}{n_h}\right)(1 - f_h)$$

where σ_h^2 is the variance of X for the hth stratum, and $f_h = n_h/N_h$ is the sampling fraction for the hth stratum. If each of the σ_h^2 is an unbiased estimate, an unbiased estimate of $\text{Var}(\bar{x}_{st})$ is obtained by substituting s_h^2 for σ_h^2:

$$\text{var}(\bar{x}_{st}) = \sum_{h=1}^{L} W_h^2 \left(\frac{s_h^2}{n_h}\right)(1 - f_h)$$

This reduces to

$$\text{var}(\bar{x}_{st}) = \sum_{h=1}^{L} W_h^2 \left(\frac{s_h^2}{n_h}\right)$$

if the sampling fractions f_h are small. To compute s_h^2, and hence $\text{var}(\bar{x}_{st})$, requires a minimum of two units to be drawn from every stratum.

In stratified random selection, the values of the sample sizes, n_h, in the respective strata are preselected by the investigator. A simple approach is to use proportional allocation $n_h = nW_h$. For example, if a given stratum is 25% of the entire target population, then 25% of the samples are allocated to that stratum. Commonly, however, it is desirable to "oversample" some strata and "undersample" others. In general, strata that have greater variability or are less expensive to sample would be oversampled.

In consideration of the use of variable sampling sizes among strata, it is important to determine whether the primary goal is to estimate the statistical characteristics of the overall population or to compare results among the strata. The former case suggests variable-size strata based on the previously mentioned considerations of the relative size of each stratum, the likely variability of the characteristic within each stratum, and the cost of sampling. On the other hand, if interest centers on comparing results among the strata, then a more equal number of samples from each stratum may be preferred.

Estimation of Proportions

Stratified random selection also can be used for estimating the proportion of units that fall into some defined class C. In this case, the idea is to select strata such that the proportion in class C varies as much as possible among the strata.

Let P_h be the proportion of units that fall in class C within the hth stratum. The sample estimate of this proportion is denoted by p_h. An estimator for the proportion P for the population as a whole is

$$p = \sum_{h=1}^{L} p_h \left(\frac{N_h}{N}\right)$$

and the variance of this sample estimate is

$$\text{Var}(p) = \sum_{h=1}^{L} W_h^2 \frac{(N_h - n_h)}{(N_h - 1)} \frac{P_h(1 - P_h)}{n_h}$$

When the sampling fraction in each stratum is negligible, this reduces to

$$\text{Var}(p) = \sum_{h=1}^{L} W_h^2 \frac{P_h(1 - P_h)}{n_h}$$

To obtain the sample estimates of the variance, one simply substitutes $p_h(1 - p_h)/(n_h - 1)$ for the unknown $P_h(1 - P_h)/n_h$ in the preceding formulas.

Selection of Statistical Strata

To apply the principles of stratified random selection, one must know the weights associated with the different strata (i.e., W_h, $h = 1, \ldots, L$). This limits the types of variables that can be used to stratify the target population to those for which information is complete enough that the relative size of the strata can be estimated.

Sometimes, stratification is established by combinations of multiple factors. For example, one might stratify by six classes that result from the combination of two intervals of depth to the water table and three land-use types. Because of limitations of sample size in most ground-water-quality investigations, the number of stratification classes should be kept small. This is particularly true if a goal is to estimate statistical properties of the strata themselves in addition to the overall population or subunit.

The method of stratification and of assigning weights differs between well-water surveys and resource assessments. For well-water surveys, ideally a complete listing of all candidate wells is available, and the value of the stratification variable(s) is known for all wells. For this situation, the process of stratifying the population is relatively straightforward, and the population weights are assigned proportional to the number of wells in each stratification class. In the absence of this information some method of approximating this ideal must be developed.

For resource assessment, the stratification variable should be available in map form over the region of interest. Examples of potential stratification variables include soil type, land use, or mapped indices of aquifer vulnerability. The assumed population weights, W_h, $h = 1, \ldots, L$, may then be based on the relative surface area covered by each stratification class.

The three-dimensional nature of groundwater systems presents special problems, because depth is an important factor in addition to areal characteristics. Depth considerations include those associated with the vadose zone and those related to thickness of the saturated zone.

Several simple measures of the depth of the vadose zone may be used as stratification variables, including depth to the top of the water-bearing unit or depth to the water table (for unconfined wells). When maps of these variables are available, the variables can be used for stratification as would any other mapped variable.

The depth associated with the saturated zone is a more complicated factor to incorporate into a stratified random design. The operational definition of depth must be related to the construction of the well. Examples include simply the depth of the well or measures of depth related to the open interval of the well such as the depth to the top, bottom, or midpoint of the open interval.

Strictly speaking, it is impossible to stratify the saturated thickness of an aquifer into independent depth strata, because any component of vertical flow into the well will circumvent the ability to draw water solely from the depth strata defined for the well. Vertical-flow components can result from heterogeneity in geologic materials, partially penetrating wells, gravel packs, and leakage into the well, among other factors.

An alternative to using depth is to stratify by position in the flow system, for example, stratifying by recharge areas, discharge areas and intermediate areas or by local and regional flow systems. This requires substantially more information than simply specifying depth, but if the information exists stratification by position in the flow system could lead to more precise results. In general, a ground-water-flow model would be required to delineate the strata. It may be difficult to find a set of existing wells with well-construction characteristics (e.g., open intervals) compatible with flow-system stratification.

Grid-Based Approaches

Random Selection within Blocks
Because potential measurement sites are commonly clustered in particular areas, simple random selection is unlikely to achieve a set of sampling sites that are areally distributed throughout the region of interest. This may lead to certain biases in the data.

A useful modification of simple random selection is to divide the region into blocks and to select randomly one or more measurement sites within each block. This procedure, referred to as random selection within blocks (see Figure 3-1C), is likely to produce a more uniform distribution of sites than does simple random selection. Another advantage of random selection within blocks and other grid-based approaches is that they tend to reduce the spatial correlations among wells. As a cautionary note, random selection within blocks may yield a biased sample for a well-water survey, because relatively isolated wells will be over-represented, unless the size of the target population within each block is accounted for in the selection process.

Intuitively, random selection within blocks seems to be more precise than simple random selection. In effect, it stratifies the population into strata represented by the blocks. Thus, if more than one site is selected per block, one can use the afore-mentioned equations for stratified random selection.

If just one site is sampled within each block, then treatment of the problem as a stratified random design is hindered, because the within-block variance cannot be estimated with just one sample. A common approach is to use the blocks for site selection, but to analyze the results as though the survey was based on simple random selection. In the context of the many other assumptions associated with collecting a "random" sample of ground-water quality, this is perhaps a reasonable approach. However, it will be strictly valid for estimating the variance of the sample mean (or proportion) for the whole population only if there are no spatial patterns or correlations in the constituent concentrations.

Systematic Grid Selection
A variation of random selection within blocks is systematic grid selection (see Figure 3-1D). Systematic grid selection consists of taking measurements at locations according to a set spatial pattern. A number of different patterns may be used for systematic grid selection. One is to simply use grid centers; another is to randomly select a starting place in one grid block and to repeat that same relative position in the other grid blocks. Gilbert (1987) discusses these and other patterns.

An advantage of systematic grid selection over random selection within blocks is that a more uniform pattern is achieved over space. For example, random selection within blocks may lead to some sites in adjacent blocks that are very close to one another, while systematic grid selection guarantees equal spacing. A disadvantage of systematic grid selection is that incorrect conclusions may be drawn if the sampling pattern corresponds to an unsuspected pattern of contamination over space, perhaps related to a pattern in land use or geology. As in the case of random selection within blocks, systematic grid selection may yield a biased sample for a well-water survey unless the size of the target population within each block is accounted for in the calculations.

An example of a systematic survey is the Iowa State-Wide Rural Well-Water Survey (Hallberg et al. 1990). The sampling frame for this survey was defined using every 5-min intersection of latitude and longitude in the State. The intersections selected for sampling were determined proportional to rural population on the basis of county-level data on rural population density.

Random Selection from Irregular Polygons
The approach to site selection for random selection within blocks is easy to apply over a single rectangular region. Unfortunately, the area of interest for a ground-water-quality sampling network commonly is not rectangular and could consist of one or more irregular polygons. As an example, Figure 3-3 shows the selection of wells for a ground-water-quality survey of alluvial and terrace deposits near Oklahoma City. The area of interest consists of several very irregular polygons that follow narrow bands along rivers traversing the area between the Cimarron and Canadian rivers.

Computerized approaches to treat situations such as shown in Figure 3-3 are described by Scott (1990) for use with geographic information systems (GIS). One of these approaches, referred to as random selection with equal-area distribution, proceeds as follows.

Let the number of blocks be denoted n_b and the number of sites to be selected per block by n_s. The value of n_s is equal to the number of primary sites, n_p, plus the number of alternative sites, n_a: $n_s = n_p + n_a$. Now, $n_p = 1$ if the complete sample is treated as a simple random sample, or $n_p = 2, \ldots$ if each block is treated as a stratum in a stratified random design. The number of alternate sites, n_a, is user-specified to provide in advance for potential difficulties that may be encountered when visiting a prospective primary site.

A set of irregular blocks is created such that each block contains an equal area of the category (subunit or statistical stratum) of interest. An example of this approach is shown in Figure 3-4.

Figure 3-4A shows a region subdivided into three categories: A, B, and C. The goal is to subdivide category A into four blocks. First, a fine-meshed square grid is overlaid on the map, and the grid boundaries are trimmed so that each includes only a single category (see Figure 3-4B). These "subareas" are then aggregated to form blocks such that each block contains approximately the same area of the category of interest. This is achieved by first dividing the map into vertical strips (see Figure 3-4C). The number of strips is computed by taking the square root of the number of blocks to be selected and rounding to the nearest integer. The vertical strips are defined so that each contains an equal area of the category.

Beginning in the lower-left corner of the map, the subareas in each horizontal row within the strip are aggregated to form a block (see Figure 3-4D). When the correct area for the block has been obtained, formation of the next block is begun. The subarea aggregation continues until the top of the vertical strip is reached and then moves to the top row of the next vertical strip to the right. The aggregation in this strip continues with each row downward until the correct area for the block has been reached. Then formation of the next block is begun. A similar process occurs when the bottom of the vertical strip is reached. Block formation continues until all subareas have been assigned to blocks.

Sites are selected from the category of interest within each block using simple random selection. The selected sites consist of either randomly selected points within the

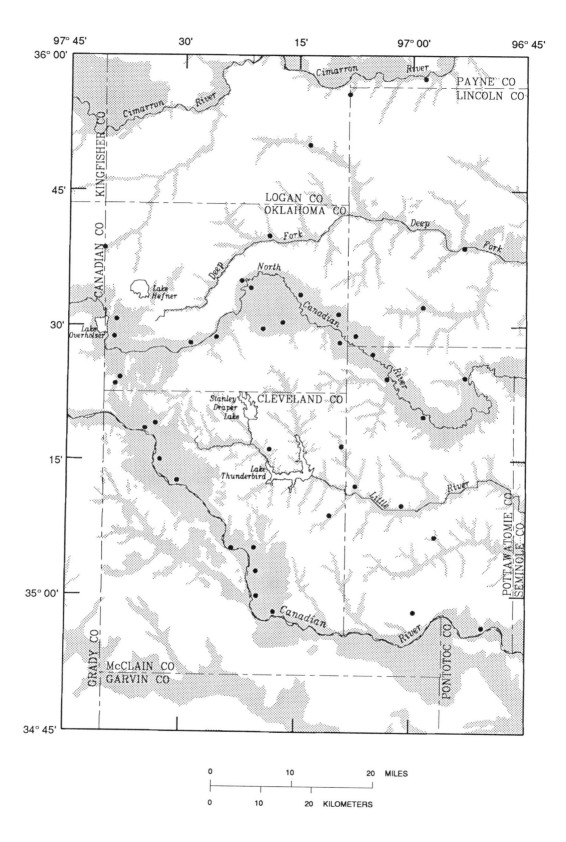

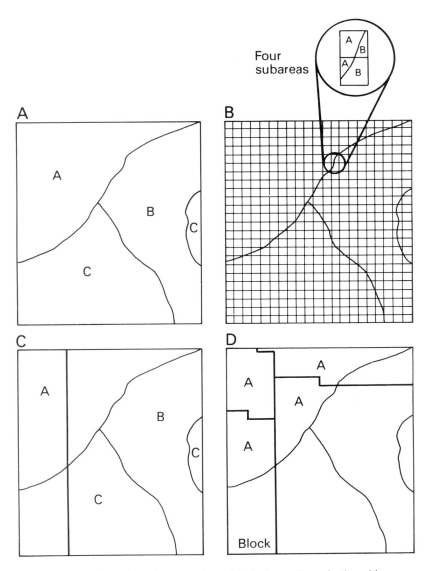

FIGURE 3-4. Example of the generation of blocks for random selection with equal-area distribution. Example is to create four blocks for areal subset A. [(A) map of areal subsets for three categories; (B) creation of subareas; (C) division into two vertical strips containing equal areas of category A; and (D) division into four blocks each having an equal area of category A]. (*After Scott 1990.*)

FIGURE 3-3. Location of wells (heavy black dots) sampled in alluvium and terrace deposits (shaded areas) in the Central Oklahoma aquifer. (*From S. C. Christenson, U.S. Geological Survey, written communication, 1991.*)

blocks or wells randomly selected from candidate wells. If random points are selected, a sampling frame of equally spaced points is first created starting from a randomly selected point below and to the left of the lower-left corner of the map.

Complications arise in the method when blocks are created that contain no potential sites or too few sites for alternative site selection. Also, the method is not completely unbiased for resource assessment, because the probability of selecting a site from a particular polygon in a block containing multiple polygons is proportional to the number of candidate sites in the polygon, not to the polygon area.

Cluster and Staged Selection

In each method discussed thus far, the sampling process selects one member of the population or stratum at a time. Cluster and two-stage selection (see Figures 3-1E, 3-1F) proceed by a somewhat different process (Cochran 1977; Gilbert 1987). In these methods, the population is subdivided into groups referred to as clusters and primary units, respectively. A subset of the clusters or primary units is then selected randomly. In cluster selection, all units in the chosen clusters are measured, while in two-stage selection each of the selected primary units is randomly subsampled. Of the two approaches, two-stage selection is the more likely choice for ground-water-quality studies.

Two-stage selection has been applied in well-water surveys when a list of wells that make up the target population is not available and is too expensive to develop for the full target population. For instance, primary units might be defined that comprise geographic units such as counties, and a subset of these counties randomly selected. The set of wells for the selected counties would be enumerated, and a random sample taken within each county. This would reduce the need for well enumeration to the selected counties only. The procedure would still require, however, an estimate of the number of wells in the full target population so that the selection probability of each chosen well is known.

Any decision to use cluster or two-stage selection must carefully consider the limited areal distribution of sampling sites that results from the selection process, as well as the potential for spatial correlation among the samples. Cluster or staged sampling may be most appropriate for surveys of large areas such as national or State well-water surveys where enumeration of the well population is a significant issue. For instance, staged selection was used in the design of the USEPA National Pesticide Survey (U.S. Environmental Protection Agency 1990) and in the final design of a survey of agricultural chemicals in rural Illinois wells (see Chapter 20). Gilbert (1987) discusses two-stage selection in the context of environmental sampling where the primary units might represent, for example, soil samples, and the individual measured units are subsamples or aliquots of these soil samples.

Random Selection with Auxiliary Variables

As discussed previously, stratification is a useful means of using supplementary information about a population to improve the precision of estimation of the mean or proportion of a population in a defined class. If the supplementary information is of a more precise kind, other applications of the information are possible.

Ratio and Regression Estimators

Assume that the following knowledge exists about an auxiliary variable, W, related to X: (1) the value of the auxiliary variable for each member of the set of observations of x_i, and (2) the average (or proportion) of the auxiliary variable for the population as a whole. In such a case, one can use this information about the auxiliary variable to improve the precision with which the population average (or proportion) of the primary variable is estimated.

One approach is to use a simple ratio estimator

$$\bar{x}' = \bar{x}\left(\frac{\mu_w}{\bar{w}}\right)$$

where $\bar{x}'$ is the adjusted estimate of the mean of X, and μ_w and $\bar{w}$ are the population and sample estimate, respectively, of the mean of the auxiliary variable. The ratio $\mu_w/\bar{w}$ thus acts as an adjustment factor on the sample mean of X.

Ratio estimators assume that the values of the primary variable are proportional to the values of the auxiliary variable. This is commonly not true. In some cases, the relation between the primary and secondary variable may be linear, but does not go through the origin. This situation suggests an estimate based on the linear regression of x_i on w_i rather than a simple ratio estimator. In this case, the adjustment to $\bar{x}$ takes the form

$$\bar{x}' = \bar{x} + b(\mu_w - \bar{w})$$

where b is the estimated slope (change in x when w is increased by unity) obtained by simple linear regression using the sample data. This relation indicates that if the sample estimate $\bar{w}$ is too low, then $\bar{x}$ also is expected to be too low and is adjusted upward by an amount $b(\mu_w - \bar{w})$. Conversely, $\bar{x}$ is adjusted downward if $\bar{w}$ is too high.

Cochran (1977) discusses ratio and regression estimators in detail, including their variance and bias. Note the assumption that the auxiliary variable has a known population mean is very restrictive. For ground-water-quality investigations, a candidate auxiliary variable that meets this restriction might be a physical attribute of the land surface (such as land use, soils, or geology) that is mapped throughout the region. However, a related water-quality constituent generally would be inappropriate as an auxiliary variable, since the population mean for any water-quality constituent usually is quite uncertain.

Double Selection

Information on another water-quality variable may be used to improve estimates of the primary water-quality variable through a somewhat different procedure, referred to as double selection. The idea of double selection is to take a large sample of a relatively inexpensive auxiliary variable that is related to the primary variable of interest, and a smaller sample of the primary variable. The sample of the auxiliary variable may be used to (1) develop ratio or regression estimators similar to those described above or (2) to obtain insight into the importance of one or more stratification variables. For ground-water-quality constituents, the auxiliary variable may be a less accurate, but relatively inexpensive measurement of the constituent of interest or a related water-quality constituent.

Double selection involves a trade-off in using some of the sampling resources that would have been used for sampling the primary variable to carry out the sampling of the auxiliary variable. It is most useful when the correlation between the auxiliary and primary variables is high, and the auxiliary variable is much less expensive to measure than the primary variable. Double selection is limited in its utility to regional ground-water-quality investigations, because the fixed costs associated with locating and documenting sites, and collecting samples from wells are a large part of the sampling costs. Hence, although the analytical costs for two variables may be quite different, the difference in total costs for sampling the two variables may be relatively small. For a detailed discussion of double selection, the reader is referred to Cochran (1977).

GRID-BASED SEARCH SAMPLING

All of the methods discussed thus far, including the grid-based methods, have involved probability sampling and random

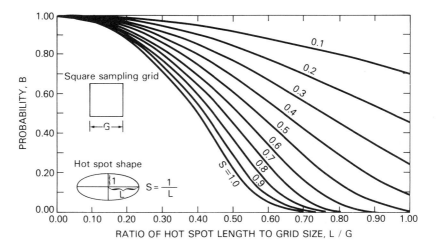

FIGURE 3-5. Probability, B, of not finding a circular or elliptical hot spot as a function of the hot-spot shape and the size of the target relative to a square grid spacing. (*After Zirschky and Gilbert 1984.*)

selection. It is worth noting that grid-based methods may be considered in an entirely different mode for search sampling. For example, suppose one has the goal of searching a two-dimensional area to determine the locations of "hot spots" of contamination, and it is desired to make statements about the spatial resolution of the sampling effort (i.e., the scale of contaminated areas that will or will not be detected).

The problem of determining the spatial resolution of target (hot spot) detection is one of chance and the problem should be specified in probabilistic terms. Gilbert (1987) lists three typical questions that might be asked: (1) What grid spacing is needed to hit a hot spot with specified confidence? (2) For a given grid spacing, what is the probability of hitting a hot spot of specified size? (3) What is the probability that a hot spot exists when no hot spots were found by sampling on a grid?

Grid-based techniques have been developed for which answers to these questions can be determined for hot spots of prespecified size and simple geometric shape. Many of these techniques have been developed by geologists and applied statisticians interested in formulating hot-spot detection schemes for mineral exploration. This work has been based, in part, on the more general studies of search theory by Koopman (1956). The methods generally require some rather restrictive assumptions, including (1) in plan view the hot spots are circular or elliptical; (2) there is no preferred orientation of the hot spots (this assumption is sometimes relaxed); (3) the distribution of hot spots does not follow a regular pattern; and (4) if a hot spot is hit with a grid point, the event can be recognized with certainty.

As an example, Zirschky and Gilbert (1984) used a computer program developed by Singer (1972) to develop nomographs for answering the three questions cited by Gilbert (1987) for a square grid pattern. These nomographs pertain to a single hot spot. Figure 3-5 is the nomograph for addressing the first question. The following must be specified to use Figure 3-5: (1) the length L of the semimajor axis of the smallest hot spot that is important to detect, (2) the expected shape S of the hot spot, where S is defined by the ratio of the short axis to the long axis, and (3) the acceptable probability B of not finding the hot spot. To use Figure 3-5, first find L/G and then solve for the required grid spacing G.

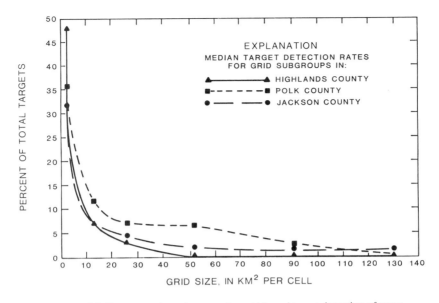

FIGURE 3-6. Median target detection rates for grid-based target detection of areas with EDB ground-water contamination in three counties in Florida. (*From Choquette and Katz 1989.*)

Procedures such as that illustrated in Figure 3-5 are simple to apply. Some applications may exist for studies of large natural geochemical anomalies, or perhaps under special circumstances, to broad-scale contamination within a relatively homogeneous area, such as shallow-ground-water contamination within an agricultural area. However, in general the procedures are likely to be inefficient for regional ground-water-quality investigations. A few of the many complications arise because (1) ground-water-quality problems are three-dimensional rather than two-dimensional in nature, (2) contamination is likely to be noncontiguous and local in scale, (3) the size and shape of the targets are not known, (4) ground water cannot be sampled at all possible points in a grid, and (5) if a hot spot is hit with a grid point, the event cannot be recognized with certainty; false positives and false negatives are quite possible given the complications in sampling and analyzing ground-water samples, particularly for trace substances.

As an illustration of some of the limitations, Choquette and Katz (1989) tested grid-based selection approaches to determine their effectiveness for target detection using data from an existing network of data on 1,2-dibromoethane (EDB). The network averaged about one well per square kilometer in each of three study areas that ranged in size from 1,040 to 2,770 km². This network represents one of the most densely sampled regional networks in the United States. On the basis of the locations of existing wells with detectable EDB, typical hydraulic properties of the aquifers sampled, and the time since the EDB was applied, Choquette and Katz (1989) estimated the location and extent of contaminated hot spots. They then performed experiments to determine the percentage of targets that would have been detected for various grid sizes.

The results are shown in Figure 3-6, which shows the median target detection rates (from three replications at each grid size) for each of the three study areas as a function of the grid size. For example, in all three study areas, less than 10% of all targets were identified with grid sizes larger than 13 km² (5 mi²). Less than half of all targets were located with the smallest grid size of 2.6 km² (1 mi²). In addition, simply detecting a set of targets at these scales provides virtually no information about their size and shape.

CONCLUDING REMARKS

To infer statistical properties from sample to population, the selection process is an integral part of the inference. Thus, if one of the objectives of a survey of ground-water quality is to make inferences about a population, the survey should adhere to certain general principles of probability sampling. These include clear statements of the survey's goals and objectives, precise definition of the target and candidate populations, and randomized well selection. A number of complexities associated with selecting and sampling wells confound the application of statistical theory to ground-water-quality surveys. In general, probability sampling theory should be used as a tool to focus objectives of the study and to minimize potential sources of error rather than to blindly dictate a set of hard-and-fast rules.

References

Blackwood, L. G. 1991. Assurance levels of standard sample size formulas. *Environmental Science and Technology* 25(8):1366–7.

Choquette, A. F., and B. G. Katz. 1980. Grid-based groundwater sampling: Lessons from an extensive regional network for 1,2-dibromoethane (EDB) in Florida. In *Regional Characterization of Water Quality*, pp. 79–86. Wallingford, United Kingdom: International Association of Hydrological Sciences Publication No. 182.

Cochran, W. G. 1977. *Sampling Techniques*. 3rd ed. New York: Wiley.

Conover, W. J. 1980. *Practical Nonparametric Statistics*. 2nd ed. New York: Wiley.

Gilbert, R. O. 1987. *Statistical Methods for Environmental Pollution Monitoring*. New York: Van Nostrand Reinhold.

Gillham, R. W., M. J. L. Robin, J. F. Barker, and J. A. Cherry. 1983. *Groundwater Monitoring and Sample Bias*. Washington, D.C.: American Petroleum Institute Publication 4367.

Hale, W. E. 1972. Sample size determination for the log-normal distribution. *Atmospheric Environment* 6:419–22.

Hallberg, G. R. et al. 1990. *The Iowa State-Wide Rural Well-Water Survey Design Report*. Iowa City, Iowa: Iowa Department of Natural Resources Technical Information Series 17.

Harris, Jane, J. C. Loftis, and R. H. Montgomery. 1987. Statistical methods for characterizing ground-water quality. *Ground Water* 25(2):185–93.

Iman, R. L., and W. J. Conover. 1983. *A Modern Approach to Statistics*. New York: Wiley.

Koopman, B. 1956. The theory of search. Part 2. Target detection. *Operations Research* 4:503–31.

Kupper, L. L., and K. B. Hafner. 1989. How appropriate are popular sample size formulas? *The American Statistician* 43(2):101–5.

Liddle, S. K., R. W. Whitmore, R. E. Mason, W. J. Alexander, and L. R. Holden. 1990. Accounting for temporal variations in large-scale retrospective studies of agricultural chemicals in ground water. *Ground Water Monitoring Review* 10(1):142–6.

National Water Well Association. 1986. *Wellfax Database*. Dublin, Ohio: National Ground Water Information Center.

Nelson, J. D., and R. C. Ward. 1981. Statistical considerations and sampling techniques for ground-water quality monitoring. *Ground Water* 19(6):617–25.

Roux, P. H., K. Balu, and R. Bennett. 1991. A large-scale retrospective ground-water monitoring study for metolachlor. *Ground Water Monitoring Review* 11(3):104–14.

Scott, J. C. 1990. *Computerized Stratified Random Site-Selection Approaches for Design of a Ground-Water-Quality Sampling Network*. Oklahoma City, Oklahoma: U. S. Geological Survey Water-Resources Investigations Report 90-4101.

Sgambat, J. P., and J. R. Stedinger. 1981. Confidence in ground-water monitoring. *Ground Water Monitoring Review* 1(1):62–69.

Singer, D. A. 1972. Elipgrid, a Fortran IV program for calculating the probability of success in locating elliptical targets with square, rectangular, and hexagonal grids.

Geocom Bulletin 5(5–6):111–26.

Spruill, T. B., and Lucila Candela. 1990. Two approaches to design of monitoring networks. *Ground Water* 28(3):430–42.

Stuart, Alan. 1976. *Basic Ideas of Scientific Sampling*. 2nd ed. New York: Hafner Press. Griffin's Statistical Monographs and Courses No. 4.

U.S. Environmental Protection Agency. 1990. *National Survey of Pesticides in Drinking Water Wells Phase I Report*. Washington, D.C.: U.S. Environmental Protection Agency. EPA 570/9-90-015.

Walker, M. J., and K. S. Porter. 1990. Assessment of pesticides in upstate New York ground water: Results of a 1985–87 sampling survey. *Ground Water Monitoring Review* 10(1):116–26.

Webster, R. 1977. *Quantitative and Numerical Methods in Soil Classification and Survey*. Oxford: Clarendon Press.

Zirschky, John, and R. O. Gilbert. 1984. Detecting hot spots at hazardous-waste sites. *Chemical Engineering* 91(14):97–100.

4

Geostatistical Models

William M. Alley

INTRODUCTION

Geostatistical theory involves the application of adaptations of classical regression techniques to problems in which the spatial dependence among data points is a key factor. The theory exploits the observation that the values of spatially distributed data commonly are spatially correlated with values recorded at close locations being more highly correlated than values recorded at widely spaced locations. Geostatistics provides a set of probabilistic techniques aimed at (1) detecting and modeling the patterns of spatial dependence of attribute(s) values in space (and perhaps in time), (2) using these models for the assessment of uncertainty about unknown values at unsampled locations, and (3) using the estimates of uncertainty to aid network design and decision making.

Geostatistical analysis may be applied to local-scale studies carried out as part of regional ground-water-quality investigations and, in some instances, to regional studies of naturally occurring constituents or perhaps, nonpoint-source contamination over broad, relatively homogeneous areas. In addition, geostatistical analysis may be useful to help determine the spatial patterns and relative uncertainty associated with other spatial data sets that are used in interpreting the water-quality data, such as ground-water levels or soil characteristics. Even in the absence of a formal geostatistical analysis, the plotting of experimental variograms described in this chapter may be a useful exercise in exploring possible spatial and directional correlation among data.

Geostatistical theory was developed by Matheron (1963) and others from earlier empirical work. The theory initially was applied for estimating concentrations of minerals in ore bodies and recoverable reserves. Geostatistical techniques subsequently have been applied to many other spatial variables, including the mapping and modeling of ground water (Delhomme 1979; Gambolati and Volpi 1979; Sophocleous, Paschetto, and Olea 1982; Kitanidis and Vomvoris 1983; Neuman and Jacobson 1984) and geochemistry and ground-water quality (Myers et al. 1982; Gilbert and Simpson 1985; Candela, Olea, and Custodio 1988; Cooper and Istok 1988; Rouhani and Hall 1988; Loaiciga 1989; Samper and Neuman 1989). Several texts devoted to geostatistical analysis are available (Journel and Huijbregts 1978; Hohn 1988; Isaaks and Srivastava 1989; Cressie 1991).

In this chapter, a brief synopsis of geostatistical analysis is provided along with a discussion of possible applications and limitations for regional ground-water-quality studies. The discussion centers on applications to mapping problems. Geostatistical analysis also has been applied to parameter estimation and network design problems in ground-water flow and solute-transport modeling. These applications are not discussed here (see ASCE Task Committee on Geostatistical Techniques in Geohydrology 1990).

SOME PRELIMINARY CONCEPTS

Any treatment of geostatistics is inherently mathematical. As background, a few key concepts of probability theory are reviewed very briefly in this section. Readers familiar with probability theory may wish to skip this section.

Distribution Functions

Two general types of random variables exist. *Discrete random variables* take on values in a discrete set of numbers, such as the set of positive integers, whereas *continuous random variables* take on values in a continuous set, such as the set of real numbers. Key concepts in summarizing the possible values of discrete or continuous random variables are those of the cumulative distribution function, the probability function, and the probability density function.

Let X denote a random variable and x a possible value of X. (Note: As in Chapter 3, uppercase letters are used for random variables, and lowercase letters are used for deterministic variables and sample values.) For any real-valued random variable X, its *cumulative distribution function* (cdf) $F_X(x)$ is the probability that the value of X is less than or equal to x; that is, $F_X(x) = \text{Prob}[X \leq x]$. It is a nondecreasing function of x bounded between 0 and 1, inclusive.

For a discrete random variable

$$F_X(x) = \sum_{x_i \leq x} P_X(x_i)$$

where the sum is taken over all values of x_i which are less than or equal to x, and $P_X(x_i)$ is the *probability function*, defined as the probability that X takes on the value x_i.

The *probability density function* (or density function), $f_X(x)$, is the continuous random variable analog to the probability function. It is always greater than or equal to zero, and the area under the density function is, by definition, equal to 1. If b and c are two constants, the probability that X is greater than b and less than or equal to c may be expressed in terms of either the density function or the cumulative distribution function as

$$\text{Prob}[b < X \leq c] = F_X(c) - F_X(b)$$
$$= \int_b^c f_X(x)\,dx$$

Mathematical Expectation

An important concept in probability and statistics is the *expectation* of a random variable. For a discrete random variable X having possible values $x_1, \ldots, x_n$, the expectation of X is the mean or first moment, defined as

$$E[X] = \sum_{i=1}^{n} x_i P_X(x_i)$$

In words, the expected value of X is a weighted average of the possible values that X can take on, each value being weighted by the probability that X assumes it.

For a continuous random variable having a density function $f_X(x)$, the expectation of X is defined as

$$E[X] = \int_{-\infty}^{+\infty} x f_X(x)\,dx$$

These principles can be readily extended to a function of X. For example, the variance, or second moment, can be defined as the expected value of the squared deviation from the mean μ. That is,

$$\sigma_X^2 = \text{Var}(X) = E[(X - \mu)^2]$$

Thus, if X is a discrete random variable,

$$\sigma_X^2 = \sum_{x_i}(x_i - \mu)^2 P_X(x_i)$$

whereas if X is a continuous random variable,

$$\sigma_X^2 = \int_{-\infty}^{+\infty}(x - \mu)^2 f_X(x)\, dx$$

The expected value "operator" has several important properties. One is that the expected value of a constant k equals that constant. Another is that the expectation of a linear function of X is a linear function of the expectation of X. For instance, if a and b are constants,

$$E[a + bX] = a + bE[X]$$

Covariance and Correlation

The *covariance* between X and Y, $C(X, Y)$, is defined as

$$C(X, Y) = \sigma_{XY} = E[(X - \mu_X)(Y - \mu_Y)]$$

Note that, by definition, the covariance of a variable with itself is equivalent to the variance.

If X and Y are independent, then $C(X, Y) = 0$. On the other hand, if X and Y are completely linearly dependent, for example when $X = a + bY$, where a and b are constants, then $C(X, Y) = \text{Var}(X)\text{Var}(Y)$.

Stationarity

In geostatistics, a set of observed data points are viewed as a single series or *realization* of a random function. To determine the properties of this random function and to make statistical inferences, one assumes the probability distribution of the random function to be the same over a specified region. This is referred to as stationarity. Under *second-order stationarity*, the mean of the random function is constant, and the covariance between two points is simply a function of their relative positions. As will be noted shortly, this type of stationarity commonly is replaced in geostatistics by a less restrictive form of stationarity, referred to as the intrinsic hypothesis.

REGIONALIZED VARIABLES AND THE INTRINSIC HYPOTHESIS

In geostatistics, spatial variables are considered as random fields with the following overall structure:

$$Z(\mathbf{x}) = m(\mathbf{x}) + R(\mathbf{x}) \quad (4\text{-}1)$$

where the vector $\mathbf{x}$ denotes the spatial coordinates in one, two, or three dimensions, $Z(\mathbf{x})$ is the actual value of the "regionalized variable" at $\mathbf{x}$, $m(\mathbf{x})$ is a slowly varying deterministic function known as the drift or trend identified as the expected value of Z at $\mathbf{x}$, and $R(\mathbf{x})$ is a spatially fluctuating random function with zero expectation and a spatial covariance structure. Regionalized variables are spatially dependent variables that have continuity from point to point, but the changes in the variable cannot be described by any "workable function," such as by a physically based model.

The first term on the right-hand side of Eq. 4-1 represents the deterministic element of the variation; the stochastic element is embodied in the second term. In many applications, the stochastic element appears to be the larger of the two sources of variation. For now, it will be assumed that an underlying deterministic spatial pattern, or drift, does not exist or has been removed. Thus, $m(\mathbf{x})$ can be replaced by a constant value, m, equal to the mean value of the "regionalized variable" Z. That is,

$$Z(\mathbf{x}) = m + R(\mathbf{x}) \quad (4\text{-}2)$$

We will return later to situations in which a drift term is included in the analysis.

As noted, certain stationarity assumptions are required to make statistical inferences. In geostatistics, a common assumption is the "intrinsic hypothesis," which assumes stationarity in the first and second moments of the *differences* between data points. The stationarity conditions of the intrinsic hypothesis can be stated as

$$E[Z(\mathbf{x}) - Z(\mathbf{x} + \mathbf{h})] = 0 \quad (4\text{-}3)$$

and

$$\begin{aligned}\text{Var}[Z(\mathbf{x}) - Z(\mathbf{x} + \mathbf{h})] \\ = E[\{Z(\mathbf{x}) - Z(\mathbf{x} + \mathbf{h})\}^2] \\ = 2\gamma(\mathbf{h})\end{aligned} \quad (4\text{-}4)$$

where $\mathbf{h}$ is a vector, the lag distance and direction, that separates the two points $\mathbf{x}$ and $\mathbf{x} + \mathbf{h}$. The function $\gamma(\mathbf{h})$ is the semi-variogram, often simply called the variogram. The variogram may be expressed as a function of distance only, or for anisotropic regionalized variables as a function of both the distance and direction of the separation between two points.

Under the more general conditions of second-order stationarity, a simple relation exists between the variogram and the more commonly used statistical property, the covariance. That is,

$$\begin{aligned}2\gamma(\mathbf{h}) &= E[\{Z(\mathbf{x}) - Z(\mathbf{x} + \mathbf{h})\}^2] \\ &= E[Z^2(\mathbf{x}) - 2Z(\mathbf{x})Z(\mathbf{x} + \mathbf{h}) \\ &\quad + Z^2(\mathbf{x} + \mathbf{h})] \\ &= C(0) - 2C(\mathbf{h}) + C(0) \\ &= 2[C(0) - C(\mathbf{h})]\end{aligned}$$

so

$$\gamma(\mathbf{h}) = C(0) - C(\mathbf{h}) \quad (4\text{-}5)$$

where $C(\mathbf{h})$ is the covariance between two data points at lag $\mathbf{h}$ and $C(0)$ is the covariance at zero lag (i.e., the variance). Note that, for simplicity and without loss in generality, the proof is shown as though the mean value m is equal to zero. The correlogram is similarly related to the variogram through

$$\rho(\mathbf{h}) = \frac{C(\mathbf{h})}{C(0)} = 1 - \frac{\gamma(\mathbf{h})}{C(0)} \quad (4\text{-}6)$$

Equations 4-5 and 4-6 apply to the regionalized variable Z only when the variable itself, not just the difference, is second-order stationary. In some instances, $2\gamma(\mathbf{h})$ appears to increase without limit as $|\mathbf{h}|$, the distance for a given direction, increases. In these cases, Eqs. 4-5 and 4-6 do not apply. The variogram can still be used, however, because of the weaker assumptions of the intrinsic hypothesis. Also, note that although $E[Z(\mathbf{x})] = m$, a constant, basing the prediction on Eq. 4-4 removes the necessity to first estimate m before estimating the spatial relationship expressed by $2\gamma(\mathbf{h})$.

VARIOGRAMS

The variogram is a key element of geostatistics; it serves to explicitly model the spatial dependence among the attribute values, and is generally an increasing function of the distance $|\mathbf{h}|$. Typically, the variogram ranges from 0, when $|\mathbf{h}|$ is equal to zero, up to a value equal to the variance $C(0)$ at some large value of $|\mathbf{h}|$. Conversely, the covariance is generally a decreasing function of $|\mathbf{h}|$. An example variogram is shown in Figure 4-1.

If the covariance function reaches zero, then at this same point, the variogram reaches a limit referred to as the *sill* value.

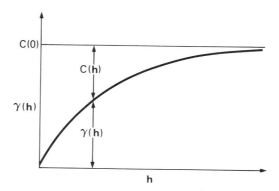

FIGURE 4-1. Example variogram showing relationship between the semivariance, $\gamma(\mathbf{h})$, and the covariance, $C(h)$, for a stationary regionalized variable.

In practice, $\gamma(\mathbf{h})$ may approach $C(0)$ asymptotically, so one may define some small value e within which $\gamma(\mathbf{h})$ is an acceptably close approximation to $C(0)$. The distance at which either the sill value is attained, or $\gamma(\mathbf{h}) + e \geq C(0)$, is called the *range*. For separation distances beyond the range, two random variables $Z(\mathbf{x})$ and $Z(\mathbf{x} + \mathbf{h})$ are considered to be uncorrelated.

Experimental Variograms

The variogram is calculated in a given direction for discrete values of $|\mathbf{h}| = kd$, $k = 1, \ldots, K$, using the existing data values, where d is the distance increment and K is the total number of increments. Since the Z values typically are not separated by exact multiples of d, the points are grouped according to the closest distance vector $\mathbf{h}$. Then the sample variogram, $\gamma^*(\mathbf{h})$, is computed as

$$\gamma^*(\mathbf{h}) = \frac{1}{2n(\mathbf{h})} \sum_{i=1}^{n(\mathbf{h})} [z(\mathbf{x}_i) - z(\mathbf{x}_i + \mathbf{h})]^2 \quad (4\text{-}7)$$

where $n(\mathbf{h})$ is the number of pairs of observations separated by the distance vector $\mathbf{h}$. The experimental variogram is obtained by plotting the values of $\gamma^*(\mathbf{h})$ for the different values of the separation distance $|\mathbf{h}|$. A simplified example of the calculation of an experimental variogram is illustrated in Figure 4-2.

Variogram Models

The estimated variogram, which commonly is quite lumpy, is then smoothed by fitting it to a chosen functional form. A number of

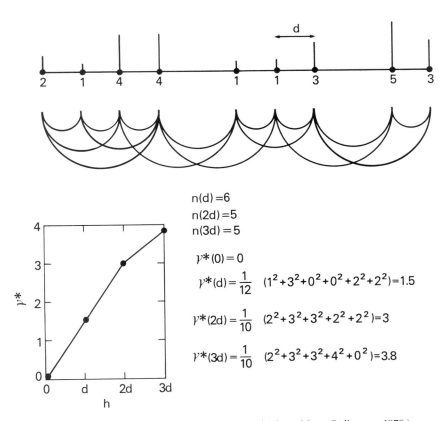

FIGURE 4-2. Example calculation of a variogram. (*Adapted from Delhomme 1978.*)

different forms are in common usage; however, the allowable functional forms must meet certain positive-definiteness conditions (the negative of any variogram must be a conditionally positive-definite function) to avoid negative prediction variances (Journel and Huijbregts 1978; Christakos 1984).

Some of the more commonly used variogram models are shown in Figure 4-3. Variogram models 1 to 3 increase without bounds. Hence, these models do not have a finite variance if the distance vector is made arbitrarily large. Only the intrinsic hypothesis can hold, and one must work with variograms rather than covariance functions. Variogram models 4 to 7, on the other hand, reach an upper bound, the sill. That bound is the variance of the random process model, and its presence means that the variable, not just its difference, is second-order stationary. Thus, in this case, one can work with the covariances as well as the variograms.

When the experimental variogram seems not to show a sill, an alternative is to fit it with a variogram model that has an arbitrarily large range and sill, but to use the model only up to distances where its fit to the experimental curve is deemed acceptable. This amounts to a second-order stationary model.

Table 4-1 gives the formulas and selected characteristics for variogram models 1 to 7. The spatial dimension shown in the penultimate column refers to the largest dimension of the Euclidean space to which the model can be applied and still satisfy the required positive-definiteness conditions. Commonly, one of the variogram models with a sill, such as the spherical or exponential models (models 5 and 6, respectively), is used in which the variogram rises and then curves to a sill value. In the case of the spherical model, the sill is reached at a range equal to the parameter a. The range of $3a$ shown in Table 4-1 for the exponential model is a practical range commonly chosen for this model. The variogram model only approaches this limit asymptotically.

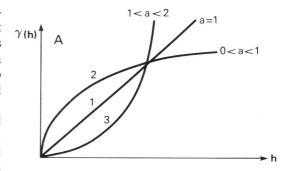

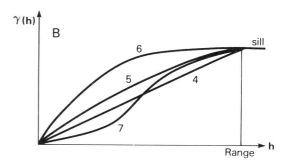

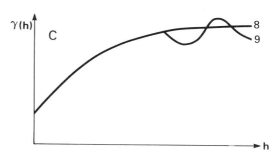

FIGURE 4-3. Examples of some commonly used variogram models: (A) power variogram models, (B) bounded variogram models with a common range and sill, and (C) variogram models with a common nugget effect. The model identification numbers are referred to in text and in Table 4-1.

Variogram models 8 and 9 in Figure 4-3 exhibit a discontinuity at the ordinate, $|\mathbf{h}| = 0$. This may occur because the variogram by definition is exactly equal to zero when $|\mathbf{h}| = 0$, but the fitted variogram may cross the y axis at a positive value c_0. It can be formalized by including a term $c_0(1 - \delta)$ in the variogram model, where δ is the Dirac func-

TABLE 4-1. Some Variogram Models in Common Use

Type	Variogram Model	Range	Sill	Spatial Dimension Allowed	Identification Number on Figure 4-3
Power	$\gamma(\mathbf{h}) = \omega|\mathbf{h}|^a \quad a = 1$	∞	∞	n	1
	$\phantom{\gamma(\mathbf{h}) = \omega|\mathbf{h}|^a \quad} 0 < a < 1$	∞	∞	n	2
	$\phantom{\gamma(\mathbf{h}) = \omega|\mathbf{h}|^a \quad} 1 < a < 2$	∞	∞	n	3
Bounded linear	$\gamma(\mathbf{h}) = c\,(|\mathbf{h}|/a) \quad \text{for } |\mathbf{h}| \le a$ $\phantom{\gamma(\mathbf{h})} = c \phantom{\,(|\mathbf{h}|/a)} \quad \text{for } |\mathbf{h}| > a$	a	c	1	4
Spherical	$\gamma(\mathbf{h}) = c\left\{\dfrac{3|\mathbf{h}|}{2a} - \dfrac{|\mathbf{h}|^3}{2a^3}\right\} \quad \text{for } |\mathbf{h}| \le a$ $\phantom{\gamma(\mathbf{h})} = c \quad \text{for } |\mathbf{h}| > a$	a	c	3	5
Exponential	$\gamma(\mathbf{h}) = c\{1 - \exp(-|\mathbf{h}|/a)\}$	$3a$	c	n	6
Gaussian	$\gamma(\mathbf{h}) = c\{1 - \exp(-|\mathbf{h}|^2/a^2)\}$	$a\sqrt{3}$	c	n	7

tion taking the value 1 when $\mathbf{h} = 0$, and zero otherwise. This kind of behavior was recognized early in the history of geostatistics in gold mining and was attributed to the chance occurrence of gold nuggets in drill cores; hence, it is referred to as the "nugget effect." The nugget effect may be due to measurement errors or to components of variability that exist at scales less than the shortest available experimental interdistance $|\mathbf{h}|$. A flat variogram is referred to as a pure nugget effect and indicates zero correlation between $Z(\mathbf{x})$ and $Z(\mathbf{x} + \mathbf{h})$ at the scale of sampling. More intensive sampling may be needed in such situations to reveal the spatially dependent variation.

Model 9 in Figure 4-3 shows a variogram model that includes a pseudoperiodic component. In practice, to adopt such a model one should have reason to suspect that such a pseudoperiodic component may actually exist, and the behavior of the experimental variogram is not simply a result of random variations and measurement errors. Otherwise, the variogram model will be overfit to the data and the model parameter values will have a very high variance. Erratic behavior of experimental variograms may result simply from the small number of sample pairs available for large values of $\mathbf{h}$.

Sometimes, two or more simple functions are combined into a single variogram model. For example, evidence may exist for two scales of variability. The first scale may correspond to very local variability and the second to broader regional variability. Such cases have been modeled by combining two models with different parameter values.

The examples thus far have been for isotropic or one-dimensional variation. Natural features, however, usually do not vary at the same rate in all directions. For example, ground-water quality in an alluvial aquifer may display greater variation perpendicular to a river course than it does parallel to the river. In such cases, an anisotropic variogram might be used. To detect anisotropy, the variogram is generally estimated in several different directions. Anisotropic models may include geometric anisotropy (the experimental variograms in all directions show the same shape and sill but different range values) or zonal anisotropy—different sill values exist in different directions (Journel and Huijbregts 1978; Isaaks and Srivastava 1989).

No completely objective criteria are available to select from the many candidate functional forms to fit to the experimental variogram. The decision to include anisotropy or to include several scales of variability in selecting a variogram model must always be carefully considered in light of the sample size available to estimate the variogram model parameters. As Journel (1987, p. 90) notes:

The general rule is that of parsimony: "Obtain the best overall fit with a parsimonious model," i.e., with a model $\gamma(\mathbf{h})$ requiring the determination of the least number of parameters.... The rule of parsimony calls for the actual modeling of such specific features as anisotropy (zonal or geometric) and pseudoperiodicity *only* if those features are clearly apparent on the experimental curves and/or they be backed by physical considerations.

Because the variogram is estimated by using squared differences in Eq. 4-7, it is susceptible to the influence of extreme data values. To detect the influences of such outliers, Journel (1987) suggests that the distributions of the $n(\mathbf{h})$ squared deviations $1/2[z(\mathbf{x}_i) - z(\mathbf{x}_i + \mathbf{h})]^2$ be plotted on the variogram using boxplots as illustrated by Figure 4-4. Note that the mean-squared deviation (i.e., the value of $\gamma^*(\mathbf{h})$ itself) at the distance kd in Figure 4-4 is located far above the interquartile range. This results from the influence of one or a very few outliers. The locations and values of the data pair(s) $z(\mathbf{x}_i)$, $z(\mathbf{x}_i + \mathbf{h})$ corresponding to these large contributions should be checked. It may be desirable to remove these data pairs in the process of estimating the variogram at the specified lag. Note that removing a data pair from the calculation of $\gamma^*(\mathbf{h})$ at a specific separation distance, $|\mathbf{h}|$, does not necessarily imply that the two data values should be removed altogether. For instance, Journel (1987) suggests that the data values should still contribute to the prediction of nearby points as part of the geostatistics process, unless one of the points is known to be unreliable.

ORDINARY KRIGING

Kriging is the generic name given to a series of geostatistical methods aimed at determining estimates of attribute values as combinations of nearby values and to estimate the uncertainty associated with these estimates. Kriging is nothing more than gen-

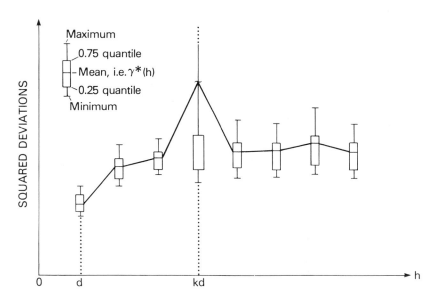

FIGURE 4-4. Example of experimental variogram plotted using boxplots of $n(\mathbf{h})$ squared deviations $1/2[z(\mathbf{x}_i) - z(\mathbf{x}_i + \mathbf{h})]^2$. Note that in this case the boxplots show the sample means rather than the sample medians. (*Modified from Journel 1987.*)

eralized least squares. The term kriging is named after D. G. Krige, who developed the method empirically for estimating gold content in South African ore bodies from fragmentary information.

Characteristics of kriging include that it yields estimates that are unbiased and have minimum estimation variances for the proposed model. Kriging is also an exact interpolator in the sense that the kriged value at a sampled point is equal to the measured value at the point.

To develop the kriging estimate, consider the random function $Z(\mathbf{x})$ whose spatial dependence is characterized by the variogram $\gamma(\mathbf{h})$. At its simplest, kriging is a method of weighted averaging of the observed values of the attribute Z using the measured values, $z(\mathbf{x}_i)$, of the attribute at n sites $\mathbf{x}_i$, $i = 1, \ldots, n$:

$$z^*(\mathbf{x}_0) = \sum_{i=1}^{n} \lambda_i z(\mathbf{x}_i) \qquad (4\text{-}8)$$

where $z^*(\mathbf{x}_0)$ is the estimated value at the location $\mathbf{x}_0$, and the λ_i's are the weights associated with the measurement points. The n sites are typically limited to a "neighborhood" of influence controlled by (1) the maximum distance of reliability of the variogram model (typically chosen as one half of the maximum interdistance, $|\mathbf{h}_{\max}|$, used to fit the model), and (2) the area under which the stationarity is assumed to hold. Unless sampling took place on a regular grid, the arrangement of surrounding measurement points differs among the different locations where the value of the variable is to be estimated, and a different set of weights must be calculated for each point. Thus, λ_i is in actuality a shortened notation for $\lambda_i(\mathbf{x}_0)$.

The λ_i's are chosen so that the estimate is unbiased and of minimum mean-square error. This is represented by the optimization problem

minimize $\sigma^2(\mathbf{x}_0) = E[(z^*(\mathbf{x}_0) - Z(\mathbf{x}_0))^2]$ (4-9)

s.t. $\sum_{i=1}^{n} \lambda_i = 1$ (4-10)

The constraint that the weights λ_i sum to 1 is used to develop unbiased estimates. Solving this optimization problem leads to a system of $n + 1$ linear equations with $n + 1$ unknowns:

$$\sum_{i=1}^{n} \lambda_i \gamma_{ik} + \mu = \gamma_{k0}, \qquad k = 1, \ldots, n \quad (4\text{-}11)$$

$$\sum_{i=1}^{n} \lambda_i = 1 \qquad (4\text{-}12)$$

The corresponding minimized error variance, or "kriging variance," is

$$\sigma_K^2(\mathbf{x}_0) = \sum_{i=1}^{n} \lambda_i \gamma_{i0} + \mu \qquad (4\text{-}13)$$

These equations show that the weights and the kriging variance depend on the geometry and density of nearby measured data points and on the statistical dependence among observations expressed by the variogram. However, neither the weights nor the kriging variance depend on the magnitude of individual values of the observations. This property allows one to use kriging in a design mode to analyze the effect of different sampling patterns and sizes on the estimation variance prior to further sampling.

Analysis of Uncertainty

Two types of maps commonly are shown from a kriging analysis. The first is a "best guess" of the configuration of the mapped variable based on the kriging model (and the use of a contouring package). The second is an "error map" expressing the relative reliability of the first map using the kriging variance (Eq. 4-13). Typical examples of these two types of maps, from an application of kriging to chloride concentrations measured at 120 wells near Barcelona, Spain, are shown in Figures 4-5 and 4-6 (Spruill and Candela 1990). The error map (Figure 4-6) shows relatively high values in

FIGURE 4-5. Estimated chloride concentrations in water from wells in Llobregat delta, Spain. (*From Spruill and Candela 1990.*)

areas of few data and relatively low values in areas of denser control.

Journel (1986a) argues that the kriging variance is not a measure of local accuracy at the specific location x_0 being estimated, but is an index of the data configuration available to estimate $Z(x_0)$. In general, the more data that are nearby and the less redundant is their distribution around a location where an attribute is being estimated (i.e., two values on different sides are more useful than two values close together), the smaller the estimation error, not necessarily at x_0, but in average over the kriging area, if the same data configuration was used throughout.

The kriging variance does not con-

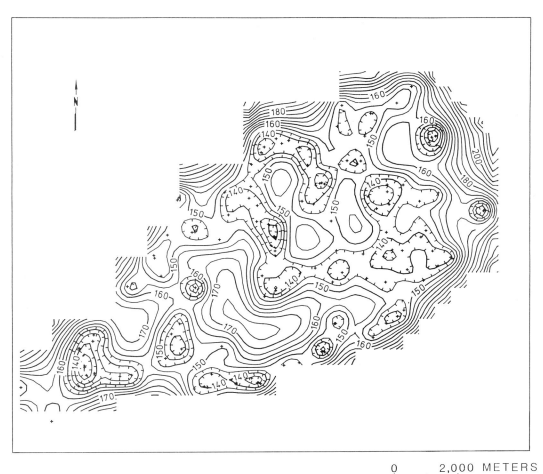

FIGURE 4-6. Standard error of estimated chloride concentrations in water from wells in Llobregat delta, Spain. (*From Spruill and Candela 1990.*)

sider the uncertainty associated with the variogram model selection and estimation. For example, Eq. 4-13 assumes that the variogram is known and correct. Estimated variograms clearly are subject to error. Their precision depends on the number of comparisons at each lag and, therefore, on the sample size. Thus, the estimates of errors should be considered only relative to one another and not in any absolute sense. Such relative errors, however, can be valuable in determining where one's knowledge of the spatial pattern of a constituent is most lacking. A common application is to direct future measurements to those areas of greatest uncertainty. However, other criteria may be used, depending on the situation.

As an example, Rouhani and Hall (1988)

examined various potential networks for ground-water quality in the shallow aquifer of the Dougherty Plain in southwestern Georgia. The purpose of the ground-water-sampling network in the shallow aquifer was to act as an early warning system for ground-water-contamination problems in the underlying principal artesian aquifer. For this reason, ground-water sampling was desired in the areas of greatest recharge to the artesian aquifer. Leakance, the ratio of vertical hydraulic conductivity to the thickness of the lower part of the shallow aquifer, was used as an indicator of potential recharge from the shallow aquifer into the principal artesian aquifer. Rouhani and Hall (1988) used geostatistical analysis to explore various network designs based on three criteria: (1) minimizing the total estimation variance of the leakance values, (2) locating sites in areas with the largest estimated magnitudes of leakance, and (3) minimizing a risk ranking. The first criterion, the traditional one, was primarily concerned with the accuracy of the estimated values, while the second considered the concern for areas with greater values of leakance. The third criterion, a risk ranking, took into account both the magnitude of the leakance and its uncertainty, and was considered by Rouhani and Hall (1988) to be the preferred criterion.

If the assumption is made that the estimation errors are normally distributed and depend only on the data configuration not the data values (i.e., are homoscedastic), then the estimation variance can be used to develop confidence intervals for each estimated value. For example, under the assumption of normally distributed errors, a 95% confidence interval about an estimated value $z^*(\mathbf{x}_0)$ will be approximately $z^*(\mathbf{x}_0) \pm 2s_K(\mathbf{x}_0)$, where $s_K(\mathbf{x}_0)$ is the estimated kriging standard deviation. Again, this confidence interval does not include many sources of error, including variogram model errors and dependence of the variance on data values. The assumption of normality may be grossly violated for many distributions of ground-water-quality constituents.

Cross Validation

Cross-validation tests should always be performed to evaluate the appropriateness of the adopted kriging model. Ideally, split-sample testing is carried out by setting aside a portion of the data during model calibration for later use to test the fitted model. Because the number of data points may be quite limited, the validation process sometimes consists of the "fictitious point" method (Delhomme 1978). This involves removing one data point at a time from the fitting process. The value for the removed point is fit from a model based on the remaining $n - 1$ points, and the estimated value is compared to the observed value for that point.

Several criteria may be used to compare alternative kriging models. Ideally, both the mean error and the error variance should be small for the model chosen from among a set of candidate models. Moreover, the average observed error variance should be close to the average estimated kriging variance from Eq. 4-13.

Journel (1987) notes that the observed error variance commonly varies little from one reasonable model to another. He suggests that several other cross-validation measures be used, including (1) plotting and contouring the observed residuals on a map, and (2) performing regressions of the observed $z(\mathbf{x}_i)$'s on the estimated $z^*(\mathbf{x}_i)$'s. The former, a plot of the residuals, should not present any pattern on the map. For example, a map that shows a drift in the kriging residuals would suggest a kriging model with drift. The regression of $z(\mathbf{x}_i)$ on $z^*(\mathbf{x}_i)$ can be used to indicate possible conditional bias and heteroscedasticity problems. For example, if the regression line deviates from a 45° line, the kriging model may be unbiased on average yet overpredict low values and underpredict high values, or vice versa. Heteroscedasticity is indicated if the amplitude of the observed residuals varies with the predicted magnitude of the attribute $z^*(\mathbf{x}_i)$. In that case, the variance of the estimate depends not only on the data configuration, as characterized by the estimated kriging variance, but also on the data values.

The results of any cross-validation analysis are limited, because model performance at the sampled sites is typically not representative of estimation at all unsampled locations. This may be particularly true if the sample data set is spatially clustered. In that case, the residuals may be more representative of only certain regions or particular ranges of values.

Block versus Point Kriging

Thus far, the discussion has focused on the estimation of point values, commonly referred to as point kriging. In some kriging applications, the linear average of the attribute Z over some subarea or block is sought. One method for obtaining such an estimate is to discretize the subarea or block into many points and then average the individual point estimates to obtain an estimate of the subarea average. Although this approach is conceptually simple, it may be computationally expensive. An alternative approach, which can significantly reduce the computational effort, is to construct and solve only one kriging system for each subarea or block. Development of these estimates, referred to as block kriging, requires some modification of the equations presented. For example, Eq. 4-11 becomes

$$\sum_{i=1}^{n} \lambda_i \gamma_{ik} + \mu = \gamma_{kB}, \qquad k = 1, \ldots, n$$

and the resulting kriging variance is

$$\sigma_K^2(\mathbf{x}_0) = \sum_{i=1}^{n} \lambda_i \gamma_{iB} + \mu - \gamma_{BB}$$

where γ_{iB} is the average variogram value between the ith sampling point and other points within block B and γ_{BB} is the average variogram value within block B. These are estimated by discretizing the block into n_B data points and computing

$$\gamma_{iB} = \frac{1}{n_B} \sum_{j=1}^{n_B} \gamma_{ij}$$

$$\gamma_{BB} = \frac{1}{n_B^2} \sum_{i=1}^{n_B} \sum_{j=1}^{n_B} \gamma_{ij}$$

Isaaks and Srivastava (1989) provide some simple examples of block kriging.

Any decision to choose block over point kriging must consider that the grouping of point estimates into block estimates will tend to decrease the observed spatial variability. Thus, for regional ground-water-quality data, the possibility exists of overly smoothing anomalous concentrations in the block kriging process. Also, some attributes do not average linearly in space. For instance, the "effective" permeability of an aquifer may be quite different from the linear average of the permeability of its subvolumes. The remaining discussion in this chapter continues to deal with point kriging.

KRIGING VARIANTS

Many variants of ordinary kriging have been developed. Six of these are discussed. Kriging in the presence of drift, lognormal kriging, disjunctive kriging, indicator kriging, probability kriging, and cokriging.

Kriging in the Presence of Drift

Thus far, we have focused on stationary models where the trend or drift, $m(\mathbf{x})$ in Eq. 4-1, is a constant. The stationarity decision is a critical one to kriging. It involves an a priori decision about the grouping of spatial data within a particular region so that average statistical properties can be estimated. In some cases, a particular region may consist of two or more subregions which have distinctly different statistical properties for the attribute of interest; hence, separate kriging equations may be developed for each subregion. The division into subregions is, of course, dependent on having a large enough data set to reliably estimate parameters within each subregion. This is a discrete or stratified approach to treating nonstationarity. A continuous approach involves modeling a spatial trend or drift.

Several techniques have been developed to include a drift or slowly varying component. One of these is the so-called universal kriging technique (Matheron 1969; Delfiner 1976) in which the drift is introduced formally into the kriging model as a low-order polynomial with undetermined coefficients. The polynomial is usually no greater than order 2. Since such a low-order approximation may not be valid over the domain of interest, universal kriging is usually applied locally over small neighborhoods. A general expression for the drift term is

$$m(\mathbf{x}) = \sum_{p=0}^{P} a_p f_p(\mathbf{x})$$

where the a_p's are unknown parameters, and the f_p's are specified functions of the coordinate(s) $\mathbf{x}$. By convention, $f_0(\mathbf{x}) = 1$. As an example, for a quadratic drift in two dimensions,

$$\begin{aligned} m(\mathbf{x}) &= m(u, v) \\ &= a_0 + a_1 u + a_2 v + a_3 u^2 \\ &\quad + a_4 v^2 + a_5 uv \end{aligned}$$

Thus, $f_0(\mathbf{x}) = 1, f_1(\mathbf{x}) = u, f_2(\mathbf{x}) = v, f_3(\mathbf{x}) = u^2, \ldots$. The amplitude parameters a_p, $p = 0, \ldots, P$, are unknown, and the order P has to be guessed. Unfortunately, the complexity of the drift model and the form of the variogram of the residual component, $R(\mathbf{x})$, are interrelated.

Closely related to universal kriging is an approach known as the use of intrinsic random functions of order k. The approach is analogous to the differencing used in time series analysis to remove nonstationarity. For example, a first-order difference (zero-order increment) in one dimension, $z(\mathbf{x} + |\mathbf{h}|) - z(\mathbf{x})$, will filter out a trend that is constant, whereas a second-order difference (first-order increment) can be used to remove linear trends. Matheron (1973) defined intrinsic functions of order k as functions for which k^{th}-order increments are weakly stationary (i.e., an intrinsic hypothesis of order k holds). A limitation of this approach is that the resulting variogram is more difficult to interpret because it applies to the differenced process and not to the original data.

Other proposed methods include an iterative procedure proposed by Neuman and Jacobson (1984) and a technique proposed by Cressie (1986) for estimating drift using median polish kriging. The latter approach requires that the data be gridded.

Determination of whether or not to include a drift term in the geostatistical analysis and, if so, to specify its form is not a trivial task. The problem is not well posed and is circular in nature; to determine the drift, one must know the underlying variogram, but to determine the variogram one must know the drift. As noted by Cressie (1986), "What is one person's nonstationarity (in mean) may be another person's random (correlated) variation."

In general, unless independent information is available that suggests the existence of a drift term and that can be used to specify its form, it may be wise to use the simpler, more parsimonious, stationary model with ordinary kriging over local search neighborhoods (Isaaks and Srivastava 1989). Journel (1987) notes that the best tools to build an argument for or against incorporation of a trend term are often the simplest—contour maps of the data values and profiles along different directions. An isopleth map whose contour lines show no definite pattern, but rather a series of bulls'-eyes, would indicate that a trend term is not needed. Conversely, a map with a clear pattern of contour lines may call for a trend component. For instance, if the distance between contour lines is constant a linear drift would be considered; if it decreases or increases linearly, a quadratic drift would be considered. Journel and Rossi (1989) suggest that kriging with a drift term often may be advantageous only when a point being estimated is beyond the correlation range of any datum.

Lognormal Kriging

As previously discussed, if the assumption is made that the estimation errors are normally distributed, then the estimation variance can be used to develop confidence intervals for each estimated value. Often, however, one would not expect the errors to be normally distributed. This is certainly the case in many regional ground-water-quality situations. One way to circumvent this problem is to assume the data are lognormally distributed and to perform kriging on log-transformed data

$$y(\mathbf{x}_i) = \ln[z(\mathbf{x}_i) + c]$$

where c is a constant sometimes added to improve the distributional fit.

The antilog back transform of the estimate is no longer a minimum-error-variance estimate, nor is it unbiased. Commonly, bias corrections are made on the basis of relations between the arithmetic mean and variance (μ_z and σ_z^2) and logarithmic mean and variance (μ_y and σ_y^2) under the assumption of multivariate lognormality

$$\mu_z = \exp(\mu_y + 0.5\sigma_y^2)$$
$$\sigma_z^2 = \mu_z^2[\exp(\sigma_y^2) - 1]$$

Bias can still result, however, if the data are not exactly multivariate lognormal. Some investigators choose to work with median predicted values which are not subject to these transformation biases. Lognormal kriging is discussed by Journel and Huijbregts (1978), Rendu (1979), Journel (1980), and Dowd (1982).

Disjunctive Kriging

Disjunctive kriging is a nonlinear estimation procedure proposed by Matheron (1976) to deal with observed data that exhibit clearly non-Gaussian characteristics even when they are log-transformed. The disjunctive kriging estimator has the form

$$z^*(\mathbf{x}_0) = \sum_{i=1}^{n} f_i[z(\mathbf{x}_i)] \quad (4\text{-}14)$$

where f_i is a function to be determined. When these functions are linear and the random function is multivariate normal, disjunctive kriging is the same as ordinary kriging. A similar relation exists between lognormal kriging and disjunctive kriging. Hence, ordinary and lognormal kriging can be considered special cases of disjunctive kriging.

The method requires the assumption that the original variable Z can be transformed into a univariate and bivariate random variable, Y, with known distribution, and it is valid only under second-order stationarity (i.e., a finite variance must exist). Usually, the bivariate normal distribution is assumed. The transform relationship between Z and Y is written in terms of a Hermite polynomial with coefficients C_k as

$$\phi[Y(\mathbf{x})] = Z(\mathbf{x}) = \sum_{k=0}^{\infty} C_k H_k[Y(\mathbf{x})] \quad (4\text{-}15)$$

The Hermite coefficients are determined by using numerical integration and the properties of orthogonality. Hermite polynomials are reviewed briefly by Yates, Warrick, and Myers (1986a) and in many texts on advanced engineering mathematics.

As in ordinary kriging, an unbiased estimator with minimum variance is sought in disjunctive kriging. The solution for the unknown functions, f_i, in Eq. 4-14 requires several steps. First, the original data are transformed into $y(\mathbf{x})$'s, and the coefficients C_k, $k = 1, \ldots, K$, are determined. A linear system of equations is then solved to provide kriging weights, b_{ik}:

$$\sum_{i=0}^{n} b_{ik}(\rho_{ij})^k = (\rho_{0j})^k, \quad j = 1, \ldots, n \quad (4\text{-}16)$$

where ρ_{ij} is the correlation function between $\mathbf{x}_i$ and $\mathbf{x}_j$. This system of equations is similar to Eqs. 4-11 and 4-12 for ordinary kriging, but it must be solved K times. For $k = 0$, Eq. 4-16 reduces to the unbiasedness condition

$$\sum_{i=1}^{n} b_{i0} = 1 \quad (4\text{-}17)$$

The kriging weights are used to estimate the Hermite polynomial, $H_k^*[y(\mathbf{x}_0)]$, at the estimation site from values of the Hermite polynomials at the sample locations

$$H_k^*[y(\mathbf{x}_0)] = \sum_{i=1}^{n} b_{ik} H_k[y(\mathbf{x}_i)]$$

Once the K values for $H_k^*[y(\mathbf{x}_0)]$ have been obtained, they are used along with the values of C_k to give the disjunctive kriging estimator

$$z^*(\mathbf{x}_0) = \sum_{k=0}^{K} C_k H_k^*[y(\mathbf{x}_0)]$$

The kriging variance for disjunctive kriging is

$$\sigma_K^2(\mathbf{x}_0) = \sum_{k=1}^{K} k! C_k^2 \left\{ 1 - \sum_{i=1}^{n} b_{ik}(\rho_{0i})^k \right\}$$

The disjunctive kriging method is more complex than ordinary kriging, and the assumptions required to use the method place additional demands on model verification. Detailed reviews of disjunctive kriging can be found in Matheron (1976), Journel and Huijbregts (1978), and Yates, Warrick, and Myers (1986a). An example application of disjunctive kriging to problems of ground-water contamination by viruses is discussed in Chapter 15.

A key attribute of disjunctive kriging is that an estimate can be obtained of the conditional probability that a random variable is above some specified cutoff or tolerance level. This type of information also is available from indicator kriging.

Indicator Kriging

Indicator kriging (Journel 1983) uses a simple transform of the data to estimate probabilities that a variable will be less than or equal to specified threshold values. The indicator function, $I(\mathbf{x}; z_q)$, is defined as

$$I(\mathbf{x}; z_q) = 1 \quad \text{if } Z(\mathbf{x}) \leq z_q$$
$$I(\mathbf{x}; z_q) = 0 \quad \text{if } Z(\mathbf{x}) > z_q$$

where z_q is the qth threshold. The random function $I(\mathbf{x}; z_q)$ appears as a Bernoulli random variable with mean and variance given by

$$E[I(\mathbf{x}; z_q)] = 1 \cdot \text{Prob}\{Z(\mathbf{x}) \leq z_q\}$$
$$+ 0 \cdot \text{Prob}\{Z(\mathbf{x}) > z_q\}$$
$$= \text{Prob}\{Z(\mathbf{x}) \leq z_q\}$$
$$= F_Z(z_q)$$

$$\text{Var}[I(\mathbf{x}; z_q)] = F_Z(z_q)[1 - F_Z(z_q)]$$

where $F_Z(z_q)$ is the cumulative distribution function of Z evaluated at z_q. The expected value of $I(\mathbf{x}; z_q)$, and hence $F_Z(z_q)$, can be estimated as the proportion of the observed values less than the threshold, z_q.

Indicator kriging has obvious utility for geostatistical analysis of ground-water-quality data that are "censored." That is, when many of the data are reported as below a laboratory analytical reporting limit. A second advantage is that an estimate of the indicator variable itself commonly is the information sought. For example, interest may center on distinguishing areas where concentrations exceed a water-quality standard from those areas that are not in exceedance. Indicator kriging also is less sensitive to the effects of outliers than the other techniques discussed thus far.

The use of multiple thresholds leads to separate variograms or covariance functions for different levels of Z. This may help resolve an inherent limitation of ordinary kriging—the assumption of homogeneous spatial variance. That is, in ordinary kriging the spatial variability of the regionalized variable is assumed to be constant and independent of the magnitude of the data values. In reality, however, the spatial variability of ground-water quality may be much different in areas of high concentrations than in areas of low concentrations. Indicator kriging may account for some of this inhomogeneity of variance through the use of different variogram models fit to the different threshold levels. (Note: In some applications, the same variogram model is used for all thresholds. Because the variogram model chosen is often based on indicator data at a threshold close to the median, this is usually referred to as median indicator kriging; cf. Isaaks and Srivastava 1989.)

Indicator kriging essentially consists of applying ordinary kriging to indicator variables. As in ordinary kriging, the kriging estimate is a weighted average of the surrounding values. In this case, the surrounding values are those of the indicator function

$$i^*(\mathbf{x}_0; z_q) = \sum_{i=1}^{n} \lambda_i(z_q) i(\mathbf{x}_i; z_q) \quad (4\text{-}18)$$

where $i^*(\mathbf{x}_0; z_q)$ can be viewed as a model for the local cdf of $Z(\mathbf{x})$ conditioned on the surrounding data values, and may be denoted $F_Z^*(z_q)$. The kriging weights are given by a linear system of equations similar to Eqs. 4-11 and 4-12:

$$\sum_{i=1}^{n} \lambda_i(z_q) \gamma_{ik}(z_q) + \mu(z_q) = \gamma_{k0}(z_q),$$
$$k = 1, \ldots, n \quad (4\text{-}19)$$

$$\sum_{i=1}^{n} \lambda_i(z_q) = 1 \quad (4\text{-}20)$$

There are as many such sets of systems of equations as there are threshold values, z_q, used to discretize the variable Z. Hence, the weights $\mu(z_q)$ and $\lambda_i(z_q)$ and the variograms $\gamma_{ij}(z_q)$ appear as a function of z_q. The results from the krigings at the different thresholds are assembled into an estimate of the complete cdf of Z for each location. As before, a different kriging system of equations applies to each different location, $\mathbf{x}_0$.

In simple indicator kriging (Solow 1986), a single threshold is used to create a binary indicator variable. In this case, the approach is no more difficult to apply than ordinary kriging and proceeds along similar lines. The approach for indicator kriging with multiple thresholds involves a few further considerations, including how to (1) choose the threshold values, z_q, $q = 1, \ldots, Q$, (2) correct for any order relations problems, and (3) interpolate to obtain the estimated probability distribution values for any threshold different from the chosen z_q's. These aspects are described in detail by Journel (1987, pp. 104–18), and in a practical sense, by Isaaks and Srivastava (1989).

The "order relations problems" referred to arise because the separate kriging equations at different thresholds may lead to estimates $i^*(\mathbf{x}_0; z_q) > i^*(\mathbf{x}_0; z_q + 1)$ for $z_q + 1 > z_q$, which suggests that the probability of being below the threshold z_q is greater than the probability of being below the larger threshold $z_q + 1$, an obvious inconsistency. There are several ways to adjust the estimates so that they satisfy the order relations (Sullivan 1984). In practice, a piecewise linear model may be fitted to the Q estimated values, $F_Z^*(z_q)$, so as to guarantee the order relations. A second problem may occur because it is possible to generate $i^*(\mathbf{x}_0; z_1) < 0$ or $i^*(\mathbf{x}_0; z_Q) > 1$. Ordinarily, this is corrected by resetting the estimate to either 0 or 1, as appropriate.

After making the above corrections, probability intervals can be derived from the cdf model. Of particular interest for water-quality applications is the probability of exceedance of a threshold b,

$$\text{Prob}[Z(\mathbf{x}) > b] = 1 - F_Z^*(b) \quad (4\text{-}21)$$

If b is not one of the values of z_q used to fit the model, interpolation will be required.

In addition to a probability statement such as Eq. 4-21, values of $z^*(\mathbf{x}_0)$ can be estimated from the fitted cdf by one of several approaches. One of the simpler ways to estimate $z^*(\mathbf{x}_0)$ is to compute the estimate based on the median of the fitted cdf. That is, $z^*(\mathbf{x}_0) = z_{0.5}$, where $z_{0.5}$ is chosen so that

$F_Z^*(z_{0.5}) = 0.5$. This estimator minimizes the expected absolute error. Journel (1989) discusses other estimators, including the traditional least-squares criterion, and consequences of different criteria.

Having developed a map showing areas where a water-quality constituent is considered to exceed a particular level, z_c, we can use indicator kriging to map the estimated risk, $\alpha(\mathbf{x})$, that an exceedance is declared when none exists (i.e., the probability of a false positive), by

$$\alpha(\mathbf{x}_0) = \text{Prob}[Z(\mathbf{x}_0) \leq z_c]$$
for all $\mathbf{x}_0$ such that $z^*(\mathbf{x}_0) > z_c$
$$\cong F_Z^*(z_c)$$

Similarly, for those locations where a water-quality constituent is considered to be less than a specified level z_c, indicator kriging can be used to map the estimated risk, $\beta(\mathbf{x})$, that an exceedance is not declared when one exists (i.e., the probability of a false negative), by

$$\beta(\mathbf{x}_0) = \text{Prob}[Z(\mathbf{x}_0) > z_c]$$
for all $\mathbf{x}_0$ such that $z^*(\mathbf{x}_0) \leq z_c$
$$\cong 1 - F_Z^*(z_c)$$

Often, additional sampling should be considered in areas with high misclassification risks α and β rather than in areas with high estimated values of the probability that the standard is exceeded.

Examples of these results are shown in Figure 4-7. (Note: The application is actually based on the probability kriging extension of indicator kriging described in the next section.) The example arises from 180 soil samples collected in the vicinity of a Dallas smelter site and analyzed for lead concentrations (Isaaks 1984).

Figure 4-7A shows the estimated contour map for soil lead concentrations (based on a least-squares criterion). The map reveals large values around the smelter site, a northeasterly trend of highs corresponding to the direction of prevailing winds, and a

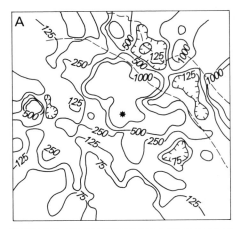
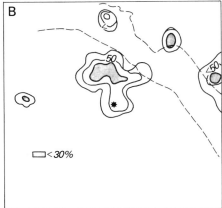
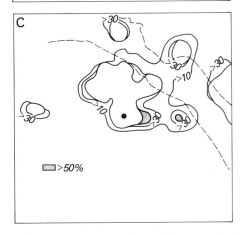

FIGURE 4-7. Examples of types of maps produced by indicator kriging. Figure A is a contour map for lead concentrations in soils near a Dallas, Texas, smelter. Figures B and C show the estimated probabilities of false positives and false negatives, respectively, for a 1,000 ppm threshold. The location of the smelter site is shown by the asterisk, and the boundaries of the Trinity River by the dashed lines. (*Modified from Isaaks 1984.*)

trough of relatively low concentrations in the upper-right corner corresponding to the flood area of the Trinity River. The high outlier area on the western side of the site was checked and found to be a junkyard with possible leakage from automobile batteries (Isaaks 1984). Figures 4-7B and 4-7C show the probabilities of false positives and false negatives, respectively, for the 1,000 ppm threshold. Note that the contours for the probabilities shown in Figures 4-7B and 4-7C are shown only within and outside, respectively, of the zones of predicted concentrations exceeding 1,000 ppm. (The boundaries of the zones of predicted concentrations exceeding 1,000 ppm are slightly different in Figure 4-7A from those in Figures 4-7B and 4-7C.)

Further examples of the application of indicator kriging are provided by Bilonick (1988) to interpret atmospheric hydrogen ion deposition data for the United States and by Johnson and Dreiss (1989) to interpret inferred relative permeability from borehole logs. Journel (1986b) discusses a procedure for including qualitative information such as information given by geologic interpretations in the interpolation process. He defines this as the soft kriging approach.

Probability Kriging

The indicator estimate uses only the indicator part of the information available about $z(\mathbf{x}_i)$. Probability kriging (Sullivan 1984; Journel 1984) is an extension of indicator kriging that, in addition to the indicator values for a particular threshold, also uses information available from a uniform transform of the original values. The kriging estimate for probability kriging is based on the model

$$i^*(\mathbf{x}_0; z_q) = \sum_{i=1}^{n} \lambda_i(z_q) \cdot i(\mathbf{x}_i; z_q)$$
$$+ \sum_{i=1}^{n} v_i(z_q) \cdot r(\mathbf{x}_i)$$

where the $v_i(z_q)$'s are additional weights which must be estimated, and $r(\mathbf{x}_i)$ is equal to $1/n$ times the rank of the datum $z(\mathbf{x}_i)$. For instance, the lowest-valued datum has rank 1 and $r(\mathbf{x}_i) = 1/n$. Conversely, $r(\mathbf{x}_i) = 1$ for the highest-valued datum. Note that the transform data $r(\mathbf{x}_i)$, $i = 1, \ldots, n$, are uniformly distributed in the interval $[0, 1]$. The rank transform is used as a scaling factor so that both the indicator variable and the rank-transformed variable $r(\mathbf{x}_i)$ have the same range in values.

The inclusion of the rank transform of the data is designed to extract more information from the data. The trade-off for this involves considerably more work in estimating additional variograms and fitting more complicated models. Altogether, probability kriging requires $2Q + 1$ variograms or covariance functions, where Q is the number of threshold values. These include Q variograms for the indicator function at the Q thresholds, the variogram of the data ranks, and Q cross variograms between the two random functions $I(\mathbf{x}; z_q)$ and $R(\mathbf{x})$. For more details on the probability kriging algorithm, the reader is referred to the works of Sullivan (1984), Isaaks (1984), or Journel (1984).

Cokriging

Cokriging is an extension of kriging in which additional variables are incorporated into the kriging model. A cokriging model for the simplest case, a single covariate, is written

$$z^*(\mathbf{x}_0) = \sum_{i=1}^{n} \lambda_i z(\mathbf{x}_i) + \sum_{j=1}^{m} v_j y(\mathbf{x}_j)$$

with the unbiasedness conditions

$$\sum_{i=1}^{n} \lambda_i = 1, \quad \sum_{j=1}^{m} v_j = 0$$

where the n coefficients λ_i and m coefficients v_j must be solved for each estimation point $\mathbf{x}_0$ (Isaaks and Srivastava 1989).

This cokriging model requires the estimation of variograms for both the variable being estimated, $Z(\mathbf{x})$, and the correlated variable, $Y(\mathbf{x})$. In addition, the spatial intercorrelation between the two variables must be modeled through a cross-covariance function or cross-variogram model. The method is typically applied when the additional variable is available at a greater number of points. The additional variable may be a separate variable (for example, a different water-quality constituent) whose values are related to the variable being kriged, or it may be a less expensive, less precise measurement of the variable of interest available at a greater number of locations. In general, the variables should have a significant number of common data points to obtain a reasonable estimation of the cross-variograms, although methods of developing "pseudo" cross-variograms (Clark, Basinger, and Harper 1989; Myers 1991) have been introduced to model situations in which the variables may not be available at common locations.

Many extensions of the simple cokriging model presented exist. These models may incorporate more than one additional variable, and they may be used to enhance the estimation of all variables simultaneously. Example applications of cokriging to groundwater data are described by Aboufirassi and Marino (1984) and Ahmed and Marsily (1987).

CONCLUDING REMARKS

Geostatistical analysis has been increasingly used in the geosciences. In this chapter, a primary objective has been to provide some intuitive grasp of geostatistics and its many options. It has been possible to provide only a cursory review of selected techniques; the literature on geostatistics is vast, and the techniques have been used in many applications. Further details can be found in the citations given; facility with the methods can be obtained only through practice.

A number of statistical packages for geostatistical analysis are commercially available. Examples include USEPA Geo-EAS (U.S. Environmental Protection Agency 1988), Geostatistical Toolbox (Froidevaux 1990), and GSLIB (Deutsch and Journel 1992). A Fortran program for disjunctive kriging can be found in Yates, Warrick, and Myers (1986b).

References

Aboufirassi, Mohamed, and M. A. Marino. 1984. Cokriging of aquifer transmissivities from field measurements of transmissivity and specific capacity. *Mathematical Geology* 16(1):19–35.

Ahmed, Shakeel, and G. de Marsily. 1987. Comparison of geostatistical methods for estimating transmissivity using data on transmissivity and specific capacity. *Water Resources Research* 23(9):1717–37.

ASCE Task Committee on Geostatistical Techniques in Geohydrology. 1990. Review of geostatistics in geohydrology. *Journal of Hydraulic Engineering* 116(5):612–58.

Bilonick, R. A. 1988. Monthly hydrogen ion deposition maps for the northeastern U.S. from July 1982 to September 1984. *Atmospheric Environment* 22(9):1909–24.

Candela, Lucila, R. O. Olea, and Emilio Custodio. 1988. Lognormal kriging for the assessment of reliability in groundwater quality control observation networks. *Journal of Hydrology* 103:67–84.

Christakos, George. 1984. On the problem of permissible covariance and variogram models. *Water Resources Research* 20(2):251–65.

Clark, I., K. Basinger, and W. Harper. 1989. MUCK—A novel approach to co-kriging. In *Proceedings of the Conference on Geostatistical, Sensitivity, and Uncertainty Methods for Ground-Water Flow and Radionuclide Transport Modeling*, ed. B. E. Buxton, pp. 473–94. Columbus, Ohio: Battelle Press.

Cooper, R. M., and J. D. Istok. 1988. Geostatistics applied to groundwater contamination. Parts I and II. *Journal of Environmental Engineering* 114(2):270–99.

Cressie, Noel. 1986. Kriging nonstationary data. *Journal of the American Statistical Association* 81(395):625–34.

Cressie, Noel. 1991. *Statistics for Spatial Data*. New York: Wiley.

Delfiner, P. 1976. Linear estimation of non-stationary spatial phenomena. In *Geostatistics for Natural Resource Characterization*, eds. G. M. Verly, M. David, A. G. Journel, and A. Marechal, pp. 49–68. Dordrecht, Netherlands: Reidel.

Delhomme, J. P. 1978. Kriging in the hydrosciences. *Advances in Water Resources* 1(5):251–66.

Delhomme, J. P. 1979. Spatial variability and uncertainty in groundwater flow parameters: A geostatistical approach. *Water Resources Research* 15(2):269–80.

Deutsch, C. V., and A. G. Journel. 1992. *GSLIB: Geostatistical Software Library and User's Guide*. Oxford: Oxford University Press.

Dowd, P. A. 1982. Lognormal kriging—The general case. *Mathematical Geology* 14:475–99.

Froidevaux, R. 1990. *Geostatistical Toolbox Primer, Version 1.30*. 10 Chemin de Drize, 1256 Troinex, Switzerland: FSS International.

Gambolati, Giuseppe, and Giampiero Volpi. 1979. Groundwater contour mapping in Venice by stochastic interpolators: 1 and 2. *Water Resources Research* 15(2):281–97.

Gilbert, R. O., and J. C. Simpson. 1985. Kriging for estimating spatial pattern of contaminants: Potential and problems. *Environmental Monitoring and Assessment* 5:113–35.

Hohn, M. E. 1988. *Geostatistics and Petroleum Geology*. New York: Van Nostrand Reinhold.

Isaaks, E. H. 1984. *Risk Qualified Mappings for Hazardous Waste Sites, A Case Study in Distribution-Free Geostatistics*. Stanford, California: Stanford University, Department of Applied Earth Sciences, MSc. thesis.

Isaaks, E. H., and R. M. Srivastava. 1989. *An Introduction to Applied Geostatistics* Oxford: Oxford University Press.

Johnson, N. M., and S. J. Dreiss. 1989. Hydrostratigraphic interpretation using indicator geostatistics. *Water Resources Research* 25(12):2501–10.

Journel, A. G. 1980. The lognormal approach to predicting local distributions of selective mining unit grades. *Mathematical Geology* 12:285–303.

Journel, A. G. 1983. Nonparametric estimation of spatial distributions. *Mathematical Geology* 15(3):445–68.

Journel, A. G. 1984. The place of non-parametric geostatistics. In *Geostatistics for Natural Resource Characterization*, eds. G. M. Verly, M. David, A. G. Journel, and A. Marechal, pp. 307–35. Dordrecht, Netherlands: Reidel.

Journel, A. G. 1986a. Geostatistics: Models and tools for the Earth sciences. *Mathematical Geology* 18(1):119–40.

Journel, A. G. 1986b. Constrained interpolation and qualitative information—The soft kriging approach. *Mathematical Geology* 18(3):269–86.

Journel, A. G. 1987. *Geostatistics for the Environmental Sciences*. Las Vegas, Nevada: U.S. Environmental Protection Agency EMSL, EPA Project No. CR 811893.

Journel, A. G. 1989. *Fundamentals of Geostatistics in Five Lessons*. Washington, D.C.: American Geophysical Union Short Course in Geology 8.

Journel, A. G., and C. J. Huijbregts. 1978. *Mining Geostatistics*. New York: Academic Press.

Journel, A. G., and M. E. Rossi. 1989. When do we need a trend model in kriging? *Mathematical Geology* 21(7):715–39.

Kitanidis, P. K., and E. G. Vomvoris. 1983. A geostatistical approach to the inverse problem in groundwater modeling (steady state) and one dimensional simulations. *Water Resources Research* 19(3):677–90.

Loaiciga, H. A. 1989. An optimization approach for groundwater quality monitoring network design. *Water Resources Research* 25(8):1771–82.

Matheron, G. 1963. Principles of geostatistics. *Economic Geology* 58:1246–66.

Matheron, G. 1969. *Le Krigeage Universel*. Fontainebleau, France: Le Cahiers du Centre de Morphologie Mathematique, Ecole des Mines de Paris, Fasc. no. 1.

Matheron, G. 1973. The intrinsic random functions and their applications. *Advances in Applied Probability* 5:439–68.

Matheron, G. 1976. A simple substitute for conditional expectation: The disjunctive kriging. In *Geostatistics for Natural Resource Characterization*, eds. G. M. Verly, M. David, A. G. Journel, and A. Marechal, pp. 221–36. Dordrecht, Netherlands: Reidel.

Myers, D. E. 1991. Pseudo-cross variograms, positive-definiteness, and cokriging. *Mathematical Geology* 23(6):805–16.

Myers, D. E., C. L. Begovitch, T. R. Butz, and V. E. Kane. 1982. Variogram models for regional groundwater geochemical data. *Mathematical Geology* 14:629–44.

Neuman, S. P., and E. A. Jacobson. 1984. Analysis of nonintrinsic spatial variability by residual kriging with application to regional groundwater levels. *Mathematical Geology* 16(5):499–521.

Rendu, J. M. 1979. Normal and lognormal estimation. *Mathematical Geology* 11:407–22.

Rouhani, Shahrokh, and T. J. Hall. 1988. Geostatistical schemes for groundwater sampling. *Journal of Hydrology* 103:85–102.

Samper, F. J., and S. P. Neuman. 1989. Estimation of spatial covariance structures by adjoint state maximum likelihood cross validation 3. Application to hydrochemical and isotopic data. *Water Resources Research* 25(3):373–84.

Solow, A. R. 1986. Mapping by simple indicator kriging. *Mathematical Geology* 18(3):335–52.

Sophocleous, M., J. E. Paschetto, and R. A. Olea. 1982. Ground-water network design for Northwest Kansas, using the theory of regionalized variables. *Ground Water* 20:48–58.

Spruill, T. B., and Lucila Candela. 1990. Two approaches to design of monitoring networks. *Ground Water* 28(3):430–42.

Sullivan, J. A. 1984. Conditional recovery estimation through probability kriging: Theory and practice. In *Geostatistics for Natural Resource Characterization*, eds. G. M. Verly, M. David, A. G. Journel, and A. Marechal, pp. 365–84. Dordrecht, Netherlands: Reidel.

U.S. Environmental Protection Agency. 1988. *Geo-EAS (Geostatistical Environmental Assessment Software)—User's Guide*. Las Vegas, Nevada: U.S. Environmental Protection Agency Report EPA 600/4-88/033.

Yates, S. R., A. W. Warrick, and D. E. Myers. 1986a. Disjunctive kriging. 1: Overview of estimation and conditional probability. *Water Resources Research* 22(5):615–21.

Yates, S. R., A. W. Warrick, and D. E. Myers. 1986b. A disjunctive kriging program for two dimensions. *Computers and Geosciences* 12:281–313.

III
Water-Quality Concepts

5

Scales in Chemical Hydrogeology: A Historical Perspective

William Back, Mary Jo Baedecker, and Warren W. Wood

INTRODUCTION

A hydrogeologic system is an energy system in which hydrogeologists identify controls on the sources, sinks, and transformation of energy in the water. This energy consists of potential and kinetic energy of ground-water flow, thermal energy, and chemical energy. Chemical hydrogeology is the study of chemical energy of the hydrologic system and can be defined simply as a study of geologic and hydrologic controls on the chemical character of ground water. The objectives of chemical hydrogeologic investigations are to determine ground-water circulation and its velocity and the sources, concentration, behavior, and fate of chemical constituents. Water quality, one of several aspects within the discipline of chemical hydrogeology, is concerned with the chemical description of water, geographic distribution of various constituents, and specific usability of water for manufacturing, agriculture, and municipal and domestic supplies.

The concept of the hydrologic cycle involves movement and distribution of water and demonstrates renewability of ground water as a resource. The hydrologic cycle is a major control on movement of gases and solutes. The concept of a hydrogeochemical cycle (Figure 5-1) helps evaluate influences of present and past climatic, geologic, hydrologic, biologic, and anthropogenic factors that affect the chemistry of ground water. These factors include the chemistry of infiltrating water, which is largely controlled by air temperature, frequency, amount and duration of precipitation, mineralogy, vegetation and soil cover, and rate of soil gas generation. Factors that additionally affect the chemistry of ground water are mineralogy and thickness of the geologic framework, and potentiometric head distribution, which is controlled by permeability distribution, type of porosity, topographic relief, and elevation. In turn, these factors control recharge rate, flow path, and residence time. The phenomena of mixing—either of water from different areas, different aquifers, or confining beds; from seawater intrusion; or from deep saline water—imposes a hydrologic control on the quality of ground water.

Superimposed on all these natural factors are the anthropogenic effects leading to chemical and physical stresses on the hydrogeologic system, which in some circumstances may dominate its characteristics, particularly in the shallow regime (Figure 5-1) In the environmental context of water

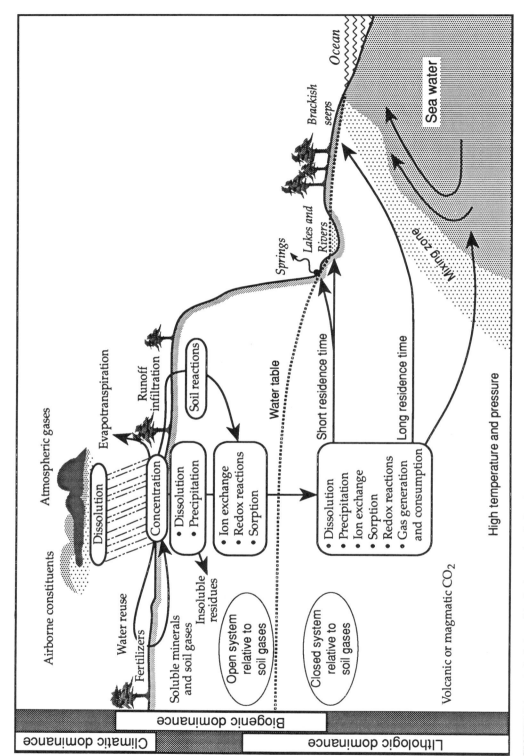

FIGURE 5-1. A depiction of the hydrogeochemical cycle.

quality, many of the areas of ground-water contamination and associated problems resulted directly from actions designed to reduce air pollution and to stop deterioration of the quality of streams and lakes. Instead of burning trash at city dumps, which generated noxious gases, smoke, and particulate matter, wastes were disposed of in "sanitary landfills," ultimately causing leaching of contaminants into the ground water. In addition, much sewage and industrial effluent that previously were discharged into streams are buried or injected underground by various methods and now enter aquifers.

Areas of saltwater encroachment (Figure 5-1) provide a good example of the relation of water quality to the rest of the environment. Ground water in coastal areas is a fragile resource, sensitive to stresses placed on the environment, not only in the immediate area, but also farther inland. The position of a saltwater–freshwater interface is controlled largely by distribution of potentiometric head in freshwater aquifers. Any effect that lowers this head will cause the saltwater to encroach farther landward. An obvious cause for encroachment is the lowering of head because of pumping freshwater for municipal and resort supplies in coastal areas. In addition, human activities that inadvertently decrease infiltration in the upgradient areas cause a decline of freshwater head and permit landward migration of the interface. Human activities that diminish or change the location of infiltration include deforestation; installation of drainage systems; paving roads, parking lots, and construction sites; and so forth. Deforestation leads to rapid runoff and reduces infiltration of water. As the demand for agriculture and development of land increases, drainage of wetlands, particularly nontidal wetlands, will increase. If this water is channeled away from the area, the water-table altitude will decline. Along the same reasoning, if roads, parking lots, and construction sites are paved in the upgradient areas, the water that previously would have infiltrated is removed via storm drains; this loss of water can contribute to head declines in coastal aquifers, thereby permitting further saltwater encroachment.

HISTORIC EVOLUTION OF SCALE IN CHEMICAL HYDROGEOLOGY

The intent of this chapter is to demonstrate the importance of scale and heterogeneity in carrying out investigations of ground-water quality. The scale of the problem, commonly referred to as "scope," and the scale of the area determines a proper approach for physical and chemical aspects of a study. The concept of scale also includes "time," which enters into studies of ground-water quality through "rates" of chemical reactions. In general, physical heterogeneity is a more dominant control in those locations where rates are most rapid. Consequently, investigations at those locations required physical measurements and chemical sampling more detailed than that required for investigations where rates of the processes and reactions are slower.

Historically, knowledge gained from detailed studies of specific areas has been generalized to understand the reactions and processes in regional systems. This generalization is analogous to the development of the science of geology. Geology has always been a site-specific science, in contrast to physics and chemistry, in which experiments can be conducted in laboratories anywhere in the world. Geology has always made great use of type sections, index fossils, stratigraphic and structural cross sections, fence diagrams, and geologic maps. Geologic investigations are made at specific sites, and the results extrapolated and transferred to other sites and larger areas.

The physical and chemical subdisciplines of hydrogeology have evolved through a series of scale changes for the systems under consideration. For example, the regional study of artesian wells by Chamberlin (1885) is generally recognized as the beginning of the science of hydrogeology in North America. This was followed by other regional mappings of extensive flow systems, such as the Dakota sandstone by Darton (1909), the San Bernardino Valley by Mendenhall (1905), the Roswell Basin by Fiedler and Nye (1933), and the Principal Artesian Aquifer of Florida by Stringfield (1936). The historical significance of these regional studies is discussed by Bredehoeft et al. (1982). Although hydrogeologic mapping of regional systems continued, beginning in the mid-1930's, most mapping was done on the much smaller scale of county-sized areas, with a density of sampling points greater than the regional mapping. Quantitative aspects of well hydraulics developed on a different scale. These began with laboratory experiments by Darcy and Schlicter, increased to site-specific field experiments based on Theis' equation, and expanded to address regional systems by use of electric analog and numerical-simulation models. For discussion of historical development of hydrogeology see Freeze and Back (1983) and Back and Freeze (1983).

The scale of studies in chemical hydrogeology developed from large to smaller areas. The basic principles of chemical hydrogeology were developed by conceptual and quantitative models of regional-scale aquifer systems; and these same principles have been applied to smaller systems, herein referred to as "local-scale" and "site-scale" systems. Other words could have been chosen for the scale terminology, but we chose "region," "local," and "site" because of their versatility and nonrigor to classify scale, which is in reality an infinite continuum with unspecified endmembers.

The term *regional scale* is used in this chapter for those systems in which a parcel of water may flow tens of kilometers or more, and in which chemical reactions *under natural conditions* may occur over hundreds to thousands of years; *local scale* is used for systems whose extent is greater than about a kilometer with reactions typically continuing for tens of years; and *site scale* is used for systems less than about a kilometer, and reactions can be expected to continue for years to a few tens of years. These spatial and temporal limits contain the implied assumption that the hydrologic regime remains unchanged. Such a classification is hydrologically arbitrary and is used only for convenience of discussion. In reality, a system is any portion of the universe that the investigator cares to define.

Today, the scales of study for water quality have, to a certain degree, come full circle with renewed interest at the regional scale motivated by concerns about nonpoint-source contaminants such as nitrate and pesticides and about general ground-water-quality conditions. Many of the present-day interests focus on shallow ground water and anthropogenic contamination. In many respects, today's "regional" water-quality studies must combine aspects of the traditional regional geochemical studies, with the newer local- and site-scale studies. This concept leads to the idea of multiple scales of study discussed in Chapter 1. Thus, regional ground-water-quality studies will build on lessons learned as chemical hydrogeology evolved through several stages of scale. This historical development is demonstrated by selected examples in this chapter.

The description and measurement of properties of aquifers are different in studies of chemical hydrogeology than in studies of physical hydrogeology. The chemical character of water is a function of the mineralogy and, therefore, the lithology of the aquifer. Thus, a minor change in mineralogy may be marked by a major change in the chemical composition of water, whereas these mineralogic changes may have little effect on the physical characteristics of the aquifer. On the other hand, even if basic lithology remains constant throughout the aquifer there can be significant changes in porosity, permeability, and thickness, resulting in changes in hydraulic conductivity, storativity, and dispersivity. These major physical changes in aquifer characteristics may be unrecognizable in the chemical composition of the water. Therefore, it has been acceptable to assume a certain degree of physical homogeneity for chemical studies of many regional aquifers, whereas such an assumption may be invalid for most physical studies of these systems.

Each decrease of scale increases the apparent complexity of both the physical and chemical characteristics. For example, it has been acceptable to assume both physical and chemical homogeneity for many regional geochemical investigations, whereas, such an assumption is usually invalid in small-scale investigations because of the great influence of physical heterogeneity. Heterogeneity becomes an increasingly important control as the scale of a system decreases. In a site-scale system, such as point-source ground-water contamination, small changes in the aquifer composition, either natural or human-induced, may have a significant effect on geochemical reactions. For example, a decrease in oxygen flux across a silty layer may cause the system to become anoxic and drastically alter the chemical composition of water and rates of reactions in the vicinity of the silty layers. In a vertical section, dissimilar reaction zones may be present within only a few centimeters of each other and may be controlled by small-scale changes in aquifer characteristics.

While considering the quality of ground water at any scale, it is instructive to consider the particular mechanisms and processes that occur at extremely small scale. The processes of transport and retardation of solutes on a small scale are not well understood; even the processes themselves that control the distribution of species may have different relative effects in regional and local environments. Several physical processes and chemical reactions control concentrations as solutes move through the aquifer. An example of chemical reactions that occur at extremely small scale is solute precipitation and dissolution at mineral surfaces. Because solutes travel with flowing water, the flow field, at all places and times, may affect the chemistry. The amount of mixing of some introduced water with the interstitial water initially present may also change the quality of water. Geologic heterogeneity causes complex distribution of permeability and, thus, ground-water velocity. An assessment of the importance of heterogeneity and scale in diverse environments is essential to modeling water quality in time and space.

During the past three decades or so, an understanding of geochemical systems has been gained from the results of detailed laboratory experiments on mechanisms of reactions, regional-field investigations, and theoretical analyses. This understanding involved applications of the principles of thermodynamics, including electrochemistry and mineral equilibria, hydrochemical facies, and use of isotopes. The most significant recent advances of a regional nature have been the expanded use of isotopes and the development of chemical modeling. Contributions from local-scale investigations include demonstration of the role of ground-water geochemistry in geologic processes such as deposition of ore minerals, diagenesis of sediments, and formation of geomorphic features. One of the important recent changes in hydrogeology has been an increased emphasis on site-scale systems because of serious problems with ground-water contamination. These investigations are resulting in the incorporation of organic geochemical and microbially mediated reactions into geochemical models.

Examples of field investigations discussed in this chapter at each scale include the following; (1) a regional-scale system is exemplified by a hypothetical coastal plain in which controlling chemical reactions are identified for the sources and sinks of ions; (2) local-scale systems are illustrated by a study of an area where overpumping of an aquifer near a lake with good-quality water caused a secondary deleterious chemical response in the well field, and by radium in ground water that discharges to an estuary; (3) site-scale systems are demonstrated by studies of chemistry of ground water resulting from a leaking gasoline tank and a broken pipeline for crude oil.

CHEMICAL HYDROGEOLOGY ON A REGIONAL SCALE

The purpose of regional geochemical studies is to identify the controlling chemical and isotopic reactions in order to explain the observed chemical character of ground water and thereby deduce more about the functioning of the physical system. In addition to major, minor, and trace solutes, these studies include evaluation of abundances of isotopes such as tritium, deuterium, nitrogen-15, oxygen-18, sulfur-34, carbon-13, carbon-14, and chlorine-36 to determine rates of reactions, sources of water, flow paths, velocity of flow, and aquifer characteristics (see Chapters 10 and 11). All of these constituents contribute to understanding the concentrations, sources, sinks, and transport of solutes that characterize water quality.

A great deal of understanding had been gained concerning the chemistry of ground water by the late 1950s, largely through areal studies of geology and ground-water resources. However, with formulation of the hypothesis that chemical thermodynamics could be used to identify controlling chemical reactions in aquifers, it was recognized that the inherent complexity of local- and smaller-scale investigations precluded acceptance of the requisite simplifying assumptions. Assumptions, such as homogeneity, simplicity of mineral assemblages, and identifiable flow paths, were more valid and realistic in regional-scale sampling than in smaller-scale systems, which require a greater sampling density. Therefore, early mineral-equilibrium studies were undertaken in regional carbonate aquifers, such as the Tertiary limestone of Florida, the Yucatan Peninsula of Mexico (Back and Hanshaw 1970; Back et al. 1979), and the Mississippian limestone in the vicinity of the Black Hills (Back et al. 1983). These regional field studies provided the basis for applying chemical thermodynamics and were sources of data for developing techniques of chemical modeling (Plummer and Back 1980; Plummer et al. 1990; Busby et al. 1991).

The selection of aquifers free of organic matter for testing methodology of chemical modeling of regional studies further simplified the conceptual model of the system and permitted the use of carbon-13 (^{13}C) as another component in the mass-balance equations. In a pristine aquifer, the carbon-isotopic composition of ground water is controlled by the composition of recharge water after it has infiltrated through the soil zone, and by the isotopic composition of carbonate minerals comprising the aquifer. Marine limestones have $\delta^{13}C$ of $0.0 \pm 2‰$ (per mil). Depending on vegetation and other factors, soil gas has a $\delta^{13}C$ of -12 to $-25‰$. Therefore, in chemical mass-balance calculations, it is possible to balance on carbon-isotopic composition without the complexities introduced by other sources of organic material with different isotopic signatures. From this relatively simple base, it has become possible to undertake increasingly complex reaction-path models in aquifers known to contain both natural and introduced organic matter.

We have chosen a hypothetical coastal plain, although one quite similar to the Atlantic Coastal Plain, to demonstrate the type, magnitude, and extent of reactions that affect water quality in time and space in regional aquifer systems. The ground-water-flow pattern is relatively simple on a regional scale in these systems, and a sufficient number of detailed studies has been completed to identify the major chemical reactions (Foster 1950; Back 1966; Lee 1985; Lee and Strickland 1988). The water quality of the regional system can be viewed as a set of small-scale systems that are temporarily in chemical equilibrium, even though the entire system is not at equilibrium over geologic time. This permits use of equilibrium thermodynamics for specific reactions without including kinetically controlled reactions that may have rate constants measured in hundreds of thousands of years.

Most regional flow systems in humid areas have a large number of local flow cells as a component near land surface. The groundwater flux through these cells is much greater than that in the regional flow systems and promotes a more rapid chemical change in the mineralogic framework of the aquifer. Water quality can change dramatically over short distances within these localized flow cells. They are truly "open" systems; in the thermodynamic sense of the word, an "open" system generally implies gain and loss of gases. This discussion of a coastal plain emphasizes intermediate and deep regimes where reactions can be somewhat separated from one another and the system is presumed "closed"; this permits viewing the system as a large reaction column. An example of the effects of local flow systems in a regional-scale investigation is provided by the case study of the Delmarva coastal-plain aquifer in Chapter 23.

The approach generally used in regional geochemical investigations is one of mass balance, in which the transfer of mass between the solid and aqueous phases is determined as water moves downgradient from recharge areas to discharge areas. The steady-state change of mass of a solute within a unit volume of the aquifer is equal to (1) the mass of that solute introduced into the unit volume by convection and diffusion, (2) minus the mass of the solute leaving the unit volume by convection and diffusion, and (3) plus or minus the chemical reactions that add or subtract the solute from the water. Mineral precipitation always removes solutes, and mineral dissolution always adds solutes to the unit volume of water. On the other hand, sorption, desorption, ion exchange, and transmutation by nuclear decay are examples of natural processes that can either remove or add solutes. In transient systems, storage of solutes may occur in the water or in the aquifer matrix.

Although most reactions identified in aquifer systems can occur in any part of the system, some tend to be dominant in certain parts (Figure 5-2). In the shallow regime of humid areas, localized short-term fast reactions and processes include mixing of infiltrating water with shallow ground water, interaction of water and gases in the unsaturated zone with infiltrating water, dissolution of minerals, and degradation of organic material. These reactions generally increase the amount of solutes in ground water, and concentrations of solutes and their isotopic composition may change rapidly in time and space. Long-term, slower, chemical reactions occur in the deeper regime where interactions of water with aquifer materials are dominant. Precipitation and dissolution of minerals, ion exchange, and sorption add or remove solutes over a long period of time in regional flow systems. In arid zones, many solutes transported by snow and rain or originating from mineral weathering are stored in the unsaturated zone and periodically reach the water table during periods of high rainfall.

These reactions and processes can be further considered in terms of sources and sinks for major cations and anions, and illustrated in a series of schematic diagrams (see Figures 5-2 to 5-7). For example, sources of calcium (Figure 5-3) are primarily from the weathering of plagioclase and other feldspars and the solution of calcite, dolomite, and gypsum. Calcium is lost from water by precipitation of calcite and by exchange with sodium and other ions on clays (for a discussion of mechanisms of these reactions see Chapter 6).

The bicarbonate ion (Figure 5-4) has sources similar to that of calcium with a few modifications. For example, bicarbonate is produced in the soil zone by introduction of atmospheric CO_2 and by biologic generation of carbon dioxide from plants in the soil. Differences in ^{13}C values indicate different sources for carbon. An isotopic composition for bicarbonate of approximately −22‰ from the soil zone in a humid area indicates decay of natural material, whereas a value 0.0‰ indicates dissolution of marine calcite without the presence of atmospheric CO_2. Sulfate-reducing bacteria use organic ma-

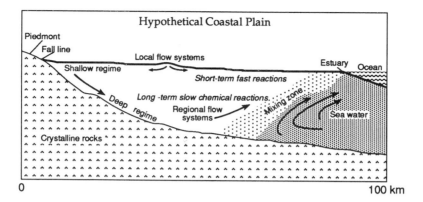

FIGURE 5-2. Hypothetical coastal-plain aquifer system indicating variations in scale of both time and space for hydrochemical processes and reactions.

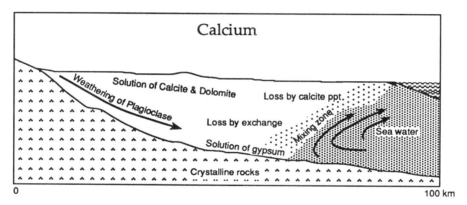

FIGURE 5-3. Typical sources and sinks for calcium in ground water in a coastal-plain aquifer.

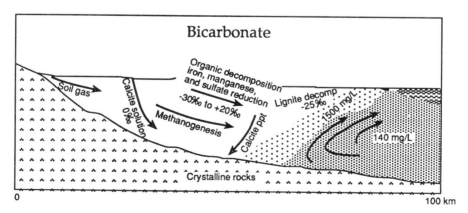

FIGURE 5-4. Sources and sinks for bicarbonate in ground water with characteristic values of $\delta^{13}C$ (of inorganic carbon) for common reactions in a coastal-plain aquifer.

terial for their energy source to produce bicarbonate which is depleted in ^{13}C and therefore has a light isotopic composition of about $-30‰$. Bicarbonate resulting from decomposition of lignite has a δ^{13}C of about $-25‰$, similar to other land plants. Anaerobic decomposition of organic material can fractionate carbon to provide bicarbonate with a wide range of isotopic compositions. This range results from the generation of methane. After methane generation, bicarbonate from the residual-carbon pool may have a value as heavy as $+20‰$ because the light carbon isotope was previously depleted in the gas phase. Chapelle and McMahon (1991) demonstrated that much of the downgradient increase in dissolved inorganic carbon in a coastal-plain aquifer can be attributed to microbial generation of carbon dioxide; this increased CO_2 causes additional dissolution of carbonate-shell material. In an earlier paper, Chapelle et al. (1987) calculated that this CO_2 has δ^{13}C values of -10 to $+5‰$ and therefore is much heavier than that of the lignite source with a value of -20 to $-24‰$. On the other hand, if the bicarbonate results from oxidation of the isotopically light biogenic methane, then the resulting bicarbonate will also be extremely light (about $-40‰$).

Bicarbonate is removed from the water in the aquifer primarily by precipitation of calcite and discharge of ground water, either directly to the ocean or into streams and estuaries. The "loss" or bicarbonate shown in Figure 5-4 by precipitation of calcite constitutes a loss from the water but not from the aquifer material. Concentrations of bicarbonate in coastal aquifers commonly are several times, and as much as 10 times, greater than the bicarbonate concentration in seawater (140 mg/L) because of combined effects of solution of calcareous material and removal of calcium by ion exchange. Ion exchange removes calcium and causes the water to remain undersaturated with calcite, resulting in the continuous dissolution of calcareous material and an abnormally high concentration of bicarbonate ion which can exceed 1,400 mg/L.

The major sources of sodium in ground water (Figure 5-5) are from maritime rainfall, hydrolysis of silicate minerals, ion exchange, seawater mixing, and certain sources of pollution, such as road salts, sewage, and possibly its release from clays by exchange with ammonium generated in landfills. Sodium is lost from the system primarily by solute discharge.

The major sources of magnesium (Figure 5-6) are weathering of ferromagnesium minerals, alteration of clays, dissolution of dolomite and magnesium calcites, and freshwater–seawater mixing. Within most ground-water regimes, the ratio of calcium to magnesium exceeds 1; the ratio is less than 1 in marine environments largely because of calcite precipitation. Losses of magnesium are primarily caused by removal by clays and by discharge of ground water to estuaries, rivers, or the ocean.

The primary sources of sulfate (Figure 5-7) in coastal-plain aquifers are rainfall, oxidation of pyrite, dissolution of gypsum, leakage from fine-grained material, and effects of mixing fresh ground water with seawater in the aquifer. The loss of sulfate from ground water is primarily by reduction of sulfate ion to hydrogen sulfide which outgasses or precipitates as a metal sulfide.

These natural processes and reactions often produce a rather standard and predictable sequence of hydrochemical facies (Figure 5-8) in humid areas (Back, 1966). Typical flow velocities within a coastal-plain aquifer are about 5 m per year, and the total dissolved solids increase downgradient along the paths of ground-water flow. In the upgradient part of the system, water is generally undersaturated with respect to calcite and other minerals; water attains equilibrium and becomes supersaturated with respect to calcite as it flows downgradient. In the recharge areas of these upgradient regimes, the typical hydrochemical facies is mixed-cation bicarbonate. The soluble minerals

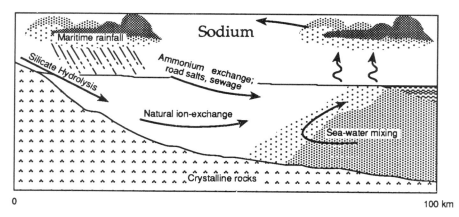

FIGURE 5-5. Typical sources and sinks for sodium in ground water in a coastal-plain aquifer.

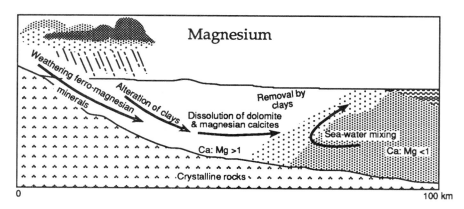

FIGURE 5-6. Typical sources and sinks for magnesium in ground water in a coastal-plain aquifer.

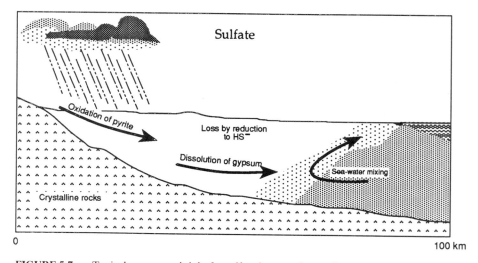

FIGURE 5-7. Typical sources and sinks for sulfate in ground water in a coastal-plain aquifer.

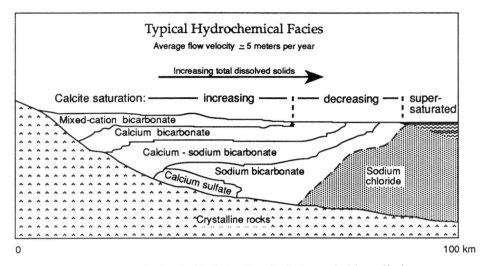

FIGURE 5-8. Typical hydrochemical facies in a hypothetical coastal-plain aquifer in a humid region.

have been leached and the low concentration of cations is derived from the comparatively resistant feldspars and other silicates. The downgradient calcium-bicarbonate facies results from dissolution of marine fossils. Sodium concentration is low because ion exchange is a faster reaction than mineral dissolution, and sodium has already been removed from the exchangeable sites. In those parts of the system in which ion exchange is still occurring, the water again becomes undersaturated with respect to calcite owing to the loss of calcium; the increased dissolution of calcareous material causes high concentrations of bicarbonate, and the characteristic facies is sodium bicarbonate. However, farther downgradient the ocean water that has intruded into the aquifer is supersaturated with respect to calcite. This body of saltwater forms a chemical boundary of the freshwater–flow system. The discharging freshwater mixes with encroaching ocean water to provide an environment for many geochemical reactions. Depending on the particular chemical composition of freshwater, the reactions in the zone of mixing can either dissolve or precipitate calcareous and other minerals (Back et al. 1979).

CHEMICAL HYDROGEOLOGY ON A LOCAL SCALE

In shifting from regional geochemical investigations to local- and smaller-scale studies, probably the most obvious distinction is the density of sampling. Increased density of sampling is required to characterize geology, determine hydrologic properties of the aquifer, and identify the nature of the chemical reactions at the level of detail required for this scale. The inherent heterogeneity of mineralogy and hydrologic properties of the aquifer become controlling factors that preclude the assumption of homogeneity. In addition, the relatively rapid chemical reactions tend to dominate the chemistry of the water in this part of the hydrogeologic regime.

An example of water-quality problems in a local-scale system is, appropriately, from the state of Rhode Island. One of the major water-quality problems associated with ground-water development in Rhode Island is a high concentration of dissolved manganese. The high-manganese water is from many heavily pumped municipal and industrial wells that tap stratified, glacial-drift aquifers. Although the drinking-water standard for manganese is 0.05 mg/L, many wells in Rhode Island yield water containing more than 0.2 mg/L, and several yield water containing from 1 to 8 mg/L. Typically, the manganese concentration in water from test wells and newly constructed wells is less than 0.05 mg/L. Generally, concentrations of manganese exceed the recommended values only after wells have been in operation for several months or, more commonly, for several years. The major aquifer of Rhode Island is stratified glacial drift, and the thickest, most transmissive parts are in valleys within a few hundred meters of streams. Consequently, high-capacity wells are close to streams, and most of them derive their water from induced infiltration of streamflow.

In a typical situation, wells adjacent to streams or ponds tap various depths in an aquifer (Figure 5-9) (Johnston and Back 1977). In this example, the pond water is of good biologic and chemical quality. Water from wells 1 and 2 has high manganese concentrations, whereas water from well 3 has yielded water with no manganese for more than 20 years. For many years, natural rainfall recharged aquifers and infiltration rate equalled the pumping rate. However, increased demands on ground water during a drought during the 1960s caused the potentiometric surface to be lowered enough to induce water to infiltrate through the bottom of the pond. The sands and gravels of the glacial deposits have manganese oxide coatings that are essentially insoluble while in contact with oxygenated water. However, depletion of oxygen in the water as it moves through organic material at the bottom of the pond generates an anoxic condition; manganese oxide is reduced to a soluble species which dissolves and enters the ground water. Similar situations occur in many areas throughout Rhode Island.

This water-quality problem results from imposition of a physical stress on the hydrologic system which causes degradation of water quality, even though the influent water is of high quality. In other parts of Rhode Island, pumpage can induce infiltration of contaminated stream water which also may mobilize manganese from glacial deposits.

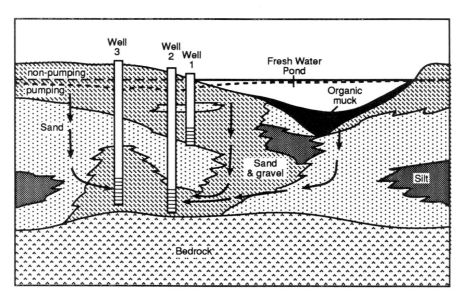

FIGURE 5-9. Schematic hydrogeologic section showing flow path of water from an oxygenated pond through an organic layer that causes ground water to become anoxic.

Thus, the presence of manganese in ground water in Rhode Island demonstrates how regional ground-water quality can be influenced by small-scale phenomena. It is through the synergistic effects of the chemical reactions occurring at a near-molecular scale superimposed on the flow path at local scale that leads to regional consequences with respect to ground-water quality.

Another example of a local-scale study that demonstrates the inherent interconnection of the geochemical cycle with the hydrologic cycle is the occurrence of radium-226 (^{226}Ra) in water (Miller, Kraemer, and McPherson 1990). Several years ago, the occurrence of radium in the environment became a topic of considerable research because of its potential adverse health effects and because of its natural availability as a tracer in the movement and mixing of ground water and surface water. Radium is a bone-seeking element in the human body and has been linked to leukemia. The USEPA has recently recommended a drinking-water limit of 20 pCi/L for each of the radium isotopes of ^{226}Ra and ^{228}Ra [1 picoCurie (pCi) = 2.22 disintegrations per minute (dpm)].

Radium concentrations are typically higher in estuarine and coastal waters than in either the ocean or river waters. These high concentrations have been attributed to (1) desorption and diffusion of radium from coastal and estuarine sediments, (2) input from salt marshes, and (3) discharge of ground water from seeps and springs. One of the coastal areas of highest concentration of ^{226}Ra is in southwestern Florida. The higher concentration in these coastal waters has been attributed to (1) enrichment as a result of circulation of water from the Gulf of Mexico through the deep, uranium-rich limestone of the Florida Peninsula, and (2) the mining and processing of phosphate ore, which contains radionuclides of the uranium-238 decay series. Radium-226 can be transported by rivers in the water and with the sediments. The water in the rivers of west Florida that flow through the phosphate deposits have concentrations of ^{226}Ra much higher than other rivers and estuaries of North America.

In order to evaluate the concentration of radium contributed from the possible sources, a study was undertaken in the upper part of the Charlotte Harbor estuary near Sarasota, Florida (Miller, Kraemer, and McPherson 1990). They found that the highest ^{226}Ra concentrations occurred in the near-shore brackish water. Among their significant conclusions is that the dominant source of ^{226}Ra is the discharge of ground water into the harbor and tidal portion of the rivers. The ground water containing high concentrations of ^{226}Ra is from the principle artesian aquifer of Florida.

This study is significant in that it clearly demonstrates the effect of regional geochemistry on a local-scale system. More commonly, under natural conditions, the local-scale systems can locally modify the chemistry of water in regional flow if there is hydrologic interconnection. The influence of the local system can become even more pronounced in contaminated systems with the infiltration and transport of organic solutes into the regional aquifer.

CHEMICAL HYDROGEOLOGY AT THE SITE SCALE

Within the terminology of this paper, "site-scale system" is generally a shallow environment with an extent of about a kilometer or less. Meaningful application of geochemical models to site-scale systems requires an understanding of geochemical reactions and aquifer characteristics in far more detail than that required for regional-scale or local-scale systems. Recognition of large changes in concentrations of constituents, and thus consequent delineation of reaction zones over short distances, adds a dimension of complexity to ground-water investigations (Back and Baedecker 1989). With each reduction in the scale of a chemical hydrogeologic study, another level of complexity is added to the system. In field investigations at the site-scale level, the assumption of homogeneity is generally invalid, and the degree of heterogeneity must be evaluated. Furthermore, the nature of the practical water-quality problems generally requires analyses of additional chemical components that increase the number of chemical reactions to be identified and evaluated.

The need to evaluate specific hydrogeologic controls imposed by physical and chemical heterogeneity resulted largely from investigations of site-scale systems associated with ground-water contamination. To gain an understanding of the controlling processes, we cannot simply generalize ground-water-flow patterns, assume homogeneity, and disregard minerals of minor abundances. Also, organic material and the microbiota in an aquifer may dominate the geochemical processes. Although the presence of biodegradable organic material may control the chemical reactions, the distribution of the reactants and products is controlled by the small-scale flow regime and the heterogeneity of the aquifer. Determination of the magnitude of controlling reactions requires a dramatic increase in frequency and density of sampling. Environments contaminated by human activities are obvious examples of site-scale systems. However, these environments may develop under natural conditions, such as where river or lake water with high concentrations of dissolved organic material locally recharges an aquifer and alters the redox potential of ground water, or where ground water interacts with peat or coal deposits.

The role of organic contaminants is generally more difficult to evaluate than inorganic wastes in ground-water environments. These problems of evaluation result from (1) complexity of field sampling, (2) difficulty of selection and identification of great numbers of possible constituents requiring analyses, (3) operation and maintenance of sophisticated instruments with delicate electronics, and (4) interpretation of massive data sets that often leads to formulation of nonunique conceptual models. The composition of organic contaminants and the mechanism by which they are deposited, accumulate, and degrade in the subsurface determines their effect on the geochemistry of an aquifer. Organic compounds occur in the subsurface from three major sources:

1. Spills and leaks of nonaqueous phase liquids (NAPL), such as gasoline that provide a source of contaminants within days of the leakage and are capable of continuous contamination for hundreds of years. In this type of contamination, the source material, in addition to the ground water, may be mobile as a separate product. The effect on the geochemistry of the aquifer is profound where these organic compounds are soluble and biodegradable.

2. Disposal of municipal and industrial wastes in landfills and lagoons that may be continuous sources of contaminants, and of composition more variable than NAPLs, which are generally of a single and known composition. These wastes and their products also have a significant effect on the geochemistry of the aquifer if the compounds are soluble in water and biodegradable.

3. Agricultural application of organic chemicals on the land surface, such as herbicides, insecticides, and fungicides that provides intermittent sources of contamination over large areas. These are incorporated into the crop biomass and can degrade or be transported in surface water and sediment.

Although these compounds can have deleterious effects on water quality for human health, their effect on the chemical reactions in aquifers is generally considered to be minimal. Whereas, with NAPLs and landfill leachates, changes in the nature and abundance of dissolved constituents alter geochemcial reactions, which proceed until the reactive materials are consumed or attenuated. The chemical heterogeneity that results from these reactions may develop in zones as thin as a few centimeters (Smith,

Harvey, and LeBlanc 1991). An obvious example is organic matter that degrades by consuming dissolved O_2. If the oxygen demand exceeds the amount of dissolved oxygen, a change from oxic to anoxic conditions will occur and persist until labile organic material is consumed. After complete degradation of organic matter, presumably oxygen may once again become available through recharge, and different geochemical reactions will be established.

A sequence of reactions occurs in contaminated aquifers, where the hydrogeologic milieu of their occurrence and products is oriented in the direction of ground-water flow (Nicholson, Cherry, and Reardon 1983; Baedecker and Back 1979). A schematic is depicted in Figure 5-10, typical of a NAPL that floats on the water table in a surficial sand-and-gravel aquifer. Note the smaller scale of this figure compared to that of the previous figures. The NAPL is a mobile phase that moves in the direction of, but at a slower rate than, ground water. Downgradient from the NAPL are zones of differing water chemistry that develop because soluble components of the NAPL are transported and degraded by microbiota. Constituents of low solubility may be transported on colloidal material moving in the aquifer.

Biochemical zonation is often difficult to delineate because advection causes mixing of reactants and products in these microenvironments. Organic compounds are degraded in biochemical zones by microbial processes that include oxidation, denitrification, reduction of sulfate, and Fe and Mn oxides. The anoxic water can become actively methanogenic. Precipitation of metal-bearing minerals or amorphous compounds occurs in and downgradient from biochemical zones. The boundaries between some of these redox zones are often not discernible, even though these reactions generally occur in distinct chemical regimes. For example, sulfate reduction and methanogenesis can occur concomitantly or in microenvironments too small to sample by conventional means in aquifers. Immediately downgradient from the NAPL (Figure 5-10) is an anoxic zone where the dissolved O_2 that infiltrates through the unsaturated zone and is transported by advection is consumed in the outer edges or mixing zone of the plume. The large concentrations of dissolved CO_2, N_2 and CH_4 from degradative

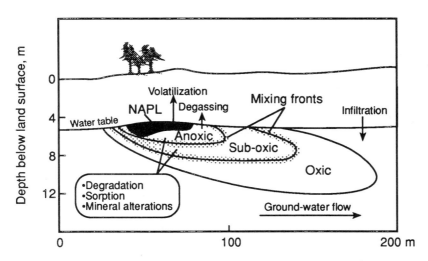

FIGURE 5-10. Schematic geochemical section depicting the zonation and characteristic reactions and processes in an aquifer contaminated with a nonaqueous phase liquid (NAPL).

processes increase the probability of outgassing to the unsaturated zone. Whereas in the saturated zone, carbonate minerals, iron and manganese minerals (Baedecker and Cozzarelli 1991), and silica (Bennett and Siegel 1987) may be dissolved and/or precipitated. In anoxic water, an increase in HCO_3^- is coupled with an increase in cations such as Fe^{2+}, Mn^{2+}, and NH_4^+ in solution.

Farther downgradient from the NAPL source is a suboxic zone where the amount of degradation of some organic compounds may increase due to the presence of O_2 in native ground water. Ferrous iron and, to a lesser extent, Mn^{2+} are oxidized to Fe^{3+} and Mn^{4+} and precipitate as hydroxides. Farther downgradient is an oxygenated zone where effects of degradative processes and chemical reactions are evident, but the concentrations of constituents are near background levels and advective transport is the major process controlling distribution of contaminants.

Examples of site-scale systems that significantly degrade the quality of ground water are the thousands of underground storage tanks that leak gasoline and other petroleum products. Attempts to reclaim a contaminated aquifer are complicated because petroleum products are mixtures of hydrocarbons that have a wide range of physical and chemical properties and considerable diversity in their geochemical fate in the environment. Some compounds are extremely soluble in water and, therefore, can move more readily with the ground water over large distances from the tanks. Other compounds are sorbed onto the aquifer material and are immobile, whereas still others degrade into gases, organic acids, or other products whose behavior and fate must be ascertained. Most of the reactions are microbially mediated oxidation-reduction reactions. The ultimate fate of the compounds is largely a function of their biodegradability in ground water.

Resolution of such environmental problems requires a great amount of data that can be obtained only by sophisticated sampling techniques and laboratory procedures. For example, a study site at a leaking gasoline tank on a farm in New Jersey is only about 600 m^2 in which more than 150 sampling points were installed at various depths. During one year of study, several thousands of analyses were made of water samples for about 20 inorganic and 50 organic constituents. Narrow, thin biochemical zones only a few centimeters thick with areas of a few square meters have quite diverse chemical character (Cozzarelli, Baedecker, and Hopple 1991). Therefore, selected physical processes, chemical reactions, and biologic activity must be considered on a small scale to determine the spatial distribution of reactants and products and to determine chemical and physical effects of organic contaminants on the aquifer. These site-scale processes will require great caution in transferring data on anthropogenic organic compounds from small-scale studies to water-quality problems at the regional scale.

CONCLUDING REMARKS

The science of chemical hydrogeology, with its principal beginnings less than 40 years ago, has matured and now provides the capability for reaction-path modeling to help explain the observed chemical character of ground water. The use of isotopes within a delineated hydrochemical framework provides for the determination of reaction rates, aquifer characteristics, and functioning of ground-water-flow systems. The application of inorganic and isotopic geochemical methods to hydrogeologic studies leads to determining the amounts of minerals that are dissolving and precipitating, velocity of ground-water flow, hydraulic conductivity, and porosity for regional-flow systems. It also is possible to determine flow paths, to assess time and rate of recharge, and to predict chemical changes that may occur in response to applied stresses. These principles have been developed largely through study of rather simple, virtually mineralogi-

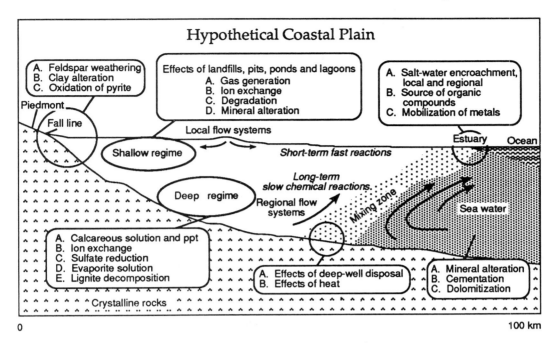

FIGURE 5-11. Dominant reactions in various regimes of a coastal ground-water system in a temperature humid region.

cally uniform, regional-aquifer systems, in which chemical and physical homogeneity was assumed. A compilation of these reactions (Figure 5-11) demonstrates their main area of influence. For example, chemically stressed environments are manifested in the shallow regime of landfills, waste-storage lagoons, deep-well waste disposal, or of salt-water from either natural encroachment or from overpumping an adjacent freshwater aquifer.

Controls imposed by heterogeneity of ground-water systems clearly demonstrate that the success of any water-quality investigation requires that scale of the sampling scheme be compatible with scale of the problem. For example, if a water-assessment program is directed toward evaluation of ground-water contamination that occurs in a site-scale system, then the sampling scheme must be at a site scale. A continuing scientific challenge is the application of conceptual models, theoretical calculations, and laboratory data to field problems. Scientific obstacles to overcome in this transfer have long been, and continue to be, the effects of time and scale of study and definition of initial and boundary conditions. The success of integrated studies depends, in part, on recognition of the appropriate scale of the problem.

Studies of ground-water contamination provide both a challenge and an opportunity to incorporate principles of organic geochemistry and microbiology into the paradigm of chemical hydrogeology. This work is now in its infancy, and the complexity of these site-specific systems requires detailed investigations, which, in turn, requires more precise determinations of mechanisms of organic-inorganic interactions and rates of reactions. Incorporation of organic reactions into chemical hydrogeology requires a continuous evaluation of their role in solute-mineral interactions, and their role in the unsaturated zone. The development of transport models is also complicated by the presence of organic compounds with a density either greater or less than that of water and with variable degrees of solubility.

When geochemical principles are applied to site-specific studies of ground-water contamination, an entire transformation of approach is required because chemical heterogeneity becomes dominant. This chemical heterogeneity develops because (1) biomediated reactions are generally more rapid than the inorganic reactions, and (2) physical heterogeneities are the dominant control on ground-water-flow paths. Resolution of the complexities introduced by chemical heterogeneity requires an understanding of the geologic environment at a detailed scale, which heretofore has been unnecessary. This understanding includes such factors as environments of deposition; diagenetic alteration of permeability and distribution of porosity; and texture, fabric, and mineralogy of the aquifer material, and its control on and transport and fate of solutes.

A key challenge to geochemists is the development of techniques to sample thin zones inherent in site-scale environments, to develop analytical capability to characterize organic and inorganic chemistry within these zones, and to interpret these chemical data in a geologic framework. Another aspect of these studies concerning mitigation of ground-water contamination is developing the rationale that permits transfer of knowledge gained at site-specific studies to regional-scale investigations. In other words, how do we transfer the understanding gained from detailed site-specific studies to a region where pesticides and fertilizers that do not have a point source are the major contaminants of ground water? And how do we move from a regional scale to a truly national scale so that we may answer such questions as, what is the chemical character of ground water now? Has it changed from the past? And, if so, at what rate is it changing? These are important questions whose resolutions are just beginning. Our activities will be guided somewhat by the cliche "study locally, but think globally." Hydrogeologists have a constant responsibility to impress upon planners, managers, and regulatory officials that the success of any water-quality investigation requires compatibility of the scale of the sampling scheme with the scale of the problem.

References

Back, William. 1966. *Hydrochemical Facies and Ground-Water Flow Patterns in Northern Part of Atlantic Coastal Plain.* U.S. Geological Survey Professional Paper 498-A.

Back, William, and M. J. Baedecker. 1989. Chemical hydrogeology in natural and contaminated environments. *Journal of Hydrology* 106:1–28.

Back, William, and R. A. Freeze. 1983. *Chemical Hydrogeology.* Benchmark Papers in Geology, volume 72. Stroudsburg, Pennsylvania: Hutchinson Ross.

Back, William, and Bruce B. Hanshaw. 1970. Comparison of chemical hydrogeology of the carbonate peninsulas of Florida and Yucatan. *Journal of Hydrology* 10:330–68.

Back, William, B. B. Hanshaw, L. N. Plummer, P. H. Rahn, C. T. Rightmire, and M. Rubin. 1983. Process and rate of dedolomitization: Mass transfer and ^{14}C dating in a regional carbonate aquifer. *Geological Society of America Bulletin* 94:1415–29.

Back, William, B. B. Hanshaw, T. E. Pyle, L. N. Plummer, and A. E. Weidie. 1979. Geochemical significance of groundwater discharge and carbonate solution to the formation of Caleta Xel Ha, Quintana Roo, Mexico. *Water Resources Research* 15: 1521–35.

Baedecker, M. J., and W. Back. 1979. Hydrogeological processes and chemical reactions at a landfill. *Ground Water* 17:429–37.

Baedecker, M. J., and I. M. Cozzarelli. 1991. Geochemical modeling of organic degradation reactions in an aquifer contaminated with crude oil. In *U.S. Geological Survey Toxic Substances Hydrology Program—Proceedings of the Technical Meeting, Monterey, California*, eds. G. E. Mallard, and D. A. Aronson, pp. 627–32. U.S. Geological Survey Water-Resources Investigations Report 91-4034.

Bennett, P. C., and D. I. Siegel. 1987. Increased solubility of quartz in water due to complexation by dissolved organic compounds. *Nature* 326:684–7.

Bredehoeft, J. D., W. Back, and B. B. Hanshaw. 1982. *Regional Ground-Water Flow Concepts in the United States: Historical Perspective, Recent Trends in Hydrogeology*, pp. 297–316. Boulder, Colorado: The Geological Society of America, Special Paper 189.

Busby, J. F., L. N. Plummer, R. W. Lee, and B. B. Hanshaw. 1991. *Geochemical Evolution of Water in the Madison Aquifer in Parts of Montana, South Dakota, and Wyoming*. U.S. Geological Survey Professional Paper 1273.

Chamberlin, T. C. 1885. *Requisite and Qualifying Conditions of Artesian Wells*. U.S. Geological Survey Annual Report 5:131–75.

Chapelle, F. H., and P. B. McMahon. 1991. Geochemistry of dissolved inorganic carbon in a coastal plain aquifer. 1: Sulfate from confining beds as an oxidant in microbial CO_2 production. *Journal of Hydrology* 127:85–108.

Chapelle, F. H., J. L. Zelibor, D. J. Grimes, and L. L. Knobel. 1987. Bacteria in deep coastal plain sediments of Maryland: A possible source of CO_2 to groundwater. *Water Resources Research* 23:1625–32.

Cozzarelli, I. M., M. J. Baedecker, and J. A. Hopple. 1991. Geochemical gradients in shallow ground water caused by the microbial degradation of hydrocarbons at Galloway Township, New Jersey. In *U.S. Geological Survey Toxic Substances Hydrology Program—Proceedings of the Technical Meeting, Monterey, California*, eds. G. E. Mallard, and D. A. Aronson, pp. 256–262. U.S. Geological Survey Water-Resources Investigations Report 91-4034.

Darton, N. H. 1909. *Geology and Underground Waters of South Dakota*. U.S. Geological Survey Water-Supply Paper 227.

Fiedler, A. G., and S. S. Nye. 1933. *Geology and Ground-Water Resources of the Roswell Artesian Basin, New Mexico*. U.S. Geological Survey Water-Supply Paper 639.

Foster, M. D. 1950. The origin of high sodium bicarbonate waters in the Atlantic and Gulf Coastal Plains. *Geochimica et Cosmochimica Acta* 1:33–48.

Freeze, R. A., and William Back. 1983. *Physical Hydrogeology. Benchmark Papers in Geology*, volume 72. Stroudsburg, Pennsylvania: Hutchinson Ross.

Johnston, H. E., and William Back. 1977. Processes of enrichment of manganese and iron in a glacial aquifer, Rhode Island. *Ground Water* 15:323.

Lee, R. W. 1985. Geochemistry of groundwater in Cretaceous sediments of the southeastern coastal plain of eastern Mississippi and western Alabama. *Water Resources Reasearch* 21(10):1545–56.

Lee, R. W., and D. J. Strickland. 1988. Geochemistry of groundwater in Tertiary and Cretaceous sediments of the southeastern coastal plain in eastern Georgia, South Carolina, and southeastern North Carolina. *Water Resources Research* 24(2):291–303.

Mendenhall, W. C. 1905. *The Hydrology of San Bernardino Valley, California*. U.S. Geological Survey Water-Supply Paper 142.

Miller, Ronald L., T. F. Kraemer, and B. F. McPherson. 1990. Radium and radon in Charlotte Harbor Estuary, Florida. *Estuarine, Coastal and Shelf Science* 31:439–57.

Nicholson, R. V., J. A. Cherry, and E. J. Reardon. 1983. Migration of contaminants in groundwater at a landfill: A case study 6. Hydrogeochemistry. *Journal of Hydrology* 63:131–76.

Plummer, L. N., and William Back. 1980. The mass balance approach: Application to interpreting the chemical evolution of hydrologic systems. *American Journal of Science* 280:130–42.

Plummer, L. N., J. F. Busby, R. W. Lee, and B. B. Hanshaw. 1990. Geochemical modeling of the Madison Aquifer in parts of Montana, Wyoming, and South Dakota. *Water Resources Research* 26(9):1981–2014.

Smith, R. L., R. W. Harvey, and D. R. LeBlanc. 1991. Importance of closely spaced vertical sampling in delineating chemical and microbiological gradients in groundwater studies. *Journal of Contaminant Hydrology* 7:285–300.

Stringfield, V. T. 1936. *Artesian Water in the Florida Peninsula*, pp. 115–95. U.S. Geological Survey Water-Supply Paper 773-C.

6

Inorganic Chemical Processes and Reactions

Donald D. Runnells

INTRODUCTION

This chapter discusses the inorganic chemical processes and reactions that determine the chemical composition of ground water in regional-aquifer systems. The first topic to be presented will be a brief historical over-view of the development of important ideas in the inorganic geochemistry of ground water. This is followed by a discussion of some relevant chemical and physical properties of water and dissolved materials, including a consideration of electrical-charge balance and concentration units. The third major topic is a brief overview of the chemical elements of major significance in ground water. The final, and longest, section is a systematic discussion of the most important inorganic chemical reactions in regional-aquifer systems.

Ground water may undergo a variety of inorganic chemical reactions as it moves through an aquifer and interacts with the solid framework materials and associated gases. Recharge water generally will be dilute, but downgradient the water will commonly evolve into higher salinities and various chemical compositions as it moves slowly through porous aquifer materials. The final composition of the water will depend chiefly on the mineralogic composition of the aquifer and the age of water; the longer the time of contact between water and aquifer solids, the greater the possible degree of chemical evolution. Because of the long distances that may be involved and the slow rate of movement in some regional aquifers, large changes can occur in the chemistry of the ground water due to inorganic reactions.

Several books discuss the geochemistry of ground water to varying degrees, including Freeze and Cherry (1979), Matthess and Harvey (1982), Lloyd and Heathcote (1985), Ward, Giger, and McCarty (1985), Drever (1988), National Research Council (1990), and Domenico and Schwartz (1990).

In recent years many studies have been published on the regional geochemistry of ground water. A few of the more significant studies are mentioned to allow the interested reader to follow the developments through time. Some additional seminal papers were reviewed in Chapter 5 in the context of scales in chemical hydrogeology.

From a chemical point of view, among the most interesting of early publications is the paper by Renick (1924) concerning cation exchange reactions, in which he showed that Ca and Mg replaced Na on smectite clays in the Lance and Ft. Union formations of

Tertiary age in Montana. The result was a rapid transformation in the cationic composition of the ground water, with essentially complete transformation in the composition of the ground water by the time a depth of about 40 m was reached. Foster (1950) also emphasized the importance of cation-exchange reactions in her study of the origin of sodium-bicarbonate waters in the regional aquifers of the Atlantic and Gulf coastal plains. She demonstrated that dissolution of limestone and cation exchange of Na for Ca can produce the high dissolved Na and HCO_3 observed in those waters.

The important concept of regional hydrochemical facies was introduced by Back (1960) in a study of ground water in the Atlantic coastal-plain aquifers of the eastern United States; he showed that the compositions and concentrations of dissolved ions in ground water respond to the lithology and flow patterns in particular regions. Chemical reactions that control the chemistry of saline ground water in deeply buried sedimentary rocks were elucidated by White in 1965, with emphasis on the idea put forth by Berry and Hanshaw (1960) that fine-grained sediments act as semipermeable membranes, permitting the passage of water while concentrating the dissolved ions. Hitchon and Friedman (1969) later used stable isotopes of hydrogen and oxygen to show that mixing of water, rock-water interaction, and ultrafiltration through shale membranes produce the chemical compositions of formation waters present in the sedimentary basin of western Canada. Runnells (1969) showed that the simple process of mixing of dissimilar ground waters could produce important and surprising chemical reactions, resulting in diagenesis and alteration on local or regional scales; the concept has subsequently been applied to a wide variety of hydrogeochemical environments.

Much important work on the regional geochemistry of ground water has focused on carbonate aquifers. Hanshaw, Back, and Rubin (1965) and Hanshaw and Back (1979) showed that the chemistry of ground water in the Floridan limestone aquifer was dependent on reactions involving carbon dioxide gas, calcite, aragonite, dolomite, and gypsum. Thrailkill (1968) and Langmuir (1971a) further developed our understanding of karst hydrology and geochemistry in regional carbonate aquifers of the eastern United States. Plummer et al. (1976) demonstrated that the chemistry of the ground water in the carbonate aquifer of Bermuda could be explained chiefly by three processes: (1) the generation of carbon dioxide gas in soils and marshes, (2) dissolution of metastable carbonate minerals, and (3) mixing with seawater.

More recent studies of regional aquifers include that by Thorstenson, Fisher, and Croft (1979) on the geochemical evolution of ground water in the Cretaceous Fox Hills–Basal Hell Creek aquifer of South Dakota. Edmunds, Bath, and Miles (1982) conducted an interesting study of the geochemistry of the ground waters in the sandstones of the Triassic East Midlands aquifer of England. A regional aquifer that has been studied extensively over a period of more than 25 years is the Milk River aquifer of southern Alberta, Canada. According to Hendry and Schwartz (1990), the reason for so much interest in the Milk River may be the regular and systematic pattern of variability in almost every geochemical parameter studied.

In 1978 the U.S. Geological Survey initiated a research program on the geology, hydrology, and geochemistry of regional aquifers, entitled the Regional Aquifer-System Analysis Program (RASA). This program, described by Sun and Weeks (1991), is designed to provide a framework of background information for important regional aquifers in the United States. One such a study is that by Henderson (1985) of the geochemistry of ground water in two sandstone aquifers in the northern Great Plains of the United States. Henderson used a variety of approaches to define the reactions which control the chemistry of the ground waters, including computer simulation modeling of rock-water interactions. A similar comprehensive study has been published for the Madison aquifer in the north-central United States by Plummer et al. (1990). As demonstrated by Henderson (1985) and Plummer et al. (1990), computer simulation modeling can be a powerful tool in interpreting the chemistry of ground water.

Studies of stable and radioactive isotopes of carbon, oxygen, sulfur, and nitrogen are critically important in interpreting chemical reactions in ground water. Two recent examples of the application of isotopic data to contamination of ground water are mentioned here. Smith, Howes, and Huff (1991) used nitrogen isotopes to study rapid denitrification in ground water in Massachusetts, showing that oxidation of organic matter to CO_2 and reduction of NO_3^- to N_2 gas are coupled chemical reactions. Bishop and Lloyd (1991) used carbon-14 modeling to determine the rates of flow, and thus the susceptibility to contamination, of ground water in different portions of the Lincolnshire Limestone of eastern England. The use of these and other environmental isotopes is addressed in Chapters 10 and 11.

Recent studies of the regional geochemistry of ground water in North America are presented in the volume on *Hydrogeology*, edited by Back, Rosenshein, and Seaber (1988), as part of the centennial series on the *Geology of North America, Decade of North American Geology*, published by the Geological Society of America. This important volume contains separate chapters for each of the 28 hydrogeologic regions of North America.

PHYSICAL AND CHEMICAL PROPERTIES

Physical Properties

Numerous articles in scientific and popular literature describe the remarkable properties of water. The same can be said about the role of water in inorganic chemical reactions. For excellent summaries of the physical and chemical properties of water, the reader is referred to Horne (1973) and Matthess and Harvey (1982). A few of the more significant properties are mentioned here.

In liquid form, the individual molecules of water seem to be clustered in a cagelike arrangement (Drost-Hansen 1966; Matthess and Harvey 1982) with a degree of structural regularity and exhibiting some properties of a polymer. Horne (1973) discusses in detail the effect of the structuring of water on its chemical behavior. Hem (1985) points out that because there are three isotopes of hydrogen and three isotopes of oxygen in nature, 18 varieties of water are possible. Numerous models have been proposed for the structure of pure water, usually for the purpose of calculating chemical and physical properties at elevated temperatures and pressures.

The molecular shape of water is that of a tetrahedron, with the oxygen atom at the center. The chemical bonds between the two hydrogens and the central oxygen are at an angle of 105° to each other. Because of the asymmetry of the molecule, with the positive hydrogen nuclei lying to the side of the strongly electronegative oxygen atom, the water molecule has a large dipole moment (charge divided by distance). The large dipole moment leads to strong dipole-dipole interactions, causing much greater internal cohesion in water than in other liquids (Matthess and Harvey 1982). The molecular dipoles also can assume resonance configurations that permit extensive hydrogen bonding among the water molecules (Drost-Hansen 1966).

The peculiar abilities of water to form several types of chemical bonds lead to exceptional physicochemical properties. Water is a member of two homologous groups of the periodic table, the hydrides of the sixth main group (H_2O, H_2S, H_2Se, H_2Te), and the hydrides of the second-period elements (BH_3, CH_4, NH_3, H_2O, and HF). However, within these series, the properties of water are anomalous in comparison to the values of many properties that one would expect by extrapolation from the other members of the series, including unexpectedly high values for the heat of fusion, heat of vaporization, surface tension, boiling point, freezing point, heat capacity, and dielectric constant.

When liquid water freezes to solid ice at low pressure, the density decreases. This property has profound implications for the history and evolution of life on Earth. If ice were to have a greater density than water, bodies of water (including the ocean) would freeze from the bottom upward. Aquatic biota that might exist in such a situation would surely be different than the forms that now inhabit the waters of the Earth.

Table 6-1 presents some physicochemical properties of liquid water that are particularly relevant in studies of ground water. All of the values in Table 6-1 refer to a total pressure of 100 kPa (1 bar); the properties of water change very slightly with pressure, even over a range of pressures far greater than can be expected in typical ground water. For example, even at total confining pressures as great as 1,000 bars (corresponding roughly to a depth of water of 10,000 m), the viscosity of water decreases by less than 3% (Horne 1973). Many important properties of water are proportionately much more sensitive to changes in temperature than to changes in pressure.

Of special interest in Table 6-1 are the values for the viscosity and the dielectric constant. The viscosity enters directly into the denominator of Darcy's equation for flow of fluids through porous media, and at constant gradient a decreasing viscosity will cause a proportional increase in the velocity of flow of the ground water. Therefore, from the values in Table 6-1, it is seen that if the temperature of ground water were to rise from 20°C to about 55°C, the effect of the change in viscosity alone would be to increase the velocity of flow of ground water by a factor of about 2 (viscosity decreases from 1,002 to about 500 µPa-s), all other factors equal. Surface tension also decreases rapidly with rising temperature and would further enhance the flow of ground water at elevated temperatures by reducing the adhesion to the surfaces of minerals in the aquifer.

From a chemical point of view, the dielectric constant controls the ability of water to act as a solvent for ionic substances. As noted earlier, the dielectric constant of water is exceptionally high, with a value of 80.20 at 20°C; in comparison, for example, the dielectric constant of benzene at 20°C is 2.284, and for methanol it is 33.62 (Lide 1991, p. 9-9). From Table 6-1 it is seen that the dielectric constant of pure water decreases with rising temperature, dropping from 87.90 at 0°C to 55.51 at 100°C, indicating that water becomes a poorer solvent for ionic substances with rising temperature if all other factors remain equal.

For purposes of investigating the geochemistry of typical ground water, it is adequate to know that the total pressure can be considered to be about 1 bar, with a temperature range from a low of about 4°C to a high of perhaps 35°C. As the temperature rises over this range, the density (Table 6-1) will decrease by less than 1%, the viscosity

TABLE 6-1. Physicochemical Properties of Liquid Water

T (°C)	Density (g/cm^3)	Viscosity (µPa-s)	Dielectric Constant	Surface Tension (mN/m)
0	0.99984	1793	87.90	75.64
10	0.99970	1307	83.96	74.23
20	0.99821	1002	80.20	72.75
30	0.99565	797.7	76.60	71.20
40	0.99222	653.2	73.17	69.60
50	0.98803	547.0	69.88	67.94
60	0.98320	466.5	66.73	66.24
70	0.97778	404.0	63.73	64.47
80	0.97182	354.4	60.86	62.67
90	0.96535	314.5	58.12	60.82
100	0.95840	281.8	55.51	58.91

Source: Adapted from Lide (1991). (Reprinted with permission from *Handbook of Chemistry and Physics*, Copyright CRC Press, Inc., Boca Raton, FL)

by nearly 50%, the dielectric constant by about 12%, and the surface tension by about 7%. The decrease in viscosity, which is the most significant physical change, may lead to doubled rates of flow through porous media at the higher temperatures.

Forms of Chemical Elements in Ground Water

Chemical elements in ground water can be present in a variety of particulate, colloidal, and dissolved forms, ranging from mechanical particles that are large enough to cause turbidity to individual molecules and hydrated ions. Colloidal particles are those that are smaller than about .1 to 1 µm (10^3–10^4 Å) in diameter, and truly dissolved substances range between about .1 and 1 nm (1–10 Å) in diameter. The broad range of forms is emphasized in Figure 6-1.

The analytical determination of the forms of chemical elements in natural waters is not a trivial task. Methods for determining the physical and chemical forms of the elements range from simple mechanical filtration, designed to separate dissolved ions from particles and colloids, to sophisticated statistical interpretation of spectrophotometric data and potentiometric titration curves. Detailed information on the techniques that are used to determine the specific forms of dissolved elements in natural waters can be found in books by Singer (1974), Leppard (1983), Kramer and Allen (1988), and Martell and Motekaitis (1988).

Dissolved Substances in Water

The exact nature of dissolved substances in water is not fully understood. However, many of the characteristics of dissolved substances can be inferred from such experimentally measurable properties of aqueous solutions as density, refractive index, electrical conductivity, absorption of sound, color and other spectroscopic properties, osmotic pressure, boiling and freezing points, viscosity, heat of dissolution, and so on. From observations of such properties, it is clear that a variety of reactions can occur when a solid or gaseous substance dissolves in water, yielding a wide variety of dissolved entities. However, in the most simplistic terms, dissolved substances in water can be divided into three categories: (1) free ions, (2) neutral molecules, and (3) ion pairs and aqueous complexes.

Free ions consist of single atoms or molecules with an electrostatic (coulombic) charge, examples of which include Ca^{2+}, Na^+, SO_4^{2-}, NO_3^-, $COOH^-$, and CH_2COO^-. Such species are always coordinated (surrounded) by water molecules and are therefore described as "hydrated" ions. The number of coordinating water molecules depends on the size, shape, and charge of the central ion, with larger numbers of water molecules coordinated to the larger and more highly charged ions. Because of the net electrostatic charge, ions can be attracted to each other in solution to form ion pairs and aqueous complexes and, if the concentrations of the ions are large enough, they can precipitate as ionic solids. Molecules of water bonded to the aqueous ions may still be retained in the solid when the ions precipitate as, for example, in gypsum ($CaSO_4 \cdot 2H_2O$) and epsom salt ($MgSO_4 \cdot 7H_2O$).

The second broad category, neutral molecules, includes aqueous species that consist of more than one atom and without a net electrostatic charge. Natural examples include dissolved $H_2CO_3^0$, H_2S^0, NH_4^0, $HCOOH^0$, CH_3COOH^0, $H_4SiO_4^0$, $Al(OH)_3^0$, and so on. Although such substances are electrically neutral, the shapes of the molecules and the spatial distributions of the electron clouds that bind the atoms together can result in significant asymmetry and polarity, giving rise to molecules, including pure water, that have strong electrical properties. Molecular substances in water can exhibit internal bonding that ranges from very weak to very strong, with the latter characterized by covalent and hydrogen bonding.

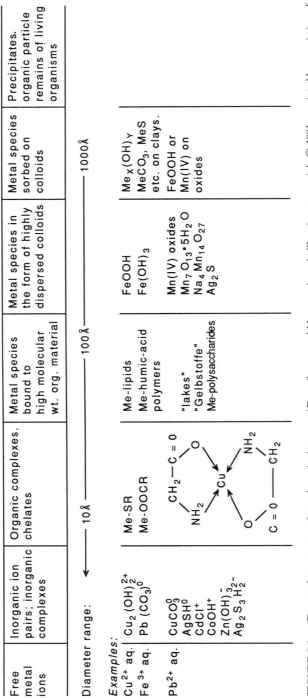

FIGURE 6-1. Forms of occurrence of metal species in water. (*From Stumm and Morgan, Aquatic Chemistry; copyright © 1981, reprinted by permission of John Wiley & Sons, Inc.*)

The third type of dissolved substance in aqueous solution is the charged ion pair or aqueous complex, in which two or more distinct dissolved species bond together into a new electrically charged entity. Examples would include Na^+ and SO_4^{2-} combined to form $NaSO_4^-{}_{(aq)}$, Ca^{2+} and Cl^- combined to form $CaCl^+{}_{(aq)}$, and Al^{3+} and formate to make $AlCOOH^{2+}$. The anionic species, SO_4^{2-}, Cl^-, and $COOH^-$ in these examples, is called the *ligand*. The terms *ion pair* and *aqueous complex* are often used interchangeably, but most researchers (cf. Stumm and Morgan 1981) draw a distinction between the weaker electrostatic (coulombic) bonding of the ion pairs and the stronger covalent or hydrogen bonding of aqueous complexes. Stumm and Morgan (1981) also specify that coordinated water molecules are retained in the ion pairs and may spatially separate the cation and the ligand, whereas in aqueous complexes the cation and ligand are immediately adjacent to each other without intervening water molecules. If the ligand is organic, the resulting ion pair or complex is termed a *chelate*.

The tendency for ions or molecules to join into ion pairs or complexes is described by an "association constant," K_A, defined for the following example for a reaction to form an aqueous complex C_mA_n, in which C and A can be organic or inorganic:

$$mC^{\alpha+}{}_{(aq)} + nA^{\beta-}{}_{(aq)} \leftrightarrow (C)_m(A)_n^{(m\alpha+)-(n\beta-)}{}_{(aq)}$$

for which the association constant is

$$K_A = \frac{[(C)_m(A)_n^{(m\alpha+)-(n\beta-)}{}_{(aq)}]}{[C^{\alpha+}{}_{(aq)}]^m[A^{\beta-}{}_{(aq)}]^n}$$

where [ ] = activity of dissolved species. Stumm and Morgan (1981) offer a fairly detailed discussion of the form and derivation of the association constants for different types of ion pairs and complexes, and the book by Singer (1974) addresses the topic of metal-organic interactions.

The larger the value of association constant, K_A, the greater the tendency for the ion pairs and complexes to form. This reaction can be quantitatively described in terms of the Gibbs energy function, in which the extent to which the new species is formed is dictated by the minimization of the Gibbs energy function of the total system.

An important aspect of the dissolution of solids and gases is the resulting disruption of the structure of the pure water. Millero and Sohn (1992) discuss this topic, giving as one interesting consequence the decrease in volume of a solution which occurs when certain salts are dissolved in water, which is the "electrostriction" effect. If the ions in solution have a smaller effective volume than in the solid state, increased pressure will tend to increase the solubility of the solids. This effect is quite important for calcite in water, causing an increase in the solubility product constant (K_{sp}) by a factor of about 1.5 at a confining pressure of 100 atm, which corresponds to a depth of about 1,000 m of water (Pytkowicz 1983).

Determination of Dissolved Species by Computer Modeling

One powerful method for determining the dissolved forms of elements in natural waters is geochemical computer modeling, as discussed in detail in Chapter 9. Briefly, geochemical modeling of aqueous systems is based on an assumption of chemical equilibrium among all of the dissolved aqueous species in a solution. The determination of the aqueous speciation results from the simultaneous solution of a set of nonlinear equations. The set of equations typically represents the equilibrium mass-action relationships among dissolved species, charge balance among positive and negative species, and mass balance between the sum of all dissolved species of each major component and the total input concentration of that component. Input data include the mass-action equilibrium constants for all reactions involving the aqueous species of interest, plus the analytically determined total concentrations of each of the major components. A measured pH and *Eh* (redox potential) may also be required, although some programs do not need these values. The algorithms typically iterate on the set of equations until an acceptable fit is obtained to the input values of the total concentrations of the major components. As the iterations take place, the concentration of each of the aqueous species is adjusted until an acceptable fit occurs between the sum of the aqueous species and the total input concentrations of each of the corresponding components.

The output from a geochemical model typically includes a listing of the equilibrium concentrations of all aqueous species of each input component, as shown in Chapter 9. This will be illustrated for a natural ground water. Table 6-2 is an analysis of a ground water from an alluvial aquifer in a mountain valley near Telluride, Colorado; the units are given in mg/dm^3, which, because the water is dilute, are numerically equal to mg/kg. The rather high concentration of SO$_4$ (140 mg/dm^3) and the moderately high (351 mg/dm^3) total dissolved solids (TDS) are probably due to the fact that the aquifer lies downgradient from significant natural occurrences of base-metal sulfides and abandoned base-metal mines and mills. The electrical-charge balance (see definition in footnote of Table 6-2) of the analysis in Table 6-2 is −2.4%, which is quite acceptable. The major composition of the water would characterize it as being a member of the calcium-sodium-bicarbonate-sulfate facies (Back 1960). The pE value of −2.16 (*Eh* = −128 mV), as measured with a platinum electrode (Langmuir 1971b; Bricker 1982), is moderately reducing, although such measurements are known to be of questionable validity (Stumm and Morgan 1981; Lindberg and Runnells 1984; Runnells and Lindberg 1990).

The analysis of Table 6-2 was used as input to the equilibrium geochemical computer code PHREEQE (Parkhurst,

TABLE 6-2. Partial Chemical Analysis of a Ground Water from an Alluvial Aquifer, Telluride, Colorado

Component	Concentration (mg/dm^3)	Concentration (mmol/dm^3)
Ca	73.6	1.84
Mg	4.0	0.16
Na	8.0	0.35
K	3.9	0.10
Alkalinity (as HCO$_3$)	112.0	1.84
Cl	<3.0	<0.08
SO$_4$	140	1.44
F	0.3	0.02
Mn	1.6	0.03
Fe	0.16	0.003
Zn	0.054	0.0008
SiO$_2$	7.6	0.13
TDS		351
pH		7.2
pE		−2.16 (*Eh* = −0.128 V)
T (K)		280
Charge balance[a]		(−)2.4%

Note: Depth approximately 30 m. All concentrations are in mg/dm^3 and mmol/dm^3 (equivalent to mg/L and mmol/L) except pH, pE, *Eh*, and *T*.
[a] Charge balance defined as [(meq cations − meq anions)/(meq cations + meq anions)] × 100.

Thorstenson, and Plummer 1980). The output from the modeling exercise includes a list of the aqueous species of the input components, as illustrated in Table 6-3. Of course, the computed distribution of aqueous species for any natural water is only an estimate, which depends on the accuracy and completeness of the thermodynamic data, as well as the accuracy and precision of the chemical analyses of the water.

It is clear from Table 6-3 that free ions (noncomplexed) are by far the predominant aqueous species in this particular ground water, even though the water has a higher concentration of TDS and SO$_4^{2-}$ than might be expected for most shallow ground waters. Other than the free ions, the only major aqueous species is CaSO$_4^0$, containing about 10 mol% of the total Ca and SO$_4$ present. About 10 mol% of the total Mg is also com-

TABLE 6-3. Modeled Concentrations of the Principal Aqueous Species of the Components Listed in Table 6-2 for a Ground Water from Near Telluride, Colorado

Species	Concentration (mmol/kg)
Ca^{2+}	1.63
$CaSO_4^0$	0.19
$CaHCO_3^+$	0.01
Mg^{2+}	0.15
$MgSO_4^0$	0.01
$MgHCO_3^+$	0.002
Na^+	0.34
$NaSO_4^-$	0.001
K^+	0.10
KSO_4^-	0.0004
HCO_3^-	1.83
CO_3^{2-}	0.001
$NaHCO_3^0$	0.0002
Total Cl	Not detected
SO_4^{2-}	1.24
$CaSO_4^0$	0.19
$MgSO_4^0$	0.01
$NaSO_4^0$	0.001
F^-	0.02
Mn^{2+}	0.03
$MnSO_4^0$	0.002
Fe^{2+}	0.003
$FeSO_4^0$	0.0002
Zn^{2+}	0.0007
$ZnSO_4^0$	0.00008
$ZnCO_3^0$	0.00007
$H_4SiO_4^0$	0.13

Note: The computer code used was PHREEQE (Parkhurst, Thorstenson, and Plummer 1980). All concentrations in mmol/kg H_2O (numerically equivalent to mmol/L and mmol/dm^3 for this dilute water).

plexed with SO_4^{2-}, as $MgSO_4^0$. The two trace metals, Fe and Mn, are each present in the form of a sulfate complex at about 10 mol% of the total dissolved concentrations. Of the total zinc, about 10% is present in the forms of $ZnSO_4^0$ and $ZnCO_3^0$. Silica is present almost entirely as nondissociated silicic acid, H_4SiO_4, and F is primarily present as free F^-.

The result of the modeling exercise summarized in Tables 6-2 and 6-3 is fairly typical for normal shallow ground waters, with most of the components being present in the form of free ions. In the case of C, S, and Si, simple ionic and neutral molecules predominate, in the form of HCO_3^-, SO_4^{2-}, and H_4SiO_4, respectively. For more saline waters, as the concentrations of dissolved species become greater, the tendency increases to form ion pairs and complexes involving those species.

The formation of chelates, involving inorganic and organic components, is also a possibility. Natural dissolved organic matter occurs chiefly in the form of complex humic and fulvic acids. Most normal ground waters contain less than 15 mg/L (15 mg/dm^3) dissolved organic carbon, with a median value of 0.7 mg/L (Thurman 1985). The fact that natural organic compounds are present at such low concentrations means that the proportions of the major inorganic components which may be complexed by dissolved organic matter will be insignificant. Thurman (1985) states that a rough rule of thumb is that 1 mg/L of dissolved organic carbon has approximately 1 µeq/L of binding capacity for metals. Thus, for cationic species in water, such as Ca and Mg, present at concentrations of a few mmol/L, less than 0.1% of the total concentration would be expected to be present in the form of organic complexes. Even for trace elements in water, such as Zn, Cd, and Pb, usually present at concentrations below 10 µg/L, less than 10% of the total metal might be expected to be complexed by dissolved organics at normal concentrations [see, for example, measurements of complexing of trace metals by dissolved organics in lake and sea water, reported by Raspor et al. (1984)]. Of course, ground waters that are exceptionally high in dissolved organic matter will show higher degrees of complexing. For example, hundreds to thousands of mg/L of dissolved organic acids (especially acetic acid) have been reported in waters from petroleum reservoirs (Barth 1991; Crossey 1991), and

Surdam, Boese, and Crossey (1984) have shown that the presence of high concentrations of such species in subsurface waters can significantly increase the solubility of common rock-forming minerals, causing enhanced diagenetic alterations of the rock. High concentrations of organic contaminants also may occur in ground water at sites that have been impacted by humans, with increased complexing between natural dissolved materials and the organic contaminants.

The transport of inorganic species in ground water in the form of colloids is being increasingly recognized as an important process. For example, Degueldre et al. (1989) found a concentration of approximately 10^{10} colloidal particles (40–1,000 nm in diameter) per liter in water from a fracture in granite at a depth of 450 m below the land surface in the Swiss Alps. The particles were made up of a mixture of organic (C, N, O, S) and inorganic (Si, Ca, Sr, Mg, Sr, and Ba) elements. Ryan and Gschwend (1990) studied colloidal mobilization in ground water of the Pine Barrens of New Jersey, defining "colloidal" as particles ranging from organic macromolecules, with diameters of 1–2 nm, up to fragments of minerals with diameters of a few micrometers. Using extremely careful methods of collection to avoid mechanical disturbance of the well, Ryan and Gschwend (1990) found total concentrations of colloids up to 60 mg/L in the ground water, primarily as particles of clay; colloidal organic carbon was also present in all samples.

A fundamental property of all aqueous solutions is that of electrical neutrality. The sum of the concentrations of positive charge must equal the sum of the concentrations of negative charge. Hem (1985) points out that the apparent charge balance between total cations and total anions, in meq/dm^3, should be between 1 and 2% for waters of moderate (250 to 1,000 mg/dm^3) total dissolved solids (TDS), with somewhat larger errors acceptable for TDS values in more dilute waters. This is demonstrated in the analysis of Table 6-2, in which the difference between the meq/dm^3 of positive and negative charges, divided by the sum of the meq/dm^3 positive and negative charges, totals −2.4%. This difference of −2.4% is an "apparent" charge balance, assuming each dissolved component is present as a free cation or a free anion with the full ionic charge. In view of the fact that the analysis in Table 6-2 is a partial analysis (not including such possible components as P, N, and dissolved organic acids), an error of 2.4% is quite acceptable and is indicative of accurate analytical determinations. If the calculation of the meq/dm^3 of positive and meq/dm^3 negative charge were to show a large difference, one would have to be skeptical about the accuracy and reliability of the analysis.

A more refined approach to the charge-balance calculation is to determine the positive-negative balance after computing the equilibrium aqueous speciation. This approach takes into account the fact that not all of the components are present in the free ionic state in the solution.

Chemical Units

The principal units to be considered in the geochemistry of ground water describe the concentrations of dissolved substances. Much of the ground-water literature prior to about 1985 reports dissolved species in concentration units of ppm (parts per million, μg/g). In reality, however, most chemical laboratories determine concentrations of dissolved materials in volumes of solution, not in masses of solution, so that most water analyses reported in the published literature as ppm are actually mg/L. The relevant relationship between ppm and mg/L is

$$\text{ppm} = \mu g/g = (A\ mg/L) \times \left[\left(\frac{1\ L}{1,000\ mL}\right)(1,000\ \mu g/mg)(1/\rho\ g/mL)\right]$$

where A represents the concentration in mg/L. If the density (ρ) is 1.00 g/cm^3 (g/mL), the concentration in ppm (μg/g) is equal to the same value in mg/L.

Geochemists have begun to report aqueous concentrations in proper SI units of moles and mass, such as mol/kg and mg/kg. The unit of moles per kilogram of water is termed *molality*. Data are also being reported in SI units of moles and volume, such as mol/dm^3 and mg/dm^3 (where dm^3 is numerically equal to liters). The unit of mmol/kg, which is an acceptable submultiple of the basic SI unit of mol/kg, is now widely used in geochemistry (see, for example, the use of mmol/kg in geochemical modeling in Chapter 9).

During this period of transition, one finds a bewildering array of concentration units in the geochemical literature. It will be some time before authors and editors in geochemistry are comfortable with true SI units. However, the non-SI units of mg/L and ppm will gradually disappear from modern geochemical literature, to be replaced by mg/dm^3 and µg/kg, respectively.

CHEMICAL ELEMENTS OF SIGNIFICANCE

Although trace amounts of every element in the periodic table are probably present in any given sample of ground water, study of the regional geochemistry of ground water usually focuses on the presence and concentration of a relatively small number of elements, less than about 25. Hem (1985) presents an excellent and comprehensive discussion of the sources and chemical reactions of many of the elements of importance in natural waters.

The significance of dissolved elements in ground water could be characterized according to many different schemes, such as source, fate, mobility, toxicity, abundance, economic value, and so on. One approach that this author has found to be particularly useful is based on the *reason* or *purpose* for analyzing the ground water for certain elements, such as geochemical modeling, economic value, environmental importance, chemical reactivity, and toxicity. This admittedly unusual approach results in at least five major categories of elements as follows:

1. Dissolved elements that are related to common rock-forming minerals
2. Elements that are economically important
3. Common elements that are involved in gaseous reactions
4. Elements that are of environmental concern because of their toxicity
5. Elements for use in isotopic studies

Clearly, as with any classification, there is significant overlap among these categories.

In the first category, elements related to common rock-forming minerals, we should include at least Na, K, Ca, Mg, Si, Al, C, S, and Fe, as well as O and H. Elements in the second category of economic importance include Cu, Pb, Zn, Mn, P, F, U, Br, I, and Au, among others. The third category, common elements that enter into gaseous reactions, includes at least Cl, H, C, S, O, and N, all of which may be present either as important dissolved gases, such as H_2, CO_2, H_2S, SO_2, and CH_4, or as ionic or molecular aqueous species, such as $Cl^-_{(aq)}$ and $NO_3^-{}_{(aq)}$. In the fourth category, elements that are commonly measured in ground water because of their environmental impact, we should include at least Pb, As, Se, Hg, Cd, N, S, U, Mn, Cu, Zn, Ba, and Cr, among others. Finally, in the fifth category, are the stable and radioactive isotopes. The stable isotopes include ^{18}O, ^{16}O, ^{2}H, ^{14}N, ^{15}N, ^{32}S, ^{34}S, ^{12}C, ^{87}Sr, and ^{13}C; these are used primarily for investigating the origin and source of water in an aquifer. Radioactive isotopes are also useful, primarily for estimating the age of ground water; important species include ^{14}C, ^{3}H, and ^{36}Cl. Isotopes are discussed in detail in Chapters 10 and 11.

Again, to emphasize the utility of this particular classification of elements, it forces an investigator, prior to sampling and analyzing the ground water, to ask a very simple question: "What is the *reason* for collecting and analyzing the water?" In the vernacular, the critical question becomes: "So what?"

TYPES OF CHEMICAL REACTIONS

In theory, thousands of different chemical reactions could conceivably occur in regional ground waters. In fact, however, studies of the regional geochemistry of ground waters indicate that a relatively small number of chemical reactions seem to control the overall chemical composition of the waters. The dominant inorganic reactions in ground water can generally be categorized as dissolution-precipitation, acid-base reactions, complexation, substitution-hydrolysis, oxidation-reduction, ion filtration and osmosis, dissolution and exsolution of gases, and sorption-desorption. In this section a brief overview of the major categories of geochemical reactions in ground water will be given, followed by a few examples from regional studies of aquifers. Additional discussion of geochemical reactions in natural waters by the author can be found in Runnells (1976, 1987) and National Research Council (1990).

Precipitation-Dissolution Reactions

The most important chemical reactions involving ground water are those of dissolution and precipitation. Dissolution is the process whereby solids dissolve into ground water, and precipitation is the formation of solid phases from the water. Most dissolved components in ground water are ultimately derived from the dissolution of rock-forming minerals in the aquifers. Generally, the longer the time of contact between the ground water and the enclosing rocks, the higher the content of dissolved materials in the water. Dissolved components in ground water may also be removed from the water through the process of precipitation. The processes of dissolution and precipitation among ground water, aquifer solids, gases, and organic matter can be summarized by the following generalized reaction (Thorstenson 1990):

initial-water composition + "reactant phases" →

final-water composition + "product phases"

The dissolution of rock-forming reactant minerals will increase the porosity and permeability of the aquifer, whereas precipitation of new product phases has the potential for decreasing the porosity and permeability. From a petrologic point of view, reactions involving ground water and solid mineral phases are classified as *diagenetic alterations*, with many well-known examples of the dissolution and precipitation of such phases as calcite, dolomite, clay minerals, pyrite, quartz, and feldspar.

Dissolution of a mineral may be either "congruent" or "incongruent." When congruent dissolution occurs, the mineral simply dissolves in water, without any change in composition of the residual portion of the solid that has not yet dissolved. In contrast, when incongruent dissolution occurs, some of the components of the mineral go into aqueous solution at a different rate, or to a different extent, than others, leaving a solid residue or forming a new precipitate which has a different composition from the original mineral. Incongruent dissolution is well known for Mg-bearing calcites (Mackenzie et al. 1983) and for many silicate minerals (Holdren and Speyer 1986).

Thermodynamically, the potential for dissolution or precipitation of solid phases can be expressed as the computed saturation index (SI):

$$SI = \log\left[\frac{IAP}{K_{sp}}\right]$$

where IAP is the empirical ion activity product for a given mineral in the water of interest, and K_{sp} is the equilibrium solubility product constant for the same mineral at the temperature and pressure of the water. Net dissolution of a solid phase should occur if the ground water is undersaturated with respect to the solid in the adjacent aquifer, as indicated by a negative value of SI. Net precipitation of a solid phase can occur only if the SI has a positive value. An SI value of zero means that there is no thermodynamic tendency for net dissolution or net precipitation of the mineral.

As mentioned, an SI value of zero for a mineral means that, thermodynamically, there should be no *net* dissolution or precipitation of the mineral. However, even for an SI value of zero, there can be a dynamic exchange of atoms and molecules between the mineral and the water at the solution-solid interface. A thermodynamic parameter, such as SI, addresses only the bulk chemical properties of substances, not the dynamic behavior of individual atoms and molecules.

The value of the saturation index is only an indicator of the thermodynamic tendency for dissolution, precipitation, or equilibrium; it does not give us any information about the rates of reaction. As discussed later in this chapter, some minerals react rapidly with water, whereas others react very slowly. Calcite is an example of an important mineral which tends to dissolve or precipitate quite rapidly in natural waters, according to whether the value of the SI is negative or positive. In contrast, both the dissolution and precipitation of quartz are very slow.

The ground water from near Telluride, Colorado (Table 6-2), can be used for the interpretation of dissolution-precipitation reactions. The saturation indices for this water, resulting from computation with the PHREEQE geochemical code (Parkhurst, Thorstenson, and Plummer 1980), are listed in Table 6-4.

From Table 6-4, the large negative SI values indicate that fluorite, gypsum, and pyrite are strongly undersaturated in the ground water and should dissolve if they are present in the aquifer. The SI values for calcite, rhodochrosite, and chalcedony are slightly negative, indicating that the water is weakly undersaturated with respect to these minerals. Quartz and barite show slightly positive SI values and are therefore potentially capable of being precipitated from the ground water. (Unfortunately, because of a lack of data for dissolved Al in the analysis of the Telluride ground water, we are unable to compute the state of saturation with respect to such important rock-forming minerals as feldspar and mica.)

TABLE 6-4. Modeled Saturation Indices for Common Minerals for a Ground-Water Sample from Near Telluride, Colorado

Mineral	Computed Saturation Index (SI)
Calcite	(−)0.60
Dolomite	(−)2.5
Fluorite	(−)1.4
Quartz	(+)0.41
Chalcedony	(−)0.16
Gypsum	(−)1.1
Siderite	(−)1.5
Pyrite	(−)7.4
Rhodochrosite	(−)0.52
Barite	(+)0.45

Note: The geochemical computer code used was PHREEQE (Parkhurst, Thorstenson, and Plummer 1980). A partial analysis of the water is given in Table 6-2. As discussed in the text, the saturation index (SI) is defined as SI = $\log(IAP/K\text{sp})$. For these Calculations, the electrical balance of the water in Table 6-2 was improved by adding approximately 7% additional total Ca (0.12 mol/kg) to the actual analysis.

Dissolution and precipitation reactions may be coupled to other reactions. For example, in the context of potential contamination of ground water, Runnells (1976) discusses precipitation of such dissolved metallic contaminants as As, Pb, Ag, Cr, Fe, Mn, Cu, Se, and Zn from ground water as a result of substitution (metathetical) reactions, hydrolysis reactions, and oxidation-reduction reactions.

The application of computer modeling to the determination of the dissolution or pre-

Acid-Base Reactions

Reactions involving the transfer of hydronium ion (H_3O^+) among dissolved aqueous species are called acid-base reactions. From a geochemical point of view, acids are substances that can give up a hydronium ion, whereas bases are able to accept a hydronium ion. The behavior of dissolved substances in water is profoundly influenced by transfers of H_3O^+. For example, below a pH of about 7, the dominant form of oxidized As in normal ground water is $H_2AsO_4^-$, whereas above a pH of about 7 oxidized As in most ground water occurs mainly as $HAsO_4^{2-}$ (Masscheleyn, Delaune, and Patrick 1991); this difference in aqueous species is one important factor in controlling the extent of absorption of dissolved As on the surfaces of minerals. Another example of the dominant role of H_3O^+ is that both the rate of dissolution and the solubility of carbonate minerals increase strongly with decreasing pH (Morse 1983). In contrast, the rate and extent of dissolution of solid SiO_2 are strongly favored by elevated pH, due to the increasing extent of formation of $H_3SiO_4^-{}_{(aq)}$ in solution (Hem 1985; Richardson and McSween 1989).

The tendency of solids to form solid precipitates or to dissolve may also be influenced by the transfer of H_3O^+. The solubilities of Fe, Mn, Al, and other metals in ground water, for example, are strongly dependent on the pH, with the precipitation of solubility-controlling metallic oxyhydroxides ($FeOOH$, $MnOOH$, $Al(OH)_3$, etc.) occurring with increasing pH (Stumm and Morgan 1981).

Because of the natural abundance of limestone and its chemical reactivity, the system $CaCO_3$-CO_2-H_2O is important in the buffering of the pH of many ground waters. Consideration of chemical reactions in regional ground-water systems should therefore always include the carbonate-buffering system.

Complexation Reactions

As discussed earlier, and as illustrated in Table 6-3, dissolved components in ground water may occur in the form of aqueous complexes. The extent of formation of the complexes depends largely on the concentrations of the reactants, with higher proportions of complex species tending to develop at higher concentrations of the complexing agents. Because of the paucity of dissolved organic compounds, inorganic aqueous complexes are generally more important in normal ground waters than are organic complexes. The solubility of elements and other components in ground water is increased by the formation of complex ions; for example, uranium is much more soluble in the presence of aqueous hydroxyl, carbonate, and phosphate complexes than in pure water (Langmuir 1978; Tutem et al. 1991).

Substitution-Hydrolysis Reactions

These reactions (sometimes called *metathetical reactions*) involve the transformation of dissolved inorganic or organic contaminants in ground water, according to the generalized reaction

$$RX + A \leftrightarrow RA + X$$

in which R represents the main portion of the dissolved molecule, X is a replaceable element or group, and A is the dissolved species that replaces X. The product RA may be aqueous or solid and either more or less reactive than the original RX.

Hydrolysis is a specific type of replacement reaction, described by

$$RX + H_2O \leftrightarrow ROH + HX$$
$$RX + H_2O \leftrightarrow RH + XOH$$

The resulting ROH or XOH may be either an aqueous or solid product, depending on the solubility of the substance and the pH of the water. Hydrolysis reactions can obviously be considered to be a type of acid-base reaction. As mentioned in the discussion of acid-base reactions, the solubility of such metals as Al, Fe, and Mn is strongly dependent on hydrolysis reactions and the pH of the water (the reader is referred especially to Hem 1985 for examples of the role of hydrolysis and pH in controlling the solubility of Al, Fe, and Mn).

Oxidation-Reduction Reactions

Oxidation-reduction (redox) reactions involve the transfer of one or more electrons among chemical elements. Implicit in this definition is the fact that the elements involved can have different oxidation (valence) numbers. The oxidation of natural organic matter is a particularly important redox reaction in ground water, involving such elements as N, S, H, and C. Dissolved species resulting from the oxidation of organic matter include NO_3^-, NH_4^+, NO_2^-, SO_4^{2-}, S^{2-}, CO_3^{2-}, CH_4, CO_2, among others.

Oxidation-reduction reactions also have profound effects on the speciation of dissolved elements and on their solubility. The reader is referred to Stumm and Morgan (1981) and Hem (1985) for detailed discussions of the role of oxidation-reduction reactions in aqueous speciation and solubility. A typical oxidation-reduction reaction would be the reduction of aqueous S(VI) (in the form of the aqueous species SO_4^{2-}) to S(II) (in the form of aqueous HS^-):

$$SO_4^{2-} + 9H^+ + 8e^- \leftrightarrow HS^- + 4H_2O$$

This reaction is known to be extremely slow at low temperatures (Stumm and Morgan 1981) unless sulfate-reducing bacteria are involved (Lindberg and Runnells 1984).

Much has been written about the problems involved in obtaining a valid measure of the redox potential of natural waters (Stumm and Morgan 1981; Langmuir 1971b; Bricker 1982). However, it is also known that the fundamental kinetics of transfer of electrons among aqueous and solid species are commonly very slow, resulting in oxidation-reduction reactions that are virtually irreversible (see Whitfield 1969; Stumm and Morgan 1981; Hostetler 1984; Lindberg and Runnells 1984; Runnells and Lindberg 1990; Kempton, Lindberg, and Runnells 1990). It is clear that microbiologic activity may be required for many potentially important oxidation-reduction reactions to take place at a geologically meaningful rates.

Ion Filtration and Osmosis

DeSitter (1947) was apparently the first person to suggest that some saline waters have been concentrated by a "salt-sieving" action of fine-grained sediments, in which water passes through fine-grained sediments but ions of dissolved salts are excluded. The physical principle involved is that the surfaces of the clay minerals which constitute the bulk of shales are negatively charged. Therefore, cations and neutral molecules, including water, can pass through the negatively charged mineral framework of the shale, but anions are rejected and are concentrated on the inlet side of the membrane.

Berry and Hanshaw (1960) carried the concept further and suggested that ion filtration and osmosis are potentially important processes in controlling both the salinity and some anomalous pressures in deep basinal ground water; as examples they gave three widely separated areas in North America (San Joaquin Valley, California; San Juan Basin, Colorado and New Mexico; Alberta Basin, Alberta, Canada) where salinities and anomalous pressures in deep aquifers may be explained by the cross-formational movement of water through semipermeable shale membranes. White (1965) elaborated on the concept, describing how semipermeable fine-grained sediments may not be equally permeable to all dissolved constituents, which could cause changes in the ionic composition of subsurface water as well as in the total salinity. Also, because of the fundamental requirement for electrical neutrality in all aqueous solutions, the exclusion of anions by a semipermeable sedimentary membrane would probably be accompanied at least by changes in pH in the ground water on both sides of the filtering membrane. The reader is referred to White (1965) for a detailed discussion of the entire concept.

Dissolution and Exsolution of Gases

Gases are intimately involved in many geochemical reactions in ground water. The solubility of a gas in water is controlled by Henry's constant, K_h (which can be expressed in many specialized forms), according to the reaction

$$[\text{gas}_{(aq)}] = K_h\,[\text{gas}]$$

in which [ ] represents the activity of the gas and its dissolved counterpart. Henry's constant is a function of temperature and pressure (Wilhelm, Battino, and Wilcock 1977; Shock, Helgeson, and Sverjensky 1989). The molal solubility (moles of gas per kilogram of solvent) of the aqueous gas is also a function of the salinity of the water because of the effect of salinity on the molal activity coefficient of the dissolved gas. At temperatures near 298 K and total pressures near 100 kPa (equal to 1 bar), the activity of the gas phase is equal to the fugacity and closely approximated by the partial pressure, and concentration is approximately equal to activity. Therefore, for many natural ground waters, the following simple relationship holds:

$$(\text{aqueous gas}) = K_h P_{\text{gas}}$$

where ( ) represents molality and P_{gas} represents the partial pressure of the gas (Hem 1985), with Henry's constant in units of moles kg^{-1} atm^{-1} (Garrels and Christ 1965). Values of Henry's constant for a representative suite of gases of importance in natural waters are given in Table 6-5. A detailed discussion of the thermodynamic and transport properties of gases in aqueous solutions, including the effect of temperature and total pressure, is given by Shock, Helgeson, and Sverjensky (1989).

The gas of most obvious geochemical significance is CO_2, which exerts a major control on the solubility of carbonate minerals and rocks. Special emphasis on the role of CO_2 gas in the geochemistry of ground water in carbonate aquifers is given by Thrailkill (1968), Langmuir (1971a), and Plummer et al. (1976). Briefly, over the range of partial pressures of CO_2 gas encountered in most normal ground waters, the solubility of carbonate minerals increases with increasing partial pressure of CO_2 gas. The details of the aqueous chemistry of CO_2 gas and solid carbonate minerals are fastidiously presented by Butler (1982) in a book entirely devoted to the subject.

Other gases in ground water are also of obvious importance. For example, the dissolution and exsolution of H_2S may control the solubility of sulfide minerals, such as pyrite (FeS_2). The presence of CO, H_2, and CH_4 are indicative of biologic activity, either in the ground water or in the recharge system (see Chapter 8). Kuivila, Murray, and Devol (1990) discuss the production of

TABLE 6-5. Values of Henry's Constant, K_h, at 298 K for Gases That Can Be Important in Ground Water

Gas	K_h
N_2	$10^{-3.19}$
CO	$10^{-3.02}$
O_2	$10^{-2.89}$
CH_4	$10^{-2.86}$
CO_2	$10^{-1.47}$
H_2S	$10^{-0.988}$
SO_2	$10^{+0.147}$

Note: Listed in order of increasing solubility. Use of K_h is as defined in Garrels and Christ (1965). Units of K_h are moles kg^{-1} atm^{-1} (convert to moles kg^{-1} kPa^{-1} by multiplying the listed Values by a conversion factor of 1.01325 × 10^{-5} atm kPa^{-1}). To obtain the approximate solubility of the gas in freshwater at 298 K and a total pressure of approximately 1 atm, multiply the listed value by the partial pressure of the gas of interest in atmospheres. Values from Wilhelm, Battino, and Wilcock (1977).

CH_4 in the interstitial water of sediments, and Lovley and Goodwin (1988) emphasize the importance of H_2 in microbiologic reactions in anaerobic sediments.

One of the factors that controls the molal solubility of a gas in water is the extent to which it reacts chemically with the water, as evidenced by a high heat of hydration. Gases such as CO_2, H_2S, and SO_2 have high solubilities in water in part because they have a strong affinity for the water molecules, as reflected by relatively large numerical values of Henry's constant for such gases (see Table 6-5). Gases such as O_2, N_2, and H_2, which do not have strong affinities to the water molecules and have lower heats of hydration, exhibit correspondingly lower aqueous solubilities, with smaller numerical values of Henry's constant (Table 6-5). A comprehensive summary of the solubility of gases in water is given by Wilhelm, Battino, and Wilcock (1977).

Berner (1981) and Lindberg and Runnells (1984) have suggested that the composition of gases in ground water should be a better indicator of the redox status of the water than traditional measurements of *Eh* using a platinum electrode. The basis for this suggestion is that the microbial populations in ground water ultimately control the redox status of the water, and that the dissolved gases reflect the degree and general type of microbial metabolism. Additional insight into the relationship between redox, microbial activity, and gases can be found in Marshall (1987) and Chapter 8 of this book.

Sorption and Desorption

Ground water is exposed to the surfaces of the rocks and minerals that constitute the solid framework of an aquifer, and the role of the surfaces of the solid framework materials is important in evaluating chemical reactions in regional ground-water systems. Ions and neutral molecules can be attracted to the surfaces of solids, either temporarily or permanently. The process by which the aqueous species are attracted to the surfaces of the solids is termed *sorption*, and the release of the sorbed species to the liquid phase is termed *desorption*. The terms *sorption* and *desorption* are used here in a broad sense to include any process involving the attraction and release of aqueous species to or from the surfaces of a solid particle. Within this broad definition, we thus include such specific processes as ion exchange, adsorption or release of protons and hydroxyl ions from the surfaces of a solid, and interactions between neutral molecules and the surfaces of solids. Sorption and desorption are important for both inorganic and organic species.

The details of the reactions that take place between aqueous species and the surfaces of solids in an aquifer have been the subject of intensive study by numerous investigators for many years. The details lie beyond the scope of this discussion, and only a brief overview will be attempted. For interesting examples of detailed investigations of sorption-desorption reactions, the reader is referred to Kavanaugh and Leckie (1980), Davis and Hayes (1986), Kramer and Allen (1988), Melchior and Bassett (1990), and Dzombak and Morel (1990).

The chemical principle which underlies sorption and desorption reactions is that of attraction between aqueous species and receptive sites on the surfaces of solids. The attraction results from chemical bonding (electrostatic, covalent, van der Waals, hydrogen, etc.) between the dissolved species and the components of the solid surface. If the bonding is electrostatic, and the reaction involves the equivalent exchange of one charged aqueous species for another that is already present on the surface site, the reaction is termed *ion exchange*. An example ion-exchange reaction of importance in ground water is

$$Na_2X_{(s)} + Ca^{2+} \leftrightarrow CaX_{(s)} + 2Na^+$$

where X may represent any solid substrate. In natural ground-water systems, X in this reaction is generally a 2:1 clay mineral. Ion-exchange reactions, such as the preceding one, can be described by an apparent equilibrium constant, K_{ex}, where

$$K_{ex} = \frac{[(CaX)[Na^+]^2]}{[(Na_2X)[Ca^{2+}]]}$$

in which ( ) represents concentrations for the solids and [ ] represents activities for the aqueous ions. In general, K_{ex} is not a constant; it varies with the type and concentration of counterions in solution, the types of ions already in exchange sites on the solid, the competing ions in solution, and the proportion of exchange sites already filled on the surface of the solid. Values of K_{ex} must be determined experimentally for each reaction and each ground water of interest.

Sorption-desorption reactions are generally fast, reaching completion over time scales of a few minutes to a few hours. However, if the adsorbed species gradually become incorporated into the crystal structure of the solid phase, the reaction can become very slow, occurring over periods of weeks to months; to distinguish the mechanisms involved, reactions in which a species is incorporated into the structure of the solid are called *absorption* rather than *adsorption*.

Most of the theoretical and predictive modeling of sorption-desorption reactions has focused on bonding reactions on the surfaces of metal oxides and hydroxides, such as SiO_2, Al_2O_3, $Al(OH)_3$, Fe_2O_3, and FeOOH (Parks 1967; Davis, James, and Leckie 1978; Davis and Leckie 1978; Dzombak and Morel 1990). The model reactions are represented as replacements of protons or hydroxyl ions from bonding sites on the surfaces, such as

$$S—O^- + H^+ \leftrightarrow S—O—H$$
$$S—O—H \leftrightarrow S^+ + OH^-$$
$$S—O—H + M^+ \leftrightarrow S—O—M + H^+$$
$$S—O^- + M^+ \leftrightarrow S—O—M$$
$$S—O—H + M^{2+} \leftrightarrow S—O—M^+ + H^+$$

and so on, where S represents the main internal structure of the substrate. In this approach, an intrinsic constant of reaction, similar to a thermodynamic equilibrium constant, is assigned to the reactions involving the surface sites. The number of sites and the strength of the electrical field adjacent to the charged surface are the remaining parameters that enter into the predictive model. One recent example of such calculations, including the parameters required, can be found in the study of sorption of organic molecules by Rea and Parks (1990). Models of surface reactions and example calculations for solids in soils, applicable to aquifer materials, are presented in a particularly clear fashion by Gast (1977).

KINETIC VERSUS EQUILIBRIUM CONSIDERATIONS

The rates of important chemical reactions in ground water vary tremendously, from extremely slow to very fast. Therefore, kinetics must not be neglected in the study of chemical processes in ground water. For example, sorption reactions and gaseous reactions tend to be quite fast, reaching completion over a period of a few hours to a few days, whereas dissolution and precipitation reactions may require months, years, or millennia to reach completion. Even within the same broad category of geochemical reaction, such as dissolutionprecipitation, rates of individual reactions vary greatly; for example, silicates tend to react with water much more slowly than carbonates.

Great effort has been devoted by many researchers to the study of the kinetics of geochemical reactions, and this area remains one of the frontiers of geochemical research. For excellent discussions and overviews see Lasaga and Kirkpatrick (1981), Colman and Dethier (1986), and Sparks (1989).

As shown in Chapter 9, numerical modeling of the chemistry of ground water is an important area of research. A great many computer codes are now available for hydrochemical modeling (see the summary of codes by Mangold and Fang, 1991), but virtually all of them are based on an assumption of complete homogeneous and heterogeneous chemical equilibrium. As time passes, kinetic considerations will certainly be incorporated into more sophisticated geochemical computer codes.

CONCLUDING REMARKS

This chapter has focused on the processes and reactions that control the chemistry of ground water. The study of chemical processes in ground water has long been an important field, for such diverse purposes as environmental contamination, exploration for new mineral deposits, age dating of ground water, and chemical relationships between the ground water and the enclosing rocks.

There are eight main categories of reactions and processes that control the chemistry of most ground water: precipitation-dissolution, acid-base, complexation, substitution-hydrolysis, oxidation-reduction, ion filtration–osmosis, dissolution and exsolution of gases, and sorption-desorption. Although thousands of theoretical chemical reactions could be written involving the water, solids, and gases in regional aquifers, the number of reactions that actually dominate the chemistry of ground water turns out to be quite small. In general, the reactions which control the major element chemistry in ground water are

1. Introduction of CO_2 gas in the unsaturated zone
2. Dissolution of calcite and dolomite, and precipitation of calcite
3. Cation exchange
4. Oxidation of pyrite and organic matter
5. Reduction of oxygen, nitrate, and sulfate, with production of sulfide
6. Reductive production of methane
7. Dissolution of gypsum, anhydrite, and halite
8. Incongruent dissolution of primary silicates with the formation of clays

The use of geochemical computer models has greatly increased our understanding of the interaction between ground water and solid aquifer materials. Current geochemical models are largely based on an assumption of complete homogeneous and heterogeneous chemical equilibrium. However, we know that the kinetics of geochemical reactions vary greatly, and as geochemical models continue to increase in sophistication, it will be helpful to incorporate considerations of reaction kinetics in order to more realistically interpret water-rock interactions.

References

Back, William. 1960. Origin of hydrochemical facies of ground water in the Atlantic Coastal Plain. *21st International Geologic Congress*, Copenhagen: Report Part I.

Back, William, J. S. Rosenshein, and P. R. Seaber, eds. 1988. *Hydrogeology*. Volume O-2. *The Geology of North America*. Boulder, Colorado: Geological Society of America.

Barth, Tanja. 1991. Organic acids and inorganic ions in waters from petroleum reservoirs, Norwegian continental shelf: a multivariate statistical analysis and comparison with American reservoir formation water. *Applied Geochemistry* 6(1):1–16.

Berner, R. A. 1981. A new geochemical classification of sedimentary environments. *Journal of Sedimentary Petrology* 51(2): 359–65.

Berry, F. A. F., and B. B. Hanshaw. 1960. Geologic field evidence suggesting membrane properties of shales. *21st International Geologic Congress*, Abstract, p. 209. Copenhagen, Denmark.

Bishop, P. K., and J. W. Lloyd. 1991. Use of ^{14}C modelling to determine the vulnerability and pollution of a carbonate aquifer: the Lincolnshire Limestone, eastern England. *Applied Geochemistry* 6(3):319–32.

Bricker, O. P. 1982. Redox potential: its measurement and importance in water systems. In *Water Analysis,* Volume 1, *Inorganic Species*, Part 1, eds. R. A. Minear and L. H. Keith, pp. 55–83. Orlando, Florida: Academic Press.

Butler, J. N. 1982. *Carbon Dioxide Equilibria and Their Applications*. Reading, Massachusetts: Addison-Wesley.

Colman, S. M., and D. P. Dethier, eds. 1986. *Rates of Chemical Weathering of Rocks and Minerals*. Orlando, Florida: Academic Press.

Crossy, L. J. 1991. Thermal degradation of aqueous oxalate species. *Geochimica et Cosmochimica Acta* 55(6):1515–27.

Davis, J. A., and K. F. Hayes, ed. 1986. *Geochemical Processes at Mineral Surfaces*. ACS Symposium Series 323. Washington, D.C.: American Chemical Society.

Davis, J. A., R. O. James, and J. O. Leckie. 1978. Surface ionization and complexation at the oxide/water interface. 1: Computation of electrical double layer properties in simple electrolytes. *Journal of Colloid and Interfacial Science* 63:480–99.

Davis, J. A., and J. O. Leckie. 1978. Surface ionization and complexation at the oxide/water interface. II: Surface properties of amorphous iron oxyhydroxide and adsorption of metal ions. *Journal of Colloid and Interfacial Science* 67:90–107.

Degueldre, C., B. Baeyens, W. Goerlich, J. Riga, J. Verbist, and P. Stadelmann. 1989. Colloids in water from a subsurface fracture in granitic rock, Grimsel Test Site, Switzerland. *Geochimica et Cosmochimica Acta* 53(3):603–10.

DeSitter, L. U. 1947. Diagenesis of oil-field brines. *American Association of Petroleum Geologists Bulletin* 31(11):2030–40.

Domenico, P. A., and F. W. Schwartz. 1990. *Physical and Chemical Hydrogeology*. New York: Wiley.

Drever, J. I. 1988. *The Geochemistry of Natural Waters*. 2nd ed. Englewood Cliffs, New Jersey: Prentice Hall.

Drost-Hansen, W. 1966. The puzzle of water. *International Science and Technology* October: 86–96.

Dzombak, D. A., and F. M. M. Morel. 1990. *Surface Complexation Modeling*. New York: Wiley-Interscience.

Edmunds, W. M., Bath, A. H., and D. L. Miles. 1982. Hydrochemical evolution of the East Midlands Triassic Sandstone aquifer. *Geochimica et Cosmochimica Acta* 46:2069–81.

Foster, M. D. 1950. The origin of high sodium bicarbonate waters in the Atlantic and Gulf Coastal Plains. *Geochimica et Cosmochimica Acta* 1:33–48.

Freeze, R. A., and J. A. Cherry. 1979. *Groudwater*. Englewood Cliffs, New Jersey: Prentice Hall.

Garrels, R. M., and C. L. Christ. 1965. *Solutions, Minerals, and Equilibria*. San Francisco: Freeman, Cooper.

Gast, R. G. 1977. Surface and colloid chemistry. In *Minerals in Soil Environments*, ed. R. C. Dinauer, pp. 27–73. Madison, Wisconsin: Soil Science Society of America.

Hanshaw, B. B., and William Back. 1979. Major geochemical processes in the evolution of carbonate aquifer systems. *Journal of Hydrology* 43:287–312.

Hanshaw, B. B., William Back, and Meyer Rubin. 1965. *Carbonate Equilibria and Radiocarbon Distribution Related to Groundwater Flow in the Floridan Aquifer, U.S.A.* Dubrovnik, Yugoslavia: International Association of Scientific Hydrology, UNESCO, October, Proceedings, vol. 1.

Hem, J. D. 1985. *Study and Interpretation of the Chemical Characteristics of Natural Water*. 3rd ed. Alexandria, Virginia: U.S. Geological Survey Water-Supply Paper 2254.

Henderson, Thomas. 1985. *Geochemistry of Ground-Water in Two Sandstone Aquifer Systems in the Northern Great Plains in Parts of Montana and Wyoming*. Alexandria, Virginia: U.S. Geological Survey Professional Paper 1402-C.

Hendry, M. J., and F. W. Schwartz. 1990. The chemical evolution of ground water in the Milk River aquifer, Canada. *Ground Water* 28(2):253–61.

Hitchon, Brian, and Irving Friedman. 1969. Geochemistry and origin of formation waters in the western Canada sedimentary basin. I: Stable isotopes of hydrogen and oxygen. *Geochimica et Cosmochimica Acta* 33:1321–49.

Holdren, G. R., Jr., and P. M. Speyer. 1986. Stoichiometry of alkali feldspar dissolution at room temperature and various pH values. In *Rates of Chemical Weathering of Rocks and Minerals*, eds. S. M. Colman and D. P. Dethier, pp. 61–81. Orlando, Florida: Academic Press.

Horne, R. A. 1973. Effects of structure and

physical characteristics of water on water chemistry. In *Water and Water Pollution*, Volume 3, ed. L. L. Ciaccio, pp. 915–47. New York: Marcel Dekker.

Hostetler, J. D. 1984. Electrode electrons, aqueous electrons, and redox potentials in natural waters. *American Journal of Science* 284:734–59.

Kavanaugh, M. C., and J. O. Leckie, eds. 1980. *Particulates in Water*. Advances in Chemistry Series 189. Washington, D.C.: American Chemical Society.

Kempton, J. H., R. D. Lindberg, and D. D. Runnells. 1990. Numerical modeling of platinum Eh measurements by using heterogeneous electron-transfer kinetics. In *Chemical Modeling of Aqueous Systems II*, eds. D. C. Melchior and R. L. Bassett, pp. 597–634. ACS Symposium Series No. 416. Washington, D.C.: American Chemical Society.

Kramer, J. R., and H. E. Allen, eds. 1988. *Metal Speciation—Theory, Analysis, and Application*. Chelsea, Michigan: Lewis.

Kuivila, K. M., J. W. Murray, and A. H. Devol. 1990. Methane production in the sulfate-depleted sediments of two marine basins. *Geochimica et Cosmochimica Acta* 53: 409–17.

Langmuir, Donald. 1971a. The geochemistry of some carbonate ground waters in central Pennsylvania. *Geochimica et Cosmochimica Acta* 33:1023–45.

Langmuir, Donald. 1971b. Eh-pH determination. In *Procedures in Sedimentary Petrology*, ed. R. E. Carver, pp. 597–634. New York: Wiley.

Langmuir, Donald. 1978. Uranium solution mineral equilibria at low temperatures with applications to sedimentary ore deposits. *Geochimica et Cosmochimica Acta* 42(60):547–69.

Lasaga, A. C., and R. J. Kirkpatick, eds. 1981. *Kinetics of Geochemical Processes*. Volume 8. *Reviews in Mineralogy*, ed. P. H. Ribbe. Washington, D.C.: Mineralogical Society of America.

Leppard, G. G. 1983. *Trace Metal Speciation in Surface Waters and its Ecological Implications*. New York: Plenum Press.

Lide, D. R., ed. 1991. *Handbook of Chemistry and Physics*. 71st ed. Boca Raton, Florida: Chemical Rubber Co.

Lindberg, R. D., and D. D. Runnells. 1984. Groundwater redox disequilibrium: applications to the measurement of Eh and computer modeling. *Science* 225:925–27.

Lloyd, J. W., and J. A. Heathcote. 1985. *Natural Inorganic Hydrochemistry in Relation to Groundwater. An Introduction*. Oxford, United Kingdom: Clarendon Press.

Lovley, D. R., and Steve Goodwin. 1988. Hydrogen concentrations as indicator of the predominant terminal electron-accepting reactions in aquatic sediments. *Geochimica et Cosmochimica Acta* 52:2993–3003.

Mackenzie, F. T., W. D. Bischoff, F. C. Bishop, M. Loijens, J. Schoonmaker, and R. Wollast. 1983. Magnesian calcites: low-temperature occurrence, solubility, and solid solution behavior. In *Carbonates: Mineralogy and Chemistry Kinetics of Geochemical Processes*, ed. R. J. Reeder, Volume 11, *Reviews in Mineralogy*, pp. 97–144. Washington, D.C.: Mineralogical Society of America.

Mangold, D. C., and Chin-Fu Tsang. 1991. A summary of subsurface hydrological and hydrochemical models. *Reviews of Geophysics* 29(1):1–79.

Marshall, K. C. 1987. *Advances in Microbial Ecology*, Volume 7. New York: Plenum Press.

Martell, A. E., and R. J. Motekaitis. 1988. *Determination and Use of Stability Constants*. New York: VCH.

Masscheleyn, P. H., R. D. Delaune, and W. H. Patrick, Jr. 1991. Effect of redox potential and pH on arsenic speciation and solubility in contaminated soil. *Environmental Science and Technology* 25(8):1414–18.

Matthess, Georg, and J. C. Harvey (translator). 1982. *The Properties of Groundwater*. New York: Wiley-Interscience.

Melchior, D. C., and R. L. Bassett, eds. 1990. *Chemical Modeling of Aqueous Systems II*. ACS Symposium Series No. 416. Washington, D.C.: American Chemical Society.

Millero, F. J., and M. L. Sohn. 1992. *Chemical Oceanography*. Boca Raton, Florida: CRC Press.

Morse, J. W. 1983. The kinetics of calcium carbonate dissolution and precipitation. In *Carbonates: Mineralogy and Chemistry*, ed. R. J. Reeder, pp. 227–64, Volume 11, *Reviews in Mineralogy*. Washington, D.C.: Mineralogical Society of America.

National Research Council. 1990. *Ground Water Models—Scientific and Regulatory Applications*. Washington, D.C.: National Academy Press.

Parkhurst, D. L., D. C. Thorstenson, and L. N. Plummer. 1980. *PHREEQE—A Computer Program for Geochemical Calculations*. Reston, Virginia: U.S. Geological Survey Water-Resources Investigations 80–96.

Parks, G. A. 1967. Aqueous surface chemistry of oxides and complex oxide minerals: isoelectric point and zero point of charge. In *Equilibrium Concepts in Natural Water Systems*, ed. Werner Stumm, pp. 121–60. Advances in Chemistry Series 67. Washington, D.C.: American Chemical Society.

Plummer, L. N., J. F. Busby, R. W. Lee, and B. B. Hanshaw. 1990. Geochemical modeling of the Madison aquifer in parts of Montana, Wyoming, and South Dakota. *Water Resources Research* 26(9):1981–2014.

Plummer, L. N., H. L. Vacher, F. T. Mackenzie, O. P. Bricker, and L. S. Land. 1976. Hydrogeochemistry of Bermuda: A case history of ground-water diagenesis of biocalcarenites. *Geological Society of America Bulletin* 87:1301–16.

Pytokowicz, R. M. 1983. *Equilibria, Nonequilibria, and Natural Waters*, Volume II. New York: Wiley.

Raspor, B., H. W. Nurnberg, P. Valenta, and M. Branica. 1984. Significance of dissolved humic substances for heavy metal speciation in natural waters. In *Complexation of Trace Metals in Natural Waters*, eds. C. I. M. Kramer and J. C. Duinker, pp. 317–328. Hingham, MA: Kluwer.

Rea, R. L., and G. A. Parks. 1990. Numerical simulation of coadsorption of ionic surfactants with inorganic ions on quartz. In *Chemical Modeling of Aqueous Systems II*, eds. D. C. Melchior and R. L. Bassett, pp. 260–71. ACS Symposium Series No. 416. Washington, D.C.: American Chemical Society.

Renick, B. C. 1924. *Base Exchange in Ground Water by Silicates as illustrated in Montana*. Washington, D.C.: U.S. Geological Survey Water-Supply Paper 520-D.

Richardson, S. M., and H. Y. McSween, Jr. 1989. *Geochemistry: Pathways and Processes*. Englewood Cliffs, New Jersey: Prentice Hall.

Runnells, D. D. 1969. Diagenesis, chemical sediments, and the mixing of natural waters. *Journal of Sedimentary Petrology* 39:1188–1201.

Runnells, D. D. 1976. Wastewater in the vadose zone of arid regions: geochemical interactions. *Ground Water* 14(6):374–385.

Runnells, D. D. 1987. Low-temperature geochemistry. In *Encyclopedia of Science and Technology*; Volume 6, pp. 36–62. San Diego, California: Academic Press.

Runnells, D. D., and R. D. Lindberg. 1990. Selenium in aqueous solution: the impossibility of obtaining a meaningful Eh using a platinum electrode, with implications for geochemical computer modeling. *Geology* 18(3):212–15.

Ryan, J. N., and P. M. Gschwend. 1990. Colloid mobilization in two Atlantic Coastal Plain aquifers: field studies. *Water Resources Research* 26(2):307–22.

Shock, E. L., H. C. Helgeson, and D. A. Sverjensky. 1989. Calculation of the thermodynamic and transport properties of aqueous species at high pressures and temperatures: standard partial molal properties of inorganic neutral species. *Geochimica et Cosmochimica Acta* 53(9):2157–83.

Singer, P. C. 1974. *Trace Metals and Metal-organic Interactions in Natural Waters*. Ann Arbor, Michigan: Ann Arbor Science.

Smith, R. L., B. L. Howes, and J. D. Huff. 1991. Denitrification in nitrate-contaminated groundwater: occurrence in steep vertical geochemical gradients. *Geochimica et Cosmochimica Acta* 55:1815–25.

Sparks, D. L. 1989. *Kinetics of Soil Chemical Processes*. San Diego, California: Academic Press.

Stumm, Werner, and J. J. Morgan. 1981. *Aquatic Chemistry*. 2nd ed. New York: Wiley-Interscience.

Sun R. J., and J. B. Weeks. 1991. *Bibliography of Regional Aquifer-System Analysis Program of the U.S. Geological Survey—Bibliography, 1978-1991*. Reston, Virginia: U.S. Geological Survey Water-Resources Investigations Report 91–4122.

Surdam, R. C., S. W. Boese, and L. J. Crossey. 1984. The chemistry of secondary porosity. In *Clastic Diagenesis*, eds. D. A. MacDonald and R. C. Surdam, pp. 127–49. Tulsa, Oklahoma: American Association of Petroleum Geologists Memoir 37.

Thorstenson, D. C. 1990. Chemical modeling of regional aquifer systems-implications for chemical modeling of low-level radioactive-waste repository sites. In *Safe Disposal of Radionuclides in Low-Level Radioactive-Waste Disposal Workshop, U.S. Geological Survey, July 11–16, 1987, Big Bear Lake, California, Proceedings*, eds. M. S. Bedinger and P. R. Stevens, pp. 110–13. Denver, Colorado: U.S. Geological Survey Circular 1036.

Thorstenson, D. C., D. W. Fisher, and M. G. Croft. 1979. The geochemistry of the Fox Hills–Basal Hell Creek aquifer in southwestern North Dakota and northwestern South Dakota. *Water Resources Research* 15(7):1479–98.

Thrailkill, John. 1968. Chemical and hydrologic factors in the excavation of limestone caves. *Geological Society of America Bulletin* 79:19–45.

Thurman, E. M. 1985. *Organic Geochemistry of Natural Waters*. Hingham, Massachusetts: Kluwer.

Tutem, Esma, Resat Apak, M. H. Turgut, and Vildan Apak. 1991. The logarithmic diagram for the uranyl-carbonate-hydroxide complex equilibria. *Journal of Chemical Education* 68(7):569–71.

Ward, C. H., Walter Giger, and P. L. McCarty, eds. 1985. *Ground Water Quality*. New York: Wiley-Interscience.

White, D. E. 1965. Saline waters of sedimentary rocks. In *American Association of Petroleum Geologists Memoir 4*, pp. 342–66. Tulsa, Oklahoma: American Association of Petroleum Geologists.

Whitfield, M. 1969. Eh as an operational parameter in estuarine studies. *Limnology and Oceanography* 14:547–58.

Wilhelm, Emmerich, Rubin Battino, and R. J. Wilcock. 1977. Low-pressure solubility of gases in liquid water. *Chemical Reviews* 77(2):219–62.

7

Organic Chemical Concepts

Paul D. Capel

INTRODUCTION

The behavior, transport, and fate of an organic chemical in ground water are controlled by the properties of the chemical and the environmental conditions of the ground water (Figure 7-1). The structure of the organic chemical determines its physical, chemical, and biological properties. A chemical's structure is determined by the three-dimensional arrangement of the various atoms of carbon (C) and hydrogen (H), and perhaps, oxygen (O), fluorine (F), chlorine (Cl), bromine (Br), nitrogen (N), sulfur (S), or phosphorus (P) atoms and the bonds that join the atoms together. The ground-water environment that surrounds the organic chemical consists of physical, chemical, and biological components (Figure 7-1). The interaction of these two, the chemical structure and the environmental conditions, controls the chemical's behavior and its effect on the environment. The environmental processes that control an organic chemical's behavior and fate in ground water can be classified into three types. Transformation processes change its chemical structure, phase transfer processes move it between the water and aquifer solids or in and out of the vapor phase in the vadose zone, and transport processes move it away from its initial location.

This chapter begins with a brief introduction to the vocabulary and concepts of organic and physical organic chemistry. Following this, each of the environmental processes will be discussed individually in relation to chemical structure and environmental conditions. The focus of this chapter is on the chemical processes affecting organic chemicals in ground water. These processes also are discussed to some extent with respect to the vadose zone. However, this topic is addressed further in Chapters 13 and 14 on organic contaminants and pesticides, respectively.

ORGANIC CHEMICALS AND THEIR NOMENCLATURE

Carbon Bonding

Carbon, the essential component of all organic molecules, is unique among the elements because of the way it bonds to other carbon atoms to form three-dimensional chemical structures. Atomic carbon has four electrons, out of a maximum of eight, in its outer electronic shell. This configuration allows carbon to share electrons with four

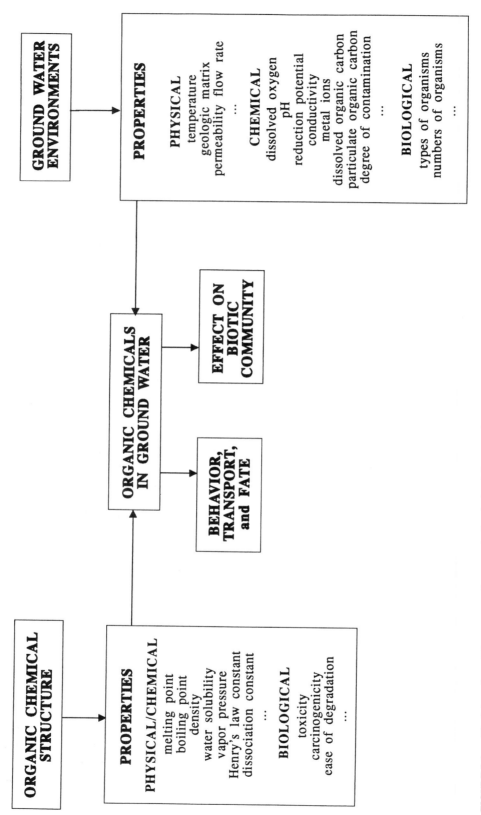

FIGURE 7-1. The interaction of the structure of an organic chemical and the ground-water environment on its behavior and effects.

other atoms, including other carbon atoms. The sharing of electrons between two atoms is termed *molecular bonding*. Covalent bonds form between two carbon atoms, where the electrons are equally shared or nearly so. The electrons between two dissimilar atoms (i.e., C—H, C—Cl, C—O, O—H) are not shared equally, giving rise to bonds that are more polar owing to a separation of electric charge. The degree of polarity in the bonds within the organic molecule affects its chemical and biological behavior. Most organic compounds have numerous C—C bonds, which form the backbone of the molecule.

Functional Groups

The simplest organic molecules are the saturated hydrocarbons. These molecules contain only carbon and hydrogen, which are all joined by single bonds (two electrons are shared between adjacent atoms). These compounds can be arranged either in linear chains, such as the normal alkanes (*n*-alkanes) (Table 7-1), as branched chains, or as rings. The suffix *ane* indicates a saturated hydrocarbon. The prefixes indicate the number of carbon atoms (*meth* = 1, *eth* = 2, *prop* = 3, etc.). These prefixes are used extensively throughout all organic chemical nomenclature.

Unsaturated hydrocarbons also contain only carbon and hydrogen, but at least one of the carbon-carbon bonds in the molecule is a double or triple bond. A double bond is the sharing of four electrons between adjacent carbon atoms; a triple bond is the sharing of six electrons between adjacent carbon atoms. The presence of double and triple bonds increases the chemical and biological reactivity of the molecule compared to the saturated hydrocarbon with the same number of carbon atoms. Organic molecules containing a double bond are termed *alkenes* and have the suffix *ene* in their name. Compounds with a triple bond are termed *alkynes* and have the suffix *yne*. Together alkanes, alkenes, and alkynes are referred to as *aliphatic hydrocarbons*.

Aromatic compounds are a second class of hydrocarbons. These molecules are ring compounds which are composed of hexagonal arrangements of carbon atoms, which share delocalized electrons. Enough electrons exist within the aromatic ring to allow, in an alternating arrangement, three single carbon-carbon bonds and three double carbon-carbon bonds. In reality, all of the carbon-carbon bonds within the aromatic ring are equivalent. The structural formula of benzene, the simplest aromatic compound, can be written as the resonance structures

However, it is commonly written as

to take into consideration the equality of the carbon-carbon bonds around the ring structure. The presence of the delocalized electrons in aromatic structures gives these compounds different physical, chemical, and biological properties relative to the aliphatic hydrocarbons. In terms of environmental behavior, one important difference is the increased water solubility of aromatic compounds compared with the aliphatic compounds with the same number of carbon atoms (compare hexane and benzene or heptane and toluene in Table 7-1).

Many organic compounds contain heteroatoms (i.e., atoms other than carbon and hydrogen, such as O, N, S, F, Cl, Br, and others) bonded to carbon. Carbon-carbon double and triple bonds, and heteroatoms are termed *functional groups* in organic molecules. A functional group is an atom or group of atoms bonded together, which, when present, determines the structural classification and reactivity of the molecule. Important functional groups are illustrated in Figure 7-2. Simple organic chemicals are named by their number of carbon atoms and their functional groups (suffixes or prefixes, Figure 7-2). As its name indicates, methanol has one carbon atom (meth-) and an alcohol group (-ol). Unfortunately, tradition and ease have allowed a number of

TABLE 7-1. Physical/Chemical Properties of Some Organic Chemical Ground-Water Contaminants (at 25°C, except 20°C for all densities and pesticide data)

Common Name	Melting Point (°C)	Density at 20°C (g/mL)	Water Solubility (moles/m^3)	Vapor Pressure (atm)	Henry's Law Constant (atm-m^3/mole)	Dissociation Constant ($-\log$)
Saturated Hydrocarbons						
Methane	−182.5	0.424	1.50	270	—	—
Ethane	−183.3	0.546	1.97	39	—	—
Propane	−189.7	0.501	1.41	9.3	—	—
Butane	−138.4	0.579	1.06	2.4	0.95	—
Pentane	−129.7	0.626	0.534	0.68	1.2	—
Hexane	−95	0.659	0.11	0.20	1.7	—
Heptane	−90.6	0.684	0.029	0.060	2.3	—
Octane	−56.2	0.703	0.0058	0.019	2.9	—
Nonane	−51	0.718	0.0017	0.0056	4.9	—
Decane	−29.7	0.730	0.00035	0.0017	6.9	—
Aromatic Hydrocarbons						
Benzene	5.5	0.879	22.8	0.13	5.4×10^{-3}	—
Toluene	−95	0.866	5.59	0.038	6.6×10^{-3}	—
ortho-Xylene	−25.2	0.897	1.61	0.0087	4.9×10^{-3}	—
meta-Xylene	−47.9	0.881	1.37	0.011	6.9×10^{-3}	—
para-Xylene	13.2	0.854	1.47	0.012	7.0×10^{-3}	—
Ethylbenzene	−95	0.870	1.52	0.013	7.9×10^{-3}	—
n-Propylbenzene	−101.6	0.862	0.46	0.0044	6.9×10^{-3}	—
Naphthalene	80.2	1.145	0.25	2.0×10^{-10}	4.2×10^{-4}	—
Anthracene	216.2	1.25	0.00023	3.2×10^{-13}	5.9×10^{-5}	—
Phenanthrene	101	1.025	0.0065	4.5×10^{-12}	3.9×10^{-5}	—
Halogenated Hydrocarbons						
Dichloromethane	−95.1	1.33	155	0.58	2.6×10^{-3}	—
Chloroform	−63.5	1.48	67	0.25	3.8×10^{-3}	—
Chloroethane	−136.4	0.90	88	0.99	2.0×10^{-3}	—
1,1-Dichloroethane	−97.0	1.18	52	0.30	5.7×10^{-3}	—
1,2-Dichloroethane	−35.4	1.23	81	0.11	1.1×10^{-3}	—
1,1,1-Trichloroethane	−30.4	1.34	67	0.16	2.8×10^{-2}	—
Vinyl chloride	−153.8	0.91	1.4	3.4	7.0×10^{-1}	—
cis-1,2-Dichloroethene	−80.5	1.28	36	0.27		—
trans-1,2-Dichloroethene	−50	1.26	65	0.43		—
1,1,2-Trichloroethene	−73	1.46	8.4	0.97	8.9×10^{-3}	—
Tetrachloroethene	−19	1.62	0.73	0.025	2.3×10^{-2}	—
Pesticides						
Alachlor	~39–41	1.133	0.48	3×10^{-8}	6×10^{-8}	—
Aldicarb	~99–100	1.195	32.	1×10^{-7}	3×10^{-9}	—
Atrazine	176	1.187	0.14	4×10^{-10}	3×10^{-9}	1.7
Cyanazine	~167–169		0.71	2×10^{-9}		—
2,4-D (acid)	140.5	1.565	1.8	1×10^{-5}	5×10^{-6}	—
Dicamba	114–116	1.57	25.	3×10^{-8}	1×10^{-9}	—
EPTC	liquid	0.955	2.0	2×10^{-5}	1×10^{-5}	—
MCPA	118–119		4.1	2×10^{-9}		3.1
metolachlor		1.12	1.87	2×10^{-8}		—
Metribuzin	~125–126	1.31	5.6			—
Picloram	215 (d)		1.8	6×10^{-10}	3×10^{-10}	3.6
Propachlor	77	1.242	2.8	3×10^{-7}	1×10^{-7}	—
Simazine	~225–227	1.302	0.025	8×10^{-11}	3×10^{-9}	1.7
2,4,5-T (acid)	~154–155	1.80	0.86	5×10^{-8}	6×10^{-8}	

Sources: Saturated, aromatic, and halogenated hydrocarbons: Mackay and Shiu (1981), Roberts, Stewart, and Caserio (1971); pesticides: Suntio et al. (1988), Hartley and Kidd (1987), Worthing and Walker (1983).

Organic Chemicals and Their Nomenclature 159

Functional Group	Structure	Prefix/Suffix
single bond	-C-C-	-ane
double bond	-C=C-	-ene
triple bond	-C≡C-	-yne
aromatic	⬡	
alcohol	-C-OH	-ol
aldehyde	H -C=O	-al
amide	O ‖ -C-NH$_2$	-amide
carboxylic acid	O ‖ -C-OH	-oic acid
acid salt	O ‖ -C-O$^-$metal$^+$	metal... -oate
acid chloride	O ‖ -C-Cl	-oyl chloride
ester	O ‖ -C-O-C-	-oate
ketone	O ‖ -C-C-C-	-one
ether	-C-O-C-	-oxy-
sulfonic acid	O ‖ -C-S-OH ‖ O	-sulfonic acid
phosphoric acid	O ‖ -C-O-P-O-C- | O | C-	...phosphoric acid
amines:		
primary	H | -N-H	amino-
secondary	H | -C-N-C-	alkylamino-
tertiary	-C- | -C-N-C-	dialkylamino-
halides:		
fluoride	-C-F	fluoro-
chloride	-C-Cl	chloro-
bromide	-C-Br	bromo-
nitro compound	-C-NO$_2$	nitro-

FIGURE 7-2. Structure and nomenclature of organic chemical functional groups. The general shorthand used in organic nomenclature is to designate the hydrocarbon portion of the molecule as "R—", which can be any structure that ends in a carbon atom, and then show the functional group in detail. As an illustration R—OH could be methanol (methyl alcohol, CH_3—OH), ethanol (ethyl alcohol, H_3C—CH_2—OH), or any organic alcohol. If there are two dissimilar hydrocarbon groups, they will be designated R and R', such as methyl ethyl ether (H_3C—O—CH_2CH_3). R will always be defined.

organic chemicals to have common names, which sometimes can make identification difficult the first time a new compound name is encountered. Examples of common names are ethylene for ethene ($H_2C=CH_2$) and chloroform for trichloromethane ($CHCl_3$).

The presence of oxygen-, nitrogen-, sulfur-, and phosphorus-containing functional groups in organic molecules changes the physical, chemical, and biological properties of the compounds. These functional groups generally tend to make organic compounds more reactive, both chemically and biologically, more water-soluble, and less volatile, as compared with hydrocarbons with the same number of carbon atoms. Many of these functional groups also can give acid or base properties to the organic chemical. The presence of halides (F, Cl, Br, I) tend to make the organic molecule less water-soluble, more volatile, and, usually less reactive and more environmentally stable.

Numerous organic chemicals, especially pesticides, which are common ground-water contaminants, have structures that contain more than one functional group. The summation of the individual effects of the multiple functional groups gives the compound its physical, chemical, and biological properties and determines its environmental fate and behavior. The usefulness of the generalizations presented in the previous paragraphs diminishes when multiple functional groups are present in a molecule. As an example, the herbicide atrazine (2-chloro-4-ethylamino-6-isopropylamino-s-triazine) contains a heterocyclic carbon and nitrogen aromatic ring, a halogen (chlorine), and two different secondary amines (ethylamino and isopropylamino). In terms of water solubility, the halogen and alkyl groups will act counter to the amino groups. The total combination of these groups, together with the heterocyclic aromatic ring, will determine the chemical's acid-base character at the amino groups. In general, different functional groups within a molecule influence, often in competing ways, every aspect of its chemical and environmental behavior. Two methods are presently available to determine the behavior of a compound with multiple functional groups. The first is simple experimentation and observation of the compound in the laboratory and in the environment. This is a direct, but slow and expensive, method of quantifying chemical behavior. The second method is through mathematical calculation of functional group contributions to chemical behavior. Each functional group and structural component (molecular fragment) is assigned a value for size, shape, and energy interactions, based on experimentation with simple molecules. The values of molecular fragments are combined for predictive estimates of chemical properties and environmental behavior (Grain 1990). The value this method provides is a fast and inexpensive means to obtain these estimates, especially for new chemicals or hard to measure parameters. Although relatively new to the environmental sciences, group contribution concepts hold an excellent promise of usefulness, especially for complex organic molecules.

Isomerism

Some organic molecules have exactly the same molecular formula, but the atoms are oriented differently in three-dimensional space. These molecules are termed *isomers*. Positional isomers have different arrangements of the atoms within the molecule. Conformational isomers have the same arrangement of atoms, but different stable configurations due to rotational restrictions of multiple bonds. The six compounds with the molecular formula C_4H_8 demonstrate both types of isomerism (Roberts, Stewart, and Caserio 1971) (Figure 7-3). The first five compounds in the figure have different arrangements of the atoms in the molecule. The first two contain cyclic arrangements; the next three are linear or branched. These five are positional isomers. The last two compounds have the same basic structure, which is $H_3C-CH=CH-CH_3$, but be-

Name and Structure	Melting Point (C)	Boiling Point (C)	Vapor Pressure (atm)	Water Solubility (mole/m3)
cyclobutane 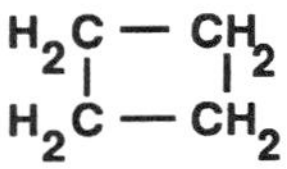 $H_2C - CH_2$ $\ \ \|\ \ \ \ \ \ \ \|$ $H_2C - CH_2$	-105	11	---	---
methylcyclopropane 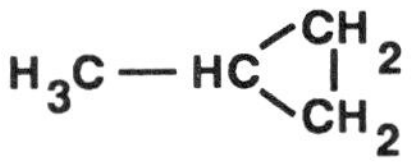 $H_3C - HC \overset{\diagup CH_2}{\underset{\diagdown CH_2}{\|}}$	---	4	---	---
1-butene $H_2C = CH - CH_2 - CH_3$	-185	-6.3	2.94	3.96
2-methylpropene 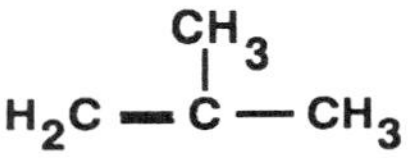 $\ \ \ \ \ \ \ \ \ \ CH_3$ $\ \ \ \ \ \ \ \ \ \ \|$ $H_2C = C - CH_3$	-140	-6.0	3.01	4.70
cis-2-butene 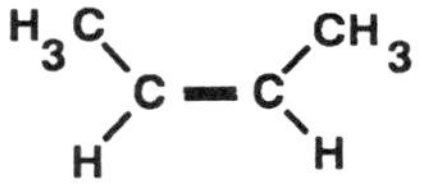 $\underset{H}{\overset{H_3C}{\diagdown}} C = C \underset{H}{\overset{CH_3}{\diagup}}$	139	3.7	---	---
trans-2-butene 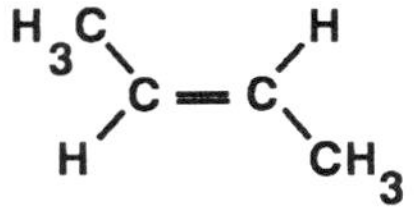 $\underset{H}{\overset{H_3C}{\diagdown}} C = C \underset{CH_3}{\overset{H}{\diagup}}$	-105	0.9	---	---

FIGURE 7-3. Effect of isomerism on the chemical properties of the six C_4H_8 compounds. (--- indicates data not available.)

cause of the double bond, the two terminal methyl groups are fixed in space either on the same side of the double bond (cis configuration) or on opposite sides of the double bond (trans configuration).

The significance of isomerism in organic chemicals is that the differences in their physical, chemical, and biological properties give rise to differences in the environmental behavior of the isomers. The differences in the physical/chemical properties for the six C_4H_8 isomers (Figure 7-3) and the three xylene isomers (ortho-, meta-, and para-; Table 7-1) illustrate the differences in properties between both positional and conformational isomers. The chemical reactivity of the compounds is also different. As an example, the four linear and branched compounds (Figure 7-3) react quickly with bromine, whereas the two cyclic compounds react slowly or not at all (Roberts, Stewart, and Caserio 1971). For other organic chemicals, biological toxicity has also been related to isomerism (Safe et al. 1983; McFarland and Clarke 1989).

PHYSICAL/CHEMICAL PROPERTIES OF ORGANIC CHEMICALS

Density

The density of an organic chemical is its mass per unit volume and is usually given in the units g/cm^3 at a specified temperature. Water has a density of 0.998 g/cm^3 at 20°C. Organic chemicals, which are present in water at concentrations above their solubilities, form a separate organic phase. These are commonly termed *nonaqueous-phase liquids* (NAPLs). NAPLs that have a density less than that of water will float (light, nonaqueous-phase liquids, LNAPL), whereas those with a density greater than that of water will sink (dense nonaqueous-phase liquids, DNAPL). Many organic chemicals, including all of the aliphatic hydrocarbons, have densities less than that of water, although halogenated organic chemicals generally have densities greater than that of water (Table 7-1). The density of the organic chemical determines its potential to penetrate beneath the water table. Whether such penetration occurs in a given situation is dependent on the volume and release rate of the organic chemical, its density, and the properties of the geologic medium (permeability, stratigraphy, depth to water table). The effects of density and immiscibility are extremely important in the fate of many organic chemicals in ground water and are discussed further in Chapter 13.

Solubility and Chemical Hydrophobicity

The solubility of an organic chemical is defined as the concentration of the chemical in a given solvent in equilibrium with excess pure chemical. Solubility is dependent on the chemical structure (molecular size, functional groups, freedom of rotation, physical state), solvent (pH, ionic strength, polarity), and environmental conditions (temperature, pressure). For environmental systems, the solubility of organic chemicals in water is of central importance. This chemical property, perhaps more than any other property, affects the behavior, transport, and fate of organic chemicals in ground water. The size (i.e., molecular surface area and/or volume) of an organic molecule partially determines its water solubility. For a homologous series, such as the saturated hydrocarbons (Table 7-1), water solubility decreases with increasing size of the molecules. Other structural aspects of molecules that affect water solubility can be discussed in the context of comparing derivatives of six-carbon organic molecules (Figure 7-4). The simplest molecule of this group, *n*-hexane (water solubility of 9.5 mg/L at 20°C), will be used as a basis of comparison. Other saturated hydrocarbon compounds, which structurally have less freedom of movement, such as cyclic and branched compounds, have greater water solubilities. The addition of a double or triple bond also increases water solubility by factors of 2 to 5. The change from the aliphatic *n*-hexane to the aromatic benzene increases the solubility by more than two orders of magnitude. This is a reflection of the ability of the delocalized electrons in the aromatic compounds to interact with the water to a greater extent than the aliphatic compounds. The presence of oxygen, nitrogen, and sulfur-containing functional groups in organic molecules, such as acids, alcohols, ketones, acid salts, amines, and thiols, can promote hydrogen bonding with water, thereby increasing their water solubilities by orders of magnitude. (Compare hexanol and hexanoic acid with hexane and phenol with benzene.) The addition of a halogen or alkyl group decreases water solubility in a systematic manner based on the size of the substituent. The series of halogen- and alkyl-substituted benzenes demonstrates this effect (Figure 7-4).

The character of the solvent also can have an effect on water solubility. An increase in ionic strength results in a decrease in water

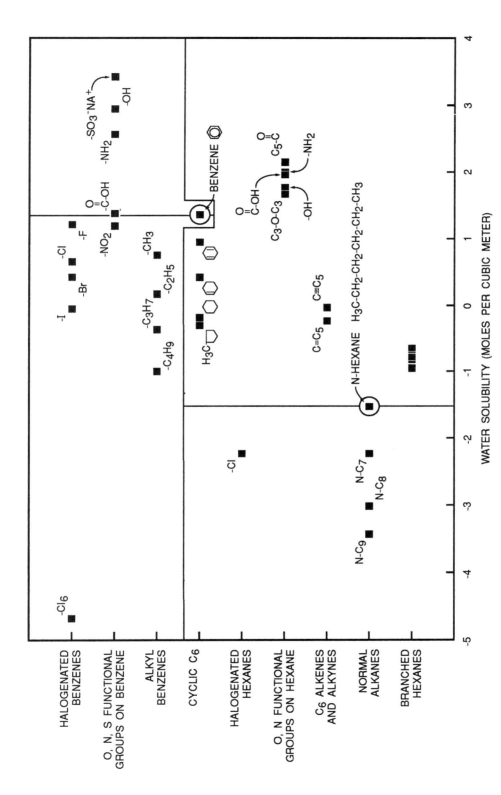

FIGURE 7-4. Effect of functional groups on the water solubility of derivatives of six-carbon molecules. *n*-Hexane and benzene are used as the basis of comparisons.

solubility (Coetzee and Ritchie 1969). An increase in solution temperature results in an increase in solubility. Generally, a two-fold solubility increase is observed for a temperature increase from 10 to 24°C for organic chemicals that are liquids or solids at ambient temperature (Seiber 1987); however, for permanent gases, an increase in temperature generally decreases water solubility. For an ionizable compound (acids, alcohols, amines), a change in the solution pH near the compound's dissociation constant (pK_a) has a very strong effect on its solubility. This will be discussed in the next section. The presence of dissolved organic carbon (DOC) (Chiou et al. 1986, 1987) and dissolved organic solvents (cosolvents), such as methanol (Morris et al. 1988), can increase the "apparent" water solubility of organic compounds, particularly for hydrophobic compounds, which have very low water solubilities. Although most non-contaminated ground waters have DOC concentrations less than 2 mg/L (Thurman 1986), ground water associated with oil shales, petroleum deposits, and chemical spills can have DOC concentrations orders of magnitude higher. High DOC and cosolvent concentrations can affect the behavior of organic chemicals in these situations.

The hydrophobicity of an organic chemical, that is, the degree to which it "dislikes" being in water, can be approximated by its water solubility or its octanol-water partition coefficient (K_{ow}). K_{ow} is a chemical's distribution between known volumes of water and the organic solvent 1-octanol. For many organic chemicals, there is a linear free-energy relation between the logarithms of water solubility and K_{ow} (Chiou et al. 1977; Miller et al. 1985). These properties are used to predict the behavior of organic chemicals in the environment.

Dissociation Constant

Some organic chemicals exhibit acid or base properties in aqueous solutions, which means they can act as either proton (H^+) donors or acceptors. Functional groups such as carboxylic acids and phenols act as proton donors

$$H_3C-\overset{\overset{O}{\|}}{C}-OH_{(aq)} + H_2O_{(l)} \leftrightarrow$$
$$H_3C-\overset{\overset{O}{\|}}{C}-O^-_{(aq)} + H_3O^+_{(aq)} \quad (7\text{-}1)$$

whereas amines and amides can act as proton acceptors

$$H_3C-NH_{2(aq)} + H_3O^+_{(l)} \leftrightarrow$$
$$H_3C-NH_3^+_{(aq)} + H_2O_{(aq)} \quad (7\text{-}2)$$

where (aq) means in the aqueous solution and (l) means in the liquid state. For Eq. 7-1, if the concentrations are assumed equal to the activities of the various organic species (i.e., the activity coefficients of all chemicals are unity), the acid-base equilibrium expression is written in the form

$$K_a = \frac{[H_3CCOO^-][H_3O^+]}{[H_3CCOOH][H_2O]} \quad (7\text{-}3)$$

where K_a is a constant at a given temperature, which describes the ratio of the ionic to molecular species of an organic acid or base as a function of pH (i.e., $[H_3O^+]$). To make an easy comparison between K_a and pH, the negative logarithm K_a, termed pK_a, is often used. Figure 7-5 demonstrates the dependence of the ionic-molecular species ratio of pentachlorophenol as a function of pH. At a pH = pK_a, the ratio of the ionic to molecular species is 1:1, that is, half of the pentachlorophenol in solution is present in the molecular (neutral) form, and half is present in the ionic (negatively charged) form. The form of the organic chemical has a significant impact on its water solubility and, therefore, its environmental behavior. The bottom part of Figure 7-5 illustrates how the organic carbon-normalized sorption coefficient (K_{oc}) of pentachlorophenol is strongly affected by its speciation. At low pH ($<pK_a$), the molecule has a sorption

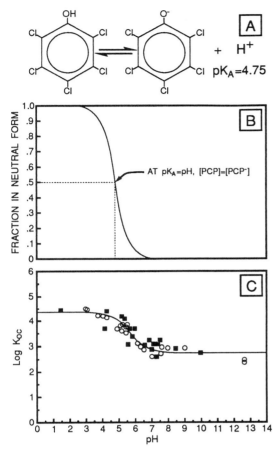

FIGURE 7-5. (a) Dissociation reaction and dissociation constant for pentachlorophenol. (b) Effect of pH on the dissociation of pentachlorophenol in aqueous solution. (c) log K_{oc} as a function of pH, showing the effect of dissociation on sorption. (*Part (c) adapted from Lee et al. 1990. Reprinted with permission from Environmental Science and Technology. Copyright 1990. American Chemical Society.* Refer to original text for the details of the data points.)

alcohols have pK_a values in the range of 9–11, and many organic amines and nitro compounds have pK_a values in the range 10–12 (Lowry and Richardson 1976). The presence of halogens in an organic molecule lowers its pK_a. As an example, phenol has a pK_a of 9.9, compared with a pK_a of 4.75 for pentachlorophenol. Most other types of organic functional groups do not have pK_a values in the typical aqueous pH range for natural water, therefore it is not a consideration in their environmental behavior.

Vapor Pressure

The vapor pressure of an organic compound is defined as the pressure, at a specific temperature, of a vapor (gas) in equilibrium with its liquid. This chemical property is a function of the size and functional groups of the organic compound. In general, small aliphatic (C < 12) and aromatic (one- and two-ring) hydrocarbons and halogenated hydrocarbons have vapor pressures greater than 10^{-4} atmosphere (atm) and are termed *volatile* organic compounds. Larger organic compounds, which include aliphatic (12 > C > 28) and aromatic (three- and four-ring) hydrocarbons, polyhalogenated aromatic hydrocarbons, and most compounds with oxygen, nitrogen, sulfur, and phosphorus functional groups, have vapor pressures between 10^{-4} and 10^{-11} atm. These compounds are termed *semivolatile*. Organic chemicals larger than the ones mentioned above, as well as organic ions and salts have extremely low vapor pressure (<10^{-11} atm) and are termed *nonvolatile* compounds.

Vapor pressure has a strong and usually nonlinear dependence on temperature. In very general terms, many organic chemicals have a twofold increase in vapor pressure for a 5°C increase in temperature. In dry soil, the vapor pressure of a chemical determines the solid-vapor distribution ratio (Chiou 1990). In moist soil and ground water, the importance of a chemical's vapor pressure often must be considered together with its water solubility, which is described in the next section on Henry's law.

coefficient about two orders of magnitude greater than at pH values above its pK_a. A sharp gradient in the sorption coefficient, K_{oc}, exists near the pK_a, reflecting the sharp changes in the ionic-molecular ratio, in this pH range. Values of pK_a of various organic chemicals are dependent on the functional groups that are present and the overall structure of the molecule.

In general, many organic acids have pK_a values in the range of 4–5, many organic

Henry's Law Constant

The Henry's law constant (H), sometimes termed the air-water partition coefficient in environmental systems, expresses the equilibrium condition between the concentration of a dissolved organic chemical in solution, [C], and the partial pressure in the air above the solution, p_C, at a given temperature:

$$H = \frac{p_C}{[C]} \qquad (7\text{-}4)$$

For organic chemicals with low water solubility, less than a few percent (i.e., grams of chemical per 100 grams of water), the Henry's law constant can be approximated as the ratio of a chemical's vapor pressure (in its liquid or supercooled liquid state) to its water solubility. As defined here, H has units of pressure divided by concentration (atm/(mole/m^3)). Both vapor pressure and solubility are temperature-dependent; thus H is temperature-dependent, generally on the order of a twofold increase in H for every increase of 8°C (Suntio et al. 1988). The Henry's law constant adequately describes the environmental air-water partition coefficient only at relatively low concentrations (<0.01 mole fraction) (Mackay and Shiu 1981), but this is usually the case for organic chemicals in the environment, except in the immediate vicinity of chemical spills. For some chemicals, depending on their vapor pressure and/or water solubility, H can be difficult to measure. The importance of the Henry's law constant to the air-water transfer of organic chemicals from ground water will be discussed in the section on volatilization.

ENVIRONMENTAL PROCESSES AFFECTING ORGANIC CHEMICALS IN GROUND WATER

Transformation Processes

The transformation of an organic chemical results in changes in its chemical structure. One or more new chemicals are produced, while the original chemical disappears. These new chemicals can be organic or inorganic molecules and ions. From an environmental point of view, the ideal fate for an organic chemical is ultimate transformation to inorganic species, such as water, carbon dioxide, and chloride ions (termed mineralization). Unfortunately, in many instances the organic chemicals formed from transformation reactions are long-lived intermediates, which themselves can have a negative impact on the environment. Chemical transformations can be mediated either by chemical, biological, or physical means. Chemically induced (abiotic) reactions include nucleophilic substitutions (hydrolysis), elimination (dehydrohalogenation), and oxidation-reduction. These will be discussed in the next sections. Biodegradation is the general term for biologically mediated reactions. Microorganisms can induce organic chemicals to undergo both hydrolytic and oxidation-reduction reactions. Photolysis is a chemical reaction induced by the energy from sunlight. Although this process does not need to be considered in ground water, it is important at the land surface for many pesticides (Miller and Hebert 1987).

Generally, first-order or pseudo-first-order kinetic expressions are adequate to describe transformation processes. The rate of disappearance of a chemical from ground water can be described as

$$\frac{d[C]}{dt} = -k[C] \qquad (7\text{-}5)$$

where k is the rate constant dependent on the specific environmental conditions and [C] is the concentration of a chemical. Unfortunately, in the environment the actual effect of the environmental condition on the kinetics of the transformation processes can be difficult to determine (Macalady, Tratnyek, and Grundl 1986). Only with very detailed laboratory and field studies can the exact transformation mechanism(s) be identified. For most purposes, transformation reactions are grouped together in a kinetic expression to describe the disappearance of an organic chemical with a lumped, pseudo-first-order reaction rate constant, k':

$$\frac{d[C]}{dt} = -k'[C] \qquad (7\text{-}6)$$

This is the approach generally used in interpreting or predicting the fate of organic chemicals in ground-water systems.

Hydrolysis and Other Nucleophilic Substitution Reactions

Hydrolysis is the chemical (or biologically mediated) reaction of an organic compound with water, usually resulting in the cleavage of the molecule into smaller, more water-soluble, portions and in the formation of new C—OH or C—H bonds. This process is important for many, but not all, organic chemicals in ground water. Hydrolysis reactions can be a result of direct attack of the water molecule (H_2O), the hydronium ion (H_3O^+), or the hydroxide ion (OH^-). These are termed neutral, acid, and base hydrolysis, respectively. Overall, the rate of hydrolysis of a chemical can be described with a first-order or pseudo-first-order rate constant, k_H:

$$k_H = k_N + k_A[H_3O^+] + k_B[OH^-] \qquad (7\text{-}7)$$

where k_N, k_A, and k_B are the rate constants for the neutral, acid-catalyzed, and base-catalyzed hydrolysis, respectively. At low solution pH, reactions are dominated by acid-catalyzed hydrolysis, while at high pH, reactions are dominated by base-catalyzed hydrolysis. At intermediate pH values, both k_N and k_A, or both k_N and k_B, can be important to the overall rate of reaction. It should be noted, however, that acid or base catalysis does not necessarily occur in all hydrolysis reactions, so that k_N alone may sometimes govern the overall rate of reaction and the rate will not depend on pH.

The structure of the organic chemical determines which of these processes, if any, are important in its hydrolysis (Mabey and Mill 1978). Organic chemicals that can undergo hydrolysis on time scales important for consideration of this process in groundwater systems (half-lives of days to years) include alkyl halides, aliphatic and aromatic esters, carbamates, phosphoric esters, and phosphoric acid esters (Vogel, Criddle, and McCarthy 1987). Some organic chemical structures undergo hydrolysis at rates too fast (half-lives of seconds to minutes) to accumulate significantly in ground water. Other organic chemicals undergo hydrolysis at rates that are too slow (half-lives of decades) to generally warrant consideration of this transformation process.

The rate of hydrolysis of a given organic compound is dependent on the characteristics of the solution. The strongest factor is pH. There is generally a two- to fourfold increase in the reaction rate for a temperature increase of 10°C. The presence of certain metal ions, humic substances, and particles can catalyze hydrolysis for some compounds (Mabey and Mill 1978; Armstrong and Chesters 1968; Burkhard and Guth 1981). The presence of cosolvents can also affect the rate of hydrolysis (Mabey and Mill 1978). The composition and concentration of ions (salts) in solution can either increase or decrease the rate of hydrolysis, depending on the organic chemical and solution conditions.

The other nucleophilic substitution reaction that can occur abiotically in groundwater systems is the reaction of sulfide species with certain halogenated aliphatic hydrocarbons. The reactive sulfide species is the sulfhydryl group (SH^-), which is more reactive than either the hydroxide ion (OH^-) or water in hydrolysis reactions. This reactive sulfide species is usually present at relatively low concentrations in most ground waters, generally making this type of reaction unimportant compared to other transformation reactions (Vogel, Criddle, and McCarthy 1987).

Elimination (Dehydrohalogenation)

Halogenated aliphatic and aromatic hydrocarbons can undergo elimination reactions, termed *dehydrohalogenation*, where a hydrogen and halide are lost from adjacent carbon atoms in the organic molecule and a double bond is formed. These types of reactions create an internal restructuring of the molecule's electrons, but there are no external electron transfers and no net change in the oxidation state of the compound. This makes them distinct from oxidation-reduction reactions, described next. Most polyhalogenated aliphatic compounds undergo dehydrohalogenation reactions, but monohalogenated compounds do not. Environmental half-lives (at 25°C) of these compounds, based on dehydrohalogenation reactions, range from 7,000 years for tetrachloromethane (carbon tetrachloride) to 0.8 years for 1,1,1-trichloroethane and 1,1,2,2-tetrachloroethane (Vogel, Criddle, and McCarthy 1987). The rates of dehydrohalogenation reactions for these compounds are difficult to predict, although generally it is recognized that the more halogens per carbon atom in an aliphatic compound, the faster the reaction rate due to dehydrohalogenation. A few aromatic hydrocarbons, such as the pesticide DDT, are also susceptible to this type of reaction.

Oxidation-Reduction
Oxidation and reduction reactions are chemical (or biologically mediated) reactions that involve a transfer of electrons. The two occur as a reaction couple; one compound or ion loses electron(s) and undergoes oxidation while another gains electron(s) and undergoes reduction. In ground water, chemical oxidation of organic contaminants generally does not occur because of a lack of oxidizing agents (chemicals to undergo reduction). However, biologically mediated oxidation does occur and will be discussed later. Certain types of ground-water contaminants have been observed to undergo reduction reactions, including organic chemicals with halogen, nitro, sulfone, and sulfoxide functional groups. It is believed that some of these reactions can occur strictly by chemical reactions, although most are still thought to be biologically mediated (Macalady, Tratnyek, and Grundl 1986; Vogel, Criddle, and McCarthy 1987; Wolfe et al. 1986).

Chemically induced reduction reactions occur in environments where organic or inorganic reducing agents exist. These reducing agents can include certain transition metals (iron, nickel, cobalt, chromium; Vogel, Criddle, and McCarthy 1987) and DOC: extracellular enzymes, iron porphyrins, chlorophylls, and other biologically derived organic chemicals (Macalady, Tratnyek, and Grundl 1986). Reduction of nitro ($-NO_2$) groups to amines ($-NH_2$) has been shown to occur. Wolfe et al. (1986) determined that the half-life of the insecticide parathion was on the order of minutes in strongly reducing environments.

Abiotically induced reduction reactions are perhaps the least understood of all environmental transformation processes. The distinction between biotic- and abiotic-mediated reactions is almost always uncertain. Herein lies the value, if not the necessity, of the lumped-transformation kinetic constants.

Biodegradation
Biodegradation is the transformation of organic chemicals mediated by living organisms using enzymes. Chemical transformation reactions can cause small structural changes in an organic chemical, but biodegradation is the only transformation process able to bring about major structural changes and completely mineralize organic chemicals (Alexander 1981). Microorganisms degrade (transform) organic chemicals as a source of energy and carbon for growth, although most of their degradative enzymes are not used directly for growth and energy processes, but rather are part of a metabolic sequence which terminates in energy release (Dagley 1983). All naturally produced organic compounds are able to be biodegraded, although for some natural products this is a slow process. On the other hand, some synthetically produced

organic chemicals have structures that are totally unfamiliar to microorganisms; thus the organisms may not have the enzymes needed for degradation of these compounds. This is the primary reason why some anthropogenic compounds are recalcitrant (very long lived) in the environment (i.e., DDE, hexachlorobenzene, mirex, etc.). However, even these synthetic compounds are observed to slowly biodegrade, probably due to a process called *cometabolism*. In cometabolism, the microorganisms are growing on other substrates (carbon sources) for growth and energy, and the unfamiliar synthetic compound enters into the process and is transformed. The microorganisms derive no particular benefit from the degradation of this compound.

The rate of biodegradation of an organic chemical is dependent on the chemical, the environmental conditions, and the microorganisms that are present. The organic chemical structure determines the types of enzymes which are able to bring about its transformation. Given this strong dependence of degradation rate on chemical structure, some progress has been made to find predictive relations between the two. The concentration of the chemical can also affect its degradation rates. At high concentrations it can be toxic; at very low concentrations it can be "overlooked" by the organisms as a potential substrate. The environmental conditions (temperature, pH, moisture, oxygen availability, salinity, concentration of other substrates) determine the species and viability of the microorganisms present. Finally, the microorganisms themselves control the rate of biodegradation depending on their species composition, spatial distribution, population density and viability, previous history with the compound of interest, and enzymatic content and activity (Scow 1990).

Phase-Transfer Processes

Phase-transfer processes involve the movement of an organic chemical from one environmental matrix to another. The important processes that can occur in ground-water environments include water-to-solid transfer (sorption), air-to-solid transfer (vapor sorption), water-to-air transfer (volatilization from water), and NAPL-to-water dissolution and NAPL-to-air vaporization (at chemical spill sites). Although the physical movement of the chemical is involved, these transfer processes should not be confused with transport processes. Transfer processes are important on the scale of molecular distances (nanometers to micrometers). Once the organic chemical has passed through the physical interface (environmental compartment boundary), it can possibly undergo transport over much larger distances. The processes of sorption and volatilization largely control the overall transport of most organic chemicals in subsurface environments.

Sorption

Organic chemicals are distributed between particle surfaces and the water to varying degrees. This process, termed *sorption*, can play a pivotal role in the environmental behavior, transport, and fate of a groundwater contaminant. An organic chemical sorbed to a particle surface will behave differently than it would in the dissolved phase. Chemicals associated with stationary aquifer particles are less available for biodegradation and not available for volatilization to the vadose zone or advective transport in the ground water. There is evidence that some particle-associated compounds undergo sorbent-catalyzed hydrolysis (Armstrong and Chesters 1968).

The extent of sorption of an organic chemical is a function of its physical-chemical properties, and particle and solution properties. Relevant aspects of the solution include pH (especially for organic chemicals which have a pK_a from 4 to 8), ionic strength, concentration of dissolved organic carbon DOC or cosolvents, and, to a lesser extent, temperature. The ionic strength of the solution affects the activity coefficient of the organic chemical in water. As the ionic strength of an aqueous solution increases, the chemical solubility decreases and the extent of sorption slightly increases. The presence of DOC and organic solvents in the water can also affect the activity coefficient of the organic chemical. This process would decrease the extent of sorption.

Because sorption is a surface process, the characteristics of the particles which have the greatest influence on the extent of sorption are surface area (sand, silt, clay, colloids) and surface coverage by organic films. In many aquifers, most of the total particle mass is in size fractions of sand and larger, but the majority of the particle surface area is found in the silt, clay, and colloidal size ranges. These fractions, although small in mass, largely control the sorptive behavior of organic chemicals in ground water. In addition to the effect of large surface area alone, the smallest particles are generally the most enriched with organic surface films. It has been shown that the organic coatings on particles, which is quantified by the fraction of organic carbon (f_{oc}) of the total particle weight, essentially control the extent of sorption for many organic chemicals to soil and aquatic particles (Karickhoff 1984; Chiou 1990). Many aquifers, which have not been highly contaminated, have very low f_{oc}, from 0.001 to 0.01 (Schwarzenbach 1986; Mackay 1990). In these cases the ultimate control on organic chemical sorption may be the inorganic surface area and surface mineralogy rather than the organic films on the particles.

The hydrophobicity of an organic chemical, which can be quantified to some extent by its water solubility or octanol-water partition coefficient, also controls the extent of sorption. The extent of sorption at equilibrium is commonly defined in terms of a linear distribution coefficient, K_d,

$$K_d = \frac{\frac{\text{mass of chemical}_{(\text{particulate})}}{\text{mass of particles}}}{\frac{\text{mass of chemical}_{(\text{aqueous})}}{\text{mass of water}}} \quad (7\text{-}8)$$

or an organic carbon-normalized distribution coefficient, K_{oc},

$$K_{oc} = \frac{K_d}{f_{oc}} \quad (7\text{-}9)$$

Tabulations of K_d and K_{oc} values for a wide range of organic chemicals can be found in Howard (1990, 1991a, 1991b) and elsewhere.

For a wide range of organic chemicals, their sorptive distribution coefficients have been shown to be strongly correlated to their water solubilities. This provides a tool for predicting the extent of sorption of a particular chemical in a particular environment where the organic coatings of the particles dominate the sorptive process. A number of structure-activity relations have been presented for these types of predictions. As an example, Schwarzenbach and Westfall (1981) have observed the relation

$$\log K_d = \log f_{oc} + 0.72 \log K_{ow} + 0.49 \quad (7\text{-}10)$$

for a series of alkylbenzenes and chlorinated hydrocarbons. In low-organic-carbon systems, however, these correlative relations lose much of their value as predictive tools. Relations based on organic-carbon content of soils and the extent of sorption in the vadose zone are discussed in Chapter 14 on pesticides.

There are a number of different kinds of "particles" present in aquifers, particularly contaminated aquifers, which can act as sorptive surfaces (Figure 7-6). In the saturated zone, an organic chemical can be truly (i.e., thermodynamically) dissolved in water, dissolved in (or present as) an organic fluid, in association with stationary particles, or in association with "mobile particles." Stationary particles are the structural components of the aquifer. "Mobile particles" include DOC, DOC-iron complexes, colloids, and droplets of an organic fluid. For an organic compound in the saturated zone, each chemical state is related to the other states through equilibrium sorption. The extents of sorp-

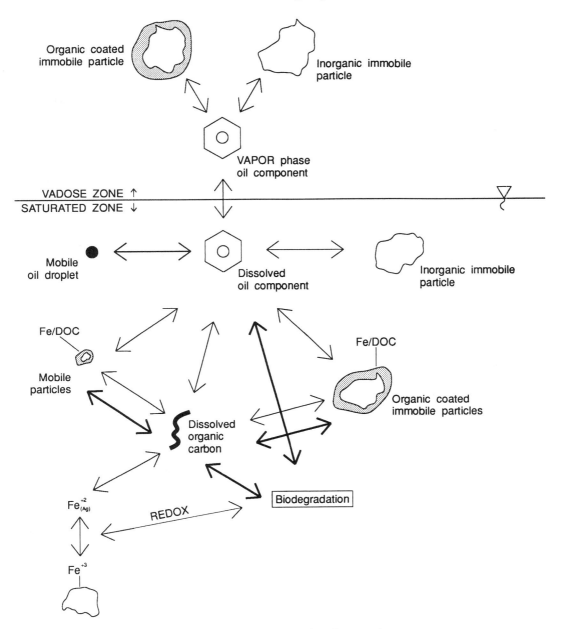

FIGURE 7-6. Conceptual model of the equilibrium distribution of an organic chemical, such as benzene, among various phases in the saturated and vadose zones. Superimposed on these distributions are the carbon and iron cycles, which in part, determine the distributions of the organic chemical at any point in time and space.

tion to the various types of "particles" determine the mechanisms and velocity of transport of a particular compound in a specific subsurface system.

Sorption is an extremely complex process. Due to a lack of complete understanding and environmental observations, sorption of organic chemicals is often assumed to be completely reversible, linear (with respect to chemical concentration), and at equilibrium in ground-water systems. Recent studies have shown that sorption and desorption are not completely reversible, at least in the laboratory (DiToro and Horzempa 1982; Karickhoff and Morris 1985) and that chemical equilibrium may not be reached, even in slowly changing ground-water environments (Mackay 1990; Gillham and Rao 1990; Brusseau and Rao 1989). For most environmental situations organic contaminants are present at concentrations low enough where a linear K_d value adequately describes its sorptive behavior. At chemical spill sites or other areas where the organic chemical concentrations are higher, at concentrations greater than about 10^{-5} moles/L or about one-half the chemical's water solubility, nonlinear sorptive behavior is observed and the use of the linear distribution coefficient is invalid (Karickhoff, Brown, and Scott 1979). Given all of the uncertainties of environmental observations of organic chemicals, the quantification of nonequilibrium, nonlinear, and nonreversible sorptive behavior has been difficult.

Vapor Sorption

Just as organic chemicals distribute themselves between the water and solid surfaces in the saturated zone, they also distribute themselves between the air and solid surfaces in the vadose zone. The extent of vapor sorption is a function of the chemical's properties, the solid's properties, and the water content of the system. Chiou (1990) has shown that in dry environments, vapor sorption interactions are stronger between the organic chemical and the inorganic surface of the solid (particularly clay surfaces) than the organic matter on the solid surface. As the water content of the system increases, the inorganic surface becomes hydrated and the water outcompetes the organic molecules for the inorganic sorption sites. The extent of vapor sorption decreases and interactions with the organic carbon surface coatings become the dominant mechanism.

Chemicals sorbed by dry solids are released by the addition of water. At about 90% relative humidity, the extent of vapor sorption is close to that of sorption in aqueous systems, if the same chemical and solid are compared.

This process governs the amount of an organic contaminant in the vapor phase in the vadose zone. Near the land surface, where a wet-dry cycle can exist, the relative humidity of the soil strongly influences the chemical's distribution. Farther from the land surface, where the relative humidity is nearly constant, the distribution of a chemical between the vapor and solid phases is more uniform and can be predicted by the water content, solid characteristics, and the vapor pressure of the chemical at the ambient temperature (Thomas 1990b). It has been suggested that the K_{oc} concept, when based on vapor concentration, can be used to described vapor sorption (Spencer 1987).

Volatilization from Ground Water

Organic chemicals can be transferred between the dissolved aqueous phase in the ground water and the vapor phase in the overlying vadose zone as a result of *volatilization* (Figure 7-6). This is controlled by the chemical nature of the air-water interface and the mass transfer (advective) rates of the chemical in water (velocity of water flow, distance of the chemical from the water-table level, etc.), its molecular diffusion in air and water, and its Henry's law constant. Figure 7-7 illustrates the relation between vapor pressure, water solubility, and Henry's law constant for a number of chemicals of interest as potential ground-water contaminants.

Thomas (1990a) has suggested that the importance of volatilization for a given chemical can be generalized based only on its Henry's law constant. For chemicals that have a Henry's law constant $<3 \times 10^{-7}$ atm-m^3/mole, volatilization from ground water is unimportant. For chemicals that have a Henry's law constant $>1 \times 10^{-5}$ atm-m^3/mole, volatilization is significant for

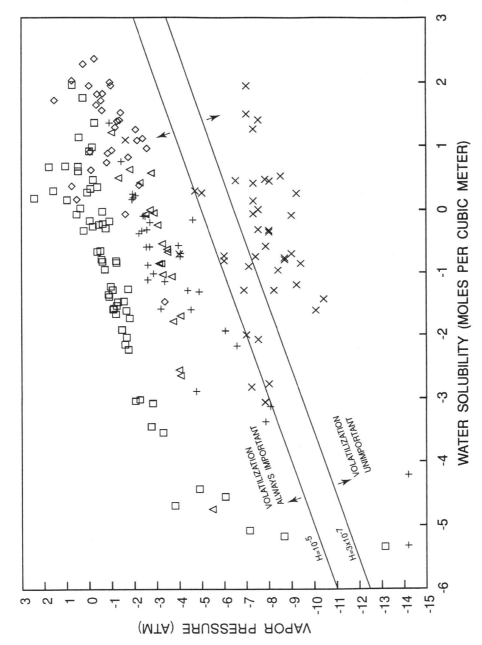

FIGURE 7-7. Henry's law constant as a function of water solubility and vapor pressure for many aliphatic hydrocarbons (□), aromatic hydrocarbons (△), chlorinated aliphatic hydrocarbons (◇), chlorinated aromatic hydrocarbons (+), and various pesticides (×).

all waters. As can be observed in Table 7-1 and Figure 7-7, many of the common ground-water contaminants, especially the hydrocarbons and halogenated hydrocarbons, have their Henry's law constant in the range where volatilization can be a significant process in their environmental behavior. In contrast to this, only a few of the pesticides commonly observed in ground water exhibit any tendency toward volatilization from ground water.

The preceding discussion has centered on volatilization (air-water transfer) of chemicals between the ground water and the vadose zone. In addition, losses to the atmosphere from plants and soils, also termed volatilization, constitute an important process limiting the leaching of many pesticides and other organic chemicals to ground water. This topic is discussed in Chapter 14.

NAPL-to-Water Dissolution and NAPL-to-Air Vaporization

At many chemical spill sites, the concentrations of organic chemicals in the ground water exceed their water solubilities and form an NAPL. These separate phases can exist as pools, droplets, or ganglia in the ground water (Mackay 1990). The transfer of an organic molecule from the NAPL to water is termed *dissolution*. The extent of this process is governed by the chemical's water solubility, but the kinetics of dissolution is determined by the rate of ground-water flow, the geometry of the NAPL, and the diffusion coefficients of the chemical in the NAPL and in the water. The transfer of an organic molecule from an NAPL (usually an LNAPL) to the air of the subsurface vadose zone is termed *vaporization*. The extent of this process is governed by the chemical's vapor pressure and Raoult's law constant (NAPL-air distribution constant; Atkins 1978), but the kinetics of vaporization is determined by the rate of convection in the vadose zone and the diffusion coefficients of the chemical in the NAPL and in the air. Although these phase-transfer processes are extremely important in the contamination of aquifers at chemical spill sites, there is presently a lack of understanding of the kinetics of these processes (Mackay 1990). The behavior of NAPLs in the subsurface is discussed further in Chapter 13 on organic contaminants and by Gillham and Rao (1990).

Transport Processes

A number of mechanisms can be responsible for the transport of organic chemicals in the subsurface. The properties of the chemical and the ground-water environment determine which transport mechanism will control the chemical's movement. An organic compound will distribute itself between a number of chemical environments (aqueous, particulate, gaseous) to reach an equilibrium condition (Figure 7-6). This distribution ultimately controls its rate of transport.

Traditionally, chemical transport in an aquifer has been viewed as a simple two-phase ("dissolved" and "sorbed") process, where the "dissolved" phase is transported by advection and dispersion and the "sorbed" phase acts to retard the process. For some organic chemicals, this is an adequate view, but for many, the situation is much more complex. As an example, migration of chemicals far beyond distances predicted on the basis of chemical properties and traditional mass transport theory has been observed for several hydrophobic organic compounds, including polychlorinated dibenzo-p-dioxins (Pereira et al. 1985), hexachlorobenzene (Schwarzenbach et al. 1983), polyaromatic hydrocarbons (Backhus 1990) and inorganic compounds, including plutonium and americium (Penrose et al. 1990). These findings have led researchers to modify conceptual models of chemical behavior in ground water from simple two-compartment systems (solid-water) to complex multicompartment systems that include "mobile particles" (solids–water–colloids–dissolved organic carbon–organic fluids) (McCarthy and Zachara 1989; Huling 1989). The transport of organic chemicals through the vapor phase, with subsequent reentry to the ground water has also been suggested as another viable mechanism (Aiken et al. 1991).

Aqueous-Phase Transport

Transport in the saturated zone can occur only when the organic chemical is in the water. While it is sorbed to a stationary particle, it is not available for transport. An organic chemical contaminant can exist in ground water in a number of chemical states, including being thermodynamically dissolved or associated with *mobile particles*, such as inorganic colloids, DOC, and DOC-iron complexes. In an aquifer contaminated with an organic fluid, the organic contaminant can also be associated with droplets of the organic fluid, which can also be a mobile "particle." Transport of organic contaminants in association with mobile particles has been termed *facilitated transport* (Huling 1989).

The role of the stationary particles (aquifer matrix), whether in the vadose or the saturated zone, is attenuation of organic chemical transport through sorption. Organic chemicals can interact directly with mineral surfaces and with organic matter sorbed onto the stationary particles. Most of the stationary-particle surfaces, especially in areas of contamination, are covered with an organic film, which changes character in space and time. Sorptive interactions between the organic chemicals and stationary particles remove the contaminant from solution (Figure 7-6) and retard its movement in the aquifer. This process can have a profound effect on the ground-water transport of organic chemicals.

One of the important components of almost all mobile particles is DOC, which is a highly transient and complex mixture of individual organic compounds with varying aqueous solubilities, chemical structures, stabilities and reactivities, and biodegradabilities. The presence of certain DOC constituents is strongly dependent on the oxidation-reduction state of the ground water. Chemical hydrophobicity, aromaticity, and molecular weight and size of the DOC determine the extent of sorption for a ground-water contaminant. A high degree of DOC-organic contaminant interaction can enhance the facilitated transport process.

The interactions between the DOC and iron in the anoxic zone of an aquifer and at the oxidation-reduction interface, where the iron is precipitating, affect the sorption and transport of organic contaminants. Interactions between organic matter and iron have been noted in soils (Antweiler and Drever 1983; Senesi 1981), in fresh surface waters (Sholkovitz and Copland 1981), and in anoxic sediments (Krom and Sholkovitz 1978). Berndt (1987) has observed that 25–48% of iron extracted from aquifer materials downgradient of a spill of crude oil is associated with organic matter. DOC-iron complexes provide additional mobile particles that enhance facilitated transport.

The transient, chemical composition of the aqueous phase and its interactions with organic contaminants suggest that concurrent, diverse transport mechanisms are at work in ground water. The thermodynamically dissolved organic contaminant will be transported by advection and dispersion on the basis of its own chemical properties. The organic contaminant associated with mobile particles will be transported "piggyback" fashion, on the basis of the properties of the mobile particle. Understanding and quantifying the role of inorganic colloids, DOC, DOC-iron complexes and droplets of organic fluids in organic chemical transport is still in its infancy.

Vapor-Phase Transport

Organic chemicals can be introduced into the vapor phase within the vadose zone by at least four routes: volatilization of chemicals leaching through the vadose zone, volatilization from a buoyant body of an organic fluid floating on the water table (LNAPL from a spill), air-solid transfer of a chemical sorbed to a particle in the vadose zone after fluctuations in the water table, and volatilization of dissolved organic chemicals from the ground water (at the water table). Once in the vadose zone, organic chemicals can be transported either in the vapor phase by diffusion and convection (often density-driven) or in an aqueous phase, if one is present. Vapor pressure, together with water solubility (i.e., Henry's law constant) and the amount of water which is present, determines the fraction of a chemical in the vapor phase in moist vadose environments. In the absence of water, vapor pressure alone determines the amount in the vapor phase (Chiou 1990). Sorptive interactions with the geologic matrix in the vadose zone attenuate the movement of the organic chemicals owing to removal from the mobile vapor and aqueous phases. In some environmental situations, organic chemical transport in the vadose zone is faster in the vapor than in the aqueous phase. It is possible that an organic molecule observed in the ground-water downgradient from a site of chemical contamination may have arrived there through vapor transport and not aqueous transport. Vapor-phase transport and its relation to ground-water transport has only recently been considered in conceptual transport models of organic compounds in unconfined aquifers (Aiken et al. 1991). Further evidence is needed to determine the importance of this vapor-phase movement and subsequent air-water exchange on the "apparent" transport of organic chemicals in ground water.

CONCLUDING REMARKS

This chapter has presented a brief introduction to the vocabulary and concepts of organic and physical organic chemistry and their relation to the environmental processing (entry, behavior, transport, and ultimate fate) of organic chemicals in ground water. The structure of the organic chemical determines its physical, chemical, and biological properties. Water solubility is perhaps the most important chemical property in determining environmental behavior. The ground-water environment, which surrounds the organic chemical, consists of physical, chemical, and biological components. The interaction of these two, chemical structure and environmental conditions, results in the chemical's behavior and its effect on the environment. Upon entry of the chemical into the ground water, the environmental processes act on the chemical until equilibrium is reached. Transformation processes will change its chemical structure. Phase-transfer processes will move it out of the water onto the aquifer solids or into the vadose zone. Transport processes will move it away from its initial location. The behavior of organic chemicals in ground water can be understood in the context of these processes.

References

Aiken, G. R., P. D. Capel, E. T. Furlong, M. F. Hult, and K. A. Thorn. 1991. Mechanisms controlling the transport of organic chemicals in subsurface environments. In *U.S. Geological Survey Toxic Substances Hydrology Program—Proceedings of the Technical Meeting, Monterey, California, March 11–15, 1991*, eds. G. E. Mallard and D. A. Aronson, pp. 633–37. Reston, VA: U.S. Geological Survey, Water-Resources Investigations Report 91-4034.

Alexander, M. 1981. Biodegradation of chemicals of environmental concern. *Science* 211:132–38.

Antweiler, R. C., and J. I. Drever. 1983. The weathering of a late Tertiary volcanic ash: Importance of organic solutes. *Geochimica et Cosmochimica Acta* 47(3):623–29.

Armstrong, D. E., and G. Chesters. 1968. Adsorption catalyzed chemical hydrolysis of atrazine. *Environmental Science and Technology* 2(9):683–89.

Atkins, P. W. 1978. *Physical Chemistry*. San Francisco: Freeman, pp. 208–15.

Backhus, D. A. 1990. *Colloids in Groundwater: Laboratory and Field Studies of their Influence on Hydrophobic Organic Compounds*. Massachusetts Institute of Technology, Boston, MA: Ph.D. Thesis.

Berndt, M. P. 1987. Metal partitioning in aquifer sediments, Bemidji, Minnesota, research site. In *U.S. Geological Survey Program on Toxic Waste—Ground-Water Contamination: Proceedings of the Third Technical Meeting,*

Pensacola, Florida, March 23–27, 1987, ed. B. J. Franks, pp. C17–C19. Reston, VA: U.S. Geological Survey Open-File Report 87–109.

Brusseau, M. L., and P. S. C. Rao. 1989. The influence of sorbate-organic matter interactions on sorption nonequilibrium. *Chemosphere* 18(9/10):1691–1706.

Burkhard, N., and J. A. Guth. 1981. Chemical hydrolysis of 2-chloro-4,6-bis(alklamino)-1,3,5-triazine herbicides and their breakdown in soil under the influence of adsorption. *Pesticide Science* 12(1):45–52.

Chiou, C. T. 1990. Roles of organic matter, minerals and moisture in sorption of nonionic compounds and pesticides by soil. In *Humic Substances in Soil and Crop Sciences*, eds. P. MacCarthy, C. E. Clapp, R. L. Malcolm, and P. R. Bloom, pp. 111–59. Madison, WI: American Society of Agronomy, Soil Science Society of America.

Chiou, C. T., V. H. Freed, D. W. Schmedding, and R. L. Kohnert. 1977. Partition coefficient and bioaccumulation of selected organic chemicals. *Environmental Science and Technology* 11(5):475–78.

Chiou, C. T., D. E. Kile, T. I. Brinton, R. L. Malcolm, J. A. Leenheer, and P. MacCarthy. 1987. A comparison of water solubility enhancements of organic solutes by aquatic humic materials and commercial humic acids. *Environmental Science and Technology* 21(12):1231–34.

Chiou, C. T., R. L. Malcolm, T. I. Brinton, and D. E. Kile. 1986. Water solubility enhancement of some organic pollutants and pesticides by dissolved humic and fulvic acids. *Environmental Science and Technology* 20(5):502–08.

Coetzee, J. F., and C. D. Ritchie. 1969. *Solute-Solvent Interactions*. New York: Marcel Dekker, pp. 329–38.

Dagley, S. 1983. Biodegradation and biotransformation of pesticides in the earth's carbon cycle. In *Residue Reviews: Residues of Pesticides and Other Contaminants in the Total Environment*, eds. F. A. Gunther and J. D. Gunther, pp. 127–37. New York: Springer-Verlag.

DiToro, D. M., and Horzempa, L. M. 1982. Reversible and resistant components of PCB adsorption-desorption: Isotherms. *Environmental Science and Technology* 16(10):594–602.

Gillham, R. W., and Rao, P. S. C. 1990. Transport, distribution, and fate of volatile organic compounds in groundwater. In *Significance and Treatment of VOC's in Water Supplies*, eds. N. M. Ram, R. F. Christman, and K. P. Cantor, pp. 141–81. Chelsea, MI: Lewis Publishers.

Grain, C. F. 1990. Activity coefficient. In *Handbook of Chemical Property Estimation Methods*, eds. W. J. Lyman, W. F. Reehl, and D. H. Rosenblatt, pp. 11.1–11.53. Washington DC: American Chemical Society.

Hartley, D., and H. Kidd. 1987. *The Agrochemicals Handbook*. Nottingham, UK: Royal Society of Chemistry.

Howard, P. H. 1990. *Handbook of Environmental Fate and Exposure Data for Organic Chemicals*. Vol. I: *Large Production and Priority Pollutants*. Chelsea, MI: Lewis Publishers.

Howard, P. H. 1991a. *Handbook of Environmental Fate and Exposure Data for Organic Chemicals*. Vol. II: *Solvents*. Chelsea, MI: Lewis Publishers.

Howard, P. H. 1991b. *Handbook of Environmental Fate and Exposure Data for Organic Chemicals*. Vol. III: *Pesticides*. Chelsea, MI: Lewis Publishers.

Huling, S. G. 1989. *Facilitated Transport, EPA Superfund Ground Water Issue*. EPA/540/4-89/003, Washington DC.

Karickhoff, S. W. 1984. Organic pollutant sorption in aquatic systems. *Journal of Hydraulic Engineering* 110(6):707–35.

Karickhoff, S. W., and K. R. Morris. 1985. Sorption dynamics of hydrophobic pollutants in sediment suspensions. *Environmental Toxicology and Chemistry* 4(3):469–79.

Karickhoff, S. W., D. S. Brown, and T. A. Scott. 1979. Sorption of hydrophobic pollutants on natural sediments. *Water Research* 13(3):241–8.

Krom, M. D., and E. R. Sholkovitz. 1978. On the association of iron and manganese with organic matter in anoxic pore waters. *Geochimica et Cosmochimica Acta* 42(6):607–11.

Lee, L. S., P. S. C. Rao, P. Nkedi-Kizza, and J. J. Delfino. 1990. Influence of solvent and sorbent characteristics on distribution of pentachlorophenol in octanol-water and soil water systems. *Environmental Science and Technology* 24(7):654–61.

Lowry, T. H., and K. S. Richardson. 1976.

Mechanism and Theory in Organic Chemistry. New York: Harper and Row, pp. 149–150.

Mabey, M., and T. Mill. 1978. Critical review of hydrolysis of organic compounds in water under environmental conditions. *Journal of Physical and Chemical Reference Data* 7(2):383–415.

Macalady, D. L., P. G. Tratnyek, and T. J. Grundl. 1986. Abiotic reduction reactions of anthropogenic organic chemicals in anaerobic systems: A critical review. *Journal of Contaminant Hydrology* 1(1):1–28.

Mackay, D. 1990. Characteristics of the distribution and behavior of contaminants in the subsurface. In *Ground Water and Soil Contamination Remediation: Toward Compatible Science, Policy, and Public Perception*, Report on a Colloquium sponsored by the Water Science and Technology Board, pp. 70–90. Washington DC: National Academy Press.

Mackay, D., and W. Y. Shiu. 1981. A critical review of Henry's law constants for chemicals of environmental interest. *Journal of Physical Chemistry Reference Data* 10(4):1175–99.

McCarthy, J. F., and J. M. Zachara. 1989. Subsurface transport of contaminants. *Environmental Science and Technology* 23(5):496–502.

McFarland, V. A., and J. U. Clarke. 1989. Environmental occurrence, abundance, and potential toxicity of polychlorinated biphenyl congeners: Considerations for a congener-specific analysis. *Environmental Health Perspectives* 81(3):225–39.

Miller, G. C., and V. R. Hebert. 1987. Environmental photodecomposition of pesticides. In *Fate of Pesticides in the Environment*, eds. J. W. Biggar and J. N. Seiber, pp. 75–86. Oakland, CA: University of California Publication 3320.

Miller, M. M., S. P. Wasik, G.-L. Huang, W. Y. Shiu, and D. Mackay. 1985. Relationships between octanol-water partition coefficient and aqueous solubility. *Environmental Science and Technology* 19(6):522–29.

Morris, K. R., R. Abramowitz, R. Pinal, P. Davis, and S. H. Yalkowsky. 1988. Solubility of aromatic pollutants in mixed solvents. *Chemosphere* 17(2):285–98.

Penrose, W. R., W. L. Polzer, E. H. Essington, D. M. Nelson, and K. A. Orlandi. 1990. Mobility of plutonium and americium through a shallow aquifer in a semiarid region. *Environmental Science and Technology* 24(2):228–34.

Pereira, W. E., C. E. Rostad, and M. E. Sisak. 1985. Geochemical investigations of polychlorinated dibenzo-*p*-dioxins in the subsurface environment at an abandoned wood-treatment facility. *Environmental Toxicology and Chemistry* 4(5):629–39.

Roberts, J. D., R. Stewart, and M. C. Caserio. 1971. *Organic Chemistry: Methane to Macromolecules.* Menlo Park, CA: W.A. Benjamin.

Safe, S., M. Mullin, L. Safe, C. Pochini, S. McCrindle, and M. Romkes. 1983. High resolution PCB analysis. In *Physical Behavior of PCBs in the Great Lakes*, eds. D. Mackay, S. Paterson, S. J. Eisenreich, and M. S. Simmons, pp. 1–14. Ann Arbor, MI: Ann Arbor Science Publishers.

Schwarzenbach, R. P. 1986. Sorption behavior of neutral and ionizable hydrophobic organic compounds. In *Organic Micropollutants in the Aquatic Environment*, eds. A. Bjorseth and G. Angeletti, pp. 168–77. Dordrecht, The Netherlands: Reidel.

Schwarzenbach, R. P., and J. Westfall. 1981. Transport of nonpolar organic compounds from surface water to groundwater. Laboratory sorption studies. *Environmental Science and Technology* 15(11):1360–67.

Schwarzenbach, R. P., W. Giger, E. Hoehn, and J. K. Schneider. 1983. Behaviour of organic compounds during infiltration of river water to groundwater: Field studies. *Environmental Science and Technology* 17(8):472–79.

Scow, K. M. 1990. Biodegradation. In *Handbook of Chemical Property Estimation Methods*, eds. W. J. Lyman, W. F. Reehl, and D. H. Rosenblatt, pp. 9.1–9.34. Washington DC: American Chemical Society.

Seiber, J. N. 1987. Solubility, partition coefficient, and bioconcentration factor. In *Fate of Pesticides in the Environment*, eds. J. W. Biggar and J. N. Seiber, pp. 53–60. Oakland, CA: University of California Publication 3320.

Senesi, N. 1981. Spectroscopic evidence on organically bound iron in natural and synthetic complexes with humic substances. *Geochimica et Cosmochimica Acta* 45(2):269–72.

Sholkovitz, E. R., and D. Copland. 1981. The coagulation, solubility and adsorption properties of Fe, Mn, Cu, Ni, Cd, Co, and humic acids in a river water. *Geochimica et Cosmochimica Acta* 45(2):181–89.

Spencer, W. F. 1987. Volatilization of pesticide residues. In *Fate of Pesticides in the Environment*, eds. J. W. Biggar, and J. N. Seiber, pp. 53–60. Oakland, CA: University of California Publication 3320.

Suntio, L. R., W. Y. Shiu, D. Mackay, J. N. Seiber, and D. Glotfelty. 1988. Critical review of Henry's law constants for pesticides. *Reviews of Environmental Contamination and Toxicology* 103(1):1–59.

Thomas, R. G. 1990a. Volatilization from water. In *Handbook of Chemical Property Estimation Methods*, eds. W. J. Lyman, W. F. Reehl, and D. H. Rosenblatt, pp. 15.1–15.34. Washington, DC: American Chemical Society.

Thomas, R. G. 1990b. Volatilization from soil. In *Handbook of Chemical Property Estimation Methods*, eds. W. J. Lyman, W. F. Reehl, and D. H. Rosenblatt, pp. 16.1–16.50. Washington DC: American Chemical Society.

Thurman, E. M. 1986. *Organic Geochemistry of Natural Waters*. Dordrecht, The Netherlands: Martinus Nijhoff/Dr. W. Junk, p. 14.

Vogel, T. M., C. S. Criddle, and P. L. McCarthy. 1987. Transformations of halogenated aliphatic compounds. *Environmental Science and Technology* 21(8):722–36.

Wolfe, N. L., B. E. Kitchems, D. L. Macalady, and T. J. Grundl. 1986. Physical and chemical factors that influence the anaerobic degradation of methyl parathion in sediment systems. *Environmental Toxicology and Chemistry* 5(11):1019–26.

Worthing, C. R., and S. B. Walker. 1983. *The Pesticide Manual*. London: British Crop Protection Council.

8

Subsurface Microbiology

Francis H. Chapelle, Paul M. Bradley, and Peter B. McMahon

INTRODUCTION

The chemical quality of ground water is a result of the combined effects of physical, chemical, and microbial processes. At first glance it may seem fairly simple to separate and identify the relative importance of these three classes of processes when dealing with ground-water quality on a regional basis. In practice, however, this can be a difficult task with many subtleties. Much of the difficulty stems from the fact that dissolved constituents commonly found in ground water can be produced by both biotic and abiotic processes. Methane in ground water, for example, may reflect either in situ microbial methanogenesis or leakage of thermally produced methane from underlying hydrocarbon reservoirs. The mere presence of methane, therefore, is equivocal and is insufficient evidence to determine the relative importance of chemical processes (thermal generation of methane), physical processes (leakage from deep reservoirs), or microbial processes (methane production by microorganisms).

Identifying the relative importance of chemical, physical, and microbial processes is important in ground-water quality assessments. Take, for example, the case of methane in ground water. The degradation of many organic contaminants in ground-water systems is dependent on ambient microbial processes. Halogenated organic solvents such as trichloroethylene (TCE) are subject to reductive dehalogenation processes under methanogenic conditions. Where aerobic microbial metabolism predominates, however, TCE is stable and does not readily degrade. Thus, determining the nature of microbial metabolism gives important information concerning the fate and transport of particular contaminants.

An understanding of subsurface microbiology is essential for determining the distribution and rates of microbially mediated processes occurring in ground-water systems. This information, when combined with appropriate hydrologic and geochemical data, allows a rational evaluation of how microbial processes influence water chemistry under pristine conditions, how contamination events have altered water chemistry in the past, and how chemistry may respond to chemical stresses in the future.

This chapter gives a brief historical overview of subsurface microbiology, describes some of the methods that are widely used, and presents some case study examples that relate regional water-quality issues to subsurface microbiology.

Historical Overview of Subsurface Microbiology

The development of microbiology has been driven by the need to solve practical problems. In the nineteenth century, the development of bacteriology by Louis Pasteur and Robert Koch was driven almost exclusively by consideration of food and wine spoilage problems and, of course, by the role of bacteria in human and animal diseases. Similarly, virology, soil microbiology, phycology (study of algae), mycology (study of fungi), and protozoology (study of protozoa) all developed in response to specific questions dealing with pathology, agriculture, and industrial concerns. In the case of subsurface microbiology, the problems at hand were related to concerns about ground-water quality.

The first such concern related to the chemical quality of ground water associated with petroleum reservoirs. In the early twentieth century, Sherburne Rogers of the U.S. Geological Survey published a comprehensive study of the Sunset-Midway oilfield in California (Rogers 1917). Rogers found that ground water associated with liquid hydrocarbons was characterized by low sulfate and high bicarbonate concentrations. In contrast, shallower ground water not associated with hydrocarbons was characterized by high sulfate and low bicarbonate concentrations. Rogers interpreted this as evidence that sulfate reduction had resulted in the removal of sulfate with the production of sulfides and bicarbonate. Furthermore, Rogers suggested that sulfate-reducing bacteria were involved in these reduction processes. At that time, however, there was no direct evidence that such microorganisms were present in deep sediments.

This evidence was soon provided by Edson Bastin, a geologist, with the help of bacteriologist Frank Greer, both of whom were from the University of Chicago. They prepared a sulfate-lactate media with a salt content similar to oil-field brines from the Illinois Basin, and inoculated the media with water produced from oil wells. In 10 days, 17 of the 19 samples exhibited growth of sulfate-reducing bacteria. Publishing his findings in the journal *Science*, Bastin (1926) observed:

From the work here reported it is evident that anaerobic bacteria of the sulphate-reducing type are present in abundance in some of the waters associated with oil in productive fields and it is very probable that they are responsible for the low sulphate content of these waters.

Throughout much of the twentieth century, the primary motivation for considering the presence of microbes in the subsurface dealt specifically with water quality. Cederstrom (1946) suggested that microbial sulfate reduction produced carbon dioxide leading to the high-bicarbonate ground water found in Atlantic coastal-plain aquifers. While conducting studies on the origins of petroleum and associated pore fluids, Claude Zobell described microbial processes in deeply buried marine sediments (Zobell 1947) and the presence of bacteria in some of the deepest parts of the ocean (Zobell 1952).

In the 1970s, the nature of subsurface microbiology-water-quality studies took a new and important direction. At this time it became apparent that significant amounts of ground water were becoming contaminated as a result of improper waste-disposal practices. Although it was known that surface-water bodies were capable of restoration from pollution, largely due to microbial processes, virtually nothing was known about the biorestoration potential of aquifer systems. Prior to 1975, it was widely assumed that abiotic processes, such as sorption and dilution, were the primary pollution attenuation mechanisms in subsurface environments. "Dilution is the solution to pollution" was the rule of thumb for many environmental scientists.

In 1973, two U.S. Environmental Protection Agency microbiologists, driven by concern about contaminated aquifers, prepared a literature search on subsurface microbiology (Dunlap and McNabb 1973). This search cited many of the early studies conducted by petroleum industry researchers and indicated that the presence of microorganisms in aquifer systems was not an impossibility. This search also indicated, however, that virtually nothing was known about the microbiology of shallow-water-table aquifers—the systems most vulnerable to contamination. Clearly, here was an important issue that required systematic investigation.

In the late 1970s and early 1980s there was a rapid increase in the rate of subsurface microbiology research, much of which was fueled by ground-water pollution issues (Leenheer, Malcolm, and White 1976; Ehrlich et al. 1982; Dunlap et al. 1977; Godsy and Ehrlich 1978). A significant effort was also mounted toward understanding the microbiology of pristine aquifers (Godsy 1980; Dockins et al. 1980; Wilson et al. 1983; Ghiorse and Balkwill 1983; White et al. 1983; Hirsch and Rades-Rohkohl 1983). Although many of these early studies were focused on relatively small-scale issues, the accumulation of numerous small-scale studies (review by Ghiorse and Wilson 1988) and the initiation of regional-scale studies (Chapelle et al. 1987; Chapelle and Lovley 1990) form a solid basis for considering microbial processes in regional evaluations of ground-water quality.

MICROBIAL ECOLOGY OF GROUND-WATER SYSTEMS

Ecology is that branch of biology dealing with relations between organisms and their physical and biological environment. An ecological community consists of all of the organisms—microbes, plants, and animals—that live in a particular area. The community, together with the physical environment to which it is tied, is called an *ecosystem*. Although a few higher animals and plants are known to inhabit some ground-water systems, ecological communities in ground-water environments consist largely of microorganisms. The major focus of subsurface ecological research is to determine how microbial communities are structured, how microorganisms within the community interact with each other and their environment, and how these interactions affect ground-water chemistry.

Pure Culture Methods

Procedures for obtaining microorganisms in pure culture and then analyzing them for their biochemical properties are deeply imbedded in the practice of microbiology. Many of the important advances in medical and agricultural microbiology were made using pure culture methods. These methods have been widely applied to studies of subsurface microbial ecology. This approach seeks to isolate as many strains of bacteria, protozoa, and fungi as possible from subsurface sediments and characterize them on the basis of colony morphology, cell morphology, and physiological-biochemical characteristics. These data are then used to assess community structure and diversity.

The first comprehensive use of culture techniques for evaluating the microbial community of an aquifer was described by Kolbel-Boelke, Anders, and Nehrkorn (1988). These investigators isolated 2,700 strains of bacteria from a shallow Pleistocene aquifer in the Lower Rhine region of Germany. In their study, samples were obtained from aseptically cored sediments as well as from water samples frow wells. Characterization of the isolates provided an interesting perspective on the ecology of microorganisms inhabiting different niches in the aquifer. For example, 61% of the isolates obtained from ground-water samples exhibited flagellar motility—the ability to move in water propelled by a flagellum. In contrast, only 46% of the sediment-bound isolates showed this feature. This implies a greater adaptation toward motility in the free-living community relative to the sediment-bound community.

One of the most important uses of pure culture methods in subsurface microbiology has been to document community diversity. For example, Balkwill, Fredrickson, and Thomas (1989) reported isolating 626 physiologically distinct types of bacteria from deep sediments of the Atlantic coastal plain. Their work demonstrated that these sediments were inhabited by an extremely diverse flora—a conclusion that was somewhat surprising to many microbiologists and geologists. In a parallel study, Fredrickson et al. (1991) used similar techniques to compare microorganisms between two hydrologically separate aquifers and showed that the indigenous microbial communities were extremely diverse, and that the flora inhabiting the hydrologically separate aquifers were distinct from one another.

Although culture methods are suitable for addressing some questions, they also have some significant disadvantages. The most serious of these is the selectivity introduced by using culture media. Many of the bacteria present in a sample may not grow on the carbon sources available in a particular media, under the culture conditions offered (pH, temperature, etc.), or with the electron acceptor (often oxygen) present in the media. For example, Ghiorse and Balkwill (1983) reported that small colonies of bacteria recovered from sediment particles often would not grow when transferred to fresh media. Another limitation of culture methods concerns the variation in the "activity" of bacteria under culture versus in situ conditions. Growth of a particular strain on culture media in the laboratory does not necessarily mean that the strain was active in the aquifer.

Perhaps the largest problem with applying pure culture methods to subsurface microbiology is that it is difficult to relate the results directly to water-quality issues. While giving an indication of the metabolic potential of microorganisms, this approach gives virtually no information on the kinds of transformations that actually occur in a particular aquifer. For this reason, in situ techniques have developed for evaluating ecological issues in subsurface environments.

Geochemical Methods in Subsurface Microbial Ecology

The basic premise for using ground-water geochemistry to evaluate the ecology of subsurface environments is that particular microbial processes impact water chemistry in characteristic ways. Thus, if particular water-chemistry patterns are observed, it is theoretically possible to deduce a great deal of information concerning the microbial ecology of hydrologic systems. In practice, however, this is rarely straightforward and a number of pitfalls are commonly encountered.

Microorganisms sustain life functions by continuously cycling organic and inorganic substrates. A very simple flowchart of these processes can be written as follows:

$$\text{substrates} \xrightarrow{\text{(microbial metabolism)}} \text{intermediate products} \xrightarrow{\text{(microbial metabolism)}} \text{final products}$$

A number of substrates, intermediate products, and final products of microbial metabolism are chemical species that are water-soluble and therefore may accumulate in and be transported by ground water. Thus, analyses of ground water for these constituents can be used to deduce ambient microbial processes by tracking (1) the consumption of particular substrates, (2) the accumulation of intermediate products, and (3) the generation of certain final products.

The consumption of particular substrates is often the simplest to interpret. For example, consumption of dissolved oxygen in an aquifer closed to the atmosphere indicates aerobic respiration according to the generalized equation

$$\underset{\text{organic carbon}}{CH_2O} + \underset{\text{dissolved oxygen}}{O_2} \rightarrow CO_2 + H_2O \quad (8\text{-}1)$$

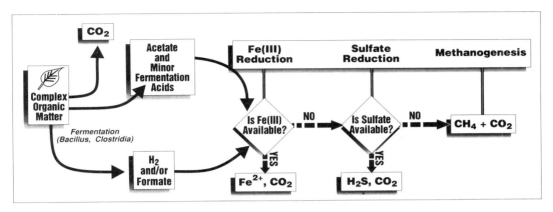

FIGURE 8-1. Diagram showing anaerobic microbial food chains that decompose complex organic matter with the accumulation of transient intermediate products as well as final products.

Similarly, consumption of dissolved nitrate or sulfate along aquifer flow lines can indicate nitrate-reducing or sulfate-reducing respiration, respectively.

The third possibility, generation of particular final products, may be fairly simple but often may present difficulties. For example, dissolved Fe(II), dissolved sulfides, and methane are products of organic matter oxidation coupled to Fe(III) reduction, sulfate reduction, and methanogenesis, respectively:

organic
carbon   ferric
(acetate)   hydroxide

$$CH_3COOH + 8Fe(OH)_3 + 16H^+ \to 8Fe(II) + 2CO_2 + 22H_2O \quad (8\text{-}2)$$

$$CH_3COOH + SO_4^{2-} + 2H^+ \to H_2S + 2CO_2 + 2H_2O \quad (8\text{-}3)$$

$$CH_3COOH \to CO_2 + CH_4 \quad (8\text{-}4)$$

Thus, the presence of these solutes may indicate the presence of these microbial processes. However, as Fe(II) may be produced by abiotic mechanisms such as dissolution of ferrous iron minerals, the simple presence of Fe(II) is not unequivocal evidence of microbial Fe(III) reduction. Alternatively, because both dissolved Fe(II) and sulfides tend to precipitate as mineral phases, the *absence* of these solutes in ground water does not rule out the possibility that these processes are ongoing in a particular hydrologic system.

Measuring concentrations of the intermediate products of microbial metabolism in ground water is perhaps the most subtle— and also perhaps the most potentially useful —indicator of microbial processes available. This method is based on the fact that anaerobic microbial metabolism proceeds via food chains in which a number of different types of microorganisms are involved in carbon cycling. This concept of microbial food chains and intermediate products of microbial metabolism is illustrated in Figure 8-1. Complex organic matter is first partially oxidized by fermentative bacteria with the production of simpler compounds such as organic acids (formate, acetate, propionate), carbon dioxide, and molecular hydrogen (H_2) as in the generalized reaction

glucose   formate
$$C_6H_{12}O_6 + 4H_2O \to 2CO_2 + 2HCOOH +$$

acetate   hydrogen
$$CH_3COOH + 6H_2 \quad (8\text{-}5)$$

These intermediate products are then completely oxidized by terminal electron-accepting respiratory processes such as Fe(III) reduction or sulfate reduction as illustrated by Eqs. 8-2 and 8-3.

The relationship between the fermentative microbial population and the respiratory population is synergistic. That is, both the fermenters and respirers work together to degrade complex carbon sources and each benefits by the activity of the other. The relationship between the terminal electron-accepting processes, on the other hand, is competitive with respect to carbon substrates. Fe(III) reducers, sulfate reducers, and methanogens compete for available formate, acetate, and hydrogen. Because Fe(III) reducers are more energetically efficient, they are able to outcompete sulfate reducers and methanogens if a suitable source of Fe(III) is available (Figure 8-1). Similarly, sulfate reducers are able to outcompete methanogens if sulfate is available (Figure 8-1). Methanogenesis, the least efficient of the anaerobic respiration processes, occurs only after Fe(III) and sulfate have been depleted in a system.

This pattern is important from a water-quality perspective. Because Fe(III) reduction is more efficient than the other terminal processes, steady-state concentrations of intermediate products such as hydrogen are lower in Fe(III)-reducing environments than in sulfate-reducing or methanogenic environments. Lovley and Goodwin (1988) tabulated hydrogen concentrations in a number of different aquatic sediments. These data (Figure 8-2) showed that hydrogen concentrations ranged from 5 to 10 nM in methanogenic sediments, 1 to 2 nM in sulfate-reducing sediments, and 0.2 to 0.4 nM in Fe(III)-reducing sediments. Nitrate-reducing sediments were characterized by hydrogen concentrations less than 0.1 nM.

Other intermediate products of microbial metabolism such as organic acids behave in a similar manner (Lovley and Klug 1986). Investigations of ground-water systems have shown similar ranges of hydrogen concen-

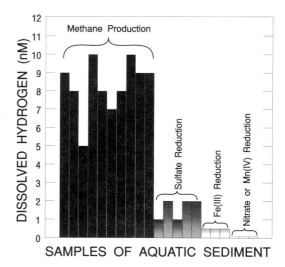

FIGURE 8-2. Measured steady-state concentrations of hydrogen for different terminal electron-accepting processes in aquatic sediments. (*From Lovley and Goodwin 1988.*)

trations associated with these microbial processes (Lovley and Goodwin 1988; Chapelle and Lovley 1990). Thus, concentrations of intermediate products, and particularly dissolved hydrogen, are potentially useful tools in deducing the presence and distribution of microbial processes in anaerobic ground-water systems.

In practice, no single water-quality parameter is unequivocally indicative of ongoing microbial processes. However, by measuring concentrations of potential substrates, intermediate products, and final products of microbial metabolism, it is often possible to deduce a great deal about the distribution of microbial processes—and thus the microbial ecology—of subsurface environments.

DISTRIBUTION OF TERMINAL ELECTRON-ACCEPTING PROCESSES (TEAPs) IN GROUND-WATER SYSTEMS

Bacteria are much more flexible in the kinds of oxidants (electron acceptors) they are capable of using than higher organisms. Most higher micro- and macroorganisms are limited to using molecular oxygen as a terminal electron acceptor. Bacteria, on the other hand, are capable of using oxygen, nitrate, Fe(III), sulfate, and carbon dioxide as electron acceptors. This flexibility allows bacteria to successfully inhabit stressful environments, such as ground-water systems, that higher organisms cannot tolerate.

The distribution of microbial terminal electron-accepting processes (TEAPs) in ground-water systems is a basic consideration in regional water-quality investigations. Many naturally occurring water-quality problems such as high concentrations of sulfides or dissolved iron are a direct result of microbial sulfate-reducing or Fe(III)-reducing respiration (Eqs. 8-2 and 8-3). Thus, knowing the geographical distribution of these processes can be of considerable practical value in locating new well fields, for example. Also, microbial degradation of many pollutants in ground-water systems are affected by the distribution of TEAPs. Using ground-water quality information to deduce ambient microbial processes is thus an important practical problem.

Many investigators have used water-quality constituents or properties as indicators of TEAPs in ground-water systems. Champ, Gulens, and Jackson (1979) suggested that platinum-electrode measurements of redox potential indicated the distribution of TEAPs in aquifer systems. However, because many microbial substrates, such as dissolved oxygen, do not react reversibly on platinum electrodes, this method has not gained wide acceptance (Lindberg and Runnells 1984). Baedecker and Back (1979) used concentrations of oxygen, nitrogen species, ferrous iron, and methane to determine the distribution of TEAPs at a municipal landfill. This method, which utilizes both the substrates and final products of microbial metabolism, has been used extensively in evaluations of microbial processes in contaminated ground-water systems (Baedecker et al. 1986, 1988) and in pristine aquifer systems (Edmunds 1973).

The first documented use of intermediate products of anaerobic metabolism as an indicator of the distribution of microbial processes was by Lovley and Goodwin (1988), who investigated the regional distribution of TEAPs in the Middendorf aquifer of South Carolina. A number of wells were sampled along a regional ground-water flow path, and concentrations of particular substrates (sulfate, nitrate), final products (ferrous iron, methane) and an intermediate product (hydrogen) were measured. The results (Figure 8-3) showed that denitrification occurred near the recharge area (which was located in an agricultural area, hence the relatively high nitrates) followed by Fe(III) reduction (as indicated by the accumulation of ferrous iron), sulfate reduction (as indicated by depletion of sulfate), and finally methanogenesis (as indicated by accumulation of methane). Concentrations of the intermediate product, hydrogen, increased systematically along the flow path from less than 0.1 nM concentration characteristic of nitrate reduction to 10 nM concentrations characteristic of methanogenesis.

The study described by Lovley and Goodwin (1988) demonstrates that by combining analyses of substrates, intermediate products, and final products of microbial metabolism, a consistent picture of the distribution of microbial processes in subsurface environments can be obtained. Furthermore, this method requires only analysis of ground water, which is relatively inexpensive to obtain, and circumvents the necessity of coring aquifer sediments for microbial analyses.

Distribution of TEAPs and the Zonation of High-Iron Ground Water in Coastal-Plain Aquifers

The TEAP method—using concentrations of substrates, intermediate products, and final products of microbial metabolism as indicators—provides a basis for deducing the geographic distribution of microbial processes in ground-water systems. The manner in which this technique is applied, however, depends largely on the water-quality issues

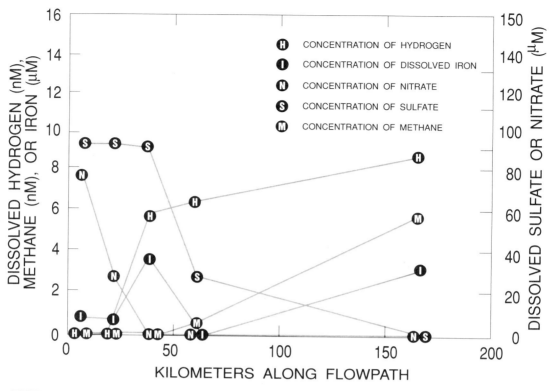

FIGURE 8-3. Concentrations of hydrogen (H), sulfate (S), nitrate (N), dissolved iron (I), and methane (M) in ground water along the flow path of the Middendorf aquifer, South Carolina. (*From Lovley and Goodwin 1988.*)

at hand. One example of how the application of this technique is problem-driven was described by Chapelle and Lovley (1992) in considering the origin of high-iron ground water in the Middendorf aquifer of South Carolina.

Coastal-plain aquifers are often characterized by discrete zones of relatively high-iron ground water. This phenomenon has been described in South Carolina (Speiran 1987), Maryland (Back and Barnes 1965), and New Jersey (Langmuir 1969). In spite of the wide occurrence of this problem and the enormous costs involved in iron-removal treatment, the fundamental causes were not known. It was generally agreed that high-iron concentrations reflected the reduction of Fe(III) oxyhydroxides to the more soluble ferrous state, but the mechanisms involved were not understood. For many years, it was assumed that these reduction processes occurred inorganically, with organic carbon compounds acting as the electron donor. However, experimental studies showed that naturally occurring organic compounds could *not* reduce Fe(III) oxyhydroxides abiotically (Lovley, Phillips, and Lonergan 1991). This finding, in turn, suggested that the zonation of high-iron ground water reflected microbial processes.

The Middendorf aquifer of South Carolina is an example of this zonation of high-iron water. This aquifer is characterized by a 40-km-wide zone in which dissolved ferrous iron concentrations commonly exceed 0.5 mg/L (Figure 8-4). To address this regional water-quality issue, Chapelle and Lovley (1992) used a combination of culture techniques and the TEAP method. Sediments from core holes were obtained and

Distribution of Terminal Electron-Accepting Processes (TEAPs) in Ground-Water Systems 189

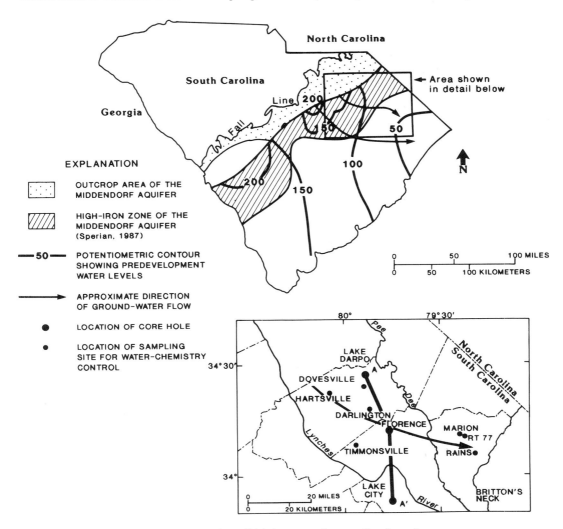

FIGURE 8-4. Map showing the distribution of high-iron ground water, directions of ground-water flow, and sampling locations for the Middendorf aquifer. (*From Chapelle and Lovley 1992.*)

microorganisms capable of oxidizing organic carbon coupled to the reduction of aquifer Fe(III) oxyhydroxides were recovered by using pure culture techniques (Lovley, Chapelle, and Phillips 1990). The simple presence of viable Fe(III)-reducing microorganisms, however, was not unequivocal evidence that they were active in situ.

Evidence supporting in situ microbial Fe(III)-reducing activity was obtained using the TEAP method. First, concentrations of potential substrates (electron acceptors such as oxygen and sulfate), final products (ferrous iron), and intermediate products (hydrogen) were measured along a flow path that was oriented normal to the observed high-iron zone (Figure 8-4). Near the recharge area of the aquifer, concentrations of oxygen were relatively high (Figure 8-5a) and ferrous iron concentrations low (Figure 8-5b). Once oxygen was consumed, however, ferrous iron concentrations increased.

Furthermore, in the zone of high-iron ground water, the isotopic composition of the dissolved inorganic carbon was identical to that of organic matter present in the aquifer (data not shown). This indicated that production of ferrous iron was related to the oxidation of organic matter.

The fact that ferrous iron was accumulating in solution was particularly important. This accumulation, in addition to causing the high-iron problem, indicated that sulfate reduction was not particularly active. Active sulfate reduction, with the production of sulfides, would tend to precipitate ferrous iron as sulfide minerals. Evidently, the *lack* of sulfate reduction was as important in the development of the high-iron problem as was the *presence* of Fe(III) reduction.

Sulfate-reducing bacteria coexisted with Fe(III)-reducing bacteria in this system. What could explain the relative activity of the Fe(III) reducers and the relative inactivity of the sulfate reducers? This question was addressed by considering concentrations of intermediate products of anaerobic microbial metabolism. Concentrations of hydrogen increased across the high iron zone from the 0.5-nM range characteristic of Fe(III) reduction to the 2-nM range characteristic of sulfate reduction (Figure 8-5d). Furthermore, concentrations of acetate (0.5–2.0 μM) and formate (0.5–6 μM) were much lower in the high-iron zone where sediments contained appreciable Fe(III) oxyhydroxides than in the downgradient Fe(III)-lacking sediments where acetate concentrations ranged from 1.5 to 3.5 μM and formate concentrations ranged from 5 to 15 μM.

Taken together, these data provide a consistent picture of the effects of microbial processes on the regional water quality of the Middendorf aquifer. In the upgradient part

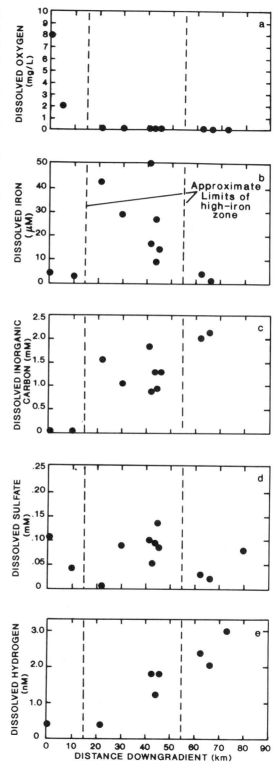

FIGURE 8-5. Concentrations of (a) dissolved oxygen, (b) dissolved iron, (c) dissolved inorganic carbon, (d) dissolved sulfate, and (e) dissolved hydrogen plotted versus distance along the flow path of the Middendorf aquifer. (*From Chapelle and Lovley 1992.*)

of the aquifer, aerobic metabolism dominates and neither Fe(III) or sulfate reduction are important processes. Once dissolved oxygen has been consumed, however, Fe(III) reduction becomes the predominant microbial process and excludes sulfate reduction by lowering hydrogen, acetate, and formate concentrations below levels required by sulfate-reducing microorganisms. This results in the delivery and accumulation of dissolved iron to the ground water. Downgradient, as the availability of Fe(III) oxyhydroxides becomes limiting, sulfate reduction becomes the predominant process. The production of sulfides causes ferrous iron to precipitate as ferrous sulfide (pyrite), and concentrations of iron decrease.

The observed zonation of high-iron ground water, an important regional water-quality problem, thus reflects the competition of microorganisms for available intermediate products of anaerobic metabolism. Stated another way, the microbial ecology of this system is a primary control on the regional water quality.

Distribution of Sulfate Reduction and Methanogenesis in a Fractured Carbonate Aquifer

The Floridan aquifer, which underlies much of the southeastern United States, is one of the most productive and economically important hydrologic systems in the world. The aquifer consists of indurated Tertiary limestones characterized by the presence of gypsum nodules and by locally extensive fracture patterns. In and near where this unit crops out at land surface, secondary dissolution features (karst topography), such as sinkholes and disappearing streams, are common. These features contribute to particular water-quality problems in certain places. One of the most interesting of these problems occurs near Valdosta, Georgia.

Near Valdosta, the channel of the Withlacoochee River intersects a series of sinkholes (Figure 8-6). These sinkholes capture much of the river's flow, which subsequently recharges the Floridan aquifer. Because the Withlacoochee is a typical South Georgia blackwater stream, characterized by high concentrations of dissolved and particulate organic carbon, a large amount of organic carbon (referred to as total organic carbon, TOC) is delivered to the Floridan aquifer at this location (Krause 1979).

The TOC delivered to the Floridan aquifer via the sinkholes draining the Withlacoochee River is a ready substrate for microbial metabolism. Microbial degradation of this TOC is indicated by its rapid decrease as ground water flows downgradient (Figure 8-6) and by the accumulation of final metabolites such as sulfide and methane. Ground water in the TOC plume is characterized by high concentrations of dissolved organic carbon (1.0–10.0 mg/L), dissolved sulfides (0.2–2.0 mg/L), and high concentrations of methane (0.2–2.0 mg/L) (James B. McConnell, U.S. Geological Survey, written communication, 1991).

The combination of this TOC, sulfide, and methane in the ground water creates several water-quality problems for Valdosta. Concentrations of sulfides ranging up to 2 mg/L, while not representing a health hazard, create a noxious odor which must be removed before people will consent to drink it. The methane, while not a direct health threat, is nevertheless a more serious problem. Chlorination of methane-bearing water, the usual water treatment for waterborne pathogens, results in the formation of trihalomethanes, a suspected carcinogen.

This raises the question as to what are the areal distributions of sulfate reduction and methanogenesis in the aquifer. This question is of scientific interest since zones of methanogenesis and sulfate reduction are not expected to coincide (Figure 8-1). Distinguishing between methanogenic and non-methanogenic zones based solely on the distribution of sulfides and methane, however, is not possible in this system. Both of these final products of microbial respiration are stable in solution and may be transported by advective ground-water flow. Thus, the

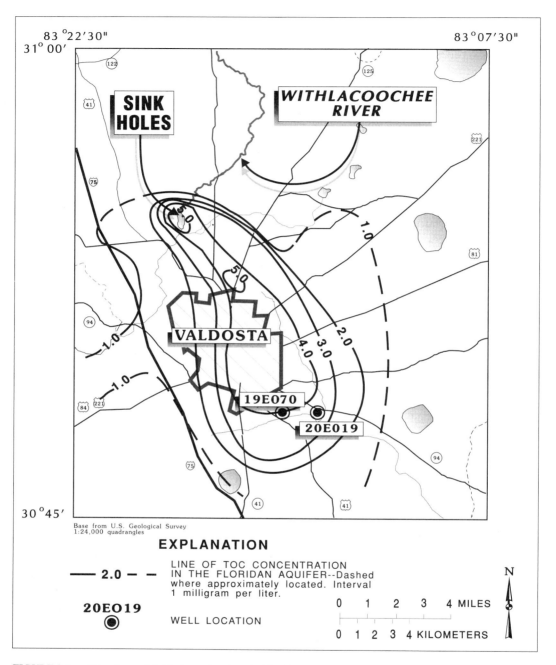

FIGURE 8-6. Locations of sinkholes recharging the Floridan aquifer, locations of selected wells, and distribution of high-TOC ground water near Valdosta, GA. (*From James B. McConnell, U.S. Geological Survey, written communication, 1991.*)

presence of methane at a particular well does not indicate that methanogenesis is the predominant process at that location.

Although the final products of microbial metabolism may be stable in solution, intermediate products are by definition *not* stable. The half-life of hydrogen in anaerobic sediments, for example, is estimated to be on the order of seconds. Because of the transient nature of intermediate products, and because their steady-state concentrations are different for methanogenesis and sulfate reduction, measurements of these compounds provide a much finer resolution of the distribution of TEAPs in this system.

An example of this is given in Figure 8-7. This figure shows concentrations of methane, sulfide, hydrogen, and acetate in water from two wells located near the tail of the poor-quality water plume (well locations shown in Figure 8-6). Because concentrations of methane and sulfide are similar for each well (Figure 8-7a), it is not possible to classify either well as being completed in a predominantly methanogenic or sulfate-reducing zone of the aquifer. However, concentrations of hydrogen and acetate in water from well 19E070 (Figure 8-7b) are clearly in the range characteristic of sulfate reduction. On the other hand, water from well 20E019 exhibits concentrations of hydrogen and acetate characteristic of methanogenic environments.

The steady-state concentrations of unstable intermediate products in water from these wells are consistent with the carbon-flow model shown in Figure 8-1. In well 19E070, which appears to tap a predominantly sulfate-reducing environment, concentrations of sulfate are sufficient (~3.3 mg/L) to support sulfate reduction. Water from well 20E019, on the other hand, has virtually no sulfate (~0.2 mg/L) which allows the initiation of methanogenesis. It may be concluded, therefore, that sulfate reduction is the predominant process in the vicinity of well 19E070, whereas methanogenesis is the predominant process in the vicinity of well 20E019.

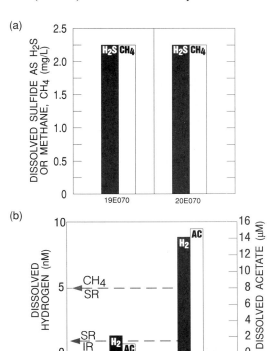

FIGURE 8-7. Concentrations of (a) methane and sulfide and (b) hydrogen and acetate in wells 19E070 and 20E019, Valdosta, GA. (SR = sulfate reduction; IR = iron reduction.)

Hydrologic systems are not simple environments either geochemically or biologically. There is no a priori reason, for example, that the entire thickness of the aquifer will be characterized by a single TEAP. Whereas the Middendorf aquifer exhibited a zonation of microbial processes definable on the scale of kilometers, the Floridan aquifer near Valdosta was more heterogeneous. Although water from some wells exhibited concentrations of intermediate products characteristic of either methanogenesis or sulfate reduction, water from other wells had intermediate-product concentrations that indicated a mixture of the two processes. Because most of the wells sampled were open boreholes as much as several hundred feet deep, these data reflect vertical zonation of TEAPs in this system. This vertical zonation apparently reflects vertical differences in the presence of gypsum nodules, which are the principal source of sulfate to the ground water. Where sulfate is abundant, sulfate reduction becomes the predominant TEAP. However, if sulfate is absent, methanogenesis becomes predominant (Figure 8-1).

The Floridan aquifer near Valdosta gives a clear example of how microbial processes (production of methane and sulfide) affect ground-water quality. Less obvious but equally important is the fact that the microbial processes depend upon external physical and chemical features of the hydrologic system. Microbial processes at this site are unusually active by the standards of many ground-water systems. This reflects the physical capture of the Withlacoochee River by sinkholes and subsequent delivery of large amounts of organic carbon to the aquifer. The complexity and variability of microbial processes in the Floridan aquifer contrasts markedly with the predictable progression exhibited by the Middendorf aquifer and reflects the variation in the abundance of gypsum in the limestone matrix. Thus, the nature of microbial processes in hydrologic systems, with their resulting water-quality effects, are strongly dependent upon external, nonbiological conditions. Knowing the nature of these nonbiological influences on microbial processes is an integral part of microbiologic investigations of regional water quality.

BIODEGRADATION OF ATRAZINE IN ALLUVIAL-AQUIFER SEDIMENTS

Microbial processes are important in the fate and transport of a variety of organic compounds in subsurface environments. Some organic compounds, such as petroleum hydrocarbons, can be directly utilized by microorganisms as substrates and are rapidly degraded in ground-water systems. Other compounds, such as chlorinated solvents, are not assimilated as easily. Nevertheless, degradation of these recalcitrant compounds via cometabolic processes—incidental reaction with microbial enzymes that does not benefit the microorganism—does occur and can substantially affect the fate of the contaminant in ground-water systems. Thus, understanding how microbial processes in ground-water systems interact with potential contaminants is an important issue.

There are two general approaches to evaluating biodegradation of particular contaminants in subsurface environments. One is to monitor the accumulation and transport of contaminants in ground water using a mass-balance approach. This has the advantage of directly observing contaminant behavior in situ, but has the disadvantage of not readily distinguishing abiotic attenuation processes (sorption, for instance) from biotransformations. Another approach is to study microbial transformations in the laboratory using sediments cored from the aquifer system under consideration. This has the advantage of separating biotic and abiotic transformations, but has the disadvantage of not representing actual in situ aquifer conditions. In practice, a combination of field and laboratory studies has proved to be the most useful approach in studying microbial degradation of organic contaminants in ground-water systems.

An example of this dual approach has recently been given by McMahon, Chapelle, and Jagucki (1992) in studying biotransformation of atrazine in an alluvial aquifer in Ohio. Atrazine is a widely used triazine herbicide that is readily biodegradable in soils. Despite this biodegradability, dissolved atrazine and its metabolites (~0.5 µg/L) were found in shallow ground water underlying field test plots to which it had been applied. This showed that atrazine degradation rates in the soils were slower than rates of atrazine transport to the water table. However, as ground water flowed downgradient of the test plots, concentrations of atrazine were observed to decrease below detectable levels, whereas concentrations of deethylatrazine (atrazine from which the ethyl side chain had been removed) persisted. The decrease in atrazine could reflect microbial degradation, abiotic sorption, or both. However, the presence and persistence of deethylatrazine suggested that microbial processes might be important. In order to evaluate the relative importance of microbial processes, it was necessary to determine experimentally if microorganisms capable of degrading atrazine were present in the aquifer sediments.

To address this issue, asceptically collected sediments were collected from the aquifer at several locations. The sediments were placed in aerobic serum vials and amended with either ring-labeled (i.e., atrazine in which the ring portion of the molecule was labeled with ^{14}C) or chain-labeled (i.e., atrazine in which the ethyl side chain was labeled with ^{14}C) atrazine. Additional vials were amended with glucose as well to see if microbial activity could be stimulated and atrazine degradation enhanced. Periodically, duplicate sets of vials were sacrificed and $^{14}CO_2$ collected in a KOH base trap. In addition, the amount of radiotracer present in cellular biomass, in the aqueous phase, and in the sediment was quantified. The accumulation of $^{14}CO_2$ would provide unequivocal evidence for oxidation of either the ring or chain portions of the atrazine molecule. Furthermore, the accumulation of radiotracer in cells would provide evidence that atrazine was being utilized for cellular growth.

There was no detectable $^{14}CO_2$ recovered from the ring-labeled incubations, suggesting that microbial degradation of the atrazine ring was small. However, there was detectable $^{14}CO_2$ recovered from the chain-labeled incubations relative to the killed controls. This showed that deethylation of atrazine in the ground-water system was microbially mediated and could account for the presence of deethylatrazine, and not atrazine, in down-gradient ground water. Furthermore, the glucose-amended treatments for the chain-labeled atrazine showed enhanced degradation (Figure 8-8). Very little of the radiotracer (<0.1%) was present as cellular

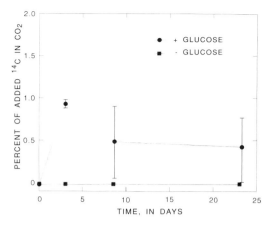

FIGURE 8-8. Oxidation of chain-labeled atrazine to CO_2, with and without glucose addition, in sediments cored from an alluvial aquifer in Ohio.

biomass in glucose-lacking or glucose-amended treatments. The lack of cellular uptake combined with the observed $^{14}CO_2$ production suggested that the degradation mechanism of atrazine was cometabolic.

The apparent rates of atrazine degradation were very slow, with about 1% of the total label being respired in one day. While this rate is relatively slow, it is significant considering the slow rate of ground-water flow. For example, the data of Figure 8-8 indicate a first-order constant for degradation of ethyl carbon of about 10^{-4} micromole $CO_2 L^{-1} d^{-1}$. This implies a degradation rate that is actually greater than required to completely deethylate atrazine between the upgradient and downgradient wells.

These experiments illustrate several points. First, while the disappearance of a compound along ground-water flow paths may suggest microbial degradation, confirmation of this requires laboratory investigation. Second, such laboratory investigations often show aspects of microbial degradation —such as the lack of cellular uptake of ethyl carbon—that are otherwise hidden. Third, and most important, it illustrates how the interpretation of laboratory degradation rates must be made in the context of the hydrologic system under study. While the degradation potential of atrazine in this system was small, it was nevertheless significant relative to ambient rates of ground-water flow.

CONCLUDING REMARKS

Microbial processes in subsurface environments are varied and depend upon the hydrologic and geochemical conditions that are present. Consequently, the effects of these processes on water quality are also varied. A number of microbiological tools, such as culture techniques and activity measurements, are available for investigating these processes. However, these methods are difficult and expensive to apply on a regional scale. Water-chemistry techniques, which track the utilization and production of microbiologically active solutes, are often more suitable.

Delineating the distribution of TEAPs provides a ready framework for describing microbial processes in ground-water systems. This method is based on the fact that microorganisms cycle substrates according to predictable patterns. Approaches to delineating the distribution of TEAPs include tracking (1) the consumption of particular substrates, (2) the production of intermediate metabolites at certain steady-state concentrations, and (3) generation of specific final products. Thus, uptake of dissolved oxygen or sulfate along aquifer flow paths indicates oxygen reduction and sulfate reduction, respectively. Generation of ferrous iron, sulfide, or methane indicate Fe(III) reduction, sulfate reduction, or methanogenesis, respectively. Finally, steady-state concentrations of metabolites such as hydrogen can be used to differentiate between Fe(III)-reducing ($H_2 \sim 0.5\,nM$), sulfate-reducing ($H_2 \sim 2-4\,nM$), or methanogenic ($H_2 \sim 10\,nM$) aquifers. Using all of these water-quality parameters in tandem, it is possible to deduce a great deal about microbial processes in a regional ground-water system.

Evaluating the biodegradation of organic compounds in ground-water systems requires a combination of (1) field sampling to track the transport and attenuation of compounds under in situ conditions, and (2) laboratory-based studies to determine mechanisms and rates of microbial degradation. In the absence of either of these components, the microbial nature of the observed degradation or its hydrologic significance will remain uncertain.

References

Back, William, and Ivan Barnes. 1965. *Relation of Electrochemical Potentials and Iron Content to Ground-Water Flow Patterns*. Washington, DC: U.S. Geological Survey Professional Paper 498-C.

Baedecker, M. J., and William Back. 1979. Hydrogeological processes and chemical reactions at a landfill. *Ground Water* 17(5):429–37.

Baedecker, M. J., B. J. Franks, D. F. Goerlitz, and J. A. Hopple. 1986. Geochemistry of a shallow aquifer contaminated with creosote products. In *U.S. Geological Survey Program on Toxic Waste—Ground-Water Contamination: Proceedings of the Second Technical Meeting, Cape Cod, Massachusetts, October 21–25, 1985*, ed. S. E. Ragone, pp. 17–20. Reston, VA: U.S. Geological Survey Open-File Report 86–481.

Baedecker, M. J., D. I. Siegel, P. C. Bennett, and I. M. Cozzarelli, 1988. The fate and effects of crude oil in a shallow aquifer. I: The distribution of chemical species and geochemical facies. In *U.S. Geological Survey Toxic Substances Hydrology Program—Proceedings of the Technical Meeting, Phoenix, Arizona, September 26–30, 1988*, eds. G. E. Mallard and S. E. Ragone, pp. 13–20. Reston, VA: U.S. Geological Survey Water-Resources Investigations Report 88-4220.

Balkwill, D. L., J. K. Fredrickson, and J. M. Thomas. 1989. Vertical and horizontal variations in the physiological diversity of the aerobic chemoheterotrophic bacterial microflora in deep southeast coastal plain sediments. *Applied and Environmental Microbiology* 55(5):1058–65.

Bastin, E. S. 1926. The presence of sulphate-reducing bacteria in oil-field waters. *Science* 63:21–24.

Cederstrom, D. J. 1946. Genesis of groundwaters in the coastal plain of Virginia. *Economic Geology* 41(3):218–45.

Champ, D. R., J. Gulens, and R. E. Jackson. 1979. Oxidation-reduction sequences in ground-water flow systems. *Canadian Journal of Earth Sciences* 16(1):12–23.

Chapelle, F. H., and D. R. Lovley. 1990. Rates of bacterial metabolism in deep coastal-plain aquifers. *Applied and Environmental Microbiology* 56:1865–74.

Chapelle, F. H., and D. R. Lovley. 1992. Competitive exclusion of sulfate-reduction by Fe(III)-reducing bacteria: A mechanism for producing discrete zones of high-iron ground water. *Ground Water* 30(1):29–36.

Chapelle, F. H., J. S. Zelibor, D. J. Grimes, and L. L. Knobel. 1987. Bacteria in deep coastal plain sediments of Maryland: A possible source of CO_2 to ground water. *Water Resources Research* 23(8):1625–32.

Dockins, W. S., G. J. Olson, G. A. McFeters, and S. C. Turbak. 1980. Dissimilatory bacterial sulfate reduction in Montana groundwaters. *Geomicrobiology Journal* 2:83–97.

Dunlap, W. J., and J. F. McNabb. 1973. U.S. Environmental Protection Agency Report No. EPA-660/2-73-014. Ada, OK: R. S. Kerr Environmental Research Laboratory.

Dunlap, W. J., J. F. McNabb, M. R. Scalf, and R. L. Cosby. 1977. *Sampling for Organic Chemicals and Microorganisms in the Subsurface*. EPA-600/2-77-176, Ada, OK: R. S. Kerr Environmental Research Laboratory.

Edmunds, W. M. 1973. Trace element variations across an oxidation-reduction barrier in a limestone aquifer. In *Proceedings of the Symposium on Hydrogeochemistry and Biochemistry, Tokyo*, Vol. 1, pp. 500–26. New York: Clarke.

Ehrlich, G. G., D. F. Goerlitz, E. M. Godsy, and M. F. Hult. 1982. Degradation of phenolic contaminants in ground water by anaerobic bacteria, St. Louis Park, Minnesota. *Ground Water* 20(6):703–10.

Fredrickson, J. F., D. L. Balkwill, J. M. Zachara, Shu-Mei W. Li, F. J. Brockman, and M. A. Simmons. 1991. Physiological diversity and distributions of heterotrophic bacteria in deep cretaceous sediments of the Atlantic coastal plain. *Applied and Environmental Microbiology* 57(2):402–11.

Ghiorse, W. C., and D. L. Balkwill. 1983. Enumeration and characterization of bacteria indigenous to subsurface environments. *Developments in Industrial Microbiology* 24:213–24.

Ghiorse, W. C., and J. L. Wilson. 1988. Microbial ecology of the terrestrial subsurface. *Advances in Applied Microbiology* 33:107–72.

Godsy, E. M. 1980. Isolation of *Methanobacterium bryantii* from a deep aquifer by using a novel broth-antibiotic disk method. *Applied and Environmental Microbiology* 39:1074–75.

Godsy, E. M., and G. G. Ehrlich. 1978. Reconnaissance for microbial activity in the Magothy aquifer, Bay Park, New York, four years after artificial recharge. *U.S. Geological Survey Journal of Research* 6(6):829–36.

Hirsch, P., and E. Rades-Rohkohl. 1983. Microbial diversity in a groundwater aquifer in northern Germany. *Developments in Industrial Microbiology* 24:183–200.

Krause, R. E. 1979. *Geohydrology of Brooks, Lowndes, and Western Echols Counties, Georgia*. U.S. Geological Survey Water-Resources Investigations Open-File Report 78–117, Doraville, GA.

Kolbel-Boelke, J., E. Anders, and A. Nehrkorn. 1988. Microbial communities in the saturated groundwater environment. II: Diversity of bacterial communities in a Pleistocene sand aquifer and their in vitro activities. *Microbial Ecology* 16:31–48.

Langmuir, Donald. 1969. *Geochemistry of Iron in a Coastal Plain Aquifer of the Camden, New Jersey Area*. Washington, DC: U.S. Geological Survey Professional Paper 650-C, pp. 224–35.

Leenheer, J. A., R. L. Malcolm, and W. R. White. 1976. *Physical, Chemical, and Biological Aspects of Subsurface Organic Waste Injection near Wilmington, North Carolina*. Denver, CO: U.S. Geological Survey Professional Paper 987.

Lindberg, R. D., and D. D. Runnells. 1984. Ground-water redox reactions: An analysis of equilibrium state applied to Eh measurements and geochemical modeling. *Science* 225:925–27.

Lovley, D. R., and Steve Goodwin. 1988. Hydrogen concentrations as an indicator of the predominant terminal electron-accepting reactions in aquatic sediments. *Geochimica et Cosmochimica Acta* 52:2993–3003.

Lovley, D. R., and M. J. Klug. 1986. Model for the distribution of methane production and sulfate reduction in freshwater sediments. *Geochimica et Cosmochimica Acta* 50:11–18.

Lovley, D. R., F. H. Chapelle, and E. J. P. Phillips. 1990. Recovery of Fe(III)-reducing bacteria from deeply buried sediments of the Atlantic coastal plain. *Geology* 18:954–57.

Lovley, D. R., E. J. P. Phillips, and D. J. Lonergan. 1991. Enzymatic versus nonenzymatic mechanisms for Fe(III) reduction in aquatic sediments. *Environmental Science and Technology* 26(6):1062–67.

McMahon, P. B., F. H. Chapelle, and M. L. Jagucki. 1992. Atrazine mineralization potential of alluvial-aquifer sediments under aerobic conditions. *Environmental Science and Technology* 26:1556–59.

Rogers, G. S. 1917. *Chemical Relations of the Oil-Field Waters in San Joaquin Valley, California*. Washington DC: U.S. Geological Survey Bulletin 653, pp. 93–99.

Speiran, G. K. 1987. Relation of aqueous geochemistry to sedimentary depositional environments. In *American Water Resources Association Monograph Series No. 9*, eds. J. Vecchioli and A. I. Johnson, pp. 79–96.

Wilson, J. T., J. F. McNabb, D. L. Balkwill, and W. C. Ghiorse. 1983. Enumeration and characterization of bacteria indigenous to a shallow water-table aquifer. *Ground Water* 21:134–42.

White, D. C., J. F. Fredrickson, M. H. Gehron, G. A. Smith, and R. F. Martz. 1983. The groundwater aquifer microbiota: Biomass, community structure, and nutritional status. *Developments in Industrial Microbiology* 24:189–99.

Zobell, C. E. 1947. Microbial transformation of molecular hydrogen in marine sediments, with particular reference to petroleum. *Bulletin of the American Association of Petroleum Geologists* 31:1709–51.

Zobell, C. E. 1952. Bacterial life at the bottom of the Philippine Trench. *Science* 63:507–08.

9

Geochemical Models

D. L. Parkhurst and L. N. Plummer

INTRODUCTION

Geochemical models are tools that aid in the interpretation of geochemical reactions. Many different geochemical models have been applied to a wide range of surface- and ground-water problems. This chapter focuses on the application of geochemical models to regional ground-water systems. Here, the models can be used for a variety of purposes, including determination of the prevailing geochemical reactions, quantification of the extent to which these reactions occur, prediction of the fate of inorganic contaminants, and estimation of the direction and rates of ground-water flow. Plummer (1984) divided geochemical modeling into two general approaches: (1) inverse modeling, which uses observed ground-water compositions to deduce geochemical reactions; and (2) forward modeling, which uses hypothesized geochemical reactions to predict ground-water compositions. Inverse modeling produces quantitative geochemical reactions that describe the chemical evolution in a ground-water system, whereas forward modeling begins at some starting composition and simulates the chemical evolution of ground water in response to sets of specified reactions. In regional ground-water studies where chemical and isotopic data are available, inverse modeling is the most efficient approach because its results always reproduce the chemistry of the available samples. The forward approach is appropriate for systems or chemical constituents as yet unstudied, where the goal is to predict the chemical composition of ground water in the absence of chemical data. The forward approach also is uniquely capable of simulating advection, dispersion, and spatial and temporal distributions of minerals and ground-water composition. In this chapter, the techniques involved in inverse and forward modeling are presented, and the utility and deficiencies of each approach are discussed. Examples of each type of modeling are presented to illustrate the interpretation of model results.

INVERSE MODELING

The purpose of inverse modeling is to determine net chemical reactions that quantitatively account for the chemical and isotopic composition of ground-water samples and are consistent with known thermodynamic constraints and mineralogic observations. The pioneering work in this approach is by Garrels and Mackenzie (1967), who quantified the amounts of specific silicate-weathering reactions that were necessary to account for the differences in chemical concentrations between two springs in the Sierra Nevadas. For ground water, the modeling typically is applied to water samples from two wells that are assumed to be on a single flow path in order to determine the net chemical reaction that has occurred between the two wells. The inverse modeling described here is a combination of speciation modeling and mass-balance modeling, with special reference to the programs WATEQF (Plummer, Jones, and Truesdell 1976) and NETPATH (Plummer, Prestemon, and Parkhurst 1991). Speciation modeling, isotopic data, and petrographic observations provide constraints on whether plausible reactant phases are dissolving, are precipitating, or are inert. Mass-balance modeling produces quantitative geochemical reactions that reproduce the compositions of the samples and are consistent with any constraints on the reactant phases.

Speciation Modeling

Speciation models calculate thermodynamic properties of aqueous solutions, including the molalities and activities of aqueous species and saturation indices of minerals. Two main approaches are currently used to calculate these properties: the ion-association approach and the specific-interaction approach. The ion-association approach attempts to account for much of the non-ideality of aqueous solutions by ion association, that is, the formation of complexes from the hydrated individual ions in solution. An example of a complexation reaction is given in Eq. 9-1.

$$Ca^{2+} + SO_4^{2-} \rightarrow CaSO_4^0 \quad (9\text{-}1)$$

The complexes have mass-action equations that relate the activities of the aqueous species; the mass-action equation for this example is

$$K_{CaSO_4^0} = \frac{[CaSO_4^0]}{[Ca^{2+}][SO_4^{2-}]} \quad (9\text{-}2)$$

where $K_{CaSO_4^0}$ is the stability constant, which is a function of temperature and pressure, and brackets indicate activity. Activities are related to molality (moles per kilogram of water) by

$$[i] = \gamma_i(i) \quad (9\text{-}3)$$

where (i) is the molality of species i and γ_i is the activity coefficient. Activity coefficients of individual ions and complexes are usually estimated using extensions of the Debye-Hückel theory.

The advantages of the ion-association approach are that it is relatively easy to modify the model to include new species and new elements. Activity coefficients for species can be estimated using a form of the Debye-Hückel equation that only requires the electrical charge of the species to be known. Extensive literature exists that identifies complexation reactions and estimates the stability constants. The major deficiencies of the models are (1) the original Debye-Hückel theory only applies at very low ionic strengths (less than 0.01), and extensions of the theory are only applicable to selected solution compositions (generally sodium chloride solutions) of moderate ionic strengths approaching seawater concentrations (ionic strength of 0.7); (2) in spite of their wide use, insufficient work has been done to ensure that the models are adequately reproducing experimental solution properties such as mineral solubilities and mean-activity coefficients; and (3) whereas attractive forces among aqueous species can be accounted for by ion association, there is no mechanism to account for repulsive forces in mixed electrolyte solutions. Some of the better known ion-association models are implemented in the codes MINTEQA2 (Allison, Brown, and Novo-Gradac 1990), GEOCHEM (Sposito and Mattigod 1980), EQ3 (Wolery 1979), WATEQF (Plummer,

Jones, and Truesdell 1976), WATEQ4F (Ball and Nordstrom 1991), and PHREEQE (Parkhurst, Thorstenson, and Plummer 1980). Characteristics of several of these and other speciation codes have been summarized by Mangold and Tsang (1991).

In contrast to the simple Debye-Hückel expressions for individual ion-activity coefficients in ion-association models, the specific-interaction approach uses a complex expansion of the Debye-Hückel expression to calculate mean-activity coefficients for chemical reactions. The expanded expression includes coefficients that account for specific interactions between every pair and triplet combination of aqueous species (see Pitzer 1979; and Harvie and Weare 1980). These mean-activity coefficients account for most of the nonideality of aqueous solutions; therefore, very few complexation reactions are included in the specific-interaction models. The advantage of this approach is that it is theoretically applicable to all mixed electrolyte solutions of any ionic strength. The specific-interaction parameters have been selected carefully to produce a consistent set of parameters and stability constants that reproduce the best available experimental data (Harvie and Weare 1980). The disadvantage of the approach is that it is currently applicable largely to elements that form strong electrolytes. In geologic contexts, the most notable omissions of elements in current implementations are aluminum and silicon as well as most trace elements. Currently, interaction parameters are available for at most a single valence state of redox elements, which means that these models are not applicable to many reactions that involve redox transitions. The specific-interaction approach has been implemented in the codes PHRQPITZ (Plummer et al. 1988) and EQ3/6 (Wolery et al. 1990), which can be used for speciation calculations.

The primary purpose of speciation modeling is to calculate mineral saturation indices, which are indicators of the saturation state of a mineral with respect to a given water composition. The saturation index is defined as

$$\text{SI} = \log_{10}\left(\frac{\text{IAP}}{K_{\text{sp}}}\right) \quad (9\text{-}4)$$

where SI is the saturation index, IAP is the ion-activity product as defined by a mass-action equation, and K_{sp} is the solubility-product constant for the mineral (also see Chapter 6). If the saturation index is less than zero, the mineral is undersaturated with respect to the solution and the mineral might dissolve. The saturation index only indicates what should happen thermodynamically; it does not indicate the rate at which the process will proceed. In many cases, a mineral having a saturation index less than zero may dissolve very slowly or not at all, depending on the kinetics of the reaction. A more definitive conclusion can be drawn if the mineral saturation index is less than zero; that is, the mineral cannot precipitate from the ground water. This conclusion is qualified only by the accuracy of the chemical analysis and the aqueous model, not by rates of reaction. By similar reasoning, if the saturation index is greater than zero, the mineral might precipitate but cannot dissolve. If the saturation index is close to zero, the mineral may not be reacting at all or may be reacting reversibly, in which case the mineral could be dissolving or precipitating.

Example of Speciation Modeling
An example of the interpretation of saturation indices is taken from a study of ground-water samples from the Madison aquifer in parts of Montana, Wyoming, and South Dakota (Plummer et al. 1990; Busby, Lee, and Hanshaw 1983). The Madison aquifer primarily is a limestone aquifer; calcite, dolomite, and anhydrite are the major minerals. Saturation indices for selected ground-water samples from this aquifer were calculated by using the speciation code WATEQF (Plummer, Jones, and Truesdell 1976). The saturation indices for calcite

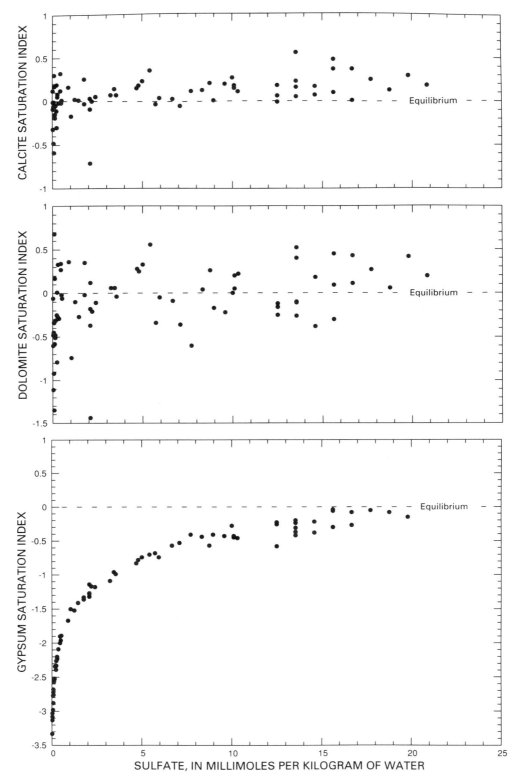

FIGURE 9-1. Comparison of calcite, dolomite, and gypsum saturation indices as a function of total dissolved-sulfate concentration for wells and springs in the Madison aquifer. (*From Plummer et al. 1990.*)

(Figure 9-1) are slightly greater than 0.0, which Plummer et al. (1990) attributed to supersaturation due to dedolomitization, pressure effects on the activities of species, variations in calcite stability caused by incorporation of sulfate into the mineral structure, or possibly CO_2 outgassing during the measurement of pH. Except in waters with very small sulfate concentrations, dolomite appears to be near saturation, if an uncertainty of ±0.5 units is accepted. Plummer et al. (1990) used saturation indices for gypsum rather than anhydrite because the stability constant for gypsum is better known as a function of temperature. However, anhydrite should be slightly more stable than gypsum at the temperatures and salinities of ground water in this aquifer, so they conclude that waters that have more than about 14 mmol/kg H_2O sulfate are in equilibrium with anhydrite. Thus, the saturation indices indicate that anhydrite is dissolving, dolomite is reacting reversibly, and calcite appears to be precipitating in the Madison aquifer. This is consistent with the process of dedolomitization, in which anhydrite dissolves and causes the dissolution of dolomite and the precipitation of calcite.

Plummer et al. (1990) also noted that the saturation indices for celestite ($SrSO_4$) are very similar to those for gypsum in that they indicate saturation at large sulfate concentrations. Strontianite ($SrCO_3$) saturation indices also are found to be nearly constant, though at a saturation index of about −1.0. Plummer et al. (1990) showed that the constant, negative saturation indices calculated for strontianite are the result of equilibria with anhydrite, calcite, and celestite and do not indicate that strontianite is reacting in the system. As long as the waters remain saturated with these three phases, the waters could never reach saturation with strontianite, even if strontianite were present in the aquifer matrix.

In the context of inverse modeling, speciation modeling is useful because the saturation indices provide thermodynamic information about which minerals could or could not be dissolving or precipitating in a ground-water system. Thus, speciation modeling can be used to constrain the set of geochemical reactions that are possible in a regional ground-water system. However, the limitations in the interpretation of saturation indices need to be considered. Sources of uncertainty include the analytical data for the ground-water sample, the completeness and thermodynamic data of the aqueous model, and the solubility-product constants for the minerals.

Uncertainties in Speciation Models

Uncertainty in the analytical data can be minimized by careful analytical procedures (also see Chapter 6). It is possible to analyze most constituents of ground water with an accuracy of 2–5%. Similar accuracy for field measurement of alkalinity also is possible in many situations. Uncertainties of this magnitude have relatively small effects on the saturation index because saturation indices are calculated from the logs of these quantities. However, the saturation indices of some minerals (carbonate, hydroxide, and aluminous phases, among others) are strongly dependent on the value of pH; thus, uncertainties in pH can cause large uncertainties in the saturation indices for these minerals.

Uncertainties in redox parameters also can produce very large uncertainties in saturation indices. Conceptually, speciation models must have a value of the redox potential, pE, whenever it is necessary to calculate the distribution of aqueous species from an analysis of the total concentration of a redox-active element; that is, when the concentrations of individual valence states of the element have not been measured. The pE may be derived from a platinum electrode measurement or from measurements of two valence states of a redox active element (a redox couple). These methods assume that redox equilibrium obtains between the solution and the electrode or between the two valence states of the redox couple, and, in addition, that this measured pE applies to other redox elements. However, redox reactions are often biologically mediated, kinetically slow, and element specific; therefore, in many aquifer systems, redox reactions are not in equilibrium. Consequently, estimates of pE based on equilibrium assumptions can be very poor, which, in turn, can cause major uncertainties in saturation-index calculations. Consider the following chemical reaction and the associated saturation-index equation:

$$FeS + 4H_2O \rightarrow Fe^{2+} + SO_4^{2-} + 8H^+ + 8e^- \quad (9\text{-}5)$$

$$SI_{FeS} = \log[Fe^{2+}] + \log[SO_4^{2-}] - 8\,pH - 8\,pE - 4\log[H_2O] - \log K_{FeS} \quad (9\text{-}6)$$

If analytical data are available only for sulfate and pE (as measured with a platinum electrode, for example), then an equation similar to Eq. 9-6 must be used to calculate the saturation index for the phase FeS. Note that any errors in the pH and pE are multiplied by a factor of 8.

In general, saturation indices of minerals containing redox elements will be most meaningful if the aqueous solution has been analyzed for the same valence states of the elements that are found in the mineral. In this example, if ferrous iron and total sulfide have been measured, then a speciation model can be used to calculate the activities of Fe^{2+} and HS^-. This calculation does not rely on a value of pE because it is not necessary to distribute iron or sulfur among different redox states. The saturation index for FeS can be calculated using the following chemical reaction and the associated saturation-index equation:

$$FeS + H^+ \rightarrow Fe^{2+} + HS^- \quad (9\text{-}7)$$

$$SI_{FeS} = \log[Fe^{2+}] + \log[HS^-] + pH - \log K^*_{FeS} \quad (9\text{-}8)$$

where the asterisk indicates that the solubility-product constant is for the reaction in Eq. 9-7 (not Eq. 9-5). Equation 9-8 does not depend on pE, and errors in pH are only multiplied by 1. Thus, Eq. 9-8 will provide more reliable estimates of the saturation index of FeS than will Eq. 9-6, but it is applicable only when the appropriate redox states of iron and sulfur have been measured.

Uncertainties in ion-association models are difficult to assess because there has been very little systematic application of the models to determine how well they reproduce experimental quantities, such as mean-activity coefficient or mineral solubility, over a wide range of solution compositions. The species included in the ion-association models and their stability constants usually are drawn from literature values and often are based on experiments that consider only a small range of solution composition. Little attempt has been made to maintain overall thermodynamic consistency. Specific-interaction models have been developed with much more concern about the accurate representation of experimental thermodynamic quantities and, therefore, are much more reliable for the chemical systems for which they have been developed, especially at higher ionic strengths (>0.7).

Stability constants for minerals may be uncertain because of lack of experimental data or because the minerals in a groundwater system have not been characterized. Thermodynamic data at low temperatures (~25°C) often are difficult to obtain because of the slow rates of equilibration. Another problem arises because minerals can have a range of stability due to differences in composition. For example, the stability of ferric oxyhydroxides can range over several orders of magnitude depending on the crystal structure and the degree of hydration (Langmuir 1971).

Compounding errors cause the uncertainty in the saturation index to be dependent on the stoichiometric formula of the mineral. A larger number of ions in the mass-action equation for a mineral causes larger uncertainties in the saturation index.

For example, the mass-action equation for calcite contains only two ions—calcium and carbonate—whereas the dolomite expression contains four ions—calcium, magnesium, and two carbonates. The larger number of ions leads to a larger variance in the saturation index of dolomite than calcite.

Thus, the interpretation of saturation indices requires knowledge of the aqueous model, the mineral stability and stoichiometry, and the kinetics of reactions. For low-temperature calculations, the number of minerals for which meaningful saturation indices can be calculated is relatively small. Saturation indices usually are meaningful for carbonate, sulfate, and chloride minerals because these minerals tend to react rapidly. Saturation indices for sulfides are useful if the environment is reducing and sulfide is measured. Under conditions where ferric iron can be measured, the saturation indices of ferric oxyhydroxides are meaningful. Saturation indices also are meaningful for aluminum oxides and hydroxides (provided accurate dissolved aluminum data are available), pure silica phases, and some minerals that contain trace elements. Except for kaolinite, saturation indices for most clays, feldspars, and other aluminosilicate minerals are at best qualitative because of uncertainties in the thermodynamic data and dissolved aluminum measurements.

Mass-Balance Modeling

Mass-balance modeling attempts to determine the nature and extent of geochemical reactions that are occurring in a groundwater system by identifying the minerals that are reacting and determining the amounts of these minerals that dissolve or precipitate. The basic conceptual model for this kind of mass-balance modeling is that two waters, for which chemical data exist, are on a single flow line, and, therefore, the downgradient (referred to as final) water composition is derived from the upgradient (referred to as initial) water composition by the occurrence of geochemical reactions. A variation of the conceptual model that is also tractable by the same approach is that two initial waters of different compositions mix as geochemical reactions occur to produce the final water composition.

The mass-balance modeling discussed here is limited to a special case of the general inverse problem that assumes steady state with respect to flow and chemical composition in the aquifer. If the water analyses have been affected by transient chemical conditions or are not within the same flow system, the techniques of mass-balance modeling may produce erroneous results. Thus, care must be taken in applying this type of mass-balance modeling to systems with changing flow conditions due to pumpage or with changing chemical compositions due to human activity. A common problem with ground-water samples is that they are collected from wells that are open to a number of intervals or to a single long interval; thus, the samples could be an aggregate of several different water compositions. Interpretation of the mass-balance results for such ground-water mixtures often requires more data than are generally available.

If the assumptions of the conceptual model are applicable, the formulation of the mass-balance problem is very simple. A linear equation can be developed for each element for which analytical data exist. For the problem with one initial and one final water composition, the equation states that the difference in the concentration of the element between the two solutions is due to the dissolution or precipitation of a set of reactant and product phases (Plummer and Back 1980; Parkhurst, Plummer, and Thorstenson 1982):

$$\Delta(i^{\text{total}}) = \sum_{p=1}^{P} b_{p,i} \alpha_p \qquad (9\text{-}9)$$

where $\Delta(i^{\text{total}})$ is the change in the total molality of element i, final concentration minus initial concentration; $b_{p,i}$ is the stoichiometric coefficient of element i in phase p; α_p is the mass transfer of phase p, in moles per kilogram water; and P is the total number of phases. If the problem entails mixing of two initial waters to form the final water, Eq. 9-10 is applicable:

$$(i_{\text{fin}}^{\text{total}}) = \alpha_{\text{init}_1}(i_{\text{init}_1}^{\text{total}}) + \alpha_{\text{init}_2}(i_{\text{init}_2}^{\text{total}})$$
$$+ \sum_{p=1}^{P} b_{p,i}\alpha_p \qquad (9\text{-}10)$$

where α_{init} refers to the fraction of each initial water that mixes to form the final water, and $(i_{\text{init}}^{\text{total}})$ refers to the total concentration of element i in the initial solutions. For this case, an additional equation is required to force the two mixing fractions to sum to 1.0; that is, $1.0 = \alpha_{\text{init}_1} + \alpha_{\text{init}_2}$.

Notice that the equations are written in terms of the total concentration of an element. The element carbon often requires special consideration because the total carbon frequently is not measured. However, a speciation model can be used to calculate the total carbon from analytical data including alkalinity and pH. The total concentration of an element includes all redox states of that element. For example, if a chemical analysis contains measurements of both sulfide and sulfate, these two quantities would be summed to calculate the total sulfur in solution and this total would be used in the mass-balance calculations.

The presence of different redox states of an element in solution or in the reactant phases indicates that redox processes may be occurring. In this case, an additional equation is required to ensure that electrons are conserved in the reactions calculated by mass-balance modeling (Parkhurst, Plummer, and Thorstenson 1982). The additional equation involves the calculation of a redox state for each solution. The redox state of a solution is defined as

$$RS = \sum_{j=1}^{J} v_j(j) \qquad (9\text{-}11)$$

where RS is the redox state of a solution, J is the total number of aqueous species, and v_j is an operational valence assigned to each aqueous species j. An operational valence u_p is also assigned to each mineral phase. One convention defines the operational valence of a species (or mineral) as the charge on the species minus the number of hydrogen atoms in the species plus two times the number of oxygen atoms in the species. For example, the operational valence of the sulfate ion SO_4^{2-} is $+6$, the bisulfide ion HS^- is -2, and the mineral $Fe(OH)_3$ is $+3$. [Parkhurst, Plummer, and Thorstenson (1982) and Plummer, Parkhurst, and Thorstenson (1983) used a similar convention except nonredox-active elements were assigned operational valences of zero.] The redox mass-balance equation for the nonmixing problem is as follows:

$$\Delta RS = \sum_{p=1}^{P} u_p \alpha_p \qquad (9\text{-}12)$$

Equations 9-9 and 9-12, or their counterparts for the mixing case, are the basic equations for mass-balance modeling. The left-hand sides of the equations are known quantities calculated from the analytical data. The right-hand sides are linear in the unknown quantities α_p. The coefficients of the unknowns are derived from the stoichiometry of the phases assumed to be reacting. Here lies the major difficulty in applying mass-balance modeling. What are the phases that are reacting? A list of all possible reactant phases and their stoichiometries must be developed. Hydrologic and isotopic evidence can determine whether the ground-water system is open to the atmosphere, in which case carbon dioxide and oxygen gas could be reactants. From geologic and mineralogic evidence, it is possible to ascertain many of the minerals that are present in an aquifer. X-ray diffraction of rock samples and optical and scanning electron microscopy of thin sections can lead to more precise estimates of the minerals present and their abundances. In addition, textural evidence from microscopic investigations can provide evidence that minerals are dissolving or precipitating. Similar evidence is derived from the saturation-index data. Some minerals, dolomite for example, are known not to form in freshwater aquifers. Combining information from all of the available sources results in a list of phases that could be reacting and constraints that some phases must only dissolve and others must only precipitate.

The stoichiometry of many minerals may be relatively well defined; however, the stoichiometry of some phases, especially phases that form solid solutions or contain trace elements, may be difficult to estimate. Uncertainty in the concentrations of trace elements in minerals and sorbed on mineral surfaces makes the mass-balance approach very difficult to apply to trace-element data. Therefore, the mass-balance approach is best suited for the predominant elements in aqueous solution or mineral compositions. For these elements, chemical, X-ray diffraction, and optical or scanning electron microscopic techniques can be used to produce quantitative or semiquantitative compositions for minerals. Frequently, compositions of minerals in the aquifer under study are estimated by analogy to other aquifers where the mineralogy is defined more accurately. Sensitivity analysis also can be used to investigate the effects of varying the mineral composition on the results of mass-balance modeling.

After the analytical data for water analyses (assumed to be on the same flow line) and the list of possible reactants with their stoichiometries are available, it is possible to proceed with the mass-balance calculations. There are a total of P reactant phases and a total of I linear equations, including one for each element and possibly one for redox. Additional constraints allow some minerals only to dissolve and some only to precipitate. Generally, there are more possible reactants than equations, or $P > I$, and there is no unique solution to the equations. Furthermore, given just I equations, it only is possible to solve for I unknowns. The program NETPATH (Plummer, Prestemon, and Parkhurst 1991) is designed to solve this type of mass-balance problem. The program exhaustively searches for sets of I phases that have mass-transfer coefficients that satisfy the linear equations and satisfy any dissolution-precipitation constraints that were imposed. The results of the search are mass-balance models that exactly account for the difference in chemical composition between the initial and final water. These mass-balance models also are consistent with any constraints that were imposed on the basis of saturation-index data or geologic or mineralogic evidence.

Example of Mass-Balance Modeling

An example of mass-balance modeling for an aquifer that receives large inputs of agricultural chemicals has been derived from a study by Denver (1989). This study investigated the ground-water composition under a field irrigated by a center-pivot irrigation system. A series of nested piezometers were installed across the field to identify horizontal and vertical variations in water chemistry. The modeling attempts to quantify the reactions that account for the evolution of rainwater to the water composition found in one of the piezometers. Although this example of mass-balance modeling is applied to a local study area, the problems involving agricultural chemicals are typical of problems found regionally throughout the Delmarva Peninsula.

The study unit was located in an unconfined coastal-plain aquifer in the Delmarva Peninsula. The geologic units are predominantly quartz sand interspersed with thin silt and clay beds. Pristine ground water in the aquifer is very dilute (specific conductance usually less than 100 µS/cm). The primary reaction accounting for the natural ground-water composition is dissolution of small amounts of feldspar and mica (Denver 1989). Ground water collected from below the field shows the effects of the use of agricultural chemicals. Specific conductance is much

TABLE 9-1. Analytical Data for Mass-Balance Modeling of Delaware Coastal-Plain Aquifer Affected by Agricultural Chemicals

Constituent or Property	Rainwater	Irrigation Water	Ground-Water Sample
Calcium	0.003	0.424	0.674
Magnesium	0.005	0.346	0.263
Sodium	0.053	0.479	0.257
Potassium	0.002	0.056	0.409
Chloride	0.054	0.536	0.508
Carbon	0.013[a]	0.896[b]	1.357[b]
Aluminum[c]	0.0	0.0	0.0
Sulfur	0.022	0.065	0.187
Nitrogen[d]	0.033	1.286	1.714
Nitrate	0.020	1.286	1.714
Ammonia	0.013	0.001	0.0[e]
Redox state[f]	1.245	10.52	16.172
Dissolved oxyen	0.25[e]	0.18[e]	0.26
pH (standard units)	4.5[e]	5.58	4.87
Temperature (°C)	20.[e]	15.	15.

Source: Denver (1989).
Note: Units of data are millimoles per kilogram of water, except as noted. Data for irrigation water are from a sample taken from the irrigation well. Data were converted from original units of milligrams per liter using NETPATH (Plummer, Prestemon, and Parkhurst 1991). The ground-water sample is from a piezometer completed at 20 to 25 ft (6.1 to 7.6 m) below land surface.

[a] Neither total carbon nor alkalinity was analyzed. Total carbon was calculated by PHREEQE (Parkhurst, Thorstenson, and Plummer 1980) assuming the pH was 4.5 and the solution was in equilibrium with atmospheric carbon dioxide.
[b] Total carbon was calculated from pH and alkalinity by NETPATH.
[c] Concentration was assumed to be negligible in all solutions.
[d] Total nitrogen is the sum of ammonium and nitrate.
[e] Estimated value.
[f] Redox state was calculated using the conventions described in NETPATH.

larger (greater than 200 µS/cm), and concentrations of nitrate, potassium, and other major ions are larger than in natural ground water. The primary agricultural chemicals used at the site (other than pesticides) include ammonia-based fertilizer, potassium chloride, lime, and ammonium thiosulfate (Denver 1989).

The sample used for the mass-balance modeling was collected at the piezometer nest closest to the irrigation well at the center pivot. Ground-water flow at this nest is nearly vertical during irrigation pumping (Denver 1989). The sample was collected from the most shallow piezometer, which was completed at 20–25 ft (6.1–7.6 m) below land surface. The analytical data for this sample are shown in Table 9-1, along with data for the composition of irrigation water and rainwater.

Two approaches were used for mass-balance modeling of these data: a mixing approach and a pure chemical-reaction approach. The mixing model used irrigation water and rainwater as the two sources of infiltrating water and then calculated the mixing fractions of each water plus additional chemical reactions. This approach is most realistic because it attempts to account for the two major sources of recharge water. However, it assumes that the composition of the irrigation water has remained constant with time, which cannot be documented with the available data. The other mass-balance modeling approach uses rainwater as the initial water and accounts for differences in chemical composition with chemical reactions alone. This approach ignores the fact that some of the water is derived from irrigation and some from rainwater by com-

TABLE 9-2. Mass-Balance Models for Delaware Coastal-Plain Aquifer Affected by Agricultural Chemicals

Reactant	Mixing		Nonmixing
	Model 1	Model 2	Model 3
Irrigation water	0.48	0.11	—
Rainwater	0.52	0.89	1.0
Dolomite [$CaMg(CO_3)_2$]	0.09	0.22	0.26
Potash (KCl)	0.22	0.40	0.45
Ammonium thiosulfate [$(NH_4)_2S_2O_3$]	0.07	0.08	0.08
Fertilizer (NH_3)	1.84	2.35	2.50
Oxygen (O_2)	2.44	3.31	3.57
Carbon dioxide (CO_2)	0.73	0.81	0.83
NH_4-Ca exchange	0.37	0.40	0.39
NH_4-Na exchange	—	0.16	0.20
NH_4-K exchange	0.16	—	—
K-Ca exchange	—	—	0.02

Note: All results are in millimoles per kilogram of water except for irrigation water and rainwater, which are mixing fractions of the two waters. Ground-water sample is from a piezometer completed at 20 to 25 ft (6.1 to 7.6 m) below land surface. For the exchange reactions, the species listed first exchanges onto the clay and the species listed second is released into solution. "-" indicates the reactant was not included in the model.

bining all of the reactions, those that were involved in making the irrigation water composition and those that occurred during infiltration, into one net reaction. This approach eliminates the need to know the composition of the irrigation water.

Many simplifying assumptions were used for this demonstration of mass-balance modeling. The naturally occurring reactions with silicate minerals (quartz and feldspars) are very slow and were assumed to be negligible. The speciation calculations indicated that carbonate minerals were orders of magnitude undersaturated and, therefore, in the mass-balance model, were assumed not to precipitate. The only redox reaction considered was biologically mediated nitrification of ammonium to nitrate with oxygen as the electron acceptor. Denitrification reactions were not considered. The only reactants considered were the agricultural chemicals—fertilizer, lime, potash, and ammonium thiosulfate; oxygen and carbon dioxide from the unsaturated zone; and clays as sites for cation exchange. The composition of the lime is not known but was assumed to be dolomite [$CaMg(CO_3)_2$]. The fertilizer was assumed to be effectively pure ammonia. The ion-exchange reactions are of the form

$$2\,NH_4^+ + CaX \rightarrow Ca^{2+} + (NH_4)_2X \quad (9\text{-}13)$$

where X represents a divalent exchange site.

Additional constraints were placed on the reactive phases based on geochemical inferences. KCl (potash) is undersaturated with respect to the ground water and was assumed only to dissolve. Ammonium thiosulfate and ammonia fertilizer also were assumed only to dissolve. Exchange reactions were assumed to occur in the following order: $NH_4 > K > Ca$, $Mg > Na$. Thus, for example, potassium was allowed to replace calcium on the clay, but calcium was not allowed to replace potassium.

The mixing mass-balance approach resulted in several models that exactly accounted for the composition of the ground-water sample. Two representative mixing models are presented in Table 9-2 (models 1 and 2). The results indicate that the ground-water composition could evolve through a mixture of irrigation water and rainwater combined with the dissolution of lime, potash, ammonium thiosulfate, and ammonia fertilizer. Oxygen is required to oxidize the ammonia to nitrate, and the reaction releases some carbon dioxide to the unsaturated zone. Some ion exchange also is necessary to account for the cation concentrations.

The second mass-balance calculations, the nonmixing approach, assumed that the ground-water composition was due to a single set of reactions; these calculations did not explicitly identify the contribution of irrigation water in the evolutionary process. Thus, the mass-balance calculations determined the net reactions necessary to create the sample composition from the rainwater composition. Several models were found using the net-reaction approach, but for simplicity only one representative model is presented in Table 9-2 (model 3). Model 3 is similar to model 2 with the addition of a small amount of K-Ca exchange and the absence of any contribution of solutes from the irrigation water.

The results of the mixing and nonmixing calculations suggest several ways that the ground water may have evolved. The amount of irrigation water that contributed to the formation of the ground water could range from none up to 50%. An avenue of investigation is suggested to determine what reasonable fractions of recharge can be expected from irrigation recharge relative to precipitation recharge. However, the differences in mixing fractions may be due to uncertainty in the irrigation water composition rather than differences in the relative amounts of recharge. It is likely that the recharge water composition varies yearly and seasonally with changes in crop type (soybeans, grown in alternate years, require no ammonia fertilizer), irrigation, application of agricultural chemicals, precipitation, and evapotranspiration. All of the models indicate that ammonium is exchanging into the clays with the release of calcium and either sodium or potassium. Experimental data on the relative abundances of exchangeable ions might determine whether ammonium can accumulate in the clays given the oxidizing environment of the aquifer, and whether calcium, sodium, or potassium are available for exchange.

Certainly this demonstration of mass-balance modeling is not definitive. The original study contains much more data that show that the spatial patterns of chemical composition are very complex (Denver 1989). The results presented in this chapter are highly simplified. The mass-balance modeling demonstrates that it is possible for the observed water compositions to evolve by reaction with chemicals known to have been used in the study area. The mass-balance calculations could be refined by investigating several topics more thoroughly, including the compositions of the lime and other agricultural chemicals, the relative rates of recharge of irrigation water and rainwater, and the composition of the exchangeable cations on clays.

Use of Stable Isotope Data in Mass-Balance Modeling

Usually, several mass-balance models can account for the observed ground-water chemistry; the question is how to narrow the number of possible models. In general, additional data of some sort are needed. Additional data can be derived from careful study of the mineralogy of the aquifer material. Other data that are widely used to refine mass-balance models are stable isotope ratios, especially of carbon and sulfur. Any valid mass-balance model must explain both the major-ion chemistry and the isotopic compositions of the observed ground water. Thus, particular mass-balance models can be eliminated if the isotopic compositions they imply are inconsistent with the observations.

The effects of a reaction on the isotopic composition of a solution can be calculated provided the mass-transfer coefficients, the isotopic compositions of the dissolving phases, and the fractionation effects for precipitating phases are known. Mass-balance modeling provides the necessary mass-transfer coefficients. It is assumed that there is no isotopic fractionation for dissolving phases; that is, the isotopic composition of the element entering the solution is the same as the isotopic composition of the element in the phase. In general, it is necessary to measure the isotopic composition of the dissolving phases. Sometimes it is possible to assume values for isotopic compositions of phases; for example, the carbon-13 ratio (relative to Peedee Belemnite, $\delta^{13}C_{PDB}$) of marine carbonates is often near 0‰, and the sulfur-34 ratio (relative to Canyon Diablo Troilite, $\delta^{34}S_{CDT}$) in sulfide minerals is usually very light, less than -30‰. Precipitation of heavy elements, like strontium, causes very little isotopic fractionation. However, precipitating phases containing lighter elements often fractionate relative to solution; that is, the isotopic composition of the precipitating phase is usually enriched in the heavier isotope relative to the isotopic composition of the solution. Equilibrium fractionation factors for carbon have been measured (Thode et al. 1965; Mook 1980; Mook, Bommerson, and Staverman 1974; Deines, Langmuir, and Harmon 1974). However, sulfate-reduction reactions generally require kinetic rather than equilibrium fraction factors. Plummer et al. (1990) published an empirical formula for the kinetic fractionation factor between dissolved sulfate and precipitated sulfide that is derived from three carbonate aquifers.

The computer program NETPATH (Plummer, Prestemon, and Parkhurst 1991) is designed to make mass-balance calculations and to calculate the effects of the mass-balance reactions on the isotopic composition of the solution. Thus, with a single program, it is possible to derive mass-balance models that are consistent with the observed chemical data and to test whether these reactions are consistent with the carbon, sulfur, and strontium isotopic data. An example from the NETPATH manual (Plummer, Prestemon, and Parkhurst 1991) based on the work of Chapelle and Knobel (1985) on the Aquia aquifer of Maryland will demonstrate the utility of isotopic data in mass-balance modeling.

The Aquia is typical of aquifers throughout the Atlantic and Gulf coastal-plain sediments in that calcium bicarbonate waters are in the upgradient areas and sodium bicarbonate waters are in the downgradient areas. The transition in the dominant cation is attributed to calcium for sodium cation exchange on marine clays, but the reaction also involves a source of carbon dioxide (Cederstrom 1946; Foster 1950). In the downgradient areas, it is unlikely that the atmosphere or soil zone is the source of carbon dioxide. The carbon dioxide is apparently produced by microbial oxidation of organic matter (Chapelle and Knobel 1985; Chapelle et al. 1987; among others). The most reasonable electron acceptor in this oxidation reaction is sulfate, which is consequently reduced to sulfide (Plummer, Prestemon, and Parkhurst 1991). Gypsum is not present within the permeable parts of the Aquia aquifer; the source of the sulfate is assumed to be the confining units (see Plummer, Prestemon, and Parkhurst 1991).

Data for two wells in the Aquia aquifer are presented in Table 9-3. The data show decreases in the calcium and magnesium concentrations and increases in the concentrations of carbon and sodium. A plausible mass-balance model that accounts for the difference is given in Table 9-4. The reaction shows that the final water can evolve from the initial water by gypsum dissolution (which may in reality be calcium and sulfate diffusing out of the confining units), oxidation of organic matter, precipitation of pyrite, dissolution of goethite, and cation exchange. This appears to be a valid mass-balance model until the $\delta^{13}C_{PDB}$ values are considered. The $\delta^{13}C_{PDB}$ of the initial water is -11.4‰ (Chapelle and Knobel 1985); the calcite is 1.1‰ (Chapelle and Knobel 1985); and the organic matter is assumed to be -22.0‰. NETPATH calculates that the isotopic composition of the final water should

TABLE 9-3. Analytical Data for Mass-Balance Modeling of the Aquia Aquifer

Constituent or Property	Initial Well	Final Well
Calcium	0.30	0.08
Magnesium	0.26	0.08
Sodium	1.13	6.09
Carbon	2.45	5.98
Iron	0.0[a]	0.0[a]
Sulfur	0.07	0.14
Redox state	10.20	24.71
$\delta^{13}C_{PDB}$ (‰)	−11.4	−6.2
Temperature (°C)	17.2	19.
pH (standard units)	8.2	8.4

Source: Chapelle and Knobel (1985).
Note: Units of data are millimoles per kilogram of water, except as noted. Data were converted from original units of milligrams per liter using NETPATH (Plummer, Prestemon, and Parkhurst 1991).
[a] Estimated value.

TABLE 9-4. Mass-Balance Model for the Aquia Aquifer

Reactant	Mass Transfer
Calcite (CaCO$_3$)	−0.15
Ca-Na exchange	2.30
Mg-Na exchange	0.18
Lignite (CH$_2$O$_{0.8}$)	3.67
Gypsum (CaSO$_4 \cdot$ 2H$_2$O)	2.22
Goethite (FeOOH)	1.08
Pyrite (FeS$_2$)	−1.08

Note: All results are in millimoles per kilogram of water. Positive mass transfers indicate dissolution, negative mass transfers indicate precipitation. For the exchange reaction, the species listed first exchanges onto the clay and the species listed second is released into solution.

be −17.8‰. However, the observed $\delta^{13}C_{PDB}$ of the final water is −6.2‰. This is too large a difference in $\delta^{13}C_{PDB}$ to consider the mass-balance model valid.

Plummer, Prestemon, and Parkhurst (1991) concluded that the only way to rectify the discrepancy between the calculated and observed $\delta^{13}C_{PDB}$ in this model is for recrystallization of aragonite or another unstable carbonate phase to occur. In this process, aragonite dissolves and is reprecipitated as calcite, which causes little change in the major-ion composition of the solution, but causes the isotopic composition of the solution to become heavier. If 9.2 mmol/kg H$_2$O of aragonite recrystallize to calcite, then the calculated $\delta^{13}C_{PDB}$ of the final water is in agreement with the observed value of −6.2‰. Thus, the mass-balance model, without consideration of the isotopic data, produced what appeared to be a plausible model. However, on consideration of the stable carbon isotopes, it was found that the model was incorrect. A new model, which includes recrystallization of a carbonate phase, now accounts for both the chemical and isotopic data. Hence, inclusion of isotopic data as part of the inverse modeling can provide a powerful tool for identifying geochemical reactions in regional ground-water systems.

Mass-Balance Modeling and Carbon-14 Dating

Another very valuable use of isotopes is to determine ground-water ages with carbon-14, ^{14}C (see Chapter 10). This radioactive isotope is generated naturally in the atmosphere and enters the ground-water system in recharge. The ^{14}C concentration in the atmosphere was relatively constant before atomic testing began at a value referred to as 100 pmc (percent modern carbon). Once in the ground-water system, the isotope decays with a half-life of 5,730 years, so if decay is the only cause for decrease in ^{14}C, then the age of the water can be determined by the following equation:

$$\text{Age} = \frac{5{,}730}{\ln(2)} \ln\left(\frac{A_0}{A}\right) \quad (9\text{-}14)$$

where A is the ^{14}C concentration of the sample and A_0 is the starting ^{14}C concentration in the ground water. However, if geochemical reactions occur that involve gain or loss of carbon from solution, then the equation must be modified to account for these carbon mass transfers. In this case, it is necessary to determine A_{nd}, which is the ^{14}C concentration that would result from the chemical reactions, excluding any changes due to radioactive decay. The appropriate equation is then

$$\text{Age} = \frac{5{,}730}{\ln(2)} \ln\left(\frac{A_{nd}}{A}\right) \quad (9\text{-}15)$$

To calculate A_{nd}, it is necessary to determine the mass transfers of carbon to and from the aqueous phases and to account for any fractionation effects. Mass-balance modeling provides quantitative mass transfers for all phases, and the equations of Wigley, Plummer, and Pearson (1978, 1979) account for the mass-transfer and fractionation effects on ^{14}C. Then Eq. 9-15 can be used to estimate ground-water ages.

Data from the Madison aquifer, which were discussed in the section on speciation modeling, will be used to demonstrate the techniques of ^{14}C dating. The predominant reactions in the aquifer are dissolution of dolomite and anhydrite with precipitation of calcite. In addition, organic-matter oxidation, sulfate reduction, iron reduction, ion exchange, and dissolution of sylvite (KCl) and halite (NaCl) occur (Plummer et al. 1990). Table 9-5 lists analytical data for a water from the recharge area and another well approximately 150 km downgradient. The following isotopic compositions of minerals were used in the calculations: $\delta^{13}C_{PDB}$ values were 4.0, 4.0, and $-25.0‰$ for dolomite, calcite, and organic matter, respectively (Plummer et al. 1990); $\delta^{34}S_{CDT}$ values for anhydrite and pyrite were 15.5 and $-22.09‰$, respectively (Plummer et al. 1990).

The mass-balance model in Table 9-6 was calculated for this analytical data in the NETPATH manual (Plummer, Prestemon, and Parkhurst 1991) and accounts for changes in chemical composition, $\delta^{13}C_{PDB}$, and $\delta^{34}S_{CDT}$. The model is consistent with dedolomitization, where irreversible dissolution of anhydrite causes the precipitation of calcite and the dissolution of dolomite. In addition, a small amount of oxidation of organic matter is coupled with the reduction of sulfate, dissolution of goethite, and pre-

TABLE 9-5. Analytical Data for Mass-Balance Modeling of the Madison Aquifer

Constituent or Property	Initial Well	Final Well
Calcium	1.2	11.28
Magnesium	1.01	4.54
Sodium	0.02	31.89
Potassium	0.02	2.54
Chloride	0.02	17.85
Carbon	4.3	6.87
Iron	0.001	0.000
Sulfur	0.16	20.12
Redox state	18.16	146.12
^{14}C (pmc)	52.0[a]	0.8
$\delta^{13}C_{PDB}$ (‰)	-6.99	-2.34
$\delta^{34}S_{CDT}$ (‰)	9.73	15.81
Temperature (°C)	9.9	63.
pH	7.55	6.61

Source: Plummer et al. (1990).
Note: Units of data are millimoles per kilogram of water, except as noted. Constituent or property: pmc, percent modern carbon. Data were converted from original units of milligrams per liter using NETPATH (Plummer, Prestemon, and Parkhurst 1991).
[a] ^{14}Carbon for the recharge water was calculated using the "mass-balance" model (see Plummer, Prestemon, and Parkhurst 1991).

TABLE 9-6. Mass-Balance Model for the Madison Aquifer

Reactant	Mass Transfer
Dolomite [CaMg(CO$_3$)$_2$]	3.53
Calcite (CaCO$_3$)	-5.32
Anhydrite (CaSO$_4$)	20.15
Organic matter (CH$_2$O)	0.87
Goethite (FeOOH)	0.09
Pyrite (FeS$_2$)	-0.09
Ca-Na exchange	8.28
Halite (NaCl)	15.31
Sylvite (KCl)	2.52
Carbon dioxide (CO$_2$)	-0.04

Note: All results are in millimoles per kilogram of water. Positive mass transfers indicate dissolution, negative mass transfers indicate precipitation. For the exchange reaction, the species listed first exchanges onto the clay and the species listed second is released into solution.

cipitation of pyrite. Halite and sylvite are assumed to dissolve from remnant evaporite deposits (Plummer et al. 1990). The mass transfer of CO_2 is approximately zero, which is consistent with a closed system for this deep flow path (see Plummer et al. 1990 for discussion). NETPATH calculates A_{nd} for this reaction model to be 12.3 pmc, assuming the ^{14}C of the recharge water was 52 pmc. This means that the chemical reactions alone decrease the ^{14}C concentration from 52 pmc in the recharge water to 12.3 pmc in the final water. Radioactive decay is assumed to cause the decrease from 12.3 pmc to the observed value of 0.8 pmc in the final water. Using these last two values in Eq. 9-15, the calculated age for the water is 22,600 years. Plummer et al. (1990) used ^{14}C ages to calculate flow velocities that range from 2.1 to 26.5 m/yr in the Madison aquifer. They also calculate hydraulic conductivities using Darcy's law, the hydraulic gradient, and these average ^{14}C flow velocities. The ^{14}C hydraulic conductivities are similar to hydraulic conductivities derived by digital simulation of the flow system.

In summary, inverse modeling combines several powerful techniques for using geochemical data to evaluate regional groundwater systems. Speciation modeling provides information on the reactant phases in the system. With knowledge of the potential reactants in the system, mass-balance modeling can identify geochemical reactions that can quantitatively account for the chemical evolution of ground water along a flow path. Isotopic data can be used to corroborate or reject the mass-balance models. And finally, mass-balance models combined with ^{14}C data can provide independent hydrologic information on the ages of ground water, rates of reaction, rates of flow, and hydraulic conductivity in a ground-water system.

FORWARD MODELING

Unlike inverse modeling, which is used to determine reactions based on observed data, forward modeling predicts the evolution of water composition based on hypothetical reactions, whether or not data are available. Forward modeling relies on a thermodynamic model of aqueous solutions and mineral phases to calculate solution composition. Two types of forward modeling are discussed here: reaction-path modeling and reaction-transport modeling. The purpose of reaction-path modeling is to predict the composition of water as it undergoes reversible and irreversible geochemical reactions. Reaction-transport modeling is similar to reaction-path modeling in that it accounts for geochemical reactions, but it also includes mechanisms to model advection and dispersion within the aquifer. In this chapter, the program PHREEQE (Parkhurst, Thorstenson, and Plummer 1980) will be used for reaction-path calculations, and results from a study by Appelo and Willemsen (1987) using a reaction-transport code will be presented.

The primary use of reaction-path modeling is to investigate geochemical reactions where analytical data are lacking. If data are available and the object of the study is to determine the geochemical reactions that are occurring in the system, then the inverse-modeling approach described earlier is the most efficient method of analyzing the data. If the results of inverse modeling are applied to reaction-path modeling—that is, the initial water is specified as the starting point and the mass transfers calculated by the mass-balance modeling are specified as an irreversible reaction—then forward modeling will reproduce the final water that was used in the mass-balance modeling. However, no new information is gained by this exercise than was available from the saturation indices of speciation modeling and the reaction calculated by mass-balance modeling. The true usefulness of reaction-path modeling is to answer hypothetical questions, such as: What is the solubility of calcite if the partial pressure of carbon dioxide is 0.1 atmospheres? or What is the first phase that becomes saturated as feldspars dissolve in rainwater? or What is the composition of the water if a specified set of reversible and irreversible reactions occur?

Reaction-transport modeling simulates

processes that are not tractable by other approaches. Reaction-transport modeling can simulate advection and dispersion, which can be difficult to simulate with inverse modeling. In addition, reaction-transport modeling can calculate ground-water and solid-phase chemical compositions as functions of time and space. Spatial and temporal resolution of chemical reactions is not possible with reaction-path modeling alone.

Reaction-Path Modeling

Reaction-path programs are based on a thermodynamic model that is very similar to those of speciation models. Speciation models have mass-balance equations for each element (except hydrogen and oxygen) and mass-action equations for each aqueous complex. The pH and redox conditions are specified explicitly for speciation calculations. Reaction-path models contain equations similar to the mass-balance and mass-action equations of speciation models, but they also include mass-action equations for phase equilibria and three of the following five equations: hydrogen mass balance, oxygen mass balance, electrical balance, redox balance, or a constant mass of water. The additional equations allow reaction-path programs to follow phase boundaries and calculate the pH, redox condition, and mass of water as a function of reaction.

All of the benefits and shortcomings of aqueous models that were discussed for speciation models are applicable to reaction-path models. Uncertainties in the theoretical formulation, the selection of aqueous species to be included, and the thermodynamic data for the aqueous species and minerals cause uncertainties in the model results.

After the aqueous model has been assembled, reaction-path calculations require only a starting water composition and a set of reversible and irreversible reactions. Reversible reactions are those that occur close to equilibrium and a mass-action phase-equilibrium equation is included in the calculations. Irreversible reactions are those that are not at equilibrium and the mass transfers involved in these must be specifically defined.

Example of Reaction-Path Modeling

As an example of reaction-path calculations, we will consider the evolution of an acid water in a mine and subsequent reactions as this water is removed from contact with the atmosphere and migrates into an aquifer. The example is based on events that occurred in the Picher mining area of northeastern Oklahoma and southeastern Kansas (Parkhurst 1987; Playton, Davis, and McClaflin 1980). However, only a few analyses are used to guide the calculations. The calculations are not intended to reproduce precisely the chemical compositions found in these studies. Rather, they are intended to be an a priori prediction of the evolution of water chemistry given a limited amount of data.

The Picher field is an extremely large lead-zinc deposit that was mined from the turn of the century until approximately 1960. The sphalerite (ZnS) and galena (PbS) ores are in a Mississipian-age limestone called the Boone Formation. Accessory minerals include pyrite and marcasite (FeS_2), calcite, and dolomite, among many other less common minerals. The mines are 30–150 m below land surface with access limited to vertical shafts. Water levels were maintained below the mine by pumpage. When mining ceased, pumping also ceased, and the mines filled with water. By about 1980, the mines and shafts were completely filled with water, and water levels in the area were about 10 m below land surface.

Three different environments were modeled in the reaction-path simulation: (1) an oxic environment corresponding to the time that the mines were partially filled with water and still in contact with the atmosphere, (2) an anoxic environment that was created when the water began to fill the mine shafts, thus essentially eliminating exchange between the mines and the atmosphere, and (3) the environment created as the anoxic

TABLE 9-7. Mass Transfers in Reaction-Path Calculations for Evolution of Mine Water in the Picher Mining Area

Reactant or Property	Reaction Step					
	1	2	3	4	5	6
Pyrite (FeS$_2$)	15.0	0.41^a	tr ppt	—	—	—
Sphalerite (ZnS)	3.0	0.08	0.0	tr ppt	−0.84^a	−0.01^a
Oxygen (O$_2$)	62.25^c	—	—	—	—	—
Carbon dioxide (CO$_2$)	−16.98^d	—	—	—	—	—
Fe(OH)$_3$	−8.63^a	—	—	—	—	—
Calcite (CaCO$_3$)	6.78	—	1.25	0.05	3.14	1.39
Dolomite [CaMg(CO$_3$)$_2$]	6.78	—	1.25	0.05	3.14	1.39^a
Gypsum (CaSO$_4$)	0.0^b	−0.10^a	−2.64^a	−0.07^a	−4.66^a	−1.60^a
Goethite (FeOOH)	—	—	—	0.0^b	−6.72^a	−0.04^a
Smithsonite (ZnCO$_3$)	—	—	—	—	0.0^b	−1.72
pH	3.09	2.46	4.39	4.88	5.94	6.30
pE	16.94	2.82	0.61	0.13	−0.96	−1.41

Note: All results are in millimoles per kilogram of water; positive mass tansfers for a mineral indicate dissolution, negative mass transfers indicate precipitation. "tr ppt" indicates a trace amount of precipitation; "—" indicates that the reactant did not react in the reaction step; a value of 0.0 indicates that the reactant is at equilibrium, but no mass transfer of the reactant has occurred in the reaction step. Irreversible reactions for the reaction steps are as follows: 1-oxidation of pyrite and sphalerite in the presence of oxygen from the atmosphere and dissolution of dolomite and calcite; 2-oxidation of pyrite and sphalerite in the absence of oxygen; 3, 4, 5, and 6-dissolution of calcite and dolomite. Initial solution for step 1 was pH 4.5 rainwater. Mass transfers for all other steps are incremental from the previous step.

^{a} Phase has reacted to equilibrium in this step.
^{b} Dissolution of calcite and dolomite has proceeded to the point where equilibrium with this phase was attained.
^{c} Partial pressure of oxygen was fixed at 0.001 atm.
^{d} Partial presure of carbon dioxide was fixed at 0.1 atm.

mine water migrates into the nonmineralized part of the Boone Formation. The analytical data show that the major ions in the mine-water environment are calcium, magnesium, iron, zinc, sulfate, and bicarbonate; for simplicity, the calculations were limited to the six elements corresponding to these ions (plus hydrogen and oxygen).

In the oxic environment, it was assumed that pyrite and sphalerite were oxidized by oxygen from the atmosphere; carbon dioxide was released to the atmosphere; calcite and dolomite dissolved; and amorphous iron hydroxide precipitated. To model these reactions, it was necessary to specify the extent to which each reaction proceeded. Pyrite and sphalerite were added irreversibly in a 5:1 ratio into a sulfuric acid solution of pH 4.5 that represented rainwater (rainwater is so dilute that the exact composition has little effect on the calculations); the total amount of pyrite dissolution was specified to be 15 mmol/kg H$_2$O. These specifications produced modeled concentrations of sulfate and zinc that approximated mine-water concentrations. Oxidation reactions were assumed to occur sufficiently fast to maintain the partial pressures of oxygen below atmospheric levels, at 0.001 atm, and carbon dioxide above atmospheric levels, at 0.1 atm. Dolomite and calcite were assumed to react in equal amounts to the point where gypsum saturation was obtained. Amorphous iron hydroxide was assumed to be in equilibrium with the mine water. The results of the reaction-path simulation of these reactions are shown in Figure 9-2 (step 1). The mass transfers of the minerals and gases are shown in Table 9-7 (step 1).

The reaction-path calculations for the

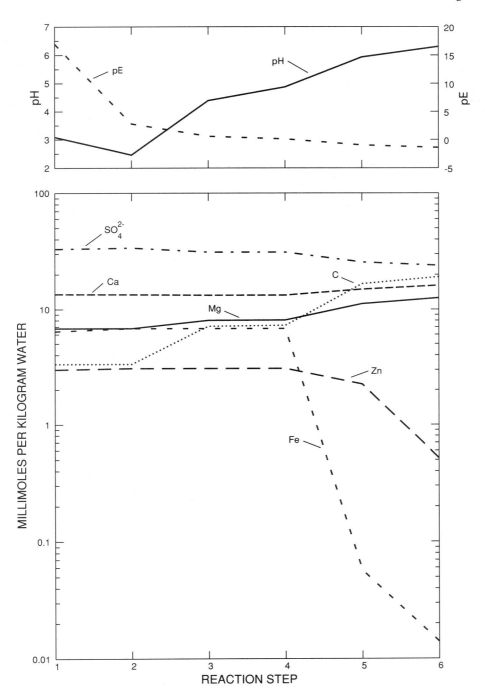

FIGURE 9-2. Hypothetical evolution of water in the mine workings and limestone aquifer of the Picher field.

mine water in contact with the atmosphere (step 1) indicate that a large amount of oxygen is necessary to oxidize the given amounts of pyrite and sphalerite. Carbon dioxide is released to the atmosphere over the mine water and amorphous iron hydroxide precipitates. Even though several millimoles of calcite and dolomite react, the pH of the mine water is near 3.0. The amount of carbonate dissolution is just sufficient to bring the solution to equilibrium with gypsum, but no gypsum has precipitated at this point in the reaction path.

The next part of the reaction path (step 2) simulated the anoxic reactions that occurred when the mine workings were cut off from the atmospheric source of oxygen. Pyrite and sphalerite were assumed to continue to react at a ratio of 5:1 to the point where pyrite reached equilibrium. The only other phase assumed to be in equilibrium during this reaction was gypsum. The results of this calculation show 0.41 and 0.08 mmol/kg H_2O of pyrite and sphalerite dissolve to reach pyrite equilibrium (Table 9-7, step 2). A small amount of gypsum precipitates to remain in equilibrium. The pE, a measure of the redox potential, drops from about 16.94 to 2.82 and the pH decreases from 3.09 to 2.46. All of the iron in solution in step 1 was ferric iron [Fe(III)], the most oxidized state for iron in natural waters. The sulfide dissolution reaction causes the reduction of all the iron to ferrous iron [Fe(II)] and a very large decrease in the redox potential. The following chemical reaction describes the reduction reaction and also explains the decrease in pH (or increase in hydrogen ion concentration).

$$15.6\,Fe^{3+} + FeS_2 + 0.2\,ZnS + 8.8\,H_2O$$
$$\to 16.6\,Fe^{2+} + 0.2\,Zn^{2+} + 2.2\,SO_4^{2-}$$
$$+ 17.6\,H^+ \qquad (9\text{-}16)$$

In the final reaction-path simulations, the mine water was assumed to migrate into the nonmineralized limestone formation. Pyrite and sphalerite were assumed to be absent in this part of the formation. The irreversible reaction was assumed to be dissolution of calcite and dolomite in equal proportions. Several phases could become supersaturated as the carbonate dissolution occurred, including pyrite, sphalerite, goethite (FeOOH), siderite ($FeCO_3$), smithsonite ($ZnCO_3$), and zinc oxyhydroxides. Several simulations were run to determine the order in which these minerals would become supersaturated.

Steps 3-6 (Figure 9-2 and Table 9-7) summarize important points along the reaction path to carbonate equilibrium. The results show that the dissolution of carbonates leads sequentially to equilibrium with sphalerite (step 3), goethite (step 4), smithsonite (step 5), and dolomite (step 6). Pyrite becomes undersaturated after step 3 and does not participate in subsequent reactions. Gypsum, sphalerite, goethite, and smithsonite precipitate in all reaction steps following the step in which they attain equilibrium. All other phases, including amorphous iron hydroxide, siderite, and zinc oxyhydroxides, are undersaturated throughout the reaction path. The reactions cause an increase in pH from 2.46 to 6.3, a decrease in iron from 6.8 to about 0.01 mmol/kg H_2O, and a decrease in zinc concentrations from 3.1 to 0.5 mmol/kg H_2O. The final concentration of iron and pH compare favorably with the secondary maximum contaminant levels (SMCLs) set by the U. S. Environmental Protection Agency (1986) of pH > 6.5 and iron < 0.005 mmol/kg H_2O (0.3 mg/L). Zinc concentrations are still considerably greater than the SMCL of 0.08 mmol/kg H_2O (5.0 mg/L).

This reaction-path simulation provides useful information about reactions involving mine water. It indicates that pyrite oxidation is sufficient to generate acid waters. Reaction of carbonates will cause gypsum and iron hydroxides to precipitate. The onset of reducing conditions tends to lower the pH further. Introduction of this acid mine water into a limestone aquifer should lead to much higher pH values and much lower concentrations of iron and zinc. All of this was derived with very little chemical data from the field. However, care must be taken not to overestimate the value of these calculations. Many assumptions about equilibrium with phases and the amounts and relative rates of reactions were used to arrive at these results. Data provide the necessary information to test assumptions, but the most efficient use of that data would be to calculate saturation indices with a speciation code to determine which minerals are indeed in equilibrium and to calculate mass transfers using the mass-balance approach to determine the amounts and relative rates of reactions.

Combining Forward and Inverse Modeling

In some cases it is useful to combine forward modeling with the results from inverse models. The mass-balance results from inverse modeling determine the net mass transfer along the flow path, but these results are only partially constrained by thermodynamics through examination of saturation indices at initial and final points on the flow path. The forward modeling can be used to prove thermodynamic consistency for the mass-balance result, provided a reaction path can be found that never precipitates a product phase from an undersaturated solution and never dissolves a reactant phase in a supersaturated solution. Recognizing that the calculated proportions of reactants and products in the mass-transfer result is only for the net reaction, it is possible that the relative reaction rates may vary throughout the reaction path between the initial and final points, provided these rates produce the net mass transfer from the inverse problem. Forward modeling can be used to test specific combinations of reaction rates to determine if the resulting reaction path is thermodynamically valid. A mass-balance model cannot be rejected if at least one thermodynamically valid reaction path can be demonstrated for the net mass transfer.

An example of combined use of inverse and forward modeling is presented by Plummer (1984) in which the mass-balance result was used to define initial relative rates of irreversible reactions (organic matter oxidation, anhydrite dissolution, halite dissolution, and temperature variations) for the Madison aquifer. Using the same recharge-water starting point and realistic thermodynamic constraints of calcite and dolomite equilibrium, the extents and relative rates of irreversible reactions were varied in forward modeling. Many thermodynamically valid reaction paths were obtained. The modeled ranges of dissolved calcium, magnesium, dissolved inorganic carbon, pH, CO_2 partial pressure, and calcite mass transfer were similar to the range of these parameters observed throughout the Madison aquifer in parts of Montana, South Dakota, and Wyoming. This suggests that the observed variations in ground-water composition in the Madison aquifer can be accounted for by a single set of reactions that vary in their relative rates and reaction extents.

After reaction models have been found for a ground-water system by using the inverse method, they may be treated in forward modeling to predict water quality at points in the aquifer where there are no analytical data. The forward modeling assumes that the relative rates of reactions derived from the inverse modeling can be applied elsewhere in the system, and that there are no other important reactions occurring that were not considered in calibrating the forward modeling from the inverse modeling.

Reaction-path modeling is a powerful tool for simulating the changes in ground-water composition in response to geochemical reactions. However, several limitations arise in applying reaction-path models to ground-water systems. The results of reaction-path modeling are functions of chemical-reaction coordinates; that is, the chemical composition of ground water is calculated as a function of the extent of specified reactions. For a ground-water study, some of the most important information is the spatial and temporal variation in water composition within the aquifer, and there is no direct way to convert the results of reaction-path modeling into the coordinates of space and time. Reaction-path modeling also lacks the capability to model the dynamic aspects of ground-water flow including the processes of advection and dispersion. These effects are especially important in systems that are affected by human-induced geochemical and hydrologic changes. Another major shortcoming is the limited ability of reaction-path modeling to account for changes in the distribution and composition of solid phases within the aquifer. Geochemical reactions frequently deplete or produce certain phases within the aquifer, resulting in geochemical fronts that propagate through the aquifer. Thus, although reaction-path modeling can simulate the chemical evolution of ground water, more complex models are needed to simulate the effects of advection and dispersion, spatial and temporal changes in water composition, and changes in the solid phases of an aquifer.

Reaction-Transport Modeling

Recent work attempts to remedy the limitations of reaction-path modeling by combining the equations of reaction-path modeling with those of single-component solute-transport modeling. Yeh and Tripathi (1989) have reviewed approaches to solving the resulting set of equations and several codes are now available that are designed to model multicomponent reaction-transport phenomena (Mangold and Tsang 1991). These codes have the capability to model both the geochemical reactions between ground water and aquifer phases and the physical processes of advection and dispersion.

Reaction-transport models are capable of modeling very complex processes, but the limitations of the approach need to be considered. The models require an aqueous model and thermodynamic data for solid phases, just as in the reaction-path and speciation models; the uncertainties involved already have been discussed. In addition, reaction-transport models must calculate the flow field for the aquifer under study. Thus, the hydraulic properties of the aquifer must be known sufficiently to calculate a realistic flow field for steady-state or, possibly, transient conditions. It may be necessary to know hydraulic properties in even more detail than for ground-water flow modeling because spatially averaged properties may not be adequate. If, for example, a contaminant is present only in one zone of an aquifer, a ground-water flow model may be adequate for quantifying ground-water flow at a resolution that does not identify the zone with the contaminant, but the reaction-transport model must consider this zone in detail. Other data needed for reaction-transport modeling include a quantitative assessment of the distribution and composition of reactive minerals within the aquifer, the initial compositions of ground water throughout the aquifer, estimates of the rates of irreversible reactions throughout the aquifer, and the history of chemical loadings and (or) ground-water withdrawals if transient conditions are modeled. Clearly a larger amount of information must be assembled to apply this kind of modeling to a specific field investigation. One other consideration for these models is computer time. A one-dimensional simulation involving a small number of nodes (less than 100) can require hours of computer time, depending on the algorithm and the complexity of the chemical system.

An example of the use of a reaction-transport model is taken from Appelo and Willemsen (1987). They developed the theory for a one-dimensional mixing-cell model that simulates advection and dispersion (or diffusion). This model used EQ3/6 (Wolery 1979) as a geochemical-equilibrium submodel [subsequent development led to the model PHREEQM which uses PHREEQE as the submodel (Appelo et al. 1990)] and a series of cells to represent the porous medium. Advection was simulated by "shifting" the contents of one cell to the next and dispersion or diffusion was simulated by mixing a fraction of the contents of a cell with the previous cell and with the next

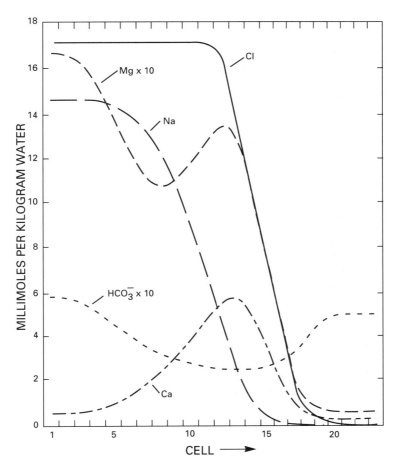

FIGURE 9-3. Reaction-transport model results for dispersive flow of saltwater into freshwater. (*From Appelo and Willemsen 1987.*)

cell in the series. This formulation of the reaction-transport model is relatively fast for one-dimensional calculations but also produces a scale dependency that must be investigated for each application of the model (Appelo and Willemsen 1987).

Appelo and Willemsen (1987) used their model to investigate seawater intrusion into freshwater in a Pleistocene aquifer in the western part of the Netherlands. One of their simulations serves as a useful example of the kinds of calculations that can be performed by reaction-transport models. In this simulation, saltwater (one-third seawater concentrations) flows into an aquifer that initially contains only freshwater. Dispersion occurs as the saltwater moves through aquifer, which tends to spread the concentration fronts in space. In addition, cation-exchange reactions affect the cation compositions of the intruding water. They assumed calcium, magnesium, potassium, and sodium are in exchange equilibrium with the preference for ions being $K^+ > Ca^{2+} > Mg^{2+} > Na^+$. Calcite equilibrium is assumed to be maintained throughout the flow system.

Figure 9-3 shows the model results for the dispersive flow of saltwater into the aquifer. The graph shows concentrations of major ions with distance (cell number) after 15 time steps. A very complex distribution of major ions is evident. Chloride is conservative, so the position of the decrease in chloride concentration is due solely to advection, and the amount of smearing of the front is due to dispersion. The sodium front is retarded relative to the chloride front because of the exchange reactions. Sodium exchanges onto the clays, releasing calcium and magnesium to solution. These reactions cause the peaks in the magnesium and calcium concentrations that are present at cells 12 and 13. The water in cells 14 through 18 is predominantly calcium chloride water because of the exchange of sodium for calcium. The increase in calcium concentration causes a simultaneous decrease in bicarbonate concentration (cells 2–14) because of precipitation of calcite. The simulation calculates the amounts of calcite that accumulate in each cell. The simulation also calculates the evolution with time of the exchanger from a low-sodium composition to a high-sodium composition as the saltwater moves through a cell. This solid-phase evolution cannot be readily simulated by inverse modeling or reaction-path modeling.

The simulation presented here is a hypothetical investigation of saltwater intrusion. Appelo and Willemsen (1987) also showed that if diffusion were the primary process accounting for the increase in salinity in the aquifer, then a much smoother distribution of ions is generated without the peaks evident in Figure 9-3. The model also was applied to a field site in order to identify how specific cation ratios could be generated. Thus, the reaction-transport model is a very versatile tool for investigating geochemical reactions in dynamic ground-water systems. This type of modeling is uniquely capable of simulating advection, dispersion, and spatial and temporal distributions of aqueous and solid-phase chemical compositions.

CONCLUDING REMARKS

Speciation, mass-balance, reaction-path, and reaction-transport models provide a powerful set of tools for the investigation of ground-water systems. The inverse-modeling approach can be used to determine the prevailing geochemical reactions occurring in an aquifer and, combined with ^{14}C data, can determine rates of flow and aquifer properties. Forward modeling can be used to investigate systems that lack sufficient data for inverse modeling and to make predictions about the possible effects of aquifer management. Appropriately used, these models provide essential information for regional ground-water assessments.

The combination of speciation and mass-balance models is described as an inverse approach because the analytical data are used to determine the chemical reactions that occurred in a ground-water system. Speciation modeling and mass-balance modeling are complementary tools for geochemical ground-water investigations. Speciation modeling provides saturation indices that indicate which reactions are thermodynamically possible, that is, which minerals may be dissolving and which may be precipitating. However, speciation modeling provides no information about reaction extents. Conversely, mass-balance modeling provides quantitative information about extents of reactions without any thermodynamic constraints. Together, these two types of models allow the identification of thermodynamically valid geochemical reaction models that quantitatively account for the chemical evolution of water along a flow path. Stable isotope data for carbon, sulfur, and strontium provide additional constraints on mass-balance models. The program NETPATH combines the capability to identify mass-balance models with the capability to calculate the ground-water isotopic composition implied by these mass-balance models. Any valid mass-balance model must account for the observed stable-isotope composition in addition to the chemical composition of ground water. NETPATH also provides the capability to use mass-balance models with ^{14}C data to calculate ground-water ages. For systems with adequate chemical, isotopic, and mineralogic data, the inverse-modeling approach of speciation and mass-balance modeling provides the most direct means of determining quantitative geochemical reaction models.

In contrast, for systems with missing or inadequate data, reaction-path modeling provides an a priori method of predicting geochemical reactions. This is a forward model because reversible and irreversible geochemical reactions must be specified before the solution composition that would result can be calculated. In this way, very complex geochemical scenarios can be modeled. However, reaction-path modeling is not as efficient as inverse modeling for determining the geochemical reactions involved in the evolution of a particular ground-water composition because the appropriate reversible and irreversible reactions must be determined by trial and error in forward modeling, whereas inverse modeling calculates the necessary reactions directly.

Recent advances in geochemical modeling have combined the capabilities of reaction-path modeling with single-component solute-transport models. The resulting multicomponent reaction-transport models are capable of simulating the physical processes of advection, dispersion, and diffusion in addition to the chemical reactions, inherent in a reaction-path model. These reaction-transport models provide the capability to model chemical reactions in time and space, which is not possible with reaction-path models alone. Reaction-transport models are suited to modeling some ground-water systems that are not tractable by inverse methods because of dispersion or non-steady-state conditions.

References

Allison, J. D., D. S. Brown, and K. J. Novo-Gradac. 1990. *MINTEQA2/PRODEFA2, A Geochemical Assessment Model for Environmental Systems: Version 3.0 User's Manual*. Athens, GA: Environmental Research Laboratory, Office of Research and Development, U. S. Environmental Protection Agency.

Appelo, C. A. J., and A. Willemsen. 1987. Geochemical calculations and observations on salt water intrusions. I: A combined geochemical/mixing cell model. *Journal of Hydrology* 94:313–30.

Appelo, C. A. J., A. Willemsen, H. E. Beekman, and J. Griffioen. 1990. Geochemical calculations and observations on salt water intrusions. II: Validation of a geochemical transport model with column experiments. *Journal of Hydrology* 120:225–50.

Ball, J. W., and D. K. Nordstrom. 1991. *User's Manual for WATEQ4F, with Revised Thermodynamic Data Base and Test Cases for Calculating Speciation of Major, Trace, and Redox Elements in Natural Waters*. U. S. Geological Survey Open-File Report 91-183.

Busby, J. F., R. W. Lee, and B. B. Hanshaw. 1983. *Major Geochemical Processes Related to the Hydrology of the Madison Aquifer System and Associated Rocks in Parts of Montana, South Dakota, and Wyoming*. U. S. Geological Survey Water-Resources Investigations 83-4093.

Cederstrom, D. J. 1946. Genesis of ground waters in the coastal plain of Virginia. *Economic Geology* 41: 218–45.

Chapelle, F. H., and L. L. Knobel. 1985. Stable carbon isotopes of HCO_3^- in the Aquia aquifer, Maryland: Evidence for an isotopically heavy source of CO_2. *Ground Water* 23(5):592–99.

Chapelle, F. H., J. L. Zelibor, Jr., D. J. Grimes, and L. L. Knobel. 1987. Bacteria in deep coastal plain sediments of Maryland: A possible source of CO_2 to groundwater. *Water Resources Research* 23(8):1625–32.

Deines, Peter, Donald Langmuir, and R. S. Harmon. 1974. Stable carbon isotope ratios and the existence of a gas phase in the evolution of carbonate ground waters. *Geochimica et Cosmochimica Acta* 38:1147–64.

Denver, J. M. 1989. *Effects of Agricultural Practices and Septic-System Effluent on the Quality of Water in the Unconfined Aquifer in Parts of Eastern Sussex County, Delaware*. Delaware Geological Survey Report of Investigations No. 45.

Foster, M. D. 1950. The origin of high sodium bicarbonate waters in the Atlantic and Gulf coastal plains. *Geochimica et Cosmochimica Acta* 1:33–48.

Garrels, R. M., and F. T. Mackenzie. 1967. Origin of the chemical compositions of some springs and lakes, In *Equilibrium Concepts in Natural Water Systems*, pp. 222–42. Advances in Chemistry Series, no. 67. Washington, DC: American Chemical Society.

Harvie, C. E., and J. H. Weare. 1980. The prediction of mineral solubilities in natural waters: The Na-K-Mg-Ca-Cl-SO_4-H_2O system from zero to high concentration at 25°C. *Geochimica et Cosmochimica Acta* 44:981–97.

Langmuir, Donald. 1971. Particle size effect on the reaction goethite = hematite + water. *American Journal of Science* 271:147–56.

Mangold, D. C., and C.-F. Tsang. 1991. A summary of subsurface hydrological and hydrochemical models. *Reviews of Geophysics* 29(1):51–79.

Mook, W. G. 1980. Carbon-14 in hydrogeological studies. In *Handbook of Environmental Isotope Geochemistry. Vol. 1: The Terrestrial Environment, A*, eds. Peter Fritz and J.-C. Fontes, pp. 49–74. New York: Elsevier.

Mook, W. G., J. C. Bommerson, and W. H. Staverman. 1974. Carbon isotope fractionation between dissolved bicarbonate and gaseous carbon dioxide. *Earth and Planetary Science Letters* 22:169–76.

Parkhurst, D. L. 1987. *Chemical Analyses of Water Samples from the Picher Mining Area, Northeast Oklahoma and Southeast Kansas*. U. S. Geological Survey Open-File Report 87–453.

Parkhurst, D. L., L. N. Plummer, and D. C. Thorstenson. 1982. *BALANCE—A Computer Program for Calculating Mass Transfer for Geochemical Reactions in Ground Water*. U.S. Geological Survey Water-Resources Investigations 82-14.

Parkhurst, D. L., D. C. Thorstenson, and L. N. Plummer. 1980. *PHREEQE—A Computer Program for Geochemical Calculations*. U.S. Geological Survey Water-Resources Investigations Report 80–96.

Pitzer, K. S. 1979. Theory: Ion interaction approach. In *Activity Coefficients in Electrolyte Solutions*, ed. R. M. Pytkowicz, pp. 157–208. Boca Raton, FL: CRC Press.

Playton, S. J., R. E. Davis, and R. G. McClaflin. 1980. *Chemical Quality of Water in Abandoned Zinc Mines in Northeastern Oklahoma and Southeastern Kansas*. Oklahoma Geological Survey Circular 82.

Plummer, L. N. 1984. Geochemical modeling: A comparison of forward and inverse methods. In *Proceedings First Canadian/American Conference on Hydrogeology: Practical Applications of Ground Water Geochemistry, Banff, Alberta, Canada*, eds., Brian Hitchon and E. I. Wallick, pp. 149–77. Worthington, OH: National Water Well Association.

Plummer, L. N., and W. Back. 1980. The mass balance approach: Application to interpreting the chemical evolution of hydrologic systems. *American Journal of Science* 280:130–42.

Plummer, L. N., J. F. Busby, R. W. Lee, and B. B. Hanshaw. 1990. Geochemical modeling of the Madison aquifer in parts of Montana, Wyoming, and South Dakota. *Water Resources Research* 26(9):1981–2014.

Plummer, L. N., B. F. Jones, and A. H. Truesdell. 1976. *WATEQF—A Fortran IV Version of WATEQ, A Computer Program for Calculating Chemical Equilibrium of Natural Waters*. U.S. Geological Survey Water-Resources Investigations 76–13.

Plummer, L. N., D. L. Parkhurst, G. W. Fleming, and S. A. Dunkle. 1988. *A Computer Program Incorporating Pitzer's Equations for Calculation of Geochemical Reactions in Brines*. U.S. Geological Survey Water-Resources Investigations Report 88–4153.

Plummer, L. N., D. L. Parkhurst, and D. C. Thorstenson. 1983. Development of reaction models for ground-water systems. *Geochimica et Cosmochimica Acta* 47:665–86.

Plummer, L.N., E. C. Prestemon, and D. L. Parkhurst. 1991. *An Interactive Code (NETPATH) for Modeling Net Geochemical Reactions along a Flow Path*. U.S. Geological Survey Water-Resources Investigations Report 91–4078.

Sposito, G., and S. V. Mattigod. 1980. *GEOCHEM: A Computer Program for the Calculation of Chemical Equilibria in Soil Solutions and Other Natural Water Systems*. Riverside and Berkeley, CA: Kearny Foundation Report, University of California.

Thode, H. G., M. Shima, C. E. Rees, and K. V. Krishnamurty. 1965. Carbon-13 isotope effects in systems containing carbon dioxide, bicarbonate, carbonate, and metal ions. *Canadian Journal of Chemistry* 43:582–95.

U. S. Environmental Protection Agency. 1986. *Secondary Maximum Contaminant Levels* (section 143.3 of part 143, national secondary drinking-water regulations). U. S. Code of Federal Regulations, Title 40, Parts 100–149, revised as of July 1, 1986, p. 374.

Wigley, T. M. L., L. N. Plummer, and F. J. Pearson, Jr. 1978. Mass transfer and carbon isotope evolution in natural water systems. *Geochimica et Cosmochimica Acta* 42:1117–39.

Wigley, T. M. L., L. N. Plummer, and F. J. Pearson, Jr. 1979. Errata. *Geochimica et Cosmochimica Acta* 43:1395.

Wolery, T. J. 1979. *Calculation of Chemical Equilibrium between Aqueous Solution and Minerals: The EQ3/6 Software Package.* Livermore, CA: Lawrence Livermore National Laboratory, Report UCRL-52658.

Wolery, T. J., K. J. Jackson, W. L. Bourcier, C. J. Bruton, B. E. Viani, K. G. Knauss, and J. M. Delany. 1990. Current status of the EQ3/6 software package for geochemical modeling. In *Chemical Modeling of Aqueous Systems II*, eds. Daniel C. Melchior and R. L. Bassett, pp. 104–16. Washington DC: American Chemical Society Symposium Series 416.

Yeh, G. T., and V. S. Tripathi. 1989. A critical evaluation of recent developments in hydrogeochemical transport models of reactive multichemical components. *Water Resources Research* 25(1):93–108.

10

Uses of Environmental Isotopes

Tyler B. Coplen

BASIC CONCEPTS

Isotopes are atoms of the same element that differ in mass because of a difference in the number of neutrons in the nucleus. The naturally occurring elements give rise to more than 1,000 stable and radioactive isotopes, commonly referred to as *environmental isotopes*. Globally distributed isotopes produced from anthropogenic sources, such as above-ground nuclear-detonation testing, also are considered to be environmental isotopes. The average terrestrial abundances of isotopes useful in hydrogeologic studies are shown in Table 10-1.

This chapter emphasizes the practical uses of selected environmental isotopes in ground-water studies, but omits a discussion of the uses of artificial isotopes, such as for tracer tests between closely spaced wells [see IAEA (1991) for a discussion of this topic]. Additionally, ^{14}C is discussed in Chapter 9, ^{3}H, ^{85}Kr, and ^{3}He isotopes are discussed in Chapter 11, and U isotopes are discussed in Chapter 17.

Stable Isotopes

The variation in the abundance of the stable isotopes of the elements hydrogen, carbon, nitrogen, oxygen, silicon, and sulfur in natural substances is small. For most studies only the relative ratio of the rare isotope to the common isotope needs to be considered. This is important because the relative ratio of two isotopes can be determined about an order of magnitude more accurately than the absolute ratio of two isotopes.

Isotopic fractionation is the fractionation or partitioning of isotopes by physical or chemical processes and is proportional to the differences in their masses. Physical isotopic fractionation processes are those in which diffusion rates are mass dependent, such as ultrafiltration or gaseous diffusion of ions or molecules. Chemical isotopic fractionation processes involve redistribution of isotopes of an element among phases or chemical species. They can be either (1) equilibrium isotopic reactions—the forward and backward reaction rates of the rare isotopic species are identical to each other, and those of the common species are identical to each other (but not to the rates of the rare species)—or (2) kinetic isotopic reactions, produced by unidirectional reactions in which reaction rates are mass dependent. In equilibrium isotopic reactions, in general, the heavy isotope will be enriched in the compound with the higher oxidation state.

TABLE 10-1. Average Terrestrial Abundance of Isotopes Used in Hydrogeologic Studies

Element	Isotope	Average Terrestrial Abundance (atom %)	Comments
Hydrogen	^{1}H	99.985	
	^{2}H	0.015	
	^{3}H	$<10^{-14}$	Radioactive, $t_{1/2} = 12.43$ years
Helium	^{3}He	0.00014	
	^{4}He	99.99986	
Carbon	^{12}C	98.90	
	^{13}C	1.10	
	^{14}C	$<10^{-10}$	Radioactive, $t_{1/2} = 5,715$ years
Nitrogen	^{14}N	99.63	
	^{15}N	0.37	
Oxygen	^{16}O	99.762	
	^{17}O	0.038[a]	
	^{18}O	0.200	
Silicon	^{28}Si	92.23	
	^{29}Si	4.67[a]	
	^{30}Si	3.10	
	^{32}Si	$<10^{-12}$	Radioactive, $t_{1/2} = 172$ years
Sulfur	^{32}S	95.02	
	^{33}S	0.75[a]	
	^{34}S	4.21	
	^{35}S	$<10^{-11}$	Radioactive, $t_{1/2} = 82$ days
	^{36}S	0.014[a]	
Chloride	^{35}Cl	75.77	
	^{36}Cl	$<10^{-12}$	Radioactive, $t_{1/2} = 300,000$ years
	^{37}Cl	24.23	
Strontium	^{84}Sr	0.56[a]	
	^{86}Sr	9.86	
	^{87}Sr	7.00	
	^{88}Sr	82.58[a]	
Uranium	^{234}U	0.0055	Radioactive, $t_{1/2} = 247,000$ years
	^{235}U	0.72	Radioactive, $t_{1/2} = 71.3 \times 10^6$ years
	^{238}U	99.27	Radioactive, $t_{1/2} = 4.51 \times 10^9$ years

Source: IUPAC (1992).
[a] These isotopes are presently not used in ground-water studies.

Thus, ^{13}C is enriched in CO_2 relative to graphite, and in graphite relative to methane. In kinetic processes, statistical mechanics predicts that the lighter (lower atomic mass) of two isotopes of an element will form the weaker and more easily broken bond. The lighter isotope is more reactive; hence, it is concentrated in reaction products, enriching reactants in the heavier isotope. Examples of kinetic isotopic reactions include biological reactions, treatment of limestone with acid to liberate carbon dioxide, and the rapid freezing of water to ice. Sulfate reduction by bacteria in respiration (Goldhaber and Kaplan 1974) is an example of a biologically mediated kinetic isotopic fractionation process. Kinetic isotopic fractionations of biological processes are variable in magnitude, making interpretation of isotopic data (primarily of nitrogen and sulfur) difficult. Kinetic isotopic fractionations may be in the direction opposite to that of equilibrium fractionations.

Isotopic equilibrium between two phases does not mean that the two phases have identical proportions of heavy and light isotopes, only that the ratio of these proportions is always constant. Water vapor in a closed container in contact with liquid water at a constant temperature is an example of two phases in oxygen and hydrogen isotopic equilibrium.

The partitioning of stable isotopes between two substances A and B is described by the *isotopic fractionation factor* α,

$$\alpha_{A-B} = \frac{R_A}{R_B} \quad (10\text{-}1)$$

where $R_x = (D/H)_x$, $(^{13}C/^{12}C)_x$, $(^{15}N/^{14}N)_x$, $(^{18}O/^{16}O)_x$, or $(^{34}S/^{32}S)_x$ of phase or species x. Note that R is the ratio of the rare isotope to the common isotope, and that the isotope ^2H is given the name deuterium and the symbol D.

If the isotopes are randomly distributed in all positions in A and B, α is related to the equilibrium constant, K, by

$$\alpha = K^{1/n} \quad (10\text{-}2)$$

where n is the number of atoms exchanged. For example, the oxygen isotope exchange reaction between sulfate and water, written for the exchange of one atom of oxygen, is

$$\tfrac{1}{4}S^{16}O_4^{2-} + H_2^{18}O$$
$$\rightleftharpoons \tfrac{1}{4}S^{18}O_4^{2-} + H_2^{16}O \quad (10\text{-}3)$$

Then the equilibrium constant is

$$K = \frac{[S^{18}O_4^{2-}]^{1/4}[H_2^{16}O]}{[S^{16}O_4^{2-}]^{1/4}[H_2^{18}O]}$$
$$= \frac{(^{18}O/^{16}O)_{SO_4^{2-}}}{(^{18}O/^{16}O)_{H_2O}} = \alpha \quad (10\text{-}4)$$

where brackets refer to activities and parentheses refer to isotope abundance ratios. Values for α are normally close to unity. For oxygen isotope exchange between dissolved sulfate and water, α is 1.020 at 88°C, and sulfate and water are species A and B in Eq. 10-1. Note that the oxygen isotopic fractionation factor between *water* and *dissolved sulfate* at 88°C is $1/1.0200 = 0.9804$. Isotopic fractionations decrease with increasing temperature—at infinite temperature all isotopic species are well mixed and all isotopic fractionation factors are 1.0000. Equilibrium isotopic fractionation factors are determined experimentally or calculated from spectroscopic data. Friedman and O'Neil (1977) and Kyser (1987) have summarized isotopic fractionation factors of hydrologic and geologic interest.

Stable isotope ratios are reported relative to a standard as δ *values* (pronounced delta, not del) in units of parts per thousand (per mil and written ‰). Thus for oxygen, we have

$$\delta^{18}O = \left[\frac{[^{18}O/^{16}O]_{sample}}{[^{18}O/^{16}O]_{standard}} - 1\right] \times 1{,}000 \quad (10\text{-}5)$$

This is equivalent to the general expression

$$\delta_x = \delta_{x-std} = \left[\frac{R_x}{R_{standard}} - 1\right] \times 1{,}000 \quad (10\text{-}6)$$

where R_x and $R_{standard}$ are D/H, $^{13}C/^{12}C$, $^{15}N/^{14}N$, $^{18}O/^{16}O$, $^{34}S/^{32}S$, or $^{87}Sr/^{86}Sr$ of the sample and standard, respectively. Thus, an oxygen sample that is +50.0‰ is enriched in ^{18}O by 5% or 50.0‰ relative to the standard; the sample is isotopically "heavy" relative to the standard. A negative δ value indicates the sample is depleted in the heavy isotope relative to the standard; the sample is isotopically "light" relative to the standard. Oxygen and hydrogen isotope values for water are reported relative to VSMOW (Vienna Standard Mean Ocean Water) or SMOW. Thus, the $\delta^{18}O$ and δD of VSMOW are both 0‰. Oxygen-isotope ratios in marine carbonates and carbon-isotope ratios are reported relative to the PDB (Peedee belemnite) or VPDB (Vienna PDB) carbonate standard (Hut 1987). Nitrogen in air is well mixed and is the nitrogen isotope standard (AIR). Sulfur isotope values are reported relative to troilite from the Canyon Diablo meteorite (CDT). Strontium isotope ratios are reported relative to seawater or as the direct ratio $^{87}Sr/^{86}Sr$ (Elderfield 1986). The precision attainable in many laboratories for sample preparation and analysis of stable isotope samples is 1‰ for δD, 0.1‰ for $\delta^{13}C$, $\delta^{15}N$, $\delta^{18}O$ and $\delta^{34}S$, and ~0.005‰ for $\delta^{87}Sr$.

A useful relationship between δ values and the isotopic fractionation factor is

$$\alpha_{A-B} = \frac{1,000 + \delta_A}{1,000 + \delta_B} \quad (10\text{-}7)$$

The difference in isotopic content, Δ, between two phases or species A and B is defined by

$$\Delta_{A-B} = \delta_A - \delta_B \simeq 1,000 \ln \alpha_{A-B} \quad (10\text{-}8)$$

because $\ln(1 + x) \simeq x$ for small values of x. Therefore, if the oxygen-isotope delta values of two minerals in isotopic equilibrium are $\delta_A = +20$‰ and $\delta_B = +5$‰, then $1,000 \ln \alpha_{A-B} \simeq 15$‰ and $\alpha_{A-B} \simeq 1.015$.

Another fractionation factor, ϵ, is occasionally used and is defined by

$$\epsilon = (\alpha - 1) \times 1,000 \quad (10\text{-}9)$$

For small values of ϵ, $\epsilon_{A-B} \simeq \delta_A - \delta_B \simeq 1,000 \ln \alpha_{A-B}$.

The isotopic composition of substances formed by combining two or more components with different isotopic compositions is additive, as is salinity, and is determined by isotope mass balance using

$$\delta_1 N_1 + \delta_2 N_2 + \delta_3 N_3 \ldots = \delta_f(N_1 + N_2 + N_3 \ldots) \quad (10\text{-}10)$$

where δ_1 is the δ value of component 1, N_1 equals the number of atoms in component 1, and δ_f is the δ value of the product. As an example, addition of 75% of water A of $\delta D = -20$‰ and 25% of water B of $\delta D = -60$‰ yields water of $\delta D = -30$‰. An important consequence of this relationship is that, on a linear plot of δD versus $\delta^{18}O$, the isotopic composition of mixtures of various proportions of the two waters will lie on a straight line connecting the δ values of the two waters. Mixing equal portions of two waters, A and B, results in a water whose δ values are midway between the delta values of A and B.

Lord Rayleigh (1896) published equations that describe the separation of gases during distillation. These *Rayleigh distillation* equations also describe the isotopic enrichment or depletion of "distillation-like processes," such as crystallization (with isolation of precipitate without further isotopic exchange with the fluid), condensation, evaporation, and passage of water through shale micropores.

Consider the hydrogen and oxygen isotopic fractionation of ground water during ultrafiltration (Coplen and Hanshaw 1973). If a solution is forced through a shale micropore, the δ value of the residual solution, δ_r, is given by

$$\delta_r = 1,000[(1 - F)^{\alpha - 1} - 1] + \delta_{ro} \quad (10\text{-}11)$$

where δ_{ro} is the initial δ value, F is the fraction of water filtered, and α is the isotopic fractionation factor. Figure 10-1 shows an example in which α, the isotopic fractionation factor between the residual fluid and the instantaneous ultrafiltrate, is 0.9990. Thus, the ultrafiltrate is depleted in the heavy isotope relative to the residual solution. As ultrafiltrate is produced, the residual solution is progressively enriched in the heavy isotope. The last remaining residual solution can be greatly enriched in the heavy isotope relative to the initial isotopic composition. When all of the solution has been forced through the shale micropore, the δ value of

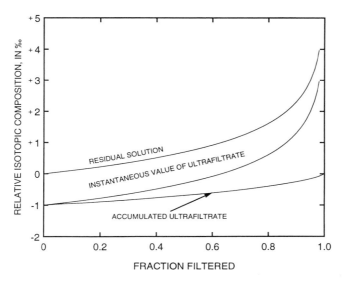

FIGURE 10-1. Relative isotopic compositions of the residual solution and ultrafiltrate during passage of water through a shale micropore—example of Rayleigh isotopic fractionation. By inverting scale on the ordinate (changing + values to −, and vice versa), the same curves illustrate the change in isotopic composition of precipitation: residual solution represents the water in clouds, instantaneous ultrafiltrate is rain, and accumulated ultrafiltrate is accumulated rain along cloud trajectory.

the accumulated ultrafiltrate must be identical to that of the initial residual solution by isotope mass balance.

Extensive use has been made of the Rayleigh distillation equation to model the isotopic content of precipitation (see Gat 1980); the isotopic composition of water vapor remaining in a cloud during precipitation is represented by δ_r in Eq. 10-11 and that of precipitation is calculated by isotope-mass balance, Eq. 10-10.

Radioactive Isotopes

The concentration of a radioactive substance decreases in a ground-water system according to the *radioactive decay law*

$$A = A_0 e^{-\lambda t} \quad (10\text{-}12)$$

where A_0 is the initial activity at the time of isolation at the surface, t is the time since isolation, A is the activity at time t, and λ equals the decay constant for that isotope (independent of chemical state). The decay constant is related to the half-life, $t_{1/2}$, by

$$t_{1/2} = \frac{\ln 2}{\lambda} = \frac{0.693}{\lambda} \quad (10\text{-}13)$$

Equation 10-12 can be rearranged to

$$t = t_{1/2}(1.44) \ln \left(\frac{A_0}{A}\right) \quad (10\text{-}14)$$

Application of Eq. 10-12 to give ages of ground-water systems often requires corrections for additional sources and sinks for the radioactive substance, such as from mixing, gas exchange, mineral dissolution or precipitation, oxidation of organic matter, etc. The correct value of A_0 is not always known and may have differed over the time scale of the application.

REVIEW OF SELECTED ISOTOPES

The following sections discuss the isotopic composition of different sources of isotopes

FIGURE 10-2. Distribution of mean $\delta^{18}O$ of precipitation, based on stations having at least two years of records. (*From Yurtsever and Gat 1981, published with permission of IAEA.*)

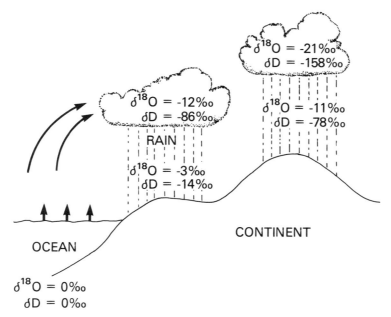

FIGURE 10-3. Hydrogen and oxygen isotopic content of ocean water and precipitation. Due to kinetic isotope fractionation during evaporation of moist oceanic vapor, the hydrogen and oxygen isotopic content of water vapor above the oceans is −86‰ and −12‰, respectively. The isotopic content of precipitation is in isotopic equilibrium with vapor in a cloud. Thus, the δD and δ^{18}O values of the first rain are −14‰ and −3‰, respectively, depleting the cloud in D and ^{18}O as precipitation occurs. Therefore, precipitation becomes isotopically lighter and lighter as it is continuously removed from clouds.

of H, C, N, O, S, Cl, and so on. These discussions of individual isotopes are followed by a review of key applications of isotopes in hydrologic investigations.

Hydrogen and Oxygen Isotopes

Because stable hydrogen and oxygen are intimately associated in the water molecule, isotopic fractionations of both are usually covariant. Consequently, the isotopic ratios of these two elements generally are discussed together. Uses of the radioactive isotope of hydrogen, ^{3}H (tritium), which has a half-life of 12.43 years, are discussed in Chapter 11.

Ground water in virtually all systems originates as precipitation. Thus, the spatial and temporal variations in isotopic content of precipitation can be used to investigate ground-water recharge. The IAEA (International Atomic Energy Agency), in cooperation with the WMO (World Meteorological Organization), has collected and analyzed global precipitation samples on a monthly basis since 1961 for D and ^{18}O content (Figure 10-2).

Factors That Control the Isotopic Content of Precipitation

The majority of the Earth's atmospheric water vapor originates from evaporation of low-latitude oceans. Precipitation is enriched in D and ^{18}O relative to atmospheric vapor. Thus, precipitation is progressively depleted in D and ^{18}O as air masses are transported to higher latitudes and lower temperatures. The isotopic composition of precipitation is related to the following factors (see also Figure 10-3).

Altitude Effect. On the windward side of a mountain ^{18}O and D abundances decrease with increasing altitude. Although

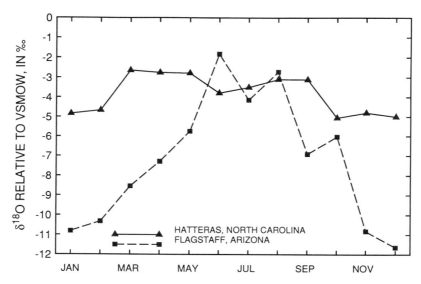

FIGURE 10-4. Monthly weighted mean $\delta^{18}O$ of a coastal (Hatteras) and an inland (Flagstaff) station. The moderating effect of the ocean upon temperature is evident in the low variability of the Hatteras $\delta^{18}O$ data. (*Data from IAEA 1981b.*)

these abundances depend upon topography and local climate, typical gradients for $\delta^{18}O$ are −0.15 to −0.5‰ per 100 m and for δD are −1.5 to −4‰ per 100 m (Yurtsever and Gat 1981). As rain falls it can evaporate, giving an additional altitude effect since evaporation is greater for rain drops that fall farther. Note that the altitude effect is often not seen (1) on mountains interior to a continent (such as in Wyoming), (2) for precipitation of snow, and (3) on the lee side of mountains.

Latitude Effect. D and ^{18}O content decreases with increasing length of storm path due to Rayleigh isotopic fractionation; thus, D and ^{18}O content decreases with increasing latitude (Figure 10-2). Over North America, the rate of $\delta^{18}O$ change is about −0.5‰ per degree of latitude.

Continental Effect. The D and ^{18}O content decreases inland from the coast.

Seasonal Variation. Winter precipitation is depleted in D and ^{18}O relative to summer rains (see Flagstaff data in Figure 10-4). This effect can be large; Fritz, Drimmie, and Render (1974) observed a 10‰ difference in oxygen isotopes in the Red River in the north central United States between winter and summer.

Amount Effect. The greater the rainfall, the more depleted the D and ^{18}O content. Evaporation increases the D and ^{18}O content of precipitation of small rainfall events more than large rainfall events. This effect does not occur for snow.

Apparent Temperature Relationship. The progressive condensation of vapor in isotopic equilibrium from a cloud during transport to higher latitudes with lower temperatures makes it difficult to separate latitude (or altitude) from temperature effects. Thus, van der Straaten and Mook (1983) observe latitudinal temperature effects of $d\delta^{18}O/dt = (0.62 \pm 0.10)$‰/°C and $d\delta D/dt = (5.1 \pm 0.9)$‰/°C, depending upon the temperature range.

Relationship between δD and $\delta^{18}O$ in Various Waters

Precipitation. The relationship between δD and $\delta^{18}O$ in meteoric water is represented (Figure 10-5) by

$$\delta D = 8\delta^{18}O + d \quad (10\text{-}15)$$

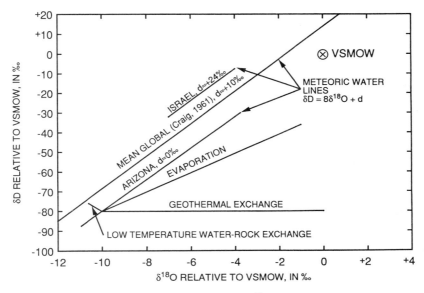

FIGURE 10-5. The relationship between δD and δ^{18}O of meteoric water, evaporating water, and water interacting with rock.

where d is the deuterium excess parameter. Craig (1961) found that the global mean value of d for freshwater sources is +10‰; however, the value of d may differ significantly from area to area and may differ over geologic time. For instance, the value for North America is +6‰ and that for the Mediterranean region is about +22‰ (Gat and Carmi 1970). The value of d may be used as an isotopic tag. For instance, during Pleistocene glaciation, cooler and more humid conditions were often characterized (probably at mid and high latitudes) by a lower deuterium excess (Merlivat and Jouzel 1979). In many studies (as discussed later), the d value of ground water differs significantly from that of modern recharge, supporting the hypothesis that the ground water is not modern water.

Evaporating Water. Evaporation from surface-water bodies is a nonequilibrium process that enriches D and ^{18}O in the water such that the δD/δ^{18}O slope is less than 8 and usually between 3 and 6 (Figure 10-5). The slope is a complicated function of humidity, temperature, salt concentration, and other factors. As salt concentration increases to high values, the isotopic content of remaining brine can decrease in D and ^{18}O abundance (Gat 1981). As water evaporates from highly saline brines, the activity of water decreases and the isotopic content of the remaining brine can actually decrease in D and ^{18}O abundance (Gat 1981) as exchange with ambient humidity becomes a dominant factor; ultimately this exchange proceeds to bring the brine into isotopic equilibrium with the ambient vapor.

Geothermal Exchange. Exchange of oxygen in water and rocks at high temperatures in hydrothermal systems increases the ^{18}O content of the water (δ^{18}O "shift") and decreases that of the rocks (Figure 10-5) in order to bring them into isotopic equilibrium. A mass transfer of only 2–4% of the oxygen can make an observable (0.5‰) shift in δ^{18}O of the water in some systems, depending upon the δ^{18}O of the rock. Because rocks contain little hydrogen relative to that in water, there is no observable shift in the δD of water.

Low-Temperature Water-Rock Exchange. A common diagenetic reaction, silicate hydrolysis, produces hydrated minerals. This

TABLE 10-2. Processes for the Production of Dissolved Carbon in Ground Water in the Closed System Model and the Effect on the Final ^{13}C and ^{14}C Content[a]

Process	Isotopic Content of Reactant		Dissolved Carbon Isotopic Content	
	$\delta^{13}C_{VPDB}$ (‰)	^{14}C (pmc)	$\delta^{13}C_{VPDB}$ (‰)	^{14}C (pmc)
A. Dissolution of marine carbonate[b]				
1. By root respiration CO$_2$ or decomposition of soil organic matter				
a. Temperate climate	−25	100	−12	50–100[c]
b. Arid climate	−15	100	−7	50–100[c]
2. By atmospheric CO$_2$ in areas of little or no vegetation (Pearson and Friedman 1970)	−7	100	−3	50–100[c]
3. By H$^+$ of humic acids (Vogel and Ehhalt 1963)	—	—	+1	0
4. By acid produced during oxidation of organic matter (Pearson and Hanshaw 1970)	−25	Variable	−12	Variable
5. By decreasing Ca concentration via Ca-Na exchange (Pearson and Swarzenki 1974)	—	—	+1	0
6. By CO$_2$ from hydrothermal or volcanic activity	Variable	0	Variable	0
7. By precipitating calcite during concurrent dissolution of dolomite (Back et al. 1983)	Variable	0	Variable	Variable
8. By CO$_2$ from coalification of lignitic detritus in deltaic or marginal marine aquifers (Winograd and Farlekas 1974)	−23	Variable	−11	Variable
B. Silicate weathering in carbonate-free environments (Pearson and Friedman 1970)				
a. Temperate climate	−25	Variable	−25	100
b. Arid climate	−15	100	−15	100
C. Sulfate reduction (Pearson and Hanshaw 1970)	Variable	0	Variable	0
D. Isotope exchange (recrystallization)	Variable	Variable	Variable	Variable[d]
E. Degradation of organic matter into methane and dissolved carbon (Baedecker and Back 1979)	−25	Variable	Up to +18	Variable

[a] In open-system environments one must also consider carbon isotopic exchange of dissolved carbon with soil CO$_2$ (Cerling 1984).
[b] $\delta^{13}C_{VPDB}$ and ^{14}C of marine carbonate are +1‰ and 0 pmc, respectively.
[c] Depending on degree of "openness" of system to atmospheric CO$_2$.
[d] ^{14}C decreases along flow path; especially important in hydrothermal systems.

reaction will increase the rock in ^{18}O content decreasing the water in ^{18}O; it may also increase the D content of water. Thus, on a plot of δD as a function of δ^{18}O, waters involved in silicate hydrolysis may plot to the left of the meteoric waterline (Figure 10-5) and the magnitude of the mass transfer needed to make an observable shift in δ^{18}O will be of the same order of magnitude as for geothermal exchange.

Carbon

^{13}C

Carbon in ground water is derived primarily from the following components of the carbon cycle: carbonate sediments, soil humus, and decay of the land-plant biomass. In less usual circumstances, carbon may be derived from atmospheric or volcanic CO_2. Table 10-2 lists typical isotopic compositions of carbon-bearing sources.

The chemical speciation of dissolved inorganic carbon is pH dependent. Bicarbonate is dominant for pH values between 6.4 and 10.3. Carbonic acid prevails below 6.4, and dissolved carbonate is dominant above 10.3. The equilibrium isotopic fractionation between some carbon species (Figure 10-6) is relatively large (~10‰). As a consequence, $\delta^{13}C$ of total dissolved inorganic carbon (DIC) in a ground-water sample, which is the parameter measured in the laboratory, is a function not only of the isotopic compositions of the individual species but also of the pH of the sample (Wigley, Plummer, and Pearson 1978, 1979). Thus, calculation of the $\delta^{13}C$ of gaseous CO_2 in carbon isotopic equilibrium with ground water requires knowledge of pH, temperature, and $\delta^{13}C$ of DIC, and is often performed using an aqueous model that takes into account activity coefficients and ion pairs.

The chemical reaction that describes the primary process for producing DIC in ground water in regions containing calcium carbonate sediments is

$$CO_2 + CaCO_3 + H_2O \rightarrow Ca^{2+} + 2HCO_3^-$$
(10-16)

The reactant CO_2 is usually root respiration CO_2 or CO_2 derived from decomposition of soil organic matter, organic matter buried in the aquifer matrix, and less often atmospheric CO_2. CO_2 reacts with carbonate sediments, which are pervasive in the lithosphere, to produce dissolved calcium bicarbonate. Note that the bicarbonate in

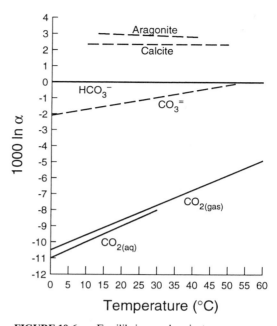

FIGURE 10-6. Equilibrium carbon isotope fractionation factors of geochemical interest. (*Data from Mook, Bommerson, and Staverman 1974; Grootes, Mook, and Vogel 1969; Turner 1982; Coplen, unpublished data.*)

Eq. 10-16 contains carbon from two different sources (reactants), gaseous CO_2, and "dead" calcium carbonate, so that the ^{13}C and ^{14}C content of the bicarbonate depends upon the ^{13}C and ^{14}C content of reactants. Table 10-2 lists reactions that affect the ^{13}C and ^{14}C content of DIC.

Generation of significant amounts of methane can occur in sanitary landfills under anaerobic conditions. On the basis of ^{13}C and ^{14}C content (Barker 1978), methane from landfills [$\delta^{13}C = -40$ to -55‰; ^{14}C content $= \sim100$ percent modern carbon (pmc)] can often be distinguished from methane from natural gas and petroleum deposits ($\delta^{13}C = -25$ to -75‰; ^{14}C activity $= \sim0$ pmc, i.e., "dead") and from methane from ground water and glacial drift gas deposits ($\delta^{13}C = -25$ to -90‰, with a median at -70‰; ^{14}C activity typically less than 5 pmc). In sanitary landfills, the degradation of organic material is a kinetic reaction that yields isotopically light methane and heavy bicarbonate [with $\delta^{13}C$ values as high as +18‰ (Baedecker and Back 1979)].

^{14}C in ground water is fractionated by the same processes that fractionate ^{13}C. Because fractionations of ^{14}C and ^{13}C are covariant, an important use of ^{13}C data is to correct ^{14}C concentration for addition of carbon by using Eq. 10-10. As discussed in the section titled "Geochemical Reaction Modeling" and Chapter 9, $\delta^{13}C$ of DIC can be used to evaluate geochemical models by calculating the $\delta^{13}C$ of DIC along flow paths.

^{14}C

As a consequence of continuous production of ^{14}C in the atmosphere by the interaction of secondary cosmic ray neutrons and nitrogen nuclei and continuous exchange in the biosphere, all living organic matter prior to 1954 contained the same concentration of ^{14}C within narrow limits ($\pm 10\%$). Since the beginning of above-ground nuclear-detonation testing in 1954, global atmospheric ^{14}C content has risen more than 50% to a maximum in the mid-1960s (IAEA 1983a).

^{14}C can enter a ground-water system by various geochemical and biogenic processes (see Table 10-2). The radioactive decay of ^{14}C can be used to date ground water (see discussion of ground-water dating in this chapter). Additionally, ^{14}C can be used to characterize a ground-water mass, as discussed later. In either case, all ^{14}C samples should be analyzed for ^{3}H content to identify input of any recent water. If ^{3}H is detectable in an "old" water sample, the sample is a mixture of young and old water, and an age is meaningless.

^{14}C activity is expressed in pmc. Occasionally, ^{14}C values are expressed in per mil using the relationship

$$A \text{ (in ‰)} = [100 - a^{14}] \times 1{,}000 \quad (10\text{-}17)$$

where a^{14} is the ^{14}C activity expressed in pmc.

Nitrogen

The variation in $\delta^{15}N$ in terrestrial materials is on the order of 100‰ because of the variation in oxidation states of N from -3 to $+5$; species and compounds include NH_3, NH_4^+, N_2, N_2O, NO_2^-, NO_2, and NO_3^-. Generally, the more oxidized the chemical species, the more positive is the $\delta^{15}N$ value. Although equilibrium isotope fractionation factors are known for many pairs of species, kinetic isotope fractionations often dominate nitrogen reactions.

Nitrogen isotope studies in ground-water systems [see general reviews by Heaton (1986) and Hübner (1986)] generally fall into one of two main categories: (1) identifying sources of nitrate contamination (e.g., synthetic fertilizers versus animal wastes) or (2) documenting oxidation and reduction (redox) reactions involving nitrogen species (e.g., denitrification). These two categories of use commonly are incompatible; the first normally requires the assumption that source characteristics are fixed in a ground-water sample in the recharge area and do not change subsequently, whereas the second deals specifically with changes that occur along ground-water flow paths. Complex situations in which source characteristics change with time and redox conditions change along flow paths may be resolved with assistance from ground-water dating, but otherwise may be difficult to interpret.

Numerous studies have demonstrated that the nitrogen isotopic composition of nitrate dissolved in ground water can indicate the dominant source of the nitrate in certain situations (Kreitler 1975, 1979; Gormly and Spalding 1979; Spalding et al. 1982; Kreitler and Browning 1983). For example, Gormly and Spalding (1979) found that the primary sources of ground-water nitrate and the corresponding $\delta^{15}N$ values in portions of Nebraska are (1) soil nitrogen ($\delta^{15}N$ of total nitrogen = +5 to +9‰ and $\delta^{15}N$ of nitrate nitrogen = +2 to +9‰); (2) commercial fertilizer (because it is produced from air, $\delta^{15}N = -2$ to $+7‰$, typically $< +3.5‰$); and (3) animal wastes ($\delta^{15}N$ of nitrate $= +10$ to $+23‰$). That study did not consider input of atmospherically derived nitrate (Moore 1977; Heaton 1987; Freyer 1978), which may be significant in many regions. The $\delta^{15}N$ of total nitrogen in fresh animal waste is $+1$ to $+6‰$. After wastes are deposited on the ground, volatilization of ^{14}N-enriched ammonia increases the ^{15}N content of remaining waste by $10‰$ or more. Subsequent nitrification occurs without major nitrogen isotopic fractionation and yields $\delta^{15}N$ values of nitrate from animal wastes of $+10$ to $+23‰$. Likewise, ammonia fertilizer undergoes nitrification without large nitrogen isotopic fractionation. In favorable situations, therefore, nitrate derived from animal wastes and from commercial fertilizers may be distinguished.

Independently of source considerations, numerous studies have shown that nitrogen isotopic fractionation can indicate chemical transformations among different nitrogen species in ground water (e.g., Vogel, Talma, and Heaton 1981; Mariotti, Landreau, and Simon 1988; Böttcher et al. 1990; Smith, Howes, and Duff 1991). Transformations that fractionate isotopes include redox reactions, most of which are mediated by microorganisms, hydrolysis, and volatilization. For example, neutralization of ammonium and volatilization of ammonia from alkaline solutions preferentially removes ^{14}N and causes residual ammonium in the solution to become isotopically heavier. Biologically mediated redox reactions generally cause residual reactant species to become isotopically heavier because of kinetic isotope effects. Examples of reactions causing kinetic fractionations of interest in ground-water studies include nitrification, denitrification, and assimilation (Hübner 1986).

Reduction of nitrate to nitrogen gas (denitrification) is an important natural mechanism for removal of nitrate from ground water. Nitrogen isotope analyses have been used in numerous studies to distinguish between dilution and denitrification as causes for decreasing nitrate concentrations along ground-water flow paths (Vogel, Talma, and Heaton 1981; Mariotti, Landreau, and Simon 1988). Because of kinetic isotope fractionation effects, ^{15}N values of residual (unreacted) nitrate increase exponentially as concentrations decrease during denitrification. Kinetic isotope fractionation factors are somewhat variable (reviewed by Hübner 1986), so that curves relating ^{15}N and concentration differ locally. Nevertheless, the effects of denitrification commonly can be distinguished from those of simple dilution (no change in ^{15}N with decreasing concentration) or mixing (many patterns of variation possible). Supporting evidence for or against denitrification or other transformations is important and may include oxygen isotope analyses (Böttcher et al. 1990), ground-water dating (Vogel, Talma, and Heaton 1981; Postma et al. 1991; Böhlke et al. 1992), as well as chemical data such as pH, DIC, sulfate, and others. Field studies have shown that denitrification and related isotope effects in ground water occur over a wide variety of time and space scales. For example, denitrification zones and flow time scales are described at scales of meters and years (Smith, Howes, and Duff 1991; Postma et al. 1991), hundreds of meters and unspecified times (Mariotti, Landreau, and Simon 1988), and tens of kilometers and thousands of years (Vogel, Talma, and Heaton 1981).

Sulfur

The largest reservoirs of sulfate are terrestrial evaporite minerals (gypsum and anhydrite) and the oceans. Other sources of sulfate include (1) oxidation of sulfide minerals or organic sulfides, (2) atmospheric precipitation of ocean-derived aerosols, and (3) atmospheric precipitation of windblown mineral sulfate dust from soils and playas. Sources of sulfides include (1) ore deposits, (2) volcanic emanations, (3) reducing environments, such as bogs, (4) organic materials within aquifers, such as coal and oil, (5) reduction of sulfates, and (6) disseminated pyrite in igneous and metamorphic rocks and marine sediments.

The existence of oxidation states of sulfur from −2 to +6 gives rise to large variations in the sulfur isotopic content of natural materials (see Goldhaber and Kaplan 1974). The $\delta^{34}S$ of dissolved sulfur in ground water (see Pearson and Rightmire 1980; Krouse 1980) may in some cases be difficult to predict due to the following factors: (1) sulfur speciation is a function of pH, with SO_4^{2-} occurring in oxidizing environments, H_2S in acidic reducing environments, and HS^- in neutral pH, reducing environments; (2) equilibrium exchange reactions such as between SO_4^{2-} and H_2S are slow; (3) bacterial reactions are kinetic and may be orders of magnitude faster than equilibrium exchange reactions; and (4) sulfur isotopic fractionation factors of bacterial reactions may be greatly different than those of eqilibrium exchange reactions. For instance, although the equilibrium isotopic fractionation factor between SO_4^{2-} and H_2S is ~1.074 at 25°C [see Figure 6-1 of Pearson and Rightmire (1980) for isotopic fractionation factors of other sulfur species], sulfate is depleted in ^{34}S during bacterial oxidation of sulfide. This is in the opposite direction to that expected for isotopic equilibrium. Therefore, in any investigation using sulfur isotopes, it is important to ensure that effects of kinetic isotopic fractionation by bacteria are either negligible or well known. Although the $\delta^{34}S$ value of contemporary dissolved oceanic sulfate and modern sulfate evaporite minerals is +20‰, that of ancient evaporites ranges from +10 to +35‰ (Claypool et al. 1980). Because there is no isotopic fractionation during dissolution, this variation in the ^{34}S content of ancient evaporites can be used, under favorable conditions, to identify the particular evaporite formation being dissolved.

The $\delta^{34}S$ of atmospheric precipitation ranges from ~+20‰ in regions containing only sea spray sulfate to ~0‰ in highly industrialized regions. Therefore, in any study involving sulfur isotopes, the $\delta^{34}S$ of atmospheric samples should be measured in situ rather than estimated using data from another location.

The ^{34}S content of dissolved sulfur in ground water can be used to test geochemical models by calculating the $\delta^{34}S$ of the chemical reaction paths (see section titled "Geochemical Reaction Modeling").

Analysis of the $\delta^{18}O$ of sulfate can be useful in isotopic geothermometry. Friedman and O'Neil (1977) show isotopic fractionation factors as a function of temperature.

Chlorine

^{36}Cl

^{36}Cl is produced in the atmosphere by spallation of heavier nuclei by cosmic rays and by slow neutron activation of ^{36}Ar. The half-life of this hydrophilic, nonreactive isotope is 300,000 years, which makes it suitable for dating in the range of 60,000 to 1 million years (60 to 1,000 ka), (see this age range in ground-water dating section later). Additionally, large amounts of ^{36}Cl were produced during above-ground nuclear-detonation testing between 1952 and 1958; thus, in soil and ground water, as an event marker of 1950s water, ^{36}Cl is useful for dating waters less than 50 years before the present (see ground-water dating section below and Chapter 11).

^{37}Cl

Long (1983) employed a technique to measure $^{37}Cl/^{35}Cl$, which shows promise as a ground-water tracer because variations of a few per mil have been observed in ground water.

Strontium

Increasing in use in ground-water studies, $^{87}Sr/^{86}Sr$ is a good tracer in the hydrologic cycle because Sr obtains its isotopic ratio by dissolution of or exchange with Sr-bearing minerals along a flow path. Higher ratios generally reflect dissolution of rocks containing ^{87}Sr produced by radioactive decay of ^{87}Rb-rich rocks. Weathering of silicates produces a spectrum of ^{87}Sr/^{86}Sr values due to variation in original Rb content. During some geologic periods ^{87}Sr/^{86}Sr values of marine carbonates were distinctive, "labeling" the carbonate. Because of the relatively high mass, no detectable isotopic fractionation accompanies precipitation of Sr within minerals. Therefore, isotopic analysis of strontium dissolved in water or precipitated in calcite can provide information on the origin of strontium in a system (Stuckless, Peterman, and Muhs 1991; Starinsky et al. 1983).

APPLICATIONS

Environmental isotopes that commonly have helped to provide solutions to or information about specific ground-water problems are summarized in Table 10-3. The reader can identify in Table 10-3 isotopes that aid in the solution of a particular problem, refer to the section on those isotopes and refer to specific applications areas below. The applications are discussed below in the order of their listing in Table 10-3. Important books on this topic include IAEA (1981a, 1983a, 1992) and Fritz and Fontes (1980).

Recharge and Flow Rate

Unsaturated Zone
Seasonal variability in D, ^{18}O, and ^{3}H content has not gained significant use in determining recharge rates in the unsaturated zone because short-term variations in isotopic abundance are masked by dispersion. However, the atomic-bomb-produced ^{3}H peak from the 1960s can be identified in vertical ^{3}H profiles in areas of low recharge. If the amount of soil water above this peak can be determined, recharge rates also can be determined (see Chapter 11).

Arid Zones
As in the foregoing case, vertical profiles of ^{3}H and ^{36}Cl/Cl abundance in wells can be used (see Table 10-3) to identify the mean depth to bomb-produced isotopes and to calculate ground-water recharge rates. Such investigations demonstrate that recharge occurs even in arid climates. Phillips et al. (1988) show that ^{36}Cl and ^{3}H from nuclear-detonation fallout have penetrated a few centimeters and 1 to 3 m, respectively, in New Mexico desert soils. Movement of ^{3}H is thought to occur by vapor transport, which does not affect ^{36}Cl.

Exchange of River and Lake Water with Ground Water
The seasonal variability of ^{18}O and D in river and lake water can be used to determine (1) relative quantities of recharge and (2) recharge velocities if dispersion does not mask these variations (see Table 10-3). McCarthy et al. (1992) showed that after five days of pumping, isotopically light Columbia River water comprised 50% of the flow of water from municipal wells located about 1 km from the river.

For optimal success in determining recharge velocities, monthly to quarterly sampling of wells located at properly spaced intervals from the river or lake is necessary. Stichler and Moser (1979) used this technique on a small lake in the Upper Rhine Valley. Samples from the lake and from about 25 nearby wells were collected monthly and analyzed for D and ^{18}O content. The D and ^{18}O content of all wells to the south were identical to each other and over time (within experimental error), indicating negligible water flow to the south. The wells to the north and east showed seasonal variation in δD and δ^{18}O. From the six-month lag in δD and δ^{18}O variations between two wells sited about 180 m apart, one can calculate a horizontal ground-water velocity of 1 m/d.

In a related subject (determining season of recharge) Winograd and Riggs (1984) used δD and δ^{18}O values to determine that summer precipitation does not contribute to ground-water recharge in Spring Mountains, Nevada.

TABLE 10-3. Ground-Water Problems Aided by Environmental Isotopes

Type of Problem	Stable Isotope Ratios						Radioactive Isotopes					
	δD	$\delta^{13}C$	$\delta^{15}N$	$\delta^{18}O$	$\delta^{34}S$	$\delta^{87}Sr$	^3H	^{14}C	^{36}Cl	^{39}Ar	^{85}Kr	U-diseq
Recharge and Flow Rate												
Unsaturated zone	C						C,a		B,b			
Arid zones							C,c		B,b			
Exchange of river and lake water with ground water	C,d			C,d			C,e	C,e				
Average ground-water flow rate in												
Systems less than 5 years old	A,f			A,f			C,f					
Systems between 5 and 30 years old							A,g	C	B,b			
Characterization of a Ground-Water Mass												
Local area less than 30 years old	C,h			C,h			C,i	C,i	B,b			
Regional systems	A,j	C,k		A,j				C,l	B,b			
Identification of Recharge Area or Source of Water												
Local area	A,m	C,n		A,m		C						
Regional systems	A,p	C		A,p		C						
Leakage between Aquifers	C,q			C,q			C,o	C,o	B,b			
Investigations of Ground-Water Flow in Fractured Rocks												
Carbonate karst rocks	A,r			A,r			A,r					
Noncarbonates	A,s			A,s			A,s	C				
Evaluation of Ground-Water Flow and Storage Characteristics												
Local system: well-mixed reservoir or piston flow							A,t					
Dispersivity investigations	C			C			C	C	C			
Separation of Stream Discharge into Ground-Water and Surface-Water Components	A,u			A,u			A,u					

Sources of Dissolved Constituents	A,v	A,y	A,w	A,v	A,aa	B,x	C,v	B,b
Geochemical Reaction Modeling					C,y		A,u	
Ground-Water Dating								
Less than 5 years	A,p	C,n	A,f				A,u	
5–50 years							A,g	C,i	B,b
50–1,000 years								C,bb	B,bb
1,000–40,000 years	A,z		A,z				A,cc	B,dd
60,000–1,200,000 years									C,ee

Key for Remarks

Frequency of Use in Ground-Water Studies
  A Has been useful in many studies
  B Has received some use; looks promising for future studies as technology improves
  C Has received some use

Example Applications
a Munnich (1983)
b Bentley, Phillips, and Davis (1986)
c IAEA (1980) and Gvirtzman and Magaritz (1986)
d Payne (1983), Stichler and Moser (1979), Carlin et al. (1975), Krabbenhoft et al. (1990), and Darling, Allen, and Armannsson (1990)
e Carlin et al. (1975)
f Stichler and Moser (1979) and Schotterer et al. (1979)
g Rauert and Stichler (1974), Fontes (1980), Phillips et al. (1988), and Solomon and Sudicky (1991)
h Payne, Quijano, and Latorre (1979)
i Fontes (1980) and Pearson et al. (1991)
j Airey et al. (1979), Gat (1971), and Pearson et al. (1991)
k Lloyd and Howard (1979)
l Fontes (1980)
m Schotterer et al. (1979), and Muir and Coplen (1981)
n Letolle and Olive (1983)
o Fontes (1980)
p Payne (1983), Gat (1971), Pearson et al. (1991), and Darling, Allen, and Armannsson (1990)
q Payne (1981)
r Davis et al. (1970) and Fontes (1983)
s Schotterer et al. (1979)
t Martinec et al. (1974)
u Sklash and Farvolden (1979 and 1982), Kennedy et al. (1986), and Stewart and McDonnell (1991)
v Payne (1983), Payne, Quijano, and Latorre (1979), and Simpson and Herczeg (1991)
w Heaton (1986) and Hübner (1986)
x Pearson et al. (1991) and Starinsky et al. (1983)
y Plummer et al. (1990) and Plummer, Prestemon, and Parkhurst (1991)
z Buchardt and Fritz (1980) and Pearson et al. (1991)
aa Claypool et al. (1980)
bb Pearson et al. (1991) and Chapter 11
cc Fontes (1983), IAEA (1983a), and Pearson et al. (1991)
dd Pearson et al. (1991), Torgersen et al. (1991), and Nolte et al. (1991)
ee Pearson et al. (1991)

Average Ground-Water Flow Rate in Systems Less than Five Years of Age

If the system under study exhibits piston flow, the seasonal variability in D, ^{18}O, or ^{3}H content can be used to determine the ground-water flow rate. The density of sampling sites along the flow path must be sufficiently dense that no annual cycles are missed. In the preceding example (Stichler and Moser 1979), well spacing was 40 m and one annual cycle stretched over 9 wells (360 m).

Average Ground-Water Flow Rate in Systems Greater than Five Years of Age

The bomb-produced ^{3}H, ^{14}C, or ^{36}Cl/Cl spike can be used to measure flow rates in systems exhibiting piston flow (see Chapter 11). Rauert and Stichler (1974) measured ^{3}H content in ground water seeping from a 7-km-long tunnel in the Austrian Central Alps. Near the ends of the tunnel, the overburden was sufficiently thin that the seepage water contained post-1952 ^{3}H. Near the center of the tunnel, the seepage water was very low in ^{3}H, indicating that post-1952 water had not yet reached the outlet.

Plummer et al. (1990) used ^{14}C to measure ground-water ages, determine flow velocities, and calculate hydraulic conductivity. The strength of the ^{14}C technique is that one calculates mean flow velocities and mean hydraulic parameters integrated over distances of tens of kilometers which is a vast improvement over pumping tests and other short-range tests in which parameters may be strongly influenced by inhomogeneous rock permeability and porosity.

Characterization of a Ground-Water Mass

Local Area Less than 30 Years of Age

The areal and stratigraphic limits of a local ground-water system can be defined by use of isotopes whose concentration in precipitation is distinguishable since the early to mid-1950s. Such isotopes include ^{3}H, ^{14}C, and ^{36}Cl, which were produced by nuclear-detonation testing (see Chapter 11). ^{18}O and D are useful in areas where the stable isotopic content of the ground-water recharge has undergone a step change during the past century. Such an example is the Mexicali Valley, Baja California, Mexico, which is discussed in the section titled "Sources of Dissolved Constituents."

Regional Systems

Regional ground-water systems can be characterized by ^{18}O, ^{14}C, and ^{36}Cl content. References to examples are given in Table 10-3.

Identification of Recharge Area or Source of Water

Local Area

The variation in δD and $\delta^{18}O$ with elevation can be used to determine the recharge area of small areas. Several examples are discussed in the section "Investigations of Ground-Water Flow in Fractured Rocks."

A novel method of determining the sources of ground water in a basin is presented by Muir and Coplen (1981). Ground-water use outstripped supply in the Santa Clara Valley of California, requiring importation of water from Lake Shasta in the mountains to the north (Figure 10-7). Lake Shasta water was recharged in a percolation pond in Upper Penitencia Creek (Figure 10-7). Because Lake Shasta water is greatly depleted in D and ^{18}O relative to the native ground water, the contribution from it to native ground water, calculated by isotope mass balance (Eq. 10-10), ranged up to 74%.

Letolle and Olive (1983) used $\delta^{13}C$ to identify the source of water flooding cellars in Viry-Chatillon, France, when all other environmental isotopes and hydrochemistry gave inconclusive results. The $\delta^{13}C$ content of total dissolved inorganic carbon in the cellar water, −11‰, matched the value of

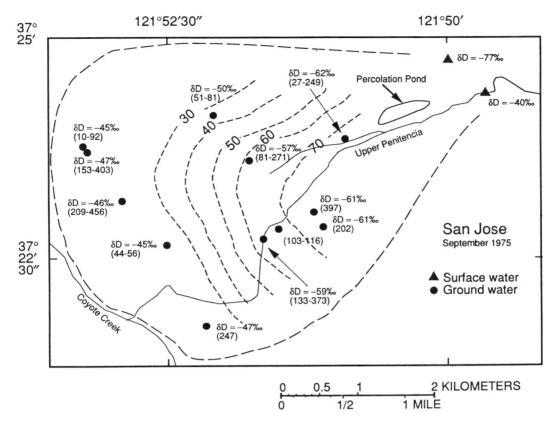

FIGURE 10-7. Sampling sites in Santa Clara Valley, California. (*From Muir and Coplen 1981.*) Values for $\delta^{18}O$ and δD are in per mil relative to VSMOW. The mean $\delta^{18}O$ and δD values of native ground water are −6.1‰ and −41‰, respectively. The mean $\delta^{18}O$ and δD values of northern California imported water which is recharged in Upper Penitencia Creek are −10.2‰ and −74‰, respectively. Wells MP11 and MP14 exhibit intermediate δD and $\delta^{18}O$ compositions and are composed of approximately 70% imported northern California water.

aqueduct water, −12‰, and was greatly different than that of local springs, −17‰.

High ^3H or ^{14}C content has been used to distinguish between "old" ground water and post-nuclear-detonation water as the source for water in basaltic rocks (Hufen, Lau, and Buddemeier 1974).

Regional Systems

The δD and $\delta^{18}O$ abundances can be used to identify the source of water (and paleowater) in a regional system if the δD and $\delta^{18}O$ values of the potential sources are isotopically distinctive due to difference in elevation, distance from the ocean, and latitude (see Table 10-3 for references to specific examples).

Leakage between Aquifers

Light stable isotope ratio and hydrochemical studies can sometimes be used to demonstrate leakage between two aquifers separated by an aquitard. Fontes (1980) and Zuppi (1983) discuss several examples. This method is useful where piezometric-gradient data are inadequate. This method requires that the isotopic or hydrochemical character of each aquifer be distinguishable. Because oxygen and hydrogen are conservative in confined low-temperature aquifers, their isotope ratios are often the most suitable tools for studying aquifer leakage. Confidence in the results of such a study can be improved through simultaneous use of oxygen and hydrogen isotope ratios; hydrochemical data also should be used if available.

Investigations of Ground-Water Flow in Fractured Rocks

Although fractured rocks frequently provide relatively low volumes for storage, they may be very important because they may be the only source of water in a region, such as large fracture-rock zones of the Mediterranean area.

In fractured-rock systems, stable isotopes (Fontes 1983) can be used to determine (1) the origin of the water by making use of the decrease in ^{18}O and D content with increasing elevation (Gonfiantini et al. 1976), (2) the degree to which ground water mixes within a fractured system by looking at spatiotemporal variation in D, ^{18}O, and ^3H content, (3) the flow rate or residence time in small watersheds by employing the seasonal variation in ^{18}O and D content (Dinçer and Payne 1971), and (4) the average age and type of flow (piston flow or well-mixed reservoir) using ^3H concentration (Pearson and Truesdell 1978). Davis et al. (1970) found that large differences in isotopic abundance in ground water on the volcanic island of Cheju in Korea could be explained by differences in flow-path lengths, storage volumes, and residence times. Acquisition of data on ^3H, ^{18}O, and D content of ground water during a flood may help distinguish a deep homogeneous ground-water flow component from a shallow seasonal flow component, both of which may combine as spring discharge (Schotterer et al. 1979). Further, one may identify isolated subsystems that discharge and recharge only during exceptionally rainy periods (Fontes 1983; Simpson and Carmi 1983).

Evaluation of Ground-Water Flow and Storage Characteristics

Local System

The use of ^3H to distinguish piston flow from flow in a well-mixed reservoir is discussed earlier and in Chapter 11. A piston flow system is indicated where seasonal variability in D and ^{18}O content can be observed in systems less than five years old.

Dispersivity Investigations

Wood (1981) has suggested that a conservative isotope tracer can be used to determine the hydrodynamic dispersivity of large-scale regional aquifer systems.

Separation of Stream Discharge into Ground-Water and Surface-Water Components

Ground-water contribution to streamflow can be determined by isotope mass balance, employing D, ^{18}O, and (or) ^3H abundances if there is a large separation in isotopic content between stream water and ground water (Sklash and Farvolden 1979 and 1982). Daily or hourly sampling may be required to obtain an adequate number of samples along the hydrograph of the stream (Turner and Macpherson 1990). Prestorm soil water may also contribute significantly to stormflow (Kennedy et al. 1986).

Sources of Dissolved Constituents

The use of nitrogen and sulfur isotopic ratios to identify the source of dissolved nitrate and sulfate in ground water is discussed in the sections titled "Nitrogen" and "Sulfur," respectively. ^{18}O, D, and ^3H content can be used to determine the mechanism of salinization (Payne 1983).

Low-temperature salinization may occur by (1) concentration of dissolved salts by evaporation, (2) ultrafiltration (Coplen and

Hanshaw 1973), (3) recharge or intrusion of seawater or other surface water, and (4) dissolution or leaching of salts in the unsaturated zone or evaporitic aquifer bedrock units by percolating ground water. Evaporative concentrations—mechanism 1—will increase the ^{18}O and D content of water as shown in Figure 10-5. On a plot of δD as a function of $\delta^{18}O$, the slope of the line for evaporating water ranges from 4 to 6 (see Gat 1981). Recharge or intrusion of major amounts of saline water—mechanism 2—can usually be identified by a linear correlation between chloride (concentration must be plotted as molality) and δD and $\delta^{18}O$. However, to quantify intrusion of small amounts (<10%) of saline water, chloride will be a more sensitive indicator and δD and $\delta^{18}O$ will be insensitive. Identification of ground-water recharge that is enriched in dissolved salt by leaching in the unsaturated zone—mechanism 3—depends upon the fact that the recharge is distinguishable from the native ground water by D or ^{18}O content. Such an example is reported by Payne, Quijano, and Latorre (1979) in the Mexicali Valley of Baja California, Mexico. They suggest that the present-day stable isotopic composition of the Colorado River when it enters the valley is more enriched in D and ^{18}O than three or four decades ago due to evaporation from reservoirs constructed on the river during this period. This allows "old" and "recent" Colorado River water that has recharged the ground-water system to be distinguished. Analysis of well waters shows a positive linear correlation between total dissolved solids and D or ^{18}O abundance, in accordance with a mixture of native ground water (pre-1930 Colorado River recharge) and saline evaporated surface water. Existence of high ^3H values only in the high-salinity zones is evidence for the recent recharge of degraded-quality river water.

^{36}Cl was used to show that chloride was concentrated by ion filtration in the Milk River aquifer, Alberta (Bentley, Phillips, and Davis 1986).

Geochemical Reaction Modeling

Geochemical reaction modeling (see Chapter 9) employs mass balance, equilibrium speciation, reaction-path calculation, Rayleigh distillation equations, and the relevant data to determine (1) the chemical reactions that have occurred in a ground-water system, (2) the extent or degree to which such reactions have occurred, (3) the conditions under which reactions occurred, such as open or closed system, (4) the effect of chemical perturbations on water quality and mineralogy of the system, and (5) the age of ground water based on ^{14}C content. For any system, several reaction models can usually be found that satisfy the data. For each model reaction path, calculations are used to predict the chemical and isotopic composition of the aqueous phase as well as the amounts of minerals dissolved or precipitated along a flow path. Sulfur and stable carbon isotope data along a flow path can be used to eliminate one or more of these reaction models (Plummer, Prestemon, and Parkhurst 1991; Plummer, Parkhurst, and Thorstenson 1983; Wigley, Plummer, and Pearson 1978 and 1979; Plummer et al. 1990).

Ground-Water Dating

The age of a ground-water sample is the length of time the sample has been isolated from the atmosphere. Due to hydrodynamic dispersion, a water sample will normally be composed of water of different ages. In some complicated systems, waters of greatly different ages may mix. Only in the simplest system can flow be compared to that in a long pipe (i.e., piston flow). On the other hand, the age of water discharging from a well-mixed reservoir of constant volume is the mean residence time of water in the system. Only in piston flow or well-mixed reservoir end-member models does apparent age have quantitative meaning. However, relative age between two sampling points in homogeneous aquifers provides information on flow rates.

Ground water is dated most commonly by one of three techniques. The first technique relies on identification of isotopes in ground water that puts limits on the minimum or maximum age of the sample, so-called event markers. For instance, high ^{3}H levels or high ^{36}Cl/Cl ratios indicate a significant proportion of post-1952 water. The second technique, of which ^{14}C is an example, relies on the measurement of the abundance in ground water of a radioactive isotope produced in the atmosphere that decays along the flow path. In the simplest case, age is calculated by the radioactive decay law (Eq. 10-14). The third technique relies on the determination of an isotope produced in the confined aquifer, for example, ^{4}He (see Pearson et al. 1991).

A common ground-water dating technique employs ^{14}C and relies upon (1) the assumption that A_0 in Eq. 10-12 was constant over tens of thousands of years, (2) the determination of the value of A_0 which depends upon the process forming DIC in the recharge area and upon the degree of atmospheric exchange with DIC, (3) the identification of sources and sinks of carbon along the flow path, (4) the use of accurate carbon isotope fractionation factors between DIC and carbonate precipitated along a flow path when such a carbon sink is present. The computer program NETPATH (Plummer, Prestemon, and Parkhurst 1991) allows one to select any of nine A_0 models and to vary carbon isotope fractionation factors. An excellent example of dating with ^{14}C is given by Plummer et al. (1990).

Less than Five Years of Age

In ground-water systems exhibiting piston flow, the seasonal variability in D, ^{18}O, and ^{3}H content of ground-water recharge can be used to date water no more than a few years old if sufficient sampling points are available along the flow path to quantify the number of annual cycles. This is discussed in the section titled "Recharge and Flow Rate." Also see Chapter 11.

Five- to 50-Year Range

^{3}H, ^{14}C, and ^{36}Cl, produced since 1952 by aboveground nuclear detonation testing and ^{85}Kr, produced by artificial nuclear fission, are useful in this range and are addressed in Chapter 11.

Fifty to 1,000-Year Range

See Chapter 11 and references in Table 10-3 for examples of this technique.

One Thousand to 40,000-Year Range

The primary dating technique in this range is ^{14}C. The best accuracy is attained by constructing a geochemical model to account for sources and sinks of carbon along the flow path (Wigley, Plummer, and Pearson 1978; 1979; Plummer, Parkhurst, and Thorstenson 1983; Plummer, Prestemon, and Parkhurst 1991; Chapter 9 herein). Plummer et al. (1990) were able to calculate ages based on ^{14}C concentrations in the Madison aquifer in Montana, South Dakota, and Wyoming, and then calculate regional hydraulic conductivities.

Sixty Thousand to 1,200,000-Year Range

The 300,000-year half-life and nonreactivity of ^{36}Cl make it potentially useful in ground-water dating (Table 10-3). Bentley, Phillips, and Davis (1986) show that ^{36}Cl ages of ground waters in the Great Artesian Basin, Australia, the Milk River aquifer (see also Nolte et al. 1991; Lehmann et al. 1991), Alberta, and the Fox Hills-Basal Bell Creek aquifer, North Dakota, are reasonable based on hydrologic data. Bentley, Phillips, and Davis (1986) show that the Carrizo aquifer in southern Texas is not appropriate for dating because ^{36}Cl input has not been constant. A fundamental difficulty of this technique is that ^{36}Cl/Cl ratios increase from a value of 20×10^{-15} in the coastal Pacific Northwest and Florida to values above 600 in the western Midwest because of the decrease in atmospheric sea salt inland from a coast, and there is no reliable method for determining initial ^{36}Cl content accurately (see Bentley, Phillips, and Davis 1986 and Nolte et al. 1991).

CONCLUDING REMARKS

Isotopic data often can be used to provide a unique solution or interpretation of ground-water systems, delineating age, flow type, hydraulic parameters, origin of salinity, and origin of water. In many instances, however, isotopic data are subject to multiple interpretations. Consequently, it cannot be stressed too strongly that any ground-water study should not be based on the results obtained with a single isotope or isotopic pair. Instead, a multidisciplinary program that integrates chemical, geologic, isotopic, and hydrologic data will yield more detailed information and may prevent erroneous conclusions.

References

Airey, P. L., G. E. Calf, B. L. Campbell, P. E. Hartley, D. Roman, and M. A. Habermehl. 1979. Aspects of the isotope hydrology of the Great Artesian Basin, Australia. In *Isotope Hydrology 1978*, Vol. I, pp. 205–17. Vienna: International Atomic Energy Agency.

Back, W., B. B. Hanshaw, L. N. Plummer, P. H. Rahn, C. T. Rightmire, and M. Rubin. 1983. Process and rate of dedolomitization: Mass transfer and ^{14}C dating in a regional carbonate aquifer. *Geological Society of America Bulletin* 94:1415–29.

Baedecker, M. J., and W. Back. 1979. Hydrogeological processes and chemical reactions at a landfill. *Ground Water* 17:429–37.

Barker, J. F. 1978. *Methane in Ground Water*. Ph.D. thesis, University of Waterloo.

Bentley, H. W., F. M. Phillips, and S. N. Davis. 1986. ^{36}Cl in the terrestrial environment. In *Handbook of Environmental Isotope Geochemistry*. Vol. 2. *The Terrestrial Environment, B*, eds. P. Fritz and J. Ch. Fontes, pp. 427–80. Amsterdam: Elsevier.

Böhlke, J. K., J. M. Denver, P. J. Phillips, C. J. Gwinn, L. N. Plummer, E. Busenberg, and S. A. Dunkle. 1992. Combined use of nitrogen isotopes and ground-water dating to document nitrate fluxes and transformations in small agricultural watersheds, Delmarva Peninsula, Maryland. *EOS* 73:140.

Böttcher, J. O. Strebel, S. Voerkelius, and H. L. Schmidt. 1990. Using isotope fractionation of nitrate-nitrogen and nitrate-oxygen for evaluation of microbial denitrification in a sandy aquifer. *Journal of Hydrology* 114:413–24.

Buchardt, B. and P. Fritz. 1980. Environmental isotopes as environmental and climatic indicators. In *Handbook of Environmental Isotope Geochemistry*, Vol 1. *The Terrestrial Environment, A*, eds. P. Fritz and J. Ch. Fontes, pp. 473–504. Amsterdam: Elsevier.

Carlin, F., G. Magri, A. Cervellati, and R. Gonfiantini. 1975. Use of environmental isotopes to investigate the interconnections between the Reno River and ground water (Northern Italy). In *Isotope Ratios as Pollutant Source and Behavior Indicators*, pp. 179–94. Vienna: International Atomic Energy Agency.

Cerling, T. E. 1984. The stable isotopic composition of modern soil carbonate and its relationship to climate. *Earth Planetary Science Letters* 71:229–40.

Claypool, G. E., W. T. Holser, I. R. Kaplan, H. Sakai, and I. Zak. 1980. The age curves of sulfur and oxygen isotopes in marine sulfate and their mutual interpretation. *Chemical Geology* 28:199–260.

Coplen, T. B., and B. B. Hanshaw. 1973. Ultrafiltration by a compacted clay membrane. I: Oxygen and hydrogen isotopic fractionation. *Geochimica et Cosmochimica Acta* 37:2295–310.

Craig, H. 1961. Isotopic variations in meteoric waters. *Science* 133:1702–3.

Darling, W. G., D. J. Allen, and H. Armannsson. 1990. Indirect detection of subsurface outflow from a rift valley lake. *Journal of Hydrology* 113:297–305.

Davis, G. H., C. K. Lee, E. Bradley, and B. R. Payne. 1970. Geohydrologic interpretation of a volcanic island from environmental isotopes. *Water Resources Research* 6:99–109.

Dinçer, T., and B. R. Payne. 1971. An environmental isotope study of the southwestern karst region of Turkey. *Journal of Hydrology* 14:307–21.

Elderfield, H. 1986. Strontium isotope stratigraphy. *Palaeogeography, Palaeoclimatology, Palaeoecology* 57:71–90.

Fontes, J. Ch. 1980. Environmental isotopes in ground water hydrology. In *Handbook of Environmental Isotope Geochemistry*. Vol. 1. *The Terrestrial Environment, A.* eds. P. Fritz and J. Ch. Fontes, pp. 75–140. Amsterdam: Elsevier.

Fontes, J. Ch. 1983. Ground water in fractured rocks. In *Guidebook on Nuclear Techniques in Hydrology, 1983 Edition*, pp. 337–50. Vienna: International Atomic Energy Agency, Technical Reports Series No. 91.

Freyer, H. D. 1978. Seasonal trends of NH_4^+ and NO_3^- nitrogen isotope composition in rain collected at Julich, Germany. *Tellus* 30:83–92.

Friedman, I., and J. R. O'Neil. 1977. Compilation of stable isotope fractionation factors of geochemical interest. In *Data of Geochemistry*, 6th ed., ed. M. Fleischer. Washington, DC: U. S. Geological Survey Professional Paper 440-KK.

Fritz, P., and J. Ch. Fontes, eds. 1980. *Handbook of Environmental Isotope Geochemistry*. Vol. 1. *The Terrestrial Environment, A*. Amsterdam: Elsevier.

Fritz, P., R. J. Drimmie, and F. W. Render. 1974. Stable isotope contents of a major prairie aquifer in central Manitoba, Canada. In *Isotope Techniques in Ground Water Hydrology*, Vol. I, pp. 379–96. Vienna: International Atomic Energy Agency.

Gat, J. R. 1971. Comments on the stable isotope method in regional investigation. *Water Resources Research* 7:980–93.

Gat, J. R. 1980. The isotopes of hydrogen and oxygen in precipitation. In *Handbook of Environmental Isotope Geochemistry*. Vol. 1. *The Terrestrial Environment, A*, eds. P. Fritz and J. Ch. Fontes, pp. 21–47. Amsterdam: Elsevier.

Gat, J. R. 1981. Isotopic fractionation. In *Stable Isotope Hydrology, Deuterium and Oxygen-18 in the Water Cycle*, eds. J. R. Gat and R. Gonfiantini, pp. 21–34. Vienna: International Atomic Energy Agency, Technical Reports Series No. 210.

Gat, J. R., and I. Carmi. 1970. Evolution of the isotopic composition of atmospheric waters in the Mediterranean Sea Area. *Journal of Geophysical Research* 75:3039–48.

Goldhaber, M. B., and I. R. Kaplan. 1974. The sulfur cycle. In *The Sea, 5, Marine Chemistry*, ed. E. D. Goldberg, pp. 569–655. New York: Wiley-Interscience.

Gonfiantini, R., G. Gallo, B. R. Payne, and C. B. Taylor. 1976. Environmental isotopes and hydrochemistry in ground water of Gran Canaria. In *Interpretation of Environmental Isotope and Hydrochemical Data in Ground Water Hydrology*, pp. 159–70. Vienna: International Atomic Energy Agency.

Gormly, J. R., and R. J. Spalding. 1979. Sources and concentrations of nitrate-nitrogen in ground water of the Central Platte region, Nebraska. *Ground Water* 17:291–301.

Grootes, P. M., W. G. Mook, and J. C. Vogel. 1969. Isotopic fractionation between gaseous and condensed carbon dioxide. *Zeitschrift für Physik* 221:257–73.

Gvirtzman, H., and M. Margaritz. 1986. Investigation of water movement in the unsaturated zone under an irrigated area using environmental tritium. *Water Resources Research* 22(5):635–42.

Heaton, T. H. E. 1986. Isotopic studies of nitrogen pollution in the hydrosphere and atmosphere—A review. *Chemical Geology* 59:87–102.

Heaton, T. H. E. 1987. $^{15}N/^{14}N$ ratios of nitrate and ammonium in rain at Pretoria, South Africa. *Atmospheric Environment* 21:843–52.

Hübner, H. 1986. Isotope effects of nitrogen in the soil and biosphere. In *Handbook of Environmental Isotope Geochemistry*, Vol. 2, eds. P. Fritz and J. Ch. Fontes, pp. 361–425. Amsterdam: Elsevier.

Hufen, T. H., L. S. Lau, and R. W. Buddemeier. 1974. Radiocarbon, ^{13}C and tritium in water samples from basaltic aquifers and carbonate aquifers on the island of Oahu, Hawaii. In *Isotope Techniques in Ground Water Hydrology 1974*, Vol. II, pp. 111–26. Vienna: International Atomic Energy Agency.

Hut, G. 1987. *Consultants' Group Meeting on Stable Isotope Reference Samples for Geochemical and Hydrological Investigations*. Vienna: International Atomic Energy Report to Director General of Meeting, 16–18 September 1985.

IAEA. 1980. *Arid-Zone Hydrology: Investigations with Isotope Techniques*. Vienna: International Atomic Energy Agency.

IAEA. 1981a. *Stable Isotope Hydrology, Deuterium and Oxygen-18 in the Water Cycle*, eds. J. R. Gat and R. Gonfiantini. Vienna: International Atomic Energy Agency, Technical Reports Series No. 210.

IAEA. 1981b. *Statistical Treatment of Environmental Isotope Data in Precipitation*. Vienna: International Atomic Energy Agency, Technical Reports Series No. 206.

IAEA. 1983a. *Guidebook on Nuclear Techniques in Hydrology, 1983 Edition*. Vienna: International Atomic Energy Agency, Technical Reports Series No. 91.

IAEA. 1983b. *Isotope Techniques in the Hydrogeological Assessment of Potential Sites for the Disposal of High-Level Radioactive Wastes*. Vienna, International Atomic Energy Agency.

IAEA. 1991. *Use of Artificial Tracers in Hydrology*. Vienna: International Atomic Energy Agency, IAEA-TECDOC-601.

IAEA. 1992. *Isotope Techniques in Waters Resources Development 1991*. Vienna: International Atomic Energy Agency.

IUPAC. 1992. Isotopic compositions of the elements—1989. *Pure and Applied Chemistry* 63:991–1002.

Kennedy, V. C., C. Kendall, G. W. Zellweger, T. A. Wyerman, and R. J. Avanzino. 1986. Determination of the components of stormflow using water chemistry and environmental isotopes, Mattole River Basin, California. *Journal of Hydrology* 84:107–40.

Krabbenhoft, D. P., C. J. Bowser, M. P. Anderson, and J. W. Valley. 1990. Estimating groundwater exchange with lakes. I: The stable isotope mass balance method. *Water Resources Research* 26:2445–53.

Kreitler, C. W. 1975. Determining the source of nitrate in ground water by nitrogen isotope studies. In *Bureau of Economic Geology Report on Investigations 83*, p. 57. Austin: University of Texas Austin.

Kreitler, C. W. 1979. Nitrogen-isotope ratio studies of soil and ground-water nitrate from alluvial fan aquifers in Texas. *Journal of Hydrology* 42:147–70.

Kreitler, C. W., and L. A. Browning. 1983. Nitrogen-isotope analysis of ground water nitrate in carbonate aquifers: Natural sources versus human pollution. In *V. T. Stringfield Symposium—Processes in Karst Hydrology*, guest eds. W. Back and P. E. LaMoreaux. *Journal of Hydrology* 61:285–301.

Krouse, H. R. 1980. Sulphur isotopes in our environment. In *Handbook of Environmental Isotope Geochemistry*, Vol. 1, *The Terrestrial Environment, A*, eds. P. Fritz and J. Ch. Fontes, pp. 435–72. Amsterdam: Elsevier.

Kyser, T. K. 1987. Equilibrium fractionation factors for stable isotopes. In *Stable Isotope Geochemistry of Low Temperature Processes*, ed. T. K. Kyser, pp. 1–84. Saskatoon: Mineralogical Association of Canada, Short Course Handbook, Vol. 13.

Lehmann, B. E., H. H. Loosli, D. Rauber, N. Thonnard, and R. D. Willis. 1991. ^{81}Kr and ^{85}Kr in groundwater, Milk River aquifer, Alberta, Canada. *Applied Geochemistry* 6:419–24.

Letolle, R., and P. Olive. 1983. Isotopes as pollution tracers. In *Guidebook on Nuclear Techniques in Hydrology, 1983 Edition*, pp. 411–22. Vienna: International Atomic Energy Agency.

Lloyd, J. W., and K. W. F. Howard. 1979. Environmental isotope studies related to ground water flow and saline encroachment in the Chalk Aquifer of Lincolnshire, England. In *Isotope Hydrology 1978*, Vol. I, pp. 311–23. Vienna: International Atomic Energy Agency.

Long, Austin. 1983. Tucson, Arizona, University of Arizona, Department of Geology, oral communication.

Mariotti, A., A. Landreau, and B. Simon. 1988. N isotope biogeochemistry and natural denitrification process in groundwater—Application to the chalk aquifer of northern France. *Geochimica et Cosmochimica Acta* 52:1869–78.

Martinec, J., U. Siegenthaler, H. Oeschger, and E. Tongiorgi. 1974. New insights into the run-off mechanism by environmental isotopes. In *Isotope Techniques in Groundwater Hydrology 1974*, Vol. I, pp. 129–43. Vienna: International Atomic Energy Agency.

McCarthy, K. A., W. D. McFarland, J. M. Wilkinson, and L. D. White. 1992. An investigation of the dynamic relationship between ground water and a river using deuterium and oxygen-18. *Journal of Hydrology* 135:1–12.

Merlivat, L., and J. Jouzel. 1979. Global climatic interpretation of the deuterium-oxygen 18 relationship for precipitation. *Journal of Geophysical Research* 84:5029–33.

Mook, W. G., J. C. Bommerson, and W. H. Staverman. 1974. Carbon isotope fractionation between dissolved bicarbonate and gaseous carbon dioxide. *Earth and Planetary Science Letters* 22:169–76.

Moore, H. 1977. The isotopic composition of ammonia, nitrogen dioxide, and nitrate in the atmosphere. *Atmospheric Environment* 11:1239–43.

Muir, K. S., and T. B. Coplen. 1981. *Tracing Ground-Water Movement by Using the Stable Isotopes of Oxygen and Hydrogen, Upper Penitencia Creek Alluvial Fan, Santa Clara Valley, California*. Washington, DC: U. S. Geological Survey Water-Supply Paper 2075.

Munnich, K.O. 1983. Moisture movement in the unsaturated zone. In *Guidebook on Nuclear Techniques in Hydrology, 1983 Edition*, pp. 203–22. Vienna: International Atomic Energy Agency.

Nolte, E., P. Krauthan, G. Korschinek, P. Maloszewski, P. Fritz, and M. Wolf. 1991. Measurements and interpretations of ^{36}Cl in groundwater, Milk River aquifer, Alberta, Canada. *Applied Geochemistry* 6:435–46.

Payne, B. R. 1981. Practical applications of stable isotopes to hydrological problems. In *Stable Isotope Hydrology, Deuterium and Oxygen-18 in the Water Cycle*, pp. 303–34. Vienna: International Atomic Energy Agency.

Payne, B. R. 1983. Ground water salinisation. In *Guidebook on Nuclear Techniques in Hydrology, 1983 Edition*, pp. 351–57. Vienna: International Atomic Energy Agency.

Payne, B. R., L. Quijano, and C. Latorre. 1979. Environmental isotopes in a study of the origin of salinity of ground water in the Mexicali Valley. *Journal of Hydrology* 41:201–15.

Pearson, F. J., Jr., and I. Friedman. 1970. Sources of dissolved carbonate in an aquifer free of carbonate minerals. *Water Resources Research* 6:1775–81.

Pearson, F. J., Jr., and B. B. Hanshaw. 1970. Sources of dissolved carbonate species in ground water and their effects on carbon-14 dating. In *Isotope Hydrology 1970*, pp. 271–85. Vienna: International Atomic Energy Agency.

Pearson, F. J., Jr., and C. T. Rightmire. 1980. Sulphur and oxygen isotopes in aqueous sulphur compounds. In *Handbook of Environmental Isotope Geochemistry, Vol. I, The Terrestrial Environment, A*, eds. P. Fritz and J. Ch. Fontes, pp. 227–58. Amsterdam: Elsevier.

Pearson, F. J., Jr., and W. V. Swarzenki. 1974. ^{14}C evidence for the origin of arid region ground water, Northeastern Province, Kenya. In *Isotope Techniques in Ground Water Hydrology 1974*, Vol. II, pp. 95–108. Vienna: International Atomic Energy Agency.

Pearson, F. J., Jr., and A. H. Truesdell. 1978. Tritium in the waters of Yellowstone National Park (Wyoming). In *Short Papers of the Fourth International Conference of Geochronology, Cosmochronology, and Isotope Geology, 1978*, ed. R. A. Zartman, pp. 327–9. U.S. Geological Survey Open-File Report 78–701.

Pearson, F. J., Jr., W. Balderer, H. H. Loosli, B. E. Lehmann, A. Matter, Tj. Peters, H. Schmassmann, and A. Gautschi. 1991. *Applied Isotope Hydrology—A Case Study in Northern Switzerland*. Amsterdam: Elsevier, Studies in Environmental Science 43.

Phillips, F. M., J. L. Mattick, T. A. Duval, D. Elmore, and P. W. Kubik. 1988. Chlorine-36 and tritium from nuclear weapons fallout as tracers for long-term liquid and vapor movement in desert soils. *Water Resources Research* 24:1877–91.

Plummer, L. N., D. L. Parkhurst, and D. C. Thorstenson. 1983. Development of reaction models for ground-water systems. *Geochimica et Cosmochimica Acta* 47:665–86.

Plummer, L. N., E. C. Prestemon, and D. L. Parkhurst, 1991. *An Interactive Code (NETPATH) for Modeling Net Geochemical Reactions along a Flow Path*. Reston, VA: U.S. Geological Survey Water-Resources Investigations Report 91–4078.

Plummer, L. N., J. F. Busby, R. W. Lee, and B. B. Hanshaw. 1990. Geochemical modeling of the Madison aquifer in parts of Montana, Wyoming, and South Dakota. *Water Resources Research* 26:1981–2014.

Postma, D., C. Boesen, H. Kristiansen, and F. Larsen. 1991. Nitrate reduction in an unconfined sandy aquifer: Water chemistry, reduction processes, and geochemical modeling. *Water Resources Research* 27:2027–45.

Rauert, W., and W. Stichler. 1974. Ground water investigations with environmental isotopes. In *Isotope Techniques in Ground Water Hydrology 1974*, Vol. I, pp. 431–41. Vienna: International Atomic Energy Agency.

Rayleigh. 1896. Theoretical considerations respecting the separation of gases by diffusion and similar processes. *Philosophy Magazine* 42:493–8.

Schotterer, U., A. Wildberger, U. Siegenthaler, W. Nabholz, and H. Oeschger. 1979. Isotope study in the alpine karst region of Rawil, Switzerland. In *Isotope Hydrology 1978*, Vol. I, pp. 351–65. Vienna: International Atomic Energy Agency.

Simpson, B., and I. Carmi. 1983. The hydrology of the Jordan tributaries (Israel): Hydrographic and isotopic investigation. *Journal of Hydrology* 62:225–42.

Simpson, H. J., and A. L. Herczeg. 1991. Salinity and evaporation in the Murray Basin, Australia. *Journal of Hydrology* 124:1–27.

Sklash, M. G., and R. N. Farvolden. 1979. The role of ground water in storm runoff. *Journal of Hydrology* 43:45–65.

Sklash, M. G., and R. N. Farvolden. 1982. The use of environmental isotopes in the study of high-runoff episodes in streams. In *Isotope Studies in Hydrologic Processes*, eds. E. C. Perry, Jr., and C. W. Montgomery, pp. 65–74. Dekalb, IL: Northern Illinois University Press.

Smith, R. L., B. L. Howes, and J. H. Duff. 1991. Denitrification in nitrate-contaminated groundwater: Occurrence in steep vertical geochemical gradients. *Geochimica et Cosmochimica Acta* 55:1815–25.

Solomon, D. K., and E. A. Sudicky. 1991. Tritium and helium 3 isotope ratios for direct estimation of spatial variations in groundwater recharge. *Water Resources Research* 27:2309–19.

Spalding, R. F., M. E. Exner, C. W. Lindau, and D. W. Eaton. 1982. Investigation of sources of groundwater nitrate contamination in the Burbank-Wallula area of Washington, U.S.A. *Journal of Hydrology* 58:307–24.

Starinsky, A., M. Bielski, A. Ecker, and G. Steinitz. 1983. Tracing the origin of salts in groundwater by Sr isotopic composition (the crystalline complex of the southern Sinai, Egypt). *Isotope Geoscience* 1:257–67.

Stewart, M. K., and J. J. McDonnell. 1991. Modeling base flow soil water residence times from deuterium concentrations. *Water Resources Research* 27:2681–93.

Stichler, W., and H. Moser. 1979. An example of exchange between lake and ground water. In *Isotopes in Lake Studies*, pp. 115–20. Vienna: International Atomic Energy Agency.

Stuckless, J. S., Z. E. Peterman, and D. R. Muhs. 1991. U and Sr isotopes in ground water and calcite, Yucca Mountain, NV: Evidence against upwelling water. *Science* 254:551–54.

Torgersen, T., M. A. Habermehl, F. M. Phillips, D. Elmore, P. Kubik, B. G. Jones, T. Hemmick, and H. E. Gove. 1991. Chlorine 36 dating of very old groundwater, 3. Further studies in the Great Artesian Basin, Australia. *Water Resources Research* 27:3201–13.

Turner, J. V. 1982. Kinetic fractionation of carbon-13 during calcium carbonate precipitation. *Geochimica et Cosmochimica Acta* 46:1183–91.

Turner, J. V., and D. K. Macpherson. 1990. Mechanisms affecting streamflow and stream water quality: An approach via stable isotope, hydrogeochemical, and time series analysis. *Water Resources Research* 26:3005–19.

van der Straaten, C. M., and W. G. Mook. 1983. Stable isotopic composition of precipitation and climatic variability. In *Palaeoclimates and Palaeowaters: A Collection of Environmental Isotope Studies*, pp. 53–64. Vienna: International Atomic Energy Agency.

Vogel, J. C., and D. Ehhalt. 1963. The use of carbon isotopes in ground water studies. In *Radioisotopes in Hydrology*, Vol. 2, pp. 383–96. Vienna: International Atomic Energy Agency.

Vogel, J. C., A. S. Talma, and T. H. E. Heaton. 1981. Gaseous nitrogen as evidence for denitrification in groundwater. *Journal of Hydrology* 50:191–200.

Wigley, T. M. L., L. N. Plummer, and F. J. Pearson, Jr. 1978. Mass transfer and carbon isotope evolution in natural water systems. *Geochimica et Cosmochimica Acta* 42:1117–39.

Wigley, T. M. L., L. N. Plummer, and F. J. Pearson, Jr. 1979. Errata. *Geochimica et Cosmochimica Acta* 43:1395.

Winograd, I. J., and G. M. Farlekas. 1974. Problems in ^{14}C dating of water from aquifers of deltaic origin. In *Isotope Techniques in Ground Water Hydrology*, Vol. II, pp. 69–91. Vienna: International Atomic Energy Agency.

Winograd, I. J., and A. C. Riggs. 1984.

Recharge to the Spring Mountains, Nevada: Isotope evidence. *Geological Society of America Abstracts with Programs 1984*, 16(6):698.

Wood, W. W. 1981. A geochemical method of determining dispersivity in regional ground water systems. *Journal of Hydrology* 54:209–24.

Yurtsever, Y., and J. R. Gat. 1981. Atmospheric water. In *Stable Isotope Hydrology, Deuterium and Oxygen-18 in the Water Cycle*, pp. 103–42. Vienna: International Atomic Energy Agency.

Zuppi, G. M. 1983. Leakage between aquifers. In *Guidebook on Nuclear Techniques in Hydrology, 1983 Edition*, pp. 327–36. Vienna: International Atomic Energy Agency.

11

Environmental Tracers for Age Dating Young Ground Water

L. N. Plummer, R. L. Michel, E. M. Thurman, and P. D. Glynn

INTRODUCTION

In studies of regional ground-water quality, it is of considerable value to supplement water-quality data with ground-water age dating. Ground-water age dating can be used to help define direction and velocity of ground-water flow. Knowledge of the spatial relations of water age can be used to recognize recharge and discharge areas in ground-water and ground-water–surface-water systems, and to estimate rates of ground-water recharge and discharge. These types of information can be quite useful in the development of hydrologic models of ground-water systems and in characterizing hydrogeologic environments on the basis of their potential for contamination.

Over the past 50 years, human activities have introduced a large number of substances into the atmosphere and hydrosphere. Some of these compounds (or their isotopes) can serve as environmental tracers (see Chapter 10), and have been of considerable use as tracers in hydrologic studies. This chapter focuses on environmental tracers that can be used to age date "young" ground water—that is, ground water recharged within approximately the past 50 years. This is a time frame of considerable relevance to those studying ground-water contamination by many pesticides and other compounds that have been manufactured only during the past 50 years or so. Longer time scales could be considered in age dating ground water, such as ground waters recharged in the past 100 or 1,000 years, but present needs are not as urgent for age dating on these scales, and few methods of age dating ground water recharged over these intervals are known (Davis and Bentley 1982; see Chapter 10). Geyh and Schleicher (1990) summarize a large number of dating methods for application in the earth sciences. Fröhlich (1990) discusses age dating of very old ground water. Pearson et al. (1991) apply a wide range of hydrochemical and isotopic dating techniques to interpret the origins of waters associated with a high-level radioactive waste site in northern Switzerland.

Two classes of environmental tracers are recognized: (1) those that mark the occurrence of a particular event, such as tritium (^3H), which is used to determine the position along a flow path of ground water recharged in 1963–64 following the 1962–63 atmospheric nuclear testing (bomb pulse), and (2) those that are present worldwide as a result of continuous, known atmospheric inputs over a period of time, such as the release of chlorofluorocarbons (CFCs) to the atmosphere. Both of these classes of environmental tracers are considered here because they can be applied over regional scales. We do not discuss applications of artificial dyes and radiotracers (Davis et al. 1980), which are usually introduced into hydrologic systems at particular sites and are applicable only on local scales.

Ground-water age usually implies duration of time since the water was recharged and, more specifically, the time since the recharge water became isolated from the atmosphere. The actual moment and position in the recharge process when water is of zero age varies somewhat with the environmental tracer used, the source term, and the recharge process. The concentrations of environmental tracers in ground water are also affected by hydrodynamic dispersion. Many standard hydrologic models can be adapted to address transient tracers, as described here for ^{3}H. These same models can be applied in studies that address other tracers, with modifications for their sources and sinks. These considerations are discussed for a variety of environmental tracers applicable to age dating young ground water. Most of the presentation treats relatively shallow ground water (generally depths less than 30 m); however, the dating methods discussed here can be applied to young ground water wherever it is present, such as in fracture systems hundreds of meters below land surface.

Figure 11-1 shows the approximate range of ages and events, in years, for dating applications of many of the more commonly recognized environmental tracers. The 0- to 50-year time scale considered here for environmental tracers includes age dating with ^{3}H, tritium/helium-3 (^{3}H/^{3}He), krypton-85 (^{85}Kr), and CFCs (F-11 and F-12). For comparison, Figure 11-1 includes the approximate dating ranges of selected other environmental tracers, including argon-39 (^{39}Ar) (Oeschger et al. 1974; Loosli and Oeschger 1978, 1980; Forster, Moser, and Loosli 1984), silicon-32 (^{32}Si) (Lal, Nuampurkar, and Rama 1970; Fröhlich et al. 1987), carbon-14 (^{14}C) (Fontes and Garnier 1979; Wigley, Plummer, and Pearson 1978; Mook 1980; Plummer, Prestemon, and Parkhurst 1991; Fontes 1983, 1992; Maloszewski and Zuber 1991), chlorine-36 (^{36}Cl) (Phillips, Mattick, and Duval 1988; Torgersen et al. 1991), and krypton-81 (^{81}Kr) (Lehmann et al. 1985; Thonnard et al. 1987).

Event markers include the 1963–64 bomb pulse, which is evident in a number of environmental tracers including ^{3}H, ^{3}H/^{3}He, ^{14}C, and ^{36}Cl produced primarily from atmospheric nuclear weapons testing over the Pacific Ocean during 1952–63; and known dates of first appearance of anthropogenic organic compounds including CFCs, surfactants, and herbicides. As Figure 11-1 shows, some tracers can be used to date ground water, both as event markers and environmental tracers. For example, the mere presence of detectable CFCs indicates the water contains at least a portion of post-1945 water, and, because of the continuous variation of CFCs in the atmosphere, the absolute CFC concentration in ground water can be used to assign a specific post-1945 recharge age. Similarly location of the 1963–64 ^{3}H bomb pulse in ground water marks the event, whereas combined measurements of ^{3}H and ^{3}He allow ^{3}H dating of the parcel of ground water.

This chapter discusses dating tools currently (1992) in use that have the greatest potential for application to young ground water. Specifically, this chapter focuses on dating applications of ^{3}H, ^{3}H/^{3}He, ^{85}Kr, CFCs, surfactants, and herbicides. No single dating method is universally applicable to all ground-water environments in the 0- to 50-year time scale. Each method has advantages and limitations. Confidence increases as overlapping dating methods and tracers are applied to the hydrochemical system and evaluated in the context of other chemical, isotopic, and hydrogeologic data.

TRITIUM

History of Tritium Input

Tritium (^{3}H), the radioactive isotope of hydrogen (H), provides an excellent tracer for

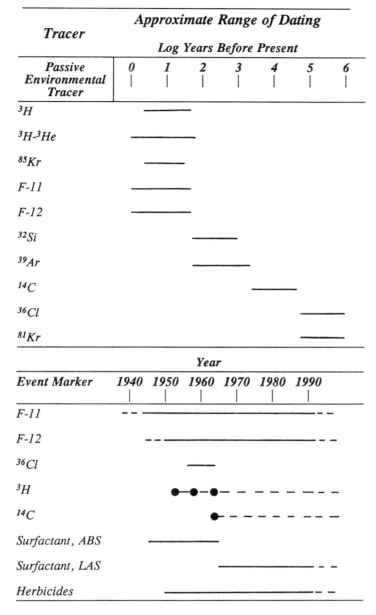

FIGURE 11-1. Approximate range (in years) of dating applications of selected environmental tracers and event markers.

determining time scales for the physical mixing and flow of ground water. Water containing a ^{3}H atom that substitutes for a H atom in the water molecule will follow the same pathway through the environment as water containing no ^{3}H atoms; only minor differences between the two waters are noticeable during phase changes. The half-life of ^{3}H [$t_{1/2}$ = 12.43 years; International Atomic Energy Agency (IAEA) 1981] and its increased production during atmospheric nuclear testing (Carter and Moghissi 1977) make ^{3}H ideally suited for studying processes that occur on a time scale of less than 100 years. Accordingly, ^{3}H is extensively used as a tracer in hydrologic studies (Brown 1961; Munnich, Roether, and Thilo 1967).

Prior to atmospheric fusion-bomb testing in the early 1950s, a steady-state ^{3}H inventory of approximately 3.5 kg was maintained on the Earth's surface as a result of cosmic-ray spallation. Beginning in 1953, ^{3}H concentrations in precipitation began to increase because of nuclear weapons testing, and the large-scale tests of 1962–63 produced ^{3}H concentrations in precipitation up to three orders of magnitude over natural concentrations during 1963–64 (Michel 1989). Tritium concentrations in precipitation are affected by latitude, distance from the ocean, and season. Water vapor over the ocean has a low ^{3}H concentration due to the molecular exchange of atmospheric water vapor with surface ocean water, which has a low ^{3}H concentration. As air masses move over the continent, input from the stratosphere and evapotranspiration become the dominant mechanisms controlling ^{3}H concentrations, and concentrations begin to rise with increasing distance from the ocean. The primary seasonal effect is caused by the breaking up of the tropopause between 30° and 60°N latitude during spring each year, resulting in an injection of water vapor with a high ^{3}H concentration into the troposphere. This injection causes a north-south gradient in ^{3}H concentrations in precipitation across the United States; concentrations decrease in the southern United States. The effects of the above processes result in the ^{3}H deposition pattern shown in Figure 11-2, whereby highest depositions are generally found in the midcontinental northern latitudes.

A realistic ^{3}H input function is required for reliable results. The seasonal, yearly, and geographic variations of ^{3}H concentrations in precipitation must be taken into account. The data base available to study the input is limited, especially for the years prior to atmospheric nuclear testing and the early years of nuclear testing. Only one station, at Ottawa, Canada, has been in continuous operation on the North American continent since 1953 (Figure 11-3). Starting in 1960, the U.S. Geological Survey began monitoring ^{3}H concentrations in rain at various locations within the continental United States (Figure 11-2). These data were supplemented by a small number of stations monitored by university laboratories. To compensate for the gaps in the data bases, particularly in the 1950s, the IAEA developed the Ottawa correlation (IAEA 1978) which correlates ^{3}H data from Ottawa with ^{3}H data from other stations where long-term data bases are available. The correlations were performed on monthly and yearly data sets and for linear correlations and log-log correlations. The correlations can be used to estimate ^{3}H concentrations in precipitation at a station for time periods when no data are available. The IAEA (1983, 1986, 1990) has continued to publish ^{3}H data in world precipitation.

For most aquifers, recharge is seasonal, and use of yearly weighted averages for ^{3}H concentrations will not be an accurate reflection of the ^{3}H concentrations for incoming waters (Rabinowitz, Gross, and Holmes 1977). Typically, recharge concentrations are calculated by assuming that most recharge water is derived from precipitation in winter or early spring when ^{3}H concentrations are lowest.

When recharge to an aquifer passes through the unsaturated zone, the ^{3}H input function will depend on transit time through the unsaturated zone. Andersen and Sevel (1974) studied movement of recharging water through the unsaturated zone above an aquifer in Denmark, and observed that the initial ^{3}H input peak was retained as it moved through the unsaturated zone, but its height and width were modified by dispersion. When precise inventories of ^{3}H are made to calculate recharge to an aquifer, ^{3}H concentrations in residual soil moisture may need to be estimated. In a study of the Chalk unsaturated zone that recharges the Chalk aquifer in England, pore-water ^{3}H content was found to be higher than could be explained by traditional hydrological balances (Foster 1975). A later study by Foster and Smith-

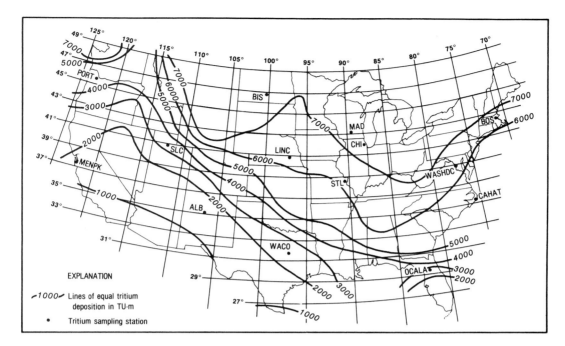

FIGURE 11-2.   Deposition in TU-meters of tritium across the continental United States from 1953 to 1983. U.S. Geological Survey sampling stations listed are Albuquerque, NM (ALB), Boston, MA (BOS), Cape Hatteras, NC (CAHAT), Chicago, IL (CHI), Lincoln, NE (LINC), Madison, WI (MAD), Menlo Park, CA (MENPK), Ocala, FL (OCALA), Portland, OR (PORT), Salt Lake City, UT (SLC), Saint Louis, MO (STL), Waco, TX (WACO), and Washington, DC (WASHDC). (*From Michel 1989.*)

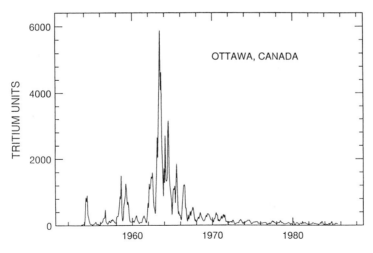

FIGURE 11-3.   Monthly tritium concentration in precipitation at Ottawa, Canada, 1953 through 1985. (*Data from IAEA 1969, 1970, 1971, 1973, 1975, 1979, 1983, 1986, 1990.*)

Carrington (1980) found that the higher ^{3}H inventories were due, in part, to the retention of summer precipitation with relatively high ^{3}H concentrations in the soil zone. The water that entered the unsaturated zone during the winter was a mixture of summer and winter precipitation, and the resulting ^{3}H concentration was higher than that found in winter precipitation alone. Thus, to perform accurate calculations of ^{3}H inventory in recharge, it was necessary to consider soil-moisture conditions and the average monthly ^{3}H concentrations in precipitation.

Few studies require a knowledge of ^{3}H input as accurate as that calculated for the Chalk aquifer study. It is important to know seasonal and yearly variations in ^{3}H concentrations in precipitation and recharge. It is also important to know where and how recharge enters the aquifer. If precipitation enters the aquifer directly with little delay during recharge events, ^{3}H concentrations in precipitation can be used directly. If recharge occurs through the unsaturated zone, then the delay and dispersion of ^{3}H before it enters the aquifer should be taken into account.

Movement of Tritium Through Aquifers

The simplest use of ^{3}H is to determine whether detectable concentrations are present in the aquifer. Tritium concentrations in precipitation prior to atmospheric nuclear testing are not well known, but probably do not exceed the estimates given by Thatcher (1962) of 2–8 tritium units (TU; 1 TU is equal to 1 ^{3}H atom in 10^{18} atoms of H, or 3.24 picocuries per liter, pCi/L). Waters derived exclusively from precipitation before nuclear-bomb testing would have maximum ^{3}H concentrations of 0.2–0.8 TU by the early 1990s. For ground waters with higher ^{3}H concentrations, some fraction of the water must have been derived since the advent of nuclear testing. In many situations, ^{3}H is a useful marker for recharge after the beginning of the fusion-bomb era. An example of this approach is found in a study conducted by Swancar and Hutchinson (1992) in the Floridan aquifer of west-central Florida. The thickness of the unit that confines this aquifer throughout the state ranges from 0 to greater than 67 m. Tritium concentrations varied from 10–12 TU where the confining unit was thin, and were near 0 TU under greater confinement (Figure 11-4). The results showed that the Floridan aquifer has received recharge in poorly confined areas, and little or no recharge in confined areas, since the advent of nuclear weapons testing.

Another possible source of ^{3}H production is from interaction of ^{6}Li and neutrons in certain aquifers (Andrews and Kay 1982). In situ production of ^{3}H is enhanced in aquifers where the rocks are enriched in uranium, thorium, and lithium, are low in boron content, and the ground water has long residence times. These conditions are rare and this possibility can be ignored in most cases of dating young ground water.

Tritium is frequently used in conjunction with ^{14}C measurements to check for possible addition of small amounts of postnuclear-testing water, which could cause errors in ^{14}C dates.

Piston-Flow Approaches

Commonly, ^{3}H measurements are used to obtain rates or directions of movement of water through an aquifer and to estimate recharge rates for the aquifer. The simplest approach used is the piston-flow method, which assumes that ^{3}H moves like a slug through the aquifer. To use the piston-flow approach, it is necessary to know the direction of the flow and to collect samples along the flow path. The appearance of detectable bomb-pulse ^{3}H will indicate water that entered the system approximately in 1953, and the maximum ^{3}H concentrations will be found in waters recharged in 1963–64. The distance from the recharge area to the 1953 and the 1963–64 bomb pulses divided by the time will yield flow velocities for those periods. The velocities obtained will be the average for the time period and will reflect any extremely dry or wet periods occurring

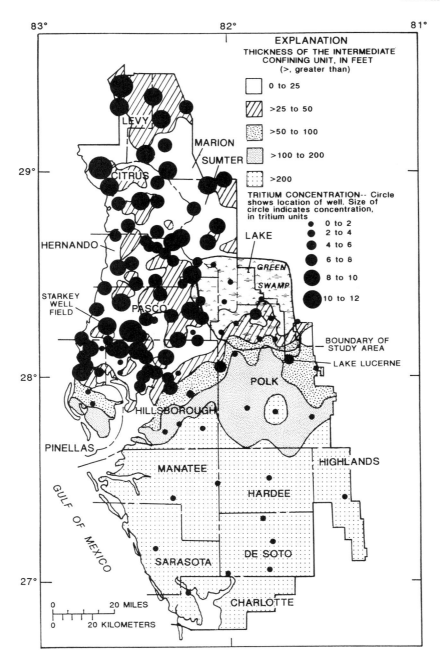

FIGURE 11-4. Distribution of tritium in the upper Floridan aquifer, central Florida, as a function of thickness of the overlying confinement. (*From Swancar and Hutchinson 1992.*)

during these years. The approach can be applied to samples collected over a single sampling period, although sample sets collected several years apart are preferable. Siegel and Jenkins (1987) demonstrate the approach for lateral flow along an alluvial fan in Nepal. Tritium concentrations increased along the flow path to a peak, and then decreased to below the detection limit of their method. Knowing the recharge area, they were able to use the location of the 1963–64 peak to obtain an average flow rate of 2–3 m/d day in the aquifer between 1963 and 1985. Knott and Olympio (1986) applied the same approach to study vertical recharge for the aquifer that supplies drinking water for Nantucket Island. For the study of vertical recharge, it is necessary to obtain a vertical profile at a ground-water divide that contains the 1963–64 ^{3}H peak to minimize problems with lateral flow.

Similar in concept to the simple piston-flow model is the method used by Fritz, Drimmie, and Fritz (1991) to study recharge in the Waterloo aquifer. The authors have two sets of measurements taken approximately one ^{3}H half-life apart from the same wells in the aquifer. The two data sets were compared to observe the decrease (or increase) in ^{3}H concentrations at each well. From these changes, the authors were able to estimate when the 1963–64 peak passed the well and make qualitative estimates for recharge in the different locations in the aquifer.

The ^{3}H interface method uses the first appearance of bomb-pulse ^{3}H to calculate recharge rates for aquifers (Andres and Egger 1985). The method requires the assumption that the flow in the aquifer is mainly vertical and downward. This assumption will be valid only in certain parts of an aquifer, such as recharge areas along ground-water-flow divides. An estimate must be made of the ^{3}H concentration that separates prenuclear and postnuclear testing ^{3}H for the area under investigation, and samples are collected from as many depths and locations in the aquifer as feasible. The ^{3}H interface is the depth below which only prenuclear testing ^{3}H concentrations are present, and above which only postnuclear testing ^{3}H concentrations are present. The depth can vary within the aquifer, depending on permeability, porosity, and differences in rate of recharge. The recharge rate can be estimated by the relation

$$\text{Recharge rate} = \frac{\text{porosity} \times \text{depth}}{\text{time}}$$

(11-1)

where porosity (dimensionless) is measured, depth refers to the depth (meters) of the ^{3}H interface below the water table, and time is years since 1953. This method has been applied to homogeneous (Larson, Delcore, and Offer 1987) and inhomogeneous (Delcore and Larson 1987) aquifers in the United States and found to give recharge rates comparable to those determined from water-budget analyses.

The piston-flow and tritium-interface methods are rapid and inexpensive ways of estimating flow and recharge rates for aquifers. Interpretations are simple, and long-term sampling series are not necessary. The main problem with this approach is the assumption that no dispersion or mixing occurs as the water moves through the aquifer. Use of the 1963–64 peak to obtain flow rates does not cause a serious problem, because the location of the peak is more important than its size. For the interface method, dispersion presents a serious problem, particularly for heterogeneous aquifers, because ^{3}H from nuclear testing will be found beyond the interface and yield recharge rates that are too high. The method also requires that the flow path be long enough or the aquifer large enough that the ^{3}H interface or 1963–64 peak is still present in the system, which is no longer the case for many aquifers.

Reservoir Models

Well-mixed reservoir or box models are the converse of the piston-flow models. Box models assume that the concentration within the system is the same everywhere, and that the mixing of the reservoir occurs on a short timescale compared to the input. The concentration of a constituent in the system for any given time period is given by

$$C(t) = -\lambda C(t-1) + k_i C_i - k_o C(t-1) \quad (11\text{-}2)$$

where $C(t)$ and $C(t-1)$ are the concentrations in the system at times t and $t-1$, C_i is the concentration in the water entering the system, λ is the decay constant ($\lambda = \ln 2/t_{1/2}$), and, k_i and k_o are rate constants for water entering and leaving the system. If the volume of water in the system is assumed to be constant, then $k_i = k_o$. This model works well for systems such as oceans, lakes, and river basins (Revelle and Suess 1957; Michel 1991), which respond like well-mixed reservoirs, but it is not applicable to most ground-water systems. It is also important to have a long-term data base, because one measurement will yield two or more possible solutions. Pearson and Truesdell (1978) compare the effectiveness of piston-flow and well-mixed reservoir models in predicting ^{3}H concentrations in outflow for a series of ground-water reservoirs in Yellowstone National Park, and show where each method is most applicable.

Compartment Models

A group of models has been developed that uses a series of compartments to study the movement of ^{3}H and water through aquifer systems. These models include the effects of dispersion and mixing as water moves through the aquifer, but retain a pistonlike nature in their flow (Przewlocki and Yurtsever 1974). The typical model is constructed of a series of compartments which are well mixed internally. Each compartment has one or more sources of incoming water from other compartments or aquifer boundaries, and one or more outputs to other compartments or discharge points. The ^{3}H concentration in a cell at a given time can be expressed as

$$C_n(t) = C_n(t-1) + \Sigma k_i C_i(t-1) \\ - \lambda C_n(t-1) - \Sigma k_i C_n(t-1) \quad (11\text{-}3)$$

where $C_n(t)$ and $C_n(t-1)$ are the concentrations in cell n at times t and $t-1$, λ is the decay constant, k_i is the fraction of water in cell n replaced from cell i in one time period, and $C_i(t-1)$ is the ^{3}H concentration in cell i at time $t-1$. Cell i can either be another cell in the system or an input from the boundary of the system. Equation 11-3 reduces to the special case of the one-reservoir model (Eq. 11-2) if only one cell has one input and output. Piston flow with no mixing occurs when the cells are connected linearly, as shown in Figure 11-5, with one input and output and all the water in the cell is replaced in each time step. Cases where only a fraction of the water in the cell is replaced results in pistonlike flow with dispersion of the ^{3}H peak.

Figure 11-5 shows the migration of the 1963–64 peak in ^{3}H concentrations of recharge water through an aquifer system consisting of six cells of equal size connected in a linear manner. The input for the system is located at cell 1 and the outflow occurs at cell 6, and k is set equal to 0.2 year^{-1} for each cell. Tritium data from Lincoln, Nebraska, are used for input concentrations, and one-year time steps are used. The 1963–64 peak begins to migrate into the aquifer within a few years as cell 1 is being diluted by lower ^{3}H concentrations in the rain. The peak continues to move down the flow path of the system while undergoing degradation caused by decay and dispersion and eventually flows out of the system. Changes in the response of the aquifer to the ^{3}H transient can be obtained by varying the value of k or the number of cells.

The application of most compartment models requires some knowledge of the hydrologic properties of the aquifer. The areas of recharge are usually specified and, frequently, the flow path for the reservoir is fixed. The model is then stepped forward at fixed time intervals from the prenuclear-

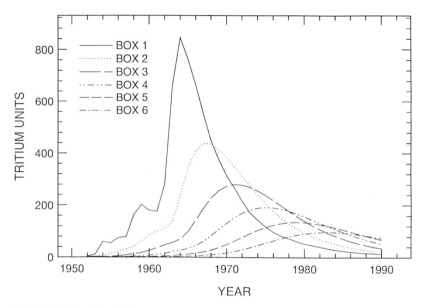

FIGURE 11-5. Modeled response of an aquifer with six compartments to the tritium transient. Input is only into box 1, and discharge is only from box 6. Precipitation data from Lincoln, Nebraska, are used for input concentrations.

bomb testing period to the time of measurement, using appropriate values for recharge and assuming values for the various exchange constants. The concentrations obtained from the model are then compared to the measured values. Values for recharge and exchange constants are varied to obtain the best fit.

Przewlocki and Yurtsever (1974) describe some of the limitations and flexibilities of the model and apply it to two aquifers where time series of ^{3}H data at the discharge area are available. By varying the size of the compartments and recharge rates so the model results match measurements, they were able to derive turnover times for the water in the aquifers and estimates of the total volume of water in the aquifer. Even if only concentrations in input to and discharge from an aquifer were available, time scales for recharge and flow could be obtained if information on other hydrologic parameters were available to limit the number of possible solutions (Yurtsever and Payne 1986).

Compartment models can be further constrained by sampling many locations in the aquifer. Allison and Hughes (1975) studied an aquifer in South Australia where recharge occurred both by flow from a local mountain range and by percolation through the unsaturated zone. Tritium data were available along the flow path, and the authors used a compartment model to study the relative importance of the two sources of recharge. Their model was able to place upper and lower limits to the recharge from each source and obtain a water budget for the aquifer. Campana and Mahin (1985) have developed a discrete compartment model where the number of compartments used is directly related to the number of locations in the aquifer where ^{3}H data are available. The recharge areas must be defined, but timescales and flow directions are determined by the model. ^{3}H data over a range of several years are needed to apply the model. The model was applied to a ^{3}H data base for the Edwards aquifer in Texas, and flow directions and turnover times for the compartments of the reservoir were determined. The flow paths matched those determined by other methods, and the turnover timescales for the compartments ranged from a few years to more than 100 years. The match of measured and modeled ^{3}H concentrations was worst for timescales of more than 100 years, and best for timescales on the order of the ^{3}H transient. The major flaw in the model as designed was that it failed to include radioactive decay, which results in concentrations that are too high.

Compartment models have been useful in describing the mixing of the ^{3}H transient with water in aquifer systems. They are able to retain the character of piston flow that is found in most aquifers, but they also account for dispersion and the degradation of the ^{3}H peak. With sufficient data, estimates of the residence times of water within the aquifer, as well as other hydrologic information, such as the size of the aquifer and the relative importance of different sources of recharge can be obtained.

Advection-Dispersion Models

The flux of ^{3}H from atmospheric nuclear-bomb testing into a homogeneous aquifer along stream paths can be modeled as an advection-dispersion process. In the one-dimensional case, the concentration in the aquifer is a function of the velocity (V), hydrodynamic dispersion (D_x), and distance from the recharge area (x). The change in concentration is given by

$$\frac{dC}{dt} = D_x \left(\frac{d^2C}{d^2x}\right) - V\left(\frac{dC}{dt}\right) - \lambda C \qquad (11\text{-}4)$$

The hydrodynamic dispersion can be expressed as

$$D_x = \alpha_L V + D^* \qquad (11\text{-}5)$$

where α_L is longitudinal dispersivity, and D^* is molecular diffusion. Molecular diffusion can be ignored in most cases. Under appropriate boundary conditions, a solution can be obtained for this equation in terms of D_x, x, t, V, and the change of concentration at the source (Cleary and Ungs 1978; van Genuch-

ten and Alves 1982). Values for D_x and V are varied to obtain a best fit between the model predictions and the actual data. The change of concentration at the source is the main uncertainty because it is not a simple function for ^{3}H. The method has been applied by Egboka et al. (1983) and Roberson and Cherry (1989).

TRITIUM–HELIUM-3

As the ^{3}H transient passes and ^{3}H concentrations decline in ground water, more interest has been directed to ^{3}H/^{3}He dating (Tolstikhin and Kamensky 1969; Torgersen, Clarke, and Jenkins 1979; Maloszewski and Zuber 1983; Zuber 1986; Weise and Moser 1987; Takaoka and Mizutani 1987; Cerling, Poreda, and Solomon 1987; Schlosser et al. 1988, 1989; Poreda, Cerling, and Solomon 1988; Solomon and Sudicky 1991; Solomon et al. 1992). Helium-3 (^{3}He) is the daughter product of ^{3}H decay ($t_{1/2}$ = 12.43 years). Therefore, in confined systems unaffected by hydrodynamic dispersion, determination of both ^{3}H and (tritiogenic) ^{3}He in ground water defines the initial ^{3}H content and allows calculation of age. The ^{3}H/^{3}He dating method has a distinct advantage over many of the ^{3}H dating applications since, with ^{3}H/^{3}He dating, the initial ^{3}H content is determined for the parcel of water being dated. Potential complications to ^{3}H/^{3}He dating include corrections for additional sources and sinks for ^{3}He in ground water, and accounting for dispersive mixing of ^{3}H and ^{3}He along the flow path.

Schlosser et al. (1989) present a mass balance for ^{3}He in ground water:

$$^3\text{He}_{\text{tot}} = {}^3\text{He}_{\text{trit}} + {}^3\text{He}_{\text{eq}} + {}^3\text{He}_{\text{exc}} + {}^3\text{He}_{\text{nuc}} \qquad (11\text{-}6)$$

where ^{3}He$_{\text{tot}}$ is the total ^{3}He content of the water sample, ^{3}He$_{\text{trit}}$ is tritiogenic ^{3}He, ^{3}He$_{\text{eq}}$ is ^{3}He derived during infiltration due to equilibration of the recharge water with the unsaturated zone air, ^{3}He$_{\text{exc}}$ is ^{3}He derived from excess air entering the ground water during recharge (Heaton and Vogel 1981), and $^3He_{nuc}$ is nucleogenic ^3He. In order to use the $^3H/^3He$ method, the tritiogenic ^3He must be separated from the total ^3He. Schlosser et al. (1989) calculated the air-derived and nucleogenic ^3He on the basis of measured concentrations of dissolved neon and ^4He (Torgersen, Clarke, and Jenkins 1979).

If a parcel of water is confined and unaffected by hydrodynamic dispersion, the quantity $^3He_{trit} + ^3H$ is conservative and the age of the water parcel is given by

$$t = \frac{t_{1/2}}{\ln 2} \ln\left[1 + \frac{^3He_{trit}}{^3H}\right] \quad (11\text{-}7)$$

where $^3He_{trit}$ is the tritiogenic ^3He in the sample, ^3H is the ^3H concentration, and $t_{1/2}$ is the ^3H half-life. The ^3He, ^4He, and neon concentrations are measured by mass spectrometry, and ^3H is more precisely determined by the ^3He in-growth method (Clarke, Jenkins, and Top 1976). Uncertainty in age because of analytic uncertainty is approximately ±0.5 year. Larger uncertainties in age result from corrections in defining the tritiogenic ^3He, the requirement that the parcel of water remain confined following infiltration, and mixing effects caused by hydrodynamic dispersion. If ^3He is lost by diffusion to the unsaturated zone air, younger ages are derived. Helium-3 can also be added to shallow ground water by dispersive transport.

Schlosser et al. (1988, 1989) reported $^3H/^3He$ dating of shallow ground water sampled from wells screened at multiple levels at Liedern/Bocholt, Germany. The ^3H from 1963–64 atmospheric nuclear-bomb testing was clearly evident in the tritiogenic ^3He at a depth of 5 to 10 m in the saturated zone. $^3H/^3He$ ages of the bomb-pulse waters were three to five years younger than the true age (1963). This difference was attributed to incomplete ^3He confinement and dispersive mixing with deeper water. From estimates of the ^3H infiltration, Schlosser et al. (1988) estimated that approximately 80% of the tritiogenic ^3He remained in the ground water at Liedern/Bocholt. Calculations based on the "Vogel" model (Vogel 1967), as applied to shallow, homogeneous sand aquifers of isotropic hydraulic conductivity, showed that the shape of the bomb pulse will be detectable in tritiogenic ^3He data for at least the next 40 years, long after the bomb pulse is lost in the ^3H data because of radioactive decay and advection-dispersion (Schlosser et al. 1989).

Helium-3 confinement has also been shown to be a function of the vertical-flow velocity (recharge rate) and dispersivity. Schlosser et al. (1989) calculated significant ^3He loss across the water table to the atmosphere at vertical flow velocities of less than 0.25 to 0.5 m/yr. Although absolute $^3H/^3He$ ages are less certain when recharge rates are small, location of the position of the bomb pulse, expressed in tritiogenic ^3He, is of great value in hydrologic studies and can be used to determine ground-water velocities as discussed earlier for ^3H.

Solomon and Sudicky (1991, 1992) used numerical simulations of simple one-dimensional and two-dimensional flow systems in hypothetical unconfined, shallow sandy aquifers to investigate the sensitivity of calculated $^3H/^3He$ ages to hydrodynamic dispersion. These authors showed that the magnitude of uncertainties in calculated $^3H/^3He$ ages depends on the ^3H input. When ^3H input is nearly constant over time, such as the ^3H input in recharge since the mid- to late 1970s, calculated $^3H/^3He$ ages tend to be within 10% of true ages. However, under transient conditions, such as for waters recharged prior to the 1960s bomb pulse, dispersion can cause more than 50% differences between calculated $^3H/^3He$ ages (Eq. 11-7) and true ages. If the vertical velocity is rapid enough to maximize ^3He confinement (Schlosser et al. 1989), $^3H/^3He$ ages determined near the water table should closely reflect the average vertical velocity.

KRYPTON-85

Krypton-85 is an inert radioactive noble gas (half-life 10.76 years) that is produced by fission of uranium and plutonium. Sources have included nuclear-bomb testing, nuclear reactors, and the release of ^{85}Kr during the reprocessing of fuel rods from nuclear reactors (Sittkus and Stockburger 1976). Because most of the nuclear industry is located in the Northern Hemisphere, most of the ^{85}Kr is released to the troposphere in the Northern Hemisphere. Because of differences in mixing rates, the ^{85}Kr activity of the Southern Hemisphere is about 20% less than that of the Northern Hemisphere (Rozanski 1979; Weiss et al. 1983). Although ^{85}Kr has a half-life similar to that of ^{3}H, ^{85}Kr has the advantage of being an environmental tracer with steadily increasing atmospheric input, whereas the ^{3}H atmospheric input is a more complex function of season and latitude and has declined since cessation of atmospheric nuclear-bomb testing in the mid-1960s. Since the beginning of the nuclear age, ^{85}Kr activity in the atmosphere has increased by about six orders of magnitude (Rozanski 1979). Smethie et al. (1992) summarize the atmospheric measurements of ^{85}Kr activity over the past 40 years (Rozanski 1979; Sittkus and Stockburger 1976; Weiss et al. 1983). The activity of ^{85}Kr in air over the Northern Hemisphere in 1990 was approximately 50 dpm/cc (disintegrations per minute per cubic centimeter) krypton at Standard Temperature and Pressure (STP). During the next 40 years, as ^{3}H and ^{3}H/^{3}He dating becomes less effective, ^{85}Kr holds considerable promise as an environmental tracer for age-dating young ground water.

Krypton-85 enters ground water by equilibration of the infiltration water with air in the unsaturated zone, which is assumed to have a ^{85}Kr activity similar to that of the atmosphere. The activity of ^{85}Kr dissolved in the infiltration water is usually reported as the specific activity of ^{85}Kr, that is, the ratio of ^{85}Kr to total dissolved krypton, which is independent of recharge temperature. The atmospheric input of ^{85}Kr has continually increased since about 1950, resulting in a continuous decrease in the ^{85}Kr specific activity in ground water with increasing distance from the recharge zone, and unambiguous definition of water age. Because ^{85}Kr is a noble gas, it is inert and not subject to microbial degradation and other chemical interactions that can alter the concentrations of organic environmental tracers.

If the effects of hydrodynamic dispersion are small, the ^{85}Kr specific activity of ground water defines the time since the infiltration water was isolated from the atmosphere. However, ^{85}Kr dating should be subject to some of the same uncertainties of confinement as pertain to ^{3}He (Schlosser et al. 1988, 1989). If not isolated quickly from the unsaturated-zone air during recharge, the ^{85}Kr in infiltrating waters can exchange with the ^{85}Kr in more recent air, resulting in an increase in specific activity of ^{85}Kr and calculated ages younger than that of the infiltration water. Smethie et al. (1992) reported a detection limit of 3 dpm/cc krypton for a 100-L water sample, thereby limiting dating to waters recharged since about 1963. Further lowering of the detection limit to, for example, 1 dpm/cc would extend ^{85}Kr dating back an additional three years to about 1960. Presently, ^{85}Kr dating is applicable over the past 30 years, but as ^{85}Kr concentrations in the atmosphere continue to increase, the maximum age of water that can be dated will increase only slightly.

The possibility of age dating ground water with ^{85}Kr was first investigated by Rozanski and Florkowski (1979) who developed a method of extracting ^{85}Kr from 200 to 300 L of ground water. From the tens of liters of dissolved gases extracted from the water, only tens of microliters of krypton gas are obtained. Special procedures for extraction, separation, and low-level counting (7 to 10 days per sample) are required (Rozanski and Florkowski 1979; Salvamoser 1983; Smethie and Mathieu 1986; Smethie et al. 1992). Rozanski and Florkowski (1979) reported 14 determinations of ^{85}Kr activity in surface water, ground water, and springs. Of these, five samples could be dated using the exponential (reservoir) model (Zuber 1983), suggesting ground-water residence times of

10, 20, and 40 years. Andrews et al. (1984) combined measurements of ^{3}H and ^{85}Kr to test for cases of mixing or leakage of shallow water into deep ground water.

Assuming air-water equilibrium during infiltration, piston flow, and radioactive decay as the only ^{85}Kr loss terms, Smethie et al. (1992) calculated the age and travel times of five water samples in the Borden aquifer of Ontario, Canada. Travel times of ground water to three sampling points in the main recharge area were nearly identical to travel times modeled using a previously developed two-dimensional steady-state model (Frind and Matanga 1985). For the other two waters, which were located farther downgradient of the main recharge area, calculated ^{85}Kr ages were younger than the modeled travel times by factors of 1.4 and 1.6, but within uncertainties in hydraulic properties of the aquifer and potential sampling uncertainties.

At the Borden landfill, Smethie et al. (1992) found that the ^{85}Kr activities in air and soil gas were nearly identical, indicating a very short residence time for air in the unsaturated zone. This means that the age of the infiltrating water at the Borden aquifer study site (unsaturated zone 1 to 6 m thick) is similar to that of the atmospheric reservoir at the moment the parcel of water was isolated from the soil air. Still, recharge rates need to be considered as discussed by Schlosser et al. (1989).

Smethie et al. (1992) also showed that the distribution of ^{85}Kr in the shallow water in the Borden aquifer was little affected by hydrodynamic dispersion. This was attributed to the nature of the ^{85}Kr input. Unlike the transients in the ^{3}H input that are significantly affected by hydrodynamic dispersion (Solomon and Sudicky 1991), the ^{85}Kr input is a generally smooth, increasing function that is less sensitive to hydrodynamic dispersion than ^{3}H (Smethie et al. 1992). Effects of hydrodynamic dispersion will generally increase for ^{85}Kr and all other transient tracers as advection-dispersion ratios decrease and travel distances increase. As we will see in the next section, there are some similarities of ^{85}Kr dating to dating with CFCs which also have a smooth, continually increasing atmospheric input function.

CHLOROFLUOROCARBONS

Chlorofluorocarbons are stable volatile organic compounds that can also be used for age-dating young ground water. The aspect of time is introduced through the recharge process and the known temporal variation of CFC concentrations in the atmosphere.

CFCs were first produced in the 1930s as the refrigerant dichlorodifluoromethane (CCl_2F_2, or F-12) followed by production of trichlorofluoromethane (CCl_3F, or F-11) in the 1940s. Annual production of CFCs now exceeds 10^9 kg. CFCs are used as refrigerants, aerosol propellants, cleaning agents, solvents, and blowing agents in the production of foam rubber and plastics. No known natural sources of CFCs exist (Lovelock 1971). All CFCs produced are eventually released to the atmosphere, and through the atmosphere they are partitioned into the hydrosphere by gas-liquid exchange equilibria. The 1990 atmospheric volume fractions of F-12 and F-11 were approximately 480 and 285 parts per trillion, respectively, and increasing at an average rate of approximately 3.7% per year.

The first measurements of CFCs in ground water were reported by Thompson (1976) and Thompson and Hayes (1979), who used a gas chromatograph with electron capture detector (ECD) to analyze on site the F-11 content of ground water from southern New Jersey, Arkansas, and south-central Texas. Their results were in good agreement with the known hydrology and ground-water ^{3}H content. Thompson and Hayes (1979) reported a plume of F-11 some 70 km in length in the Edwards aquifer of Texas, which is evidence of the persistence of F-11 and its resistence to microbial degradation in the hydrologic environment. Schultz (1979) confirmed the findings of Thompson and Hayes (1979) and used the F-11 observations to model the hydrology of the Edwards aquifer.

Tracer Applications

CFCs are invaluable in a qualitative sense as tracers of recent recharge. Considering the atmospheric concentrations of F-11 and F-12 since 1940, and assuming an average recharge temperature of 10°C, the analytical detection limit of 1 pg/L for F-11 and F-12 in water allows identification of post-1940 water using F-12 and of post-1947 water using F-11. Because most ground-water samples are regarded as mixtures, the presence of detectable F-11 or F-12 indicates that the water contains at least some post-1940 or post-1947 water, respectively. Because of the extremely low detection limit for CFCs, it is possible to detect mixtures of as little as 0.01% modern water in pre-1940 ground water, and, if the source of recent water has been contaminated with CFCs from anthropogenic sources in addition to the atmosphere, as is often observed in urban areas, some modern waters containing CFC concentrations at least several orders of magnitude less than 0.01% often can be detected. In a study of the purging of deep (250 m) municipal supply wells in central Oklahoma, Busenberg and Plummer (1992) measured the discharge of CFCs over a period of more than 3 h of continuous pumping; the change in CFCs concentration over time indicated mixing of approximately 1% shallow water with deep formation water. Because the ^{14}C concentration in water in the aquifer was very low, the presence of CFCs was attributed to drawdown of small portions of shallow ground water around the grout and gravel pack surrounding the well casing.

Important uses of CFCs as hydrologic tracers are demonstrated by Thompson, Hayes, and Davis (1974), Schultz et al. (1976), Randall and Schultz (1976), Davis et al. (1980), and Davis and Bentley (1982). Schultz et al. (1976) and Busenberg and Plummer (1991) show that CFCs are excellent tracers of sewage. Busenberg and Plummer (1992) traced CFC anomalies in shallow ground water of the alluvium and terrace deposits along the North Canadian River and its tributaries in central Oklahoma to CFC anomalies in the river located at and downstream from discharge points of sewage-treatment plants.

Water samples transported to laboratories for CFC analysis in conventional bottles and containers usually become contaminated by diffusion of CFCs through walls of plastic bottles or through plastic caps prior to analysis. To overcome this sampling problem, Busenberg and Plummer (1992) designed a sampling apparatus for collecting water samples in the field that excludes contact with air. The samples are then welded into borosilicate-glass ampules for transport and laboratory storage until they are analyzed. Using this procedure, water samples free of CFCs have been collected in the field, and after months of storage remained free of CFCs. The collection, transport, and storage of water samples without contamination is a critical step to reliable age dating with CFCs, particularly for old waters that may contain CFC concentrations of a few to tens of picograms per liter.

Age Dating with CFCs

Northern latitude tropospheric CFC concentrations are reconstructed for the continental U.S. from 1940 to the present (1992) in Figure 11-6 (Busenberg and Plummer 1992). This atmospheric CFC input function is derived, in part, from CFC production records and atmospheric measurements that began in the mid-1970s. Exceptions to Figure 11-6 are urban areas where CFC concentrations can be considerably higher, and southern latitudes where CFC concentrations are approximately 15% lower than those in northern latitudes.

The equilibrium solubility of the ith CFC compound in water is defined by Henry's law:

$$C_{CFC_i} = K_{CFC_i, T}\, P_{CFC_i} \qquad (11\text{-}8)$$

where K is the Henry's law constant for the ith CFC compound at temperature T, and P

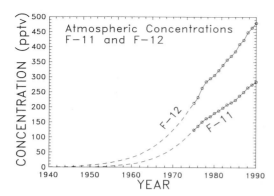

FIGURE 11-6. Comparison of atmospheric concentrations of F-11 and F-12 (parts per trillion by volume); dashed where reconstructed from manufacturing data. (*After Busenberg and Plummer 1992.*)

is the atmospheric partial pressure of the ith CFC compound. The Henry's law constants for F-12 and F-11 in water and seawater were determined by Warner and Weiss (1985) between 0°C and 40°C and 0 and 40 parts per thousand salinity. Although the present atmospheric concentration of F-12 is nearly twice that of F-11, their solubilities are reversed. The solubility of F-11 in water in equilibrium with the 1990 atmosphere is approximately 830 pg/L, compared with 320 pg/L for F-12 (both calculated at 10°C).

From inspection of Figure 11-6 and Eq. 11-8, it is apparent that CFC concentrations of meteoric water recharged in equilibrium with air will, to a first approximation, vary with the corresponding atmospheric age and the equilibration temperature. As new water is recharged, concentrations of CFCs in ground water will increase continuously. This forms the basis of CFC age dating of ground water.

Calculations combining the atmospheric CFC growth curves of Figure 11-6 and the Henry's law solubilities of Warner and Weiss (1985) allow estimation of the expected concentrations of F-12 and F-11 in water recharged in equilibrium with air between 1940 and 1990 at temperatures of 0 to 20°C (Figure 11-7). If the recharge temperature is known, and there are no further hydrologic or chemical complications, analyses of concentrations of F-11 and F-12 in a ground-water sample should each independently indicate the same water age on Figure 11-7. However, the recharge temperature is usually not known precisely, and there are usually a number of hydrologic and chemical complications (discussed below) that must be carefully considered in CFC age dating.

Interpretation of CFC Age

All processes and mechanisms that can alter the CFC solubility governed by air-water equilibria (Figure 11-7) must be considered in interpreting CFC age from the CFC concentrations in ground water. Many of these processes cannot be fully evaluated, resulting in assignment of uncertainties to the dating or, in some cases, determination that a particular water sample is not suited for CFC age dating. The least of the uncertainties in CFC age dating is the uncertainty associated with the actual analysis. Purge-and-trap GC technology with ECD (Bullister and Weiss 1983) usually defines the CFC concentrations within 1 pg/L, representing age uncertainties of about one year. However, when dating with CFCs, careful attention needs to be given to other factors, including (1) determination of recharge temperatures, (2) recognition of waters whose CFC concentrations may have been altered by additional sources or sinks of CFCs, and (3) recognition of physical processes that can alter CFC concentrations.

Recharge Temperature

The recharge temperature is the temperature at the base of the unsaturated zone where final equilibration of recharge water and unsaturated zone air occurs. The recharge temperature affects the solubilities of all gases in the infiltrating water. Uncertainties related to recharge temperature almost completely cancel out in the case of ^{85}Kr dating, which is based on the specific activity of ^{85}Kr—that is, the ratio of the concentrations of ^{85}Kr to total dissolved Kr in the water. No such relation exists for CFCs. Although

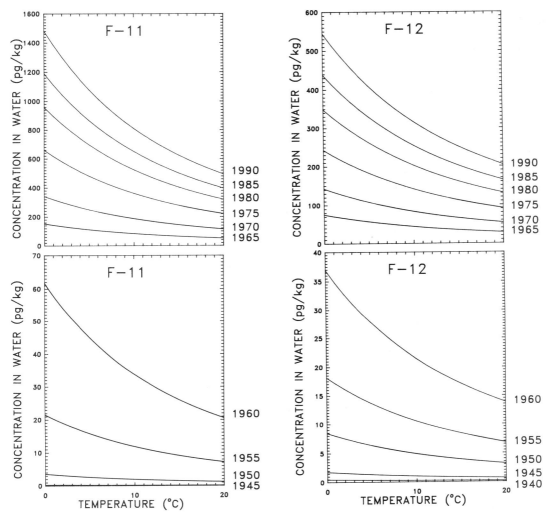

FIGURE 11-7. Equilibrium air-water concentrations of F-11 and F-12 in ground water calculated using the solubility data of Warner and Weiss (1985) and the atmospheric F-11 and F-12 data of Figure 11-6. Contours show solubilities for a particular year as a function of recharge temperature.

the F-11:F-12 ratio has been used successfully in dating ocean water masses (Smethie et al. 1988), the ratio is less reliable in ground-water systems because of a variety of interactions, including sources and sinks that can affect the ratio after the initial air-water equilibration step (see discussion following). Therefore, determination of ground-water ages by use of CFCs should be done independently for F-11 and F-12 using data for the recharge temperature.

In deep unsaturated zones, the recharge temperature corresponds to the mean annual temperature (Mazor 1972; Herzberg and Mazor 1979; Heaton and Vogel 1981; Andrews and Lee 1979). In shallow unsaturated zones, however, temperatures vary with season. In North America during the late winter-spring rains when most recharge typically occurs, the temperature of the unsaturated zone can be cooler than the mean annual temperature. Ideally, seasonal studies of unsaturated-zone temperatures should be done when age dating shallow ground-water systems by use of CFCs. Alternatively, noble gases and N_2/Ar solubilities are excellent indicators of recharge temperature (Heaton 1981; Heaton and Vogel 1981; Andrews et al. 1985).

Inspection of Figure 11-7 indicates the range of errors in CFC ages resulting from uncertainities in recharge temperature. A maximum age error of several years results from uncertainties of several degrees Celsius in recharge temperature when using both F-11 and F-12 in the youngest waters. CFC dating of older waters is less sensitive to uncertainties in recharge temperature.

Additional Sources and Sinks of CFCs

Water containing F-11 and F-12 concentrations greater than those corresponding to equilibrium of water with modern air is clearly contaminated and cannot be dated with CFCs. Water recharged in urban areas often contains CFC concentrations greater than can be in equilibrium with tropospheric air. The excess CFCs partly result from local releases of CFCs to urban air and other human-made sources, such as plastic containers, air conditioners, propellants in aerosol cans, and so forth. Some sewage effluents contain CFC concentrations orders of magnitude greater than air-water-equilibrium concentrations. Consequently, it may not be possible to age-date waters by use of CFCs in many urban or industrialized areas. If the water is contaminated with one CFC compound but not another, dating may be possible if it is assumed that the concentration of the other CFC compound is determined by air-water equilibrium. In this case, the possibility exists that the water sample may contain CFCs from air-water equilibration and an unknown contaminant source; therefore, the CFC age should be regarded as a minimum age.

In addition to recharge of contaminated water, several other processes are known to add excess CFCs to ground water. Studies of N_2 and noble gas concentrations in ground water have demonstrated the presence of "excess air" (Heaton 1981; Heaton and Vogel 1981; Andrews et al. 1985). Apparently, small volumes of air are trapped in pore spaces of the capillary fringe as water tables decline. As the water table rises, following, for example, barometric pressure changes or recharge events, air becomes trapped below the water table. Under the increased hydrostatic pressure, additional air dissolves in the water. Excess quantities of all atmospheric gases are introduced into ground water by this mechanism, including CFCs. The largest reported content of excess air in ground water is 30 cc/L (Heaton and Vogel 1981) but is typically less than 5 cc/L (Busenberg and Plummer 1992). Addition of excess air to ground water results in calculated CFC ages that are younger than actual ages. In a quantitative evaluation of the effect of excess air on CFC age dating, Busenberg and Plummer (1992) found that CFC ages based on F-12 were more significantly affected than CFC ages based on F-11. The CFC age error caused by excess air increases with increasing recharge temperatures, increasing volumes of excess air, and is more significant for recently recharged waters that contain increased concentrations of CFCs. For example, addition of 10 cc/L of excess air at approximately 30°C to recent water (1986) can result in F-12 ages approximately three years younger than actual. The error is less than one year for F-11 under these conditions. If the recharge temperature is less than 10°C, the F-12 error is less than two years for 10 cc/L excess air. These errors are small and can normally be ignored, especially for typically small volumes of excess air that are encountered.

In arid climates where the shallow unsaturated zone can dry between recharge events, Russell and Thompson (1983) present evidence for the sorption of CFCs on dry organic matter in soils. The extent of halomethane adsorption by soils decreases with increasing fluorine content ($CCl_4 > CCl_3F > CCl_2F_2$) (Weeks, Earp, and Thompson 1982; Khalil and Rasmussen 1989; Brown 1980). Russell and Thompson (1983) found that CFCs are subsequently released to the soil air as moisture is introduced, as has been observed for many volatile organic compounds (Chiou and Shoup 1985). Water recharged through previously dried unsaturated zones may contain CFC concentrations in excess of atmospheric air-water-equilibrium concentrations. Because F-11 is more strongly adsorbed by soils than is F-12, F-11 can be released during recharge in excess of F-12. In arid zones, this sorption-desorption mechanism can lead to erroneously younger ages, particularly for F-11. Because of the competition of CFCs and water vapor for sorption sites on organic matter, excess CFCs are probably not recharged through this sorption-desorption process when the unsaturated zone remains moist, such as occurs in temperate zones in the east and southeast United States.

CFCs are extremely stable compounds as shown by their persistence in aquifers for tens of years (Thompson and Hayes 1979; Busenberg and Plummer 1991, 1992). Nevertheless, most of the ground-water environments investigated have been aerobic. In recent studies conducted by the U.S. Geological Survey (Plummer and Busenberg, U.S. Geological Survey, written communication, 1991), no evidence has been found for microbial degradation of CFCs in aerobic ground water. Little is known of the possibility of anaerobic degradation of CFCs. Semprini et al. (1990) reported the microbial degradation of CFCs in an anaerobic, confined aquifer after biostimulation by addition of acetate. Degradation of CCl_4, 1,1,1-trichloroethane, F-11, and F-113 was observed mainly after denitrification reactions were completed. The rate of microbial degradation decreased with increasing fluorine content of the halocarbon. Lovley and Woodward (1992) reported anaerobic degradation of F-11 and F-12 in laboratory microcosm experiments. Field studies are presently being conducted by U.S. Geological Survey personnel to investigate microbial degradation of CFCs in natural, anaerobic ground water.

Physical Processes

Because of the generally smooth, increasing nature of the atmospheric CFC input function (Figure 11-6), it is anticipated that CFC concentrations in shallow ground water will not be greatly affected by hydrodynamic dispersion. A series of numerical simulations were made to test this hypothesis using an analytical solution to the one-dimensional advection-dispersion equation in a steady-state flow field (van Genuchten and Alves 1982). The model simulates the flow and transport of $^3H+^3He$, F-11, and F-12 through a porous medium of semi-infinite length at an average interstitial velocity of 1 m/yr. The ^3H input function was constructed for a latitude of approximately 39°N along the east coast of the United States using data of Michel (1989) and is shown in Figure 11-8A. The input concentrations of F-11 and F-12 were calculated for air-water equilibrium at 9°C (solid line on Figure 11-8B shows the F-11 input function). In order to compare hydrodynamic dispersion effects caused by the nature of the CFC and ^3H input functions, the transport of the conservative component $^3H+^3He$ was modeled. A constant-flux boundary condition (Type 3, van Genuchten and Alves 1982) was used. The ^3H and CFC concentrations were maintained constant during any single year. The simulation spans the time interval 1940 through 1991.

The modeled distribution of $^3H+^3He$ and F-11 for the final year of simulation (1991) assuming dispersivities of 0.1, 0.5, 1.0, and 2.0 m is shown in Figure 11-8. Simulated input of ^3H and CFCs began in 1940; after 51 years of flow, the 1963 bomb pulse has traveled approximately 28 m into the aquifer. The $^3H+^3He$ and F-11 input concentrations are shown as solid lines in Figure 11-8 and compared with F-11 and $^3H+^3He$ concentrations modeled with different dispersivities. Because of the nature of the two input functions, dispersion has different effects on $^3H+^3He$ and CFC concentrations and, therefore, on the water ages calculated from these concentrations. For example, dispersion causes an increase in $^3H+^3He$ and a decrease in CFC concentrations in waters younger than 1970. Dispersion also spreads the bomb pulse into waters older than approximately 1963 and increases CFC concentrations in those waters.

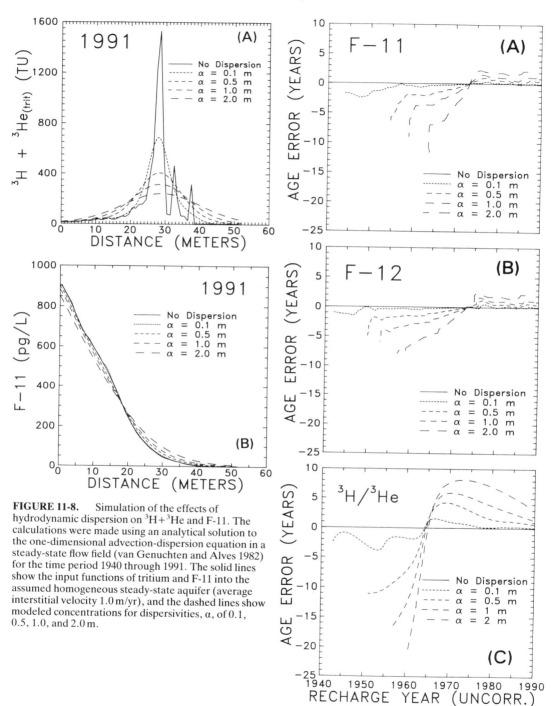

FIGURE 11-8. Simulation of the effects of hydrodynamic dispersion on ^{3}H+^{3}He and F-11. The calculations were made using an analytical solution to the one-dimensional advection-dispersion equation in a steady-state flow field (van Genuchten and Alves 1982) for the time period 1940 through 1991. The solid lines show the input functions of tritium and F-11 into the assumed homogeneous steady-state aquifer (average interstitial velocity 1.0 m/yr), and the dashed lines show modeled concentrations for dispersivities, α, of 0.1, 0.5, 1.0, and 2.0 m.

FIGURE 11-9. F-11, F-12 and ^{3}H/^{3}He age-dating errors as a function of measured age and dispersivity. Advective recharge year is the recharge year (uncorrected for hydrodynamic dispersion) plus the age error (in years).

The modeling results of Figure 11-8 can be used to calculate the errors in ^{3}H/^{3}He and CFC age dating caused by dispersion in the hypothetical one-dimensional system. F-11 and F-12 age errors were calculated by comparing the positions of the piston-flow concentrations with the positions of the modeled dispersed concentrations. Dispersivities of 1 to 2 m can cause measured F-11 and F-12 ages (uncorrected for hydrodynamic dispersion) to be 1 to 2 years older than the actual advective year during the period 1973 to present (1991) (Figures 11-9A,B). CFC age errors caused by dispersion are approximately zero for water recharged in 1973. Large dispersion-related errors in CFC dating are evident in waters recharged before 1973, causing measured F-11 and F-12 ages (uncorrected for hydrodynamic dispersion) to be younger than the actual recharge year (Figure 11-9A,B). The lower solubility of F-12 relative to F-11 in water (Figure 11-7) results in a lower rate of increase in input of F-12 relative to F-11 in recharge. Because the resulting F-12 concentration gradient in ground water is lower than that of F-11, dispersion causes smaller age errors for F-12 than for F-11 (Figure 11-9). For example, Figure 11-9 shows that a water with a measured F-11 recharge year of 1960, in an aquifer with a dispersivity of 1 m, would have actually been recharged in the year 1956. Similarly a water with a measured F-12 recharge year of 1960 would have an actual recharge year of approximately 1957.

^{3}H/^{3}He age-dating errors were calculated by an application of Eq. 11-6 to modeled ^{3}H and ^{3}H+^{3}He profiles. Figure 11-9 shows that hydrodynamic dispersion has a greater effect on ^{3}H/^{3}He age dating than on F-11 and F-12 age-dating. The ^{3}H/^{3}He ages of waters recharged in 1964–65 are little affected by hydrodynamic dispersion. For waters recharged after 1964 the measured recharge year (uncorrected for hydrodynamic dispersion) will be older than the actual (advective) recharge year. The opposite effect occurs for waters recharged before 1964. For example a water with a measured ^{3}H/^{3}He recharge year of 1973 would have an actual recharge date of 1979 to 1981 for dispersivities of 1 to 2 m. A water with a measured ^{3}H/^{3}He recharge year of 1960 would actually be 13 to 21 years older than observed for dispersivities of 1 to 2 m (Figure 11-9C). The modeled ^{3}H/^{3}He dating errors are similar to those of Solomon and Sudicky (1991, 1992).

Ground-water age dating is greatly enhanced when more than one dating tracer is measured on the same water sample. Figure 11-10 compares measured (uncorrected for hydrodynamic dispersion) F-12 and ^{3}H/^{3}He recharge years for dispersivities of 0 to 2 m in the hypothetical one-dimensional flow tube. The calculations of Figure 11-10 show that F-12 and ^{3}H/^{3}He recharge ages should be strongly correlated, but with greater departures from linearity with increasing dispersivity. Waters with equal F-12 and ^{3}H/^{3}He dispersion errors, or with zero dispersivity are linearly correlated. The measured age equals the advective age for both tracers in cases of zero dispersivity and, for F-12, the years 1972–74, and for ^{3}H/^{3}He, the years 1964–65. The isochrons (solid lines connecting dots on Figure 11-10) give the actual recharge year. In principle, Figure 11-10 indicates the possibility of determining the actual recharge age of a single water sample as well as the dispersivity that has affected it along its flow path, within the assumptions of the one-dimensional model. Potential problems in CFC dating are indicated by arrows on Figure 11-10, indicating older apparent CFC ages for waters affected by CFC sorption or degradation, and younger apparent ages for waters contaminated by CFCs.

The actual age of a water sample is further complicated in the sampling process where waters of different ages are drawn into the well screen and mix. The uncertainty in ground-water age increases with larger sampling intervals. In spite of these limitations, reliable ground-water ages can often be assigned, particularly in relatively homogeneous aquifers, or where water samples are obtained over narrow sampling intervals.

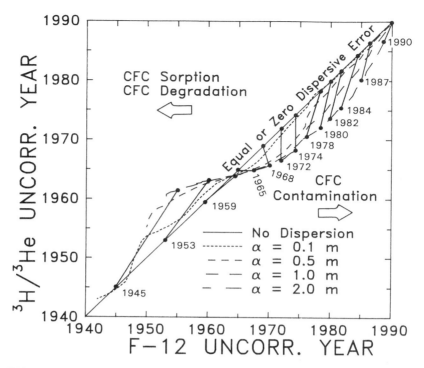

FIGURE 11-10. Comparison of measured (uncorrected for hydrodynamic dispersion) F-12 and $^3H/^3He$ recharge years as a function of dispersivity. Solid tie-lines indicate the advective recharge year.

The CFC-based age is determined at the moment the water in the recharge zone becomes isolated from the unsaturated-zone air. If the air in the unsaturated zone is old, the recharge water will have a CFC concentration and age consistent with that of the air. In areas with deep unsaturated zones, recharge water may come in contact with old, deep air, thereby increasing the apparent age of the recharge water. Weeks, Earp, and Thompson (1982) found that F-11 and F-12 concentrations in the upper 5 to 10 m of the unsaturated zone are similar to that of modern air. Variations in barometric pressure and temperature apparently promote convective mixing of air in shallow unsaturated zones (Weeks, Earp, and Thompson 1982). Therefore, in CFC age dating of shallow surficial aquifers, it is usually assumed that the unsaturated zone air and atmosphere have similar compositions. Alternatively, in deep unsaturated zones, precipitation may be tens of years old before it reaches the water table. If air circulates more rapidly through the unsaturated zone, the water recharged could appear to be younger than that of the actual precipitation. Younger ages also can occur with shallow unsaturated zones if the recharge rate is small, resulting in less efficient confinement of infiltration water. As CFC concentrations in the unsaturated zone air increase, previously recharged waters are not completely isolated from the air and the shallow water equilibrates with younger air. In this case the flux of CFCs is into the water table, but otherwise this process is analogous to the ^3He confinement problem discussed by Schlosser et al. (1988, 1989). As we learn more about age dating shallow ground water and begin to compare

CFC ages with those based on other dating methods, such as ^{3}H/^{3}He and ^{85}Kr age dating, detailed understanding of the recharge process will likely be necessary to account for observed age differences.

Environments and Conditions Best Suited for CFC Age Dating

Not all aquifers and ground-water environments are well suited for CFC dating. The most likely candidate aquifers for CFC dating have the following properties:

- Located in rural settings where CFC concentrations are more likely to be at natural concentrations, as determined by air-water equilibrium only.
- Relatively thin unsaturated zone, probably less than 10 m, where unsaturated-zone air is mixed rapidly and is similar to atmospheric air composition.
- Oxic ground water where there is no evidence for microbial degradation of CFCs. After denitrification, microbial degradation of CFCs may begin.
- Unsaturated zones with minimal or no organic matter content which minimizes the sorption of CFCs on soil organic matter during dry periods.
- Temperate climates where unsaturated zones have less opportunity to dry between successive recharge events.
- Shallow waters that contain detectable concentrations of CFCs (that is, post-1940 recharge).

Precautions in sample collection also can improve the reliability of CFC dating. Ground waters should be sampled over narrow depth intervals, such as through piezometers installed with narrow screens. CFC contamination during the sampling process can be minimized by use of CFC-free materials that contact the water. Water samples should be collected in air-free environments and preserved in welded borosilicate glass ampules. Finally, more representative samples are obtained by use of low pumping rates to minimize drawdown around the screen or casing.

Additional data also should be obtained in CFC age-dating studies. These include (1) measurement of dissolved gases to define recharge temperatures, and (2) determination of the ^{3}H content of the water. The most reliable recharge temperatures are obtained from noble gas data. However in aerobic ground water, where denitrification has not occurred, reliable recharge temperatures have been obtained from dissolved N_2 and Ar concentrations. Tritium data are particularly useful in recognizing old waters that have mixed with small portions of CFC-contaminated water, and in general, for partial validation of the CFC dating process. Frequent monitoring of water-quality parameters, such as dissolved oxygen, temperature, pH, and conductivity during well purging, is useful in determining when to begin CFC sampling and in recognizing possible hydraulic anomalies during sampling. Normally three to six successive CFC samples are taken from a single well to confirm representative concentrations. Present research on age dating of shallow ground water with CFCs being conducted by the U.S. Geological Survey suggests that under optimal conditions, such as these listed here, CFC ages may be determined within a few years.

COMPARISON AMONG ENVIRONMENTAL TRACERS

During the past decade, several environmental tracers, specifically ^{3}H/^{3}He, ^{85}Kr, and CFCs, have seen relatively wide use in oceanographic studies to age-date parcels of water, trace water masses, and define mixing proportions. These techniques are only recently receiving greater attention in groundwater investigations. In age dating ground water with ^{3}H/^{3}He, ^{85}Kr, and CFCs, additional physical and chemical processes that were not important in oceanographic studies must be considered. The effects of hydrodynamic dispersion as water flows through porous media, and various sources and sinks for these tracers are presently being carefully evaluated in applications to the ground-water environment. The advantages and limitations of environmental tracers must be carefully considered in relation to the specific hydrogeologic environment under study. There are also practical considerations, including specialized collection and analytical procedures necessary for age dating with $^3H/^3He$, ^{85}Kr, and CFCs. These practical aspects are discussed more fully in the original sources. We compare and contrast some of the advantages and limitations of $^3H/^3He$, ^{85}Kr, and CFC age dating of young ground water.

The input function must be known for use of all three environmental tracers. For ^{85}Kr and CFC dating, only average atmospheric input functions are known for the activity of ^{85}Kr, and the F-11 and F-12 concentrations. The average atmospheric input functions are based partly on worldwide atmospheric measurements and partly on production records and atmospheric models. The input functions for ^{85}Kr and CFCs may vary locally from the average atmospheric functions due to local sources, such as excess CFCs in urban air, and excess ^{85}Kr near nuclear fuel processing plants. A major advantage in $^3H/^3He$ dating is that the initial input ^3H value is actually measured and reconstructed for each water sample through determination of the ^3H and tritiogenic ^3He content. However, due to the variable nature of the ^3H input, $^3H/^3He$ dating becomes less certain for waters older than the mid-1960s bomb pulse due to dispersive mixing. Consequently, $^3H/^3He$ dating is most reliable only for the past 20 to 25 years (Solomon et al. 1992) when ^3H input has been relatively constant, and therefore, influenced to a lesser extent by hydrodynamic dispersion (Solomon and Sudicky 1991). In contrast to $^3H/^3He$ dating, ^{85}Kr and CFCs have relatively smooth, increasing atmospheric input functions and to a good approximation, appear to transport under approximately piston-flow conditions over much of the dating range of the tracer.

The ranges of $^3H/^3He$, ^{85}Kr, and CFC dating are not identical. Due to counting limitations, ^{85}Kr dating presently is limited to post-1962. CFC dating has potentially the longest dating period of the three environmental tracers, extending to 1940 for F-12 and approximately 1947 for F-11. Recent applications of $^3H/^3He$ dating have recognized waters recharged over the past 30 years (Schlosser et al. 1988). It is expected that the usefulness of $^3H/^3He$ dating will decrease in the future as the mid-1960s bomb pulse dissipates due to decay and dispersive mixing (Schlosser et al. 1989). The ^{85}Kr input function is expected to increase in the future, but without significant improvements in extraction and/or counting procedures, the dating range of ^{85}Kr will remain approximately 30 years. As long as CFC concentrations continue to increase in the atmosphere, the potential lifetime of CFC dating should increase linearly with time. However, as production in F-11 and F-12 declines worldwide, their atmospheric concentrations will also decrease. This will result in some ambiguity in CFC ages for younger waters that can only be resolved by a greater sampling density to locate the expected CFC maximum in ground water.

Although CFC dating is attractive in terms of its dating lifetime and prognosis for future applications, its usefulness can be limited by a variety of chemical interactions, such as sorption and microbial degradation, or contamination from additional anthropogenic sources beyond that of air-water equilibrium. These problems may limit the environments suited for CFC dating to relatively rural, pristine environments with aerobic ground water and relatively shallow unsaturated zones low in particulate organic matter. Recently, Smethie et al. (1992) reported ^{85}Kr dating near a landfill, an environment very likely to be contaminated with CFCs and unsuitable for CFC dating. $^3H/^3He$ and ^{85}Kr dating applications are not limited by the chemical interactions that affect CFCs.

As all three dating methods involve a gas phase, each is subject to uncertainties caused by gas exchange between the unsaturated zone and the water table. If recharge rates are slow, ^{3}He$_{trit}$ can escape by diffusion into the unsaturated zone resulting in apparently younger ^{3}H/^{3}He ages. Similarly, shallow waters can take up additional ^{85}Kr or CFCs before recharge is sufficient to confine the parcel of water, again resulting in apparently younger ages. CFCs have the added potential uncertainty of gaining additional CFCs during recharge by exchange of water vapor with CFCs sorbed on soil organic matter. This again results in apparently younger ages.

Only CFC dating is subject to uncertainties in the recharge temperature. As explained earlier, recharge temperatures are needed to define the solubility of CFCs at the water table. The recharge temperatures are best determined from dissolved gas measurements. The need to know recharge temperature is eliminated in ^{85}Kr dating when the observations are expressed in terms of specific activity. Tritium input is independent of the recharge temperature, but, in ^{3}H/^{3}He dating, there is a minor dependency on recharge temperature in defining the amount of ^{4}He dissolved in recharge water. The ^{4}He concentration is used in determining the ^{3}He$_{trit}$ content. In ^{3}H/^{3}He dating, the recharge temperature is usually taken as the mean annual temperature, and uncertainties of several degrees Celsius in recharge temperature can be ignored in most cases of ^{3}H/^{3}He dating.

All three environmental tracers require special sampling procedures in the field. Collection of ^{3}H/^{3}He samples requires relatively simple filling and crimping of copper tubing. CFC samples have been collected using a special sampling apparatus that excludes contact with air and allows welding into borosilicate glass ampules (Busenberg and Plummer 1992). Sampling for ^{85}Kr is the most time-intensive, requiring vacuum-stripping of gas from large (more than 100 L) volumes of water. There are relatively few laboratories presently capable of analyzing the water and gas samples for ^{3}He, ^{85}Kr, and CFCs.

Environmental tracers have the potential capability to date any given parcel of water, provided the sample contains natural and measurable concentrations of the tracer. Placement of time in aquifers using event markers requires not only measurable quantities of the tracer, but sufficient spatial sampling to locate the position of the tracer event in the flow system.

SELECTED NUCLEAR EVENT MARKERS

Several other radiochemical tracers have been applied to specific geochemical problems. These tracers can be divided into event markers or steady-state tracers. The event markers are generally produced by anthropogenic activity, typically nuclear weapons testing. They include the transuranics, ^{137}Cs, ^{133}Cs, ^{89}Sr, ^{90}Sr, and ^{36}Cl. These tracers have been used extensively for studies on sediment and glacial deposition (Koide et al. 1979; Fabryka-Martin and Davis 1987) or to trace movement of solutes through the unsaturated zone (Phillips, Mattick, and Duval 1988). They make poor tracers for water movement due to the variety of chemical interactions that they undergo, and require more effort in sampling than ^{3}H. They have been of limited use in the study of ground-water hydrology. The ^{14}C bomb pulse has been used in studies of young ground waters, but it generally confirms results obtained from ^{3}H or other tracers (Munnich, Roether, and Thilo 1967).

Another group of potential tracers are the radioisotopes produced in a steady state by cosmic-ray spallation (Lal and Peters 1967). Many of the isotopes, while useful for studying certain surface processes (Cooper et al. 1991), have too short a half-life to furnish information on ground-water aquifers. Other isotopes which have longer half-lives (^{32}Si, ^{22}Na, etc.) present analytical and geochemical difficulties and have received limited use in regional ground-water studies (Lal, Nuampurkar, and Rama 1970; Fröhlich et al. 1987).

ORGANIC COMPOUNDS AS EVENT MARKERS: SELECTED EXAMPLES

If data are available on time of application, rates of degradation, and extent of sorption to soil and aquifer sediments, a variety of anthropogenic organic compounds could be used as tracers of shallow ground-water flow through both time and space. The degradation products of organic compounds also may be used as a further indicator of flow path and rates of transport. Background information and field examples are presented to demonstrate two applications of surfactants and herbicides as event markers and tracers of shallow ground water.

Surfactants

Anionic surfactants used in detergents are of three major types: branched-chain alkylarylsulfonates (ABS), linear alkylarylsulfonates (LAS), and sodium dodecylsulfates (NaLS). Anionic surfactants are collectively referred to as methylene-blue active substances (MBAS). The MBAS test is a standard method (American Public Health Association 1981) that is used to measure concentrations of surfactants in water and wastewater.

ABS surfactants are known in the detergent industry as "hard" detergents, because these substances are difficult to degrade biologically. The ABS surfactants were introduced into the marketplace in 1946, and their use was continued for nearly 20 years until 1965. On the other hand, LAS and NaLS are "soft" detergents, because they are easily degraded, probably to aromatic sulfonic acids and carbon dioxide (Layman 1984). The detergent industry switched almost exclusively to LAS detergents in 1965 after Weaver and Coughlin (1964) reported that biodegradation of LAS surfactants occurred more rapidly than biodegradation of ABS surfactants and that streams and rivers were being polluted by ABS surfactants. When in 1965 the detergent manufacturers changed to LAS, the foaming problem on rivers nearly disappeared. The combination of time of use (1946 to 1965), the lack of degradation, and the poor adsorption of surfactants on aquifer sediments (Thurman, Barber, and LeBlanc 1986) make the ABS surfactant a useful tracer of shallow ground-water flow.

As an example, in a study of sewage contamination of ground water on Cape Cod (Thurman, Barber, and LeBlanc 1986) it was hypothesized that two types of surfactants were present in ground water, and that the surfactants formed two distinct plumes, an ABS plume and a LAS plume. The ABS plume represented time lines from 1945 to 1965, and the LAS plume represented time lines from 1965 to 1985, when the study was completed. This hypothesis was substantiated in a later paper (Thurman et al. 1987) by analyzing the surfactants with ^{13}C-NMR (nuclear magnetic resonance). Both LAS and ABS were measured in ground water at 0.1 mg/L, a lower limit of detection of MBAS. Surfactants may be isolated and purified by XAD resins at the same concentrations (Thurman et al. 1987). The results of this study of the two MBAS plumes and the utility of these results in establishing the approximate age of trichloroethylene (TCE) and tetrachloroethylene (PCE) in the sewage-contaminated ground water are discussed later.

The Otis Air Force Base sewage-treatment facility is located 10 km north of Nantucket Sound on a broad glacial outwash plain of Pleistocene age near Falmouth, Massachusetts. The aquifer that receives the treated sewage is composed of Pleistocene glacial deposits of sand, gravel, silt, and clay that overlie crystalline bedrock. Ground water in the aquifer is unconfined, and the water table slopes uniformly to the south except where it is distorted by Ashumet and Coonamessett ponds (Figure 11-11). The only natural source of water to the aquifer is recharge from precipitation. Recharge occurs over most of the study area, and direct surface runoff is negligible because the sandy soils are permeable. Ground water

Organic Compounds as Event Markers: Selected Examples 281

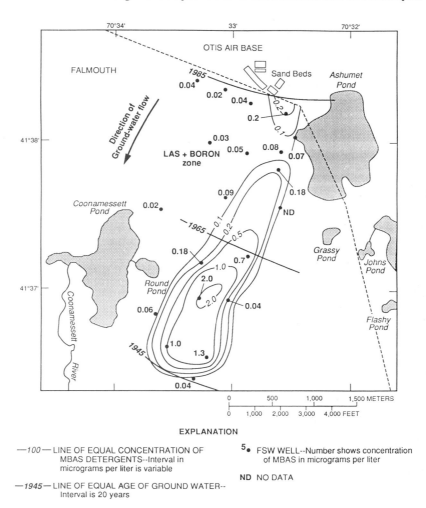

FIGURE 11-11. Distribution of ABS surfactants as determined by MBAS and C13 NMR on ground water samples near Otis Air Force Base, Falmouth, Massachusetts. (*After Thurman, Barber, and LeBlanc 1986.*)

in the study area generally flows south as inferred from the water-table map and the path of the plume (LeBlanc 1982, 1984; Thurman, Barber, and LeBlanc 1986). Flow is nearly horizontal except near the ponds and infiltration beds.

The plume of contaminated ground water has been delineated from measurements of conductance and boron, which show the general location of the contaminated water (LeBlanc 1982, 1984; Thurman, Barber, and LeBlanc 1986). The plume extends 3–4 km

downgradient from the sand beds. Based on the location of the plume, hydraulic conductivity measurements, and digital modeling of the plume (LeBlanc 1984), an average velocity of 0.3 m/d was estimated (Thurman, Barber, and LeBlanc 1986).

The boron concentration is relatively constant in the effluent because of its use as a bleaching additive in soaps and detergents. Boric acid is the likely chemical species present. Boron also extends 3–4 km downgradient (Figure 11-11), similar to the conductance plume that is presumed to be conservative. Although it is not certain that conductance and boron are moving conservatively in ground water, it appears that retardation is small (if at all), based on ground-water velocities from modeling. The center of mass of boron extends 3 km downgradient, quite similar to the center of mass of the ABS surfactants (Figure 11-11).

Figure 11-11 shows the distribution of ABS surfactants, as determined by MBAS. The core concentrations of MBAS at 2.0 mg/L occurs at 3.0 km downgradient from the sand beds. This core of MBAS was hypothesized as ABS detergents (Thurman, Barber, and LeBlanc 1986) that were disposed of between 1945 and 1965, and this was supported by the ^{13}C-NMR evidence, which shows that the alkylaryl benzene sulfonates are the branched-type ABS (Thurman et al. 1987).

At 0.5–1.5 km from the sand beds LAS surfactants were identified by ^{13}C-NMR. These surfactants were introduced in 1965 to the time of study, 1985. However, the observed concentrations in ground water were low because of biological degradation. Thus, these two types of surfactants map out the distribution of several time lines, a 1945 to 1965 time zone and a 1965 to 1985 time zone (Figure 11-11).

These zones may be used to evaluate other contaminate zones in the plume. For example, Figure 11-12 shows the distribution of TCE plus PCE in ground water in 1983. Based solely on the time lines established by the ABS surfactant zone, the core of the volatile organic compounds lies in the time zone from approximately 1970 to 1975. However, one must also consider possible retardation due to sorption of TCE and PCE onto the aquifer solids with subsequent alteration of the time of entry. For example, if the apparent velocity of the TCE and PCE plume was reduced to one half of the groundwater velocity, then the time of entry would be much earlier, approximately 1960 to 1965. To establish the sorption properties of the compounds they may be compared to other compounds present in the plume. For example, Barber et al. (1988) found that dichlorobenzene (DCB) was a major chlorinated aromatic hydrocarbon in the sewage that was transported at a rate nearly coincident with boron and is following the plume of the ABS surfactant (Figure 11-13). They also showed that sorption by hydrophobic or partitioning processes was low for compounds as water-soluble as dichlorobenzene. The volatile plume of TCE and PCE is only slightly sorbed, if at all, and the time of entry for these compounds is between 1970 to 1975 based on the time lines by ABS and LAS surfactants. Although TCE, PCE, and DCB may have retardation coefficients sufficient to retard these compounds during transport, the field data support conservative transport for these species. The low organic carbon content of the aquifer solids (<0.2%) is probably responsible for the lack of retardation. The time of entry (1970–75) is consistent with records maintained at the treatment plant that show a solvent recovery system being used at the treatment plant in the early 1970s (LeBlanc, U.S. Geological Survey 1985, written communication), which may be a source for the TCE and PCE plume.

Surfactants have been used above as indicators of time frames of point-source contamination. Surfactants may also be used for nonpoint-source contamination studies. An example is the application of surfactants as inert ingredients in pesticide formulations, which are added to emulsify pesticides and to enhance movement of pesticides into surface layers of soil. These surfactants may be measured in soil waters and ground waters receiving recharge from agricultural fields. Surfactants as indicators of inert ingredients is a topic for future research.

Herbicides

Preemergent herbicides have been used since the mid-1950s to control weeds in corn and soybeans. The use is most extensive in the midwestern United States, where triazine herbicides have been detected in the sur-

Organic Compounds as Event Markers: Selected Examples 283

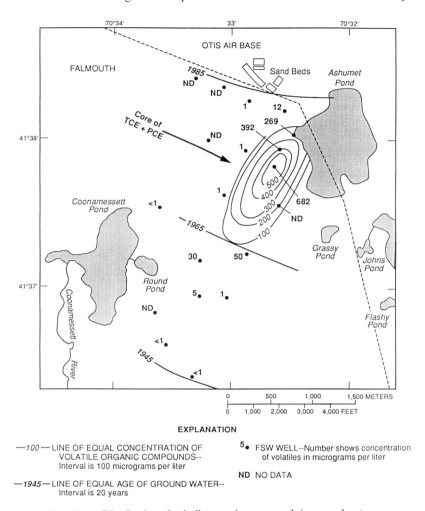

FIGURE 11-12. Distribution of volatile organic compounds in ground water samples near Otis Air Force Base, Falmouth, Massachusetts. (*After Thurman, Barber, and LeBlanc 1986.*)

face water and ground water of this region (Thurman et al. 1991; Hallberg 1989). The triazines are used primarily on corn, and today are still one of the most widely used preemergent herbicides in the United States.

Of the triazines, atrazine is the singly most used compound in the 12-state Corn Belt region of the midwestern United States. Atrazine degrades in soil to two major degradation products, deethylatrazine and deisopropylatrazine (Figure 11-14). In both cases the decomposition products are the result of dealkylation reactions. These two degradation products are produced in soil and carried by recharge into ground water. Because the two compounds are produced in the soil they may be used as tracers of ground-water contamination in agricultural regions (Thurman et al. 1991; Adams and Thurman 1991).

The concept of using the degradation products of atrazine to distinguish point-source and nonpoint-source contamination by herbicides was recently introduced by

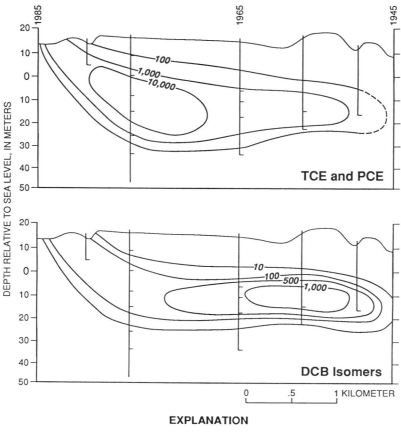

FIGURE 11-13. Distribution of TCE, PCE, and DCB in ground waters at Otis Air Force Base, Falmouth, Massachusetts. (*Adapted from Barber et al. 1988.*)

Adams and Thurman (1991), who coined the term DAR to denote the deethylatrazine-to-atrazine ratio. They found that as atrazine is dissipated in the soil under normal field conditions in the Corn Belt region, the major degradation product was deethylatrazine, not deisopropylatrazine. A similar result was reported by Mills (1991). Furthermore, she found that deisopropylatrazine was a significant metabolite in runoff waters from cornfields, but was not found in significant amounts in soil waters. Mills (1991) proposed that deisopropylatrazine degraded much more rapidly in soil waters than did deethylatrazine to the metabolite didealkylatrazine (Figure 11-14). The reason for the rapid degradation of deisopropylatrazine is the loss of an ethyl group rather than the more recalcitrant isopropyl group.

This difference in pathway and rate of degradation for atrazine's two degradation products makes atrazine and its degradation products good candidates for tracers of agricultural contamination. Furthermore, the ratios of the degradation products to parent compound may be used to distinguish ground

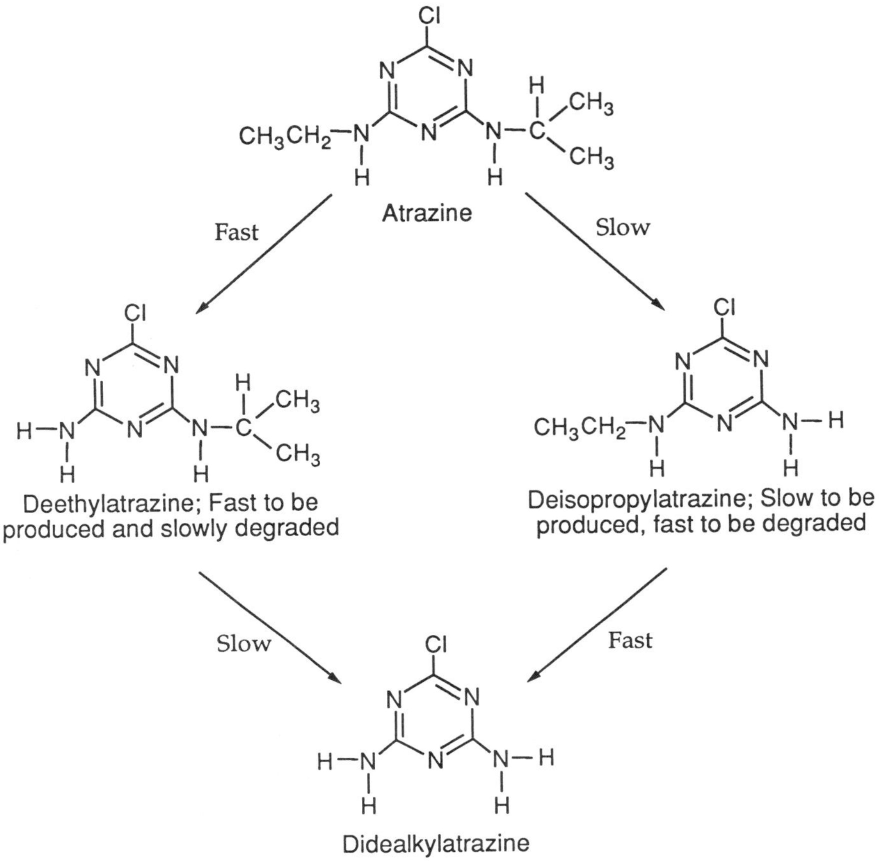

FIGURE 11-14. Degradation pathway of atrazine to deethylatrazine, deisopropylatrazine, and didealkylatrazine in unsaturated zone waters and runoff waters. (*From Mills 1991.*)

water that has taken a normal flow path through the unsaturated zone to groundwater from water that has been quickly transported to the aquifer, perhaps due to point-source contamination or natural preferential flow paths. Scenarios that would generate both nonpoint- and point-source contamination and how these three compounds (atrazine, deethylatrazine, and deisopropylatrazine) might be used to determine source or flow path are illustrated in Figure 11-15.

In the case of nonpoint-source contamination (scenario A in Figure 11-15), the flow path is percolation through the unsaturated zone. The herbicide is applied in the spring at the time of planting and degrades slowly in the soil over a period of several months (Nash 1988). During this time atrazine is partially degraded to both deethylatrazine and deisopropylatrazine. The deisopropylatrazine further degrades rapidly to either didealkylatrazine, or other products. On the other hand, both atrazine and deethylatrazine may persist in the soil waters for more than two growing seasons (Mills 1991) at concentrations from 0.2 to 1.0 µg/L in soil waters. Commonly, the concentration of deethylatrazine will equal or exceed the concentration of atrazine. Thus, values for the DAR will be from 0.5 to greater than 1.0 (Adams and Thurman 1991; Mills 1991).

Runoff from field plots, on the other hand,

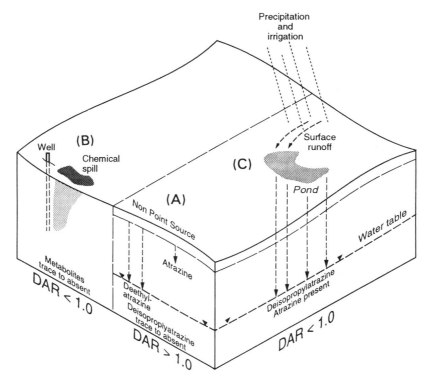

FIGURE 11-15. Scenario of higher concentrations of atrazine, deethylatrazine and deisopropylatrazine entering ground water by (A) nonpoint-source percolation through the soil, (B) point-source mechanisms, and (C) direct interaction of surface water and ground water.

has much greater concentrations of parent compounds and degradation products (in the hundreds of µg/L concentration range), and the ratio of deisopropylatrazine to deethylatrazine may be as high as 1:3. Deisopropylatrazine is a major degradation product found in surface water, which is not usually present in ground waters. It may be a useful tracer of surface-water impact on ground water or of point-source contamination.

There are many possible pathways for ground water to bypass the normal degradation pathways of herbicides in soil. For example scenario B in Figure 11-15 shows a chemical spill, such as washing herbicide tanks in the vicinity of a well and the leaking of herbicide around the gravel pack of the well to later contaminate the aquifer. Such a scenario has been shown to occur (Perry 1990) with high concentrations of parent herbicides and a low DAR value (Adams and Thurman 1991). Other examples include macropore flow, surface-water injection into wells to prime pumps, and incorrect placement of check valves in center-pivot irrigation. All of these examples could result in point-source contamination of ground water with a result of low DAR values and detectable quantities of deisopropylatrazine.

Higher concentrations of deisopropylatrazine than would be normally expected may also occur due to direct interaction between surface and ground waters. For example, scenario C in Figure 11-15 shows diagrammatically how surface-water interaction with ground water by holding ponds may lead to higher concentrations of deisopropylatrazine in ground water. These ponds capture runoff from fields that contain high concentrations of atrazine and its degradation products, including deisopropylatrazine. These waters may rapidly penetrate to the aquifer.

Future Studies of Organic Tracers

The previous two examples show how both surfactants and herbicides may be used to either date or trace ground-water flow. These two examples point out that an understanding of transport mechanisms is needed. In particular, one needs to know the history of use of a chemical, its degradation pathways, and its sorption characteristics on soil or aquifer solids. These factors allow one to carefully monitor the fate of the compound and its use for predictive purposes in ground water. Furthermore, the presence of other compounds is helpful in understanding history, degradation, and sorption rates and may be used in conjunction with the compounds of interest. Other compounds that could be considered for ground-water tracers include sewage-related organic compounds, such as chlorinated phenols and alkylbenzenes (Barber et al. 1988).

In agricultural areas the various herbicide classes should be considered as tracers of surface-water and ground-water flow, especially with respect to relative degradation rates of the compounds. For example, the differences in degradation rates and the types of products formed from atrazine, metolachlor, alachlor, and cyanazine could be examined as tracers. These herbicides are the major herbicides found in streams and rivers of the midwestern United States (Thurman et al. 1991), and the relative losses of each of these compounds makes them excellent potential tracers of ground water.

CONCLUDING REMARKS

There are relatively few environmental tracers suitable for age-dating ground water recharged over the past 50 years. The recognized dating techniques can be subdivided into environmental tracers, which can yield a continuous record of water age with distance along a flow path, and event markers, which, when located in a ground-water system, assign an age to a specific point in the flow system. Because environmental tracers are dissolved solutes which are transported along with the ground water, it is usually necessary to consider effects of hydrodynamic dispersion on the modeled age.

Of the environmental tracers discussed here, only ^3H is widely accepted and used in hydrologic studies, both as an event marker of the mid-1960s atmospheric nuclear testing and as a tool to interpret ground-water residence times. Tritium data have been applied to the study of aquifers by the use of a variety of models ranging from the simple piston-flow approach to numerical simulations. Information provided by these models include flow paths and velocities, residence times, aquifer size, and dispersivity coefficients. A long-term ^3H data base is desirable for all approaches; for reservoir models and some compartment models, it is essential. For most techniques, preexisting knowledge of some aquifer parameters (e.g., direction of flow, recharge area) must exist. Caution should be used to ensure that the conditions for applying the model are met. Zuber (1986) reviews many of the mathematical models that have been applied to environmental tracers in ground water. The major problem in applying many of the models is obtaining an accurate ^3H input function. Methods have been developed for both compartment models and advection-dispersion models that treat the input realistically. Tritium has found considerable use as a marker of the 1963–64 bomb pulse.

For use of ^3H as an environmental tracer, the initial ^3H input must be known for a given parcel of water. This is usually accomplished by determination of both ^3H and its decay product, ^3He, in a given water sample. Reliable ages can be obtained by making combined $^3H/^3He$ measurements on ground water, particularly for waters recharged during the past 20 to 25 years when ^{3}H input has been uniformly declining. However, due to dispersive mixing of the ^{3}H peak from atmospheric nuclear testing, ^{3}H/^{3}He ages uncorrected for hydrodynamic dispersion can be highly unreliable for waters recharged prior to the mid-1960s. Schlosser et al. (1989) showed that ^{3}He confinement can be as much as 80%, which is sufficient for most ^{3}H/^{3}He dating applications. These authors also showed that ^{3}H/^{3}He dating should be possible for at least the next 40 years before the bomb-pulse signature is lost to dispersive mixing effects.

Dating with ^{85}Kr is potentially the most attractive of environmental tracers, especially for waters recharged over the past 30 years. Because the input function is smooth and increasing, the ^{85}Kr specific activity is not significantly affected by hydrodynamic dispersion (Smethie et al. 1992). Due to continued reprocessing of nuclear fuels, the atmospheric ^{85}Kr input function is expected to increase. Like ^{3}He, ^{85}Kr is an inert dissolved gas that is not affected by chemical interactions (other than radioactive decay) in ground water. Because the ^{85}Kr specific activity (rather than concentration) is measured, ^{85}Kr input to ground water is independent of recharge temperature. To date, relatively few measurements of ^{85}Kr have been made in ground water, mainly because of time-intensive field sampling and laboratory analytical procedures.

In contrast to ^{85}Kr dating, CFC samples can be collected relatively quickly (Busenberg and Plummer 1992) and analyzed quickly by using purge-and-trap GC-ECD instrumentation. Consequently, it may be more practical to obtain large numbers of CFC samples from ground-water systems, and to check the CFC dating with a fewer number of ^{3}H/^{3}He and/or ^{85}Kr ages. Unfortunately, CFCs cannot be applied to all ground-water environments. CFCs can be introduced to ground water from a variety of contaminant sources, in addition to equilibration of recharge water with the unsaturated zone air. Although they are highly stable organic compounds in aerobic systems, CFCs can be degraded by microbial activity in anaerobic systems. Dissolved-gas analyses are usually required in CFC dating to define the recharge temperature, and ^{3}H data are useful in evaluating potentially contaminated or mixed waters. Because the CFC atmospheric input functions are smooth and increasing, CFC dating may be relatively independent of dispersive mixing.

Event markers are useful in recognizing pre- or postevent waters. In addition to ^{3}H several other event markers of nuclear detonation have application to hydrologic studies, including ^{36}Cl and ^{14}C, but they are not widely used for this purpose. Surfactants prove to be valuable event markers in sewage effluent plumes, where CFC dating is useless owing to contaminant sources of CFCs, and where ^{85}Kr and ^{3}H/^{3}He dating is less certain owing to mixed input from treated water and natural precipitation sources. Here the surfactants ABS and LAS can be used to recognize pre- and post-1965 sewage effluent waters. Herbicides and their degradation products are beginning to be recognized as tracers of agricultural waters and as a tool in recognizing runoff and recharge waters. Atrazine and its metabolites may also be used as event markers of flow paths rather than time, especially for point- and nonpoint-source contamination of ground water.

References

Adams, C. D., and E. M. Thurman. 1991. Formation and transport of deethylatrazine in the soil and vadose zone. *Journal of Environmental Quality* 20:540–47.

Allison, G. B., and M. W. Hughes. 1975. The use of environmental tritium to estimate recharge to a South-Australian aquifer. *Journal of Hydrology* 26(3/4):245–54.

American Public Health Association. 1981. Methylene blue method for methylene-blue-active substances (MBAS). In *Standard Methods for the Examination of Water and Wastewater*, pp. 530–32, Washington DC: American Public Health Association.

Andersen, L. J., and T. Sevel. 1974. Six years environmental tritium profiles in the unsaturated and saturated zones, Gronhoj, Denmark. In *Symposium on Isotope Techniques in Groundwater Hydrology*. II, pp. 3–20. Vienna: International Atomic Energy Agency.

Andres, R., and R. Egger, 1985. A new tritium interface method for determining the recharge rate of deep groundwater in the Bavarian Molasse Basin. *Journal of Hydrology* 82(1/2):27–38.

Andrews, J. N., and R. L. F. Kay. 1982. Natural production of tritium in permeable rocks. *Nature* 298:361–63.

Andrews, J. D., and D. J. Lee. 1979. Inert gases in groundwater from the Bunter sandstone of England as indicators of age and paleoclimatic trends. *Journal of Hydrology* 41:233–52.

Andrews, J. D., J. E. Goldbrunner, W. G. Darling, P. J. Hooker, G. B. Wilson, M. J. Youngman, L. Eichinger, W. Rauert, and W. Stichler. 1985. A radiochemical, hydrochemical and dissolved gas study of groundwaters in the Molasse Basin of Upper Austria. *Earth Planetary Science Letters* 73:317–32.

Andrews, J. N., W. Balderer, A. H. Bath, H. B. Clausen, G. V. Evans, T. Florkowski, J. E. Goldbrunner, M. Ivanovich, H. Loosli, and H. Zojer. 1984. Environmental isotope studies in two aquifer systems. In *Isotope Hydrology 1983*, pp. 535–76. Vienna: International Atomic Energy Agency, IAEA-SM-270/93.

Barber, L. B., Jr., E. M. Thurman, M. P. Schroeder, and D. R. LeBlanc. 1988. Long-term fate of organic micropollutants in sewage-contaminated ground water. *Environmental Science and Technology* 22:205–11.

Brown, J. D. 1980. *Evolution of fluorocarbon compounds as ground water tracers—Soil column studies*, M.S. Thesis. Tucson, AZ: Department of Hydrology and Water Resources, University of Arizona.

Brown, R. M. 1961. Hydrology of tritium in the Ottawa Valley. *Geochimica et Cosmochimica Acta* 21(3/4):199–216.

Bullister, J. L., and R. F. Weiss. 1983. Anthropogenic chlorofluoromethanes in the Greenland and Norwegian Seas. *Science* 221:265–68.

Busenberg, E., and L. N. Plummer. 1991. Chlorofluorocarbons (CCl_3F and CCl_2F_2): Use as an age-dating tool and hydrologic tracer in shallow ground-water systems. In *Proceedings U.S. Geological Survey Toxic Substances Meeting*, March 11–15, 1991 at Monterey, CA, eds. G. L. Mallard and D. A. Aronson, pp. 1–6. U.S. Geological Survey Water-Resources Investigations Report 91-4034.

Busenberg, E., and L. N. Plummer. 1992. Use of chlorofluoromethanes (CCl_3F and CCl_2F_2) as hydrologic tracers and age-dating tools: The alluvium and terrace system of central Oklahoma. *Water Resources Research* 28:2257–83.

Campana, M. E., and D. A. Mahin. 1985. Model-derived estimates of groundwater mean ages, recharge rates, effective porosities and storage in a limestone aquifer. *Journal of Hydrology* 76(3/4):247–64.

Carter, M. W., and A. A. Moghissi. 1977. Three decades of nuclear testing. *Health Physics* 33(1):55–71.

Cerling, T. E., R. J. Poreda, and D. K. Solomon. 1987. Use of tritium and helium isotopes in the study of a shallow unconfined aquifer. Abstract, *EOS* 68:300.

Chiou, C. T., and T. D. Shoup. 1985. Soil sorption of organic vapors and effects of humidity on sorptive mechanism and capacity. *Environmental Science and Technology* 19:1196–1200.

Clarke, W. B., W. B. Jenkins, and Z. Top. 1976. Determination of tritium by mass spectrometric measurement of ^{3}He. *International Journal of Applied Radiation and Isotopes* 27:217–25.

Cleary, R. W., and M. T. Ungs. 1978. *Analytical Models for Groundwater Pollution and Hydrology*. Princeton, NJ: Princeton University Water Resources Program Report 78-15.

Cooper, L. W., C. R. Olsen, D. K. Solomon, I. L. Larsen, R. B. Cook, and J. M. Grebmeier. 1991. Stable isotopes of oxygen and natural and fallout radionuclides used for tracing runoff during snowmelt in an Arctic watershed. *Water Resources Research* 27(9):2171–9.

Davis, S. N., and H. W. Bentley. 1982. Dating groundwater, a short review. In *Nuclear and Chemical Dating Techniques: Interpreting the Environmental Record*, ed. L. A. Cutrie, pp. 187–222. Washington, DC: American

Chemical Society Symposium Series, No. 176.

Davis, S. N., G. M. Thompson, H. W. Bentley, and G. Stiles. 1980. Ground-water tracers—A short review. *Ground Water* 18:14–23.

Delcore, M. R., and G. J. Larson. 1987. Application of the tritium interface method for determining recharge rates to unconfined drift aquifers. II: Non-homogeneous case. *Journal of Hydrology* 91(1/2):73–81.

Egboka, B. C. E., J. A. Cherry, R. N. Farvolden, and E. O. Frind. 1983. Migration of contaminants in groundwater at a landfill: a case study. 3. Tritium as an indicator of dispersion and recharge. *Journal of Hydrology* 63(1/2):51–80.

Fabryka, J., and S. N. Davis. 1987. Applications of ^{129}I and ^{36}Cl in hydrology. *Nuclear Instruments and Methods in Physics Research B* 29:361–71.

Fontes, J.-Ch. 1983. Dating of groundwater. In *Guidebook on Nuclear Techniques in Hydrology*, pp. 285–317. Vienna: International Atomic Energy Agency, Technical Report Series 91.

Fontes, J.-Ch. 1992. Chemical and isotopic constraints on ^{14}C dating of groundwater. In *Radiocarbon After Four Decades*, eds. R. E. Taylor, A. Long, and R. S. Kra, pp. 242–61. New York: Springer-Verlag.

Fontes, J.-Ch., and J.-M. Garnier. 1979. Determination of the initial ^{14}C activity of the total dissolved carbon: A review of the existing models and a new approach. *Water Resources Research* 15:399–413.

Forster, M., H. Moser, and H. Loosli. 1984. Isotope hydrological study with carbon-14 and argon-39 in the Bunter Sandstone of the Saar region. In *Isotope Hydrology 1983*, IAEA-SM-27/93, pp. 515–33. Vienna: International Atomic Energy Agency.

Foster, S. S. D. 1975. The Chalk groundwater tritium anomaly—a possible explanation. *Journal of Hydrology* 25(1/2):159–65.

Foster, S. S. D., and A. Smith-Carrington. 1980. The interpretation of tritium in the Chalk unsaturated zone. *Journal of Hydrology* 46(3/4):343–64.

Frind, E. O., and G. B. Matanga. 1985. The dual formulation of flow for contaminant transport modeling. 1: Review of theory and accuracy aspects. *Water Resources Research* 21:159–69.

Fritz, S. J., R. J. Drimmie, and P. Fritz. 1991. Characterizing shallow aquifers using tritium and ^{14}C: Periodic sampling based on the tritium half-life. *Applied Geochemistry* 6(1):17–34.

Fröhlich, K. 1990. On dating old groundwater. *Isotopenpraxis* 26(12):557–560.

Fröhlich, K., T. Franke, R. Gellermann, D. Hebert, and H. Jordan. 1987. Silicon-32 in different aquifer typers and implications for groundwater dating. In *Isotope Techniques in Water Resources Development*, pp. 149–63. Vienna: International Atomic Energy Agency.

Geyh, M. A., and H. Schleicher. 1990. *Absolute Age Determination*. Berlin: Springer-Verlag.

Hallberg, G. R. 1989. Pesticide pollution of ground water in the humid United States. *Agriculture Ecosystems and Environment* 26:299–367.

Heaton, T. H. E. 1981. Dissolved gases: Some application to groundwater research. *Transactions of the Geological Society South Africa* 84:91–97.

Heaton, T. H. E., and J. C. Vogel. 1981. Excess air in groundwater. *Journal of Hydrology* 50:201–16.

Herzberg, O., and E. Mazor. 1979. Hydrological applications of noble gases and temperature measurements in groundwater in underground water systems: Example from Israel. *Journal of Hydrology* 41:217–31.

International Atomic Energy Agency. 1969. *Environmental Isotope Data No. 1: World Survey of Isotope Concentration in Precipitation (1953–1963)*. Vienna: IAEA Technical Report Series, No. 96.

International Atomic Energy Agency. 1970. *Environmental Isotope Data No. 2: World Survey of Isotope Concentration in Precipitation (1964–1965)*. Vienna: IAEA Technical Report Series, No. 117.

International Atomic Energy Agency. 1971. *Environmental Isotope Data No. 3: World Survey of Isotope Concentration in Precipitation (1966–1967)*. Vienna: IAEA Technical Report Series, No. 129.

International Atomic Energy Agency. 1973. *Environmental Isotope Data No. 4: World Survey of Isotope Concentration in Precipitation (1968–1969)*. Vienna: IAEA Technical Report Series, No. 147.

International Atomic Energy Agency. 1975. *Environmental Isotope Data No. 5: World Survey of Isotope Concentration in*

Precipitation (1970–1971). Vienna: IAEA Technical Report Series, No. 165.

International Atomic Energy Agency. 1978. *Statistical Treatment of Environmental Isotope Data in Precipitation*. Vienna: IAEA Technical Report Series, No. 206.

International Atomic Energy Agency. 1979. *Environmental Isotope Data No. 6: World Survey of Isotope Concentration in Precipitation (1972–1975)*. Vienna: IAEA Technical Report Series, No. 192.

International Atomic Energy Agency. 1981. *Statistical Treatment of Environmental Isotope Data in Precipitation*. Vienna: IAEA Technical Report Series, No. 206.

International Atomic Energy Agency. 1983. *Environmental Isotope Data No. 7: World Survey of Isotope Concentration in Precipitation (1976–1979)*. Vienna: IAEA Technical Report Series, No. 226.

International Atomic Energy Agency. 1986. *Environmental Isotope Data No. 8: World Survey of Isotope Concentration in Precipitation (1980–1983)*. Vienna: IAEA Technical Report Series, No. 264.

International Atomic Energy Agency. 1990. *Environmental Isotope Data No. 9: World Survey of Isotope Concentration in Precipitation (1983–1987)*. Vienna: IAEA Technical Report Series, No. 311.

Khalil, M. A. K., and R. A. Rasmussen. 1989. The potential of soils as a sink of chlorofluorocarbons and other man-made chlorocarbons. *Geophysical Research Letters* 16:679–82.

Knott, J. F., and J. C. Olympio. 1986. *Estimation of Recharge Rates to the Sand and Gravel Aquifer Using Environmental Tritium, Nantucket Island, Massachussetts*. Washington, DC: U.S. Geological Survey Water-Supply Paper 2297.

Koide, M., R. Michel, E. D. Goldberg, M. M. Herron, and C. C. Langway. 1979. Depositional history of artificial radionuclides in the Ross Ice Shelf, Antarctica. *Earth and Planetary Science Letters* 44:205–23.

Lal, D., V. N. Nuampurkar, and S. Rama. 1970. Silicon-32 hydrology. In *Isotope Hydrology*, pp. 847–68. Vienna: International Atomic Energy Agency.

Lal, D., and B. Peters. 1967. Cosmic ray produced radioactivity on the earth. *Handbook of Physics* 46:551–612.

Larson, G. J., M. R. Delcore, and S. Offer. 1987. Application of the tritium interface method for determining recharge rates to unconfined drift aquifers. I: Homogeneous case. *Journal of Hydrology* 91(1/2):59–72.

Layman, P. L. 1984. Detergent report—Brisk detergent activity changes picture for chemical suppliers. *Chemical Engineering News* 62:17–49.

LeBlanc, D. R. 1982. *Sewage Plume in a Sand and Gravel Aquifer, Cape Cod, Massachusetts*. Boston, MA: U.S. Geological Survey Open-File Report 82–274.

LeBlanc, D. R. 1984. Description of the hazardous-waste research. In *Movement and Fate of Solutes in a Plume of Sewage-Contaminated Ground Water, Cape Cod, Massachusetts*, ed. D. R. LeBlanc, Chap. A, pp. 1–10. Boston, MA: U.S. Geological Survey, Open-File Report 84–475.

Lehmann, B. E., H. Oeschger, H. H. Loolsi, G. S. Hurst, S. L. Allman, C. H. Chen, S. D. Kramer, M. G. Payne, and R. C. Phillips. 1985. Counting ^{81}Kr atoms for analysis of groundwater. *Journal of Geophysical Research* 90(B13):11, 547–51.

Loosli, H. H., and H. Oeschger. 1978. Argon-39, carbon-14 and krypton-85 measurements in groundwater samples. In *Isotope Hydrology 1978*, pp. 931–45. Vienna: International Atomic Energy Agency.

Loosli, H. H., and H. Oeschger. 1980. Use of ^{39}Ar and ^{14}C for groundwater dating. *Radiocarbon* 22(3):863–70.

Lovelock, J. E. 1971. Atmospheric fluorine compounds as indicators of air movements. *Nature* 230:379.

Lovley, D. R., and J. C. Woodward. 1992. Consumption of Freons CFC-11 and CFC-12 by anaerobic sediments and soils. *Environmental Science and Technology* 26:925–29.

Maloszewski, P., and A. Zuber. 1983. The theoretical possibilities of the ^{3}H-^{3}He method in investigations of groundwater systems. *Catena* 10:189–98.

Maloszewski, P., and A. Zuber. 1991. Influence of matrix diffusion and exchange reactions on radiocarbon ages in fissured carbonate aquifers. *Water Resources Research* 27(8):1937–45.

Michel, R. L. 1989. Tritium deposition over the continental United States, 1953–1983. In

Atmospheric Deposition, pp. 109–115. Oxfordshire, UK: International Association of Hydrological Sciences.

Michel, R. L. 1991. Residence times in river basins as determined by analysis of long-term tritium records. *Journal of Hydrology* 130:367–78.

Mills, M. S. 1991. *Field Dissipation of Encapsulated Herbicides: Geochemistry and Degradation*. Master's thesis, Lawrence, KS: Department of Geology, University of Kansas.

Mook, W. G. 1980. Carbon-14 in hydrogeological studies. In *Handbook of Environmental Isotope Geochemistry*. Vol. 1. *The Terrestrial Environment, A*, Chap. 2, eds. P. Fritz and J.-Ch. Fontes, pp. 49–74. New York: Elsevier.

Munnich, K. O., W. Roether, and L. Thilo. 1967. Dating of groundwater with tritium and 14C. In *Isotopes in Hydrology*, pp. 305–20. Vienna: International Atomic Energy Agency, IAEA-SM-83/21.

Nash, R. G. 1988. Dissipation in soil. In *Environmental Chemistry of Herbicides*, ed. R. Grover, pp. 131–69. Boca Raton, FL: CRC Press.

Oeschger, H., L. H. Gugelmann, U. Schotterer, U. Siegenfhaler, and A. Wiest. 1974. Ar dating of groundwater. In *Isotope Techniques in Groundwater Hydrology 1974*, pp. 179–89. Vienna: International Atomic Energy Agency.

Pearson, F. J., Jr., W. Balderer, H. H. Loosli, B. E. Lehmann, A. Matter, Tj. Peters, H. Schmassmann, and A. Gautschi. 1991. *Applied Isotope Hydrogeology. A Case Study in Northern Switzerland*. New York: Elsevier, Studies in Environmental Sciences 43.

Pearson, F. J., Jr., and A. H. Truesdell. 1978. Tritium in the waters of Yellowstone National Park. In *Short Papers of the Fourth International Conference, Geochronology, Cosmology, Isotope Geology*, pp. 327–29. Washington, DC: U.S. Geological Survey Open-File Report 78-701.

Perry, C. A. 1990. *Source, Extent, and Degradation of Herbicides in a Shallow Aquifer Near Hesston, Kansas*. Lawrence, KS: U.S. Geological Survey Water-Resources Investigation Report 90–4019.

Phillips, F. M., J. L. Mattick, and T. A. Duval. 1988. Chlorine-36 and tritium from nuclear weapons fallout as tracers for long-term liquid and vapor movement in desert soils. *Water Resources Research* 24(11):1877–91.

Plummer, L. N., E. C. Prestemon, and D. L. Parkhurst. 1991. *An Interactive Code (NETPATH) for Modeling NET Geochemical Reactions along a Flow PATH*. Reston, VA: U.S. Geological Survey Water-Resources Investigations Report 91–4078.

Poreda, R. J., T. E. Cerling, and D. K. Solomon. 1988. Tritium and helium isotopes as hydrologic tracers in a shallow unconfined aquifer. *Journal of Hydrology* 103:1–9.

Przewlocki, K., and Y. Yurtsever. 1974. Some conceptual mathmatical models and digital simulation approach in the use of tracers in hydrological systems. In *Symposium on Isotope Techniques in Groundwater Hydrology*, pp. 425–50. Vienna: International Atomic Energy Agency.

Rabinowitz, D. D., G. W. Gross, and C. R. Holmes. 1977. Environmental tritium as a hydrometeorologic tool in the Roswell Basin, New Mexico. I: Tritium input function and precipitation-recharge relation. *Journal of Hydrology* 31(1/2):3–17.

Randall, J. H., and T. R. Shultz. 1976. Chlorofluorocarbons as hydrologic tracers: A new technology. *Hydrology Water Resources Arizona Southwest* 6:189–95.

Revelle, R., and H. E. Suess. 1957. Carbon dioxide exchange between atmosphere and ocean and the question of an increase of atmospheric CO_2 during the past decades. *Tellus* 9(1):18–27.

Robertson, W. D., and J. A. Cherry. 1989. Tritium as an indicator of recharge and dispersion in a groundwater system in central Ohio. *Water Resources Research* 25(6):1097–109.

Rozanski, K. 1979. Krypton-85 in the atmosphere 1950–1977, A data review. *Environmental International* 2:139–43.

Rozanski, K., and T. Florkowski. 1979. Krypton-85 dating of groundwater. In *Isotope Hydrology 1978*, Vol. II, pp. 949–61. Vienna: International Atomic Energy Agency, IAEA-SM-228/51.

Russell, A. D., and G. M. Thompson. 1983. Mechanism leading to enrichment of the atmospheric fluorocarbons CCl_3F and CCl_2F_2 in groundwater. *Water Resources Research* 19:57–60.

Salvamoser, J. 1983. ^{85}Kr for groundwater dating: Measurement-models-applications. In *International Symposium on Isotope Hydrology in Water Resources Development, Extended Abstracts*, p. 128. Vienna: International Atomic Energy Agency, IAEA-SM-270.

Schlosser P., M. Stute, H. Dorr, C. Sonntag, and K. O. Munnich. 1988. Tritium/^{3}He dating of shallow groundwater. *Earth and Planetary Science Letters* 89:353–62.

Schlosser P., M. Stute, C. Sonntag, and K. O. Munnich. 1989. Tritiogenic ^{3}He in shallow groundwater. *Earth and Planetary Science Letters* 94:245–56.

Semprini, L., G. D. Hopkins, P. V. Roberts, and P. L. McCarty. 1990. In-situ biotransformation of carbon tetrachloride, 1,1,1,-trichloroethane, Freon-11, and Freon-12 under anoxic conditions. Abstract, *EOS* 71(43):1324.

Schultz, T. R. 1979. *Trichlorofluoromethane as a Ground-Water Tracer for Finite-State Models*. Ph.D. dissertation, Tucson, AZ: University of Arizona.

Schultz, T. R., J. H. Randall, L. G. Wilson, and S. N. Davis. 1976. Tracing sewage effluent recharge—Tucson, Arizona. *Ground Water* 14:463–70.

Siegel, D. I., and D. T. Jenkins. 1987. Isotopic analysis of groundwater flow systems in a wet alluvial fan, southern Nepal. In *Isotope Techniques in Water Resources Development*, pp. 475–82. Vienna: International Atomic Energy Agency.

Sittkus, A., and H. Stockburger. 1976. Krypton-85 als indikator des kernbrennstoffverbrauchs. *Naturwissenschaften* 63:266–72.

Smethie, W. M., Jr., and G. Mathieu. 1986. Measurement of krypton-85 in the ocean. *Marine Chemistry* 18:17–33.

Smethie, W. M., Jr., D. W. Chapman, J. H. Swift, and K. P. Koltermann. 1988. Chlorofluoromethanes in the Arctic Mediterranean Seas: Evidence for formation of bottom water in the Eurasian Basin and deep-water exchange through Fram Strait. *Deep-Sea Research* 35(3):347–69.

Smethie, W. M., Jr., D. K. Solomon, S. L. Schiff, and G. G. Mathieu. 1992. Tracing groundwater flow in the Bordon aquifer using krypton-85. *Journal of Hydrology* 130:279–97.

Solomon, D. K., and E. A. Sudicky. 1991. Tritium and helium 3 isotope ratios for direct estimation of spatial variations in groundwater recharge. *Water Resources Research* 27(9):2309–19.

Solomon, D. K., and E. A. Sudicky. 1992. Correction to "Tritium and helium 3 isotope ratios for direct estimation of spatial variations in groundwater recharge." *Water Resources Research* 28(4):1197.

Solomon, D. K., R. J. Poreda, S. L. Schiff, and J. A. Cherry. 1992. Tritium and helium 3 as groundwater age tracers in the Borden aquifer. *Water Resources Research* 28(3):741–55.

Swancar, A., and C. B. Hutchinson. 1992. *Chemical and Isotopic Composition and Potential for Contamination of the Upper Floridan Aquifer, West-Central Florida, 1985–1989*. Tampa FL: U.S. Geological Survey Open-File Report 92–47.

Takaoka, N., and Y. Mizutani. 1987. Tritiogenic ^{3}He in groundwater in Takaoka. *Earth and Planetary Science Letters* 85:74–78.

Thatcher, L. L. 1962. The distribution of tritium fallout in precipitation over North America. *Bulletin of the International Association of Scientific Hydrology* 7(2):48–58.

Thompson, G. M. 1976. *Trichloromethane, a New Hydrologic Tool for Tracing and Dating Ground Water*. Ph.D. dissertation, Bloomington, IN: Department of Geology, Indiana University.

Thompson, G. M., and J. M. Hayes. 1979. Trichloromethane in groundwater—A possible tracer and indicator of groundwater age. *Water Resources Research* 15:546–54.

Thompson, G. M., J. M. Hayes, and S. N. Davis. 1974. Fluorocarbon tracers in hydrology. *Geophysical Research Letters* 1:177–80.

Thonnard, N., R. D. Willis, M. C. Wright, and W. A. Davis. 1987. Resonance ionization spectroscopy and the detection of ^{81}Kr. *Nuclear Instruments and Methods in Physics Research B* 29:398–406.

Thurman, E. M., L. B. Barber, Jr., and D. LeBlanc. 1986. Movement and fate of detergents in ground water: A field study. *Journal of Contaminant Hydrology* 1:143–61.

Thurman, E. M., D. A. Goolsby, M. T. Meyer, and D. W. Kolpin. 1991. Herbicides in surface waters of the midwestern United States: The effect of spring flush. *Environmental Science and Technology* 25(10):1794–6.

Thurman, E. M., T. Willoughby, L. B. Barber, Jr., and K. A. Thorn. 1987. Determination of alkylbenzenesulfonate surfactants in ground water using macroreticular resins and carbon-13 nuclear magnetic resonance spectrometry. *Analytical Chemistry* 59:1798–1802.

Tolstikhin, I. N., and I. L. Kamensky. 1969. Determination of groundwater ages by the T-^{3}H method. *Geochemistry International* 6:810–11.

Torgersen T., W. B. Clarke, and W. J. Jenkins. 1979. The tritium/helium-3 method in hydrology. In *Isotope Hydrology 1978*, pp. 917–30. Vienna: International Atomic Energy Agency, IAEA-SM-228/49.

Torgersen T., M. A. Habermehl, F. M. Phillips, D. Elmore, P. Kubik, B. G. Jones, T. Hemmick, and H. E. Gove. 1991. Chlorine 36 dating of very old groundwater 3. Further studies in the Great Artesian Basin, Australia. *Water Resources Research* 27(12):3201–13.

van Genuchten, M. Th., and W. J. Alves. 1982. *Analytical Solutions of the One-Dimensional Convective-Dispersive Solute Transport Equation*. Riverside, CA: U.S. Department of Agriculture, Technical Bulletin 1661.

Vogel, J. C. 1967. Investigation of groundwater flow with radiocarbon. In *Isotopes in Hydrology*, pp. 355–69. Vienna: International Atomic Energy Agency, IAEA-SM-83/24.

Warner, M. J., and R. F. Weiss. 1985. Solubilities of chlorofluorocarbons 11 and 12 in water and seawater. *Deep-Sea Research* 32:1485–97.

Weaver, P. J., and F. J. Coughlin. 1964. Measurement of biodegradability. *Journal of the American Oil Chemists Society* 41:738–41.

Weeks, E. P., D. E. Earp, and G. M. Thompson. 1982. Use of atmospheric fluorocarbons F-11 and F-12 to determine the diffusion parameters of the unsaturated zone in the southern high plains of Texas. *Journal of Geophysical Research* 18:1365–78.

Weise, S. M., and H. Moser. 1987. Groundwater dating with helium isotopes. In *Isotope Techniques in Water Resource Development*, pp. 105–26. Vienna: International Atomic Energy Agency, IAEA-SM-299/44.

Weiss, W., A. Sittkus, H. Stockburger, and H. Sartorius. 1982. Large-scale atmospheric mixing derived from meridional profiles of krypton-85. *Journal of Geophysical Research* 88:8574–78.

Wigley, T. M. L., L. N. Plummer, and F. J. Pearson, Jr. 1978. Mass transfer and carbon isotope evolution in natural water systems. *Geochimica et Cosmochimica Acta* 42:1117–39.

Yurtsever, Y., and B. R. Payne. 1986. Mathematical models based on compartmental simulation approach for quantitative interpretation of tracer data hydrological systems. In *Proceedings of the 5th International Symposium on Underground Water Tracing*, pp. 341–53. Athens: Institute of Geology and Mineral Exploration.

Zuber, A. 1986. Mathematical models for the interpretation of environmental radioisotopes in groundwater systems. In *Handbook of Environmental Isotope Geochemistry*. Vol. 2. *The Terrestrial Environment, B.*, eds. P. Fritz and J.-Ch. Fontes, pp. 1–59. Amsterdam: Elsevier.

IV

Selected Water-Quality Issues

12

Nitrate

G. R. Hallberg and D. R. Keeney

INTRODUCTION

Nitrate is perhaps the most widespread contaminant of ground water. Nitrogen is ubiquitous in the environment, and its conversion to nitrate is part of the natural functioning of any ecosystem. Nitrogen is one of the most important plant nutrients; nitrate is highly soluble and very mobile, which facilitates plant uptake, but also makes it highly susceptible to leaching through the soil with infiltrating water. Nitrogen is an important component of all proteins and, hence, is found in all foodstuffs and animal wastes. Consequently, there are many sources of nitrate, natural and anthropogenic, that can contribute to ground-water contamination.

The surficial activities of modern society, primarily through disposal of organic wastes and through agricultural and horticultural practices, are the primary contributors to the widespread contamination of ground water by nitrate. Efforts to understand and mitigate nitrate loading to ground water must deal with the complex interplay of numerous land uses and a host of potential nitrogen inputs and sinks. Nitrogen is continually cycled in the environment, and this processing must be recognized to understand the distribution of nitrate in water resources.

NITROGEN CYCLING

The nitrogen (N) cycle has been the subject of numerous texts, monographs, and symposia (see Keeney 1986 for a comprehensive list of references). The N cycle (Figure 12-1) illustrates the complex interactions possible, with numerous sources, transformations, storage pools, and loss mechanisms of N. Many of the transformations are carried out by microbial processing or catalysis.

Mineralization-Immobilization

Mineralization-immobilization processes are of critical importance in the N cycle. The largest pool of N in the soil occurs as organic N. Nitrogen entering the system in inorganic forms can move into the large soil organic-N pool (immobilization), while organic forms can move out of the pool (mineralization), converted into inorganic or mineral N. These opposing processes occur simultaneously (Jansson and Persson 1982). The direction of the transformations are affected by the ratio of organic carbon and nitrogen (C/N ratio). Organic N may be mineralized to ammonium, but some of this ammonium may rapidly recycle through microbial biomass back into the organic pool. Some of the organic C is simultaneously mineralized to

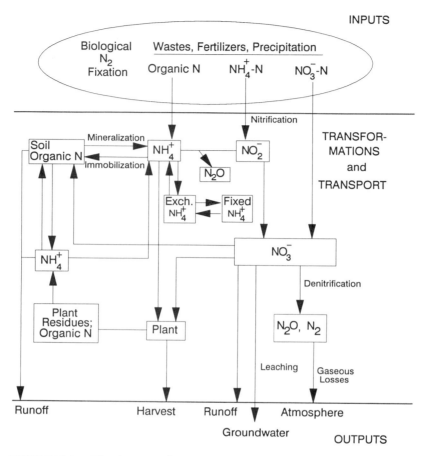

FIGURE 12-1. The nitrogen cycle.

CO_2 while some goes to form fresh biomass. The net result over time is a lowering in the amount of C available for heterotrophic bacteria growth and an increase in mineral nitrogen (ammonium). Some of the ammonium will be taken up by plants, some will be nitrified, and a small portion will be incorporated into recalcitrant soil organic matter that is only very slowly mineralized. Any source of ammonium (e.g., fertilizer, plant residues, wastes, or that absorbed from the atmosphere) can undergo nitrification, immobilization, and subsequent remineralization. The identity of the initial source of N is often quickly lost in this cycling (see Chapter 10 for a discussion of nitrogen isotopes). Nitrate also enters into the microbial immobilization phase of the N cycle but heterotrophs strongly prefer ammonium if available (Jansson and Persson 1982). In contrast, nitrate is more rapidly assimilated by some higher plants than is ammonium.

Nitrification

Nitrification is the microbial oxidation of ammonium to nitrite (NO_2) and further to nitrate (NO_3). With the exception of some atmospheric reactions, nitrification is the sole natural source of nitrate to the biosphere. These reactions are important relative to N losses, since they transform the relatively immobile ammonium ion into nitrate, which is highly mobile in water and can readily be leached or denitrified. Nitrification is considered to be carried out almost exclusively by the Gram-negative, chemosynthetic, autotrophic bacteria of the family Nitrobacteriaceae. Five genera of the ammonium oxidizers have been recognized (Schmidt 1982), although most pure culture studies have been done with the genus *Nitrosomonas*. Similarly, *Nitrobacter* is considered the dominant NO_2 oxidizer.

Denitrification

Combined N may be returned to the atmosphere as gaseous N (N_2 or N_2O) through denitrification. Many consider biological denitrification to be a major pathway of N loss and hence fertilizer N inefficiency in agriculture. Recent research on denitrification has been stimulated by interest in groundwater pollution problems and the concern over the destruction of stratospheric ozone by N_2O.

Denitrification generally refers to the microbial respiratory process in which nitrogen oxides serve as terminal electron acceptors. During respiratory electron transport, electrons move from a reduced electron donating substrate (e.g., organic matter), through numerous electron carriers, to the more oxidized nitrogen oxides (Firestone 1982). Denitrifying bacteria are capable of normal respiratory growth in the presence of oxygen, but under microaerobic and anaerobic conditions they use nitrate, nitrite, or nitrous oxide as the terminal electron acceptors. Nitrate can also be reduced to ammonium under anaerobic conditions (Korom 1992), but this requires a more highly reduced environment and usually excess organic C.

At least 14 genera of denitrifying bacteria have been identified, and they are present in most soil and aquatic environments. In addition to an anaerobic environment, organic matter or reduced sulfur compounds are essential as a source of electrons and energy. Some studies suggest chemical denitrification, involving oxidation of ferrous iron, can take place (Strebel et al. 1989), but this probably involves complex biological catalysis in the oxidation of sulfide minerals (Postma et al. 1991). The exact reactions and rates depend on the thermodynamic setting, the reaction kinetics, and the capacity for biological mediation (Korom 1992).

Denitrification is an important source of N losses from the shallow soil-water system, particularly in wetlands, lakes, and streams, and intermittently in poorly drained soils (e.g., Keeney 1986; Trudell et al. 1986; Chen and Keeney 1974; Volz et al. 1975; Lund et al. 1974). Its magnitude may have been overestimated in many soil studies, however. Many older plot studies attributed much of the N that could not be accounted for to losses through denitrification as opposed to leaching.

Spatial Complexity

In the same soil, nitrification and denitrification processes may alternate, related to changing soil moisture conditions. Anaerobic conditions and denitrification may dominate when the soil is saturated, but when water drains from the soil, oxidizing conditions and nitrification resume. Also, these processes may go on simultaneously, varying on a small scale within the soil or rock matrix (cf. Christensen et al. 1990a, 1990b). The porosity (and hydraulic conductivity) of soil and rock materials is complex and is now commonly being recognized and described in terms of a "dual-porosity" or "multi-domain" concept (Beven and Germann 1982; Shaffer et al. 1979; Hallberg et al. 1986). This affects the hydraulic conductivity and the movement of solutes through the soil, with relatively slow movement through the small pores in the relatively low conductivity matrix (Darcian or displacement flow), and more rapid, preferential flow through larger pores, macropores, and along structural planes, which are secondary features providing discontinuities through the matrix. Denitrification can go on locally within soil peds or small aggregates in the matrix, where limited porosity maintains reducing conditions even though the surrounding, more porous matrix has drained and returned to oxidizing conditions (Myrold and Tiedje 1985; Sexstone et al. 1985).

Denitrification and Ground Water

Generalizations about denitrification in subsoils and aquifers are difficult. Strong evidence is shown for areally extensive denitrification in poorly drained subsoils, on parts of the coastal plain of the southeastern United States, where substantial fertilizer N is leached below the root zone of agricultural areas, but little nitrate reaches ground water (Daniels et al. 1975; Gambrell et al. 1975; Gilliam et al. 1974; Jacobs and Gilliam 1985). Trudell et al. (1986) document denitrification in an unconfined sand aquifer in Ontario, Canada, while in a similar sand aquifer in Ontario, Hill (1986) found that denitrification did not have significant impact on NO_3 losses under fertilized potato cropping.

Local spatial variability in denitrification also occurs. Thompson et al. (1986), in a contiguous alluvial aquifer, found local areas where denitrification was apparent, amidst other areas where significant nitrate leaching continued. The areas related to different age alluvial deposits; denitrification occurred in areas of young alluvium which contained substantial organic matter, while continued leaching occurred in older alluvium that was well oxidized, containing little organic matter.

The N Cycle and the Hydrologic Cycle

In addition to the N cycling in the soil environment, many factors affect the resultant concentrations of nitrate that reach ground water. Most studies indicate that, over the long term, there are several primary controlling factors: the amount of N available, the amount of infiltrating or percolating water (and the hydraulic conductivity of the material), depth to the water table, and the potential for denitrification. The seasonal interaction of rainfall and recharge, mineralization and nitrification, and application of nitrogenous materials also can affect the resultant leaching of nitrate.

The water that runs over the soil surface during a rainfall or snowmelt event, by rill or sheet flow, or even high-order channelized flow, may have relatively high concentrations of ammonium and organic nitrogen attached to suspended particulate matter. But strict overland flow is typically quite low in nitrate concentration. Nitrate forms in the soil (by nitrification) and hence is primarily moved in the water infiltrating the soil (i.e., "soil" water or ground water), not the water moving over it (i.e., runoff or surface water). Thus, the high nitrate flux that may occur in streams draining agricultural land is derived primarily from the ground-water contributions to streamflow, including shallow interflow ("subsurface runoff") during discharge events, tile-drainage effluent, bank-storage return flow, and deeper base-flow (e.g., Baker and Johnson 1977; Baker and Laflen 1983; Hallberg 1987b, 1987c).

SOURCES OF NITRATE TO GROUND WATER

There are many sources of nitrate, natural and anthropogenic, that can contribute to ground-water contamination. Rainfall, for example, is an important contributor of nitrogen and nitrate to the soil-plant system in natural ecosystems and pastoral settings, and undoubtedly contributes some nitrate to ground water. Most natural ecosystems maintain a relative nutrient balance that minimizes such losses, however. There are isolated areas where natural geologic deposits have contributed to high concentrations. Many of the activities of modern society overload the soil-water-plant system with nitrogen, resulting in nitrate contamination of water resources. Such practices as the disposal of sewage and some industrial wastes, the handling of animal manures, and the intensive use of fertilizers all contribute.

Geological Nitrogen

There are areas where natural geologic deposits have contributed to high concentrations of nitrate in ground water, and areas where the mineralization of soil organic nitrogen is an important source. However, these occurrences commonly result from anthropogenic disturbance. For example, Boyce et al. (1976) found substantial quantities of nitrate under never-fertilized rangeland on Pleistocene age loess of semiarid southwestern Nebraska. They concluded that, with the development of irrigation, nitrate is being leached out of the deposits. The nitrate likely was derived from the degradation of organic matter from vegetation that was buried during loess deposition. The transformed nitrogen may exist throughout the dry loess deposits that were seldom or never saturated below the soil solum, in the recent geologic past, until irrigation was introduced.

High concentrations of nitrate have been reported associated with some volcanic rocks and related alluvium in the eastern Mojave Desert and the San Joaquin Valley, California (Marrett et al. 1990; Strathouse et al. 1980). As in Nebraska, some of this nitrate has been leached to the ground water after the advent of irrigation (Strathouse et al. 1980). Natural soil nitrogen also has been identified as a major nitrate source to the ground waters of Runnels County, Texas (Kreitler and Jones 1975). Here, the nitrate probably arose from land disturbance from earlier dryland farming (i.e., tillage), but it was not mobilized until the water table rose after the introduction of water-conserving terraces.

Forest Land

Forested areas, and other natural ecosystems, are in relative balance between nutrient inputs and uptake, and allow little nitrate to escape the root zone (e.g., Johnson 1992). The general consensus is that unmanaged (natural), mature forests are N-conserving but that disturbances that affect N cycling can lead to large N losses, usually as nitrate, to ground water (Keeney 1980; Frazer et al. 1990; Dillon et al. 1991). Vitousek and Millilo (1979) reviewed literature on nitrate losses from clear-cutting and other forest disturbance. Losses varied widely depending on the system and management. However, the increase in nitrate-N loss following disturbance from most systems is less than 10 kg/ha per year, and soil solution nitrate-N concentrations are seldom greater than 10 mg/L.

While nitrate leaching from forests is a potential threat to ground water, this source is small compared with agricultural sources. Many local comparative studies show nitrate concentrations in ground water are often 5–10 times greater under agricultural land than adjacent forested areas (e.g., Beck et al. 1985; Pionke and Urban 1985). Significantly, Omernik (1976, 1977) summarized data from 904 watersheds from throughout the United States to evaluate nonpoint-source water-quality impacts. He noted that N concentrations in streams draining agricultural watersheds were fivefold greater than forested watersheds (Figure 12-2).

The greatest increase in the progression from natural areas to agricultural watersheds is for inorganic N, which is dominantly NO_3-N (Omernik 1977). In natural systems, the major N load is as organic N, much of it in the particulate fraction (related to organic matter), but the major load in many streams in agricultural areas is as soluble NO_3-N.

Forested areas also have potential for improving water quality. Riparian forests, and buffer strips, situated between croplands and streams, can remove some of the N from shallow ground water that originates in upgradient cropland, reducing the nitrate concentration of the ground water delivered to the stream as baseflow (Peterjohn and Correll 1984; Jacobs and Gilliam 1985). Riparian areas can also provide conditions conducive to denitrification as well

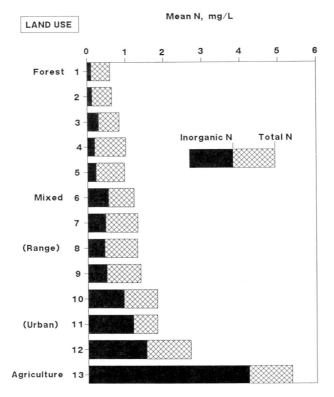

FIGURE 12-2. Land use and mean inorganic and total nitrogen concentrations from stream sample data from 904 nonpoint-source-type watersheds distributed throughout the United States (*After Omernik 1977.*) Land-use categories: 1. ≥90% forest ($n = 68$); 2. ≥75% forest ($n = 77$); 3. ≥50% forest ($n = 295$); 4. ≥75% cleared, unproductive ($n = 5$); 5. ≥50% cleared, unproductive ($n = 16$); 6. mixed ($n = 103$); 7. ≥50% range, remainder predominantly forest ($n = 12$); 8. ≥75% range ($n = 17$); 9. ≥50% range, remainder predominantly agriculture ($n = 10$); 10. ≥50% agriculture ($n = 144$); 11. ≥40% urban ($n = 11$); 12. ≥75% agriculture ($n = 72$); 13. ≥90% agriculture ($n = 74$).

(Groffman et al. 1991). Ground-water seepage to streamflow that is largely vertical may not be available to riparian vegetation for nitrate removal, but can still undergo significant denitrification through interaction with organic-matter-rich stream sediments.

Grasslands, Forage, and Pastoral Agriculture

There is a wide range of grassland ecosystems. Nitrogen cycling will vary greatly in these systems, but little research has been conducted regarding nitrate leaching losses. Since most native and extensive managed grasslands are N-deficient, very little nitrate will be available for leaching; also, many of these grasslands are in semiarid regions where little leaching may occur. Limited studies in natural prairie areas have shown nitrate concentrations in shallow ground water and surface water are an order of magnitude lower than agricultural areas (McArthur et al. 1985).

On the other hand, intensively managed forage and especially grazed grasslands can be the source of considerable nitrate to ground water (e.g., Russelle 1992). Grasslands, like croplands, have annual aboveground biomass cycles that leave nitrate in the soil profile susceptible to leaching at times of the year when plant uptake is minimal, usually spring and autumn. Animal wastes, particularly urine, are in concentrated patches in grazed pastures, leading to inefficiency of waste N use and potential for N losses leading to ground-water contamination by nitrate. This has been observed in intensive grazing areas in New Zealand, for example (Ball et al. 1979). In Europe, grasslands for grazing are often more intensively managed and fertilized than in the United States, resulting in excess nitrate that can be leached to ground water (Strebel et al. 1989).

Leguminous crops (which fix nitrogen), such as alfalfa, have been used to scavenge nitrate from the soil profile, but a few studies have shown that significant nitrate leaching can occur during, and particularly following, the plowing of alfalfa, grasslands, or other green manure crops. However, large nitrate losses in these situations were associated with additions of fertilizer, fallowing the land, or following the plowed grass or alfalfa with crops that were not significant nitrogen users (e.g., Cameron and Wild 1984; Robbins and Carter 1980; Strebel et al. 1989). Nitrate leaching in relation to fertilization practices and overwatering has also been documented for turf grass, golf courses, and even home lawn use (e.g., Exner et al. 1991; Morton et al. 1988; Keeney 1986).

Waste Materials

Many local sources of waste materials can contribute nitrate to ground water, including sites used for disposal of human and animal sewage; industrial wastes related to food processing, munitions, and some polyresin facilities (Vomocil 1987); and sites where handling and accidental spills of nitrogenous materials may accumulate. Hallberg (1986a) discussed cases around mixing and rinsing facilities of local agricultural chemical dealerships, where nitrate-N levels were substantially increased, likely from many small-scale spills and rinse-water discharge over time.

Organic Wastes

Organic wastes, including farm manures, sewage sludges, food-processing wastes, and crop residues, are often land applied. The soil is an ideal medium to receive these materials for recycling through the nitrogen transformations that take place (Sommers and Giordano 1984). Such organic material is often considered a desirable N source because the N is in the mineralization-immobilization cycle longer and thus is more slowly available. As a result this N is not subject to as rapid a loss as are inorganic N sources. Organic wastes have disadvantages as sources of N for crops, however, including that (1) they usually are low in N content and thus expensive to transport and handle; (2) they are variable in composition and quality; (3) the extended time for the N to be mineralized may be out of phase with the high rate of N uptake needed by growing crops; (4) they may be high in ammonia, and much of the N can be easily lost by volatilization if the material is not immediately incorporated; and (5) they sometimes contain undesirable contaminants, such as heavy metals or toxic organic compounds, as in municipal sewage sludges, for example (Smith and Peterson 1982; Sommers and Giordano 1984). One of the main problems with use of organic wastes, especially sludges and manures, is obtaining an accurate estimate of the rate of N mineralization and of net N availability for calculation of loading rates (Bouldin et al. 1984). Hence, in practice they have often been applied in excess.

Animal Wastes

Animal wastes comprise a large potential source of nitrate. Various estimates suggest that animal manure N returned to the soil may equal 10-20% of the fertilizer N consumed in the United States (Follett et al. 1987; Olson 1985; CAST 1975), equaling approximately 1 million metric tons of N. Animal manures often are concentrated in large commercial poultry, dairy, hog, and beef operations. Even a small farm will often have feedlot or barnyard areas with a high density of manure. Some older farm and household wells are located in proximity to animal holding areas, and nitrate leached from manures may contaminate the well, obscuring other effects of land use on water quality.

The obvious preferred method of disposal of animal wastes is to recycle these materials on the soil as a replacement for fertilizers. However, in the case of concentrated animal feeding operations some waste accumulation in the feedlot cannot be avoided. Concentrated operations have generated many localized problems, because the large mass of manure generated may be excessive for the amount of land locally available for its disposal. Furthermore, manures are difficult to handle and often the manure is disposed of, rather than recycled, by applying it to croplands at rates far in excess of fertilizer N needs.

Sewage Sludge and Effluent

Land-application programs using either municipal sewage sludge or effluent on cropland, forests, or other areas, are generally under close scrutiny and regulation with regard to their potential damaging effects on the environment, including their effects on ground-water quality. Assuming that the wastes are applied by approved methods, nitrate contamination of the ground water from municipal wastes should be minimal (Sommers and Giordano 1984). Locally such materials can constitute a substantial N source.

Septic Tanks

The disposal of human and household wastes in nonsewered areas is almost exclusively by use of the septic tank–soil seepage field. About one third of the individual households in the United States, as well as numerous rural commercial establishments, utilize such systems. In most cases, an economic, environmentally acceptable alternative approach does not exist. Significant ground-water contamination from septic tanks has been reported in many areas of the United States and elsewhere (Keeney 1986).

As reviewed by Bicki et al. (1984), the concentration of home sewage disposal in subdivisions has clearly created problems. Where ground-water contamination occurs, it is generally related to the density of the septic systems. Septic systems can be a major local source of nitrate in densely populated settings. Because nitrate sinks are limited (Magdoff and Keeney 1976), this may provide a local input similar to row-crop agriculture (Walker et al. 1973a, 1973b). However, several studies have found no general or regional relationship between nitrate in individual well water and septic tanks in more sparsely populated areas (Hallberg 1987a; Libra et al. 1984; Kross et al. 1990; Mancl and Beer 1982).

Agriculture

The most extensive source of nitrate delivered to ground water is agriculture (see reviews: Follett 1989; Follett et al. 1991; Hallberg 1986a, 1986b, 1987a, 1989; Keeney 1982, 1986; OECD 1986; Pratt 1984). Numerous studies on various scales, from controlled plot studies to basin-size inventories, have shown that nitrate concentrations in ground water (in shallow, freshwater aquifers) can be related directly to agricultural land use. Many of these studies show a range from a 3-fold to a 60-fold increase in nitrate concentrations in ground water moving from forested, grassland, or even pasture areas (generally <2 mg/L NO_3-N) to agricultural areas. In Germany, for instance, it is estimated that diffuse agricultural sources contribute over 70% of the nitrate load to water; the most affected areas are those with high fertilization rates and shallow aquifers (OECD 1986).

As summarized by Keeney (1986), the greatest problems with nitrate in the United States arise with heavy fertilization in the intensive row-cropping practices in rain-fed grain production (such as corn), in intensive irrigated grain agriculture, in the irrigation and fertilization of shallow-rooted vegetable crops (e.g., potatoes) on sandy soils, and locally in intensive animal-feeding and handling operations. In Europe, findings are similar, but also include regions of intensively managed grasslands. Grasslands, manures, and organic wastes were discussed in the preceding sections. This section will focus on crop production and related nitrogen management.

Nitrate leaching in relation to fertilization-nitrogen management has been well documented for many crops—from citrus and onions to tobacco and turf grass, as well as other vegetable and specialty crops (e.g., Embleton et al. 1986; Bruck 1986; De Roo 1980; Morton et al. 1988; Keeney 1986). Hill (1982), for example, studied land use and water quality in a water-table aquifer setting and found that the nitrate concentration beneath forested and pasture areas was typically <1 mg/L but that nitrate often was >10 mg/L beneath potato-growing areas. He found significant positive correlations between nitrate concentrations and the percentage area of fertilized crops (potatoes, corn, sod, and asparagus) and with N fertilizer application rates in the vicinity of the sampling sites. Chloride concentrations in ground water also correlated positively with KCl fertilizer application rates.

In particular, many studies show a direct relationship between nitrate in ground water and nitrogen fertilization rates and/or total nitrogen management and fertilization history in row-crop production. Rather direct impacts on ground water have been illustrated by controlled field studies monitoring tile-drainage water, which is shallow ground water (e.g., Baker and Johnson 1981; Hallberg 1987a; Hallberg et al. 1986). Such studies, utilizing different fertilizer N rates for corn on replicated plots, commonly show a rapid increase in NO_3-N in the shallow drainage water, and typically in direct proportion to the rates of fertilizer N applied (e.g., Spalding and Kitchen 1988). In a review, Baker and Laflen (1983) note that "NO_3-N losses with subsurface drainage related in nearly linear fashion to N application for rates exceeding 50 kilograms per hectare."

Row-Crop Production

Row crops occupy large land areas in many developed regions. In the United States, for example, around 70–80 million ha typically are harvested in the major crops—corn, cotton, soybeans, and wheat. Such vast areas provide large nonpoint sources relative to other nitrate sources. Several factors contribute to the importance of row-crop agriculture as a source of nitrate. First, significant external N inputs (except for soybeans) from commercial fertilizer, manure, and/or N wastes are typically associated with row crops. Second, these farming systems produce annual crops and often leave the land bare for a significant portion of the season. This not only promotes soil erosion losses by the actions of water and wind, but also inefficient N use since no plant uptake of N occurs during most of the year. Third, row crops typically involve some form of tillage or seedbed preparation that promotes mineralization of soil N. Under certain conditions (e.g., crop fallow) more N can be mineralized than utilized by the crop, leading to nitrate losses by leaching, even without application of other N (Lamb et al. 1985).

Fertilizer N has become the largest annual input of N into most agricultural systems. In the United States, consumption of nitrogen fertilizer has grown from a negligible amount prior to 1945 to over 10^7 Mg (as N). In Great Britain, between 1938 and 1976 the total N availability in the soil (from fertilizer, rainfall, manure, fixation, etc.) increased more than 50%, and nearly 75% of the increase was in fertilizer N (OECD 1986). The intensity of fertilizer N application also has increased significantly. Across the United States, average fertilizer N rates on corn increased from 72 kg/ha in 1965 to more than 150 kg/ha in the 1980s (Hallberg et al. 1991). Corn accounts for about 20–25% of all U.S. cropland, but also accounts for 40–

43% of the U.S. fertilizer N use, which highlights the concerns with nitrate contamination in the extensive midwestern Corn Belt region of the United States.

In the Corn Belt in Iowa, for example, statewide surveys have shown that water from more than 35% of *all* private wells exhibited nitrate N concentrations >3 mg/L, and more than 18% had >10 mg/L. High nitrate concentrations occur in the water-table alluvial aquifers in many areas of Nebraska related to fertilization rates in irrigated corn production (e.g., Schepers et al. 1991).

The humid, southeastern United States has high leaching rates in winter when the soils are wet but not frozen, while in the summer evapotranspiration minimizes leaching. The North Carolina coastal plain consists of varying permeability surficial sediments underlain by relatively impermeable, highly reduced sediments. Nitrate accumulates in the soil and may reach shallow ground water in response to agricultural activities, but because of the zones of reducing conditions (up to 15 m thick) nitrate does not typically move to depth nor reach underlying aquifers.

Such conditions can be altered by other impacts on the environment. In the coastal-plain areas of Georgia, rapid development of center-pivot irrigation systems has greatly increased the potential for nitrate contamination of shallow ground water, especially where the subsoils are more permeable (Hubbard et al. 1984). A study of shallow ground water found that nitrate concentrations averaged 20 mg/L NO_3-N under center-pivot irrigated areas of intensive multiple-cropping systems, but under nearby forest NO_3-N averaged <1 mg/L (Hubbard et al. 1984).

Vegetable Crops

Some vegetable crops, particularly shallow-rooted species, are difficult to manage with respect to fertilizer and water. Potatoes are a particular challenge. They have a high N requirement, are often grown on coarse-textured soils and irrigated, and must not be subjected to long periods of moisture stress because of the potential formation of misshapen (and thus unmarketable) tubers (Saffigna and Keeney 1977). The roots intercept only a portion of the volume in the root zone, and nitrate leached below about 15–20 cm is not recovered by the crop.

Irrigated Agriculture

Irrigation of croplands is now widely practiced in many areas, particularly arid regions where it is required for consistent crop production and where economic returns are greatest (Pratt 1984). In the arid regions, irrigation supplies essentially all of the water required for crop growth, while in other regions, irrigation is supplemental to rainfall. In regions where supplemental irrigation is practiced, irrigated soils often are droughty soils with low water-holding capacity. Since irrigation is capital and energy-intensive, crops are usually high value and often receive high rates of fertilizer. In arid regions, the rooting zone must be periodically leached to remove salts so that soils do not become saline and unproductive. In the more humid areas, rainfall is unpredictable and leaching can occur even with the use of good irrigation practices. The high probability of leaching, combined with large N inputs, results in irrigated agriculture being a major potential source of nitrate to ground water (Pratt 1984).

Irrigation practices have enhanced the leaching of nitrate from croplands in many areas (Hubbard et al. 1984, 1986; Timmons and Dylla 1981). However, the better water management afforded with irrigation offers significant potential for reducing leaching problems in many settings. With careful water management, the volume of water moving below the root zone can be reduced, soil-water conditions can be optimized for plant growth, and fertilizer N use in irrigated systems can be more efficient than in areas of rain-fed agriculture. Various California studies (reviewed in Pratt 1984) have demonstrated that, with careful management of moisture and N in the upper soil profile, nitrate is not leached unless excess fertilizer N is added or the soils are overirrigated. However, even in the arid agricultural situation where water is easier to manage, nitrate will be leached when salts are leached out of the profile.

Crop and Livestock Production

As described, areas of intensive livestock feeding and handling can be major sources of nitrate. Particularly for cattle and hog production, these areas often coincide with areas of row-crop grain production which is used for feed. The combination of heavily fertilized grain areas with excess livestock wastes can amplify these problem areas. Further, in some cattle production areas, and particularly in dairy cattle regions, corn and other grains are often rotated with legumes, primarily alfalfa, which provides high-protein forage. The high nitrogen fixation rates further compound the possible sources of excessive N in these regions if careful N management is not practiced.

N Management

The efficient agronomic management of N to minimize nitrate leaching losses involves minimizing excess nitrate in the root zone at times when the soil is vulnerable to substantial rainfall or excessive irrigation. This means that all sources of plant-available N are accounted for and are just sufficient to provide economic crop yields. However, given our lack of knowledge about all the site-specific factors that affect N availability, the imprecise nature of our understanding of the availability of N from soil organic matter, crop residues, and wastes, and the impossibility of predicting yearly weather patterns, this concept is very difficult to execute. There are sound economic reasons to meet the goal of maximum efficiency and environmental protection (Hallberg et al. 1991), but this has not been the driving force behind N fertilizer recommendations (Bock 1984). Rather, to prevent large yield reductions and loss in profit, fertilizer use has often aimed to maintain sufficient available N to provide maximum yields as opposed to optimal economic yields.

NITRATE DISTRIBUTION AND VARIABILITY

The preceding sections have outlined many factors that affect the occurrence and variability of nitrate in ground water: the nitrogen cycle and biogeochemical processing, varied sources of N in the environment, and temporal changes in source and delivery. Particularly when looking at regional water quality, it is imperative to appreciate the interaction of these factors, for variability is the norm that must be sorted through.

Figure 12-3 schematically summarizes some typical patterns of NO_3 variability in an unconfined, alluvial aquifer. Even in this relatively simple hydrogeologic setting some of the multidimensional complexities are evident. Spatial variability of loading of nitrate to the ground water is apparent, related to land use (N sources). Little N is lost from the forested areas; significant N may be lost from the cultivated, farmed land. Variations in NO_3-N concentrations typically would be evident related to the hydrologic regime. The nitrate delivered to the ground water from the source areas will move in "plumes" to depth and downflow. Nitrate concentrations will typically be greater at the top of the water table and will decline with depth and downflow as dispersion takes place as well as some mixing and dilution with water moving from low-nitrate source areas.

Further depth and lateral variations may occur related to biogeochemical processing in the system. The "redox" boundary reflects a change in the oxidation-reduction potential, and across that boundary denitrification becomes significant, resulting in a further, pronounced depth stratification of nitrate. In the riparian area adjacent to the stream, in this example, some N uptake occurs, and conditions also occur to promote denitrification. Hence, as the plumes move laterally into the area, further processing and removal of nitrate takes place. Often this boundary is

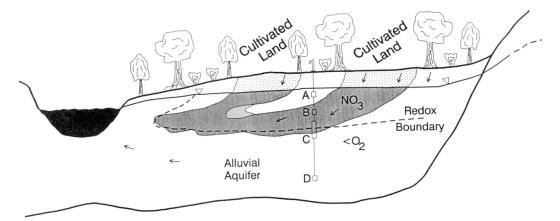

FIGURE 12-3. Schematic diagram of an alluvial aquifer and a contaminant plume of nitrate. Varying land use delivers variable nitrate concentrations to the ground water. Below the redox boundary, dissolved oxygen is depleted ($<O_2$) and the nitrate is reduced (denitrified) and escapes the system as N_2 (or N_2O) gas. These sources of variability result in different concentrations of nitrate in samples derived from wells open to various depths (A–D) in the aquifer.

at the immediate sediment interface with the surface-water body (e.g., Chen and Keeney 1974).

Temporal variability must also be considered. Concentrations will change seasonally, related to recharge periods and to the amount of water flux through the soil that can mobilize the nitrogen. Recharge will not only vary seasonally but from year to year. Further, one of the cultivated areas may be in a continuous row crop, receiving annual applications of nitrogen, while the other is in a crop rotation with very different N application needs and potential for loss from year to year. The result may be differential N loading to the ground water, by season, by year, and by site that will result in different concentrations pulsing through the plume over time.

This exemplifies part of the difficulty in ground-water studies where samples are derived from wells. In Figure 12-3, four possible screened or open well intervals are shown as A through D. Samples derived from these discrete intervals would provide a very different picture of local conditions. A and D would exhibit no, or very low, NO_3-N, but for different reasons. Well B would produce relatively high NO_3-N concentrations, whereas C would produce a mixed sample. Wells B and C would certainly change over time, related to the temporal factors discussed, but also because the flow system changes related to changing recharge and head conditions. When this is extended to a more complex aquifer and hydrologic regime, it becomes apparent why understanding results from water-quality monitoring is not always easy or straightforward.

Depth Distribution

An additional feature of nitrate contamination of ground water that has been noted in studies worldwide is the general, inverse relationship between nitrate concentration and depth below the land surface or within an aquifer (e.g., Freeze and Cherry 1979; Singh and Sekhon 1978; Hallberg 1989; Jacks and Sharma 1983; Ritter and Chirnside 1984). In general, this relationship is a function of time and rates. In many areas, as noted, nitrate in ground water has only become a problem in recent decades. In many aquifers lateral and vertical transport rates are low, and surface-derived solutes such as

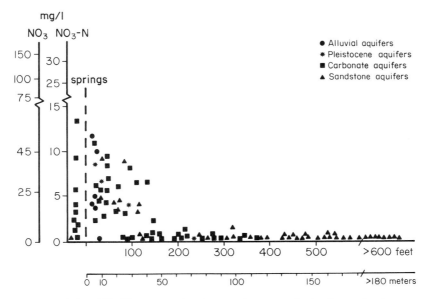

FIGURE 12-4. NO₃-N concentrations in ground water vs. depth of ground water tapped by well, from Iowa. (*After Hallberg 1986a; Hallberg et al. 1983a.*)

NO$_3$ may accumulate in the upper zone of the aquifer, with a slow gradual movement downward, by mass flux, diffusion, and hydrodynamic dispersion. In many settings there has simply not been sufficient time for nitrate to move great distances into an aquifer. Denitrification may also deplete the nitrate as it is transported to greater depth in the system.

This general depth relationship is illustrated in Figure 12-4 (see also Figure 1-5). The data are from aquifers of various lithologies to illustrate the general aspect of this trend. This depth distribution trend is typical of all surficially derived contaminants in ground water.

Nearly all large nitrate sampling data sets from complex situations show skewed, often multimodal distributions, as the data in Figure 12-4 suggest. Typically there is a major mode at low concentrations, in the local "background" range, often <1 mg/L, and in areas (or at depths) exhibiting contamination, a secondary mode at higher concentrations. Hence, the use of the median and nonparametric statistical techniques are more appropriate than statistics assuming normality, though logarithmic transformations can also be employed.

Areal Variability and Depth Distribution

One controlling factor in this general relationship between depth and nitrate concentration is the depth to the water table or, conversely, the thickness of the unsaturated zone, and the properties of the soils and surficial geologic materials that may act as confining beds (or aquitards). Where the unsaturated zone is of variable thickness, or where the aquifer is overlain by a variable thickness of an aquitard (e.g., fine-textured soils, shales), this depth relationship may not always be clear because of the different rates of transmission of recharge water and nitrate into the underlying aquifer. In such cases, the depth to the aquifer becomes another factor, and differences in nitrate concentrations may be reflected in areal variability.

Table 12-1 is illustrative. In eastern Iowa, bedrock aquifers are overlain by glacial de-

TABLE 12-1. Median Nitrate-N Concentration Data, Summarized by Geologic Region and Well Depth

Well Depth (m)	Median NO_3-N Concentration (mg/L) Hydrogeologic Region (depth to regional aquifer)			
	Deep Bedrock (>15 m)	Shallow Bedrock (<15 m)	Karst Region (<10 m)	Total Area
<15	7.3a	5.8a	6.2a	6.2
15–29	1.3a	4.2b	7.6c	4.0
30–44	<1.0a	3.6b	5.1b	1.6
45–59	<1.0a	1.3b	2.2b	1.6
60–74	<1.0a	<1.0a	<1.0a	<1.0
75–89	<1.0	1.0	<1.0	<1.0
90–149	<1.0	<1.0	<1.0	<1.0
>150	<1.0	<1.0	<1.0	<1.0
All well depths	<1.0a	2.0b	4.2c	1.3

Source: After Hallberg and Hoyer (1982); Libra et al. (1987).
Notes: Data from 6,039 Water Well Samples, from 22 Counties in Iowa, Analyzed during 1977–1980.
Medians within row followed by different letter indicate statistically significant differences at $p \leq .001$.

posits and, locally, by shales, which collectively form an aquitard. Regional mapping shows that the thickness of this aquitard varies from being negligible (where the bedrock aquifer outcrops in the root zone) to more than 100 m (Hallberg and Hoyer 1982). In various studies, some simple, yet effective hydrogeologic regions have been defined in the area (Hallberg and Hoyer 1982; Hoyer and Hallberg 1991; Libra et al. 1984) to summarize water-quality findings and to define the susceptibility or vulnerability of the aquifers to surficial contamination. These areas are (1) "deep-bedrock" aquifer areas, where 15 m or more of aquitard overlies the aquifer; (2) "shallow-bedrock" aquifer areas, where <15 m of aquitard materials overlie the aquifer; and (3) karst areas, shallow carbonate bedrock (limestone and dolomite) areas, typically with <10 m of aquitard cover, where karst solutional development has affected the subjacent carbonate aquifer and is expressed locally at the land surface by clusters of sinkholes.

Regional data, summarizing over 6,000 nitrate analyses from 22 counties, approximately 35,000 km^2 in area, illustrate these relationships (Table 12-1). The data are summarized by well-depth classes, as well as by region. They exhibit a systematic decline in NO_3-N concentration with increasing depth below the land surface, as well as a significant difference in the aggregate median NO_3-N concentration among the areas. In the susceptible areas, elevated nitrate concentrations extend to 30 meters greater depth than in the confined, deep-bedrock region. Also, the wells in the <15-m depth class in the deep-bedrock region do not penetrate into the bedrock aquifer; they are finished in glacial or local alluvial deposits and reflect shallow, water-table conditions. (The thickness of the aquitard cover, equivalent to the depth to the bedrock aquifer, ranges from 15 m to more than 100 m; so even some of the wells in deeper categories would not be finished in the bedrock aquifer.) Hence, in the deep-bedrock region only relatively low nitrate concentrations may occur in the bedrock aquifer (e.g., median of 1.3 mg/L; 15–29-m depth class), but high nitrate concentrations still occur in the more shallow part of the ground-water system. There is no statistically significant difference in NO_3-N among the regions in the <15-m depth range. However, in the susceptible regions where the nitrate has been delivered directly into the bedrock aquifer, the greater rates of water and solute transport in the aquifer have dispersed the nitrate to significantly greater depth. (The possible role of denitrification will be discussed in a later section.)

Temporal Variations

Seasonal variations in nitrate concentrations, in shallow, responsive ground-water systems, have been well documented (e.g., Hallberg 1986a, 1987a, 1987b), with nitrate concentrations often rising when recharge

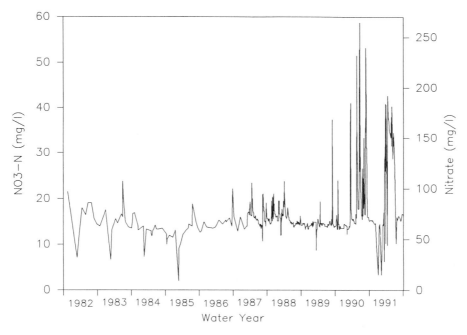

FIGURE 12-5. Nitrate-N concentrations over time from water-table monitoring beneath an agricultural field in the Big Spring Basin, Iowa. (Site L22-T of Littke and Hallberg 1991; Hallberg et al. 1984.)

through the soil occurs. Figure 12-5 illustrates temporal variations in nitrate concentration from monitoring at the water table under an agricultural field. Except during 1984–85, nitrogen-fertilized corn has been produced in the field. Seasonal variations are readily apparent: annual peak concentrations coincide with the spring–early summer and/or fall recharge periods; annual minima invariably coincide with spring snowmelt recharge. These early spring snowmelt minimum concentrations are related to the seasonally lower nitrate concentrations in the soil, and to preferential flow through the soil with nitrate-free snowmelt waters. Some of the sharply defined short-term changes, from low to high values, reflect changes on a scale of hours, and are caused by preferential flow through the soil.

Other sources of variability are also apparent. The general decline and lower concentrations in 1984–85 coincide with management changes. During this time only a portion of the contributing field was in fertilized corn with large N inputs; portions of the field were set aside and were in cover crops of grasses and some clover. The low variability in 1988–89 followed by the extreme high concentrations of 1990–91 follow climatic patterns: drought conditions in 1988–89 limited recharge and crop production; this was followed by very wet conditions with high recharge rates which also mobilized the excess nitrogen that was left as residual from the drought. Figure 1-6 also illustrated climatically induced annual variations in nitrate concentrations in relation to a longer-term trend of increasing nitrate in ground water from shallow aquifers.

Studies from the Big Spring ground-water basin in northeastern Iowa have been widely cited as illustrative of the long-term trend relationship between agricultural inputs and ground-water quality. The hydrogeology of the area has been intensively studied, and the quantity and quality of surface and ground

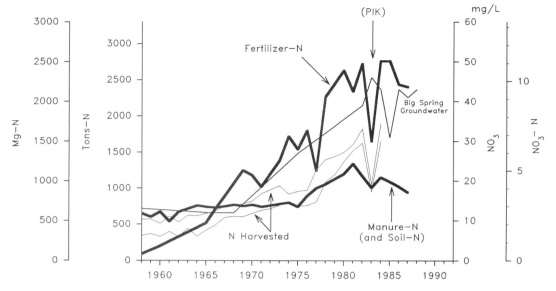

FIGURE 12-6. Mass of fertilizer N and manure N applied in the Big Spring Basin, Iowa and annual average nitrate concentration (medium-weight line, right axis for scale) in ground water at Big Spring, and mass of N harvested in corn grain (two lines shown indicating range of estimates). PIK is Payment-In-Kind, a U.S. government commodity program that removed land from corn production during 1983, hence reducing nitrogen applied. (*After Hallberg et al. 1984; Hallberg 1986a, 1986b; Libra et al. 1986.*)

waters have been measured. Ground-water discharges from a large spring that integrates the water flux from a 270-km² basin that is essentially all agricultural. Land treatment, farming practices, and agricultural chemical use are inventoried annually, and various historic records have been compiled which are illustrative of the long-term changes in agricultural practices and in ground-water quality (e.g., Hallberg et al. 1983, 1984, 1985; Hallberg 1987a, 1987b, 1989; Libra et al. 1986, 1991; Littke and Hallberg 1991; Kalkhoff and Kuzniar 1991; Kapp 1986; Padgitt 1986, Rowden and Libra 1990).

In the 1930s, nitrate concentrations in the ground-water aquifer in the Big Spring Basin area were <1 mg/L NO_3-N. During the 1950s and 1960s, the nitrate concentrations in the ground water averaged about 3 mg/L NO_3-N. By the 1980s, the NO_3 had increased three times to an annual average of 10 mg/L NO_3-N in 1983. Nitrogen inputs and balance data have been reviewed by Hallberg (1987a) and Hallberg et al. (1984). Estimates and measurements of the nitrogen contributed from rainfall, crop-rotation effects, and soil mineralization from increasing areas of cultivation were all included. The two greatest sources of added N in this basin are manure and fertilizer. Between the late 1950s and the 1980s, manure N increased 0.3 times, while fertilizer N applied increased nearly threefold, from a minor source, to the largest N input to the system, as a function of increasing rate of application and the greater corn acreage. The increase in nitrate in ground water directly paralleled these increases (Figure 12-6). The estimates of N removed with harvested grain amplify the prior relationships noted; from the late 1960s to the present, as the difference between the amount of N applied and the amount of N removed by crops increased, the nitrate concentration in the ground water increased.

Similar trends have been noted elsewhere. In Merrick County, Nebraska, Spalding et al. (1978) note that the average nitrate concentration in ground water increased from about 2.5 mg/L NO_3-N, in the late 1940s, to 11–12 mg/L in the mid-1970s. In Holt County, Nebraska, Exner and Spalding noted (1974) that nitrate concentrations were increasing at about 1.1 mg/L per year beneath N-fertilized and irrigated areas. In areas of fine-textured soils in Nebraska, even where the vadose zone is >30 m thick, NO_3-N in ground water has experienced increases of 0.1–0.2 mg/L per year in agricultural areas (Spalding and Kitchen 1988).

Such trends also have been documented in Europe (Hallberg 1986a, 1987b; OECD 1986; Strebel, Duynisveld, and Bottcher 1989). In England, for example, rates of nitrate increases in ground water ranging from 0.2 to about 1.0 mg/L per year have been noted for the last 10–20 years in many agricultural areas, and these rates are projected to increase over the next 40 years because of the slow transit time into some aquifers (Carey and Lloyd 1985; Howard 1985). In the former USSR, in an agricultural basin over a 10-year period, the fertilizer N input increased five times and manure N increased 1.5 times. This resulted in increased nitrate leaching and an eightfold increase in NO_3-N in ground water; average concentrations rose from 0.8 to 6.5 mg/L (Kudeyarov and Bashkin 1980).

Depth and Time

A major difference between ground water and surface water is the time involved for water and solutes to travel and disperse. As inferred, there is clearly a time component to the depth distribution and spatial variability of nitrate concentrations that appear in ground water at a particular monitoring point. Figure 1-5 illustrated this relationship. Over a 10-year period, nitrate increases were noted at all well depths, but the greatest increase occurred at shallower depths.

Hallberg (1989) noted differences in the timing (on a scale of years to decades) for increases in nitrate concentration in ground water related to differences in the thickness of the soil mantle over different aquifers in the same region. This aspect of temporal change has also been noted or inferred in other studies (Olsen et al. 1970; Spalding and Kitchen 1988). Even in sandy soils, if the water table is deep, some studies suggest it will take 10–50 years for the NO_3-N to reach ground water (Pratt et al. 1972; Adriano et al. 1972; Adelman et al. 1985). Carey and Lloyd (1985) modeled areas in Britain with a moderate thickness of glacial deposits, and noted that the full impact of current excess N availability will not be noted in the subjacent aquifers for possibly 30 to 40 years.

Denitrification and Time

In areas affected by nitrate contamination, the lack of nitrate in an aquifer, or the variability of nitrate among wells, within aquifers, or between aquifers, may relate either to the age of the water or to removal by denitrification. As discussed, in many areas the nitrate concentration in recharge water has increased considerably since the 1960s; hence, "older" waters would not contain significant nitrate. As discussed, there is little doubt that denitrification goes on in soils and shallow-ground-water settings locally, though its magnitude may have been overestimated in many settings. It is not clear, however, whether or not the groundwater environment can "remove" (by denitrification) the greater concentrations of nitrate now being delivered into shallow environments as they move into deeper ground-water aquifers. Although the requisite anaerobic, reducing conditions may exist in confined and/or deeper aquifers, there is likely to be less substrate to sustain significant bacteriological denitrification. Thus, in relation to long-term impacts of nitrate in ground water the efficacy of denitrification in deeper ground-water systems needs considerable research.

Furthermore, as discussed, nitrification-denitrification processes may vary on a small scale within the soil or rock matrix. In the Chalk aquifer regions of Britain extensive nitrate leaching and ground-water contamination has been shown. In studying recharge and nitrate leaching in the aquifer, Foster et al. (1984) concluded that denitrification was occurring in the relatively immobile pore waters in the matrix of the rock of the aquifer, while nitrate moved into/through the aquifer in the mobile water, moving through the secondary pore space formed by fractures and bedding planes and solution openings. Nitrate concentrations in the aquifer declined to very low concentrations downgradient in the direction of flow. Foster et al. (1984) also noted that the disappearance of nitrate, going downdip, may be related to denitrification. Smith-Carrington et al. (1983) suggested that the downdip disappearance of NO_3 may be related to dispersion and dilution, and that the apparent downdip advance of a front of sulfate and increasing Eh (redox potential) with time might indicate that biochemical activity was too limited to sustain reducing activity (such as denitrification). Howard (1985), using additional major-ion data and isotopes (tritium and carbon-14) to date the relative age of the water, concluded that the NO_3 changes were simply related to different-age ground waters and that denitrification could not be relied upon to reduce the nitrate in modern recharge waters.

The relationship between nitrate concentration and ground-water age was also explored in a series of studies in the Devonian carbonate aquifers in Iowa (Libra et al. 1984, 1987; Hallberg et al. 1987). Tritium (see Chapter 11) was used to "date" the ground water. The presence (>5 TU, tritium units) or relative absence (<5 TU) of tritium in ground water was used for a simple age classification for the water into "old," pre-1953 water, or "modern," post-1953 water. The potential for denitrification was also evaluated by analyzing the ground-water samples for dissolved oxygen (DO) concentrations. The absence of significant DO (i.e.,

TABLE 12-2. Nitrate-N Concentration Data from 184 Wells in Upper Devonian Aquifer in Two Counties in Northern Iowa, Summarized by Hydrogeologic Regions

Hydrogeologic Regions	NO_3-N Concentration (mg/L)	
	Mean	Median
Deep bedrock	0.7	<0.1
Shallow bedrock	5.3	2.4
Very shallow bedrock	6.9	6.9
Karst	10.4	9.6

Source: After Libra et al. (1984, 1987); Hallberg et al. (1987).

<1.0 mg/L) indicates that the requisite reducing conditions for denitrification have been established.

The study sampled 51 private-water-supply wells, a subset of wells from previous statistically designed, systematic sampling, representing 10% of farm wells in the two counties (Hallberg et al. 1987; Libra et al. 1987). The area was classified into the hydrogeologic regions discussed with Table 12-1.

The nitrate data (Table 12-2, from the total data set) show that in the deep-bedrock aquifer area, where the aquifer is confined, very little nitrate is present in the aquifer, but where the aquitard thins and the aquifer becomes unconfined, in the shallow-bedrock and karst areas, NO_3-N has been delivered into the aquifer from surficial recharge from the agricultural areas. About 90% of the wells from the deep-bedrock areas show <1 mg/L nitrate, as expected from presumptions about the relative age of the ground water. In the areas susceptible to contamination, the mean or median value for all wells was 5–10 mg/L NO_3-N, but 25–40% of the wells in these areas also show <1 mg/L NO_3-N. Research piezometers (unaffected by pumping gradients) showed that modern (post-1953), tritiated water with nitrate has permeated the upper-Devonian aquifer to depths greater than 30 m under natural gradients in the areas susceptible to contamination (i.e., the shallow bedrock and karst areas, Table 12-2). Lower aquifers exhibited "old" water.

Three types of significantly different ground water were identified. Type 1 comprises modern water, which is aerobic, and all samples contained >3 mg/L NO_3-N. Type 1 water occurs only in the susceptible areas and is distributed throughout the upper aquifer areally and with depth. Type 2 is anaerobic, had <1.0 mg/L nitrate, but was also modern water (median of 20 TU). Type 2 waters occur throughout the susceptible areas at various depths, although the majority are relatively deep in the upper aquifer. These samples suggest that denitrification is taking place within the aquifer. Type 3 is pre-1953 water (median of 1 TU), anaerobic, with <0.05 mg/L NO_3-N. The occurrence of Type 3 water is divided into two groups, one group which represents wells from the deep-bedrock aquifer area, and one which documents the presence of old water within the uppermost Devonian aquifer, in the susceptible areas, from various depths. Within the deep-bedrock areas, nearly 80% of the wells produced old water, with 20% producing modern water but with little nitrate (Type 2). Some of these wells have had intermittent detections of NO_3-N >1 mg/L.

These data indicate that within the susceptible areas (i.e., exclusive of the deep-bedrock area), the majority of water within the aquifer is of quite recent origin (i.e., modern, post-1953) and contains elevated nitrate concentrations. Within just the susceptible areas approximately 35% of the wells show very low nitrate concentrations (<1 mg/L); 15% produce old water from within the aquifer, and 20% produce modern water but suggest denitrification.

The data suggest that in susceptible areas the majority of water (about 90%) within the upper aquifer is modern, and most of it contains elevated nitrate concentrations. The data indicate that, even in a carbonate aquifer, rather complete oxygen consumption and denitrification can take place in less than 35 years, at least within parts of the ground-water-flow system. However, the exact age and initial nitrate concentration of the denitrified water cannot be determined. The nitrate concentration in recharge water in this region, during the 1953–65 period likely averaged about 3 mg/L NO_3-N (Hallberg 1989). It is not clear whether the system is capable of removing the greater nitrate concentrations contained in more recent recharge waters.

These examples also indicate other complexities in the ground-water system that contribute to the variability in nitrate distribution. The data suggest that the bulk of the aquifer has been permeated by modern waters, with relatively high concentrations of nitrate. While there is some depth stratification, nitrate does occur to the base of the aquifer. But, locally, there appear to be pockets of "older" water preserved that have not yet been displaced by modern water.

Nitrate Variability among Wells

The preceding sections have discussed a number of significant factors that affect the resultant concentrations of nitrate that may be measured from an individual well. The interaction of these factors produces substantial spatial variability in nitrate concentrations among individual wells, on many scales. Some studies have suggested that such heterogeneity must indicate that local "point-source" or well-construction factors are the principal "cause" of nitrate contamination. Because large variability in nitrate concentrations in ground water is the rule, not the exception, particularly in areas where variable thickness of an aquitard overlies an aquifer, such a priori conclusions based on variability are not warranted. Furthermore, many of the older and more poorly constructed wells are located in rural areas with the longest and most intensive agriculture. While nearly all field studies find that rural, private well construction and well placement are factors that may allow local nitrate contamination to occur, it must be clearly noted that these are factors, not the *cause* or *source* of contamination.

The nitrate concentration noted at any well will reflect a complex interaction of the land uses and N sources in areas of differing recharge characteristics, the nature and thickness of material over the aquifer, the hydraulic properties of the surface materials and aquifer, the three-dimensional ground-water-flow system, and possible related stratification of solutes (either because of chemical properties, flow-system effects, denitrification, or temporal changes in loading)—all in relationship to the very small portion of the aquifer tapped by that well. With stratification of solutes and/or the nonuniform, preferential flow of water and contaminants, even subtle differences in the depth of the open portion of a well can make a major difference in nitrate concentrations (e.g., Childs et al. 1974; Libra et al. 1984; Piskin 1973). For example, Ronen et al. (1987) document variations of 10–15 mg/L NO_3-N over less than a meter difference in depth. The aforementioned factors affect the rate and timing of the appearance of a particular parcel of water (and solutes) at a particular well. Even nearby wells may show seasonal variations that are out of phase, because of differences in the time of arrival of the "seasonal changes" to a well. Given these effects, the time frame over which sampling is carried out, or repeated, in a given area can also affect the apparent nature of these variations.

CONCLUDING REMARKS

Many practices have contributed to nitrate contamination: disposal of human sewage, handling of animal manures, and industrial waste disposal. Case studies document that all these activities are sources of nitrate contamination of ground water in local areas. On a regional basis, however, agricultural crop production has become recognized as the major source of nitrate delivered into water resources.

Background areas such as forest, prairie and grasslands, and often even pasture and rangeland, commonly show <2–3 mg/L NO_3-N in ground water under stable conditions. Many agricultural areas exhibit over 10 mg/L in shallow ground water, at least seasonally. Many studies show a direct relationship between nitrate concentrations in ground water and nitrogen fertilization rates and/or fertilization history in row-crop production. While these relationships are clear in many areas, nitrogen management is the key component: nitrogen derived from the effects of crop rotation and from animal and municipal wastes also contributes to the "overloading" that has taken place, and often all occur in the same region.

Various data and reviews show similar results, pointing out some of the more widespread problems. Major areas exhibiting problems with nitrate contamination of ground water include (1) rain-fed grain production (particularly corn), marked by intensive row cropping and heavy fertilization; (2) intensive irrigated grain agriculture; (3) locally intensive animal feeding and handling operations, particularly where intensive poultry and other livestock production intermingle with other crop production, and where relatively dense rural populations occur; and (4) irrigation and fertilization of vegetable and specialty crops, particularly shallow-rooted vegetable crops (e.g., potatoes) on sandy soils.

Many of the activities that contribute to regional nitrate contamination are relatively recent developments (e.g., intensive fertilization, concentrated waste disposal). Various studies in responsive ground-water environments have shown increases in nitrate concentrations over the past few decades, with rates commonly ranging from 0.1 to 1.0 mg NO_3-N per liter per year. At present, nitrate contamination (as well as other contaminants from surface activities) is only pronounced in the more shallow and responsive portions of the ground-water-flow system. This is, in part, a function of time. The high loading of nitrate into the soil-water system has only been taking place for a short time, relative to the response time and dimensions of the ground-water system.

In less responsive settings, only slight increases have occurred, and in many areas, particularly deeper portions of the ground-water system, no effects are yet obvious. Several studies have noted, because of the time involved for ground-water flux in many areas, that the full impact of current excess N availability will not be noted in some widely used ground-water aquifers for possibly 30 to 40 years. Thus, the depth distribution of NO_3-N currently noted is only a temporary status quo. For some ground-water supplies these problems are just beginning to appear. This is particularly true if the ground-water environment has a limited capacity for denitrification.

Monitoring studies for nitrate in ground water typically show a great deal of variability, except in local, more uniform water-table aquifer settings. Nitrate concentrations in ground water vary in many dimensions—areally, with depth, and over time—and the variation in these dimensions are often interrelated in a complex manner. An understanding of this variability is important to understand the extent of the problem and to design programs to mitigate it.

References

Adelman, D. D., W. J. Schroeder, R. J. Smaus, and G. R. Wallin. 1985. Overview of nitrate in Nebraska's ground water. *Transactions of the Nebraska Academy of Science* 13:75–81.

Adriano, D. C., P. F. Pratt, and F. H. Takatori. 1972. Nitrate in the unsaturated zone of an alluvial soil in relation to fertilizer nitrogen rate and irrigation level. *Journal of Environmental Quality* 1:418–22.

Baker, J. L., and H. P. Johnson. 1977. Impact of subsurface drainage on water quality. In *Proceedings of the Third National Drainage Symposium*. St. Joseph, MO: American Society of Agricultural Engineering.

Baker, J. L., and H. P. Johnson. 1981. Nitrate-nitrogen in tile drainage as affected by fertilization. *Journal of Environmental Quality* 10:519–22.

Baker, J. L., and J. M. Laflen. 1983. Water quality consequences of conservation tillage. *Journal of Soil and Water Conservation* 38:186–93.

Ball, R., D. R. Keeney, P. W. Theobald, and P. Nes. 1979. Nitrogen balance in urine-affected areas of a New Zealand pasture. *Agronomy Journal* 71:309–14.

Beck, B. F., L. Asmussen, and R. Leonard. 1985. Relationship of geology, physiography, agricultural land use, and ground-water quality in southwest Georgia. *Ground Water* 23:627–34.

Beven, K., and P. Germann. 1982. Macropores and water flow in soils. *Water Resources Research* 18(5):1311–25.

Bicki, T. J., R. B. Brown, M. E. Collins, R. S. Mansell, and D. F. Rothwell. 1984. *Impact of On-site Sewage Disposal Systems on Surface and Ground Water Quality*. Florida Department of Health and Rehabilitation Services, Environmental Health Programs, Tallahassee, FL.

Bock, B. R. 1984. Efficient use of nitrogen in cropping systems. In *Nitrogen in Crop Production*, ed. R. D. Hauck, pp. 273–94. Madison, WI: American Society of Agronomy.

Bouldin, D. R., S. D. Klausner, and W. S. Reid. 1984. Use of nitrogen in manure. In *Nitrogen in Crop Production*, ed. R. D. Hauck, pp. 224–44. Madison, WI: American Society of Agronomy.

Boyce, J. S., J. Muir, A. P. Edwards, E. C. Seim, and R. A. Olson. 1976. Geologic nitrogen in Pleistocene loess of Nebraska. *Journal of Environmental Quality* 5:93–6.

Bruck, G. R. 1986. Pesticide and nitrate contamination of ground water near Ontario, Oregon. In *Proceedings of the Agricultural Impacts on Ground Water Conference*, pp. 597–612. Worthington, OH: National Water Well Association.

Cameron, K. C., and A. Wild. 1984. Potential aquifer pollution from nitrate leaching following the plowing of temporary grasslands. *Journal of Environmental Quality* 13:274–78.

Carey, M. A., and J. W. Lloyd. 1985. Modelling non-point sources of nitrate pollution of groundwater in the Great Ouse Chalk, U.K. *Journal of Hydrology* 78:83–106.

CAST. 1975. *Utilization of Animal Manures and Sewage Sludges in Food and Fiber Production*. Ames, Iowa: Council of Agricultural Science and Technology Report 11.

Chen, R. L., and D. R. Keeney. 1974. The fate of nitrate in lake sediment columns. *American*

Water Works Association, Water Resources Bulletin 10:1162–71.

Childs, K. E., S. B. Upchurch, and B. Ellis. 1974. Sampling of variable, waste-migration patterns in ground water. *Ground Water* 12:369–75.

Christensen, S., S. Simkins, and J. M. Tiedje. 1990a. Spatial variation in denitrification: Dependency of activity centers on the soil environment. *Soil Science Society of America Journal* 54:1608–13.

Christensen, S., S. Simkins, and J. M. Tiedje. 1990b. Temporal patterns of soil denitrification: Their stability and causes. *Soil Science Society of America Journal* 54:1614–18.

Daniels, R. B., J. W. Gilliam, E. E. Gamble, and R. W. Skaggs. 1975. Nitrogen movement in a shallow aquifer system of the North Carolina coastal plain. *Water Resources Research* 11:1121–30.

De Roo, H. C. 1980. *Nitrate Fluctuations in Ground Water as Influenced by Use of Fertilizer*. New Haven, CT: Connecticut Agricultural Experiment Station Bulletin 779.

Dillon, P. J., L. A. Molot, and W. A. Schneider. 1991. Phosphorus and nitrogen export from forested stream catchments in central Ontario. *Journal of Environmental Quality* 20:857–64.

Embleton, T. W., M. Matsumura, L. H. Stolzy, D. A. Devitt, W. W. Jones, R. El-Motaium, and L. L. Summers. 1986. Citrus nitrogen fertilizer management, groundwater pollution, soil salinity, and nitrogen balance. *Applied Agricultural Research* 1:57–64.

Exner, M. E., M. E. Burbach, D. G. Watts, R. C. Shearman, and R. F. Spalding. 1991. Deep nitrate movement in the unsaturated zone of a simulated urban lawn. *Journal of Environmental Quality* 20:658–62.

Exner, M. E., and R. F. Spaulding. 1974. *Groundwater Quality of the Central Platte Region, 1974*. Resource Atlas No. 2, Lincoln, NE: Conservation and Survey Division, University of Nebraska-Lincoln.

Firestone, M. K. 1982. Biological denitrification. In *Nitrogen in Agricultural Soils*, ed. F. J. Stevenson, Agronomy (Monograph) Series, no. 22, pp. 289–326. Madison, WI: American Society of Agronomy.

Follett, R. F., ed. 1989. *Nitrogen Management and Ground Water Protection. Developments in Agriculture and Managed-Forest Ecology* 21. Amsterdam: Elsevier.

Follett, R. F., S. C. Gupta, and P. G. Hunt. 1987. Conservation practices: Relation to the management of plant nutrients for crop production. In *Soil Fertility and Organic Matter as Critical Components of Production Systems*, pp. 19–51. Soil Science Society of America Special Publication No. 19.

Follett, R. F., D. R. Keeney, and R. M. Cruse, eds. 1991. *Managing Nitrogen for Groundwater Quality and Farm Profitability*. Madison, WI: Soil Science Society of America Inc.

Foster, S. S. D., D. P. Kelley, and R. James. 1984. The evidence for zones of biodenitrification in British aquifers. In *Planetary Ecology*, eds. D. E. Cladwell, J. A. Brierly, and C. L. Brierly, pp. 356–69. New York: Van Nostrand Reinhold.

Frazer, D. W., J. G. McColl, and R. F. Powers. 1990. Soil nitrogen mineralization in a clearcutting chronosequence in a northern California conifer forest. *Soil Science Society of America Journal* 54:1145–52.

Freeze, R. A., and J. A. Cherry. 1979. *Groundwater*. Englewood Cliffs, NJ: Prentice-Hall.

Gambrell, R. P., J. W. Gilliam, and S. B. Weed. 1975. Denitrification in subsoils of the North Carolina coastal plain as affected by soil drainage. *Journal of Environmental Quality* 4:311–16.

Gilliam, J. W., R. B. Daniels, and J. F. Lutz. 1974. Nitrogen content of shallow ground water in the North Carolina coastal plain. *Journal of Environmental Quality* 3:147–51.

Groffman, P. M., E. A. Axelrod, J. L. Lemunyon, and W. M. Sullivan. 1991. Denitrification in grass and vegetated filter strips. *Journal of Environmental Quality* 20:671–74.

Hallberg, G. R. 1986a. Overview of agricultural chemicals in ground water. In *Agricultural Impacts on Ground Water*, pp. 1–63. Worthington, OH: National Water Well Association.

Hallberg, G. R. 1986b. From hoes to herbicides: agriculture and groundwater quality. *Journal of Soil and Water Conservation* 41:357–64.

Hallberg, G. R. 1987a. Nitrates in groundwater in Iowa. In *Rural Groundwater Contamination*, eds. F. M. D'Itri and L. G.

Wolfson, pp. 23–68. Chelsea, MI: Lewis Publishers.

Hallberg, G. R. 1987b. The impacts of agricultural chemicals on groundwater quality. *Geo Journal* 15:283–95.

Hallberg, G. R. 1987c. Agricultural chemicals in groundwater: Extent and implications. *American Journal of Alternative Agriculture* 2:3–15.

Hallberg, G. R. 1989. Nitrate in groundwater in the United States. In *Nitrogen Management and Groundwater Protection*, ed. R. F. Follett, Chap. 3, pp. 35–74. Amsterdam: Elsevier.

Hallberg, G. R., and B. E. Hoyer. 1982. *Sinkholes, Hydrogeology, and Groundwater Quality in Northeast Iowa*. Iowa City, IA: Iowa Geological Survey Open-File Report 82-3.

Hallberg, G. R., J. L. Baker, and G. W. Randall. 1986. Utility of tile-line effluent studies to evaluate the impact of agricultural practices on groundwater. In *Agricultural Impacts on Groundwater*, pp. 298–328. Worthington, OH: National Water Well Association.

Hallberg, G. R., C. K. Contant, C. A. Chase, G. A. Miller et al., 1991. *A Progress Review of Iowa's Agricultural-Energy-Environmental Initiatives: Nitrogen Management in Iowa*. Iowa City, IA: Iowa Department of Natural Resources, Geological Survey Bureau, Technical Information Series 22.

Hallberg, G. R., B. E. Hoyer, E. A. Bettis III, and R. D. Libra. 1983. *Hydrogeology, Water Quality, and Land Management in the Big Spring Basin, Clayton County, Iowa*. Iowa City, IA: Iowa Geological Survey Report 83-3.

Hallberg, G. R., R. D. Libra, E. A. Bettis III, and B. E. Hoyer. 1984. *Hydrogeologic and Water-Quality Investigations in the Big Spring Basin, Clayton County, Iowa: 1983 Water-Year*. Iowa City, IA: Iowa Geological Survey Report 84-4.

Hallberg, G. R., R. D. Libra, and B. E. Hoyer. 1985. Nonpoint source contamination of groundwater in karst-carbonate aquifers in Iowa. In *Perspectives on Nonpoint Source Pollution*, pp. 109–14. Washington, DC: U.S. Environmental Protection Agency, EPA 440/5 85-001.

Hallberg, G. R., R. D. Libra, K. Long, and R. Splinter. 1987. Pesticides, groundwater, and rural drinking-water quality in Iowa. In *Pesticides and Groundwater: A Health Concern for the Midwest*, pp. 83–104. Navarre, MN: The Freshwater Foundation and the USEPA.

Hill, A. R. 1982. Nitrate distribution in the ground water of the Alliston region of Ontario, Canada. *Ground Water* 20(6):696–702.

Hill, A. R. 1986. Nitrate and chloride distribution and balance under continuous potato cropping. *Agriculture, Ecosystems, Environment* 15:267–80.

Howard, K. W. F. 1985. Denitrification in a major limestone aquifer. *Journal of Hydrology* 76:265–80.

Hoyer, B. E., and G. R. Hallberg. 1991. *Groundwater Vulnerability Regions of Iowa*. Iowa City, IA: Iowa Department of Natural Resources, Geological Survey Bureau, Special Map Series 11.

Hubbard, R. K., L. E. Asmussen, and H. D. Allison. 1984. Shallow groundwater quality beneath an intensive multiple cropping system using center pivot irrigation. *Journal of Environmental Quality* 13:156–61.

Hubbard, R. K., G. J. Gascho, J. E. Hook, and W. G. Knisel. 1986. Nitrate movement into shallow ground water through a coastal plain sand. *Transactions of the American Society of Agricultural Engineers* 29(6):1564–71.

Jacks, G., and V. P. Sharma. 1983. Nitrogen circulation and nitrate in groundwater in an agricultural catchment in southern India. *Environmental Geology* 5(2):61–64.

Jacobs, T. C., and J. W. Gilliam. 1985. Riparian losses of nitrate from agricultural drainage waters. *Journal of Environmental Quality* 14:472–78.

Jansson. S. L., and J. Persson. 1982. Mineralization and immobilization of soil nitrogen. In *Nitrogen in Agricultural Soils*, ed. F. J. Stevenson. *Agronomy* 22:229–52.

Johnson, D. W. 1992. Nitrogen retention in forest soils. *Journal of Environmental Quality* 21:1–12.

Kalkhoff, S. J., and R. L. Kuzniar. 1991. *Hydrologic Data for the Big Spring Basin, Clayton County, Iowa, Water Year 1989*. Iowa City, IA: U.S. Geological Survey, Open-File Report 91-63.

Kapp, J. D. 1986. Implementing best

management practices to reduce nitrate levels in northeast Iowa groundwater. In *Agricultural Impacts on Groundwater*, pp. 412–27. Worthington, OH: National Water Well Association.

Keeney, D. R. 1980. Prediction of soil nitrogen availability in forest ecosystems: A literature review. *Forest Science* 26:159–71.

Keeney, D. R. 1982. Nitrogen management for maximum efficiency and minimum pollution. In *Nitrogen in Agricultural Soils*, ed. F. J. Stevenson. *Agronomy Monographs* 22:605–49.

Keeney, D. R. 1986. Sources of nitrate to ground water. *CRC Critical Reviews in Environmental Control* 16(3):257–304.

Korom, S. F. 1992. Natural denitrification in the saturated zone: A review. *Water Resources Research* 28:1657–68.

Kreitler, C. W., and D. C. Jones. 1975. Natural soil nitrate: The cause of the nitrate contamination of ground water in Runnels County, Texas. *Ground Water* 15:53–58.

Kross, B. C., G. R. Hallberg, D. R. Bruner, R. D. Libra, K. D. Rex, et al. 1990. *The Iowa State-Wide Rural Well-Water Survey, Water-Quality Data: Initial Analysis*. Iowa City, IA: Iowa Department of Natural Resources, Geological Survey Bureau, Technical Information Series 19.

Kudeyarov, V. N., and V. N. Bashkin. 1980. Nitrogen balance in small river basins under agricultural and forestry use. *Water, Air, and Soil Pollution* 14:23–27.

Lamb, J. A., G. A. Peterson, and C. R. Fenster. 1985. Wheat fallow tillage systems-effect on a newly cultivated grassland soils nitrogen budget. *Soil Science Society of America Journal* 49:352–56.

Libra, R. D., G. R. Hallberg, B. E. Hoyer. 1987. Impacts of agricultural chemicals on groundwater quality in Iowa. In *Ground Water Quality and Agricultural Practices*, ed. D. M. Fairchild, pp. 185–217. Chelsea, MI: Lewis Publishers.

Libra, R. D., G. R. Hallberg, B. E. Hoyer, and L. G. Johnson. 1986. Agricultural impacts on groundwater quality: The Big Spring Basin study. In *Agricultural Impacts on Ground Water*, pp. 253–73. Worthington, OH: National Water Well Association.

Libra, R. D., G. R. Hallberg, J. P. Littke, B. K. Nations, D. J. Quade, and R. D. Rowden. 1991. *Groundwater Monitoring in the Big Spring Basin 1988–1989: A Summary Review*. Iowa City, IA: Iowa Department of Natural Resources, Geological Survey Bureau, Technical Information Series 21.

Libra, R. D., G. R. Hallberg, G. R. Ressmeyer, and B. E. Hoyer. 1984. *Groundwater Quality and Hydrogeology of Devonian-Carbonate Aquifers in Floyd and Mitchell Counties, Iowa*. Iowa City, IA: Iowa Geological Survey Report 84-2.

Littke, J. P., and G. R. Hallberg. 1991. *Big Spring Basin Water-Quality Monitoring Program: Design and Implementation*. Iowa City, IA: Iowa Department of Natural Resources, Geological Survey Bureau, Open-File Report 91-1.

Lund, L. J., D. E. Adriano, and P. F. Pratt. 1974. Nitrate concentrations in deep soil profiles as related to soil profile characteristics. *Journal of Environmental Quality* 3:78–82.

Magdoff, F. R., and D. R. Keeney. 1976. Nutrient mass balance in columns representing fill systems for disposal of septic tank effluents. *Environmental Letters* 10:285–94.

Mancl, K., and C. Beer. 1982. High-density use of septic systems. Avon Lake, Iowa. *Proceedings of the Iowa Academy of Science* 89:1–6.

Marrett, D. J., R. A. Khattak, A. A. Elseewi, and A. L. Page. 1990. Elevated nitrate levels in soils of the eastern Mojave desert. *Journal of Environmental Quality* 19:658–63.

McArthur, J. V., M. E. Gurtz, C. M. Tate, and F. S. Gilliam. 1985. The interaction of biological and hydrological phenomena that mediate the qualities of water draining native tallgrass prairie on the konza Prairie Research Natural Area. In *Perspectives on Nonpoint Source Pollution*, pp. 478–82. Washington, DC: U.S. Environmental Protection Agency, EPA-440/5-85-001.

Morton, T. G., A. J. Gold, and W. M. Sullivan. 1988. Influence of overwatering and fertilization on nitrogen losses from home lawns. *Journal of Environmental Quality* 17:124–30.

Myrold, D. D., and J. M. Tiedje. 1985. Diffusional constraints on denitrification in soil. *Soil Science Society of America Journal* 49:651–57.

OECD. 1986. *Water Pollution by Fertilizers and Pesticides*. Paris: Organization for Economic Co-operation and Development (OECD).

Olsen, R. J., R. F. Hensler, O. J. Attoe, S. A. Witzel, and L. A. Peterson. 1970. Fertilizer nitrogen and crop rotation in relation to movement of nitrate nitrogen through soil profiles. *Soil Science Society of America Proceedings* 34:448–52.

Olson, R. A. 1985. Nitrogen problems. In *Plant Nutrient Use and the Environment*, pp. 115–38. Washington, DC: The Fertilizer Institute.

Omernik, J. M. 1976. *The Influence of Land Use on Stream Nutrient Levels*. Corvallis, OR: Environmental Research Laboratory, U.S. Environmental Protection Agency, EPA-600/3-76-014.

Omernik, J. M. 1977. *Nonpoint Source-Stream Nutrient Level Relationships: A Nationwide Study* (3 maps and text). Washington DC: U.S. Government Printing Office, U.S. Environmental Protection Agency, EPA-600/3-77-105.

Padgitt, S. 1986. Agriculture and ground water quality as a social issue: Assessing farming practices and potential for change. In *Agricultural Impacts on Ground Water*, pp. 134–44. Worthington, OH: National Water Well Association.

Peterjohn, W. T., and D. L. Correll. 1984. Nutrient dynamics in an agricultural watershed: Observations on the role of a riparian forest. *Ecology* 65:1466–75.

Pionke, H. B., and J. B. Urban. 1985. Effect of agricultural land use on ground-water quality in a small Pennsylvania watershed. *Ground Water* 23:68–80.

Piskin, R. 1973. Evaluation of nitrate content of ground water in Hall County, Nebraska. *Ground Water* 11:4–13.

Postma, D., C. Boesen, H. Kristiansen, and F. Larsen. 1991. Nitrate reduction in an unconfined sandy aquifer: Water chemistry, reduction processes, and geochemical modeling. *Water Resources Research* 27:2027–45.

Pratt, P. F. 1984. Nitrogen use and nitrate leaching in irrigated agriculture. In *Nitrogen in Crop Production*, ed. R. D. Hauck, pp. 319–33. Madison, WI: American Society of Agronomy.

Pratt, P. F., W. W. Jones, and V. E. Hunsakes. 1972. Nitrate in deep soil profiles in relation to fertilizer rates and leaching volumes. *Journal of Environmental Quality* 1:97–102.

Ritter, W. F., and A. E. M. Chirnside. 1984. Impact of land use on groundwater quality in southern Delaware. *Ground Water* 22:38–47.

Robbins, C. W., and D. L. Carter. 1980. Nitrate-nitrogen leached below the root zone during and following alfalfa. *Journal of Environmental Quality* 9:447–50.

Ronen, D., M. Magaritz, H. Gvirtzman, and W. Garner, 1987. Microscale chemical heterogeneity in groundwater. *Journal of Hydrology* 92:173–8.

Rowden, R. D., and R. D. Libra. 1990. *Hydrogeologic Observations from Bedrock Monitoring Well Nests in the Big Spring Basin*. Iowa City, IA: Iowa Department of Natural Resources, Geological Survey Bureau, Open-File Report 90-1.

Russelle, M. P. 1992. Nitrogen cycling in pasture and range. *Journal of Production Agriculture* 5:13–22.

Saffigna, P. G., and D. R. Keeney. 1977. Nitrate and chloride in ground water under irrigated agriculture in central Wisconsin. *Ground Water* 15:170–7.

Schepers, J. S., M. G. Moravek, E. E. Alberts, and K. D. Frank. 1991. Maize production impacts of groundwater quality. *Journal of Environmental Quality* 20:12–16.

Schmidt, E. L. 1982. Nitrification in soil. In *Nitrogen in Agricultural Soils*, ed. F. J. Stevenson. *Agronomy* 22:253–88.

Sexstone, A. J., N. P. Revsbech, T. B. Parkin, and J. M. Tiedje. 1985. Direct measurement of oxygen profiles and denitrification rates in soil aggregates. *Soil Science Society of America Journal* 49:645–51.

Shaffer, K. A., D. D. Fritton, and D. E. Baker. 1979. Drainage water sampling in a wet, dual-pore soil system. *Journal of Environmental Quality* 8:241–6.

Singh, B., and G. S. Sekhon. 1978. Nitrate pollution of groundwater from farm use of nitrogen fertilizers—A review. *Agriculture and Environment* 4:207–25.

Smith, J. H., and J. R. Peterson. 1982. Recycling of nitrogen through land application of agricultural, food processing, and municipal wastes. In *Nitrogen in Agricultural Soils*, ed. F. J. Stevenson. *Agronomy* 22:791–832.

Smith-Carrington, A. K., L. R. Bridge, A.S. Robertson, S. D. Foster. 1983. *The Nitrate*

Pollution Problem in Groundwater Supplies from Jurassic Limestones in Central Lincolnshire. Institute Geological Sciences Report 83-3, London, UK: Natural Environmental Research Council.

Sommers, L. E., and P. M. Giordano. 1984. Use of nitrogen from agricultural, industrial, and municipal wastes. In *Nitrogen in Crop Production*, ed. R. D. Hauck, pp. 207–20. Madison, WI: American Society of Agronomy, Crop Science Society of America, Soil Science Society of America.

Spalding, R. F., J. R. Gormly, B. H. Curtiss, and M. E. Exner. 1978. Nonpoint nitrate contamination of ground water in Merrick County, Nebraska. *Ground Water* 16:86–95.

Spalding, R. F., and L. A. Kitchen. 1988. Nitrate in the intermediate vadose zone beneath irrigated cropland. *Ground Water Monitoring Review* 7:89–95.

Strathouse, S. M., G. Sposito, P. J. Sullivan, and L. J. Lund. 1980. Geologic nitrogen; a potential geochemical hazard in the San Joaquin Valley, California. *Journal of Environmental Quality* 9:54–60.

Strebel, O., W. H. M. Duynisveld, and J. Bottcher. 1989. Nitrate pollution of groundwater in western Europe. *Agriculture, Ecosystems, and Environment 26* (3–4):189–214.

Thompson, C. A., R. D. Libra, and G. R. Hallberg. 1986. Water quality related to ag-chemicals in alluvial aquifers in Iowa. In *Agricultural Impacts on Ground Water*, pp. 224–42. Worthington, OH: National Water Well Association.

Timmons, D. R., and A. S. Dylla. 1981. Nitrogen leaching as influenced by nitrogen management and supplemental irrigation level. *Journal of Environmental Quality* 10:421–6.

Trudell, M. R., R. W. Gillham, and J. A. Cherry. 1986. An in-situ study of the occurrence and rate of denitrification in a shallow unconfined sand aquifer. *Journal of Hydrology* 83:251–68.

Vitousek, P. M., and J. M. Millilo. 1979. Nitrate losses from disturbed forests: Patterns and mechanisms. *Forest Science* 25:605–10.

Volz, M. G., L. W. Belser, M. S. Ardakani, and A. D. McLaren. 1975. Nitrate reduction and associated microbial population in a ponded Hanford Sandy Loam. *Journal of Environmental Quality* 4:99–102.

Vomocil, J. A. 1987. Fertilizers: Best management practices to control nutrients. In *Proceedings of the Northwest Nonpoint Source Pollution Conference*, pp. 88–97. Olympia, WA: Department Social and Health Services, State of Washington LD-11.

Walker, W. G., J. Bouma, D. R. Keeney, and F. R. Magdoff. 1973a. Nitrogen transformations during subsurface disposal of septic tank effluent in sands. I: Soil transformations. *Journal of Environmental Quality* 2:475–80.

Walker, W. G., J. Bouma, D. R. Keeney, and P. G. Olcott. 1973b. Nitrogen transformations during subsurface disposal of septic tank effluent in sands. II: Ground water quality. *Journal of Environmental Quality* 2:521–5.

13

Organic Contaminants

Douglas M. Mackay and Lynda A. Smith

INTRODUCTION

Contamination of ground water by the inadvertent release or improper disposal of organic chemicals has occurred throughout the world. This problem came forcefully to the public attention in the mid- to late 1970s when a particular subset of organic contaminants, the volatile organic chemicals (VOCs), were widely detected in ground waters extracted for public drinking-water supplies in the United States. However, other organic chemicals in addition to VOCs had been and continue to be detected in soils and ground waters. Since the late 1970s, a substantial amount has been learned about ground-water contamination by various types of organic chemicals, but it is also clear that more information is required in order to assess the severity of the problems and to design the most efficient response to them.

The goal of this chapter is to establish a framework for understanding, monitoring, and responding to ground-water contamination by organic chemicals. We begin with a review of current understanding of how organic chemicals enter and distribute themselves in the subsurface, which requires that we consider the complexities of the subsurface environment as well as the wide variety of chemicals of potential concern. We then examine what is known about organic contamination of ground water, starting with a review of the results of past and ongoing monitoring efforts designed to determine how widespread the problem is. We then evaluate the findings of these studies and conclude the chapter by illustrating a range of issues that are currently not well understood. This approach allows identification of matters that should be considered in future surveys of organic contamination of ground water. It is also possible to gain insights into the future for VOC and other organic contaminant detection in ground-water supplies and for remediation of contaminated ground water. The material presented herein is drawn in large part from a series of papers addressing organic contamination of ground water in California (Mackay, Gold, and Leson 1988; Smith, Green, and Mackay 1990; Smith, Lashgari, and Mackay 1991), amended as appropriate with information drawn from studies in Japan, the United States as a whole, or other states. In this chapter, we often draw on examples specific to California; although the details (hydrogeology, contaminant type, etc.) will differ in other parts of the world, the same general issues will require careful consideration in

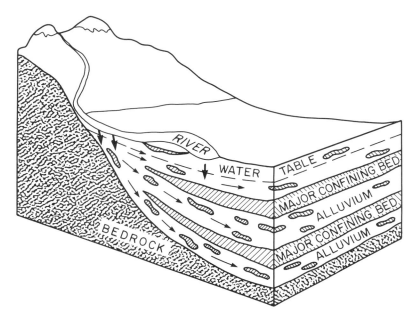

FIGURE 13-1. Geologic heterogeneities typical of alluvial ground-water basins. Bold arrows denote water flow in recharge areas, and fine arrows denote flow in aquifers.

the design and interpretation of monitoring studies elsewhere.

HYDROGEOLOGY

In order to define the current and prospective quality of ground water, we must have a thorough understanding of the physical environment occupied by the resource, i.e., the hydrogeologic setting. This is a major part of the overall challenge. California, for example, has more than 400 ground-water basins in which the water is contained within porous media ranging in composition from granular alluvial material to fractured rock. In California, the majority of the usable ground water is located within large alluvial systems, which are characterized by three-dimensional heterogeneity such as illustrated in Figure 13-1. In part because of geologic heterogeneities, the direction and rate of ground water flow may vary with depth or horizontal location in these systems. Furthermore, rates and directions of ground-water flow may vary with time in responses to rainfall, snowmelt, pumping of supply wells, and so on. Heterogeneous structure and spatially and temporally variable water flow will also characterize ground-water resources in other geologic environments around the world (e.g., karst aquifers, fractured-rock aquifers, etc.).

Strictly speaking, assessing the present and especially the future quality of ground water in a region would require a fully three-dimensional understanding of the distribution of all contaminants in all of the subsurface strata within the region in addition to a complete understanding of the geologic heterogeneity and water flow. The challenges inherent in the latter two tasks are discussed in Chapter 2. In the following section, we consider the types, sources, and properties of organic contaminants which may impact the subsurface, and illustrate how their subsurface distribution and fate depend on geologic heterogeneities and water flow.

TYPES OF ORGANIC POLLUTANTS

As described in Chapter 7, there are many classes of organic chemicals that may present threats to ground-water resources. One group of chemicals that has been widely implicated in ground-water contamination are the volatile organic chemicals. In general, the VOCs most widely implicated in ground-water contamination fall into two main subclasses: (1) halogenated organic chemicals, including the synthetic industrial solvents trichloroethene (TCE), tetrachloroethene (PCE), and so forth, and (2) aromatic, generally petroleum-derived compounds including benzene, toluene, and xylene (collectively referred to as the BTX contaminants). Many of these acronyms and even some of the chemical names are now a part of the public vernacular because of their common detection in ground-water supplies.

Despite the considerable attention paid to the VOCs, however, a large number of other organic chemicals have been detected in, or are suspected to present threats to, ground water. Appendix IX to the Resource Conservation and Recovery Act (RCRA) regulations, which governs ground-water monitoring near hazardous-waste-disposal areas, lists numerous organic chemicals in addition to VOCs (Appendix IX in parts 264 and 270, 40CFR 25942, July 9, 1987). Furthermore, there are more organic chemicals that may potentially contaminate ground water than are found on such lists. The reason for this is that the regulatory lists generally address only those compounds for which approved analytical methods are available, while other organic contaminants of concern may not yet be detectable by such methods. For example, in establishing the final Appendix IX list, the U.S. Environmental Protection Agency deleted a number of organic chemicals from its original list (Appendix VIII in part 261, 45CFR 33107, May 19, 1980). One of the major reasons cited for the deletions was the general inability of commercial laboratories to analyze for the compounds or compound classes. Many of the deleted compounds were aromatic amines (e.g., 1-aminonaphthalene, 2,4-diaminotoluene, benzidine, etc.) and organic compounds containing sulfur (sulfonates, sulfonic acids, sulfoxides, sulfones, etc.). We will return later to give examples of such organic contaminants which are not on the typical target lists for monitoring but which have been detected in contaminated ground water.

POLLUTANT SOURCES

Aside from intentional distribution during agricultural, landscaping, and similar efforts, organic chemicals are inadvertently released to the subsurface environment by many activities, such as manufacturing, transport, storage, and waste disposal. Depending on the properties of the chemicals and the nature of the source, they may be released in some or all of the following phases: (1) single or multicomponent nonaqueous-phase liquids (NAPLs), such as petroleum hydrocarbons, creosotes, PCB oils, and solvents, which are not miscible with water; (2) vapors, such as arise from release of fumigant pesticides or volatile solvents; (3) aqueous solutions, in which the pollutants are dissolved in water, such as in landfill leachates and wastewater from municipalities or industries; (4) mixed solvent solutions, in which the pollutants are dissolved in a mixture of miscible solvents such as water and acetone, water and alcohols, and so on; and (5) polluted solids, such as discarded containers, spent materials from treatment systems (e.g., activated carbon), sludges, and soils. Once released to the subsurface, the mass of each pollutant may be redistributed among these phases; for example, some constituents of NAPLs may partially dissolve in the native ground water or vaporize into the soil gas, vapors may partially dissolve in ground water, and dissolved constituents may become partially sorbed to the geologic medium (gravel, sand, silt, clays, etc.) or volatilize into the soil gas.

326 Organic Contaminants

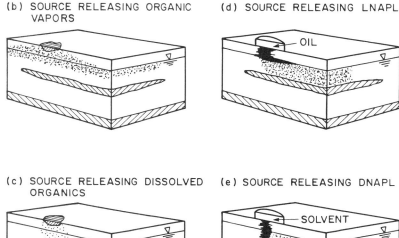

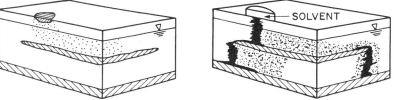

FIGURE 13-2. Organic contaminant distribution resulting from interactions between geologic heterogeneities and types of sources. Stippling represents dissolved or vapor-phase contaminants. Solid black represents nonaqueous-phase liquids (NAPLs).

The ability of an organic chemical to migrate from the point of release, and therefore to present a significant threat to groundwater supplies, is strongly dependent on its physical and chemical characteristics, the phases in which it is released and/or is migrating, the amount and rate of mass released, and the characteristics and conditions of the hydrogeologic environment. Examples of potential migration routes for the release of organic chemicals into a geologic environment similar to that sketched in Figure 13-1 are given in Figure 13-2. Figure 13-2a illustrates the relatively simple three-dimensional geologic heterogeneity of the assumed case and the relatively simple ground-water flow directions. Infiltration of water downward through the vadose zone is assumed to be negligible. Also, it is assumed that no in situ transformations of the chemicals occur. This simple example is used to allow clear differentiation between the shapes and extents of plumes generated by different types of sources. Of course, the subsurface at a real site may often be much more heterogeneous, which could either amplify these differences or in some cases make the differentiation more difficult. Furthermore, it may often be that ground-water plumes would be significantly affected by downward infiltration of water from the surface (recharge), either locally due to specific sources of water (rivers, ponds, leaking sewers, etc.) or areally due to rainfall. Such recharge would result in the plumes being displaced downwards compared to the illustrations.

Figure 13-2b assumes that only contaminant vapors are free to migrate from some source in the unsaturated zone, illustrating that horizontally broad but vertically thin plumes in the upper portions of the saturated zone may be produced by such sources. The concentrations at and immediately below the water table may be quite high; however, in the absence of significant downward flow of infiltrating water, the plumes generally remain quite thin (centimeters or decimeters). Field experiments and computer modeling have confirmed that such behavior is expected (Conant and Gillham 1992; Mendoza and McAlary 1989).

Figure 13-2c assumes that only dissolved contaminants emerge from the source zone and that the infiltrating water is sufficient in quantity to penetrate to the water table. In sand and gravel aquifers, when there is little or no water-table mounding at the source, the resultant ground-water plumes tend to be relatively narrow in the horizontal direction yet occupy significant vertical intervals in the aquifer. A long (>3.5 km), narrow (1 km) plume of a variety of organic contaminants was identified to have arisen from sewage infiltration lagoons at Otis Air Force Base on Cape Cod, Massachusetts (Barber et al. 1988). A sewage lagoon in Regina, Saskatchewan, produced a narrow plume of elevated chloride and sodium 5 miles long (Luba 1992). Plumes arising from septic tank leach fields have also been shown to be narrow in the horizontal direction (Robertson, Cherry, and Sudicky 1991). Controlled field experiments involving the injection of dissolved organic pulses have confirmed that the migrating contaminants do not spread much in the horizontal direction when the ground-water flow is not diverging—in other words, when the ground-water flow lines are essentially parallel (Mackay et al. 1986; Barker, Patrick, and Major 1987).

Figure 13-2d illustrates the case of a source which leaks LNAPLs, that is, NAPLs which are less dense than water, such as oils and gasoline. As is well known, if sufficient LNAPL is released in relatively permeable material, it will penetrate to the water table and float upon it; the soluble constituents will then dissolve into the ground water passing beneath the floating LNAPL, forming a plume. The plume produced by an LNAPL source may be similar in extent to that produced by a source of dissolved contaminants, except that the concentrations are likely to be much higher (denser stippling in the figure). A long narrow plume with very high concentrations along the centerline was generated by dissolution of a spill of aviation gasoline at the Traverse City Air National Guard Base, Michigan (Rifai et al. 1988). Note that the movement of the LNAPL toward the water table may be more complex than illustrated in Figure 13-2d if there are stratigraphic features that direct it laterally.

This potential lateral movement of NAPLs is illustrated more graphically for the next case in Figure 13-2e, in which the source is releasing DNAPLs, that is, NAPLs which are denser than water, such as halogenated solvents. Here, sufficient DNAPL is released to penetrate below the water table and run laterally along a submerged impermeable layer in at least two directions, one opposite to the ground-water flow on the left face of the block and one diagonally across the direction of water flow which emerges on the right face of the block. This results because the DNAPL flow is driven primarily by gravity and is therefore controlled by the topography of the impermeable layer rather than the ground-water flow (Feenstra and Cherry 1988; Poulsen and Kueper 1992). Because the released volume is sufficiently great, both DNAPL flows then spill over the edge of the impermeable layer and plunge deeper, reaching and forming pools on the continuous impermeable layer at the bottom of the block. If the lower impermeable layer were sloped, then the DNAPL might migrate to some extent along it. From each part of the widely distributed DNAPL, constituents dissolve into the ground water flowing past it, leading to a complex distribution of plumes. Figure 13-2e illustrates that the pathways and subsurface distribution of DNAPLs may be quite complex, leading to contaminated ground water appearing to emanate from areas at some distance from the actual surface source. In essence, the surface source has created a number of subsurface sources capable of contaminating ground water. Other conceptual examples of the potential complexity of subsurface migration of DNAPLs are given by Schwille (1988) and Feenstra and Cherry (1988).

Poulsen and Kueper (1992) present the results of two small-scale field experiments which explored the influence of stratigraphy on vertical penetration of a DNAPL. In their work, spills of 6 L of tetrachloroethylene (PCE) were found to penetrate on the order of 10 ft into a sandy unsaturated zone, having following a complex set of paths strongly affected by subtle variations in media properties. On the basis of their results, they estimate that slow leaks of a drum's worth of such DNAPLs may penetrate to 100 m in such unsaturated sandy media. Rivett et al. (1991) conducted a controlled field experiment with a small DNAPL source (18 L of mixed halogenated solvents) of known geometry (1 m high, 0.5 m in direction of groundwater flow, 1.5 m in width cross-gradient to the flow) emplaced beneath the water table in a sand aquifer. Over the course of several years, they monitored the development of a long narrow plume from dissolution of the DNAPL; at 322 days, for example, the plume was approximately 50 m long but only 4–5 m wide. Examples of ground-water plumes that are likely to have arisen from spills of DNAPL at industrial and other sites are presented by Mackay and Cherry (1989). In such cases subsurface migration of DNAPLs is often not recognized since it is rarely the case that immiscible-phase liquids are actually detected below the water table. However, a panel of experts has recently concluded that plumes of halogenated solvents containing concentrations that exceed even 1% of the water solubility of the contaminants should be assumed to have arisen from subsurface DNAPL penetration unless a strong case can be made to reject this hypothesis (U.S. Environmental Protection Agency 1992).

APPROACHES TO MONITORING OF THE SUBSURFACE

Taken together, the frames of Figure 13-2 illustrate the challenge of monitoring the subsurface for organic contaminants. The contaminants may be present in relatively limited portions of the subsurface or they may be more widely distributed, depending on geologic heterogeneity, water flow, contaminant properties, and source characteristics. Some insight into ground-water quality can be gained by monitoring water-supply wells, and indeed these have generally served as the primary monitoring tool for assessing regional ground-water quality. However, as illustrated in Figure 13-3, water-supply wells generally are designed to maximize water delivery, not information on contaminant distribution. They tend to have intakes (screened sections) which span large vertical intervals, sometimes including more than one permeable stratum. In many cases, illustrated by well A in Figure 13-3, this will lead to dilution of the contaminant in the well; in some cases the dilution may be so great that the contaminant is below the analytical detection limit and the existing

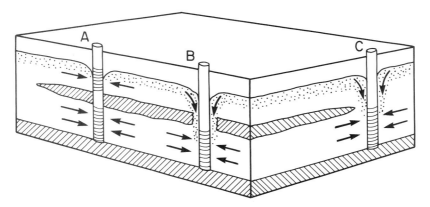

FIGURE 13-3. Dilution or distortion of contaminant plumes by large supply wells. See text for discussion.

contamination will go unrecognized until the concentration increases or the detection limit decreases. Furthermore, when contamination is noted, the incorrect conclusion may be drawn that contamination is uniform throughout the vertical interval of the well intake. In other cases in which wells are installed through impermeable layers but without competent seals in those layers (e.g., well B), the contaminants may be drawn into aquifers that would otherwise have been unpolluted. If the contaminants were at high enough concentrations to be detected, this latter case would give the impression that contamination was present in and migrating through the lower aquifer rather than being short-circuited to it along the well casing. A similar misinterpretation might occur if a supply well happened to be located in an area in which there was a natural discontinuity in an impermeable layer which was generally assumed to separate two aquifers (well C).

The point of these examples is that monitoring of supply wells provides information on the quality of the extracted water, but rather limited understanding of the subsurface distribution of contaminants. Gaining insight into the latter requires other approaches. A common approach is the use of permanently installed monitoring wells with relatively short screened sections such as illustrated in Figure 13-4. However, the insight provided by monitoring this type of well may also be limited. We have illustrated some special aspects and problems inherent in monitoring in the vicinity of a large spill of DNAPL in Figure 13-4. Eight clusters of monitoring wells are illustrated. Monitoring clusters 1 and 2 leads to the correct conclusion that there is no contamination in the upper or lower aquifers in the area somewhat upgradient of the source. If, as often occurs in practice, a small portion of the DNAPL pool is smeared into the lower aquifer during installation of cluster 3, monitoring this cluster may lead to the incorrect conclusion that contaminants have entered the lower aquifer upgradient of the cluster. A more critical problem is illustrated for cluster 4, whose inappropriate installation has allowed significant downward leakage of the DNAPL which otherwise would have been confined to the upper aquifer. A small consolation is that cluster 4 ought to allow detection of the problem it created, that is, the introduction of contamination in the lower aquifer. Cluster 5 will certainly allow determination that the contamination extends throughout the entire thickness of the upper aquifer. However, monitoring the lower of the two monitoring wells in cluster 5 may not necessarily lead to detection of the DNAPL phase. For various reasons de-

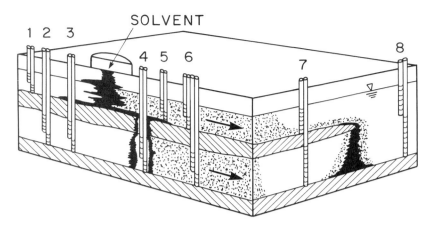

FIGURE 13-4. Illustration that monitoring wells may yield good resolution on contaminant distribution but may also fail to detect more migrating contaminants or may even cause DNAPL leakage through low-permeability layers. See text for discussion.

scribed by Cherry (1992), the DNAPL phase may not flow into the well screen, or if it does it may not be collected into typical sampling equipment. Monitoring wells in cluster 6 should correctly detect contamination throughout both aquifers. Clusters 7 and 8 are meant to illustrate the typically sparse monitoring conducted at some distance from the source. If the dissolved plumes are narrow in the horizontal direction, it is conceivable that such clusters will miss much of the contamination migrating away from the source (out the right end of the box). In this example, the upper well in cluster 7 detects contaminants that are moving along the low-permeability layer, having originated from dissolution of the DNAPL running diagonally along that layer from the source to the right face of the box. The lower well in cluster 7 misses both the plume emanating from the DNAPL which leaked down cluster 4 and the plume emanating from the DNAPL which has emerged on the right face of the box. Similarly, monitoring wells in cluster 8 miss the nearby subsurface DNAPL source and plume.

The point is that monitoring well installation will not necessarily guarantee the unambiguous location of organic contaminants in the subsurface. Part of the problem is that monitoring budgets tend to constrain the number of monitoring wells that can be installed. This problem is exacerbated by the fact that such wells are not pumped continuously, but are typically sampled after only slight purging; thus they cannot capture contaminants migrating nearby (as do water-supply wells) and tend to allow detection of only those contaminants present along the well screen at the time of sampling. If hydrogeology and contaminant source type conspire to produce long thin plumes, such plumes may in some cases thread their way past all or a portion of monitoring networks. Examples of contaminated sites in which such elusive plumes have been documented are presented by Kopania (1991) and Osiensky, Winter, and Williams (1984).

Cherry (1992) discusses new tools and approaches which offer promise of improving our insight into subsurface organic contaminant distribution. One approach that is being increasingly applied in the early stages of site investigation is sampling liquids and/or solids by means of devices temporarily pushed or drilled into the subsurface. For example, hollow-stem augers with screened sections at their end may be used to collect water samples during drilling. Since rapid

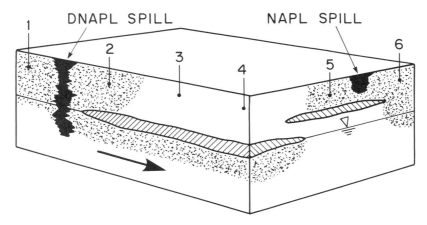

FIGURE 13-5. Illustration of some advantages and disadvantages of vadose-zone gas monitoring. See text for discussion of numbered sampling locations.

and reliable field analysis of the samples is now possible using mobile laboratories or portable gas chromatography equipment, having such samples can allow much more insightful installation of permanent monitoring wells with screen lengths appropriate to the detected contaminant distribution. Also, Cherry (1992) points out the advantage of recent advances in piezocone technology. With this technology, devices mounted on the tip of a narrow rod may be pushed to depths on the order of 20 m, generating information on geologic stratigraphy while also allowing collection of solid and liquid samples. After sampling, the rods are retracted, decontaminated, and pushed into a series of new locations. This technology is generally more rapid than traditional drilling and thus can generate more spatial information in shorter times, which in turn can be used to design more efficient networks of conventional monitoring wells for subsequent installation. A disadvantage of the conventional piezocone technology is that it produces a hole through the subsurface which, at least until sealed, may allow downward migration of DNAPLs. This problem is alleviated considerably in some new versions of the device which have the capability to seal the bore as the cone is withdrawn.

Lastly, in recent years there has been considerable interest in and application of methods to sample the vadose-zone gas, often under the assumption that detected volatile organic contaminants are likely to arise from or be indicative of underlying ground-water contamination. The gas may be actively sampled through driven miniwells, by using a small vacuum pump, and then analyzed on-site or in the laboratory. An alternative, passive approach is to place sorbing devices in the subsurface for a period of time which are capable of concentrating volatile organics from the gas; these devices are then exhumed and sent to a laboratory where the contaminants are desorbed from them as the first step in standard chemical analyses. Figure 13-5, a compilation of concepts explored by Lin (1990) and Rivett and Cherry (1991), illustrates some of the factors that may affect the success or failure of such an approach. First, if a source of volatile contaminants remains in the vadose zone, it may yield a vapor plume which will mask any underlying ground-water plume, and not necessarily be areally coincident with it. For example, samplers at positions 1, 2, and 5 in Figure 13-5 are detecting contaminants which are migrating in the vapor phase. Below sampler 2 there is in fact some ground-water contamination, but it is not "detected" by the soil gas monitoring, per se, since upward migration of the contaminants from the ground-water plume is not occurring. Second, if there are low-permeability layers above the water table, they may impede significant upward migration of vapors from the plume, thereby preventing plume detection via soil gas monitoring; samplers 3 and 4 illustrate this problem. In other cases, illustrated by sampler 5, a vapor source in the vadose zone may create an easily detectable vapor plume which is prevented in some areas from downward migration by an impermeable layer. In such areas, detection of vapors does not imply the presence of underlying ground-water contamination. Lastly, when vapors are free to migrate to the water table from a vadose-zone source, as in the vicinity of sampler 6, they may produce ground-water contamination. Rivett and Cherry (1991) described a field experiment which resulted in ground-water contamination caused by vapors moving from a source in the unsaturated zone. They also described another experiment in which the bulk of the DNAPL source for a ground-water plume was intentionally restricted to a vertical interval a meter or so below the water table. In that case, the authors concluded that soil gas monitoring was not effective at identifying the plume location, even though the ground-water VOC concentrations were extremely high in the experiment. Thus it appears that while soil gas monitoring may in some cases provide rapid and inexpensive information on subsurface distribution of volatile organic contaminants, the results may remain ambiguous or even misleading regarding the presence of significant ground-water contamination.

The foregoing discussion implies that the three-dimensional distribution of organic contaminants in typically heterogeneous subsurface environments can only be understood imperfectly, regardless of which tool or combination of tools is used in any particular case. This is an inevitable consequence of the complexity of the problem. However, the examples presented also suggest that the clearest understanding of contaminant distribution can be gained only if the monitoring and data analysis is conducted with the proper *conceptual* model or models in mind, that is, considering the phases in which significant masses of contaminants are migrating (in vapor, water, or NAPL) and considering the unique hydrogeology of the site or region.

NATIONAL SURVEYS OF VOC CONTAMINATION

United States

The earliest detections of VOCs in ground waters, which resulted from monitoring conducted or required by state or local agencies, were not included in the compilations of the various national surveys that were conducted later. Nevertheless, it is known that VOCs were found at relatively high concentrations in many wells nationwide, most of which were removed from service. Pye and Kelley (1984) cite sources for such information in a few specific regions: South Brunswick, New Jersey (Geraghty and Miller, Inc. 1979); New Castle County, Delaware (Frick and Shaffer, undated); and Long Island, New York, and California (Council on Environmental Quality, 1981). The Commonwealth of Massachussets (1986) detailed VOC detections and well closings. As of July 1986, 41 communities had abandoned water supplies, most commonly because of VOC contamination (the most frequently detected VOCs were TCE, 1,1,1-TCA, and PCE). The VOC levels in the community supplies ranged from a few to thousands of parts per billion (micrograms per liter). Thousands of wells in California were found to be contaminated with one VOC (dibromochloropropane or DBCP, a soil fumigant) from 1979 to 1983 (Litwin, Hantzsche, and George 1983) and, although documentation is elusive, it has been estimated that hundreds of wells were closed in California prior to the mid-1970s due to VOC contamination before the organized statewide monitoring

TABLE 13-1. Selected Results from the National Ground Water Supply Survey

	Small Systems (serving <10,000)	Large Systems (serving >10,000)
Random Sampling		
Systems sampled	280	186
Systems positive[a]	47 (17%)	52 (28%)
10 most frequent VOCs	Tetrachloroethene	Tetrachloroethene
	1,1,1-Trichloroethane	Trichloroethene
	1,1-Dichloroethane	1,1,1-Trichloroethane
	Trichloroethene	c,t-1,2-Dichloroethene
	m-Xylenes	Carbon tetrachloride
	o + p-Xylenes	1,1-Dichloroethane
	Carbon tetrachloride	1,1-Dichloroethene
	1,1-Dichloroethene	1,2-Dichloropropane
	Toluene	1,2-Dichloroethane
	c,t-1,2-Dichloroethene	p-Dichlorobenzene
Nonrandom Sampling		
Systems sampled	321	158
Systems positive[a]	72 (22%)	59 (37%)
10 most frequent VOCs	Tetrachloroethene	Trichloroethene
	1,1,1-Trichloroethane	c,t-1,2-Dichloroethene
	Trichloroethene	1,1,1-Trichloroethane
	c,t-1,2-Dichloroethene	Tetrachloroethene
	m-Xylene	1,1-Dichloroethane
	Carbon tetrachloride	1,1-Dichloroethene
	o + p-Xylene	Carbon tetrachloride
	1,1-Dichloroethane	Vinyl chloride
	1,1-Dichloroethene	1,2-Dichloroethane
	Benzene	Benzene

Source: Adapted from data presented by Westrick (1990).
[a] "Positive" means that one or more VOC was detected.

efforts began. The point of this brief discussion is that many of the more severe organic contamination problems may have been identified and wells removed from service before the more recent and more organized ground-water surveys were begun.

Westrick (1990) has reviewed in detail the sequence, design, and results of the series of national surveys of VOC contamination of public water supplies. In this section, we examine the key conclusions of one of the most recent efforts, the Ground Water Supply Survey (GWSS). The earlier surveys had served to identify VOC contamination as a nationwide problem and had highlighted the halogenated VOCs as the most frequently detected in ground water. These and other findings served as the basis for the design of the GWSS.

The GWSS was undertaken by the U.S. Environmental Protection Agency to determine the fraction of public-drinking-water supplies that were contaminated with VOCs. The survey was designed to sample several hundred supplies selected randomly and several hundred other supplies identified by state or local officials as likely to be impacted by VOCs. Selected results from the GWSS are presented in Table 13-1, adapted from data presented by Westrick (1990). Interestingly, the nonrandom selection process was only slightly more successful at finding contaminated supplies than the random process, which illustrates that relatively little was known at that time about VOC sources, transport paths, and transport rates. Halogenated VOCs were found in generally greater frequencies in large water-supply systems than in the smaller systems, a finding considered to reflect the generally more industrialized nature of the communities with large water-supply systems. Aromatic VOCs (benzene, toluene, xylene, etc.) were infrequently found compared to halogenated VOCs and were more of a problem for the smaller water-supply systems.

It is interesting to consider why, out of all of the many organic chemicals that are certain to have been released to the subsurface over the last few decades, only a few VOCs are commonly detected. Certainly a part of the answer is that such contaminants are relatively mobile and persistent when they are migrating in the dissolved phase (e.g., Mackay, Roberts, and Cherry 1985; Mackay 1990). But only recently has the nature of the source been recognized as equally important. The VOCs detected in water-supply wells may in large part have derived from NAPL spills, either DNAPL halogenated solvents or LNAPL petroleum products. As described previously, NAPLs which penetrate to or into ground water result in high concentrations near their sources. These high initial dissolved concentrations combine with relatively high mobility and persistence in the dissolved state to produce a situation in which significant concentrations may extend for great distances from the source areas. Examples of extensive, but typical, plumes of VOCs, whose lengths are often measured in kilometers and whose volumes are measured in billions of liters, are given by Mackay and Cherry (1989). As described by Mackay and Cherry (1989), many of these large plumes may have arisen from what would previously have been thought to be relatively small spills or releases of NAPLs.

Approximately half of the contaminated ground-water supplies were affected by more than one VOC, and there was evidence that some halogenated VOCs often co-occurred with other halogenated VOCs (Westrick 1990). One reason for the co-occurrence could have been release of the co-occurring compounds to the subsurface. Another reason could have been in situ transformation of some of the compounds leading to formation of the co-occurring VOCs. For example, chloroethene (vinyl chloride) was never found as the sole contaminant, but generally was found in samples that also included *cis/trans*-1,2-dichloroethene, trichloroethene, and tetrachloroethene. This would be consistent with expectations if PCE and/or TCE were undergoing in situ anaerobic biotransformation (Vogel and McCarty 1985). Similarly, 1,1-dichloroethene was never found as the sole contaminant, suggesting it may have originated at least in part from in situ transformation; indeed, mechanisms have been reported for the production of 1,1-dichloroethene by the abiotic transformation of 1,1,1-trichloroethane (Vogel and McCarty 1987) and possibly from microbiological transformation of TCE and PCE (Vogel and McCarty 1985). During the 1980s, investigations of specific contaminated sites and their plumes of ground-water contamination have confirmed that in situ anaerobic transformation of the halogenated solvents occurs relatively often, in some cases yielding substantial plumes of the dichloro and monochloro daughter products.

Japan

Organic contamination of ground water similar to that found in the United States has also been identified in Japan. The Environmental Agency in Japan began a nationwide ground-water-quality survey in 1982 after TCE and PCE were detected in drinking water derived from ground water during compliance monitoring for trihalomethanes (Magara and Furuichi 1986). In the ground-water quality survey, shallow wells and deep wells were sampled in 15 cities throughout Japan. The most frequently detected VOCs were TCE, PCE, chloroform, 1,1,1-trichloroethane, carbon tetrachloride, and

TABLE 13-2. Selected Results from Monitoring Programs for Public Water Systems in California

	Small Systems (5–199 connections)	Large Systems (>200 connections)
Systems sampled	4,476	807
Systems positive[a]	364 (8%)	184 (23%)
10 most frequently detected organic contaminants	Dibromochloropropane Tetrachloroethene Chloroform 1,1,1-Trichloroethane Trichloroethene 1,2-Dichloropropane 1,2-Dichloroethane Bromodichloromethane Dibromochloromethane Trichlorotrifluoroethane	Tetrachloroethene Trichloroethene Dibromochloropropane Chloroform 1,1-Dichloroethene 1,1,1-Trichloroethane Carbon tetrachloride Atrazine 1,2-Dichloroethene Simazine

Source: Adapted from data presented by California Department of Health Services (1986, 1990).
[a] "Positive" means that one or more organic contaminant was detected.

cis-1,2-dichloroethene. The TCE and PCE occurred in 28% and 27%, respectively, of the monitored supply wells, with concentrations ranging from 0.5 to 4,800 µg/L for TCE and 0.2 to 23,000 µg/L for PCE. The experience of Japan and the United States suggests that similarly extensive VOC contamination of ground water is to be expected in other industrialized nations.

STATE SURVEYS OF GROUND-WATER CONTAMINATION

California

Since 1984, a large number of water-supply wells in large and small public water systems in California have been monitored for organic contaminants (California Department of Health Services 1986, 1990). Selected results from these monitoring programs are presented in Table 13-2. It is clear from the table that many supply systems have been impacted by a wide variety of organic chemicals; generally the concentrations of the chemicals detected in the supplies have been low, but, in a significant number of cases, the concentrations have exceeded one or more of the applicable regulatory standards. The frequently detected organic chemicals are mostly VOCs, including some industrial solvents and a fumigant pesticide (DBCP). Overall, approximately 2,700 water-supply wells have been affected by VOC contamination (California State Water Resources Control Board 1990a).

The frequent occurrence of trihalomethanes (chloroform, bromodichloromethane, and dibromochloromethane) in small system wells has been attributed to chlorination of well water (California Department of Health Services 1990). The rest of the organic contaminants appear to have been released to the subsurface from a variety of industrial and agricultural activities. The most commonly detected pesticide, dibromochloropropane (DBCP), has appeared and continues to appear in thousands of wells at levels exceeding California's drinking-water standard of 0.2 µg/L; many of these wells are in the productive agricultural areas of the San Joaquin Valley. This continued appearance is important to note in light of the fact that the use of DBCP in California was suspended in 1979. Similarly, ethylene dibromide (EDB) and 1,2-dichloropropane (1,2-D) have been prohibited in pesticide formulations in California since 1984 (Smith,

Green, and Mackay 1990), yet they continue to be detected in California ground water. The continued detection in ground water of these organic chemicals, which have not been used for 6 to 12 years, suggests that a large reservoir of contaminants develops by the time the contamination is noted in water-supply wells, and, furthermore, that the impacts of the historic contamination may persist long after the surface sources have been eliminated. It should be noted that this lesson may well apply to other types of contaminants detected in water-supply wells.

The most commonly detected solvents—1,1,1-trichloroethane (TCA), trichloroethene (TCE), and tetrachloroethene (PCE)—generally have appeared at low concentrations, but, in approximately 350 water-supply wells, solvent concentrations have exceeded the drinking-water standards (California State Water Resources Control Board 1990a). Many of the large supply systems which have detected organic contamination are located in Los Angeles County, the most densely populated metropolitan area in the State (California Department of Health Services 1986), or in the Santa Clara Valley, an area well known for its electronics industries. The most common sources of the widespread contamination in Los Angeles County have been identified as poor storage and handling of industrial solvents, leaking underground storage tanks, and aboveground sumps and clarifiers. Similar sources have been implicated in the Santa Clara Valley. There is mounting evidence that DNAPL has penetrated into the subsurface at many of these sites.

Unlike the national surveys, components of petroleum hydrocarbons (i.e., the BTX contaminants benzene, toluene, and xylene) have rarely been detected in the public-water-supply monitoring programs in California. This finding at first seems odd considering the known leakage of large quantities of petroleum hydrocarbons from underground storage tanks throughout California (California State Water Resources Control Board 1989). The minimal impact of the BTX contaminants on California water supplies appears to be, at least in part, a consequence of their natural biodegradation. It is known from field studies elsewhere that the BTX compounds degrade relatively quickly in the subsurface if sufficient oxygen is present (Barker, Patrick, and Major 1987; Rifai et al. 1988). Apparently the conditions in the vadose and shallow saturated zones in California are such that aerobic degradation of the BTX contaminants proceeds rapidly enough to protect most water-supply wells (Smith, Green, and Mackay 1990; Hadley and Armstrong 1991). In some cases, the reasons that BTX contaminants are not detected in supply wells may be of a purely physical nature. For example, when supply wells are screened far below the water table, there may be significant dilution of the BTX plume as it is drawn downwards toward the intake (e.g., well B in Figure 13-3). Although some BTX contaminants may reach the intake, the concentration in the extracted water may be diluted below the detection limit of the analytical method.

Reviewers of California's ground-water monitoring efforts have repeatedly identified the inert ingredients of pesticide formulations as potential ground-water contaminants about which little is known (Cohen and Bowes 1984; Mackay and Smith 1990). Inert ingredients are those ingredients which do not contribute to the intended pesticidal activity of the formulation but which are used for a variety of other purposes. The U.S. Environmental Protection Agency (1987) determined that some of the inert ingredients used in pesticide formulations nationwide are of known or potential toxicological concern; examples of the compounds include VOCs such as chloroform, carbon tetrachloride, tetrachlorethene, and trichloroethene. The U.S. Environmental Protection Agency list makes it clear that numerous VOCs and other organic contaminants often associated with industrial activity may also be released to the environment in significant quantities by agriculture.

Many of the inert ingredients might affect ground water to an extent similar to that of the active ingredients detected in ground water because they have similar mobility and persistence characteristics (Mackay 1988). Cohen and Bowes (1984) roughly estimated the amount of inert ingredients released to land in California as 200 million pounds during the period 1971–81. To date, no better estimates have been possible since the composition of the inert fraction of pesticide formulations has been considered proprietary. Generally, the monitoring efforts in agricultural areas of California have focused on a selected subset of active ingredients known to have been used or known to be of public health significance. Thus, few data exist to assess whether inert ingredients such as the VOCs listed by the U.S. EPA (1987) accompany the active ingredients as they migrate in the subsurface in California. This issue appears to warrant more attention in regional efforts to characterize organic contamination of ground water. In essence, even if no industrial activity has occurred in an otherwise agricultural area, the potential for VOC contamination cannot be automatically discounted. Furthermore, even though such inert ingredients may have been or are now being voluntarily removed from pesticide formulations, if they were used in significant quantities in the past, they may have left a legacy similar to that of the now banned active ingredients such as DBCP, EDB, and 1,2-D.

The California monitoring data have been collected over the years through a variety of efforts spearheaded by a number of regulatory agencies. Aside from the now routine monitoring of water supply wells and the more detailed monitoring of specific sites with known problems, the state has conducted exploratory monitoring in areas judged to be susceptible to contamination. This "Hot Spots" program (California State Water Resources Control Board 1990b) has targeted areas on the basis of proximity of industries, disposal areas, etc., and has been relatively successful at pinpointing sources and finding ground-water plumes through the use of soil gas monitoring coupled with conventional monitoring well installation or ground-water sampling using temporarily driven sampling points (modifications of the piezocone technology). On the basis of this success and common sense, it would seem that such detective work should be a part of any thorough regional effort to define the principal locations of organic contamination of soils and ground water.

Other States

The Arizona Department of Health Services has summarized ground-water monitoring data, detailing the occurrence of VOCs in monitoring wells and water-supply wells at 30 contaminated sites (Graf 1986). To illustrate the extent of ground-water contamination, Arizona has reported that VOC-contaminated wells are located in 122 one-mile-square sections of land at these contaminated sites. Most of the identified sites are located in major ground-water basins in and around the two principal metropolitan areas of Phoenix and Tucson. The primary sources of the ground-water contamination include dry wells used for injection of waste solvents, waste impoundments and leach fields, leaking storage tanks, and landfills. Like California, the most frequently detected VOCs are TCE, PCE, 1,1-DCE, 1,1,1-TCA, t-1,2-DCE, and 1,1-DCA. In several cases, TCE and 1,1,1-TCA have been detected at levels close to limits for water solubility, which can be taken as confirmation that DNAPL has penetrated into the aquifer at those sites. It is expected that additional VOC-contaminated sites will be discovered as ground-water-monitoring programs progress.

A statewide monitoring program of drinking-water-supply wells was undertaken in Wisconsin to analyze for VOC contamination (Krill and Sonzogni 1986). Both community and private wells were included in the monitoring program, the latter being selected based upon their proximity to suspected sources of VOCs and the soil characteristics in the area. The study found 6% of the community supply wells and 13% of the private wells had detectable levels of VOCs. The most frequently detected contaminants were TCE, PCE, 1,1,1-TCA, toluene, 1,1,2-TCA, benzene, 1,2-DCE, and ethylbenzene. Future monitoring efforts for water-supply wells will be focused near probable sources of contaminants.

The Wisconsin Department of Natural Resources monitored the occurrence of VOCs in ground water at 39 municipal solid-waste landfills and 12 industrial solid-waste landfills (Friedman 1988; Battista and Connelly 1989). Samples were collected from upgradient and downgradient monitoring wells and from leachate collection systems at the landfills. VOC-contaminated ground water occurred at 67% of the municipal landfills and 25% of the industrial landfills. Municipal landfills without clay liners comprised the majority of landfills where VOCs were detected in ground water. The detected compounds included 1,1-DCA, TCE, PCE, 1,1,1-TCA, benzene, trichlorofluoromethane, toluene, xylene, chloroethane, and both isomers of 1,2-DCE. The sources of VOC contamination at municipal landfills are assumed to be household wastes such as small-sized solvent, fuel, paint, and refinishing containers (Battista and Connelly 1989). The Department of Natural Resources has recommended continued VOC monitoring for ground water and leachates at Wisconsin landfills.

State agencies in Nebraska started monitoring ground water for the presence of organic chemicals in 1982 (Goodenkauf and Atkinson 1986). Even though Nebraska was not expected to have significant VOC contamination due to the low population density and limited industrial development, results of the statewide ground-water-monitoring programs indicated VOC occurrence comparable to other national and state surveys. As of 1984, 16.5% of the sampled public-water-supply wells and 15.9% of the sampled private wells contained detectable levels of VOCs. The most frequently detected VOCs in public-water-supply wells were carbon tetrachloride, TCE, and chloroform, and the most frequently detected VOCs in private wells were *cis/trans*-1,2-dichloroethene, PCE, and 1,1,1-trichloroethane. The occurrence of VOCs in Nebraska's ground water was attributed to generally permeable soils and shallow aquifers, poor storage and handling of organic chemicals, and some bias in the selection of ground-water sampling sites in areas at risk for ground-water contamination.

Although monitoring data from only a few states besides California are reviewed here, it is clear that the halogenated solvents are the most commonly detected organic ground-water contaminants. While this fact is inescapable and should strongly influence the target list of analytes for future ground-water monitoring efforts, the following information reminds us not to ignore other organic contaminants in attempts to define current and potential risks to public health.

LIMITATIONS OF EXISTING GROUND-WATER QUALITY DATA

Based upon their evaluation of ground-water-monitoring data collected in California over the past decade, Smith, Lashgari, and Mackay (1991) noted a number of limitations to existing ground-water-quality data which affect the conclusions that may be drawn to data and which should be considered in the design of future monitoring efforts. These limitations are likely to be applicable to previous and ongoing monitoring efforts worldwide.

These potential or actual limitations of existing ground-water-quality data fall into two categories: (1) accuracy of the data and (2) completeness of the understanding of the types, distribution, and environmental behavior of all organic contaminants of public health concern.

Overall, data accuracy has certainly improved through the years, because of the marked advances in analytical instrumentation and the creation and expanding conformance to quality-assurance and quality-control procedures (QA/QC). Nevertheless, despite considerable (and costly) efforts, data quality problems persist in some monitoring efforts. In California, at least, state officials still find occasional evidence of faulty sampling techniques (which introduce negative bias for VOCs, resulting in lower reported contamination levels), incorrect preparation and use of blanks (which may cause low concentrations detected in samples to be considered an artifact and thus ignored), and inaccurate reporting of final results (due to a variety of transcription and other errors). Most analytical laboratories have good intralaboratory QA/QC programs and use standard, reliable protocols for analyses. However, few interlaboratory studies have been conducted. Some variability in analytical results is to be expected, an inevitable consequence of the often complex procedures. However, anecdotal reports from California and elsewhere suggest that the results for analyses of splits from a given sample may vary, sometimes widely, from one laboratory to another and even for one laboratory over time.

Despite the problems outlined, it is likely that most of the analytical data, especially from the more recent monitoring studies, are of high enough quality for general monitoring purposes, that is, to establish the approximate concentrations of the targeted analytes in given samples. Of more potential importance to the interpretation of data from monitoring programs, however, are the following issues: (1) whether the subsurface environment is adequately characterized, (2) whether all significant organic contaminant sources are sufficiently understood, (3) whether sampling of ground water yields a complete understanding of the subsurface distribution of contaminants, and (4) whether all of the organic contaminants of public health significance have been detected. The challenges inherent in answering the first three questions have already been addressed. In the following we show, by example, the problems which may in some cases be encountered in detecting all organic contaminants significantly impacting a ground-water resource.

In a sense, the first problem is to discard the common monofocus on the VOCs. Although it is true that they are the most frequently detected ground-water contaminants, it may be a mistake to aim all resources at determining their distribution while ignoring the possibility that other organic contaminants are present. In some cases, of course, there may be other organic contaminants present at the source which are unlikely to pose a significant threat to underlying ground water. For example, PAHs and PCBs are of concern at many industrial and hazardous waste sites because of their presence in wastes or in the soils. These compounds, however, are rarely found to contaminate significant volumes of ground water, presumably since they are strongly sorbed by geologic media and therefore are generally quite immobile. The question that must be considered in such cases is whether there are other organic contaminants present which, while yet to acquire the attention of the involved public or regulators, are actually able to migrate and impact ground water.

Consider the case of a large plume of contaminants emanating from sewage infiltration lagoons at Otis Air Force Base in Cape Cod, Massachusetts. Aside from a wide range of VOCs (Barber et al. 1988), portions of the plume contained relatively high concentrations of relatively nonbiodegradable alkylbenzenesulfonic acid surfactants (Thurman et al. 1986). Such surfactants were used in detergent formulations worldwide during the period 1950–65, after which they were replaced by more biodegradable surfactants in many countries (see Chapter 11). Similar plumes of detergents have been reported from investigations of septic tank leach fields (Robertson, Cherry, and Sudicky 1991) and should be anticipated in many situations in which domestic wastes have been disposed through leach fields or infiltration lagoons.

Another case is that of the Stringfellow Acid Pits, a hazardous-waste-disposal site in Southern California. Much of the regulatory and monitoring attention at this site had originally been directed to VOCs, particularly TCE, because the TCE plume extended relatively far from the site impacting a ground-water resource used for domestic supply. However, the sum of the concentrations of detected halogenated organic contaminants was noted by the California Department of Health Services to be much less than the total organic halogen concentration (TOX is determined by a technique which detects halogen atoms such as chlorine, bromine, etc., which are present in organic compounds in the sample). This suggested the presence of substantial concentrations of previously undetected organic contaminants, and, indeed, after considerable analytical work, it was determined that high concentrations (mg/L range) of 4-chlorobenzenesulfonic acid were present in portions of the plume (Kim et al. 1990). This and other detected chlorinated aromatic sulfonic acids were presumed to be byproducts of DDT manufacture, probably present in the "sulfuric acid" wastes disposed at Stringfellow (Kim et al. 1990). Similar compounds were found in ground water beneath BKK hazardous waste landfill, also in Southern California (Kim et al. 1990). In both cases, the VOCs accounted for only a few percent of the total halogenated organic contaminants.

There is evidence from other hazardous waste sites that the organic contaminants present in leachates and sometimes in ground water can be numerous and poorly characterized. For example, Bramlett et al. (1987) attempted to characterize leachates from 13 hazardous waste sites. On average, they found that only a few percent of the total organic carbon content of the leachates could be identified. In the identified fraction, only about 20% were VOCs. In part because the VOCs are so easily identified by comparison to the myriad other contaminants present, they are generally targeted for use as analytes in monitoring intended for screening purposes, that is, to scan for the probable presence of organic contamination. While this makes considerable economic sense, our point here is that it must not be forgotten that the VOCs detected in ground water may often be the "tip of the iceberg" when complex contaminant sources are nearby, and may in fact not represent the most significant public health threat.

CONCLUDING REMARKS

The examples and discussions illustrate that it is crucial, during investigations of organic chemical contamination of ground water, to formulate conceptual models which incorporate the best understanding of hydrogeologic features of significance as well as the most reasonable assumptions regarding the known or probable contaminant sources. Although volatile organic chemicals have been the most commonly detected contaminants, it is important to explore whether other classes of chemicals may also be presenting significant threats to ground water. In some cases, these other organic chemicals may present greater health risks than the VOCs. In other cases, particularly in which significant DNAPL penetration of the saturated zone has occurred, the VOCs may present by far the most persistent impact.

In designing a monitoring effort, the first conceptual model may have to be based on rather scant site-specific information. However, revisions to the model should be made consistent with information gathered during the investigation itself and, of course, with the rapidly expanding base of scientific knowledge. Only in this way is it likely that a reasonably complete understanding of the three-dimensional distribution and probable future impact of organic contaminants will be obtained. In less favorable cases, in which monitoring is undertaken with only a vague conceptual model in mind, it is likely that time and resources will be wasted in generating a data base of uncertain completeness. There is little excuse for the latter case any more, given the recent advances in monitoring technologies and in the scientific understanding of the types and behavior of organic contaminants which threaten our ground-water resources.

ACKNOWLEDGMENTS

Much of the material presented herein was drawn from work funded in part by the University of California Toxic Substances Control Program through its support of a project entitled "Nonconventional Pollutants in Raw and Treated Groundwater" (D. Mackay, principal investigator), the William and Flora Hewlett Foundation through a grant for scientific and public policy studies to the UCLA Environmental Science and Engineering Program (D. Mackay, former faculty member), the University of California Water Resources Center through various grants to D. Mackay, and the University of Waterloo Centre for Groundwater Research through its support of research and teaching by D. Mackay.

References

Barber, L. B., II, E. M. Thurman, M. P. Schroeder, and D. R. LeBlanc. 1988. Long-term fate of organic micropollutants in sewage-contaminated groundwater. *Environmental Science and Technology* 22(2):205–11.

Barker, J. F., G. C. Patrick, and D. Major. 1987. Natural attenuation of aromatic hydrocarbons in a shallow sand aquifer. *Ground Water Monitoring Review* Winter: 64–71.

Battista, J. R., and J. P. Connelly. 1989. *VOC Contamination at Selected Wisconsin Landfills—Sampling Results and Policy Implications*. Wisconsin Department of Natural Resources, PUBL-SW-094 89, June.

Bramlett, J., C. Furman, A. Johnson, W. D. Ellis, H. Nelson, and W. H. Vick. 1987. *Composition of Leachates from Actual Hazardous Waste Sites*. Cincinnati, OH: U.S. Environmental Protection Agency Technical Report EPA/600/2-87/043.

California Department of Health Services. 1986. *Organic Chemical Contamination of Large Public Water Systems in California*. April.

California Department of Health Services. 1990. *Organic Chemical Contamination of Small Public Water Systems; AB 1803 Final Status Report*. Office of Drinking Water. June.

California State Water Resources Control Board. 1989. *Leaking Underground Storage Tank Information System—Quarterly Report*. Division of Loans and Grants. January.

California State Water Resources Control Board. 1990a. *Well Investigation Program, Volatile Organic Chemicals in Public Water Supply Wells*. December.

California State Water Resources Control Board. 1990b. *Ground Water Hot Spots Phase II—Toxic Pollutant Identification, Correction and Prevention*. Regulatory and Monitoring Branch, Division of Water Quality. April.

Cherry, J. A. 1992. Groundwater monitoring: Some deficiencies and opportunities. In *Hazardous Waste Site Investigations; Toward Better Decisions*. Proceedings of the 10th ORNL Life Sciences Symposium, Gatlinburg, TN, eds. R. B. Gammage and B. A. Berven, pp. 119–34. Chelsea, MI: Lewis Publishers.

Cohen, D. B., and G. W. Bowes. 1984. *Water Quality and Pesticides: A California Risk Assessment Program*. Sacramento, CA: California State Water Resources Control Board Report No. 84-6SP.

Commonwealth of Massachusetts. 1986. *Contamination in Municipal Water Supplies*. Special Legislative Commission on Water Supply, State House, Boston, MA. December.

Conant, B. H., and R. W. Gillham. 1992. Field experiments and modeling of vapour transport of trichloroethylene in the unsaturated zone (abstract). *EOS Transactions, American Geophysical Union* 73(14):127.

Council on Environmental Quality. 1981. *Contamination of Ground Water by Toxic Organic Chemicals*. Washington, DC.

Feenstra, S., and J. Cherry. 1988. Subsurface contamination by dense nonaqueous-phase liquid (DNAPL) chemicals. In *Proceedings of the International Groundwater Symposium*, Halifax, Nova Scotia, May 1–4. International Association of Hydrogeologists.

Frick, D., and L. Shaffer. Undated. *Assessment*

of the Availability, Utilization, and Contamination of Water Resources in New Castle County, Delaware. New Castle County, Delaware: Department of Public Works, Office of Water and Sewer Management (for U.S. EPA Office of Solid Waste Management Programs, Contract No. WA-6-99-2061-J).

Friedman, M. A. 1988. *Volatile Organic Compounds in Groundwater and Leachate at Wisconsin Landfills.* Wisconsin Department of Natural Resources, PUBL-WR-192 88. February.

Geraghty and Miller, Inc. 1979. *Investigations of Ground Water Contamination in South Brunswick Township, N. J.* Syosset, New York: Geraghty and Miller, Inc.

Goodenkauf, O., and J. C. Atkinson. 1986. Occurrence of volatile organic chemicals in Nebraska ground water. *Ground Water* 24(2):231–3.

Graf, C. G. 1986. VOCs in Arizona's ground water: A status report. In *Proceedings of the Conference on Southwestern Ground Water Issues,* October 20–22, 1986, Tempe, Arizona, pp. 269–87. Dublin, OH: National Water Well Association.

Hadley, P. W., and R. Armstrong. 1991. "Where's the benzene?"—Examining California ground-water quality surveys. *Ground Water* 29(1):35–40.

Kim, I. S., F. I. Sasinos, R. D. Stephens, and M. A. Brown. 1990. Anion-exchange chromatography particle beam mass spectrometry for the characterization of aromatic sulfonic acids as the major organic pollutants in leachates from Stringfellow, California. *Environmental Science and Technology* 24(12):1832–6.

Kopania, A. A. 1991. *Geologic Heterogeneity and Groundwater Contamination: Guidelines for Determination of Significance to Investigation and Remediation.* D. Env. thesis, UCLA School of Public Health, Environmental Science and Engineering Program.

Krill, R. M., and W. C. Sonzogni. 1986. Chemical monitoring of Wisconsin's groundwater. *Journal of the American Water Works Association* 78(9):70–5.

Lin, S-F. 1990. *Soil Gas Analysis for the Investigation of Groundwater Contamination by Volatile Organic Chemicals.* D. Env. thesis. UCLA School of Public Health, Environmental Science and Engineering Program.

Litwin, Y. J., N. N. Hantzsche, and N. A. George. 1983. *Groundwater Contamination by Pesticides: A California Assessment.* Sacramento, CA: California State Water Resources Control Board.

Luba, L. 1992. *Study of an Extensive Contaminant Plume near Regina Saskatchewan.* M. Sc. Report, Department of Earth Sciences, University of Waterloo, Waterloo, Ontario.

Mackay, D. M. 1990. Characterization of the distribution and behavior of contaminants in the subsurface. In *Ground Water and Soil Contamination Remediation: Toward Compatible Science, Policy and Public Perception,* pp. 70–90. Washington, DC: National Academy Press, Water Science and Technology Board.

Mackay, D. M., and J. A. Cherry. 1989. Groundwater contamination: pump-and-treat remediation. *Environmental Science and Technology* 23(6):630–6.

Mackay, D. M., and L. A. Smith. 1990. Agricultural chemicals in groundwater: Monitoring and management in California. *Journal of Soil and Water Conservation* 45(2):253–5.

Mackay, D. M., D. L. Freyberg, P. L. McCarty, P. V. Roberts, and J. A. Cherry. 1986. A natural gradient experiment on solute transport in a sand aquifer. I: Approach and overview of plume movement. *Water Resources Research* 22(13):2017–30.

Mackay, D., M. Gold, and G. Leson. 1988. Current and prospective quality of California's ground water. In *Proceedings of the 16th Biennial Conference on Ground Water,* ed. J. J. Devries, pp. 97–110. Water Resources Center Report No. 66.

Mackay, D. M., P. V. Roberts, and J. A. Cherry. 1985. Groundwater transport of organic contaminants in sand and gravel aquifers. *Environmental Science and Technology* 19(5):384–92.

Magara, Y., and T. Furuichi. 1986. Environmental pollution by trichloroethylene and tetrachloroethylene: A nationwide survey. In *New Concepts and Developments in Toxicology,* eds. P. L. Chambers, P. Gehring, and F. Sakai, pp. 231–43. New York: Elsevier.

Mendoza, C. A., and T. A. McAlary. 1989. Modeling of ground-water contamination caused by organic solvent vapors. *Ground Water* 28(2):199–206.

Osiensky, J. L., G. V. Winter, and R. W. Williams. 1984. Monitoring and mathematical modeling of contaminated ground-water plumes in fluvial environments. *Ground Water* 22(3):298–306.

Poulsen, M. M., and B. H. Kueper. 1992. A field experiment to study the behaviour of tetrachloroethylene in unsaturated porous media. *Environmental Science and Technology* 26(5):889–95.

Pye, V. I., and J. Kelley. 1984. The extent of groundwater contamination in the United States. In *Groundwater Contamination*, ed. Geophysics Study Committee, National Research Council, pp. 23–33. Washington, DC: National Academy Press.

Rifai, H. S., P. B. Bedient, J. T. Wilson, K. M. Miller, and J. M. Armstrong. 1988. Biodegradation modeling at aviation fuel spill site. *Journal of Environmental Engineering* 114(5):1007–29.

Rivett, M. O., and J. A. Cherry. 1991. The effectiveness of soil gas surveys in delineation of groundwater contamination: Controlled experiments at the Borden site. In *Proceedings of the Conference on Petroleum Hydrocarbons and Organic Chemicals in Ground Water*, pp. 107–24. Worthington, OH: National Water Well Association.

Rivett, M. O., S. Feenstra, and J. A. Cherry. 1991. Field experimental studies of a residual solvent source emplaced in the groundwater zone. In *Proceedings of the Conference on Petroleum Hydrocarbons and Organic Chemicals in Ground Water*, pp. 283–99. Worthington, OH: National Water Well Association.

Robertson, W. D., J. A. Cherry, and E. A. Sudicky. 1991. Groundwater contamination from two small septic systems on sand aquifers. *Ground Water* 29(1):82–92.

Schwille, F. 1988. *Dense Chlorinated Solvents in Porous and Fractured Media: Model Experiments*. Chelsea, MI: Lewis Publishers.

Smith, L. A., K. P. Green, and D. M. Mackay. 1990. Quality of ground water in California: Overview and implications. In *Proceedings of the 17th Biennial Conference on Ground Water*, ed. J. J. Devries, pp. 93–107. University of California Water Resources Center, Report No. 72.

Smith, L. A., A. Lashgari, and D. M. Mackay. 1991. The health of California's groundwater: Implications for drinking water supplies. In *Proceedings of the Conference on Protecting Drinking Water at the Source*, pp. 13–23. University of California Water Resources Center, Davis, CA, Report No. 76.

Thurman, E. M., L. B. Barber II, and D. L. LeBlanc. 1986. Movement and fate of detergents in groundwater: A field study. *Contaminant Hydrology* 1:143–62.

U.S. Environmental Protection Agency. 1987. Inert ingredients in pesticide products; Policy statement. *Federal Register*, 52(77), Notices: 13305–09, April 22.

U.S. Environmental Protection Agency. 1992. *Estimating Potential for Occurrence of DNAPL at Superfund Sites—Quick Reference Fact Sheet*. Ada, OK. R. S. Kerr Environmental Research Laboratory, Office of Solid Waste and Emergency Response Publication 9355, 4-07FS.

Vogel, T. M., and P. L. McCarty. 1985. Biotransformation of tetrachloroethylene to trichloroethylene, dichloroethylene, vinyl chloride and carbon dioxide under methanogenic conditions. *Applied and Environmental Microbiology* 49(5):1080–3.

Vogel, T. M., and P. L. McCarty. 1987. Rate of abiotic formation of 1,1-dichloroethylene from 1,1,1-trichloroethane in groundwater. *Journal of Contaminant Hydrology* 1(3):299–308.

Westrick, J. J. 1990. National surveys of volatile organic compounds in ground and surface waters. In *Significance and Treatment of Volatile Organic Compounds in Water Supplies*, eds. Neil M. Ram, Russell F. Christman, and Kenneth P. Cantor, pp. 103–38. Chelsea, MI: Lewis Publishers.

14

Pesticides

P. S. C. Rao and William M. Alley

INTRODUCTION

Pesticide contamination of ground water has become a major environmental issue. Present-day concerns began largely with the 1979 discoveries of aldicarb in ground water in New York, and dibromochloropropane (DBCP) in California, the subsequent finding of these pesticides in ground water in other states, and the widespread detection of another pesticide, ethylene dibromide (EDB), in ground water in the early 1980s in California, Florida, Georgia, and Hawaii. Pesticides were also reported in ground water prior to these findings (cf. Richard et al. 1975). Hallberg (1989) and Ritter (1990) review reported occurrences of pesticide contamination of ground water in the United States.

The purpose of this chapter is to review knowledge of pesticides in ground water in the context of regional-scale studies. The chapter is divided into four sections. The first, and longest, section focuses on how pesticides may be transported *to* ground water. This is necessary, because to understand the problems of pesticides in ground water, one needs to understand the mechanisms whereby they enter ground water.

The second section continues this idea with an overview of methods for assessing regional vulnerability of ground water to pesticide contamination. The third section reviews selected studies of the distribution of pesticides in ground water, and the chapter closes with some implications for regional-scale investigations of pesticides. Although some inorganic constituents, such as arsenic and sulfur, are used as pesticides, this chapter will focus exclusively on organic chemicals.

A number of recent texts can be consulted for further information on pesticide chemistry, processes in the subsurface, modeling, and assessment of impacts, including those by Biggar and Seiber (1987), Grover (1988), Grover and Cessna (1988), and Cheng (1990).

FACTORS INFLUENCING CONTAMINATION POTENTIAL

Many factors affect the spatial and temporal variability of the delivery of pesticides to ground water. These are grouped for discussion purposes under three topics: (1) pesticide behavior and fate in the vadose zone, (2) point-source contamination, and (3) agricultural practices and pesticide use.

Pesticide Behavior and Fate in the Vadose Zone

Processes affecting the behavior and fate of pesticides in the vadose zone include phase-transfer processes, transformation processes, and transport processes (see Chapter 7). *Phase-transfer processes* involve the movement of organic chemical from one environmental media to another, including water-to-solid transfer (sorption), air-to-solid transfer (vapor sorption), and water-to-air transfer (volatilization). Phase transfer is mainly considered to be a physical process. *Transformation processes*, on the other hand, result in changes in the chemical structure of a pesticide. Transformation processes involve both abiotic (chemical) and biotic (microbiological) pathways by which a pesticide is transformed to metabolites and carbon dioxide. *Transport processes* move the contaminant through the vadose zone, and include aqueous- and vapor-phase transport. Aqueous-phase transport is generally viewed as the primary contributor to downward leaching to ground water, but also can cause upward movement due to evaporation and plant uptake. Likewise, vapor-phase transport can lead to downward movement or to upward dispersal of pesticide residues into the atmosphere. As discussed in Chapter 7, chemical transport in association with water in the subsurface can involve complex mixtures of water, dissolved constituents, colloids, and organic fluids.

Further complicating these interacting processes are the different rates at which the individual processes take place. Differences in chemical structures, soil properties and conditions, weather and climatic conditions, microbiological conditions, plant types and growth stages, and soil-crop-pesticide management practices all affect the rates of the processes and the ultimate fate of pesticides.

Ideally, an assessment of the behavior and fate of a pesticide in the vadose zone would incorporate a complete assessment of all of the aforementioned processes. Unfortunately, because the processes and their interactions are so complex and difficult to characterize, greatly simplified representations must be used and some processes ignored. This is particularly true at the regional scale in comparison to a detailed site investigation.

A comprehensive discussion of the processes and factors controlling the behavior and fate of pesticides in the vadose zone is beyond the scope of this chapter. Instead, the discussion will first focus on two factors —sorption and degradation rate—and how simple representations of these two factors can be utilized to evaluate the likelihood of ground-water contamination with pesticides. Next, complicating factors related to transport processes; namely, the episodic nature of contaminant transport through the vadose zone and the presence of preferential flow paths will be discussed. Finally, losses from surface runoff, volatilization, and plant uptake will be considered briefly.

Sorption

The retention of pesticides and other organic chemicals by natural sorbents (soils, clays, and sediments) is usually referred to as sorption, without specifying the nature of the interaction between the solute and the sorbent. Sorption includes both *ad*sorption at the solid-liquid interfaces (e.g., on clay mineral surfaces) and *ab*sorption into the interior of the sorbent matrix (e.g., uptake into organic matter). The latter process is generally thought to be more dominant than the former for many nonpolar and nonionic organic chemicals (Chiou, Porter, and Freed 1983). However, sorption on mineral surfaces plays an important role, especially in soils and sediments with low-organic-carbon content, and for ionic or ionizable organic chemicals (Mingelgrin and Gerstl 1983). Soil organic-carbon content usually decreases with depth, whereas clay content often increases with depth. Thus, sorption on clay and other mineral surfaces may increase in relative importance in deeper parts of the soil profile and in the intermediate vadose zone.

Sorption of pesticides is commonly quantified using a sorption coefficient (K_d; typical units of mL/g), which is the ratio of the sorbed-phase concentration (units of μg/g) to the solution-phase concentration (units of μg/mL) at equilibrium. Use of a sorption coefficient requires assumptions that the sorption isotherm is linear and reversible, and that instantaneous equilibrium is achieved. Although data collected on pesticide sorption raise questions about each of these simplifying assumptions (cf. Rao and Davidson 1980; Rao, Berkheiser, and Ou 1984; Green and Karickhoff 1990), the approximation appears to be adequate for many practical applications.

The value of the sorption coefficient, K_d, can vary over several orders of magnitude, as determined by a number of physical and chemical properties of both the sorbent (soil, sediment, or organic carbon) and the sorbate (pesticide). For many pesticides, the variations in K_d among different sorbents can be reduced considerably by normalizing K_d to the organic-carbon content of the sorbent. This organic-carbon-normalized sorption coefficient is referred to as K_{oc} (units of mL/g). The two sorption coefficients are related by $K_d = K_{oc} f_{oc}$, where f_{oc} is the fraction of soil organic carbon on a mass basis; that is

$$f_{oc} = \frac{\text{mass of organic carbon}}{\text{mass of dry soil or sediment}}$$

Pesticides with a larger K_{oc} are expected to be retarded to a greater extent, hence, have longer residence times in the vadose zone. For a given pesticide, the soil-specific sorption coefficient, K_d, increases with increasing soil organic-carbon content.

A number of assumptions are inherent in the use of K_{oc} to represent sorption of pesticides. First, normalizing K_d to the organic-carbon content of the sorbent will not always reduce the variation in K_d, especially for low-organic-matter-containing materials, such as are commonly present in the intermediate vadose zone (and in the saturated zone). Furthermore, the approach is most appropriate for nonionic pesticides. Sorption of ionizable pesticides (e.g., 2,4-D, picloram) is strongly affected by soil pH. For example, anions are repelled by the generally negative charge balance at soil surfaces; thus, sorption of weak acids decreases as the proportion of chemical present in the anionic form increases with increasing pH (see Figure 7-5 and accompanying text in Chapter 7). Conversely, sorption of organic bases (e.g., atrazine, quinoline) increases dramatically with decreasing pH as the protonated species is sorbed via cation-exchange mechanisms.

Use of K_{oc} requires the assumption that sorption processes are at equilibrium, which will occur only if water flow is sufficiently slow and sorbing materials are uniformly distributed throughout the porous medium. Various factors control the rate at which sorption equilibrium is achieved for organic chemicals. Mathematical models to describe nonequilibrium sorption have been reviewed by Brusseau and Rao (1989, 1991). The assumption of equilibrium sorption is most likely to be violated in cases of preferential flow in solution channels and macropores.

Compilations of K_{oc} values for a large number of pesticides have been published and updated (Hamaker and Thompson 1972; Rao and Davidson 1980; Lyman 1990; Wauchope et al. 1991). Some typical values of K_{oc} for pesticides that are applied to Florida are listed in Table 14-1. These data are used later in an example for the use of K_{oc} values. The pesticides listed do not include some common pesticides that are presently banned, such as DBCP and EDB.

Considerable variation exists among reported values of K_{oc} for a given pesticide. Coefficients of variation (CV) in K_{oc} values compiled from the literature for specific pesticides range from 44% to 256% (Gerstl 1990). Values of K_{oc} for a pesticide often approximate a lognormal distribution, such that the CVs for log K_{oc} values are

TABLE 14-1. Sorption and Persistence in Soils for Selected Pesticides Used in Florida

Number on Figure 14-1	Common Name	Trade Name(s)[a]	K_{oc} (mL/g)	$t_{1/2}$ (days)
1	Dalapon	Basfapon, Dowpon	1	30
2	Dicamba	Banvel	2	14
3	Chloramben	Amiben	15	14
4	Picloram	Tordon	16	90
5	Metalaxyl	Ridomil	16	21
6	Carbofuran	Furadan, Curaterr	22	50
7	Oxamyl	Vydate	25	4
8	Aldicarb	Temik	30	30
9	Bromacil	Hyvar, Bromax	32	60
10	Metsulfuron-Methyl	Ally, Escort	35	30
11	Terbacil	Sinbar	55	120
12	Fomesafen	Flex	60	100
13	Ethoprop	Mocap	70	25
14	2,4,5-T	Dacamine 4T, Trioxone	80	24
15	Fensulfothion	Dasanit	89	33
16	Atrazine	Attrex	100	60
17	Fluometuron	Cotoran, Lanex	100	85
18	Chlorimuron-Ethyl	Classic	110	40
19	Simazine	Princep	130	60
20	Prometon	Pramitol	150	500
21	Propazine	Milogard, Primatol P	154	135
22	Alachlor	Alanex	170	15
23	Cyanazine	Bladex	190	14
24	Metolachlor	Bicep	200	90
25	Propham	Ban-Hoe, Chem-Hoe	200	10
26	Captan	Orthocide, Captanex	200	3
27	Diphenamid	Enide, Rideon	210	30
28	Monolinuron	Aresin, Afesin	284	321
29	Ametryn	Evik	300	60
30	Carbaryl	Sevin	300	10
31	Dichlobenil	Casoron	400	60
32	Linuron	Lorox, Aflon	400	60
33	Prometryn	Caparol, Primatol Q	400	60
34	Chlorpropham	Beet-Kleen, Furloe	400	30
35	Diuron	Karmex	480	90
36	Isofenphos	Oftanol	600	150
37	Fonofos	Dyfonate	870	40
38	Chlorbromuron	Maloran	996	45
39	Iprodione	Rovral	1,000	14
40	Phorate	Thimet	1,000	60
41	Azinophos-Methyl	Guthion	1,000	10
42	Fluridone	Sonar	1,000	21
43	Cacodylic Acid	Bolate, Bolls-Eye	1,000	50
44	Diazinon	Basudin, Spectracide	1,000	40
45	Lindane	Isotox	1,100	400
46	Cyhexatin	Plictran	1,380	180
47	Procymidone	Sumilex	1,650	120
48	Chloroneb	Terraneb	1,650	130
49	Malathion	Cythion	1,800	1
50	Benomyl	Benlate	1,900	67
51	Chloroxuron	Tenoran, Norex	3,000	60
52	Ethalfluralin	Solanan	4,000	60
53	Parathion Ethyl	Thiophos, Bladan	5,000	14
54	Methyl Parathion	Penncap-M, Metacide	5,100	5

TABLE 14-1. *Continued*

Number on Figure 14-1	Common Name	Trade Name(s)[a]	K_{oc} (mL/g)	$t_{1/2}$ (days)
55	Fenvalerate	Extrin, Sumitox	5,300	35
56	Esfenvalerate	Asana	5,300	35
57	Chlorpyrifos	Lorsban, Dursban	6,070	30
58	Trifluralin	Treflan	8,000	60
59	Ethion	Ethion	10,000	150
60	Endosulfan	Thiodan, Endosan	12,400	50
61	Glyphosate	Roundup	24,000	47
62	Fluvalinate	Mavrik, Spur	1,000,000	7

Source: Data compiled from laboratory studies; after Rao and Hornsby (1989), updated with values from Wauchope et al. (1991).

[a] The use of trade or product names is for identification purposes only, and does not constitute endorsement by the University of Florida or the U.S. Geological Survey

on the order of 5% to 34% (Rao and Davidson 1980; Gerstl 1990). The variations in K_{oc} values are attributed to Green and Karickhoff (1990): (1) variations in the nature of organic carbon (e.g., functional groups, aromaticity, origin, degree of humification); (2) the contribution from other sorbent surfaces and the contribution of other soil factors (e.g., clay type and content, cation exchange capacity, pH, and surface area); and (3) differences in procedures used to measure K_{oc}.

Various attempts have been made to correlate K_{oc} values to physical-chemical properties of pesticides. Among the most successful of these is the inverse relation between log K_{oc} and log S_w (aqueous solubility) or a direct correlation between log K_{oc} and log K_{ow} (the octanol-water partition coefficient; see Chapter 7). Based on analysis of K_{oc}, K_{ow}, and S_w data compiled for more than 400 pesticides, Gerstl (1990) concluded that regression equations of log K_{oc} versus log S_w (or versus log K_{ow}) for individual groups of chemicals were preferred over a single equation for all pesticide classes. Gerstl noted, however, that the differences among groups of chemicals could be taken into account through the addition of a "group correction term" to the single regression equation. Green and Karickhoff (1990) proposed a detailed scheme for estimating sorption coefficients that recognizes the need for alternative procedures for nonhydrophobic chemicals.

The uncertainty associated with estimates of sorption coefficients must be recognized in regional-scale assessments of pesticide behavior in the vadose zone (cf. Loague et al. 1989). Green and Karickhoff (1990, p. 96) note that

The principal justification for adopting simplified approaches that are known to contribute additional uncertainty to modeling results is that a reasonable estimate of sorption is better than no estimate at all. Additionally, modelers may be inclined to rationalize that errors in sorption estimates are relatively small in comparison to parameter estimates for other processes involved in the modeling of pesticide fate in soils and water.... We have encouraged model users to evaluate for each application the impact of uncertainty arising from errors in sorption estimates.

Degradation

Pesticide degradation in the vadose zone involves both abiotic and biotic processes. While chemical pathways can lead to a transformation of the parent compound and the production of intermediate metabolites, only biotic pathways result in complete mineralization of pesticides to carbon dioxide, water, and other simple products. Thus, two types of degradation rates can be defined: one based on the loss of parent compound (the "primary degradation" or "disappearance" rate), and another based on CO_2 evolution (the "ultimate degradation" or "mineralization" rate). The difference between these two rates is an indicator of the rate of accumulation of intermediate metabolites, some of which may be of environmental concern.

Most of the literature on biodegradation of pesticides (and other organic chemicals) is descriptive and deals with identification of the microorganisms, enzyme systems, and metabolic pathways (cf. Smith 1988; Bollag and Liu 1990; Wolfe, Mingelgrin, and Miller 1990). According to Scow (1990), quantitative data on degradation rates are limited, because (1) standard experimental protocols for measuring biodegradation are lacking; (2) the results are not comparable and apply only to a particular set of experimental conditions; and (3) the variables that control degradation rates are not well understood and they have not been examined across a broad spectrum of chemical classes.

Although a number of models have been proposed for describing pesticide degradation kinetics, the most widely used is the pseudo-first-order kinetic model. Values for first-order rate constants (k; typical units of day^{-1}) or half-lives ($t_{1/2} = .693/k$) have been reported for degradation of several pesticides in soils. Some compilations of degradation rate constants may be found in Rao and Davidson (1980), Rao, Berkheiser, and Ou (1984), Nash (1988), and Scow (1990). More recently, extensive compilations of degradation rate constants for pesticides (Wauchope et al. 1991) and a broad range of organic compounds (Howard et al. 1991) have been published. Half-lives for pesticide degradation in soils range from a few days or less to many years. Some typical values of half-lives for pesticide degradation in surface soils are shown in Table 14-1.

The more careful documentations of rate constants differentiate between k values for abiotic and biotic pathways, and distinguish between k values for biotic processes under aerobic and anaerobic conditions. Generally, however, compiled values for k and $t_{1/2}$ (including those values listed in Table 14-1) reflect overall kinetics of parent chemical loss, and are commonly averaged over data reported for diverse soils and environmental conditions. Thus, published compilations of k and $t_{1/2}$ values must be used with considerable caution, and only as initial estimates in the absence of site-specific and compound-specific data.

Many of the published data for pesticide degradation are based on laboratory studies. Problems are encountered in extrapolating these laboratory-measured degradation rates to predict field-scale degradation of pesticides. Among the reasons cited to explain observed differences in laboratory- and field-measured degradation rates (Scow 1990) are that (1) neither the microbial populations nor their activity is likely to be similar between field and controlled laboratory studies; and (2) high substrate and nutrient concentrations used in laboratory studies for experimental expedience do not reflect conditions likely to be encountered in nature. Such arguments are presented to suggest that pesticide loss via degradation under field conditions might be smaller than that predicted from laboratory studies. On the other hand, additional pathways of loss under field conditions (e.g., volatilization, plant uptake, and so forth) can result in field-measured dissipation rates (i.e., loss of parent compound) that exceed degradation rates measured under laboratory conditions.

Many of the published degradation rates for pesticides are based on studies using surface soils (0–30-cm depth). Very limited data are available for pesticide degradation in subsoils and aquifers. A common assumption is that degradation rate constants decrease with increasing depth because microbial activity and diversity are likely to decrease at greater depths. Thus, the relative importance of chemical pathways is assumed to increase with depth, such that the rate constant for chemical hydrolysis is assumed to represent the lower limit for k in subsoils and aquifers. The validity of this simplifying assumption is being questioned with accumulating evidence that both the diversity and numbers of microbial populations in subsoils and aquifers are greater than those often previously assumed to exist (see Chapter 8).

Published degradation rates are for particular values of temperature and soil-water content. Biodegradation rates may decrease rapidly as soils dry and as the temperature of the soil water decreases; abiotic degradation rates might be faster in drier and hotter conditions. As the soil-water content approaches saturation, and anoxic conditions prevail, the biodegradation rates and intermediate metabolites can be different. Soil pH also will influence the rates of chemical hydrolysis, if the reactions are acid- or base-catalyzed. Mass-transfer processes can have a significant effect on biodegradation by determining the "available" concentration and the rate of pesticide diffusion to microsites containing active microbial populations (Scow and Hutson 1992; Scow and Alexander 1992).

Simple Indices of Residence Time and Degradation Rate

A ratio of the half-life ($t_{1/2}$) to the organic-carbon-normalized sorption coefficient (K_{oc}) for a pesticide can serve as a simple index of its leaching potential, and these two parameters can be used to group pesticides in terms of their relative potentials for ground-water contamination. An example is discussed based on the *attenuation factor* (AF), an index developed by Rao, Hornsby, and Jessup (1985).

The AF index is defined as

$$\text{AF} = \exp\left[\frac{-0.693 d \, \text{RF} \, \theta_{FC}}{q t_{1/2}}\right] \quad (14\text{-}1)$$

where d is the distance from the land surface to ground water (or to some other reference location); θ_{FC} is the volumetric water content at field capacity (where field capacity is the idealized water content after a soil has been thoroughly wetted and allowed to drain freely for several days); q is the net ground-water recharge rate; and RF is the *retardation factor* accounting for pesticide sorption effects. RF is defined as

$$\text{RF} = \left[1 + \frac{\rho_b f_{oc} K_{oc}}{\theta_{FC}}\right] \quad (14\text{-}2)$$

where ρ_b is the soil bulk density (g/cm^3).

The AF index is equivalent to the fraction of the applied pesticide mass that is likely to leach past the chosen reference depth, d. The AF value varies between 0 and 1, with larger values indicating a greater contamination potential. Simplifying assumptions made in deriving Eqs. 14-1 and 14-2 are discussed by Rao, Jessup, and Hornsby (1985).

Substitution of Eq. 14-2 into Eq. 14-1, and after algebraic manipulation and taking logarithms on both sides yields the following expression (Rao, Jessup, and Alley 1992):

$$\text{Log } t_{1/2} = \text{Log } (\theta_{FC} + \rho_b f_{oc} K_{oc}) + F \quad (14\text{-}3)$$

where

$$F = -\text{Log}\left[\frac{2.303(\text{Log } 1/\text{AF})}{0.693(d/q)}\right] \quad (14\text{-}4)$$

By choosing the numerical values for reference depth (d), recharge rate (q), and the extent of leaching (AF), the term F can be calculated from Eq. 14-4. Given soil-specific parameters required in the first term (θ_{FC}, ρ_b, and f_{oc}), Eq. 14-3 can be plotted and superimposed on a plot of published values for $t_{1/2}$ and K_{oc} for several pesticides.

A plot generated in this manner is shown as Figure 14-1. Published data for properties of four major soil types found in Florida and data for the 62 pesticides listed in Table 14-1 were used to generate Figure 14-1. The soils used in this example cover a wide range in relevant properties for major soil types found in Florida (Table 14-2), and the 62

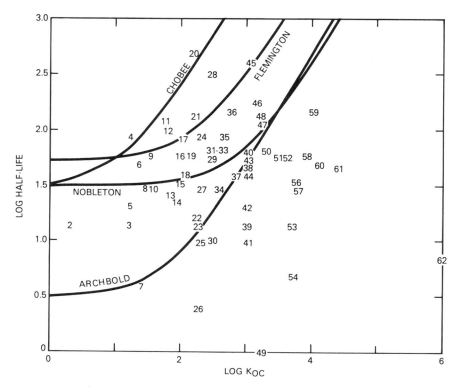

FIGURE 14-1. Log-log plot for four major soil types in Florida of the combinations of half-life ($t_{1/2}$; days) and organic-carbon-normalized sorption coefficient (K_{oc}; mL/g of organic carbon) that result in 0.1% leaching past a depth of 2 m under a recharge rate of 0.2 cm/d. The soil properties used are listed in Table 14-2. The values of $t_{1/2}$ and K_{oc} for 62 common pesticides used in Florida also are plotted on this diagram (the plotted numbers correspond to pesticides listed in Table 14-1).

TABLE 14-2. Depth-Weighted Average Values For Soil Properties Used in Eq. 14-3 for Generating the Four Lines Shown in Figure 14-1

Soil Type	θ_{FC}	$\rho_b f_{oc}$
Archbold sand	0.032	0.0005
Nobleton fine sand	0.326	0.0004
Flemington fine sandy loam	0.546	0.0029
Chobee loamy sand	0.312	0.0239

pesticides selected span a broad range of K_{oc} and $t_{1/2}$ values. The net recharge value (q) was set at 0.2 cm/day (a typical value for Florida soils), and the value of d was set at 2 m (a typical maximum depth for the crop root zone and deepest sampling depth in soil-characterization studies conducted by the Soil Conservation Service). The value of Log(1/AF) was set at 3.0, which is equivalent to setting AF = 10^{-3}; that is, 0.1% of the applied pesticide mass is assumed to leach past the depth of interest (d).

For each soil, those pesticides lying to the right of the solid line have relatively low potential for contaminating ground water, because they have sufficiently long residence times and/or short half-lives. Those pesticides that lie above the line have relatively large contamination potential, because they degrade slowly and/or leach rapidly—that is, for a particular soil, pesticides with data values lying to the right of the curve for that soil are predicted to have less than 0.1% of the applied pesticide mass leach past the 2-m depth, and pesticides with data values lying to the left of the curve for that soil are predicted to have greater than 0.1% of the applied pesticide mass leach past the 2-m depth.

Some pesticides lie to the right of all curves and, thus, have relatively little potential for leaching through any of the soils. Other pesticides lie to the left of all curves and thus have relatively high potential for leaching through any of the soils. The region between the Chobee and Archbold soils in Figure 14-1 covers essentially all mineral soils occurring in Florida. Pesticides with data values falling between these two curves, may or may not have a high leaching potential dependent upon the local soil type. A large number of data points fall within this region. Ideally, data points plotted for each pesticide in a graph such as Figure 14-1 would include error bars to indicate uncertainty. Likewise, the curves plotted for different soils would ideally be plotted as bands to indicate uncertainty.

Similar relations to those shown in Figure 14-1 have been produced by several authors. For example, Jury, Focht, and Farmer (1987) used a similar model, but with depth-dependent biodegradation. Gustafson (1989) developed an empirical index of K_{oc} and $t_{1/2}$ to identify which pesticides would leach and which would not. Boesten and van der Linden (1991) examined effects of temperature dependence of the degradation rate and time of pesticide application on development of relations such as those shown in Figure 14-1.

In summary, the likelihood for groundwater contamination can be viewed as a competition between how fast the pesticide leaches and how fast it can be degraded. The net result of these two processes determines what fraction of soil-applied pesticides potentially reaches ground water. The effects of shorter residence times can be tempered by rapid degradation rates, while longer residence times can compensate for slower degradation rates. Given site-specific soil and climatic characteristics and pesticide properties, the approaches discussed above offer a means of ranking pesticides in terms of their relative contamination potentials.

Equations 14-1 to 14-4 are highly simplified, and are intended to be used as screening tools rather than as predictive models of actual contamination of ground water with pesticides. They strictly apply only under steady, unsaturated water-flow conditions and to porous media that approximate a uniform homogeneous ideal medium. In natural settings, these conditions rarely occur because of (1) the episodic nature of contaminant transport through the vadose zone, (2) the presence of preferential flow paths and heterogeneity of the porous media. These complications are discussed next.

Effects of Episodic Water Flow and Solute Transport

The assumption of steady, unsaturated flow conditions may hold reasonably well in the intermediate vadose zone, at least for situations where the water table is sufficiently deep. Water flow in the root zone, however, is usually far from steady. Here, outputs from the Pesticide Root-Zone Model (PRZM; Carsel et al. 1984) are used to illustrate some significant features of flow and transport in the root zone that should be considered in interpreting regional pesticide data and in modeling pesticide transport in the subsurface.

The PRZM simulations were performed for a deep, well-drained sandy soil under irrigation and were carried out for a one-year period, commencing February 1. A single application of 100 kilograms per hectare (kg/ha) of a nonsorbed tracer (e.g., chloride) and 10 kg/ha of a pesticide was assumed to occur on day 1 of the simulation period, a week before corn was planted. Soil-water extraction by a growing root system (maximum rooting depth of 1 m) was considered in the simulations, but tracer and pesticide losses by means of plant uptake or degradation were not considered.

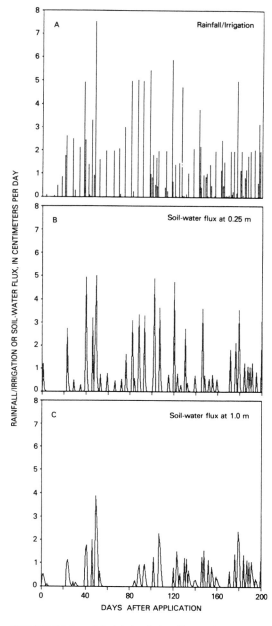

FIGURE 14-2. Model simulation illustrating episodic soil-water fluxes at two soil depths under multiple rainfall and irrigation events.

Simulated rainfall and irrigation events, varying in intensity and duration and in the time interval between events, resulted in a complex, time-varying boundary condition at the ground surface (see Figure 14-2A). As a direct consequence, soil-water flux was episodic at all depths within the root zone, as well as in the upper portion of the intermediate vadose zone (Figure 14-2B, C). Soil-water depletion above the depth of interest must be satisfied before excess water is available for recharge past that depth; thus, the time sequence of the rainfall and irrigation events and the "recharge" events at subsurface depths did not match. In examining solute transport (Figure 14-3), the time for first arrival of the nonsorbed solute at a given depth depended on the sequence of rainfall and irrigation events from the date of application and varied by depth (compare Figure 14-3A,B). Sorption further affected the episodic nature of the pesticide fluxes at subsurface depths. The time of first arrival of the pesticide at a given depth was delayed due to retardation, and the timing and magnitude of the pesticide "leaching" episodes were different compared to the nonsorbed tracer (compare Figure 14-3C, D with A and B).

The episodic nature of water and solute fluxes at the bottom of the crop root zone clearly can lead to complex input boundary conditions for the ground-water-flow system. Usually, however, much simpler, time-invariant boundary conditions are employed in modeling studies. Ground-water simulation studies that have considered time-varying outputs from root-zone or vadose-zone models have shown that contamination of ground water by pesticides can be characterized by highly transient loadings with rainfall events dominating the timing and magnitude of the pesticide leaching (cf. Jones et al. 1987; Dean and Atwood 1987; Huyakorn, Kool, and Wadsworth 1988).

The effects of episodic leaching of pesticides to ground water will be dampened with increasing thickness of the vadose zone and with greater depths below the water table. Nonetheless, for shallow ground water, episodic delivery of pesticides to ground water can complicate interpretation of water-quality samples. Furthermore, the mean net

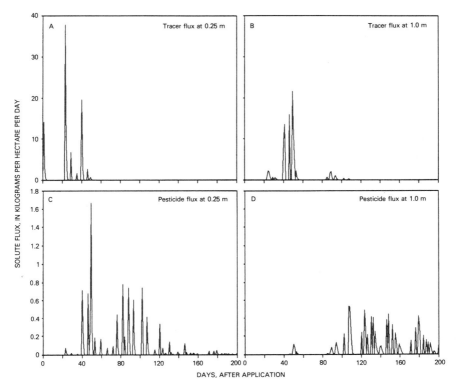

FIGURE 14-3. Model simulation illustrating episodic fluxes of a nonsorbing tracer and a sorbing pesticide at two soil depths resulting from the multiple rainfall and irrigation events shown in Figure 14-2.

recharge value used in simple indices of pesticide leaching, such as the AF, may be unrelated to vulnerability, if a few large storms are the principal events that cause the leaching of pesticides to ground water.

Effects of Preferential Flow

Many soils have a high probability of developing so-called preferential flow in the vadose zone. Preferential flow is taken to include all forms of rapid downward movement by means of high-conductivity zones. During an infiltration event, water and solutes can preferentially flow into and through these zones and bypass a major portion of the soil matrix. As a consequence, water and solute fronts penetrate to greater depths, and much faster, than might be anticipated from conventional modeling approaches.

Preferential flow in the vadose zone can be caused by numerous factors. Beven (1991) distinguishes the following two broad phenomena leading to preferential flow: (1) water and solute flow in continuous, noncapillary structural voids or in zones of capillary pores that have locally high hydraulic conductivities; and (2) instability in wetting fronts under certain conditions, even in soils lacking obvious structural voids. Preferential flow in macropores (cracks, root channels, worm holes, etc.) are examples of the first category, while "fingering" arising from pore-scale or grain-scale heterogeneities is an example of the latter category. In many situations, both categories of preferential flow are likely to occur.

The occurrence of preferential flow in macropores depends on antecedent soil-water conditions, the contrast in hydraulic properties between the soil matrix and structural voids, the spatial distribution and pore geometry of the preferential flow paths, and nonuniform boundary conditions for water and solute input at the ground surface. The impact of preferential water flow on solute transport depends on the time lag between solute application and the sequence of events causing preferential flow, and on the sorptive characteristics of the pore walls of the cracks, worm holes, and so on. A further confounding factor is that both the structure and function of preferential paths may not be spatially or temporally stable. For example, wetting-drying or freezing-thawing cycles can produce cracks or fissures whose geometry and spatial distribution are unpredictable.

In addition to its widely reported occurrence in fine-textured soils, preferential flow also can occur in sandy materials having high matrix permeability and containing discrete lenses of coarse or fine material. These lenses can act as barriers to downward flow and cause focusing or funneling of water (Kung 1990a, 1990b). In such situations, physical properties of the preferential-flow region may be similar to those of the surrounding matrix, but the flow velocity is quite different.

A major effect of preferential flow is accelerated movement of surface-applied solutes (such as pesticides) through the vadose zone. In field studies, the average behavior of the bulk of the applied pesticides may be consistent with the advective-dispersive transport and sorption-equilibrium assumptions common in pesticide leaching models (e.g., PRZM) or indices such as the AF, but some of the applied chemicals is often found to undergo rapid and deep movement (cf. Jury, Elabd, and Resketo 1986). Macropore flow is at least partially responsible for the frequent reports that field-measured dispersion coefficients are much larger (by an order of magnitude or more) than those measured in packed laboratory columns. An important anomaly with respect to subsurface pesticide mobility is that strongly sorbed compounds can be transported in macropores in the sorbed phase, if attached to colloidal particles (McDowell-Boyer, Hunt, and Sitar 1986).

Although it has been recognized for well over a century that preferential flow occurs in soils, it is only during the past 10–15 years that experimental investigations and modeling of macropore flow and transport have been a major focus of investigations by soil scientists and subsurface hydrologists (Germann 1988; van Genuchten, Ralston, and Germann 1990). Beven (1991) reviewed various attempts to model preferential flow, and grouped the modeling approaches into the following categories: (1) mechanistic models, where the porous medium is represented as having a biomodal pore-size distribution with Darcy-type flow in matrix pores and Poiseuille-type flow in macropores, (2) statistical models, typified by the transfer function models, which utilize a statistical representation of flow velocities or residence times without specifying the geometric or hydrologic features of the medium, and (3) statistical-mechanistic models, which attempt to combine the best features of the two approaches by using mechanistic models to represent matrix flow, but use statistical distributions to represent spatial or geometric variability of preferential-flow paths.

Beven (1991) concluded that much more needs to be learned about preferential flow. Inclusion of such flow effects in models will require complex representation, and involves multiple parameters that cannot be easily and independently calibrated. Guidance on how to use simplified models, especially for regional-scale analyses, is lacking. The likely importance of macropores and other preferential-flow paths suggests that conventional models for pesticide leaching are primarily useful for ranking pesticides in terms of their relative leaching potential through the major portion of the soil matrix, rather than as absolute predictors of pesticide movement in the subsurface.

Other Processes and Factors

Other processes and factors beyond those discussed affect the amount of pesticides potentially leached to ground water. For instance, surface runoff can carry off pesticides before they leach to ground water (Leonard 1990), and photodecomposition may be an important loss mechanism for pesticides applied to plant canopy or at the soil surface (Miller and Hebert 1987).

Volatilization can result in significant loss of some pesticides from soils (and plant surfaces) to the atmosphere. Volatilization losses from plants or moist soil surfaces can be very high, with losses approaching 80–90% within a few days after application for some highly volatile pesticides (Taylor and Spencer 1990). Volatilization losses of soil-incorporated pesticides are dependent upon the rate at which the pesticide moves to the soil surface by gaseous and liquid diffusion or by convection in evaporating water.

In a series of papers, Jury and co-workers (Jury et al. 1983, 1984a, 1984b, 1984c) describe and apply a screening model for assessing the relative roles of three major pathways of soil-applied organic chemicals: leaching, degradation, and volatilization to the atmosphere. The model assumes linear, equilibrium partitioning among vapor, liquid, and sorbed chemical phases, and first-order degradation in a manner similar to that of the AF index. Chemical movement to the atmosphere by volatilization loss is through a stagnant air boundary layer at the soil surface. A steady-state upward or downward flow of water is assumed by the model. The screening model requires knowledge of Henry's constant (K_H; see Chapter 7) in addition to the values of K_{oc} and half-life for the chemical. Using this screening model, Jury et al. (1984a) classified chemicals into three categories that differ distinctly in their behavior. Category I compounds have high Henry's constants and high fluxes to the atmosphere, but flux to the atmosphere decreases after application whether water is evaporating or not. Category III chemicals have low Henry's constants and tend to move to the surface in evaporating water faster than they can diffuse to the atmosphere through the boundary layer. Consequently, under evaporative conditions, the chemical concentration increases at the soil surface and the flux to the atmosphere increases with time. Category III compounds are predicted to have insignificant losses to the atmosphere without water evaporation. Category II chemicals are intermediate in their behavior between Categories I and III.

Finally, pesticide uptake by plants (Nash 1974; O'Connor, Chaney, and Ryan 1991) has not been considered in most modeling efforts or in any of the pesticide leaching indices. This is primarily a result of the lack of quantitative experimental information, and the presumption that the mass of pesticide absorbed by the plant is a small fraction of that applied to the system. While this assumption is unproven for pesticide uptake, it is important to recognize that plant extraction of water greatly influences water fluxes, which, in turn, affect pesticide transport through the vadose zone.

Point-Source Contamination

The discussion thus far has focused on non-point sources of pesticide contamination. Point sources of pesticides can be associated with commercial or manufacturing facilities that handle pesticides or with on-farm usage. For example, accidental back-siphoning and spills of concentrated product or formulation, particularly at or near the wellhead, can serve as a local point source. On-farm point sources might also arise from simply washing equipment near a wellhead. Point sources can thus contribute to ground-water contamination through direct movement of pesticides down wells or their annuli (outside of casings) or from localized, high concentration inputs to the vadose zone.

The extent to which pesticides detected in ground water arise from point sources, as opposed to leaching after nonpoint agricultural applications, is a key issue in determining how to prevent ground-water contamination from pesticides. If nonpoint agricultural applications of pesticides lead to subsequent contamination of ground water, then the best preventive measures might include developing new soil-crop-pesticide management practices, and, in some cases, restricting the use of certain pesticides. On the other hand, if pesticide contamination of ground water results from point sources, then the best management strategy might be education and enforcement to ensure that approved storage and handling practices are followed.

Agricultural chemicals from point sources can reach ground water within short times, rather than the months or years which may be required for leaching after normal applications. Thus, the levels detected in ground water from point sources could reflect the highest levels found, and point sources can introduce some compounds that would not otherwise be leached to ground water (Hallberg 1986). However, flood irrigation or a large storm can also result in rapid transport of pesticides from nonpoint sources to ground water.

Hallberg (1989) discusses the many nuances associated with defining "normal" agricultural applications. He notes that normal agricultural applications can include such practices as discharging of excess formulation or rinse water across a field in a dispersed manner, or applying a little stronger mix than the label direction "just to be sure" or because of pest-resistance problems. The higher concentrations of pesticides introduced to the vadose zone by these mechanisms, or by spills and other point sources, may lead to a greater mass flux of pesticides to ground water than would be predicted by simple models or indices such as AF. This can occur when the pesticide sorption isotherms are nonlinear. In this case, sorption coefficients may be lower at higher pesticide concentrations, and retardation effects on leaching are overpredicted by using standard published values of K_{oc} (cf. Davidson et al. 1980). In addition, pesticide degradation rates might be much smaller in the presence of high pesticide concentrations (cf. Ou et al. 1978), causing predicted losses to be overestimated using published values of $t_{1/2}$.

Many recent regional-scale studies attempt to infer whether pesticide detections in ground water result from point or nonpoint sources. Empirical evidence may be obtained, for example, through site investigations and interviews with landowners and farmers about pesticide management practices near each sampled well. Also, evidence that pesticides are entering ground water from nonpoint sources is provided indirectly if high correlations are found between the detection of pesticides in ground water and natural factors expected to control pesticide movement through the vadose zone, such as soil type and depth to water. In addition, the type of wells in which the pesticides are detected may provide further evidence about the sources of the pesticides. For instance, detection of pesticides in new wells established far away from pesticide storage and handling areas suggests that pesticides are entering ground water from agricultural applications. On the other hand, frequent detection of pesticides in existing farm wells, but not in specially constructed wells in farmed areas, suggests the pesticides arise from pesticide handling practices near wells. Profiles of pesticide occurrences in the unsaturated zone can provide additional supportive evidence for or against the leaching of pesticides. Furthermore, if the vadose zone is bypassed or pesticides are rapidly transported through the vadose zone as a result of point sources, a different suite of pesticide degradation products might occur (Adams and Thurman 1991; see Chapter 11).

Seasonal patterns in the concentrations of a pesticide in a well might suggest nonpoint sources, whereas a large concentration that occurred once, with perhaps some time decay in concentrations, may indicate point-source contamination. Such temporal patterns of pesticide detection in individual wells should be interpreted cautiously, however, because long-term storage and episodic leaching of pesticides in the vadose zone can greatly complicate the interpretations.

Typically, evidence about the likelihood of point sources as opposed to nonpoint sources of pesticides in ground water is inconclusive. For example, wells located near point sources of pesticides also are typically close to nonpoint sources, and vice versa, making it difficult to isolate a single source. It also is possible that pesticides detected in a well result from a combination of point and nonpoint sources. Thus, conclusions drawn about the sources of pesticides in wells should be made with great caution, and the supporting evidence and its limitations carefully documented.

Agricultural Practices and Pesticide Use

Agricultural practices that affect leaching include the amount of pesticide applied, the pesticide formulation (e.g., the surfactants added), the timing of the application relative to seasonal rainfall or temperature extremes, the method of application (e.g., direct application to soil or foliar spraying), the cultivation practices, and the amount of irrigation water used (U.S. Environmental Protection Agency 1987). A review of agricultural approaches to reduce contamination of ground water by pesticides and other agrichemicals was published by the U.S. Office of Technology Assessment (1990).

The occurrence of pesticides in ground water is perhaps most directly controlled by pesticide use; if a pesticide is not applied to the contributing area of a well, then it would not be expected in ground-water samples from that well. Unfortunately, despite their potential utility, data on pesticide use are limited. Some states, such as California, publish pesticide-use data on an annual or periodic basis. Recently, the National Agricultural Statistics Service of the U.S. Department of Agriculture began publishing on-farm agricultural chemical use statistics for selected crops and states on a biannual basis. As of the end of 1991, two reports were available (National Agricultural Statistics Service 1991a, 1991b).

A national data base on herbicide use in the United States has been developed by Resources for the Future (Gianessi and Puffer 1990). The estimates of herbicide use are calculated using two coefficients: the percent of land area that is treated and the average annual application rate per unit treated area. The land-area estimates come from the Census of Agriculture conducted by the U.S. Department of Agriculture. The usage coefficients come from a variety of sources.

Some limitations associated with many of the pesticide-use data bases are that (1) they are available for only a limited suite of pesticides, usually herbicides (although, a national insecticide-fungicide data base is under development by Resources for the Future; L.P. Gianessi, oral commun., 1992); (2) the resolution of the data is usually coarse (e.g., at the state or county level) making it difficult to estimate pesticide usage within the assumed contributing area to a well, (3) the usage coefficients may be transferred across areas with very different climates, geography, and agricultural practices, because of a lack of site-specific data; (4) the data represent only a limited time period; (5) the usage data may be based solely on commercial licensed applications (perhaps of restricted-use pesticides only); (6) many of the usage estimates are based on expert opinions without independent surveys to verify these opinions; and (7) the data typically do not account for nonagricultural uses, or perhaps even noncropland uses.

Nonagricultural uses of pesticides include the elimination of unwanted vegetation from roadsides, railways, airfields, forestry breaks, and other areas, as well as urban and suburban applications by homeowners. Although more limited than the agricultural use data, some data on nonagricultural uses do exist. For example, pesticide-use data for selected pesticide active ingredients used on golf courses and by urban applicators are published by Resources for the Future (Gianessi and Puffer 1991).

Limitations in existing pesticide-use data suggest consideration should be given to field examination of agricultural practices in the vicinity of sampled wells and to discussions of pesticide use with local farmers and extension agents. The information obtained should include not just current agricultural practices and pesticide applications, but also those of previous years. Unfortunately, crop types change and new pesticides are frequently introduced; hence, a given contributing area to a well may have numerous pesticides applied to it over several years. Furthermore, available records of pesticide use are likely to be for an entire farm rather than individual fields. This suggests the importance of maintaining careful up-to-date records of pesticide use and cropping patterns for wells selected for long-term studies.

REGIONAL ASSESSMENT OF GROUND-WATER VULNERABILITY

Ground-water vulnerability to contamination can be defined as the propensity or likelihood for contaminants to reach some specified position ("reference location") in the ground-water system after their introduction at some point above the top of the uppermost aquifer. Methodologies for vulnerability assessment range from the use of empirical techniques to those based on simulation models (U.S. Environmental Protection Agency 1992b).

Methods currently available for assessment of ground-water vulnerability can be grouped into three categories: (1) overlay and index methods, (2) methods based on simulation models, and (3) statistical methods. These methods have been applied at scales ranging from a few hectares to the national scale, and they draw to varying degrees on knowledge of the aforementioned processes and factors affecting pesticide behavior in the vadose zone.

In general, the overlay and index methods and statistical methods have been used for assessments at map scales smaller than 1:50,000 (i.e., large study areas), while methods based on simulation models have been used at much larger map scales (i.e., smaller study areas). Early attempts at vulnerability assessment involved simple manual overlay of paper maps; however, over the past decade, the introduction of Geographic Information Systems (GIS) has facilitated and popularized the use of computers and digitized maps and data bases. GIS-based approaches also enable the coupling of simulation models to spatial data bases, and facilitate graphical display of model outputs.

Overlay and Index Methods

Overlay and index methods utilize various spatial attributes of a region (e.g., depth to ground water, soil type, geology, net recharge). In the simplest methods, maps of attributes are overlaid, and areas with a combination of certain characteristics (e.g., shallow water tables with high recharge) are designated as having greater vulnerability. Recent examples of this approach exist for Illinois (McKenna and Keefer 1991), Iowa (Hoyer and Hallberg 1991), and the 48 conterminous United States (Pettyjohn, Savoca, and Self 1991). The Illinois vulnerability map is shown in Chapter 20 as Figure 20-1. An example of the state aquifer vulnerability maps developed by Pettyjohn, Savoca, and Self (1991) is shown in Figure 14-4 for Idaho. Their method was developed specifically for the USEPA Underground Injection Control Program, but the authors suggest that "the products are equally valuable to assess the potential for ground-water contamination from other surface or near surface sources" (Pettyjohn, Savoca, and Self 1991).

Many recent overlay and index techniques attempt to improve upon the simple overlay approaches by assigning numerical weights and scores to each of the attributes in recognition of their relative importance. The best-known example of this approach is the DRASTIC index (Aller et al. 1987). A number of state regulatory agencies have

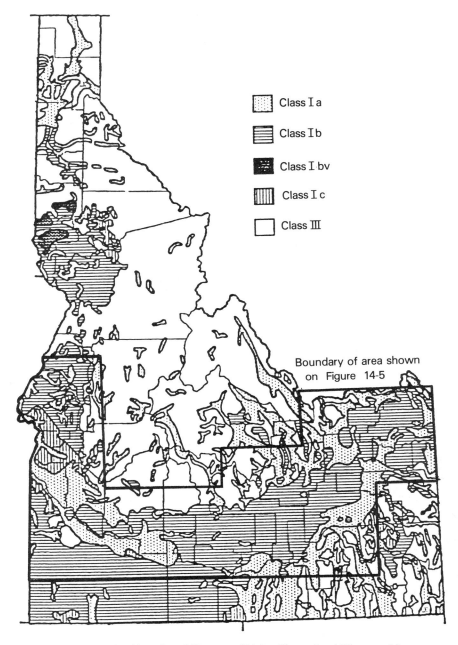

FIGURE 14-4. Aquifer vulnerability map of Idaho. Classes I and III are used for high- and low-vulnerability areas, respectively: a = unconsolidated aquifers; b = soluble and fractured bedrock aquifers; c = semiconsolidated aquifers; v = variably covered aquifers. (*After Pettyjohn, Savoca, and Self 1991.*)

developed assessment methods similar to DRASTIC. For example, the Idaho Department of Health and Welfare recently assessed vulnerability (see Figure 14-5) on the basis of a depth-to-water map (developed through kriging), several recharge classes (based on land cover and the type of irrigation), and several soils attributes.

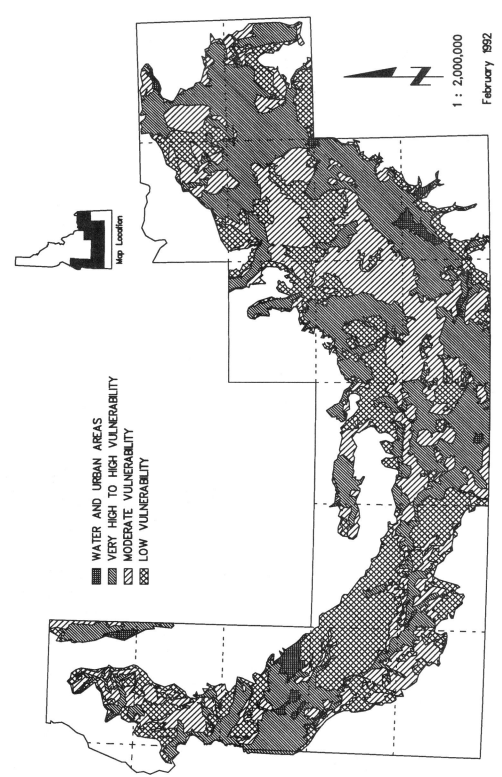

FIGURE 14-5. Relative ground-water vulnerability map for the Snake River Plain in Idaho. (*After Rupert et al. 1991.*)

Some overlay and index methods, such as DRASTIC, attempt to be universally applicable and incorporate parameters that should be available to some degree virtually everywhere; other methods are adjusted to the setting and information data base available in a particular area. An advantage of the latter approach is that geologic and geographic features unique to a particular area can be taken into account. For example, the Illinois method (McKenna and Keefer 1991) is based on an intensive examination of stratigraphy and the identification of low-and high-permeability units in a three-dimensional context throughout the state.

As noted in the discussion of pesticide transport in the vadose zone, ground-water vulnerability to contamination is likely to differ considerably among pesticides with differing properties. Hence, the lack of consideration of specific contaminants or contaminant groups is potentially a significant limitation of many overlay and index vulnerability assessment methods.

It is instructive to compare vulnerability maps where separate assessments have been made of the same region to illustrate some of the uncertainties and effects of scale associated with these maps. For example, consider the Snake River Plain in Idaho. In the vulnerability assessment map for Idaho shown in Figure 14-4, three rock types (classes Ia, Ib, and Ic) in the Snake River Plain are distinguished, but a high-vulnerability class is assigned to all three rock types, and hence, to most of the area. A different pattern of relative vulnerabilities in the Snake River Plain is exhibited by Figure 14-5, which shows more areas of low vulnerability. Some of the differences between the two maps result from differences in the factors considered. For example, Figure 14-4 is based strictly on geology, whereas Figure 14-5 incorporates recharge and soils information. Many of the areas assigned a low vulnerability rating in Figure 14-5 are in areas of raw lava flows with deep ground water and without agriculture or irrigation. Other differences between the two maps result from scale effects. For example, Figure 14-4 is one of 48 maps that together delineate vulnerability for the entire United States, whereas Figure 14-5 was based on a more detailed analysis of a particular part of Idaho.

Further comparisons can be made in other areas. For example, in a state map for Maryland, Pettyjohn, Savoca, and Self (1991) assign a high-vulnerability rating to the entire coastal plain in Maryland. A more detailed examination of a large part of this coastal plain in Chapter 23 (see Figure 23-3) delineates hydrogeomorphic regions of differing vulnerability based on variations in soils and landscape characteristics, including some areas with confining units overlying the uppermost utilized aquifer.

Methods Based on Simulation Models

Simulation models of varying conceptual complexity have been used for vulnerability assessment, depending on the availability of input data, spatial scale of the region, and intended use of the assessments. Values of the input parameters required in these models are obtained by estimation of extrapolation from published maps and from existing data bases (e.g., soil maps and soil characterization data published by the Soil Conservation Service; weather records available from the National Weather Service). A listing of key parameters that have been included in pesticide simulation models is shown in Table 14-3.

Wagenet and Rao (1990) present a comprehensive review of current modeling approaches. Pennell et al. (1990) compared predictions of five simulation models at the field scale with a common data set. They concluded that the models tested were able to predict the location of the center of mass of the solute and amount of mass remaining within the soil profile within approximately 50% of the actual value; however, the distribution of solute concentrations in the soil profile was much more poorly simulated. Examples of regional vulnerability assessments based on simulation models coupled to GIS include use of (1) the LEACHM

TABLE 14-3. A Listing of Some Key Parameters in Models of Pesticide Transport in Soils

Pesticide Parameters
   Organic-carbon-normalized sorption coefficient (K_{oc})
   Distribution coefficient (K_d)
   Aqueous solubility
   Henry's constant
   Saturated vapor density
   Gas phase diffusion coefficient
   Half-life
   Hydrolysis half-life
   Oxidation half-life
   Foliar decay rate

Soil Parameters
   Dispersion coefficient
   Saturated water content
   Field-capacity water content (θ_{FC})
   Wilting-point water content
   Hydraulic properties
   Bulk density (ρ_b)
   Organic-carbon content (f_{oc})
   pH
   Cation-exchange capacity
   Heat flow parameters

Crop Parameters
   Root density distribution
   Maximum rooting depth
   Pesticide uptake rates

Climatological Parameters
   Rainfall or irrigation rates
   Pan evaporation rates
   Daily maximum and minimum temperature
   Snowmelt
   Hours of sunlight

Management Parameters
   Pesticide application rate and timing
   Pesticide application method and formulation
   Crop production-system variables
   Soil-management variables

Source: Adapted from Wagenet and Rao (1990).

model (Wagenet and Hutson 1989) to assess a 70-km² region in New York (Petach, Wagenet, and DeGloria 1991), and (2) the CMLS model (Nofziger and Hornsby 1987) to map a 107-km² region in Florida (Hornsby et al. 1990).

The data requirements for the simulation models increase with increasing level of conceptual complexity and tend to quickly outstrip data availability on a regional scale. This has prompted the use of indices such as the attenuation factor (AF; Eq. 14-1) and retardation factor (RF; Eq. 14-2) for generating regional vulnerability maps. These indices have been developed on the basis of process-oriented models, but with very simple process representations. Despite their simplicity, these process-based indices should provide more appropriate combinations and weighting of relevant attributes than empirical indices such as the DRASTIC index.

Khan and Liang (1989) used the RF and AF indices and a GIS to produce vulnerability maps for several Hawaiian islands. Meeks and Dean (1990) used a leaching potential index (LPI) to map an area of about 1,000 km² in the San Joaquin Valley in central California. The conceptual basis for the LPI is essentially equivalent to that for the AF.

Statistical Methods

The third category of assessment methods is based on statistical relations between water-quality data and variables such as depth to the water table, soil properties, and pesticide usage. Examples include multiple regression (Chen and Druliner 1988) and discriminant analysis (Teso et al. 1988). These methods attempt to provide a statistical basis for the numerical weights assigned to various spatial attributes. The methods require groundwater monitoring data for their development.

Uncertainty in Vulnerability Assessment

A major missing element in most vulnerability assessments is evaluation of the uncertainty that is pervasive in essentially all of the data bases (or maps) used. The level of certainty portrayed in vulnerability maps (or implied by not specifically recognizing errors of various types) usually cannot be defended. For example, the subtle distinctions in vulnerability claimed for areas within a region may not be justified given the uncertainty in the supporting data bases. Loague and coworkers (Loague et al. 1989, 1990; Loague 1991; Kleveno, Loague, and Green 1992) have used first-order uncertainty analysis

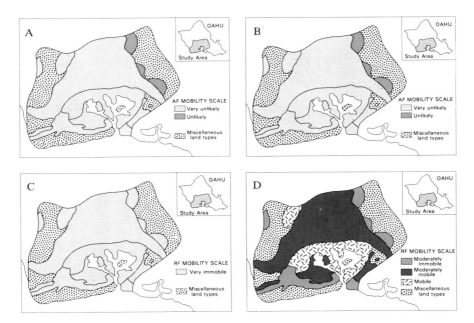

FIGURE 14-6. AF and RF rating map for the herbicide diuron for the Pearl Harbor basin, Hawaii. Maps show ratings of ground-water vulnerability on the basis of (A) AF values, (B) AF values plus one standard deviation, (C) RF values, and (D) RF values minus one standard deviation. (*After Loague et al. 1990, Figures 7, 8, 11, and 12.*)

to illustrate the need for recognizing the "fuzziness" of the boundaries delineating areas of different levels of vulnerability. An example representation of map uncertainty derived in this manner is shown in Figure 14-6, which shows vulnerability ratings for the herbicide diuron in the Pearl Harbor basin of Hawaii. These ratings were developed using the AF (Figure 14-6A) and RF (Figure 14-6C) indices. Alongside these maps are rating maps with one standard deviation added to the AF (Figure 14-6B) and one standard deviation subtracted from the RF (Figure 14-6D); these are directions of greater simulated vulnerability. Estimates of uncertainty were determined by first-order uncertainty analysis of hydrologic parameters (e.g., recharge rate), pesticide parameters (e.g., K_{oc} and $t_{1/2}$), and soil parameters (e.g., ρ_b, f_{oc}, and θ_{FC}). Note that no change occurs in the AF map rating classes, but a large change occurs in the RF map rating classes, mostly resulting from uncertainty in f_{oc} and K_{oc}. Loague et al. (1990) note that, with either larger recharge rates or longer half-lives, the uncertainties in the AF map might also become pronounced.

Another major limitation in vulnerability assessments is the absence of objective and quantitative protocols for judging their validity. The existing measures of performance that have been used for site-specific evaluations of simulation models cannot be readily transferred to regional-scale vulnerability assessments. An evaluation of vulnerability assessments should consist of multiple lines of evidence that either support or contradict the relative vulnerability ranking of different areas. One source of evidence might come from the use of age-dating isotopes (see Chapters 10 and 11) to evaluate "intrinsic" vulnerability, defined by the time of travel of water from the point of contaminant entry to the reference location in the ground-water system. Contaminant-specific vulnerability is much more difficult to check, because chemicals are not applied at the land surface uniformly and the loads are unknown spatially and temporally. Thus, simple comparison of chemical concentrations or detections among wells in areas assigned to different vulnerability classes should be approached with caution as an evaluation method. The differences can result from actual differences in vulnerability or from different chemical loadings, different well depths sampled, movement of contaminants down wells or their annuli, and so forth. The reference location of a vulnerability method should also correspond to the production zone of the wells used to test it. This suggests the use of data from wells screened near the water table for testing vulnerability methods that use the water table as the reference location. Unfortunately, as indicated in the section "Effects of the Open Interval" in Chapter 1, it is particularly difficult to obtain representative samples of ground-water quality at the water table.

Clearly, current vulnerability assessments can be highly uncertain. Existing vulnerability assessment methods do not have the predictive capability to forecast the actual occurrences or magnitudes of ground-water contamination, nor should they be used to make site-specific management or policy decisions. They should be used only to evaluate relative vulnerability within a region or among different pesticides used over a region. A recent National Research Council report (1993) discusses in greater detail uncertainty in vulnerability assessments and its implications to management and policy issues.

SELECTED STUDIES OF PESTICIDES IN GROUND WATER

Studies investigating the occurrence of pesticides in ground water range in size from field plot studies to the USEPA National Survey of Pesticides in Drinking-Water Wells (U.S. Environmental Protection Agency 1990). A brief review of a few studies is provided here, progressing from more localized to very broad scale studies. These selected examples illustrate the combined effects of pesticide transport to ground water and characteristics of ground-water-flow systems. The examples are also indicative of the level of detail achievable at different scales of investigation. It is noteworthy that detailed field studies to determine pesticide properties in ground water are less prevalent than for some other organic contaminants such as many organic solvents. Part of the reason for this is the dispersed nature of pesticide delivery to ground water and the rapid degradation of many pesticides.

Aldicarb in the Central Sand Plain, Wisconsin

Some of the earliest findings of pesticides in ground water were residues of the pesticide aldicarb, in particular, its oxidative degradation products aldicarb sulfoxide and aldicarb sulfone. One of the areas in which aldicarb residues have been investigated is the Central Sand Plain of Wisconsin, an area of glacial outwash deposits that form highly transmissive and important aquifers in the region. The results of early field studies of aldicarb residues (simply referred to as "aldicarb" here) in the Central Sand Plain are reported by Rothschild, Manser, and Anderson (1982). They monitored water levels and concentrations of aldicarb beneath five fields with known histories of aldicarb treatment, and sampled 25 private wells and seven irrigation wells. The measurements indicated that most aldicarb beneath the fields studied was concentrated in roughly a 1.5-m-thick zone. The absence of strong vertical gradients in hydraulic potential in the aquifer caused the aldicarb to remain concentrated near the water table. No aldicarb was detected in any of the deep monitoring wells beneath the fields (those approximately 18 m below the water table). Aldicarb was found, however, in some of the irrigation wells finished at approximately the same depth. It was suggested that downward hydraulic gradients created during pumping of ground water for irrigation uses caused aldicarb to move into the irrigation wells and hence to the deeper part of the aquifer near these wells. After a typical irrigation cycle of 30 h, water levels in the irrigation wells were about 3 m below the static water level, but recovery of the water levels took only a couple of hours. Aldicarb was not detected in samples from the irrigation wells after the water levels recovered from pumping.

In a subsequent investigation of aldicarb in nine irrigated fields (Harkin et al. 1986), the pattern of aldicarb leaching through the soil and the levels detectable in ground water were "highly erratic," particularly in light of the uniform application rate across the treated part of each field and the relatively uniform soils. The pesticide concentrations in ground water decreased at rates faster than could be explained by dilution and dispersion from ground-water movement. The decline in concentrations was apparently accelerated by chemical hydrolysis and microbial degradation. The investigators suggested that important variables affecting the rates of breakdown likely included seasonal fluctuations of ground-water temperatures and variations in pH and alkalinity with location and depth.

Additional studies have documented the presence of aldicarb in the Central Sand Plain of Wisconsin and explored the factors affecting its fate (e.g., Anderson 1986; Holden 1986; Fathulla et al. 1988; Kraft and Helmke 1992). These studies confirm that the persistence of aldicarb residues in ground water is highly variable. Degradation of aldicarb in ground water is faster with greater depth, higher pH, and lower dissolved-oxygen concentrations (Kraft and Helmke 1992). Longer half-lives for aldicarb residues in Wisconsin (and Long Island) ground water than in other places, such as Florida, have been attributed to colder ground-water temperatures at these locales (Jones 1986).

River-Aquifer Interactions in Iowa and Nebraska

Two recent local-scale studies of the herbicide atrazine illustrate that water flow and solute transport of pesticides between rivers and ground water can be dynamic with reversals in flow and transport from one water source to the other.

The first study investigated the sources of atrazine and one of its degradation products, deethylatrazine, to the Cedar River in Iowa (Liszewski and Squillace 1991; Squillace, Thurman, and Furlong 1993). Alluvial ground water adjacent to the Cedar River was the major source of atrazine and deethylatrazine to the river during baseflow conditions. Atrazine and deethylatrazine detected in the alluvium resulted from bank storage of river water and from ground-water recharge in areas distant from the river. Atrazine was never applied to the soil surface near the wells.

The effects of river-aquifer interactions on the movement of atrazine between the Cedar River and its alluvium are illustrated in Figure 14-7 by the cross-sectional distribution shown for three sampling periods in 1990. (Note: The results for deethylatrazine were similar to those for atrazine, but are not shown here.) Atrazine was detected in the river and alluvial aquifer during the preapplication period for the herbicide (February 20–22; Figure 14-7A). During the second sampling period (March 20–22; Figure 14-7B), the river stage peaked almost 2 m above its baseflow level, and a reverse hydraulic gradient between the river and the alluvial aquifer resulted in the movement of pesticide-laden river water into the aquifer as bank storage. As a result, the concentrations of atrazine in the alluvial aquifer increased as far as 30 m from the river's edge. A couple of weeks later, the hydraulic gradient between the river and the alluvial aquifer had reversed again, and atrazine was being discharged from bank storage in the alluvium to the Cedar River (April

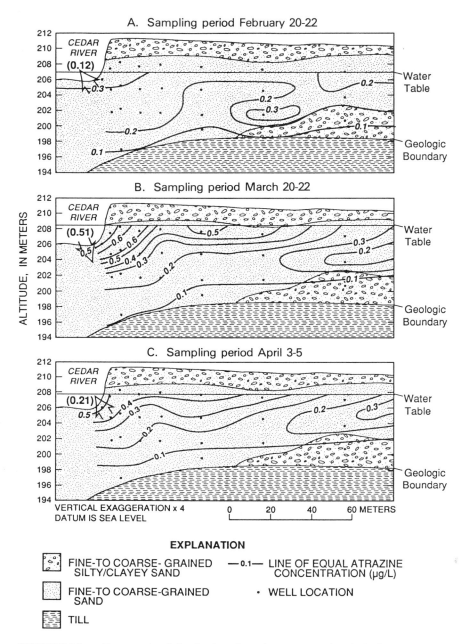

FIGURE 14-7. Cross-sectional distribution of atrazine concentrations in Cedar River alluvium near Cedar Rapids, Iowa, February through April 1990. The direction of seepage between the river and alluvium is shown by the arrows. (*After Squillace, Thurman, and Furlong 1993.*)

3–5; Figure 14-7). After depletion of bank storage, persistent lower concentrations of atrazine continued to be released from the alluvial aquifer into the river.

A second study of river-aquifer interactions by Duncan et al. (1991) is associated with withdrawals of ground water by the city of Lincoln, Nebraska, from a well field near

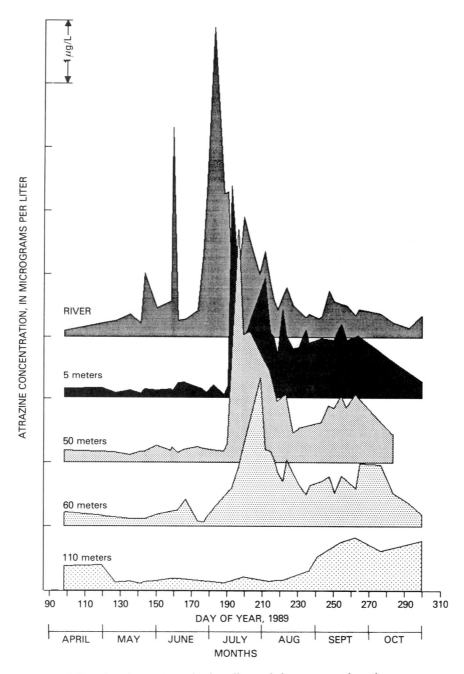

FIGURE 14-8. Atrazine concentration in wells sampled on a transect from the Platte River to a well field near Lincoln, Nebraska. Approximate distances of wells from the river are indicated at left-hand side of each graph. (*From Duncan et al. 1991, Figure 4, Ground Water Monitoring Review,* © *1991.*)

the Platte River. The focus of this study was on atrazine movement from the Platte River into the Lincoln well field. Water samples were collected from monitoring wells located at different distances from the river. Atrazine detected in samples from these wells was attributed to induced river recharge, since the herbicide is not applied in the immediate vicinity of the well field. The temporal patterns of high and low concentrations in the river water were traceable to some extent in the wells (see Figure 14-8). Those wells closer to the river showed a response sooner than those farther from the river. Peak concentrations of pesticides in river water during late-May and early-June, however, did not have obvious corresponding peaks in any of the monitoring wells, probably because of minimal induced recharge. Only small increases in river stage occurred at these times, and heavy midsummer pumping had not yet begun.

Insecticides on Long Island

Ground-water samples were collected on Long Island, New York, from 90 shallow wells screened in the unconfined, upper glacial aquifer beneath five different land-use areas (Eckhardt, Siwiec, and Cauller 1989; Cauller and Eckhardt 1990). The areas, which range from 57 to 114 km^2, represent suburban land sewered more than 22 years (long-term sewered), suburban land sewered less than 8 years (recently sewered), unsewered suburban land, agricultural land, and undeveloped (forested) land. All five areas are on the regional ground-water divide, where downward-moving water recharges deeper parts of the Long Island aquifer system.

Three classes of insecticides were detected in ground water—organochlorine, organophosphorous, and carbamate insecticides. The results for the carbamate and organochlorine insecticide residues are shown in Figure 14-9. Carbamate insecticide residues, mainly aldicarb metabolites and carbofuran, were detected almost exclusively within the agricultural areas and at the highest concentrations of any pesticide class. On the other hand, organochlorine insecticides, mainly dieldrin and chlordane, were commonly detected in samples from suburban areas. The results indicate that urban, as well as agricultural areas may have significant con-

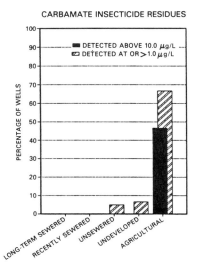
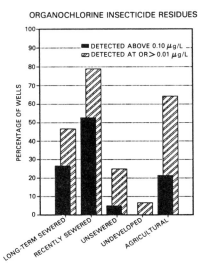

FIGURE 14-9. Concentrations of carbamate and organochlorine insecticide residues in shallow ground water underlying five land-use types on Long Island, New York. (*After Cauller and Eckhardt 1990.*)

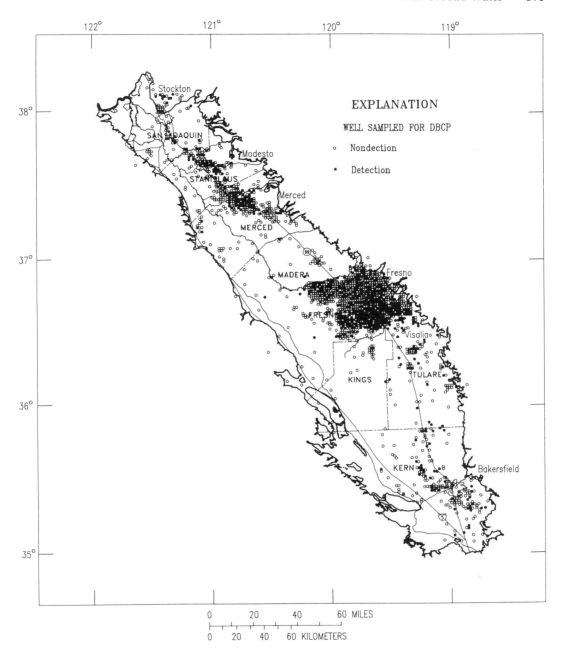

FIGURE 14-10. Areal distribution of wells sampled for dibromochloropropane (DBCP) in the San Joaquin Valley, California, 1975–88. (*After Domagalski and Dubrovsky 1991; data from California Department of Food and Agriculture [Cardozo et al. 1988].*)

centrations of certain pesticides. This is also evident from the case study for the Oklahoma City area discussed in Chapter 24.

DBCP in the San Joaquin Valley, California

Dibromochloropropane (DBCP) has been one of the most commonly detected pesticides in ground water. DBCP was banned in 1979, but continues to be detected in ground water in many areas today.

Results from sampling for DBCP from 1975 to 1988 in the San Joaquin Valley, California, are shown in Figure 14-10. DBCP was detected in 1,280 of 3,016 wells sampled, illustrating the potential widespread nature of some pesticides in ground water. Most wells sampled and most DBCP detections were near the cities of Fresno, Merced, and Modesto. These were also areas of heavy use of DBCP. Studies of DBCP in California ground water have shown that principal factors affecting leaching include the application pattern, texture, and permeability of the soil, and location of subsurface features such as fine-grained layers (Schmidt 1986; Meeks and Dean 1990). For example, DBCP contamination apparently is less extensive in the western San Joaquin Valley, despite heavy use in some areas. This is likely a result of the greater presence of fine-grained sediments in the western part of the valley (Domagalski and Dubrovsky 1991).

Surveys of Selected Minnesota Wells

Between July 1985, and June 1987, the Minnesota Departments of Health (MDH) and Agriculture (MDA) conducted cooperative surveys of water wells for selected pesticides (Klaseus, Buzicky, and Schneider 1988; Klaseus and Hines 1989). Altogether 725 wells and five drain tiles were sampled. In general, the sampling sites were selected from agricultural areas with an emphasis on areas vulnerable to pesticide contamination due to soils and hydrogeologic conditions.

One or more pesticides were detected in 38% of the wells sampled and in three drain tiles. Pesticides were detected in wells in 51 counties, but were most commonly found in the karst settings of southeastern Minnesota and in shallow sand and gravel outwash and alluvial aquifers in central and southwestern Minnesota. In general, pesticide detections decreased with increasing well depth and casing depth. For example, in the bedrock aquifer areas, pesticides were detected in 41% of the wells where bedrock was less than 30 m deep but only in 10% of wells where bedrock was deeper (Klaseus, Buzicky, and Schneider 1988). Pesticides were detected more frequently in observation wells and private-drinking-water wells than in public-water-supply wells. The authors attribute this to the shallower depths of many of the observation and private-drinking-water wells and to their close proximity to fields receiving pesticide applications.

At nine of the MDA sites, two observation wells within 0.7 m of one another were screened at two depths in unconfined sand and gravel aquifers. At three of the nine sites, pesticides were detected more prevalently in the deeper of the two wells.

Repetitive monitoring showed varying pesticide concentrations. Some pesticides, including a few with rapid degradation rates, were observed only for short periods, while other pesticides, particularly atrazine, tended to persist throughout the year. For some wells, the pesticide concentration was at or near its maximum observed concentration just prior to the spring pesticide application, suggesting that these pesticides may be leaching into ground water with the renewed water flux of late winter, early spring recharge.

USEPA National Pesticide Survey

The U.S. Environmental Protection Agency sampled approximately 1,300 community water system wells and rural domestic wells for the presence of 101 pesticides, 25 pesticide metabolites, and nitrate as part of its National Pesticide Survey (U.S. Environmental Protection Agency 1990). Twelve of the 126 pesticides and pesticide metabolites included in the survey were found in one or more sampled wells at or above the minimum reporting limits of the survey. The two most commonly detected pesticide analytes were DCPA metabolites and atrazine. DCPA metabolites were reported in 6.4% and 2.5% of the community water system and rural domestic wells, respectively. Two key observations about the high detection frequencies of the DCPA metabolites are that (1) the parent compound, DCPA, was not reported in any well by the survey, despite its lower reporting limit than the metabolites (0.06 µg/L for DCPA versus 0.10 µg/L for DCPA metabolites), and (2) the health advisory for DCPA metabolites (4,000 µg/L) is well above the maximum measured concentration of 7.2 µg/L. Other pesticide analytes detected include alachlor, bentazon, DBCP, dinoseb, EDB, ethylene thiourea, hexachlorobenzene, lindane, prometon, and simazine.

From the National Pesticide Survey (NPS), USEPA estimated that approximately 10% of the nation's community water system wells and 4% of the nation's rural domestic wells contain at least one pesticide or pesticide metabolite (of the 126 pesticide analytes tested) at or above the minimum reporting limits used in the survey. The finding that a larger percentage of community water system wells than rural domestic wells were estimated to contain pesticides was unanticipated. Most detections of pesticides and pesticide metabolites were below established or proposed USEPA life-time health advisory levels (HALs) or maximum contaminant levels (MCLs). The USEPA estimated that at most 0.8% of the nation's community water system wells and 0.6% of the nation's rural domestic wells contain at least one pesticide above the MCL or HAL.

In a follow-up analysis, USEPA investigated possible correlations or associations between detections of pesticides (or nitrate) in drinking-water wells and a number of factors that might be related to their presence in those wells (U.S. Environmental Protection Agency 1992a). A number of associations were found, although these results should be interpreted with caution because of the large number of variables examined. An interesting result was that the NPS county-level and subcounty-level DRASTIC assessments "did not prove to be a useful means of locating drinking water wells containing pesticides or nitrate." Similar findings were reported by the National Alachlor Well Water Survey (Holden et al. 1992). The lack of association between contamination and the DRASTIC scores in these studies may reflect limitations in DRASTIC as a vulnerability assessment method and (or) may have resulted because the screened intervals for the sampled wells in these studies are at a wide range of depths from the water table. It is interesting to note, however, that in the National Alachlor Well Water Survey, a simple measure of vulnerability (based on the most likely aquifer tapped) was associated with pesticide contamination despite a "less clear" relation of the well-specific DRASTIC score to pesticide occurrence.

Pesticide-use data developed from questionnaires about pesticide use near sampled wells did not display strong correlations with detections of pesticides in the NPS, and pesticides were detected in well water in areas where they were not reported as used. USEPA concluded that factors contributing to the latter result include (1) the limited time period (three to five years) covered by the questionnaire, (2) a lack of correspondence between the contributing area to the well and the geographic area from which the pesticide-use information was obtained, and (3) inaccurate information provided by respondents.

IMPLICATIONS FOR REGIONAL-SCALE STUDIES

The results from these cited studies, and many others conducted in recent years, demonstrate large variability in the extent of pesticide contamination and in the factors that relate to pesticide contamination at the regional scale. However, several general statements can be made.

Pesticide occurrence. It is clear that pesticides are detected far more commonly in ground water than the preconceptions of a decade or so ago would have predicted. Pesticide concentrations measured in ground water are generally low, however, compared with those commonly measured in soils and surface water (Hallberg 1989); the majority of pesticide concentrations reported in ground water are less than 1 µg/L.

Pesticide analytes. Investigations of pesticides are complicated by the wide range of chemical classes of compounds in common use. The pesticides most frequently detected in ground water include mobile and persistent soil fumigants and nematicides such as EDB and DBCP; insecticides such as aldicarb, carbofuran, and chlordane; and commonly used herbicides such as alachlor and atrazine (Moody et al. 1988; Hallberg 1989). In designing future studies, it is important to recognize that other pesticides may be discovered to be prevalent in ground water as new pesticides are used, pesticide application practices change, and analytical methods improve. Some of the pesticides reported in ground water today have either been banned or are no longer in significant use in the United States (e.g., DBCP).

Pesticide metabolites. It is increasingly clear that ground-water-monitoring studies of pesticides should, in many cases, include analyses for the metabolites of pesticides in addition to the parent compounds. Some of these metabolites are as toxic or more toxic than the parent compounds. Metabolites of some pesticides, such as aldicarb, have been studied for many years. Little is known, however, about the occurrence or behavior of metabolites for many pesticides. In some cases, pesticide metabolites are sorbed to a lesser extent, and, as a result, are more mobile compared to the corresponding parent compounds. The degradation half-lives for the metabolites can also be different. Findings about the spatial and temporal distribution of pesticides may turn out to be misleading in the absence of data on the metabolites. Pesticide metabolites may be found in samples without the parent compound, and analysis of metabolites may indicate more widespread contamination by pesticide residues than analysis of parent compounds alone. For example, Denver and Sandstrom (1991) found desethylatrazine further along a ground-water-flow path than the atrazine parent compound in an agricultural area in Delaware.

Analyte detection limits. The detection or reporting limits used by different studies have a major effect on the reported prevalence of particular pesticides, as well as perceptions about pesticide contamination in general. The continued lowering of these limits with advances in analytical technology presents somewhat of a double-edged sword. On the one hand, the lowering of reporting limits requires stricter quality-assurance and quality-control procedures and considerable care in communicating results to the public. Greatly different statistics on the frequency of "occurrence" of pesticides can result from using different reporting limits. This can lead to undue cause for alarm in the case of using very low reporting limits or to a false sense of security in the case of using high reporting limits. On the other hand, the high frequency of samples with concentrations "below the reporting level" severely constrains the extent to which the spatial distribution of pesticides and the causes of this distribution can be discerned.

At present, many pesticides commonly are reported at 5% or less of the sites sampled in a regional study; even the most prevalent pesticides commonly are reported much less than half the time. Furthermore, there has been a trend toward using pesticides that have increased efficacy, and as a result, application rates have declined from tens of kg/ha to as little as 0.01 kg/ha for such newer types of pesticides as sulfonylurea compounds (Nash et al. 1991). This trend makes it more difficult to detect the newer classes of pesticides.

Field monitoring techniques. Future regional studies of pesticides are likely to benefit increasingly from advanced monitoring techniques that can be applied in the field. For example, immunoassay analysis can provide sensitive, rapid, and low-cost screening of water samples for a number of pesticides (Vanderlaan et al. 1990). The principle of immunoassay analysis involves the attachment or binding of the compound of interest (atrazine for example) to an antibody developed for that compound. Only the target compound and other compounds with similar chemical structure will react. The tests require only a few drops of water and results can be obtained in as few as 15 minutes. Immunoassay tests are becoming widely used for water-quality monitoring and research (cf. Thurman et al. 1991). The sensitivity and low cost of the tests make it possible to screen numerous samples for selected pesticides and confirm results for a small fraction of the samples by more expensive conventional laboratory techniques.

Comparison of regional studies. At present, it is difficult to compare the results from any two regional studies, because investigations of pesticides in ground water in different localities have used different designs, pesticide analytes, reporting limits, sample-collection methodologies, and so forth. Thus, generalizations about large-scale effects such as climate and regional physiographic differences are very difficult to make at this time and may be misleading. A similar statement could be made about reports of the number and types of pesticides detected in different parts of the United States. Statistics on the proportion or number of occurrences of pesticides in each state are likely to be largely a function of the field and laboratory procedures used and the degree to which highly vulnerable areas to pesticide contamination have been targeted for sampling.

Temporal variability. Many studies report marked seasonal or episodic fluctuations in the concentrations of pesticides in ground water, but questions remain about how to generalize the results from any given study. The extent of seasonal or episodic fluctuations is a complex function of relations among the timing of pesticide applications, periods of recharge (and runoff), persistence of the pesticide, properties of the vadose zone, and position within the flow system of the producing zone of a well. A number of scenarios are possible from different combinations of these factors. For example, rapidly degraded pesticides may be detected in shallow ground water only if significant recharge occurs shortly after application. On the other hand, more persistent pesticides may display large concentrations in shallow ground water during spring recharge a year or more after their application. Many situations between these two extremes can occur. In addition to seasonal and episodic fluctuations, substantial differences in pesticide occurrence have been attributed to climatic variations by studies encountering differences in weather conditions (Keim et al. 1989; Kross et al. 1990); these results emphasize the value of long-term studies.

Local-scale studies. More focused sampling of ground water, directed along flow paths through the subsurface, is needed to better understand both the depth variations and temporal fluctuations of pesticides in ground water. Data from isolated well clusters are difficult to interpret with respect to the causes of observed variations with depth. Likewise, temporal sampling from individual wells provides little insight on the causes of temporal variability. Future progress in understanding pesticide contamination of ground water may be best made through studies that bridge the gap between traditional field-plot studies of the vadose zone and broad surveys of pesticides in ground water. The scales of study must better match the scales of pesticide contamination, while at the same time maintaining a regional perspective. Unfortunately, a notable lack of integrated studies of ground water and the vadose zone exist that examine pesticide behavior throughout the subsurface.

Well type. The type of well and general siting practices for wells with respect to areas of pesticide applications may have a significant effect on reported detections of pesticides (Cohen, Eiden, and Lorber 1986).

Association of nitrate and pesticide contamination. Many studies have attempted to relate pesticide detections and concentrations to nitrate (cf. Spalding, Junk, and Richard 1980; Pionke et al. 1988; U.S. Environmental Protection Agency 1992a). These efforts have met with varying degrees of success. Complications in relating pesticide detections or concentrations to nitrate concentrations in ground water include the following: (1) the sources of nitrate, such as septic systems and animal feedlots, may be quite different from the sources of pesticides; (2) the relative application rates of pesticides and fertilizers will differ among agricultural fields; (3) denitrification can remove the nitrate derived from cropped areas, but the pesticides could still be present; (4) pesticides can be sorbed or degraded in areas with little denitrification; and (5) the reporting levels of nitrate and pesticide analytical methods can differ relative to fertilizer and pesticide application rates (i.e., represent different dilution ratios). Thus, the relation between nitrate and pesticides must be evaluated in each area based on the commonalities of sources, the potential for denitrification, and the properties of specific pesticides relative to those of nitrate. Moreover, it may be necessary to account for pesticide metabolites in searching for relations between nitrate and pesticides. In some cases, other constituents associated with agricultural practices, such as potassium and chloride from potash fertilizers, may be better correlated with pesticide occurrence.

ACKNOWLEDGMENTS

Approved for publication as Florida Agricultural Experiment Station Journal Series No. R-02355. The assistance of Denie Augustijn and Ron Jessup in performing the PRZM model simulations shown in Figures 14-2 and 14-3, and the assistance of Molly Maupin in preparing the relative vulnerability map for the Snake River Plain shown in Figure 14-5 are greatly appreciated.

References

Adams, C. D., and E. M. Thurman. 1991. Formation and transport of dethylatrazine in the soil and vadose zone. *Journal of Environmental Quality* 20:540–7.

Aller, Linda, T. Bennett, J. H. Lehr, R. J. Petty, and G. Hackett. 1987. *DRASTIC-A Standardized System for Evaluating Ground-Water Pollution Potential Using Hydrogeologic Settings*. Ada, OK: U.S. Environmental Protection Agency Report 600/2-87/035.

Anderson, M. P. 1986. Field validation of ground water models. In *Evaluation of Pesticides in Ground Water*, eds. W. Y. Garner, R. C. Honeycutt, and H. N. Nigg, pp. 396–412. Washington, DC: American Chemical Society Symposium Series 315.

Beven, K. 1991. Modeling preferential flow: An uncertain future? In *Preferential Flow*, eds. T. J. Gish and A. Shirmohammadi, pp. 1–11. St. Joseph, MI: American Society of Agricultural Engineers, Proceedings National Symposium, Dec. 16–17, 1991.

Biggar, J. W., and J. N. Seiber, eds. 1987. *Fate of Pesticides in the Environment*. Oakland, CA: Agricultural Experiment Station, University of California Publication 3320.

Boesten J. J. T. I., and A. M. A. van der Linden. 1991. Modeling the influence of sorption and transformation on pesticide leaching and persistence. *Journal of Environmental Quality* 20:425–35.

Bollag, J.-M., and S.-Y. Liu. 1990. Biological transformation processes of pesticides. In *Pesticides in the Soil Environment: Processes, Impacts, and Modeling*, ed. H. H. Cheng, pp. 169–211. Madison, WI: Soil Science Society of America.

Brusseau, M. L., and P. S. C. Rao. 1989. Sorption nonideality during organic contaminant transport in porous media. *CRC Critical Reviews in Environmental Control* 19:33–99.

Brusseau, M. L., and P. S. C. Rao. 1991. Sorption kinetics of organic chemicals: Methods, models, and mechanisms. In *Rates of Soil Chemical Processes*, eds. D. L. Sparks and D. L. Suarez, pp. 281–302. Madison, WI: Soil Science Society of America Special Publication 27.

Cardozo, C., M. Pepple, J. Troiano, D. Weaver, B. Fabre, S. Ali, and S. Brown. 1988. *Sampling for Pesticide Residues in California Well Water: 1988 Update—Well Inventory Data Base*. Sacramento, CA: California Department of Food and Agriculture.

Carsel, R. F., C. N. Smith, L. A. Mulkey, J. D. Dean, and P. P. Jowsie. 1984. *User's Manual for the Pesticide Root Zone Model (PRZM): Release 1*. Washington, DC: U.S. Environmental Protection Agency EPA-600/3-84-109.

Cauller, S. J., and D. A. V. Eckhardt. 1990. Relation between land use and quality of shallow ground water in central and eastern Long Island, New York. (abs.) *Ground Water* 28(5):792–3.

Chen, H., and A. D. Druliner. 1988. Agricultural chemical contamination of ground water in six areas of the High Plains aquifer, Nebraska. *National Water Summary 1986—Hydrologic Events and Ground-Water Quality*, pp. 103–8. Reston, VA: U.S. Geological Survey Water-Supply Paper 2325.

Cheng, H. H., ed. 1990. *Pesticides in the Soil Environment: Processes, Impacts, and Modeling*. Madison, WI: Soil Science Society of America, SSSA Book Series No. 2.

Chiou, C. T., P. E. Porter, and V. H. Freed. 1983. Partition equilibria of nonionic organic compounds between soil organic matter and water. *Environmental Science and Technology* 17:227–31.

Cohen, S. Z., C. Eiden, and M. N. Lorber. 1986. Monitoring ground water for pesticides. In *Evaluation of Pesticides in Ground Water*, eds. W. Y. Garner, R. C. Honeycutt, and H. N. Nigg, pp. 170–96. Washington, DC: American Chemical Society Symposium Series 315.

Davidson, J. M., P. S. C. Rao, L. T. Ou, W. B. Wheeler, and D. F. Rothwell. 1980. Adsorption, movement, and biological degradation of large concentrations of selected pesticides in soils. Cincinnati, OH: U.S. Environmental Protection Agency EPA-600/2-80-124.

Dean, J. D., and A. F. Atwood. 1987. *Exposure Assessment Modelling of Aldicarb in Florida*. Washington, DC: U.S. Environmental Protection Agency EPA-600/3-85/051.

Denver, J. M., and M. W. Sandstrom. 1991. Distribution of dissolved atrazine and two metabolites in the unconfined aquifer, southeastern Delaware. In *U.S. Geological Survey Toxic Substances Hydrology Program —Proceedings of the Technical Meeting, Monterey, California*, March 11–15, 1991, eds. G. E. Mallard and D. A. Aronson, pp. 314–18. Reston, VA: U.S. Geological Survey Water-Resources Investigations Report 91-4034.

Domagalski, J. L., and N. M. Dubrovsky. 1991. *Regional Assessment of Nonpoint-Source Pesticide Residues in Ground Water, San Joaquin Valley, California*. Sacramento, CA: U.S. Geological Survey Water-Resources Investigations Report 91-4027.

Duncan, D., D. T. Pederson, T. R. Sheperd, and J. D. Carr. 1991. Atrazine used as a tracer of induced recharge. *Ground Water Monitoring Review* 11(4):144–50.

Eckhardt, D. A., S. F. Siwiec, and S. J. Cauller. 1989. Regional appraisal of ground-water quality in five different land-use areas, Long Island, New York. In *U.S. Geological Survey Toxic Substances Hydrology Program—Proceedings of the Technical Meeting, Phoenix, Arizona*, September 26–30, 1988, pp. 397–403. Reston, VA: U.S. Geological Survey Water-Resources Investigations Report 88-4220.

Fathulla, R. N., R. A. Jones, J. M. Harkin, and G. Chesters. 1988. Distribution of aldicarb residues in the sand and gravel aquifer of central Wisconsin. 1: Relationship between aldicarb residue concentration and groundwater chemistry. In *Advances in Environmental Modelling*, ed. A. Marani, pp. 59–84. Amsterdam: Elsevier.

Germann, P. F., ed. 1988. Rapid and far-reaching hydrologic processes in the vadose zone (15 papers). *Journal of Contaminant Hydrology* 3:115–380.

Gerstl, Z. 1990. Estimation of organic chemical sorption by soils. *Journal of Contaminant Hydrology* 6(4):357–75.

Gianessi, L. P., and C. A. Puffer. 1990. *Herbicide Use in the United States: National Summary Report*. Washington, DC: Resources for the Future.

Gianessi, L. P., and C. A. Puffer. 1991. *Estimation of County Pesticide Use on Golf Courses and by Urban Applicators*. Washington, DC: Resources for the Future.

Green, R. E., and S. W. Karickhoff. 1990. Sorption estimates for modeling. In *Pesticides in the Soil Environment: Processes, Impacts, and Modeling*, ed. H. H. Cheng, pp. 79–101. Madison, WI: Soil Science Society of America.

Grover, R., ed. 1988. *Environmental Chemistry of Herbicides*, Vol. I. Boca Raton, FL: CRC Press.

Grover, R., and A. J. Cessna, eds. 1988. *Environmental Chemistry of Herbicides*, Vol. II. Boca Raton, FL: CRC Press.

Gustafson, D. I. 1989. Ground-water ubiquity score: A simple method for assessing pesticide leachability. *Environmental Toxicology and Chemistry* 8:339–57.

Hallberg, G. R. 1986. From hoes to herbicides: Agriculture and ground-water quality. *Journal of Soil and Water Conservation* 41(6):357–64.

Hallberg, G. R. 1989. Pesticide pollution of groundwater in the humid United States. *Agriculture, Ecosystems, and Environment* 26:299–367.

Hamaker, J. W. and J. M. Thompson. 1972. Adsorption. In *Organic Chemicals in the Soil Environment*, Vol. 1, eds. C. A. I. Goring and J. W. Hamaker, pp. 49–143. New York: Marcel Dekker.

Harkin, J. M., F. A. Jones, R. N. Fathulla, E. K. Dzantor, and D. G. Kroll. 1986. Fate of aldicarb in Wisconsin ground water. In *Evaluation of Pesticides in Ground Water*, eds. W. Y. Garner, R. C. Honeycutt, and H. N. Nigg, pp. 219–55. Washington, DC: American Chemical Society Symposium Series 315.

Holden, L. R., J. A. Graham, R. W. Whitmore, W. J. Alexander, R. W. Pratt, S. K. Liddle, and L. L. Piper. 1992. Results of the National Alachlor Well Water Survey. *Environmental Science and Technology* 26(5):935–43.

Holden, P. W. 1986. *Pesticides and Groundwater Quality: Issues and Problems in Four States*. Washington, DC: National Academy Press.

Hornsby, A. G., P. S. C. Rao, J. G. Booth, P. V. Rao, K. D. Pennell, R. E. Jessup, and G. D. Means. 1990. *Evaluation of Models for Predicting Fate of Pesticides*. Project Completion Report submitted to Florida Department of Environmental Regulation. Gainesville, FL: Soil Science Department, University of Florida.

Howard, P. H., R. S. Boethling, W. F. Jarvis, W. M. Meylan, and E. M. Michalenko. 1991. *Handbook of Environmental Degradation Rates*. Boca Raton, FL: Lewis Publishers.

Hoyer, B. E., and G. R. Hallberg. 1991. *Groundwater Vulnerability Regions of Iowa*. Iowa City, IA: Iowa Department of Natural Resources Special Map 11.

Huyakorn, P. S., J. B. Kool, and T. D. Wadsworth. 1988. A comprehensive simulation of aldicarb transport at the Wickham Site on Long Island. In *International Conference and Workshop on the Validation of Flow and Transport Models for the Unsaturated Zone*, Ruidoso, New Mexico, May 23–26, 1988, eds. P. J. Wierenga and D. Bachelet, pp. 176–86. New Mexico State University.

Jones, R. L. 1986. Field, laboratory, and modeling studies on the degradation and transport of aldicarb residues in soil and groundwater. In *Evaluation of Pesticides in Ground Water*, eds. W. Y. Garner, R. C. Honeycutt, and H. N. Nigg, pp. 197–218. Washington, DC: American Chemical Society Symposium Series 315.

Jones, R. L., A. G. Hornsby, P. S. C. Rao, and M. P. Anderson. 1987. Movement and degradation of aldicarb residues in the saturated zone under citrus groves on the Florida ridge. *Journal of Contaminant Hydrology* 1:265–85.

Jury, W. A., H. Elabd, and M. Resketo. 1986. Field study of napropamide movement through unsaturated soil. *Water Resources Research* 22(5):749–55.

Jury, W. A., D. D. Focht, W. J. Farmer. 1987. Evaluation of pesticide groundwater pollution potential from standard indices of soil-chemical adsorption and biodegradation. *Journal of Environmental Quality* 16(4):422–8.

Jury, W. A., W. F. Spencer, and W. J. Farmer. 1983. Behavior assessment model for trace organics in soil. I: Model description. *Journal of Environmental Quality* 12(4):558–64.

Jury, W. A., W. J. Farmer, and W. F. Spencer. 1984a. Behavior assessment model for trace organics in soil. II: Chemical classification and parameter sensitivity. *Journal of Environmental Quality* 13(4):567–72.

Jury, W. A., W. F. Spencer, and W. J. Farmer.

1984b. Behavior assessment model for trace organics in soil. III: Application of screening model. *Journal of Environmental Quality* 13(4):573–9.

Jury, W. A., W. F. Spencer, and W. J. Farmer. 1984c. Behavior assessment model for trace organics in soil. IV: Review of experimental evidence. *Journal of Environmental Quality* 13(4):580–6.

Keim, A. M., L. C. Ruedisili, D. B. Baker, and R. E. Gallagher. 1989. Herbicide monitoring of tile drainage and shallow groundwater in northwestern Ohio farm fields—a case study. In *Pesticides in Terrestrial and Aquatic Environments*, ed. D. L. Weigmann, pp. 62–78. Blacksburg, VA: Virginia Water Resources Research Center and Virginia Polytechnic Institute and State University.

Khan, M. A., and T. Liang. 1989. Mapping pesticide contamination potential. *Environmental Management* 13(2):233–42.

Klaseus, T. G., and J. W. Hines. 1989. *Pesticides and Groundwater: A Survey of Selected Private Wells in Minnesota*. Minneapolis, MN: Minnesota Department of Health, Report prepared for the U.S. Environmental Protection Agency, Office of Ground Water, Region 5.

Klaseus, T. G., G. C. Buzicky, and E. C. Schneider. 1988. *Pesticides and Groundwater: Surveys of Selected Minnesota Wells*. Minneapolis, MN: Minnesota Department of Health and Minnesota Department of Agriculture, Report Prepared for the Legislative Commission on Minnesota Resources.

Kleveno, J. J., K. M. Loague, and R. E. Green. 1992. Evaluation of a pesticide mobility index: Impact of recharge variation and soil profile heterogeneity. *Journal of Contaminant Hydrology* 11:83–99.

Kraft, G. J., and P. A. Helmke. 1992. Dependence of aldicarb residue degradation rates on groundwater chemistry in the Wisconsin central sands. *Journal of Environmental Quality* 21:368–72.

Kross, B. C., and twenty others. 1990. *The Iowa State-Wide Rural Well-Water Survey Water-Quality Data: Initial Analysis*. Iowa City, IA: Iowa Department of Natural Resources Technical Information Series 19.

Kung, K-J. S. 1990a. Preferential flow in a sandy vadose zone. 1: Field observation. *Geoderma* 46:51–58.

Kung, K-J. S. 1990b. Preferential flow in a sandy vadose zone. 2: Mechanism and implications. *Geoderma* 46:59–71.

Leonard, R. A. 1990. Movement of pesticides into surface waters. In *Pesticides in the Soil Environment: Processes, Impacts, and Modeling*, ed. H. H. Cheng, pp. 303–49. Madison, WI: Soil Science Society of America.

Liszewski, M. J., and P. J. Squillace. 1991. The effect of surface-water and ground-water exchange on the transport and storage of atrazine in the Cedar River, Iowa. In *U.S. Geological Survey Toxic Substances Hydrology Program—Proceedings of the Technical Meeting, Monterey, California, March 11–15, 1991*, eds. G. E. Mallard and D. A. Aronson, pp. 195–202. Reston, VA: U.S. Geological Survey Water-Resources Investigations Report 91-4034.

Loague, K. M. 1991. The impact of land use on estimates of pesticide leaching potential: Assessments and uncertainties. *Journal of Contaminant Hydrology* 8(2):157–75.

Loague, K. M., R. E. Green, T. W. Giambelluca, T. C. Liang, and R. S. Yost. 1990. Impact of uncertainty in soil, climate, and chemical information in a pesticide leaching assessment. *Journal of Contaminant Hydrology* 5(2):171–94.

Loague, K. M., R. S. Yost, R. E. Green, and T. C. Liang. 1989. Uncertainty in a pesticide leaching assessment for Hawaii. *Journal of Contaminant Hydrology* 4:139–61.

Lyman, W. J. 1990. Adsorption coefficients for soils and sediments. In *Handbook of Chemical Property Estimation Methods*, eds. W. J. Lyman, W. F. Reehl, and D. H. Rosenblatt, pp. 4-1–4-33. Washington, DC: American Chemical Society.

McDowell-Boyer, L. M., J. R. Hunt, and N. Sitar. 1986. Pesticide transport through porous media. *Water Resources Research* 22(13):1901–21.

McKenna, D. P., and D. A. Keefer. 1991. *Potential for Agricultural Chemical Contamination of Aquifers in Illinois*. Champaign, IL: Illinois State Water Survey Open File 1991-7R.

Meeks, Y. J., and J. D. Dean. 1990. Evaluating ground-water vulnerability to pesticides. *Journal of Water Resources Planning and Management* 116(5):693–707.

Miller, G. C., and V. R. Hebert. 1987.

Environmental photodecomposition of pesticides. In *Fate of Pesticides in the Environment*, eds. J. W. Biggar and J. N. Seiber, pp. 75–86. Oakland, CA: Agricultural Experiment Station, University of California Publication 3320.

Mingelgrin, U., and Z. Gerstl. 1983. Reevaluation of partitioning as a mechanism of nonionic chemicals adsorption in soils. *Journal of Environmental Quality* 12:1–11.

Moody, D. W., J. E. Carr, E. B. Chase, and R. W. Paulson, eds. 1988. *National Water Summary 1986—Hydrologic Events and Ground-Water Quality*. Reston, VA: U.S. Geological Survey Water-Supply Paper 2325.

Nash, R. G. 1974. Plant uptake of insecticides, fungicides, and fumigants from soils. In *Pesticides in Soils and Water*, ed. W. G. Guenzi, pp. 257–313. Madison, WI: Soil Science Society of America.

Nash, R. G. 1988. Dissipation from soil. In *Environmental Chemistry of Herbicides*, Vol. I, ed. R. Grover, pp. 131–170. Boca Raton, FL: CRC Press.

Nash, R. G., C. S. Helling, S. E. Ragone, and A. R. Leslie. 1991. Groundwater residue sampling—Overview of the approach taken by government agencies. In *Groundwater Residue Sampling Design*, eds. R. G. Nash and A. R. Leslie, pp. 1–13. Washington, DC: American Chemical Society ACS Symposium Series 465.

National Agricultural Statistics Service. 1991a. *Agricultural Chemical Usage: 1990 Field Crops Summary*. Washington, DC: U.S. Department of Agriculture.

National Agricultural Statistics Service. 1991b. *Agricultural Chemical Usage: 1990 Vegetables Summary*. Washington, DC: U.S. Department of Agriculture.

National Research Council. 1993. *Ground-Water Vulnerability Assessment: Predicting Relative Contamination Potential Under Conditions of Uncertainty*. Washington, DC: National Academy Press (in Press).

Nofziger, D. L., and A. G. Hornsby. 1987. *CMLS: Interactive Simulation of Chemical Movement in Layered Soils*. Gainesville, FL: Institute of Food and Agricultural Sciences, University of Florida Circular 780.

O'Connor, G. A., R. L. Chaney, and J. A. Ryan. 1991. Bioavailability to plants of sludge-borne toxic organics. *Reviews of Environmental Contamination and Toxicology* 121:129–55.

Ou, L. T., D. F. Rothwell, W. B. Wheeler, and J. M. Davidson. 1978. The effect of high 2,4-D concentrations on degradation and carbon dioxide evolution in soils. *Journal of Environmental Quality* 7:241–6.

Pennell, K. D., A. G. Hornsby, R. E. Jessup, and P. S. C. Rao. 1990. Evaluation of five simulation models for predicting aldicarb and bromide behavior under field conditions. *Water Resources Research* 26(11):2679–93.

Petach, M. C., R. J. Wagenet, and S. D. DeGloria. 1991. Regional water flow and pesticide leaching using simulations with spatially distributed data. *Geoderma* 48:245–69.

Pettyjohn, W. A., M. Savoca, and D. Self. 1991. *Regional Assessment of Aquifer Vulnerability and Sensitivity in the Conterminous United States*. Ada, OK: U.S. Environmental Protection Agency Report EPA/600/2-91/043.

Pionke, H. B., D. E. Glotfelty, A. D. Lucas, and J. D. Urban. 1988. Pesticide contamination of groundwaters in the Mahantango Creek watershed. *Journal of Environmental Quality* 17:76–84.

Rao, P. S. C., V. E. Berkheiser, and L. T. Ou. 1984. *Estimation of Parameters for Modelling the Behavior of Selected Pesticides and Orthophosphate*. Washington, DC: U.S. Environmental Protection Agency EPA-600/3-84-019.

Rao, P. S. C., and J. M. Davidson. 1980. Estimation of pesticide retention and transformation parameters required in nonpoint source pollution models. In *Environmental Impacts of Nonpoint Source Pollution*, eds. M. R. Overcash and J. M. Davidson, pp. 23–67. Ann Arbor, MI: Ann Arbor Science Publishers.

Rao, P. S. C., and A. G. Hornsby. 1989. *Behavior of Pesticides in Soils and Water*. Gainesville, FL: Institute of Food and Agricultural Sciences, University of Florida Soil Science Fact Sheet SL-40 (revised).

Rao, P. S. C., A. G. Hornsby, and R. E. Jessup. 1985. Indices for ranking the potential for pesticide contamination of groundwater. *Proceedings, Soil and Crop Science Society of Florida* 44:1–8.

Rao, P. S. C., R. E. Jessup, and W. M. Alley. 1992. Empirical and model-based approaches for regional-scale assessments of groundwater contamination: Illusion of certainty. Paper presented at Beltsville Symposium XVII: *Agricultural Water Quality Priorities, A Team Approach to Conserving Natural Resources*, May 4–8, 1992. Beltsville, MD.

Richard, J. J., G. A. Junk, M. J. Avery, N. L. Nehring, J. S. Fritz, and H. J. Svec. 1975. Analysis of various Iowa waters for selected pesticides: Atrazine, DDE, and dieldrin-1974. *Pesticide Monitoring Review* 9:117–23.

Ritter, W. F. 1990. Pesticide contamination of ground water in the United States—A review. *Journal of Environmental Science and Health B* 25(1):1–29.

Rothschild, E. R., R. J. Manser, and M. P. Anderson. 1982. Investigation of aldicarb in ground water in selected areas of the Central Sand Plain of Wisconsin. *Ground Water* 20(4):437–45.

Rupert, M., T. Dace, M. Maupin, and B. Wicherski. 1991. *Ground Water Vulnerability Assessment: Snake River Plain, Southern Idaho*. Boise, ID: Idaho Department of Health and Welfare.

Schmidt, K. D. 1986. DBCP in ground water of the Fresno-Dinuba area, California. In *Proceedings of the Agricultural Impacts on Ground Water Conference*, pp. 511–29. Dublin, OH: National Water Well Association.

Scow, K. M. 1990. Rate of biodegradation. In *Handbook of Chemical Property Estimation Methods*, eds. W. J. Lyman, W. F. Reehl, and D. H. Rosenblatt, pp. 9-1–9-85. Washington, DC: American Chemical Society.

Scow, K. M., and M. Alexander. 1992. Effect of diffusion on the kinetics of biodegradation: Experimental results with synthetic aggregates. *Soil Science Society of America Journal* 56:128–34.

Scow, K. M., and J. Hutson. 1992. Effect of diffusion and sorption on the kinetics of biodegradation: Theoretical considerations. *Soil Science Society of America Journal* 56:119–27.

Smith, A. E. 1988. Transformations in soil. In *Environmental Chemistry of Herbicides*, Vol. I, ed. R. Grover, pp. 171–200. Boca Raton, FL: CRC Press.

Spalding, R. F., G. A. Junk, and J. J. Richard. 1980. Pesticides in ground water beneath irrigated farming in Nebraska, August 1978. *Pesticides Monitoring Journal* 14(2):70–73.

Squillace, P. J., E. M. Thurman, and E. Furlong. 1993. Groundwater as a nonpoint source of atrazine and deethylatrazine in a river during base-flow conditions. *Water Resources Research* (in press).

Taylor, A. W., and W. F. Spencer. 1990. Volatilization and vapor transport processes. In *Pesticides in the Soil Environment: Processes, Impacts, and Modeling*, ed. H. H. Cheng, pp. 213–69. Madison, WI: Soil Science Society of America.

Teso, R. R., T. Younglove, M. R. Peterson, D. L. Sheeks III, and R. E. Gallavan. 1988. Soil taxonomy and surveys: Classification of areal sensitivity to pesticide contamination of groundwater. *Journal of Soil and Water Conservation* 43(4):348–52.

Thurman, E. M., D.A. Goolsby, M. T. Meyer, and D. W. Kolpin. 1991. Herbicides in surface waters of the midwestern United States: The effect of spring flush. *Environmental Science and Technology* 25(10):1794–6.

U.S. Environmental Protection Agency. 1987. *Agricultural Chemicals in Ground Water: Proposed Pesticide Strategy*. Washington, DC: U.S. Environmental Protection Agency EPA 440/6-88-001.

U.S. Environmental Protection Agency. 1990. *National Survey of Pesticides in Drinking Water Wells Phase I Report*. Washington, DC: U.S. Environmental Protection Agency EPA 570/9-90-015.

U.S. Environmental Protection Agency. 1992a. *Another Look: National Survey of Pesticides in Drinking Water Wells Phase II Report*. Washington, DC: U.S. Environmental Protection Agency EPA 579/09-91-020.

U.S. Environmental Protection Agency. 1992b. *A Review of Methods for Assessing the Sensitivity/Vulnerability of Aquifers to Pesticide Contamination*. Washington, DC: U.S. Environmental Protection Agency.

U.S. Office of Technology Assessment. 1990. *Beneath the Bottom Line: Agricultural Approaches to Reduce Agrichemical Contamination of Groundwater*. Washington, DC: U.S. Government Printing Office, OTA-F-418.

Vanderlaan, Martin, L. H. Stanker, B. E. Watkins, and D. W. Roberts, eds. 1990. *Immunoassays for Trace Chemical Analysis*.

Washington, DC: American Chemical Society Symposium Series 451.

van Genuchten, M. Th., D. E. Ralston, and P. F. Germann, eds. 1990. Transport of water and solutes in macropores (special issue). *Geoderma* 46:1–297.

Wagenet, R. J., and J. L. Hutson. 1989. *LEACHM: A Finite-Difference Model for Simulating Water, Salt, and Pesticide Movement in the Plant Root Zone*. Ithaca, NY: New York State Water Resources Research Institute, Cornell University, Continuum Vol. 2.

Wagenet, R. J., and P. S. C. Rao. 1990. Modeling pesticide fate in soils. In *Pesticides in the Soil Environment: Processes, Impacts, and Modeling*, ed. H. H. Cheng, pp. 351–99. Madison, WI: Soil Science Society of America.

Wauchope, R. D., T. M. Butler, A. G. Hornsby, P. W. M. Augustijn-Beckers, and J. P. Burt. 1991. The SCS/ARS/CES pesticides properties database for environmental decision-making. *Reviews in Environmental Contaminant Toxicology* 123:1–155.

Wolfe, N. L., U. Mingelgrin, and G. C. Miller. 1990. Abiotic transformations in water, sediments, and soil. In *Pesticides in the Soil Environment: Processes, Impacts, and Modeling*, ed. H. H. Cheng, pp. 103–68. Madison, WI: Soil Science Society of America.

15

Pathogens

Marylynn V. Yates and Scott R. Yates

INTRODUCTION

Ground water supplies over 100 million Americans with their drinking water; in rural areas there is an even greater reliance on ground water as it constitutes up to 95% of the water used (Bitton and Gerba 1984). It has been assumed traditionally that ground water is safe for consumption without treatment because the soil acts as a filter to remove pollutants. As a result, private wells generally do not receive treatment (DiNovo and Jaffe 1984), nor do a large number of public water-supply systems. The U.S. Environmental Protection Agency (USEPA) has estimated that approximately 72% of the public water-supply systems in the United States that use ground water do not disinfect (USEPA 1990). However, use of contaminated, untreated, or inadequately treated ground water has been the cause of approximately 50% of the waterborne disease outbreaks in this country since 1920 (Craun 1986a, 1986b, 1991; Herwaldt et al. 1992). The majority of the outbreaks were caused by pathogenic (disease-causing) microorganisms.

Characteristics of Microorganisms

Bacteria are microscopic organisms, ranging from approximately 0.2 to 10 µm in length. They are distributed ubiquitously in nature and have a wide variety of nutritional requirements. Many types of harmless bacteria colonize the human intestinal tract and are routinely shed in the feces. One group of intestinal bacteria, the coliform bacteria, has historically been used as an indication that an environment has been contaminated by human sewage. In addition, pathogenic bacteria, such as *Salmonella* and *Shigella*, are present in the feces of infected individuals. Thus, a wide variety of bacteria is introduced into septic tanks. Many of these bacteria can survive and grow in septic tanks, and are present in the effluent when it moves to the soil absorption field.

Viruses are obligate intracellular parasites; that is, they are incapable of replication outside of a host organism. They are very small, ranging in size from approximately 20 to 200 nm. Viruses that replicate in the intestinal tract of man are referred to as human enteric viruses. These viruses are shed in the fecal material of individuals who are infected either purposely (i.e., by vaccination) or inadvertently through consumption of contaminated food or water, by swimming in contaminated water, or by personal contact with an infected individual. More than 100 different enteric viruses may be excreted in human fecal material (Melnick and Gerba

1980). As many as 10^6 plaque-forming units (pfu) of enteroviruses (a subgroup of the enteric viruses) per gram and 10^{10} rotaviruses per gram may be present in the feces of an infected individual (Tyrrell and Kapikian 1982). Thus, viruses are present in domestic sewage and, depending on the type of treatment process(es) used, between 50 and 99.9999998% (i.e., from less than one to greater than eight orders of magnitude) of the viruses are inactivated during sewage treatment (Stewart 1990).

A third group of microorganisms of concern in domestic sewage is the protozoan parasites. In general, protozoan parasite cysts (the resting stage of the organism which is found in sewage) are larger than bacteria, although they can range in size from 2 µm to over 60 µm. Protozoan parasites are present in the feces of infected persons; however, they also can be excreted by healthy carriers. Cysts are similar to viruses in that they do not reproduce in the environment, but are capable of surviving in the soil for months or even years, depending on environmental conditions.

Sources of Microorganisms

Microorganisms may be introduced into the subsurface environment in a variety of ways. In general, any practice that involves the application of domestic wastewater to the soil has the potential to cause microbiological contamination of ground water. This is because the treatment processes to which the wastewater are subjected do not effect complete removal or inactivation of the disease-causing microorganisms present. Goyal, Keswick, and Gerba (1984) isolated viruses from the ground water beneath cropland being irrigated with sewage effluent. Viruses have been detected in the ground water at several sites practicing land treatment of wastewater; these cases were reviewed by Keswick and Gerba (1980). The burial of disposable diapers in sanitary landfills is a means by which pathogenic microorganisms in untreated human waste may be introduced into the subsurface. Vaughn et al. (1978) detected viruses as far as 408 m downgradient of a landfill site in New York. Land application of treated sewage effluent for the purposes of ground-water recharge has also resulted in the introduction of viruses to the underlying ground water (Vaughn and Landry 1977, 1978).

Septic-tank effluent may be the most significant source of pathogenic bacteria and viruses in the subsurface environment. Septic tanks are the source of approximately 1 trillion gallons of waste disposed to the subsurface every year (Office of Technology Assessment 1984), and are the most frequently reported source of ground-water contamination (U.S. Environmental Protection Agency 1977). The overflow or seepage of sewage, primarily from septic tanks and cesspools, was responsible for 25% of the reported outbreaks caused by the use of untreated ground water from 1971 to 1985 (Table 15-1).

There have been several waterborne disease outbreaks documented to have been caused by the contamination of ground water with septic-tank effluent; many of these have been reviewed by Yates (1985). Twelve hundred people in a town of 6,500 developed acute gastroenteritis after consuming tap water which had been contaminated by septic-tank effluent (Craun 1981). A dye tracer was used to show that the source of

TABLE 15-1. Causes of Waterborne Disease Outbreaks in Untreated Ground-Water Systems, 1971–1985

Cause	Outbreaks (%)
Overflow or seepage of sewage into wells and springs	25
Surface runoff or flooding from contaminated streams	13
Chemical contamination	8
Contamination through limestone or fissured rock	5
Improper system construction	3
Insufficient data to classify	46

Source: From Craun (1990).

contamination was a septic tank located 49 m (150 ft) from the city's drinking-water well. Effluent from a septic tank serving a household which had recently had infectious hepatitis contaminated a well used to make commercial ice, resulting in a 98-person outbreak of hepatitis (Craun 1979). A drinking-water spring contaminated with septic-tank effluent was responsible for over 400 persons developing gastroenteritis caused by a Norwalk-virus-like agent (Craun 1984). More recently, 900 persons developed gastroenteritis caused by a Norwalk-like virus after consuming well water that had been contaminated by an onsite sewage-treatment facility (Herwaldt et al. 1992).

Another source of microorganisms in the subsurface is municipal sludge. Land application of municipal sludge is becoming a more common practice as alternatives are sought for the disposal of the ever-increasing amounts of sludge produced in this country.

The sludge that is produced during the process of treating domestic sewage contains high levels of nitrogen and other nutrients that are required by plant materials. However, it also contains pathogenic microorganisms at concentrations sufficient to cause disease in exposed individuals.

Several studies conducted in the late 1970s suggested that viruses are tightly bound to sewage solids and are not easily released into the soil. However, in a more recent study, viruses were detected in a 3-m deep well at a site where anaerobically digested sludge was applied to a sandy soil; this occurred 11 weeks after sludge application (Jorgensen and Lund 1985).

Waterborne Disease in the United States

Between 1920 and 1990, 1,674 outbreaks of waterborne disease were reported in the United States, involving over 450,000 people

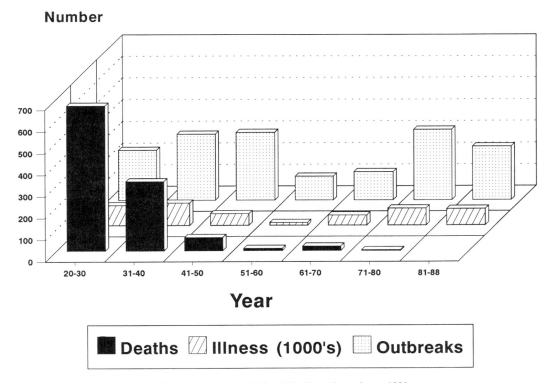

FIGURE 15-1. Waterborne-disease outbreaks, 1920–1988. (*Data from Craun 1990, 1991.*)

and resulting in 1,083 deaths (Craun 1991; Herwaldt et al. 1992). The data are summarized for 10-year periods in Figure 15-1 (Craun 1991). The number of reported outbreaks and the number of associated cases of illness have risen dramatically since 1971, as compared with the period 1951–70. From 1971 to 1980, an average of 32.6 outbreaks per year were reported. From 1981 through 1988, the average was 31, as compared with averages of 11 and 13 for 1951 to 1960 and from 1961 through 1970, respectively (Craun 1991). The increase in reported numbers of outbreaks may be due to an improved reporting system implemented in 1971 (Craun 1985); however, it is still believed that only a fraction of the total number of outbreaks is reported (Lippy and Waltrip 1984). Based on survey data from the Centers for Disease Control, it has been estimated that waterborne infections affect 940,000 people and are responsible for 900 deaths every year in the United States (Bennett et al. 1987).

Causative agents of illness were identified in approximately one-half the disease outbreaks from 1971 to 1990 (Table 15-2). The most commonly identified causative agents were *Giardia*, chemicals, and *Shigella* spp. *Giardia lamblia* caused over 18% of the illness associated with waterborne-disease outbreaks. Enteric viruses (viral gastroenteritis and hepatitis A) were identified as the causative agents of disease in 8.8% of the outbreaks during this period.

In the 1980s the use of untreated or inadequately treated ground water was responsible for 44% of the outbreaks that occurred in the United States (Figure 15-2) (Craun 1991). When considering outbreaks that have occurred due to the consumption of contaminated, untreated ground water, from 1971 to 1985, sewage was most often identified as the contamination source (Table 15-1). In ground-water systems, etiologic (disease-causing) agents were identified in only 38% of the outbreaks, with *Shigella* sp. and hepatitis A virus being the most commonly identified pathogens (Craun 1990). In more than one half of the outbreaks, no etiologic agent could be identified, and the illness was simply listed as gastroenteritis of unknown etiology. However, retrospective serologic studies of outbreaks of acute nonbacterial gastroenteritis from 1976 through 1980 indicated that

TABLE 15-2. Causative Agents of Waterborne Disease Outbreaks, 1971–1990

Disease	Number of Outbreaks (%)	Cases of Illness (%)
Gastroenteritis, unknown cause	293 (49.66)	67,367 (47.26)
Giardiasis	110 (18.64)	26,531 (18.61)
Chemical poisoning	55 (9.32)	3,877 (2.72)
Shigellosis	40 (6.78)	8,806 (6.18)
Viral gastroenteritis	27 (4.58)	12,699 (8.91)
Hepatitis A	25 (4.24)	762 (<1)
Salmonellosis	12 (2.03)	2,370 (1.66)
Campylobacterosis	12 (2.03)	5,233 (3.67)
Typhoid fever	5 (<1)	282 (<1)
Yersiniosis	2 (<1)	103 (<1)
Cryptosporidosis	2 (<1)	13,117 (9.20)
Chronic gastroenteritis	1 (<1)	72 (<1)
Toxigenic *E. coli*	2 (<1)	1,243 (<1)
Cholera	1 (<1)	17 (<1)
Dermatitis	1 (<1)	31 (<1)
Amebiasis	1 (<1)	4 (<1)
Cyanobacteria-like bodies	1 (<1)	21 (<1)
Total	564 (100.)	138,247 (100.)

Source: Data from Craun (1991) and Herwaldt et al. (1992).

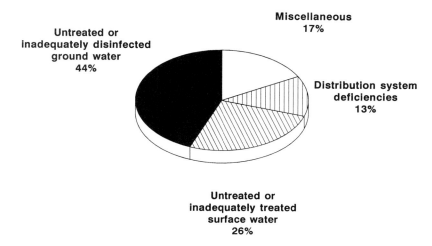

FIGURE 15-2. Causes of waterborne-disease outbreaks, 1981–1988. (*From Craun 1991.*)

42% of these outbreaks (i.e., the 62% for which no etiologic agent was identified) were caused by the Norwalk virus (Kaplan et al. 1982). Thus, it has been suggested that the Norwalk virus is responsible for approximately 23% of all reported waterborne outbreaks in the United States (Keswick et al. 1985).

The difficulty in the isolation of many enteric viruses from clinical and environmental samples probably accounts for the limited number of viruses identified as causes of waterborne disease. As methods for the detection of enteric viruses have improved, so has the percentage of waterborne disease identified as having a viral etiology (Gerba 1984a).

The fact that microorganisms are responsible for numerous waterborne disease outbreaks every year led the U.S. Environmental Protection Agency to reexamine the coliform standard, which has been used to indicate the microbiologic quality of drinking water in the United States for more than 75 years. Increasing amounts of evidence collected during the past 15 to 20 years suggest that the coliform group may not be an adequate indicator of the presence of pathogenic viruses and possibly protozoan parasites in water. For example, in a study of the removal of viruses and indicator bacteria at seven drinking-water treatment plants in Canada, none of the bacterial indicators were correlated with the concentration of viruses in the finished water (Payment, Trudel, and Plante 1985). In 1985, the USEPA proposed maximum contaminant level goals (MCLGs) for viruses and *Giardia*, a protozoan parasite (USEPA 1985). These standards are in addition to the standard for the indicator microorganism, total coliform. Rather than require public water systems to monitor the water for the presence of these pathogenic microorganisms, the USEPA (1989) proposed treatment requirements with the goal that the level of pathogenic viruses and *Giardia* in the treated water would result in a risk of less than 1 in 10,000 infections per person per year.

Detecting Pathogens in Ground Water

Detecting pathogens, especially viruses and parasites, in ground water requires the collection of very large samples; a typical sample volume is hundreds to thousands of liters. The large sample volume is necessitated by the fact that the concentration of these microorganisms is expected to be very low in ground water. However, due to the fact that one virus or parasite particle may be sufficient to cause infection, even very low concentrations can pose a threat to public health. For example, the World Health Organization (1984) has suggested that an acceptable virus concentration in drinking water is zero per 100 to 1,000 L. In the United States, using the USEPA's goal of limiting the risk of increased infections to 1 in 10,000 persons per year would result in an acceptable virus concentration of two viruses per 10^6 L of water (Regli et al. 1991).

In order to sample such large volumes, the water is passed through an electropositive or electronegative filter which adsorbs viruses present in the water. The viruses are then removed from the filter with a small volume of proteinaceous liquid such as beef extract (American Public Health Association 1985). The concentrated sample may then be analyzed for the presence of viruses by using various techniques. The traditional method involves inoculating the appropriate living cells and waiting for evidence of infection. This process may take up to three weeks to complete. However, more rapid techniques exist for detecting viruses in environmental samples, including radioimmunoassay (RIA) and enzyme-linked immunosorbent assay (ELISA). These methods require the use of a specific antibody against the virus of interest. The drawback of such methods is that very high numbers (10,000–100,000) of virus particles are required. The number of virus particles that would be expected in a concentrated ground-water sample would be orders of magnitude lower.

Recent advances in recombinant DNA technology have provided new tools that can be used to detect viruses in ground water. Nucleic-acid probes can be made to be specific to one virus or a specific group of viruses. The probes have the advantage that results are obtained rapidly; however, they are unable to differentiate between infective and noninfective virus particles. Another recent development that should make virus detection more sensitive is polymerase chain reaction (PCR) technology. PCR enables one to make copies of the nucleic acid from virus particles present in a sample, thereby amplifying one's ability to detect viruses. These new methods are currently being used for research purposes, and many improvements will have to be made before they could be generally used for detecting viruses and other pathogens in environmental samples. However, the development of this technology will undoubtedly have a major impact on the field of water microbiology.

FACTORS AFFECTING MICROBIAL FATE AND TRANSPORT

The fact that microorganisms remain infective long enough and can travel far enough in the subsurface to contaminate drinking water and cause waterborne-disease outbreaks has led to attempts to develop predictive models of microbial fate in the subsurface. In order to model the survival and transport of microorganisms in the subsurface, it is necessary to determine the factors which influence them. In addition to identifying these factors, it is necessary to quantify these effects in some way so that they can be used in the development of predictive models. Once in the subsurface, there are two major factors which control microbial fate: survival and transport. The longer a microorganism persists, the greater the chance that it will still be capable of causing infection when it reaches the ground water after being transported through the soil.

In general, both survival and transport are controlled by the specific microorganism type, the nature of the subsurface environment, and the climate of the environment (Gerba 1983). The susceptibility of microorganisms to different environmental factors varies considerably among different species as well as strains. The size and chemical composition of different microorganisms influence the extent to which they can travel or be transported in the subsurface. Soil properties play a major role in the survival and transport of bacteria, viruses and parasites. The texture of the soil, its pH, organic matter content, and moisture content all influence how long microorganisms can survive and how far they can be transported in the subsurface. Two aspects of climate are particularly important in determining microbial fate: temperature and rainfall. Microorganisms can survive for extended periods of time at low temperatures. Rainfall is important in that it can mobilize attached microorganisms and promote their migration to ground water.

More specifically, the factors that control the fate of bacteria and viruses in the subsurface are listed in Tables 15-3 and 15-4, respectively. The exact mechanism(s) whereby these factors influence the inactivation or protection of microorganisms is unknown in many instances. In some cases, it is difficult to consider the factors separately, as interactions among them undoubtedly occur.

Inactivation

The inactivation of microorganisms in water and soil has usually been described as a first-order reaction (Yates, Gerba, and Kelley 1985; Hurst, Gerba, and Cech 1980; Reddy, Khaleel, and Overcash 1981; Vilker 1981a; Hejkal and Gerba 1982). Nonlinear survival curves may result if viral aggregates are present or a significant variation exists in sensitivities among the microbial population to the factors causing inactivation. The inactivation rate described by a first-order reaction rate expression is

$$\text{Inactivation rate} = \frac{\partial C}{\partial t} = -kC \quad (15\text{-}1)$$

where C is the infective microorganism concentration at time t and k is the first-order

TABLE 15-3. Factors Influencing Bacterial Fate in the Subsurface

Factor	Influence on	
	Survival	Transport
Temperature	Bacteria survive longer at low temperatures.	Unknown
Microbial activity	Increased survival time in sterile soil	Unknown
Moisture content	Greater survival time in moist soils and during times of high rainfall.	Generally, transport increases under saturated flow conditions.
pH	Increased survival time in alkaline soils (pH > 5) than in acid soils.	Low pH enhances bacterial retention.
Salt species and concentration		Generally, increasing the concentration of ionic salts and increasing cation valences enhances bacterial adsorption.
Soil properties		Greater bacterial transport in coarse-textured soils; bacteria are retained by the clay fraction of soil.
Bacterium type	Different bacteria vary in their susceptibility to inactivation by physical, chemical, and biological factors.	Filtration and adsorption are affected by the physical and chemical characteristics of the bacterium.
Organic matter	Increased survival and possible regrowth when sufficient amounts of organic matter are present.	The accumulation of organic matter can aid in the filtration process.
Hydraulic conditions		Generally, bacterial transport increases with increasing hydraulic loads and flow rates.

Source: From Yates and Yates (1988).

TABLE 15-4. Factors Influencing Virus Fate in the Subsurface

Factor	Influence on Survival	Influence on Transport
Temperature	Viruses survive longer at lower temperatures.	Unknown
Microbial activity	Some viruses are inactivated more readily in the presence of certain microorganisms; however, adsorption to the surface of bacteria can be protective.	Unknown
Moisture content	Some viruses persist longer in moist soils than dry soils.	Generally, virus transport increases under saturated flow conditions.
pH	Most enteric viruses are stable over a pH range of 3 to 9; survival may be prolonged at near-neutral pH values.	Generally, low pH favors virus adsorption and high pH results in virus desorption from soil particles.
Salt species and concentration	Some viruses are protected from inactivation by certain cations; the reverse is also true.	Generally, increasing the concentration of ionic and increasing cation valences enhances virus adsorption.
Virus association with soil	In many cases, survival is prolonged by adsorption to soil; however, the opposite has also been observed.	Virus movement through the soil is slowed or prevented by association with soil.
Virus aggregation	Enhances survival.	Retards movement
Soil properties	Effects on survival are probably related to the degree of virus adsorption.	Greater virus transport in coarse-textured soils; there is a high degree of virus retention by the clay fraction of soil.
Virus type	Different virus types vary in their susceptibility to inactivation by physical, chemical, and biological factors.	Virus adsorption to soils is probably related to physicochemical differences in virus capsid surfaces.
Organic matter	Presence of organic matter may protect viruses from inactivation; others have found that it may reversibly retard virus infectivity.	Soluble organic matter competes with viruses for adsorption sites on soil particles.
Hydraulic conditions	Unknown	Generally, virus transport increases with increasing hydraulic loads and flow rates.

Source: From Yates and Yates (1988).

inactivation constant (t^{-1}). In Eq. 15-1, k is an expression of the sum total of all factors which influence microorganism survival. Measurement of virus inactivation has been conducted on a wide variety of surface waters, but such information on soils and ground water has been limited until recently. Reddy, Khaleel, and Overcash (1981) obtained values from virus inactivation studies during anaerobic digestion or from studies with soil columns flooded with sewage. Matthess and Pekdeger (1981) used values obtained for surface waters by Akin, Benton, and Hill (1971). Grosser (1984) conducted a survey of virus inactivation in a number of environments, and used a variety of values for virus inactivation in his studies. Vilker (1981a) discussed in detail virus inactivation observed in various environments, but values for ground water had not been determined at the time. In general, most of the previous work has lacked experimentally determined values for k in ground water and soil.

Literature reports which provide information on the inactivation rates for viruses in ground water are provided by Keswick et al. (1982), Bitton et al. (1983); Yates, Gerba, and Kelley (1985), Sobsey et al. (1986), and Pancorbo et al. (1987). The reported inactivation rates were 0.19 $\log_{10}$

day^{-1} for coxsackievirus B3 and 0.21 log$_{10}$ day^{-1} for poliovirus in an 84-m-deep water well in Houston, Texas; 0.0456 log$_{10}$ day^{-1} inactivation rate for poliovirus 1 in Florida, and a mean inactivation rate of 0.1615 log$_{10}$ day^{-1} in 11 ground-water samples collected from around the United States. Hepatitis A virus was found to have an inactivation rate of 0.0357 log$_{10}$ day^{-1} in sterile ground water and 0.143 log$_{10}$ day^{-1} in nonsterile samples (Sobsey et al. 1986). Pancorbo et al. (1987) reported an inactivation rate of 0.158 log$_{10}$ day^{-1} for human rotavirus, strain Wa.

Such information, while providing an idea of inactivation rates for a particular virus which could be used in the development of a model, does not provide information which can be applied generally to specific locations where environmental factors may dictate widely varying inactivation rates from those already reported. Yates, Gerba, and Kelley (1985) studied the influence of various factors likely to be useful in predicting virus inactivation in ground water. They found that ground-water temperature was the single most important predictor of virus inactivation. A linear regression analysis gave a correlation coefficient of 0.88, which was significant at the 0.01 level. The coefficient of determination, r^2, was 0.775, meaning that 77.5% of the variation in inactivation rates among samples could be explained by temperature. The inactivation rate for coliphage MS2 as a function of temperature was expressed by the following equation:

$$\text{Inactivation rate } (\log_{10}\#/\text{day}) = -0.181 + 0.0214T \, (°C) \quad (15\text{-}2)$$

However, viruses persisted for longer periods of time in well-water samples than had been found in experiments using surface waters incubated at similar temperatures (Yates, Gerba, and Kelley 1985). For example, Hajkal and Gerba (1982) analyzed all the published literature (143 cases) on the effect of temperature on virus inactivation in seawater and found that it varied with temperature according to the following equation:

$$\text{Inactivation rate } (\log_{10}\#/\text{day}) = -0.184 + 0.0335T \, (°C) \quad (15\text{-}3)$$

Examination of the equation developed by Yates, Gerba, and Kelley (1985) indicates that as ground-water temperatures approach about 8°C virus inactivation becomes negligible. It is probable that virus inactivation occurs at these temperatures but over much greater time periods than could be observed in laboratory experiments covering a time period of three to four months. Field studies by Stramer (1984) on the persistence of viruses leached into ground water from a septic tank appear to confirm prolonged virus survival at ground-water temperatures of 9 to 20°C. She observed persistence of poliovirus type 1 in a contaminated aquifer for over 105 days after being released from the septic tank into the aquifer.

Hurst, Gerba, and Cech (1980) studied the survival of seven viruses (poliovirus 1, echovirus 1, coxsackieviruses A9 and B3, rotavirus SA11, and bacteriophage T2 and MS2) in nine different soils to determine what soil characteristics affect virus inactivation rate. They found that soil temperature and virus adsorption to soil were the most important factors affecting virus survival. An equation was developed using stepwise multiple regression to predict virus survival in soil:

$$y = 0.1005 + 0.0025x_1 - 0.0008x_2 - 0.0007x_3 - 0.0510x_4 \quad (15\text{-}4)$$

where y is the average of the survival rate values for the three viruses under the conditions of the experiment, x_1 is the average percent adsorption of all three viruses to the soil, x_2 is the resin-extractable phosphorus value (parts per million) for the given soil, x_3 is the exchangeable aluminum value (parts per million) for the given soil, and x_4 is the saturation pH value for the given soil. The authors acknowledged that it would be difficult to use this equation to predict virus survival under natural field conditions of constantly changing temperature and soil moisture. They suggest that the equation might best be used to estimate relative virus survival in different soils based on their known physical properties.

Reddy, Khaleel, and Overcash (1981) also attempted to model microbial survival in soil. They stated that the concentration of microorganisms in the soil at time t could be described using the first-order rate expression

$$M_t = M_0 \exp[(K_B - K_D)t]$$
$$= M_0 \exp(-Kt) \quad (15\text{-}5)$$

where M_t is the microbial concentration at time t, M_0 is the initial microbial concentration after the waste application, K_B is the rate coefficient for the rate of division of the microorganism, K_D is the rate coefficient for the die-off rate of the microorganism, and $K = K_D - K_B$.

These investigators felt that temperature, pH, moisture, and the method of waste application were the most important factors controlling microbial inactivation. They reviewed the literature to find data which could be used to develop equations quantifying the influence of these factors on inactivation and developed a functional relationship for each of the four factors. These were incorporated into the following expression for the die-off rate constant, K_2:

$$K_2 = K_1 F_T F_M F_{pH} F_{ma}$$
and $$t_{1/2} = 0.693/K_2 \quad (5\text{-}6)$$

where the correction coefficients are F_T, temperature; F_M, soil moisture content; F_{pH}, soil pH; F_{ma}, method of application, and $t_{1/2}$ is the half-life for the survival of the organisms. Reddy et al. (1981) present a table which lists the $t_{1/2}$ values for fecal coliform under various environmental conditions. These results are not easily generalized to other systems, as most of the equations used by these investigators were developed from one set or a limited set of experimental data, usually involving only one or a few microorganisms.

Physical Filtration

One factor which affects the transport of microorganisms through porous media is filtration. Cookson (1970) states that the filtration mechanism includes straining, sedimentation, inertial impingement, and diffusion. The straining mechanism occurs where the particle in suspension in the porous matrix cannot pass through a smaller pore opening or constriction (i.e., the wedge between two soil particles) and thus its transport is halted (Corapcioglu and Haridas 1984, 1985). The relative magnitude of the effect of this process depends on many soil, water, and microbial factors (Wollum and Cassel 1978). For small microbial particles (e.g., viruses) in coarse-grained material that are not aggregated or solids-associated, filtration is probably negligible (Gerba and Bitton 1984; Sobsey 1983; Bitton 1980; Corapcioglu and Haridas 1984, 1985). For large bacteria and parasites, on the other hand, physical straining may be an important consideration (Gerba and Bitton 1984; Corapcioglu and Haridas 1984; Wollum and Cassel 1978; Smith et al. 1985). In general, under high flow velocities the amount of bacteria filtered is less than for low-flow velocities (Bouma 1979; Butler, Orlob, and McGauhey 1954; Wollum and Cassel 1978; White 1985), probably because a larger amount of the total flow quantity is derived from the larger pores, which will transmit a greater portion of the total number of bacteria present.

Smith et al. (1985) found that more bacteria (*E. coli*) were filtered by disturbed soil than by undisturbed soils. It was postulated that this was a result of the fact that there was a larger component of flow through macropores in the undisturbed soils than in the disturbed soils. In particular, it was found that 21–78% of the bacteria applied to three undisturbed soils was retained, whereas for disturbed samples of the same three soils, 93–99.8% of the bacteria was retained. This finding suggests that if one desires to filter bacteria from contaminated water, the soil should be disturbed to close any macropores or preferential paths that may be present. Although undisturbed samples filter fewer bacteria, in general, than disturbed samples, the difference between two undisturbed samples can also be large (White 1985).

Wollum and Cassel (1978) investigated the transport of streptomycetes conidia and an unidentified bacterium in a saturated sand. They found that at higher pore-water velocities, the maximum concentration of conidia in the effluent occurred at about 0.85 pore volume. It was postulated that for high flow rates the conidia were preferentially transported in the interconnected higher velocity pores. This same reasoning was used to explain the apparent increased maximum concentration found for the higher pore-water velocities, because higher velocities would help prevent conidia from becoming entrapped by the soil particles. The entrapment of microorganisms by the smaller pores was also demonstrated by comparing the concentration profiles for two columns of different length and water content. It was found that a larger quantity of conidia was retained for the longer column which had a lower water content, since for this case there is a reduced likelihood of continuous pathways. A final comparison was made between the streptomycete conidia and the bacterium. It was found that a greater quantity of the bacterium was retained than conidia. The investigators suggested that this was due to the fact that the bacterium produced a sticky extracellular polysaccharide. The authors suggest that pore-water velocity, number of organisms, morphology of the organisms, and soil-water properties, such as water content, bulk density, and clay type and quantity, affect the entrainment of microorganisms. They also found that more microorganisms were filtered out near the surface of the soil.

Sedimentation in the pores occurs where there is a density difference between the microorganism and water. If the microorganism is denser than water and the flow properties are such that the tendency for gravitational settling is greater than the tendency to be resuspended into the flow stream, the bacteria may settle into quiescent parts of the porous matrix. Corapcioglu and Haridas (1984, 1985) analyze the gravitation settling using Stokes' law (Sears and Zemansky 1955). A disadvantage of using this method is that Stokes' law was derived for fluids at rest, which is generally not the case in ground-water systems.

Adsorption

Numerous studies have demonstrated that soils can effectively remove viruses from water. Filtration is not believed to be a significant mechanism of virus removal in coarse-textured soils such as sands, but may be of some importance in fine-textured soils where the pore sizes of the soil matrix are of the same magnitude as the virus (approximately 20–200 nm) (Matthess and Pekdeger 1981) or when viruses are aggregated or solids-associated. One of the most important physicochemical processes for virus removal in soils is adsorption of viruses on solid surfaces. This mechanism is also important in the removal of bacteria in soils.

Suspended virus particles have been dealt with as dispersions of colloidal particles, and interactions between the suspended viruses and solid surfaces often are described using physical equilibrium adsorption isotherms. Gerba (1984b) has summarized both the theoretical and applied aspects of virus adsorption to surfaces. For the purposes of developing a quantitative relationship for this removal process, Langmuir and Freundlich isotherms and kinetic adsorption models have been used.

The Langmuir equilibrium isotherm is based on the assumption that every adsorption site is of equal strength, that there is no interaction between adsorbed molecules on the surface, and that maximum adsorption corresponds to a saturated monolayer of molecules on the adsorbent surface. It can be written as (Bohn, McNeal, and O'Connor 1979; Hendricks 1972; Enfield, Phan, and Walters 1981)

$$C_S = \frac{K_L C C_{sm}}{1 + K_L C} \qquad (15\text{-}7)$$

where C_s is the concentration of adsorbed virus on the solid phase, C_{sm} is the maximum concentration when all of the active surface sites are occupied, and C is the concentration of viruses in suspension. The term K_L is a constant related to the bonding energy.

Moore et al. (1981) found that a Langmuir isotherm fit their data for poliovirus type 2 adsorption to minerals and soils under near-saturation conditions quite well. Vilker et al. (1978) also used a Langmuir isotherm to describe the Cookson and North (1967) batch equilibrium measurements of T4 phage adsorption on activated carbon.

As Vilker (1981b) points out, when the adsorbed state is only weakly favored and the liquid phase concentration is small so that $K_L C \ll 1$, the Langmuir isotherm reduces to

$$C_s = K_L C C_{sm} \qquad (15\text{-}8)$$

which is essentially a linear isotherm. Whether the assumption that the adsorbed state is only weakly favored is valid is highly dependent upon the soil properties and the specific virus of concern, as discussed previously. Under such environmental conditions where these assumptions are valid, the equations which describe the virus transport can be simplified (i.e., linearized) and allow the solution to be written in analytical form for situations with simplified geometries and initial and boundary conditions.

Freundlich isotherms have also been used quite successfully to describe the adsorption of viruses in a variety of virus-soil-water systems (Gerba 1984b). The isotherm, which does not assume homogeneity among active sites for adsorption, is expressed as (Bohn, McNeal, and O'Connor 1979; Freeze, and Cherry 1979; Rao and Jessup 1983)

$$C_s = K_F C^n \qquad (15\text{-}9)$$

where K_F and n are constants. For many systems, the empirical "constant" n is not significantly different from unity, and the Freundlich isotherm reduces to the linear form

$$C_s = K_F C \qquad (15\text{-}10)$$

which is indistinguishable from the linear form of the Langmuir isotherm (Eq. 15-8).

For the purposes of quantifying adsorption data, the choice of an equilibrium isotherm may simply be one of convenience in fitting the isotherm constants, particularly when the isotherm appears linear. For a fluid of arbitrary concentration, however, the use of a Langmuir type of isotherm (Eq. 15-7) connotes saturation-limited adsorption. Adsorption data can indicate a linear relationship if virus adsorption is characterized by a large number of active sites and an equilibrium which strongly favors the suspended phase over the adsorbed phase (Vilker 1981a). Laboratory experiments have shown that even when high concentrations of viruses were applied to soils, less than 1% of the soil-particle surface was covered with viruses (Moore et al. 1981). These investigators suggested that it is unlikely that even coarse-textured materials would become saturated under natural conditions.

For the description of virus movement through soils, the choice of an isotherm may have important consequences. The isotherm parameters K_L and C_{sm}, or K_F and n, are macroscopic variables which reflect the integrated effects of molecular or ionic interactions between viruses and adsorption sites on the soil matrix. This interaction is largely electrostatic in nature. Thus, divalent cations can be particularly effective in increasing adsorption by compressing the Gouy layers around both virus and soil particles. This phenomenon could explain the desorption and enhanced movement of viruses observed by Duboise, Sagik, and Malina, Jr. (1974) and Lance, Gerba, and Melnick (1976) when deionized water was cycled with secondarily treated wastewater percolates in sandy soil columns. Thus, both the adsorption and desorption isotherms may be important in predicting virus transport.

Kinetic Adsorption

Another method for characterizing the adsorption-desorption process is with nonequilibrium adsorption models. A fairly detailed discussion of nonequilibrium adsorption is given by Rao and Jessup (1983) along with a list of review articles concerning the adsorption process. Conceptually, two types of sorption are described by Rao and Jessup (1983), chemically controlled and physically controlled adsorption. In the chemically controlled adsorption process, the soil is viewed as being composed of two types of adsorption sites, one type which behaves in an equilibrium-adsorption manner, the other in a first-order reversible manner. For the physically controlled adsorption, the adsorption process is viewed as a two-step transport process whereby a contaminant is transported by diffusion through a quiescent layer, followed by instantaneous adsorption to the soil surface on the soil side of the water film (quiescent layer). Since this latter conceptualization has been used to describe the adsorption process in the transport of microorganisms, it will be described in more detail.

Using the notation of Vilker (1981a), the nonequilibrium rate of adsorption is mathematically described as

$$\rho_b \frac{\partial M}{\partial t} = k_{f,c}(C - C^*) \qquad (15\text{-}11)$$

where ρ_b is the bulk density of the soil, $k_{f,c}$ is an adsorption rate coefficient which describes the diffusive transport of a virus particle from the bulk solution across a quiescent water film toward the soil surface, M is the mass of virus adsorbed to the soil surface, C is the concentration of viruses, and C^* is the equilibrium concentration in the fluid in immediate contact with the soil particles. In this case, compared with equilibrium sorption, the sorption process is time-dependent.

MODELING MICROBIAL TRANSPORT

At the present time, the physical processes related to the bulk movement of water through saturated and unsaturated soils are fairly well understood. A variety of methods and models are available to simulate the flow of water in soils, especially saturated soils. Within the context of predicting the movement and attenuation of microorganisms in the subsurface, the flow models currently available are probably adequate for predicting the bulk movement of water. The major exceptions are in fractured rocks and karst systems which are difficult to characterize, both structurally and hydrogeologically. For the most part, current modeling techniques are probably as reliable as, if not more reliable than, estimates of the physical and hydraulic properties required to describe the flow system. The major impediments to predicting concentrations of microorganisms in ground water lie in quantifying the transport processes, and in particular, determining the actual pathways a microbial particle takes.

Applying Models to Regional Microbial Transport

Few studies report the use of models to describe microbial transport on a regional basis. Most efforts to model the transport of microorganisms use simulation for purely hypothetical purposes or to describe transport in laboratory column experiments. An example of a hypothetical modeling study is provided by Grosser (1984), who used a one-dimensional form of the advection-dispersion equation to model virus transport between septic tanks and private water-supply wells, where it was assumed that the adsorption process could be described by a linear adsorption isotherm and that adsorption occurs instantaneously. From these simulations, Grosser (1984) found that there was an 80% removal of virus within the first 1 m of flow and that about 15 m were required before the virus concentration was below prescribed limits of 1 pfu/100 gal or 1 pfu/1,000 gal. A sensitivity analysis demonstrated that the inactivation rate was the most important parameter affecting the concentration profile at steady-state.

Using laboratory data, Grondin and Gerba (1986) investigated whether viruses will disperse in a similar manner to solutes and produce a concentration distribution which can be described using a conventional solute transport theory. They were also interested in determining whether available analytical solutions could be adequately fitted to column effluent data for use in describing the virus concentration in laboratory column experiments. The laboratory experiment was conducted using a column 1.15 m in length and 5.04 cm in diameter, containing a saturated gravelly sand soil. MS2 bacteriophage (10^3–10^4 pfu/mL) were introduced at a constant flow rate. Grondin and Gerba (1986) found that dispersion of MS2 did occur and could be described by using existing solutions.

An interesting result of this study was that the average virus velocity (from the calibrated solution) was found to be 1.6 to 1.9 times that of the average velocity for the water. To account for this behavior, a value for the retardation factor, R, of less than unity was used in the analytical solution. An alternative to reducing R to a value less than unity (which implies that the viruses are subjected to anion exclusion) would be to use the average virus velocity instead of the average pore-water velocity. One possible explanation for the observed behavior is pore-size exclusion. Pore-size exclusion would cause the viruses (and aggregates of viruses) to be restricted to the larger pores which have, on average, a pore-water velocity greater than the average pore-water velocity of all the pores taken together. Similar results have been demonstrated, using macromolecules, by Enfield and Bengtsson (1988) and streptomycete conidia by Wollum and Cassel (1978). Among other results of this study, viruses were found not to adsorb to the gravelly sandy soil, and there was no appreciable inactivation over the time period of the experiments.

A Geostatistical Approach

For problems on a regional scale, typical modeling approaches become less useful. This is due to a combination of inadequate data characterizing the flow and transport domain over the scale of interest, as well as the possibility of processes that occur at large scales which are not included in the classical approaches. For example, in many field studies the dispersivity has been observed to increase with the scale of the experiment. However, classical approaches to modeling the solute transport (derived from column (laboratory) experiments) assume that the dispersivity is a constant value for the entire domain. Using models with this assumption will lead to erroneous results.

There are currently no proven methods for modeling the transport of solutes, microorganisms, and so on, in porous media at the regional scale. In theory, assuming that complete knowledge of the material characteristics of the entire transport domain is known and that computing capabilities exist which would allow a fully coupled, three-dimensional model for flow, heat, and solute transport, then the most accurate approach would probably be a deterministic numerical solution to the problem. However, since the assumptions are currently impossible to satisfy, and will likely remain so into the future, other approaches are sought which require less information and allow for a simpler solution. As a method becomes simpler (i.e., that simplifying assumptions are employed) and less input data are required, the uncertainty of the results of the simulation increases.

Geostatistical methods offer an alternative method for characterizing the spatial heterogeneity observed at a variety of scales. Geostatistical methods have been used in the analysis of a variety of agricultural and hydrologic problems (see Chapter 4).

A technique which is relatively new to soil and hydrologic sciences—disjunctive kriging —offers several advantages over linear estimation methods. First, since disjunctive kriging is a nonlinear estimator, it may provide a more accurate estimate of the property of interest. Second, and more importantly, an estimate of the conditional probability that the variable of interest has exceeded a specified value can be easily obtained. The conditional probability can be used as an input to management decision-making models providing a quantitative method for determining whether management actions are necessary.

Two pieces of information must be available to use disjunctive kriging for management decision making. The first is the hazard level for a property of interest. This value is called the cutoff or critical level, and values of the property which are larger than this level represent the undesirable event. Another necessary piece of information is the probability level that causes a management corrective action. This is the critical probability level where the undesirable levels of the property being investigated will no longer be tolerated. Whenever the value of a property in a region is larger than the cutoff level at a probability equal to or greater than the critical probability level, some corrective action will be applied to lower the value of the property to acceptable levels.

In this manner, disjunctive kriging can be used to provide a quantitative means for aiding in making management decisions. With other geostatistical techniques such as ordinary kriging, the crucial piece of information—the conditional probability—is not generally available. Without this information, regions where the level of a property is greater than a cutoff level can be found, but since the region is demarked only from estimates it remains unknown whether the region is significantly over the cutoff level.

To demonstrate how disjunctive kriging can be used as a management decision-making tool, we give an example of the optimization of septic-tank placement to protect drinking water from viral contamination. This work is described in more detail by Yates and Yates (1989). Since a description of disjunctive kriging is provided in Chapter 4 and further descriptions can be found in a number of sources (i.e., Matheron 1976; Journal and Huijbregts 1978; Yates, Warrick, and Myers 1986a, 1986b; Yates 1986; Webster 1991) it will be omitted here. For illustrating how geostatistics can be used to minimize the risks of drinking virus-contaminated water, a model and supporting data are required. Since the purpose of this section is to determine the probability that the septic-tank setback distance is appropriate at any specified location, a model which will allow the determination of the setback distance is required.

Setback distance is defined as the distance between a water-supply well and a potential source of contamination such as a septic tank. A very simple model for obtaining this distance uses Darcy's law (Freeze and Cherry 1979); that is,

$$\text{Setback distance} = \frac{tKi}{\eta_e} \quad (15\text{-}12)$$

where t, K, i, and η_e, respectively, are the travel time (day), the hydraulic conductivity (m day^{-1}), the hydraulic gradient (m/m), and the effective porosity (m^3 m^{-3}). Once the virus decay rate is known and the amount of removal of virus organisms is specified, the necessary travel time in the aquifer can be determined.

To obtain the virus decay rate, 71 water samples were collected from municipal drinking-water wells in the Tucson Basin. These samples were used to determine the virus inactivation rates. Viruses were added to the water samples, which were then incubated at the prevailing ground-water temperature. Periodically, subsamples were analyzed to determine the number of viruses remaining in the water samples. Virus inactivation rates were calculated based on these analyses. The procedures used to obtain the inactivation rates are described in greater detail by Yates et al. (1986) and Yates and Yates (1987).

To obtain a value for the travel time in Eq. 15-12, we assumed that 10^7 viruses must be removed from the ground water during the time interval t. Also assumed is that the travel time depends only on the regional flow characteristics, that only horizontal flow occurs, and that the local values for the gradient, hydraulic conductivity, and effective porosity remain constant near a well. Although the assumptions used to determine the setback distances are somewhat restrictive, the proposed management decision-making method can be illustrated equally well with this simple model as with more complex models. However, if the proposed method is to be used by a municipality for regulatory purposes it is suggested that a more comprehensive technique for determining the setback distances be used, one which includes all the important factors affecting the travel time and, therefore, setback distance.

Hydraulic gradients were calculated from a water-table elevation map obtained from the City of Tucson. Hydraulic conductivity values were calculated on the basis of transmissivity values provided by the State of Arizona Department of Water Resources. Travel times were calculated by using the virus inactivation rate as the number of days required to achieve a 7-log decrease in virus numbers. These values were used in Eq. 15-12 to obtain a setback distance at each of the sample locations.

Shown in Figure 15-3A–D are the estimates, kriging variances, and conditional probability levels that the estimated setback distance is greater than the critical setback distance of 15 m and 30 m, respectively. In this example, if the maximum allowed probability that the setback distance is greater than the cutoff level (i.e., 15 or 30 m) is known, then Figure 15-3C or 15-3D can be used to delineate the areas where septic tanks should not be allowed, which provides a means for ground-water managers to make more informed decisions. In this way, allowing or not allowing the placement of a septic tank or a water-supply well (if septic tanks are already present) can be based on quantitative information which accounts for the environmental conditions. Figure 15-3C, D would be useful in aiding management decision making for the situation where zoning laws specify the setback distances. In most parts of the United States the mandatory setback distance is about 15–30 m (Plews 1977). However, Figure 15-3 shows that in many parts of the Tucson Basin a setback distance greater than the 30 m required by law would be necessary to ensure 7 logs of virus inactivation. Therefore, for locations where the setback distance is regulated, it would be possible to use disjunctive kriging and the conditional probability to more accurately determine whether it would be safe to place a septic tank at a given location. It would also be possible to use a zonal approach in a management decision-making model. For example, a regulation might state that if the probability is less than 0.25, no restrictions are necessary in that area. Between 0.25 and 0.5, septic tanks could be placed on lots meeting certain requirements, such as soil percolation values and minimum lot size. Above 0.5, an even larger lot size could be required, or septic tanks could be prohibited in those areas. The method can easily be manipulated to allow such flexibility.

Another approach to regulate septic-tank placement would be to require that the probability be high that a prescribed level (i.e., seven orders of magnitude) of virus inactivation will occur in the time that the water moved from the septic tank to the well. For this case, the regulations would provide a

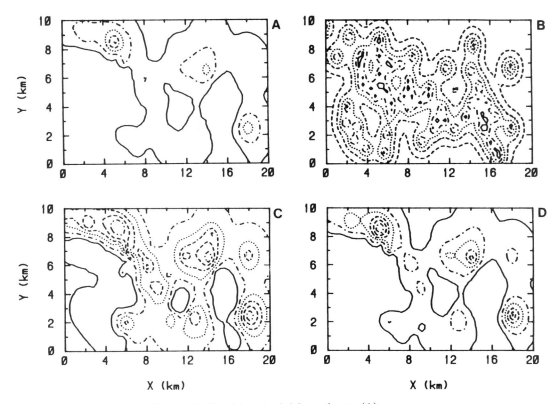

FIGURE 15-3. Contour diagrams for the disjunctive kriging estimates (A), estimation variance (B), and conditional probabilities that the setback distances are greater than 15 m (C) and 30 m (D). Contour line types, ———, — · — · —, · · · · ·, — — · — —, — — — — — and — · · — · · are (A) 15, 30, 45, 60, and 75 m; (B) 100, 150, 200, 250, and 300 m²; (C and D) 0.15, 0.30, 0.45, 0.60, 0.75, and 0.90 m, respectively.

minimum probability level which must be achieved. Then disjunctive kriging could be used to determine the critical setback distances, that is, the minimum allowed setback distance associated with the required probability level. This has been done for the Tucson Basin for four probability (of contamination) levels, 0.20, 0.10, 0.05, and 0.01, and is shown in Figure 15-4A–D, respectively. For this situation, and given a minimum allowed probability level, the contour levels denote the minimum setback distance necessary to ensure adequate distance to achieve seven orders of magnitude of virus inactivation as the ground water travels that distance. As expected, the setback distances associated with the higher probability levels are greater than would be required in the same location compared with those calculated for a lower level. For example, at (18,1), a 40-m setback distance would be required to be 80% certain that the ground water would be free of viral contamination at a well (i.e., 1 − probability of contamination). However, 100 m would be required at the same location if a 99% probability of freedom from contamination is required. Thus, the probability level chosen will have a profound effect on the calculated setback distance.

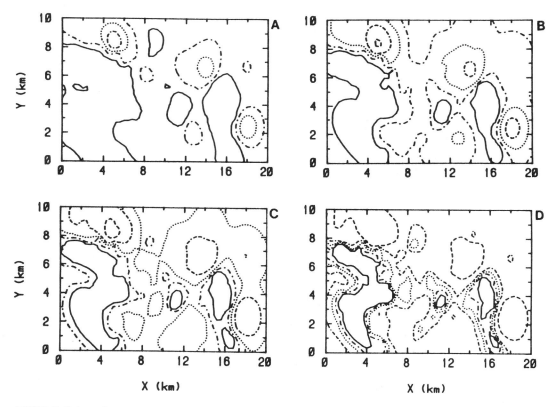

FIGURE 15-4. Contour diagram for the critical setback distances given a minimum allowed probability level using disjunctive kriging. In A, B, C, and D, the conditional probability used to determine the critical setback distance is 0.20, 0.10, 0.05, and 0.01, respectively. The contour levels are marked as ——— 20.0 m, —·—·— 40.0 m, ····· 60.0 m, ——·—— 80.0 m, and ———— 100.0 m.

CONCLUDING REMARKS

This chapter has been devoted to a discussion of the biological, chemical, and physical factors that influence virus and bacterial survival and transport in the subsurface (i.e., temperature, pH, microbial activity, etc.). Well-designed and -conducted field studies are needed to determine how closely laboratory studies represent what is occurring in the field. In addition, many of the available data have been determined in laboratory studies by using a few "model" microorganisms. It has become apparent that it is unlikely that there is a good bacterial indicator of virus behavior, and "model viruses" such as poliovirus may not mimic the behavior of other viruses such as hepatitis A or Norwalk. As methods are developed for the cultivation and detection of other viruses known to cause waterborne-disease outbreaks, such as Norwalk and Norwalk-like viruses, these viruses will also have to be evaluated in terms of their ability to survive and be transported in the subsurface environment.

References

Akin, E. W., W. H. Benton, and W. F. Hill. 1971. Enteric viruses in ground and surface waters: A review of their occurrence and survival. In *Proceedings of the 13th Water Quality Conference on Virus and Water Quality: Occurrence and Control*, ed.

V. Snoeyink, pp. 59–73. Urbana, IL: The University of Illinois.

American Public Health Association. 1985. *Standard Methods for the Examination of Water and Wastewater*, 16th ed. Washington, DC: American Public Health Association.

Bennett, J. V., S. D. Holmberg, M. F. Rogers, and S. L. Solomon. 1987. Infectious and parasitic disease. In *Closing the Gap: The Burden of Unnecessary Illness*, eds. R. W. Amler and H. B. Dull. New York: Oxford University Press.

Bitton, G. 1980. *Introduction to Environmental Virology*. New York: John Wiley.

Bitton, G., and C. P. Gerba. 1984. Groundwater pollution microbiology: The emerging issue. In *Groundwater Pollution Microbiology*, eds. G. Bitton and C. P. Gerba, pp. 1–7. New York: Wiley.

Bitton, G., S. R. Farrah, R. H. Ruskin, J. Butner, and Y. J. Chou. 1983. Survival of pathogenic and indicator organisms in ground water. *Ground Water* 21:405–10.

Bohn, H., B. McNeal, and G. O'Connor. 1979. *Soil Chemistry*. New York: Interscience.

Bouma, J. 1979. Subsurface applications of sewage effluent. In *Planning the Uses and Management of Land*, p. 665. Madison, WI: American Society of Agronomy.

Butler, R. G., G. T. Orlob, and P. H. McGauhey. 1954. Underground movement of bacterial and chemical pollutants. *Journal of American Water Works Association* 46:97.

Cookson, J. T. 1970. Removal of submicron particles in packed beds. *Environmental Science and Technology* 4:128–34.

Cookson, J. T., and W. J. North. 1967. Adsorption of viruses on activated carbon. *Environmental Science and Technology* 1:46–52.

Corapcioglu, M. Y., and A. Haridas. 1984. Transport and fate of microorganisms in porous media: A theoretical investigation. *Journal of Hydrology* 72:149–69.

Corapcioglu, M. Y., and A. Haridas. 1985. Microbial transport in soils and groundwater: A numerical model. *Advances in Water Resources* 8:188–200.

Craun, G. F. 1979. Waterborne disease—status report emphasizing outbreaks in groundwater systems. *Ground Water* 17:183–91.

Craun, G. F. 1981. Outbreaks of waterborne disease in the United States: 1971–1978. *Journal of American Water Works Association* 73:360–69.

Craun, G. F. 1984. Health aspects of groundwater pollution. In *Groundwater Pollution Microbiology*, eds. G. Bitton and C. P. Gerba, pp. 135–79. New York: Wiley.

Craun, G. F. 1985. A summary of waterborne illness transmitted through contaminated groundwater. *Journal of Environmental Health* 48:122–27.

Craun, G. F. 1986a. Statistics of waterborne outbreaks in the U.S. (1920–1980). In *Waterborne Diseases in the United States*, ed. G. F. Craun, pp. 73–159. Boca Raton, FL: CRC Press.

Craun, G. F. 1986b. Recent statistics of waterborne disease outbreaks (1981–1983). In *Waterborne Diseases in the United States*, ed. G. F. Craun, pp. 43–69. Boca Raton, FL: CRC Press.

Craun, G. F. 1990. *Methods for Investigation and Prevention of Waterborne Disease Outbreaks*. U.S. Environmental Protection Agency, Office of Research and Development, Report No. EPA-600/1-90/005a.

Craun, G. F. 1991. Causes of waterborne outbreaks in the United States, *Water Science and Technology* 24:17–20.

DiNovo, F., and M. Jaffe. 1984. *Local Groundwater Protection, Midwest Region*. Chicago, IL: American Planning Association.

Duboise, S. M., B. P. Sagik, and J. F. Malina, Jr. 1974. Virus migration through soils. In *Virus Survival in Water and Wastewater Systems*, eds. J. F. Malina and B. P. Sagik, pp. 233–40. Austin, TX: Center for Research in Water Resources.

Enfield, C. G., and G. Bengtsson. 1988. Macromolecular transport of hydrophobic contaminants in aqueous environments. *Ground Water* 26:64–70.

Enfield, C., T. Phan, and D. Walters. 1981. Kinetic model for phosphate transport and transformation in calcareous soils. II: Laboratory and field transport. *Soil Science Society of America Journal* 45:1064–70.

Freeze, R. A., and J. A. Cherry. 1979. *Groundwater*. Englewood Cliffs, NJ: Prentice-Hall.

Gerba, C. P. 1983. Virus survival and transport in groundwater. *Developments in Industrial Microbiology* 24:247–51.

Gerba, C. P. 1984a. *Strategies for the Control of*

Viruses in Drinking Water. Report to the American Association for the Advancement of Science, Washington, DC.

Gerba, C. P. 1984b. Applied and theoretical aspects of virus adsorption to surfaces. *Advances in Applied Microbiology* 30:133–68.

Gerba, C. P., and G. Bitton. 1984. Microbial pollutants: Their survival and transport pattern to groundwater. In *Groundwater Pollution Microbiology*, eds. G. Bitton and C. P. Gerba, pp. 65–88. New York: Wiley.

Goyal, S. M., B. H. Keswick, and C. P. Gerba. 1984. Viruses in groundwater beneath sewage irrigated cropland. *Water Research* 18:299–302.

Grondin, G. H., and C. P. Gerba. 1986. Virus dispersion in a coarse porous medium. *Hydrology and Water Resources in Arizona and Southwest* 16:11–15.

Grosser, P. W. 1984. *A One-Dimensional Mathematical Model of Virus Transport*. Paper presented at the Second International Conference on Ground Water Quality Research, Tulsa, OK, pp. 105–7.

Hejkal, T. W., and C. P. Gerba. 1982. *A Model of Virus Die-off in Seawater*. Rockville, MD: NOAA Office of Marine Pollution Assessment.

Hendricks, D. W. 1972. Sorption in flow through porous media. In *Developments in Soil Science*, Vol. 2: *Fundamentals of Transport Phenomena in Porous Media, International Association of Hydraulic Research*, pp. 384–92. New York: Elsevier.

Herwaldt, B. L., G. F. Craun, S. L. Stokes, and D. D. Juranek. 1992. Outbreaks of waterborne disease in the United States: 1989–1990. *Journal of American Water Works Association* 84:129–35.

Hurst, C. J., C. P. Gerba, and I. Cech. 1980. Effects of environmental variables and soil characteristics on virus survival in soil. *Applied Environmental Microbiology* 40:1067–79.

Jorgensen, P. H., and E. Lund. 1985. Detection and stability of enteric viruses in sludge, soil and ground water. *Water Science and Technology* 17:185–95.

Journel, A. G., and Ch. J. Huijbregts. 1978. *Mining Geostatistics*. New York: Academic Press.

Kaplan, J. E., G. W. Gary, R. C. Baron, W. Singh, L. B. Schonberger, R. Feldman, and H. Greenberg. 1982. Epidemiology of Norwalk gastroenteritis and the role of Norwalk virus in outbreaks of acute nonbacterial gastroenteritis. *Annals of Internal Medicine* 96:756–61.

Keswick, B. H., and C. P. Gerba. 1980. Viruses in groundwater. *Environmental Science and Technology* 14:1290–97.

Keswick, B. H., C. P. Gerba, S. L. Secor, and I. Cech. 1982. Survival of enteric viruses and indicator bacteria in groundwater. *Journal of Environmental Science and Health* A17:903–12.

Keswick, B. H., T. K. Satterwhite, P. C. Johnson, H. L. DuPont, S. L. Secor, J. A. Bitsura, G. W. Gary, and J. C. Hoff. 1985. Inactivation of Norwalk virus in drinking water by chlorine. *Applied Environmental Microbiology* 50:261–4.

Lance, J. C., C. P. Gerba, and J. L. Melnick. 1976. Virus movement in soil columns flooded with secondary sewage effluent. *Applied Environmental Microbiology* 32:520–6.

Lippy, E. C., and S. C. Waltrip. 1984. Waterborne disease outbreaks—1946–1980: A thirty-five-year perspective. *Journal of American Water Works Association* 76:60–7.

Matheron, G. 1976. A simple substitute for conditional expectation: The disjunctive kriging. In *Proceedings of NATO Advanced Study Institute Series "Geostat 75"*. Hingham, MA: Reidel.

Matthess, G., and A. Pekdeger. 1981. Concepts of a survival and transport model of pathogenic bacteria and viruses in groundwater. *Science of the Total Environment* 21:149–59.

Melnick, J. L., and C. P. Gerba. 1980. The ecology of enteroviruses in natural waters. *CRC Critical Reviews in Environmental Control* 10:65–93.

Moore, R. S., D. H. Taylor, L. S. Sturman, M. M. Reddy, and G. W. Fuhs. 1981. Poliovirus adsorption by 34 minerals and soils, *Applied Environmental Microbiology* 42:963–75.

Office of Technology Assessment. 1984. *Protecting the Nation's Groundwater from Contamination*, Vol. 1. Washington, DC: Office of Technology Assessment OTA-O-233.

Pancorbo, O. C., B. G. Evanshen, W. F.

Campbell, S. Lambert, S. K. Curtis, and T. W. Wolley. 1987. Infectivity and antigenicity reduction rates of human rotovirus strain Wa in fresh waters. *Applied Environmental Microbiology* 53:1803–11.

Payment, P., M. Trudel, and R. Plante. 1985. Elimination of viruses and indicator bacteria at each step of treatment during preparation of drinking water at seven water treatment plants. *Applied Environmental Microbiology* 49:1418–28.

Plews, G. 1977. Management guidelines for conventional and alternative onsite sewage systems—Washington State. In *Individual Onsite Wastewater Systems, Proceedings of the Third National Conference*, ed. N. I. McClelland, pp. 187–93. Ann Arbor, MI: Ann Arbor Science.

Rao, P. S. C., and R. E. Jessup. 1983. Sorption and movement of pesticides and other toxic substances in soil. In *Chemical Mobility and Reactivity in Soil Systems*, pp. 183–99. Madison, WI: American Society of Agronomy.

Reddy, K. R., R. Khaleel, and M. R. Overcash. 1981. Behavior and transport of microbial pathogens in soils treated with organic wastes. *Journal of Environmental Quality* 10:255–66.

Regli, S., J. B. Rose, C. N. Haas, and C. P. Gerba. 1991. Modeling the risk from *Giardia* and viruses in drinking water. *Journal of American Water Works Association* 83:76–84.

Sears. F. W., and M. W. Zemansky. 1955. *University Physics*. Reading, MA: Addison-Wesley.

Smith, M. S., G. W. Thomas, R. E. White, and D. Ritonga. 1985. Transport of *Escherichia coli* through intact and disturbed soil columns. *Journal of Environmental Quality* 14:87–91.

Sobsey, M. D. 1983. Transport and fate of viruses in soils. In *Microbial Health Considerations of Soil Disposal of Domestic Wastewaters*, pp. 174–91. Cincinnati, OH: U.S.EPA Report No. EPA-600/9-83-017.

Sobsey, M. D., P. A. Shields, F. H. Hauchman, R. L. Hazard, and L. W. Caton III. 1986. Survival and transport of hepatitis A virus in soils, groundwater and wastewater. *Water Science and Technology* 18:97–106.

Stewart, M. 1990. *Pathogen Removal in Treated Sewage*, Metropolitan Water District of Southern California, Water Quality Laboratory, La Verne, CA, Internal Report.

Stramer, S. L. 1984. *Fates of Poliovirus and Enteric Indicator Bacteria during Treatment in a Septic Tank System Including Septage Disinfection*. Ph.D. dissertation, Madison, WI: University of Wisconsin.

Tyrrell, D. A., and A. Z. Kapikian. 1982. *Virus Infections of the Gastrointestinal Tract*. New York: Marcel Dekker.

U.S. Environmental Protection Agency. 1977. *Waste Disposal Practices and Their Effects on Ground Water*. The Report to Congress, Washington, DC.

U.S. Environmental Protection Agency. 1985. National Primary Drinking Water Regulations: Synthetic Organic Chemicals, Inorganic Chemicals, and Microorganisms. Proposed rule. *Federal Register* 50:46936–47022.

U.S. Environmental Protection Agency. 1989. National Primary Drinking Water Regulations: Filtration; Disinfection; Turbidity; *Giardia lamblia*; Viruses; *Legionella*; and Heterotrophic Bacteria. Final rule. *Federal Register* 54:27486.

U.S. Environmental Protection Agency. 1990. Strawman Regulation for Ground Water Disinfection Requirements (GWDR). Washington, DC: Office of Drinking Water.

Vaughn, J. M., and E. F. Landry. 1977. *Data Report: An Assessment of the Occurrence of Human Viruses in Long Island Aquatic Systems*, Department of Energy and Environment. Upton, NY: Brookhaven National Laboratory.

Vaughn, J. M., and E. F. Landry. 1978. The occurrence of human enteroviruses in a Long Island groundwater aquifer recharged with tertiary wastewater effluents. In *State of Knowledge in Land Treatment of Wastewater*, Vol. 2, pp. 233–45. Washington, DC: U.S. Government Printing Office.

Vaughn, J. M., E. F. Landry, L. J. Baranosky, C. A. Beckwith, M. C. Dahl, and N. C. Delihas. 1978. Survey of human virus occurrence in wastewater-recharged groundwater on Long Island. *Applied Environmental Microbiology* 36:47–51.

Vilker, V. L. 1981a. Simulating virus movement in soils. In *Modeling Wastewater Renovation: Land Treatment*, ed. I. K. Iskandar, pp. 223–53. New York: Wiley.

Vilker, V. L. 1981b. Virus transport through percolating beds. Technical Completion

Report, *California Office of Water Research and Technology Project No. B-184-CAL and California Water Resources Center Project UCAL-WRC-W-523, NTIS PB81-21240-9.*

Vilker, V. L., L. H. Frommhagen, R. Kamdar, and S. Sundarum. 1978. Applications of ion exchange/adsorption models to virus transport in percolating beds. *AICHE Symposium Series* 74:84–92.

Webster, R. 1991. Local disjunctive kriging of soil properties with change of support. *Journal of Soil Science* 42:301–18.

White, R. E. 1985. Transport of chloride and non-diffusible solutes through soil. *Irrigation Science* 6:3–10.

Wollum A. C., and D. K. Cassel. 1978. Transport of microorganisms in sand columns. *Soil Science Society of America Journal* 42:72–76.

World Health Organization. 1984. *Guidelines for Drinking Water Quality*, Vol. 1. *Recommendations*. Geneva, Switzerland: World Health Organization.

Yates, M. V. 1985. Septic tank density and ground water contamination. *Ground Water* 23:586–91.

Yates, M. V., and S. R. Yates. 1987. A comparison of geostatistical methods for predicting virus inactivation rates in groundwater. *Water Research* 21:1119–25.

Yates, M. V., and S. R. Yates. 1988. Modeling microbial fate in subsurface environments. *CRC Critical Reviews in Environmental Control* 17:307–44.

Yates, M. V., and S. R. Yates. 1989. Septic tank setback distances: A way to minimize virus contamination of drinking water. *Ground Water* 27:202–8.

Yates, M. V., C. P. Gerba, and L. M. Kelley. 1985. Virus persistence in ground water. *Applied Environmental Microbiology* 49:778–81.

Yates, M. V., S. R. Yates, A. W. Warrick, and C. P. Gerba. 1986. Use of geostatistics to predict virus decay rates for determination of septic tank setback distances. *Applied Environmental Microbiology* 52:479–83.

Yates, S. R. 1986. Disjunctive kriging. III: Cokriging. *Water Resources Research* 22:1371–76.

Yates, S. R., A. W. Warrick, and D. E. Myers. 1986a. Disjunctive kriging. I: Overview of estimation and conditional probability. *Water Resources Research* 22:615–22.

Yates, S. R., A. W. Warrick, and D. E. Myers. 1986b. Disjunctive kriging. II: Examples. *Water Resources Research* 22:623–30.

16

Acid Precipitation

Gunnar Jacks

INTRODUCTION

Environmental acidification caused by anthropogenic acidic deposition is an old phenomenon. The famous biologist Carl von Linné commented on the deplorable status of the environment at the ancient Falu copper mine in central Sweden by saying: "Never has a theologian described a Hell so dreadful as what is seen here." The London air was in a poor state already in the eighteenth century (Evelyn 1661; Graunt 1662). However, until World War II, the effects were essentially local. With the increase in use of fossil fuels during the past four decades, environmental acidification has become a regional phenomenon in central and northern Europe, in the northeastern United States (Likens et al. 1977), and in the neighboring parts of Canada (Altshuller and McBean 1979). Large emissions of acidifying substances are causing increasing concern in China and in southern and southeastern Asia (Galloway and Rodhe 1991).

Acidification of ground water may be caused by internal processes in the soil zone or by external factors imposed on the soil-water system. Oxidation processes such as the oxidation of reduced sulfur and nitrogen compounds in the soil zone are examples of internal factors (Grass, Aronovici, and Muckel 1962). These processes have been recognized for a long time and have been extensively studied. The effect of acid deposition on ground water, a rather new phenomenon, is an example of an external factor (Holmberg 1986). Effects on surface waters were evident in the late 1950s (Gorham 1957), but concern about effects on ground water did not arise until about 20 years later. Initial work focused on shallow ground water, particularly ephemeral springs associated with snowmelt in springtime. However, on closer scrutiny of old data it became evident that widespread changes had occurred in the chemical composition of ground water, especially the anion composition. While under pristine conditions most ground waters are of the calcium bicarbonate type, acid sulfur deposition has caused SO_4^{2-} to be a major anion. The study of chemical trends in ground water, however, has met with considerable difficulties as the degree of spatial and temporal variation is far greater than in surface-water systems. Lakes and streams integrate a larger area than ground water and there is a considerable integration of the chemical composition with time.

In this chapter, the acidification of ground water will primarily be treated with regard to the effect of anthropogenic acidifying sulfur and nitrogen compounds. The main problems connected with ground-water acidification are the potential release of toxic metals from soils and pipe systems and corrosion causing economic losses. The release of toxic metals from soils is of great concern for surface-water organisms, but it seems that only in extreme cases do metal concentrations in wells exceed the health criteria set for metals in drinking water. Once in contact with piping materials, however, metal concentrations in acidic or poorly buffered water can easily exceed those limits.

The discovery of environmental acidification occurred in reverse order; it was first discovered in surface water, later in ground water, but clear evidence for soil acidification became apparent only in the early 1980s. As most water does pass through the soil before entering streams and lakes, one would anticipate that soil acidification should precede water acidification. The belated proof of soil acidification is due to the scarcity of representative old data and the very large spatial variation in soil conditions. Changes in soil chemical analytical procedures also contributed to the difficulties confronting researchers.

SOIL ACIDIFICATION

Most soil investigations have concerned the surface layers of soils, often the top 20 cm where root penetration is intense and most of the nutrients are found. Little or no changes were found with time in these top sections of forest soils and many of the changes that were observed could be attributed to effects of vegetation (e.g., Likens et al. 1977; Johnson et al. 1991). Hallbäcken and Tamm (1986) made a careful resampling of an old soil investigation plot in southwestern Sweden which was meticulously documented in 1927. Their investigation disclosed a deep acidification at the same localities that were initially tested. Several other investigations have strengthened the picture that deep acidification has occurred in areas with high deposition and/or sensitive soils (Johnston et al. 1986; Falkengren-Grerup 1987; Eriksson, Karltun, and Lundmark 1992).

Most soil processes involve the production and consumption of protons, and thus are acidifying or neutralizing. The following reactions are common acid-base reactions that occur in soils.

Neutralizing Reactions

Reduction

$$4NO_3^- + 5CH_2O + 4H^+ \rightarrow 2N_2 + 5CO_2 + 7H_2O$$

$$SO_4^{2-} + 2CH_2O + 2H^+ \rightarrow H_2S + 2CO_2 + H_2O$$

Weathering

$$CaCO_3 + H_2O + CO_2 \rightarrow Ca^{2+} + 2HCO_3^-$$

$$CaAl_2Si_2O_8 + 3H_2O + 2CO_2 \rightarrow Ca^{2+} + 2HCO_3^- + Al_2Si_2O_5(OH)_4$$

Anion-exchange and nutrient uptake of anions

$$SO_4^{2-} + Al(OH)_3 \rightarrow AlOHSO_4 + 2OH^-$$

$$NO_3^- + Root\text{-}OH^- \rightarrow Root\text{-}NO_3^- + OH^-$$

Acidifying Reactions

Oxidation

$$NH_4^+ + 2O_2 \rightarrow NO_3^- + 2H^+ + H_2O$$

$$FeS_2 + 3.5O_2 + H_2O \rightarrow Fe^{2+} + 2SO_4^{2-} + 2H^+$$

$$Fe^{2+} + 0.25O_2 + 2.5H_2O \rightarrow Fe(OH)_3 + 2H^+$$

Hydrolysis

$$Al^{3+} + 3H_2O \rightarrow Al(OH)_3 + 3H^+$$

The reaction with oxidation of ferrous iron also contains an element of hydrolysis.

Soil Acidification

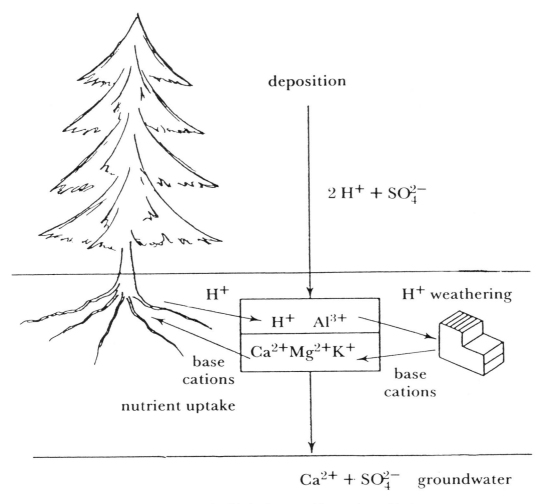

FIGURE 16-1. Cation fluxes in a highly simplified soil system with an exchangeable storage.

Cation exchange and nutrient uptake of cations

$$Ca^{2+} + R\text{-}(COOH)_2$$
$$\rightarrow R\text{-}(COO)_2\text{-}Ca^{2+} + 2H^+$$
$$K^+ + Root\text{-}H^+ \rightarrow Root\text{-}K^+ + H^+$$

Some soil acidifying processes are of a transient nature in relation to climatic fluctuations or changing land-use activities. A number of dry years may cause the lowering of ground-water levels and oxidation of reduced sulfur compounds such as pyrite or organically bound sulfur (Nevell and Wainwright 1986). Fertilization with acidifying compounds such as ammonium-containing nitrogenous fertilizers is another reason for a transient effect. Effects on soil water and surface water may be observed in connection with precipitation from storm events containing unusually high amounts of seasalts, mainly sodium chloride. This may cause an acid surge in water due to ion exchange (Rosenqvist 1977; Sullivan et al. 1988).

If we disregard transient processes of durations up to a few years, the acid-base budget can be simplified to that of Figure 16-1. The figure illustrates the effect on cations in the soil zone of the major processes of deposition, nutrient uptake, and weathering. In this case, sulfate is regarded as a fully mobile anion in the soil zone. This may be the case in the young glacial soils that occur in Scandinavia, Canada, and the northeastern United States. If there is a mature rather acid soil profile with large amounts of ferric oxides or hydroxides and aluminum hydroxides, soil acidification will continue with no effect on the surface and ground water until the sulfate adsorption capacity is saturated with respect to the ambient sulfate concentrations in the soil water (Fuller, David, and Driscoll 1985). A good example of the difference in the development of water acidification reflecting sulfate adsorption capacity is seen in the northeastern United States; surface-water acidification is common in areas with glacial soils while it is not seen further south where there has been more mature soil development (Rochelle, Church, and David 1987).

In areas with nitrogen deposition in excess of nutrient uptake, nitrification and nitrate leaching can cause rapid soil acidification. Nitrification may take place even in very acid soils, further enhancing base-cation and aluminum leaching (Stams et al. 1991).

WEATHERING

As seen in Figure 16-1, weathering is a key neutralizing process, actually more or less the only long-term process being capable of buffering acidic inputs to soils. Weathering and weathering processes have been studied for many years, for example the description of laterite formation by Buchanan from 1807. Tamm (1930) performed a series of weathering experiments in which ground minerals were slurried in water for various periods of time to examine the effects of mineral dissolution. Tamm also estimated the release of base cations that had occurred during Holocene time, about 10,000 years B.P., and calculated a historic weathering rate. Such weathering rate estimates have been used by other researchers as well (April et al. 1986; Olsson and Melkerud 1990). The current weathering rate has been difficult to estimate, because the fluxes of base cations in the soil comprise base cations deposited from the atmosphere, released from the exchangeable stores and root exudates, as well as those released during mineral weathering.

The weathering rate is determined by a number of factors such as the mineralogic composition, exposed surface area, and the acidity of the soil solution. Sverdrup (1990) classified the weathering rates of minerals into groups as follows:

- The dissolving group: *calcite, dolomite*
- The fast weathering group: *anorthite, olivine, garnet, diopside*
- The intermediate weathering group: *epidote, pyroxenes, amphiboles, chlorite, biotite*
- The slow weathering group: *sodic plagioclase feldspars*, clay minerals like *vermiculite* and *montmorillonite*
- The very slow weathering group: *K-feldspar, muscovite*
- The inert group: *quartz, rutile, zircon*

The intermediate and slow weathering groups include minerals common in areas sensitive to acidification, in soils derived from granites and gneisses. The laboratory weathering rates differ between the groups by about one exponential unit, which means that, for instance, 3% of biotite has approximately the same neutralizing effect as 30% of oligoclase, the most common plagioclase feldspar. Rutile and zircon are accessory minerals, but they are important because they can be used in estimating historic losses of material in a soil profile and thus historic weathering rates.

The importance of the acidity of the soil solution has been extensively discussed (Reuss and Johnson 1985, 1986). If the weathering rates of common minerals were directly proportional to the concentration of protons there would be no acidification until the weatherable minerals were dissolved. For most feldspars, the dissolution under laboratory conditions is independent of pH between 4.5 and 8 (Holdren and Speyer 1985, 1987). At higher acidity, the dissolution rate increases, but not proportionally to the increase in acidity (Wollast and Chou 1985; Mast and Drever 1987). Minerals such as hornblende and biotite dissolve faster and respond to increases in acidity already below a pH of about 7 (Lin and Clemency 1981). However, increasing concentrations of aluminum decrease the dissolution rate of these minerals (Sverdrup 1990).

It is not only the acidity of the soil solution that matters; weathering rates are also enhanced by the presence of proton donors and complexing agents such as organic acids (Berthelin 1985). The influence of fulvic acids in a natural soil profile is likely to make a difference between laboratory and field mechanisms and rates. Experiments with organic acids will to some extent give answers on these uncertainties. However, organic matter in nature has a wide range of different acidic and complexing groups. Lundström (1990), using moor extracts in laboratory mineral dissolution experiments, found a threefold enhancement in rate compared to a strong acid.

A number of approaches have been used to estimate the current weathering rate. Input-output budgets can be used if steady-state or near-steady-state conditions exist. Input-output budgets can also be used if transient processes are considered. Paces (1986) has done this by assuming that sodium and silica are not subject to ion exchange and are not considerably involved in the formation of secondary minerals; thus there is no soil sink for these elements. Still another way of calculating the current weathering rate is to synthesize the runoff chemistry from weathering reactions as deduced from soil mineral compositions (Cleaves, Godfrey, and Bricker 1970; Velbel 1985). This approach was first tested by Garrels (1967) in his well-known paper on the Sierra Nevada waters. Under certain conditions, it is possible to trace the fluxes of calcium through the soil zone by using strontium as an analogue (Åberg et al. 1989). Calcium and strontium behave very similarly in a simple ecosystem, and strontium of different origin can be separated by its $^{87}Sr/^{86}Sr$ ratio. Old granitic and gneissic rocks, and soils derived from them, are enriched in ^{87}Sr through a slow decay of ^{87}Rb. Finally, there are several modeling approaches. Most models like MAGIC (Cosby et al. 1985) arrive at the weathering rate by a calibration procedure, adjusting the output of the model to a measured composition of runoff or ground water. PROFILE (Sverdrup and Warfvinge 1988) calculates weathering rates by assemblying weathering reactions of a set of soil minerals, using either laboratory determined weathering rates for individual minerals or field rates determined by some of the aforementioned methods.

Weathering rates as estimated by these methods vary greatly, depending on the soil environment. A major determinant is the mineral composition of the soil. Till soils in Scandinavia have been thoroughly investigated using different methods (Sverdrup 1990; Jacks 1990). A till terrain covered with 1–5 m of till shows a base cation weathering rate of 20–60 millimol/m^2 · a. If the till is derived from acid granites and gneisses, the rate will be in the lower part of the range. It will approach the upper limit if derived from intermediate granites and gneisses. Weathering of calcareous soils would produce a considerably higher proportion of base cations. A 1-m-thick soil containing 0.3% of calcium carbonate if weathered would produce 300 millimol of base cations/m^2 · a (Sverdrup, de Vries, and Henriksen 1990). Forest soils in North America and central Europe commonly show base cation weathering rates of about 200 millimol/m^2 · a (de Vries 1990). In the Adirondack region of the United States, April, Newton, and Treuttner Coles (1986) found base cation weathering rates of 20 and 170 millimol/m^2 · a, respectively, in two neighboring watersheds primarily as a result of differences in till thickness. Sand and gravel would be expected to be highly sensitive to acidification due to low weathering rates. Eriksson

(1986) estimated the base cation weathering rate in a 3-m-thick sand layer to be 12 millimol/m$^2 \cdot$ a.

PLANT UPTAKE OF NUTRIENTS

The uptake of nutrients occurs both in the form of cations and anions and gases (Cole and Rapp 1981; Johnson and Henderson 1989). Thus, nitrogen can be taken up as ammonium or as nitrate. Some nutrients such as sulfur are almost solely taken up as an anion. However, most nutrients, and especially those required in larger amounts, are cations, such as potassium, magnesium, and calcium. Cations are taken up in exchange for protons to maintain the electroneutrality both inside the root as well as in the soil solution. Root uptake is thus predominantly an acidifying process (Nilsson et al. 1982). The rate of acidification depends on the rate of growth and on the extent and frequency of harvesting. In a virgin forest where no harvest takes place, cations are returned to the soil again after the death and decay of trees. However, in modern agriculture, harvest occurs yearly and the amount of nutrients contained in matter removed is considerable.

The uptake of nutrients in a forest varies with tree species and age of the trees, being highest in a middle-aged stand (Alban 1982). To illustrate the differences between forest and agricultural environments, some rough figures are presented in Table 16-1.

As would be expected, the plant uptake is closely related to the release of nutrients, in fact to the weathering rate of the soil. In agricultural lands it is a common practice to add fertilizers including lime or limestone to match the effects of nutrient uptake. On noncalcareous soils a dose of 2–3 tons of limestone per hectare is usually added every five years. This is more or less equivalent to the acidification created by harvesting as given in the table.

RELATION BETWEEN GROUND-WATER AND SURFACE-WATER ACIDIFICATION

As mentioned in the introduction, surface-water acidification was discovered first, although most surface water has passed through the soil zone and even through the groundwater zone. With the rain intensities common in the humid zone, infiltration capacities for most soils are seldom exceeded; thus overland flow should be a rare phenomenon. It is, however, observed that streamflow may be rather immediately affected by rainfall. This has been explained by Dunne and Black (1970) as the result of overland flow in areas where the water table is at or very near the soil surface. This would also explain acid surges in otherwise well-buffered streams (Jacks et al. 1986). Hewlett and Hibbert (1967) suggested an active role of soil and ground water in runoff generation during storm events. The use of environmental isotopes has confirmed that rainwater can be just a small fraction of a flood event, and that most of the water is often pre-event soil or ground water (Sklash and Farvolden 1979; Rodhe 1987). Studies of runoff processes have shown that the upper portion of the soil zone, especially tills, are very permeable and can conduct rapid flow and dis-

TABLE 16-1. Acidification from Plant Uptake and Harvest

Vegetation	Soil	Acidification from Plant Uptake and Harvest (millimol/m$^2 \cdot$ a)
Norway spruce	Fairly productive	20–70
Scotch pine	Poor sandy site	10
Agricultural crop	Good	500–1,000

Source: Data from Nilsson, Miller, and Miller (1982).

charge to streams under saturated conditions (Lundin 1982; Newton and Driscoll 1987). The permeability decreases rapidly with depth due to a number of factors. Root penetration may increase the surface permeability and root channels may act as highly conductive macropores under saturated conditions. Frost heaving may also increase the hydraulic conductivity. In tills, the surface soil may be melt-out material loosely deposited on a denser lodgement till. It thus seems that many soils have a near-surface, ground-water zone which is rather actively exchanged in response to recharge. This ground-water zone is partly ephemeral. The deeper ground water is more stagnant. Ground water in hard rocks at a few hundred meters depth may have ages of thousands of years (Pettersson and Allard 1991).

Figure 16-2 gives a conceptual picture of the turnover rates of different types of ground water in a typical glaciated terrain along with rough estimates of pH (Jacks et al. 1984). The ground water used for consumption is usually extracted from below the depth where recharge from single rain events is occurring. In order to avoid bacterial pollution and coloring by humus from the soil zone, it is customary to isolate the well from the topmost soil layers. Thus, it is rather logical that surface water is acidified before any effect is seen in wells or springs used for water supply, which are usually tapping water with a longer residence time in the ground.

Lake water acidification is largely determined by the watershed characteristics. Eilers et al. (1983) could distinguish three different types of lakes in the north-central United States with decreasing susceptibility to acidification by atmospheric deposition: precipitation lakes, surface-runoff lakes, and ground-water lakes. The influence of the ground-water flux to lakes is clearly demonstrated by April and Newton (1985) for three neighboring watersheds with similar mineralogy but different soil depth. The influence of the mineralogy is, on the other hand, the strongest factor in a case study on the prediction of the acidification sensitivity of streams made by Bricker and Rice (1989). A direct flux of well-buffered ground water into lakes may be important even though only minor amounts of water are involved (Vanek 1987).

OBSERVED TRENDS IN GROUND-WATER COMPOSITION AS INFLUENCED BY ACIDIC DEPOSITION

The pH is usually not a good parameter to detect changes in ground water caused by environmental acidification. It is difficult to measure pH in the poorly buffered waters that are likely to be affected by acidic deposition (Galloway, Cosby, and Likens 1979). The loss of carbon dioxide before and during measurement is another factor that tends to give dubious pH values.

Acidification by sulfur deposition will initially result in increased base cation losses from the soil to the water with sulfate as an anion (Reuss and Johnson 1986). This is a rather easily detectable change in hydrochemistry of ground waters. In older sets of analyses, sulfate was not always analyzed as it was not considered to be important either from health aspects or from technical points of view at the ambient levels normally found. However, hardness was almost always measured, and analyses are usually reliable as is the case for alkalinity. The ratio of hardness to alkalinity is often a good indicator for detecting trends in ground water affected by environmental acidification. Alkalinity may change in either direction, depending on the soil conditions. If there is a calcareous soil, alkalinity may increase because of dissolution of calcite, while alkalinity is likely to decrease in carbonate-free soils when the base cation stores in the soil are being depleted as a source for neutralization.

A trend of increasing hardness to alkalinity ratio was seen by Jacks et al. (1984) over a period of 30 years in small wells and springs in till in southeastern Sweden. The wells were situated in meadows and forest

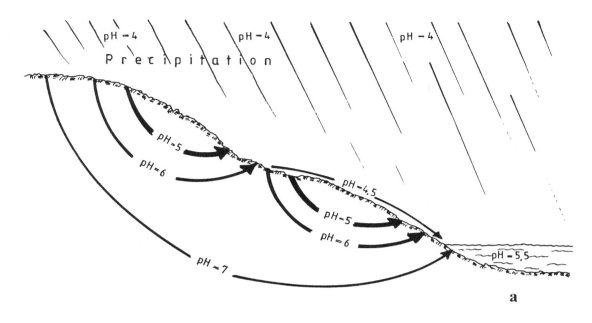

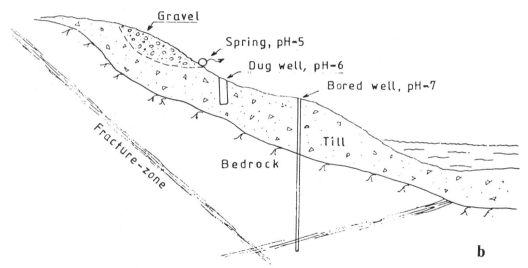

FIGURE 16-2. Ground-water flow pattern in a glaciated terrain. (a) An idealized sketch with thickness of flow lines representing fluxes of water. (b) A geologic situation on which the flow pattern applies. The gravel is a washed-out till formed by wave action during the Holocene periods with higher seawater level.

lands. There was little likelihood for the increase of any other anion than sulfate. The grazing by cows and other domestic animals in these areas decreased considerably over the period of investigation thus making increased nitrate leaching unlikely.

The water was originally a calcium bicarbonate water, other components were small. Gradually, it has become dominated by calcium sulfate.

A similar trend in the hardness to alkalinity ratio in data sets from 1952, 1962,

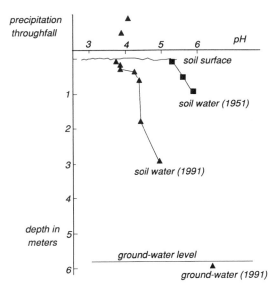

FIGURE 16-3. Soil-water acidification observed over a period of 40 years in southwestern Sweden. (*After Knutsson et al. 1993.*)

1968, and 1980 for 150 public ground-water supplies in Sweden was found by Grimvall et al. (1986). There were indications of an alkalinity decrease but no significant change in hardness. These wells represent larger aquifers than those in the previous example.

A striking decrease in alkalinity has been observed in several populations of well-water analyses, each numbering 2 to 3,000, in some counties in southwestern Sweden (Jonasson, Lång, and Swedberg 1985). The well analyses represent water from newly constructed wells, which were analyzed by local laboratories. It is likely that the well design has been improved over the period, eliminating poor well design as causing the trend.

A deep progressing soil-water acidification has been observed in a site in southwestern Sweden by Knutsson et al. (1993) (Figure 16-3).

HEALTH EFFECTS OF ACIDIC GROUND WATER

Heavy metals and base cations are mobilized by increasing acidity in soil and ground water. The potentially toxic soil metals of concern in connection with human health are aluminum and cadmium. When acidic ground water is fed to a pipe system other metals, such as lead and copper, may be elevated to toxic levels as well. Many investigations show that a sharp increase in the concentration of most metals occurs when the pH drops below about 4.5. These metals mobilized from the soil have not been a problem in ground water used for water supply, because such a low pH is seldom encountered. However, Edmunds, Kinniburgh, and Moss (1990) have found interstitial waters in sandstones with a pH of 4.0–4.5 to several meters in depth. These waters have high contents of aluminum along with several other trace metals. Aluminum may constitute a definite health risk. There is a suspicion that increased levels of aluminum may cause Alzheimer's disease, which manifests itself as a premature dementia (Martyn et al. 1989). Epidemiologic investigations are, however, not as yet conclusive on this question.

Although ground water of low pH may carry cadmium in 10-fold concentrations as compared with background levels of more neutral water, the present health limit of 5–10 µg/L adopted in most countries should seldom be exceeded. Any increased body burden of cadmium is of course unwanted, as it is found that the kidney cortex in present-day man contains about 10-fold more than that of the past century's population (Drasch 1983).

While it is still quite uncertain whether there are any health risks connected with acidic well water, it is well documented that acidic water in contact with piping systems can cause health problems. The most serious problems arise in areas where it has been customary to use lead pipes, as in Anglo-Saxon countries (WHO Working Group 1986). Several milligrams per liter of lead has been detected in Scottish tap water, and lead poisoning is a well-known problem to most Scottish doctors (Moore et al. 1981).

Copper, a safer pipe material, is abundantly used in many countries. Copper corrosion will increase rapidly when pH drops below 6. This is not only a function of the acidity but also of the relative proportions of bicarbonate and sulfate in the water. The formation of a protective copper-hydroxycarbonate layer in pipes is inhibited by sulfate ions (Mattson and Fredriksson 1968). It has been suspected that infant diarrhea can be caused by tap water having in excess of about 1 mg/L of copper. So far, it has not been possible to confirm this suspicion. However, a far more serious effect of excess copper intake via drinking water, fatal liver cirrhosis, has been observed in several countries (Bhave, Pandit, and Tanner 1987; Muller-Höcker, Meyer, and Wiebecke 1988).

MODELING OF SOIL AND GROUND-WATER ACIDIFICATION

Many of the models developed for surface-water modeling have been adapted for modeling ground water as well. Most models use two sets of routines, one for describing the hydrologic pathways and another for the hydrochemical reactions. These routines require a number of empirical parameters whose ranges are roughly known, but the value of which, for each case, are defined by calibration of the models versus a measured record of runoff and runoff chemistry.

Some of the models can be run over long periods of time, creating long-term scenarios, while others can be used only for short-time simulations. For the calibrated models, it is of utmost importance to keep the number of parameters as low as possible and to test the sensitivity of the output to a range of parameter values. The calibrations should always be verified by using an independent period.

One of the first models to consider ground water was discussed by Eriksson (1986) in connection with the definition of critical loads. Assuming a steady-state system, he

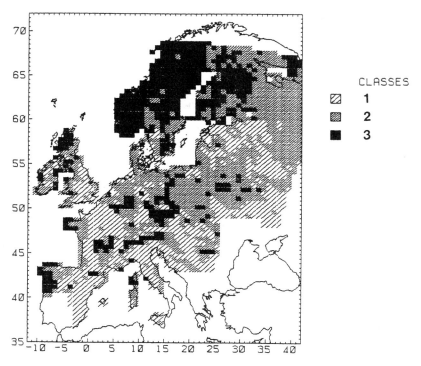

FIGURE 16-4. Aquifer sensitivity to acidification in Europe. (*After Holmberg, Johnston, and Maxe 1990.*)

used hydrochemical budgets for catchments to estimate weathering rates in different kinds of soils, particularly considering the texture and absence or presence of calcium carbonate. Using knowledge of the mean depth of the unsaturated zone, he was able to estimate critical loads for different kinds of common aquifers in Scandinavia. The critical loads were in fair agreement with what has later been found with more elaborate tools.

Aquifer sensitivity has been modeled on a European scale by Holmberg, Johnston, and Maxe (1990). The result is seen in Figure 16-4. They also evaluated the risk for acidification by combining the sensitivity map with a map of the current sulfur deposition.

MAGIC has been quite extensively used to generate scenarios for acidification of surface and ground water (Cosby et al. 1985). It can be run on yearly input data for runoff, or data with higher resolution. The hydrochemical portion of the model contains subroutines for sulfate retention, cation exchange, aluminum solubility, weathering of minerals, and dissociation of carbonic acid.

Other models that have been used for ground-water modeling are TRICKLE DOWN (Schnoor, Palmer, and Glass 1984) and ILWAS (Gherini et al. 1985). The latter has a very detailed hydrologic routine, while the former is more elaborate on the chemical side.

A recent development in modeling is PROFILE (Sverdrup and Warfvinge 1988; Sandén and Warfvinge 1992). The PROFILE model is a soil-profile model that takes into account acid deposition, plant uptake, cation exchange, sulfate retention, and weathering (Figure 16-5). It has been run on hydrologic data of different resolution, but most commonly on a yearly basis. This model is unique in that it is not calibrated. It calculates weathering rates by using dissolution rates for individual minerals derived from laboratory tests or from field data. It requires a detailed description of the mineralogy of the soil profile. This model has been tested on quite a number of soil profiles, especially

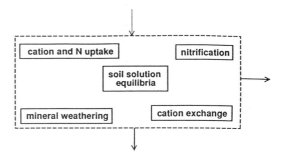
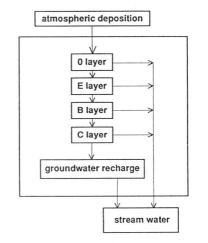

FIGURE 16-5. Structure of a soil-water chemical model. (*After Sandén and Warfvinge 1992.*)

in Scandinavia where other independent weathering rate estimates have been made. The agreement has generally been good. The model has also been run on a national scale by using soil data from a large number of forest soil inventories.

The SAFE model has been used for the reconstruction of the soil chemistry history and the description of future scenarios with different atmospheric loadings (Sverdrup et al. 1990). Another model for dynamic simulation is RESAM (de Vries 1990) which calculates weathering rates from a total soil base cation content correlation.

As noted, the calcium and magnesium content in soil minerals is likely to be an important factor with regard to the neutralizing capacity of a soil. Calcium and mag-

TABLE 16-2. Mineralogic and Petrologic Classification of Soil Material

Class	Minerals Controlling Weathering	Assumed Parent Rock
1	Quartz, K-feldspar	Quartzite, granite
2	Muscovite, plagioclase, biotite (<5%)	Granite, gneiss
3	Biotite, amphibole (<5%)	Granodiorite, gabbro, schist, greywacke
4	Pyroxene, epidote, olivine (<5%)	Gabbro, basalt
5	Carbonates	Limestone, marlstone

Source: Adapted from Nordic Council of Ministers 1988.

nesium are cations found in all fast or rather fast weathering minerals such as calcite, hornblende, biotite, and the more calcic plagioclases. In addition to mineralogy, temperature of the locality is of importance. Based on these assumptions, Olsson and Melkerud (1990) have constructed a simple weathering model which compares well with independent data on weathering rates.

A comprehensive description of acidification models applicable to soils and ground waters is given by Sverdrup, de Vries, and Henriksen (1990).

CRITICAL LOADS OF SULFUR AND NITROGEN FOR GROUND WATER

As the processes and mechanisms of acidification have become more clarified and arguments raised for action to lower emissions of acidifying substances, an immediate need for quantification of permissible deposition has arisen. The critical loads of acidifying substances on an ecosystem have been developed as such a concept. Critical load is defined as "a quantitative estimate of an exposure to one or more pollutants below which significant harmful effects on specified sensitive elements of the environment do not occur according to our present knowledge" (Nordic Council of Ministers 1988).

The first estimate of a critical load was for sulfur in surface water and was set at the Stockholm Conference on Acidification in 1982 at $0.5\,g\text{-}S\,m^{-2}\,a^{-1}$ for Scandinavian conditions.

In the late 1980s the work was intensified in Europe with two workshops held in Norway and Sweden in 1986 and 1988 (Nordic Council of Ministers 1986 and 1988). These workshops saw the participation of North American as well as European experts. In short, the critical load for sulfur was set equal to the weathering rate plus the removal of base cations in harvested biomass in soils with small sulfate adsorption capacity, i.e., for glacial soils. For nitrogen, the critical load is governed by the biological uptake and subsequent harvest in the ecosystem. No acidification of the soil or water takes place until there is nitrate leaching. Nitrogen is presently an acidifying factor in restricted areas in central European forests, especially in Holland and Germany. In Holland, nitrate leaching causes concentrations in ground water to exceed the permissible health limit or 50–200 mg/L of nitrate (de Vries 1988).

The classification that was made for soils will be equally applicable for ground water. As the mineral weathering rate is the fundamental governing factor, the classification of soil can be done on the basis of the soil mineralogy (Table 16-2).

By integrating available weathering rate data, the expert committee was able to set critical sulfur loads for forest soils and for ground water (Table 16-3).

The critical load can be modeled by using some of the models mentioned. The model to be used depends on the input data available. The calibrated models need a record of hydrochemistry or soil chemistry. The noncalibrated models like PROFILE and

TABLE 16-3. Critical Loads for Forest Soil 0–50 cm

Class	Total Acidity (millimol/m$^2 \cdot$ a)	Equivalent Amount of Sulfur (g/m$^2 \cdot$ a)
1	<20	<0.3
2	20–50	0.3–0.8
3	50–100	0.8–1.6
4	100–200	1.6–3.2
5	>200	>3.2

Source: Adapted from Nordic Council of Ministers 1988.

SAFE need a good mineralogic description for the area concerned.

The critical load for nitrogen is the deposition which can be accepted without causing increased nitrate leaching. Excess nitrate leaching will be accompanied by base cations and may cause acidification of the soil in cases where weathering rates are insufficient to buffer cation losses. The critical load will be the long-term uptake by the harvested biomass plus the background nitrogen leaching, mostly in the form of organically bound nitrogen. In addition, it may also be considered that in many boreal forests the carbon storage is increasing due to better forest management. For coniferous forests in Scandinavia and in the Scottish highlands the uptake is from 5 to 15 kg-N ha^{-1} year^{-1}, the background leaching is about 1–2 kg-N ha^{-1} year^{-1}, and the increased nitrogen storage affected by the increase in carbon stores is of the same order (Nordic Council of Ministers 1988). Thus, a sum of about 7–20 kg N ha^{-1} year^{-1} is considered to be a critical load for these ecosystems. The risk of acidification on agricultural land is not as great because fields are limed for production purposes. On the other hand, excess nitrate leaching may exceed the health limit of 45 to 50 mg/L commonly adopted in many countries (de Vries 1988).

The mapping of critical loads is currently being done on a European scale using a combination of different models and other approaches (Kämäri 1991; Sverdrup, de Vries, and Henriksen 1990). Maps of critical loads for ground water will soon be published.

AMENDMENTS FOR ACIDIC GROUND WATER

Acidic ground water has mainly been a problem from the technical-economical aspect due to accelerated corrosion in pipe systems. The most common countermeasure is to install a neutralizing filter in the pipe system. This being a technical matter unrelated to the text, it is not treated here. Other countermeasures have been the liming of the recharge areas of wells and springs and the recirculation of water through limestone beds (Sundlöf 1988). Soil liming has been modeled by Warfvinge (1988).

Liming of the recharge area of wells and springs has been tested on small household wells and springs. In general, the effect on well water has been marginal. Investigations of soil profiles show that migration of the liming effect through the soil is very slow, it will take about five years to reach a depth of 0.5 m (Norrström and Jacks 1993).

Installation of limestone beds in the pathway of ground water, upstream of wells or springs, has been used and found effective where a considerable portion of the flow has been intercepted. Recirculation of the water through limestone beds has also been used and found to be a viable alternative to filters in the pipe system.

CONCLUDING REMARKS

Acidification of ground water caused by anthropogenic emissions of acidifying substances is a problem where these emissions are high and/or where the buffering capacity of the soil is low. The weathering rates of soils is a key factor in neutralizing acidic inputs. In areas with soils derived from granites and gneisses, ground-water acidification may occur even where the deposition is moderately acidic. The critical load for acid sulfur deposition is a direct function of the weathering rate, while the critical load for nitrogen deposition depends on the biological uptake of nitrogen and the harvesting of biomass.

Acidification of surface water will generally occur before acidification of ground water used for water supply due to the near-surface pathways followed by the bulk of the runoff water.

The most serious problem with acidified ground-water supply is the increased corrosion of pipe systems. In addition to increased maintenance costs, there is a definite risk of release of metals from the pipe systems which may be detrimental to human health. Not only increased acidity, but also an increased sulfate content of ground water, may cause corrosion due to the inhibition of formation of protective layers in copper pipes.

A number of models have been developed which have the potential for modeling the effect of acid deposition of ground water at regional scales. Preliminary results from such work indicate that large-scale ground-water acidification is likely unless sulfur and nitrogen emissions are reduced.

References

Åberg, G., G. Jacks, and P. J. Hamilton. 1989. Weathering rates and 87Sr/86Sr ratios: An isotopic approach. *Journal of Hydrology* 109:65–78.

Alban, D. H. 1982. Effects of nutrient accumulation by aspen, spruce and pine on soil properties. *Soil Science Society of America Journal* 46:853–61.

Altshuller, A. P., and G. A. McBean. 1979. *The LRTAP Problem in North America*. The U.S.-Canada Bilateral Research Group on the Long Range Transport of Air Pollutants. Atmospheric Environmental Service. Downsview, Ontario.

April, R., and R. Newton. 1985. Influence of geology on lake acidification in the ILWAS watersheds. *Water, Air, and Soil Pollution* 26:373–86.

April, R., R. Newton, and L. Truettner Coles. 1986. Chemical weathering in two Adirondack watersheds: Past and present-day rates. *Geological Society of America Bulletin* 97:1232–8.

Berthelin, J. 1985. Microbial weathering processes in natural environments. In *Physical and Chemical Weathering in Geochemical Cycles*, eds. A. Lerman and M. Meybeck. NATO ASI Series, Vol. 251-33-60.

Bhave, S. A., A. N. Pandit, and A. N. Tanner. 1987. Comparison of feeding history of children with Indian childhood cirrhosis. *Journal of Pediatrics, Gastroenterology and Nutrition* 6:562–7.

Bricker, O. P., and K. C. Rice. 1989. Acidic deposition to streams. A geology-based method predicts their sensitivity. *Environmental Science and Technology* 23:379–85.

Cleaves, E. T., A. E. Godfrey, and O. P. Bricker. 1970. Geochemical balance of a small watershed and its geomorphic implications. *Geological Society of America Bulletin* 81:3015–32.

Cole, D. W., and M. Rapp. 1981. Elemental cycling in forest cycling. In *Dynamic Properties of Forest Ecosystems*, ed. D. E. Reichle, pp. 341–409. London: Cambridge University Press.

Cosby, B. J., R. F. Wright, G. M. Hernberger, and J. N. Galloway. 1985. Modelling the effects of acid deposition: Assessment of a lumped parameter model of soil water and streamwater chemistry. *Water Resources Research* 21:51–63.

de Vries, W. 1988. Critical deposition levels for nitrogen and sulphur on Dutch forest ecosystems. *Water, Air, and Soil Pollution* 42:221–39.

de Vries, W. 1990. Philosophy, structure and application methodology of a soil acidification model for the Netherlands. In *Impact Models to Assess Regional Acidification*, ed. J. Kämäri, pp. 3–21. Dordrecht: Kluwer.

Drasch, G. A. 1983. An increase of cadmium body burdens for this century—an investigation on human tissues. *Science of the Total Environment*. 26:111–19.

Dunne, T., and R. D. Black. 1970. Partial area contributions to storm runoff in a small New England watershed. *Water Resources Research* 6:1296–1311.

Edmunds, W. M., D. G. Kinniburgh, and P. D. Moss. 1990. Trace metals in interstitial waters from sandstones-acidic inputs to shallow groundwaters. In *international Conference on Acidic Deposition*, Conference Abstracts, p. 60. Royal Society of Edinburgh.

Eilers, J. M., G. E. Glass, K. E. Webster, and J. A. Rogalla. 1983. Hydrologic control of lake susceptibility to acidification. *Canadian Journal of Fisheries and Aquatic Science* 40:1896–904.

Evelyn, J. 1661. *Fumifugium*. London: Bedel and Collins.

Eriksson, E. 1986. Critical loads for acid deposition on groundwater. In *Critical Loads for Nitrogen and Sulphur*, pp. 71–86. Nordic Council of Ministers, Environmental Report 1986:11.

Eriksson, E., E. Karltun, and J. E. Lundmark. 1992. Acidification of forest soils in Sweden. *Ambio* 21:148–54.

Falkengren-Grerup, U. 1987. Long term changes in pH of forest soils in southern Sweden. *Environmental Pollution* 43:79–90.

Fuller, R. D., M. B. David, and C. T. Driscoll. 1985. Sulfate adsorption relationships in forested spodosols of the Northeastern USA. *Soil Science Society of America Journal* 49:1034–40.

Galloway, J. N., B. J. Cosby, and G. E. Likens. 1979. Acid precipitation: Measurement of pH and alkalinity. *Limnology and Oceanography* 24:1161–65.

Galloway, J. N., and H. Rodhe. 1991. Regional atmospheric budgets of S and N fluxes: How well can they be quantified? In *Acidic Deposition, Its Nature and Impacts*, eds. F. T. Last and R. Watling, pp. 61–80. Edinburgh: The Royal Society of Edinburgh.

Garrels, R. M. 1967. Genesis of some ground waters from igneous rocks. In *Researches in Geochemistry*, ed. P. H. Abelson, pp. 405–20. New York: Wiley.

Gherini, S. A., L. Mok, R. J. Hudson, G. F. Davis, C. W. Chen, and R. A. Goldstein. 1985. The ILWAS model: Formulation and application. *Water, Air, and Soil Pollution* 26:425–59.

Gorham, E. 1957. The chemical composition of lake waters in Halifax County, Nova Scotia. *Limnology and Oceanography* 2:12.

Grass, L. B., V. S. Aronovici, and D. C. Muckel. 1962. Some chemical characteristics of submerged and reclaimed sediments of the San Francisco Bay system. *Soil Science Society of America Journal* 26:453–55.

Graunt, J. 1662. *Natural and Political Observations Mentioned in a Following Index, and Made upon the Bills of Mortality*. London: Martin, Allestry and Dicas.

Grimvall, A., C. A. Cole, B. Allard, and P. Sandén. 1986. Quality trends in public water supplies in Sweden. *Water Quality Bulletin* 11:6–11.

Hallbäcken, L., and C. O. Tamm. 1986. Changes in soil acidity from 1927 to 1982–84 in a forest area of southwest Sweden. *Scandinavian Journal of Forestry Research* 1:219–32.

Hewlett, J. D., and A. R. Hibbert. 1967. Factors affecting the response of small watersheds to precipitation in humid areas. In *International Symposium on Forest Hydrology 1965*, Pennsylvania State University, eds. W. E. Sopper, and H. W. Lull, pp. 275–90. New York: Pergamon Press.

Holdren, G. R., and P. M. Speyer. 1985. Reaction rate-surface area relationships during early stages of weathering. I. *Geochemica et Cosmochimica Acta* 49:675–81.

Holdren, G. R., and P. M. Speyer. 1987. Reaction rate-surface area relationships during early stages of weathering. II. *Geochemica et Cosmochimica Acta* 51:2311–18.

Holmberg, M. 1986. *The Impact of Acid Deposition on Groundwater: A Review*. Laxenburg, Austria: International Institute of Applied Systems Analysis, WP-86-31.

Holmberg, M., J. Johnston, and L. Maxe. 1990. Mapping groundwater sensitivity to acidification in Europe. In *Impact Models to Assess Regional Acidification*, ed. J. Kämäri, pp. 51–64. Dordrecht: Kluwer.

Jacks, G. 1990. Mineral weathering studies in Scandinavia. In *The Surface Water Acidification Programme*, ed. B. J. Mason, pp. 215–22. Cambridge: Cambridge University Press.

Jacks, G., G. Knutsson, L. Maxe, and A. Fylkner. 1984. Effect of acid rain on soil and groundwater in Sweden. In *Pollutants in Porous Media. Ecological Studies 47*, eds. B. Yaron, G. Dagan, and J. Goldshmid, pp. 94–114. Berlin: Springer-Verlag.

Jacks, G., G. Werme, and E. Olofsson. 1986. An acid surge in a well-buffered stream. *Ambio* 15:282–5.

Johnson, D. W., M. S. Cresser, S. I. Nilsson, J. Turner, B. Ulrich, D. Blinkley, and D. W. Cole. 1991. Soil changes in forest ecosystems: Evidence for and probable causes. In *Acidic Deposition, Its Nature and Impacts*, eds. F. T. Last, and R. Watling. Royal Society of

Edinburgh, Ser. B., Vol. 97.

Johnson, D. W., and G. S. Henderson. 1989. Terrestrial nutrient cycling. In *Analysis of Biogeochemical Cycling Processes in Walker Branch Watershed*, eds. D. W. Johnson, and R. I. van Hook, pp. 233–300. New York: Springer-Verlag.

Johnston, A. E., K. W. T. Goulding, and R. P. Poulton. 1986. Soil acidification during more than 100 years under permanent grassland and woodland at Rothamsted. *Soil Use and Management* 2:3–10.

Jonasson, S., L. O. Lång, and S. Swedberg. 1985. *Factors Affecting pH and Alkalinity—An Analysis of Acid Well Waters in Southwest Sweden*. Swedish National Environmental Protection Board, Report 3021.

Kämäri, J., ed. 1991. *Impact Models to Assess Regional Acidification*. IIASA. Dordrecht: Kluwer.

Knutsson, G., L. Maxe, L. Lundin, and G. Jacks. 1993. *Field Investigations of Groundwater Acidification*. Stockholm, Sweden: Department of Land and Water Resources, Royal Institute of Technology, Research Report.

Likens, G. E., F. H. Borman, R. S. Pierce, J. S. Eaton, and N. M. Johnson. 1977. *Biogeochemistry of a Forested Ecosystem*. Berlin: Springer Verlag.

Lin, F. C., and C. Clemency. 1981. Dissolution of phologopite. II: Open systems. *Clay and Clay Minerals* 29:107–12.

Lundin, L. 1982. *Soil Water and Ground Water in Till*. Uppsala University, Department of Natural Geography Report 56.

Lundström, U. 1990. Laboratory and lysimeter studies of chemical weathering. In *The Surface Waters Acidification Programme*, ed. B. J. Mason, pp. 267–74. Cambridge: Cambridge University Press.

Martyn, C. N., C. Osmond, J. A. Edwardson, D. J. P. Baker, E. C. Harris, and R. F. Lacey. 1989. Geographical relation between Alzheimer's disease and aluminum in drinking water. *Lancet* 1:59–62.

Mast, M. A., and J. I. Drever. 1987. The effect of oxalate on the dissolution rates of oligoclase and tremolite. *Geochemica et Cosmochimica Acta* 51:2559–68.

Mattsson, E., and A. M. Fredriksson. 1968. Pitting corrosion in copper tubes—causes of corrosion and countermeasures. *British Corrosion Journal* 3:246–57.

Moore, M. R., A. Goldberg, W. M. Fyfe, and W. N. Richards. 1981. Maternal lead levels after alteration to water supply. *Lancet* 2:661–62.

Muller-Höcker, J., U. Meyer, and B. Wiebecke. 1988. Copper storage disease of the liver and chronic intoxication in two further German infants mimicking Indian childhood cirrhosis. *Pathology Research Practice* 183:39–45.

Nevell, W., and M. Wainwright. 1986. Increases in extractable sulphate following submergence with water, dilute sulphuric acid or acid rain. *Environmental Pollution* 12B:301–11.

Newton, R. M., and C. T. Driscoll. 1987. The role of flow paths in controlling stream water chemistry at Pancake Creek in the Adirondack region of New York State. In GEOMON, *International Workshop on Geochemistry and Monitoring in Representative Basins*, eds. B. Moldan and T. Paces, pp. 255–7. Prague: Geological Survey.

Nilsson, S. I., H. G. Miller, and J. D. Miller. 1982. Forest growth as a possible cause of soil and water acidification: An examination of the concepts. *Oikos* 39:40–9.

Nordic Council of Ministers. 1986. *Critical Loads for Nitrogen and Sulphur*, ed. J. Nilsson. Report from a Nordic workshop.

Nordic Council of Ministers. 1988. *Critical Loads for Sulphur and Nitrogen*, eds. J. Nilsson and P. Grennfelt. Report from a workshop held at Skokloster, Sweden, 19–24 March 1988.

Norrström, A. C., and G. Jacks. 1993. Soil liming as a measure to improve acid groundwater. *Environmental Technology* 14:125–34.

Olsson, M., and P. A. Melkerud. 1990. Determination of weathering rates based on geochemical properties of the soil. In *Environmental Geochemistry in Northern Europe*, ed. E. Pulkinen, pp. 69–78. Rovaniemi, Finland: Geological Survey of Finland.

Paces, T. 1986. Weathering rates of granites and depletion of exchangeable cations in soils under environmental acidification. *Journal of the Geological Society of London* 143:673–77.

Pettersson, C., and B. Allard. 1991. Dating of groundwaters from ^{14}C-analysis of dissolved humic substances. In *Humic Substances in the Aquatic and Terrestrial Environment, Lecture Notes in Earth Sciences*, pp. 135–41. Heidelberg: Springer-Verlag.

Reuss, J. O., and D. W. Johnson. 1985. Effect of

soil processes on the acidification of water by acid deposition. *Journal of Environmental Quality* 14:26–31.

Reuss, J. O., and D. W. Johnson. 1986. *Acid Deposition and the Acidification of Soils and Waters. Ecological Studies 59*. New York: Springer-Verlag.

Rochelle, B. P., M. R. Church, and M. B. David. 1987. Sulfur retention in intensively studied watersheds in the U.S. and Canada. *Water, Air, and Soil Pollution* 33:73–83.

Rodhe, A. 1987. *The Origin of Streamwater Traced by Oxygen-18*. Uppsala University, Department of Natural Geography, Report Series A 41.

Rosenqvist, I. Th. 1977. *Acid Soil, Acid Water*. Oslo: Ingenjörforlaget.

Sandén, P., and P. Warfvinge. 1992. *Modelling Groundwater Response to Acidification*. Swedish Meteorological and Hydrological Survey, Report in Hydrology, April 1992.

Schnoor, J. L., W. D. Palmer, Jr., and G. E. Glass. 1984. Modelling impacts of acid precipitation for northeastern Minneosta. In *Modelling of Total Acid Precipiation Impacts*, Vol. 9, ed. J. L. Schnoor. Boston: Butterworth.

Sklash, M. G., and R. N. Farvolden. 1979. The role of groundwater in storm runoff. *Journal of Hydrology* 43:45–65.

Stams, A. J. M., H. W. G. Booltink, I. J. Lutke-Schipholt, B. Beemsterboer, J. R. W. Woittiez, and N. van Breemen. 1991. A field study on the fate of ^{15}N-ammonium to demonstate nitrification of atmospheric ammonium in an acid forest soil. *Plant and Soil* 129:241–55.

Sullivan, T. J., C. T. Driscoll, J. M. Ellers, and D. H. Landers. 1988. Evaluation of the role of sea salt inputs in the long-term acidification of coastal New England lakes. *Environmental Science and Technology* 22:185–90.

Sundlöf, B. 1988. *In Situ Treatment of Acid Groundwater*. Stockholm: Royal Institute of Technology, Report Trita-Kut 1051.

Sverdrup, H. U. 1990. *The Kinetics of Base Cation Release due to Chemical Weathering*. Lund: Lund University Press.

Sverdrup, H. U., W. deVries, and A. Henriksen. 1990. Mapping critical loads. *Nordic Council of Ministers, Environmental Report* 1990:14.

Sverdrup, H. U., and P. Warfvinge. 1988. Weathering of primary silicate minerals in the natural soil environment in relation to a chemical weathering model. *Water, Air, and Soil Pollution* 38:387–408.

Tamm, O. 1930. Experimentelle Studien über die Verwitterung und Tonbildung von Feldspaten. *Chemie der Erde* 4:420–30.

Vanek, V. 1987. The interactions between lake and groundwater and their ecological significance. *Stygologia* 3:1–23.

Velbel, M. A. 1985. Geochemical mass balances and weathering rates in forested watersheds in southern Blue Ridge. *American Journal of Science* 285:904–30.

Warfvinge, P. 1988. *Modeling Acidification Mitigation in Watersheds*. Thesis, Lund: Lund University, Department of Chemical Engineering II.

WHO Working Group. 1986. Health impacts of acidic deposition. *Science of the Total Environment* 52:157–87.

Wollast, R., and L. Chou. 1985. Kinetic study of the dissolution of albite with a continuous flow-through fluidized bed reactor. In *The Chemistry of Weathering*, ed. J. I. Drever, pp. 75–96. Dordrecht, The Netherlands: Reidel.

17

Natural Radionuclides

Richard B. Wanty and D. Kirk Nordstrom

INTRODUCTION

This chapter reviews nuclear and chemical processes that control the abundance and behavior of natural radionuclides in ground water. Principles of radiochemistry and geochemistry will be used to explain the mobilities of individual radionuclides. Some examples will be presented as appropriate, and the reader is encouraged to pursue literature cited in this chapter for more in-depth treatments of various applications.

Radionuclides exist throughout Earth's atmosphere and geosphere. The natural radionuclides ^{238}U, ^{235}U, ^{232}Th, ^{87}Rb, and ^{40}K originated with the solar system. The radioactive series of ^{238}U, ^{235}U, and ^{232}Th have received much attention from an environmental standpoint. Within these series, daughter radionuclides are produced as decay products by spontaneous fission. Radionuclides outside these series may be produced from neutron or cosmic-ray bombardment of nonradioactive elements (Lal 1988; Andrews et al. 1989a).

Radionuclides have been used for dating rocks or ground water (Brownlow 1979; Faure 1986), and for petrogenesis and rock provenance studies (Henderson 1982). They also have been used to determine geochemical and hydrogeologic properties of natural systems (Dyck 1975; Key, Guinasso, and Schink 1979; Gruebel and Martens 1984; Fleischer 1988; Andrews et al. 1989a; 1989b; Wanty et al. 1991). This chapter focuses on radionuclides in the ^{238}U series because of their importance in environmental issues, although the general concepts apply to other radioactive isotopes.

Potential health hazards attributed to natural radiation include accumulations of airborne ^{222}Rn in homes and other structures, large concentrations of radionuclides in drinking-water supplies, and occupational exposures such as in mining and mineral processing. Serious environmental concerns about synthetic radionuclides, especially in regard to disposal of radioactive waste, have focused considerable international research on studies of repository suitability and stability. Many of these studies examined the migration and fixation of natural radionuclides in the proposed repository environment and hence contribute to our understanding of natural radionuclide mobility.

Radionuclide Abundances in Rocks and Ground Water

Uranium and thorium, the original elements in the three natural decay series, are found in all rocks and therefore may be leached into

TABLE 17-1. Typical Abundances of U and Th in Various Rocks and Minerals[a] and Concentrations of U and Th in Natural Waters[b]

Residence of Element	Uranium	Thorium
Earth's crust (continental)	2.7 ppm	9.6 ppm
Granite	4.4 ppm	16. ppm
Basalt	0.8 ppm	2.7 ppm
Sandstone	0.45 ppm	1.7 ppm
Shale	3.8 ppm	12. ppm
Limestone and dolostone	2.2 ppm	1.7 ppm
Phyllosilicates (biotite, muscovite)	20. ppm	25. ppm
Potassium feldspar	1.5 ppm	5.0 ppm
Plagioclase	2.5 ppm	1.5 ppm
Zircon	2,500 ppm	2,000 ppm
Seawater	3.3 ppb	1.5×10^{-3} ppb
Chemically oxidizing ground water	0.1–100 ppb	<1 ppb
Chemically reducing ground water	<0.1 ppb	

[a] Data from Krauskopf (1979), Brownlow (1979), and Henderson (1982).
[b] Data from Brownlow (1979), Wanty et al. (1991), and Langmuir and Herman (1980).

ground water. Abundances of U and Th vary greatly according to rock type, crystallization conditions, depositional and diagenetic environment, and extent of weathering and alteration. Average abundances in some rocks and minerals, in surface seawater, and in typical ground waters are given in Table 17-1. Both U and Th are highly lithophile and are concentrated in the Earth's crust by factors of 100 to 500 over their average cosmic abundances.

Anomalous radionuclide concentrations in ground water are related to rock composition, mineralogy, geologic structure and ground-water chemistry (Asikainen and Kahlos 1979; Coveney et al. 1988; Gascoyne 1989; Gundersen 1989). Natural radionuclide mobilities also are influenced by their atomic location and chemical bonding in minerals and rocks, pressure and temperature conditions, and composition and flow regime of the ground water. If a radionuclide substitutes for another atom during crystal growth, such as a U atom for Zr in relatively insoluble zircon, it may be effectively isolated from contact with ground water. Conversely, a radionuclide adsorbed on a mineral surface is in intimate contact with ground water and may be desorbed depending on changes in ground-water chemistry.

In addition to the crystallographic residence of radionuclides, macroscopic properties of the aquifer system such as water-rock ratio, specific surface area of the rock (Paĉes 1973, 1983), ground-water-flow rates, and total volume of ground-water flux also may affect radionuclide concentrations. A porous sandstone aquifer probably has a greater surface area per unit volume of rock than a fractured crystalline rock. On the other hand, fractured crystalline rock may have a greater surface area per unit volume of entrained ground water because of the extremely low porosity of many crystalline aquifers. The effects of these variables on radionuclide mobility are not yet completely understood.

Ground-water geochemical properties may control radionuclide concentrations through oxidation-reduction conditions, pH, and presence or absence of complexing agents. The solubility and mobility of radionuclides also are affected by temperature and, to a lesser extent, by pressure conditions. Summaries of T and P dependence of mineral solubilities are available (Kern and Weisbrod 1967; Nordstrom and Munoz 1986). However, predictions based on thermodynamics may be approximate, and experimental data should be sought if extrap-

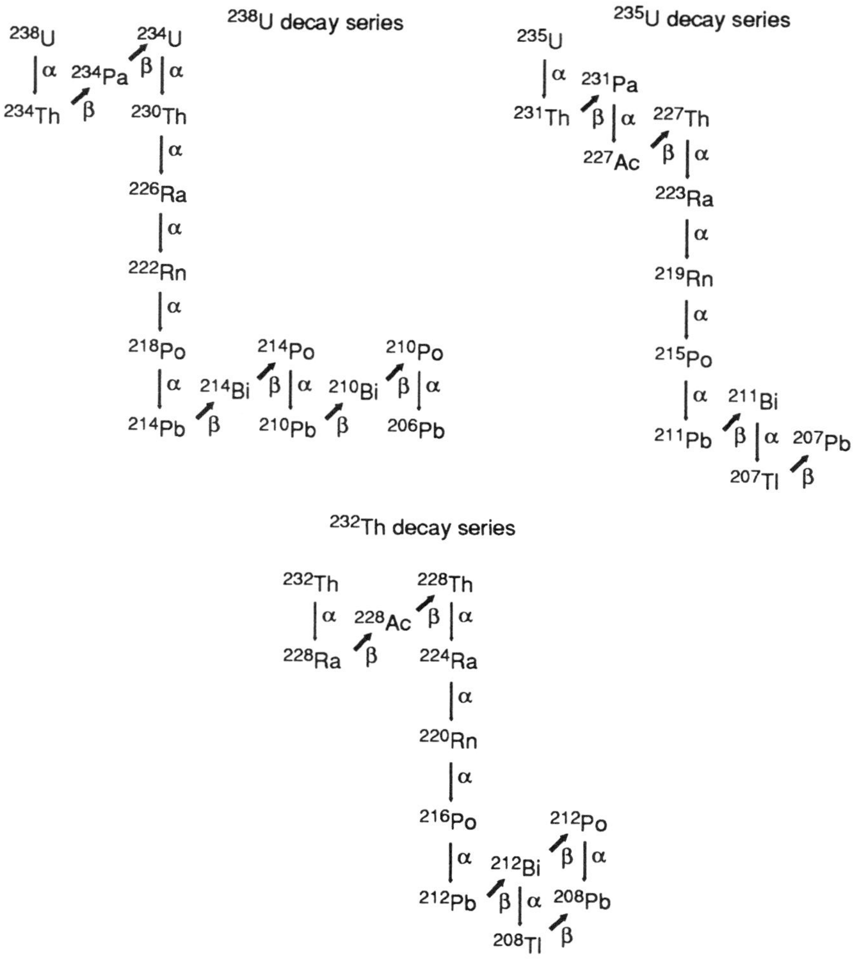

FIGURE 17-1. The ^{238}U, ^{235}U, and ^{232}Th decay series, showing the principal mode of decay of each radionuclide. Other data for radionuclides in these series are given in Table 17-2.

olations are made to high temperature or pressure (cf. Lemire and Tremaine 1980; Davina 1983).

RADIOCHEMISTRY

A greater understanding of radionuclides is attained through a combined understanding of the radiochemical and geochemical processes involved. This section presents the mathematical aspects of radioactivity, and shows how the principles of radioactivity can be applied to natural systems.

Radioactive Decay

Radioactive decay can be described mathematically by the equation $N_t = N_0 e^{-\lambda t}$, where N_t and N_0 refer to the number or concentration of atoms at time t and initially, respectively, λ is the decay constant (time^{-1}), and t is the elapsed time. For a simple system of a radionuclide decaying to produce a single nonradioactive daughter, such as ^{87}Rb decaying to form ^{87}Sr, the abundance of the daughter atom would be simply $N_0 - N_t$.

The mathematics are more complicated

TABLE 17-2. Properties of Radionuclides in the ^{238}U, ^{235}U, and ^{232}Th Series

Radionuclide	Decay Mode	Half-Life (years)	λ (years^{-1})	SpA (pCi/gm)
\multicolumn{5}{c}{^{238}U Decay Series}				
^{238}U	α	4.51E + 09	1.54E − 10	3.33E + 05
^{234}Th	β	6.60E − 02	1.05E + 01	2.32E + 16
^{234}Pa	β	7.70E − 04	9.00E + 02	1.99E + 18
^{234}U	α	2.47E + 05	2.81E − 06	6.19E + 09
^{230}Th	α	8.00E + 04	8.66E − 06	1.94E + 10
^{226}Ra	α	1.60E + 03	4.33E − 04	9.89E + 11
^{222}Rn	α	1.05E − 02	6.60E + 01	1.53E + 17
^{218}Po	α (99.98%)	5.80E − 06	1.20E + 05	2.83E + 20
^{214}Pb	β	5.10E − 05	1.36E + 04	3.28E + 19
^{214}Bi	β (97%)	3.75E − 05	1.85E + 04	4.46E + 19
^{214}Po	α	5.20E − 12	1.33E + 11	3.22E + 26
^{210}Pb	β (>99.99%)	2.10E + 01	3.30E − 02	8.11E + 13
^{210}Bi	β (99%)	1.37E − 02	5.06E + 01	1.24E + 17
^{210}Po	α	3.79E − 01	1.83E + 00	4.50E + 15
\multicolumn{5}{c}{^{235}U Decay Series}				
^{235}U	α	7.1E + 08	9.76E − 10	2.14E + 06
^{231}Th	β	2.91E − 03	2.38E + 02	5.32E + 17
^{231}Pa	α	3.25E + 04	2.13E − 05	4.77E + 10
^{227}Ac	β (98.6%)	2.16E + 01	3.21E − 02	7.30E + 13
^{227}Th	α	5.07E − 02	1.37E + 01	3.11E + 16
^{223}Ra	α	3.13E − 02	2.21E + 01	5.13E + 16
^{219}Rn	α	1.27E − 07	5.46E + 06	1.29E + 22
^{215}Po	α	5.64E − 11	1.23E + 10	2.95E + 25
^{211}Pb	β	6.86E − 05	1.01E + 04	2.47E + 19
^{211}Bi	α (99.72%)	4.09E − 06	1.69E + 05	4.15E + 20
^{207}Tl	β	9.09E − 06	7.63E + 04	1.90E + 20
\multicolumn{5}{c}{^{232}Th Decay Series}				
^{232}Th	α	1.41E + 10	4.92E − 11	1.09E + 05
^{228}Ra	β	5.77E + 00	1.20E − 01	2.72E + 14
^{228}Ac	β	6.99E − 04	9.92E + 02	2.25E + 18
^{228}Th	α	1.91E + 00	3.63E − 01	8.22E + 14
^{224}Ra	α	9.97E − 03	6.95E + 01	1.60E + 17
^{220}Rn	α	1.74E − 06	3.98E + 05	9.35E + 20
^{216}Po	α	4.75E − 09	1.46E + 08	3.49E + 23
^{212}Pb	β	1.21E − 03	5.73E + 02	1.39E + 18
*^{212}Bi	β (63.4%)	1.15E − 04	6.03E + 03	1.47E + 19
^{212}Po	α	9.63E − 15	7.20E + 13	1.75E + 29
*^{212}Bi	α (36%)	1.15E − 04	6.03E + 03	1.47E + 19
^{208}Tl	β	5.89E − 06	1.18E + 05	2.92E + 20

Source: Half-life data from Weast and Astle (1980).
Note: * indicates branch in the chain (see Figure 17-1). Decay constant (λ) and specific activity (SpA) calculated using equations given in the text.

for chains of parents and daughters that are linked in radioactive decay series such as those in Figure 17-1. Nevertheless, relatively simple computer programs can be used to calculate the buildup of daughter products as a function of time in a decay series (Harvey 1962). This decay-generated increase in daughter concentration is referred to as *ingrowth*.

The *half-life* ($t_{1/2}$) of a radionuclide is the amount of time it takes for half of the original atoms to undergo radioactive decay. In the preceding notation, $N_t = 0.5N_0$ after one half-life. By substituting this relation in the exponential decay equation, one can solve for the decay constant as $\lambda = \ln(2)/t_{1/2}$. When applying decay constants to problems in radiochemistry, care must be taken that time units are consistent, as half-lives can vary by many orders of magnitude within a decay series (Table 17-2). Half-lives, decay constants, and other parameters are given in Table 17-2 for radionuclides in the ^{238}U, ^{235}U, and ^{232}Th series.

Radionuclides all undergo some form of nuclear disintegration, so the unit for reporting radioactivity is disintegrations per unit time. The becquerel (Bq) is defined as one disintegration per second (not to be confused with counts per second, which depends on the configuration and calibration of a counter). In SI units, radioactivity is reported as Bq/m^3. The Curie (Ci) is another common radioactivity unit, defined as 3.7×10^{10} Bq; 1 pCi/L = 37 Bq/m^3. Radioactivity values in ground water commonly are reported in units of kilobecquerels per cubic meter or picocuries per liter (pCi/L).

Analyses for radionuclides typically are reported in radioactivity units, but it is useful to be able to convert these results to mass units such as grams or moles per liter. This is accomplished through the *specific activity*, which expresses the radioactivity of a nuclide in terms of its mass (Table 17-2). Specific activity (SpA; pCi/gm) is calculated from the half-life ($t_{1/2}$; years), and atomic mass (m) by the formula

$$\text{SpA} = \frac{3.578 \times 10^{17}}{m t_{1/2}}$$

where the factor 3.578×10^{17} converts from disintegrations to picoCuries. For example, a ^{222}Rn concentration of 1,000 pCi/L is equivalent to 6.51×10^{-15} gm/L or 2.93×10^{-17} mole/L. Thus, extremely small mass quantities of certain radionuclides are detectable through analytical techniques that measure radioactive properties.

Radioactive Equilibria

Radioactive decay series consist of a succession of radionuclides (decay products) each with different decay rates. Depending on the relative decay rates, various steady states may be achieved. For example, ^{226}Ra and ^{222}Rn in a closed system are related by the expression (Lapp and Andrews 1963; Atkins 1978)

$$N_{\text{Rn}} = N_{0,\text{Ra}} \frac{\lambda_{\text{Ra}}}{\lambda_{\text{Rn}} - \lambda_{\text{Ra}}} (e^{-t\lambda_{\text{Ra}}} - e^{-t\lambda_{\text{Rn}}})$$

where λ_i refers to the decay constant, N_{Rn} is the ^{222}Rn concentration at time t, and $N_{0,\text{Ra}}$ is the initial ^{226}Ra concentration. Given that $\lambda_{\text{Ra}} \ll \lambda_{\text{Rn}}$ (Table 17-2), for values of t greater than six or seven ^{222}Rn half-lives the equation reduces to $N_{\text{Rn}}\lambda_{\text{Rn}} = N_{\text{Ra}}\lambda_{\text{Ra}}$. This expression describes the steady-state condition, known as *secular equilibrium*. It requires a closed system and the relationship $\lambda_{\text{parent}} \ll \lambda_{\text{daughter}}$. The ingrowth of ^{222}Rn is shown graphically in Figure 17-2A. At secular equilibrium, the radioactivity of each daughter is maintained, or *supported*, by its parents. Disequilibrium is commonly observed in natural systems when a daughter is *unsupported* by sufficient parent to maintain a constant daughter activity. Excess daughter is thus subject to radioactive decay according to its half-life. *Transient equilibrium* applies when the parent decays somewhat more slowly than the daughter. In this case, parent

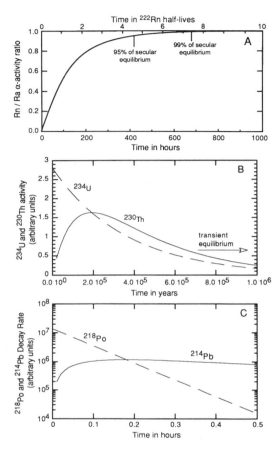

FIGURE 17-2. Ingrowth of ^{222}Rn from parent ^{226}Ra in a closed system. No ^{222}Rn is present initially. (A) Secular equilibrium. Note the time scale on the x-axis is long relative to the half-life of ^{222}Rn, but short relative to the half-life of ^{226}Ra. (B) Transient equilibrium between ^{234}U and ^{230}Th in the ^{238}U-decay series. (C) Nonequilibrium between ^{218}Po and ^{214}Pb in the ^{238}U- decay series.

and daughter decay at steadily decreasing rates as the supply of parent is exhausted (Figure 17-2B). If the parent decays more rapidly than the daughter, equilibrium is never attained (Figure 17-2C; Harvey 1962).

Secular equilibrium is attained in a closed system to within 1% after seven half-lives of the daughter, about four weeks for ^{226}Ra and ^{222}Rn. An entire decay series attains secular equilibrium to within 1% after seven half-lives of the longest-lived daughter in the decay series. In the ^{238}U series, ^{234}U is the longest-lived daughter and about 1.7 million years is required. Secular or transient equilibria may be attained within stepwise parts of a decay series even though the entire series is not at secular equilibrium.

If secular equilibrium is attained in a decay series, three conditions are met. First, each radionuclide decays at the rate at which it is produced. For elapsed times that are short relative to the half-life of the parent, the radioactivity of the daughter is essentially constant. This condition is analogous to a steady state in the terminology of chemical kinetics. Second, the radioactivity of parent and daughter is equal at secular equilibrium, so concentrations of every radionuclide in the series, measured in terms of radioactivity units such as Ci or Bq, will be equal. Third, the ratio of concentrations of a parent and daughter, expressed in terms of number of atoms or mass, will be in proportion to the ratio of their half-lives. For the ^{226}Ra-^{222}Rn system, 1.52×10^5 atoms of ^{226}Ra will be present for each atom of ^{222}Rn (Table 17-2).

The preceding discussion assumes zero mass transfer into and out of the system containing the radionuclides, i.e., a thermodynamically closed system (cf. Kern and Weisbrod 1967; Nordstrom and Munoz 1986). Attainment of radioactive equilibrium infers, but does not prove, closed-system behavior because there may be compensating fluxes of individual radionuclides into and out of a system with a resultant net flux of zero. Radioactive disequilibrium indicates insufficient time since closed-system conditions were established, or true open-system behavior. By determining the particular radionuclides being transported, net direction of the flux, ground-water chemistry, and so forth, many aspects of a rock-water system may be characterized.

Principles of radioactive equilibria (or disequilibria) in ground water are applied in many geologic techniques. Radiometric dating requires, among other things, a closed system with respect to the specific parent and daughter radionuclides being measured and any intermediate daughters (Brownlow 1979; Henderson 1982). Open-system behavior is often documented by examining relative abundances of ^{238}U-series daughters. The nature of U-series disequilibria in U ore deposits can suggest chemical conditions that produced fractionation of chemically dissimilar daughters and also may indicate the timing of remobilization of ore (Rosholt et al. 1965; Rosholt, Tatsumoto, and Dooley 1965). Moreira-Nordemann (1980) used ^{234}U/^{238}U ratios in waters, rocks, and soils in a river basin to estimate weathering rates of rocks.

Another application of secular equilibrium concepts is in the routine measurement of γ-ray energy attributed to ^{214}Bi in γ-spectral surveys (Duval 1990). The concentration of ^{214}Bi is directly measured, and this result is used to calculate the equivalent U concentration, eU, assuming secular equilibrium in the ^{238}U decay series. Considering the number of decay steps (Figure 17-1) and the various chemical behaviors of the elements involved, this assumption is not always justified, although comparison of eU concentrations with total U (U_T) in the same area can be informative. Ratios of eU:U_T greater than unity indicate migration of a U daughter to the site of measurement or leaching of U from that site. Ratios less than unity indicate accumulation of U that has not had sufficient time to reach secular equilibrium with its daughters or migration of the daughters away from the site of the measurement.

A final but important distinction must be made between chemical and radioactive equilibria. Chemical equilibria among aqueous species and solid phases are defined according to the laws of chemical thermodynamics. Chemical reactions naturally drive the system toward favorable energy states. On the other hand, secular or transient equilibria are attained as a function of time and may be aided, hindered, or prevented by the progress of chemical reactions such as precipitation-dissolution, adsorption-desorption, and so forth.

Nuclear versus Chemical Properties

Radioactive properties are a function of nuclear configuration while chemical properties are a function of electron configuration. Nuclear instability leads to spontaneous radioactive decay at a fixed rate (half-life), whereas the chemical stability of each radionuclide is determined by its electron configuration and by the interaction of its electrons with those of surrounding atoms.

In addition to the obvious transmutation of elements accompanying radioactive decay, the chemical behavior may be affected by alpha recoil, the "kickback" that a daughter atom undergoes as the alpha particle is ejected. Alpha recoil may disrupt crystal lattices, leaving the daughter in a more leachable position (Fleischer and Raabe 1978). It may lead to direct ejection of the daughter from a solid to adjacent pore space (Kigoshi 1971; Fleischer 1983; Semkow 1990), or it may place the daughter deeper in the crystal lattice, farther removed from solution. Alpha-recoil effects have been used to explain radioactive disequilibria in a variety of systems (for example, Kigoshi 1971; Fleischer and Raabe 1978; Fleischer 1980; Petit, Langevin, and Dran 1985).

Alpha recoil, rather than diffusion, is probably the most important mechanism for transferring ^{222}Rn produced in a rock to the entrained ground water (Flügge and Zimens 1939). This transfer is quantitatively referred to as emanation efficiency. Emanation models presented by Fleischer (1983, 1988), and Semkow (1990) describe recoil of Rn into pores and consider aspects of surface geometry and ^{226}Ra distribution in the solid. Despite their rigorous treatment of ^{222}Rn emanation, these models include properties that are impossible to measure, such as specific surface area of the rock, surface roughness, and inhomogeneous ^{226}Ra distribution in the solid.

The models of Fleischer (1983, 1988) and Semkow (1990) predict greater ^{222}Rn emanation efficiencies if the mineral surfaces are wet. Apparently, water is a much more effective "stopping" medium for alpha particles than air, and wet surfaces limit the recoil of ^{222}Rn across pores into adjacent mineral grains (Semkow 1990). Experiments by Thamer, Nielson, and Felthauser (1981) show that emanation efficiencies increased by 1.5 to 4 times in wet versus dry rock samples. In addition to the stopping power of water, Fleischer (1983) credits leaching effects of water for increasing ^{222}Rn emanation.

GEOCHEMISTRY OF U, Th, Ra, AND Rn

In this section, the geochemistry of environmentally important radionuclides in the ^{238}U-decay series will be briefly reviewed to allow the reader to gain a process-oriented understanding of natural radionuclide occurrences. This discussion is based largely on thermodynamics and so considers individual species of the elements under equilibrium conditions. Kinetics of chemical reactions are not considered. In general, aqueous complexation reactions are very fast, adsorption-desorption and oxidation-reduction reactions are somewhat slower, and precipitation-dissolution reactions are very slow. Compared to most ground-water-flow rates, complexation, adsorption, and redox reactions should attain equilibrium, although solubility reactions may not. Despite this apparent shortcoming, mobility of radionuclides can be better understood in terms of the thermodynamic principles outlined here.

Uranium

Uranium may be present in natural waters with formal oxidation states of IV, V, or VI. Precipitation of U(IV) minerals limits dissolved U concentrations to µg/L levels or lower in reducing waters. The intermediate oxidation state, U(V), is stable only under a narrow range of conditions seldom found in nature. Despite its thermodynamic instability, U(V) plays an important role in the mechanism of reduction of U(VI) to U(IV) (Mohagheghi 1985). Thus, U(V) may be present in natural systems as an intermediate product. In oxidizing water, factors such as pH and dissolved inorganic or organic carbon can influence the amount of dissolved U(VI). Uranium concentrations in oxidizing waters usually are less than 10 µg/L (4.2×10^{-8} molal). The highest observed natural U concentrations are on the order of several mg/L.

Dissolved U(VI), the most soluble oxidation state of U, takes the form of the uranyl ion (UO_2^{2+}) or some complex thereof. Uranyl is complexed by carbonate and biphosphate above pH 4 (Hostetler and Garrels 1962; Langmuir 1978; Dongarra and Langmuir 1980). Uranyl also is complexed by soluble humic acids at pH values between 3.5 and 7 (Kribek and Podlaha 1980), although Shanbag and Choppin (1981) show that most uranyl–humic acid complexes are relatively insoluble and that precipitated humic acid should scavenge uranyl from solution.

Uranium (VI) concentrations up to several milligrams per liter are predicted on the basis of solubility calculations for U(VI) minerals, but these levels are rarely realized. In oxidizing waters, uranyl is strongly adsorbed onto Fe oxides (Hsi 1981; Ames et al. 1983; Hsi and Langmuir 1985), organic material (Moore 1954; Breger, Deul, and Rubinstein 1955; Schmidt-Collerus 1967; Leventhal, Daws, and Frye 1986), clay minerals (Tsunashima, Brindley, and Bastovanov 1981; Ames, McGarrah, and Walker 1983a), and micaceous minerals (Ames, McGarrah, and Walker 1983c). Adsorption of uranyl is strong at near-neutral pH and weaker at higher and lower values. Below pH 4–5, hydrogen ions compete for surface sites (Hsi and Langmuir 1985); thus, high U concentrations may be found in acid mine waters. At pH greater than 8 or 9, carbonate complexes of uranyl favor desorption. Consistent with carbonate complexation, higher U concentrations have been found in slightly basic, carbonate-rich ground waters than in slightly acidic, carbonate-poor ground waters (Wanty et al. 1991). In strongly acidic waters (pH <

4), such as in acid mine drainage, U concentrations also may be quite high, often exceeding 1 ppm (Plumlee et al. 1992; Wanty et al. in press).

Uranyl is stable in oxidizing systems and is reduced through U(V) to U(IV) under progressively reducing conditions. Uranyl is readily reduced by H_2S (Mohagheghi 1985); the reaction is strongly catalyzed by mineral surfaces. The end product of this reaction is the reduced U(IV) oxide, uraninite (UO_2), or some less crystalline and more oxidized form with a stoichiometry approaching U_3O_8, commonly referred to as pitchblende. In natural waters with high silica concentrations, coffinite ($U[SiO_4]_{4-x}[OH]_{4x}$) may precipitate (Goldhaber et al. 1987).

Thorium

Natural Th has one oxidation state, Th[IV] (Zebroski, Alter, and Heumann 1951). Aqueous Th[IV] is strongly hydrolyzed in the series Th $(OH)_n^{4-n}$, where $n = 0$ to 4 (Langmuir and Herman 1980); the species Th $(OH)_4^0$ dominates above a pH of about 4.5. Strong complexes between Th and biphosphate, oxalate, and EDTA exist at pH values commonly found in nature (Zebroski, Alter, and Heumann 1951; Waggener and Stoughton 1952; Langmuir and Herman 1980).

Thorium is highly insoluble in natural waters; even in solutions with high concentrations of complexing agents, calculated solubilities for thorianite (ThO_2) rarely exceed 1 ppb Th above pH 5 (Langmuir and Herman 1980). Minor amounts of Th substitute as in phosphate, oxide, or silicate minerals. Thorium minerals such as monazite ([Ce, La, Nd, Th]PO_4) may be concentrated in stream sediments and reworked beach sands (Rankama and Sahama 1950). Thorium is the major component in the rare minerals thorianite and thorite ($ThSiO_4$).

Thorium is adsorbed strongly onto kaolinite (Riese 1982) and α-quartz (James and Healy 1972; Riese 1982) at pH values above about 4. Thorium adsorption is inhibited by high sulfate concentrations (Riese 1982; Hunter, Hawke, and Choo 1988), but still adsorption is virtually complete at pH values above 5.5.

Radium

Radium has only one oxidation state in natural systems, Ra(II). Radium behaves in many respects like other alkaline earth elements. It forms strong complexes with sulfate and carbonate and weak complexes with nitrate and chloride (Langmuir and Riese 1985). Complexing of Ra by organic materials is not well characterized, although weak 1:1 complexes with several organic acid species have been detected (Schubert, Russell, and Myers 1950).

Radium mobility in natural systems is enhanced by α-recoil effects and limited by coprecipitation in barite (Doerner and Hoskins 1925) or calcite (Tanner 1964), by adsorption, and by ion exchange. Discrete precipitates of $RaSO_4$ and $RaCO_3$ rarely form because Ra concentrations are so restricted by other processes that natural waters seldom are oversaturated with respect to these compounds (Langmuir and Riese 1985). Radium is strongly adsorbed by minerals, including quartz, clay minerals, and Fe(III) oxyhydroxides (Riese 1982; Ames, McGarrah, and Walker 1983b; Ames et al. 1983). Levinson (1980) suggests that Ra that is initially *ad*sorbed on precipitating ferric oxyhydroxide may be *ab*sorbed by the mineral as it grows. Weigel (1977) provides a detailed review of the chemisty of Ra.

Radon

Radon mobility in natural water is affected by physical rather than chemical processes. A noble gas, Rn is not ionized in solution, nor does it precipitate in solid phases. It may be adsorbed by organic matter, especially charcoal (Lowry and Brandow 1985), but there is little evidence to suggest that it is adsorbed to any degree by most minerals.

The solubility of Rn in water is sensitive to temperature. At 10°C, the solubility of Rn is approximately 61 times that of oxygen on a mass basis (Lowry and Brandow 1985). Radon readily degasses from water exposed to air because its ambient partial pressure is extremely low. Thus, ^{222}Rn concentrations in surface waters are in general less than 100 pCi/L, while ground-water ^{222}Rn concentrations commonly are on the order of hundreds to thousands of pCi/L. A thorough review of Rn chemistry is provided by Weigel (1978).

In ground water, dissolved ^{222}Rn concentrations are rarely supported by aqueous ^{226}Ra, usually exceeding the concentration of its parent by several orders of magnitude. The inescapable conclusion is that ^{226}Ra is in the solid, but near enough to the water-rock interface to provide the source for dissolved ^{222}Rn. Radium adsorbed on mineral surfaces fulfills this requirement. However, usually less than 30% of the Rn generated in a rock is transferred to ground water (Wanty, Lawrence, and Gundersen, 1992).

Environmental Screening for Gross-α and Gross-β

In addition to analyses for individual radionuclides, measurements often are made of gross-α and gross-β activity of dried residues of water samples. The USEPA has recommended maximum levels for these two parameters as 15 pCi/L and 4 mrem/yr effective dose equivalent, respectively. Caution should be applied when interpreting results of these tests, as pointed out by Janzer (1980). First, the calibration standard used is critical. For instance, analytical errors up to 100% arise if gross-α is measured against a ^{241}Am standard as opposed to a natural U standard. Further, because these measurements are made on dried residues of water samples, a potential problem arises from the phenomenon of self-absorption. Greater residue weights tend to absorb emitted radiation before it escapes to the counter, decreasing sample counting efficiency and producing artificially low results. At any rate, the results of gross-α and gross-β analyses cannot be used to determine the concentrations of individual radionuclides in a sample.

CASE STUDIES OF NATURAL RADIONUCLIDES

Radionuclides in Drinking-Water Supplies

If radionuclides such as ^{238}U, ^{226}Ra, or ^{222}Rn are dissolved in sufficient concentrations in drinking water, health hazards may result. Current drinking-water standards are established for Ra (5 pCi/L of ^{226}Ra plus ^{228}Ra), but new revised standards have been proposed by the US EPA to increase the limit to 20 pCi/L of either ^{226}Ra or ^{228}Ra. Standards have been proposed for U (20 μg/L) and ^{222}Rn (300 pCi/L) (U.S. Environmental Protection Agency 1991).

Ingested Ra and U are identified with a variety of health problems (Aieta et al. 1987; Cothern 1987). Because of its chemical similarity to calcium, Ra collects in bone tissue. Inducement of bone cancers appears to be the greatest health threat posed by ingested Ra (National Research Council 1988). The toxicity of U is twofold: as a heavy metal, U may cause kidney damage with extreme concentrations leading to renal failure; as a radionuclide, U may induce bone cancers. The former effect appears more important (National Research Council 1988).

Direct ingestion of ^{222}Rn from drinking water appears not to be a serious health risk (Aieta et al. 1987; Cothern 1987; Crawford-Brown and Cothern 1987; Crawford-Brown 1990). Rather, ^{222}Rn introduced in domestic-water supplies may degas and remain in indoor air, where it poses a more serious risk. The health risk of airborne ^{222}Rn arises from inhalation of ^{222}Rn and its α-emitting daughters, all of which may deposit on the lung tissue. Once deposited in the lung, up to three α-decays per ^{222}Rn atom may contact the lung tissue, and prolonged respiration of ^{222}Rn-rich air can lead to carcinogenesis in the lungs (National Research Council 1988).

Natural Radionuclides in Ground Water

Variations of hydrogeologic properties within an aquifer may affect the ^{222}Rn concentration of ground water. For instance, inverse correlations between dissolved ^{222}Rn and well yield have been found (Rumbaugh 1983; Hall, Boudette, and Olszewski 1987). Rumbaugh (1983) proposed that increased average fracture aperture leads to lower aqueous ^{222}Rn concentrations because the ^{222}Rn produced in the rock is effectively diluted by the greater volume of transmitted water. Lawrence (1990) found this to be true for high-yield wells in his study area but found variable ^{222}Rn concentrations in low-transmissivity wells, possibly because of variable amounts of ^{226}Ra adsorbed to mineral surfaces.

Significant differences in dissolved ^{222}Rn concentrations between fracture-controlled aquifers and porous aquifers may be attributed to aquifer properties. Ground-water-flow rates, tortuosities, water-rock ratios, and specific surface areas vary greatly within and between aquifer types. Comparisons between porous and fractured aquifers in North and South Carolina (King, Michel, and Moore 1982; Loomis 1987) show average dissolved ^{222}Rn activities to be lower in the porous sedimentary aquifers by one to three orders of magnitude. The higher porosity of the sedimentary aquifers compared to fractured ones may help explain this difference (Wanty and Gundersen 1987).

Studies of ^{238}U-series radionuclides in ground water show that ^{222}Rn concentrations are rarely supported by dissolved parent radionuclides. Such is the case in ground waters near Great Salt Lake, Utah, (Tanner 1964), in two-mica granites of Maine and New Hampshire (Wathen 1986), and in the Reading Prong in eastern Pennsylvania (Wanty and Gundersen 1987; 1988). Tanner (1964) proposed that the required additional ^{226}Ra was adsorbed on Fe oxide surfaces. Autoradiographs by Wathen (1986) showed secondary enrichment of U on rims of oxidized biotite grains and in sericitized plagioclase grains. In both cases, ^{226}Ra concentrated on or near grain boundaries was proposed as the source of ^{222}Rn to the water. Wanty and Gundersen (1987, 1988) and Gundersen, Reimer, and Agard (1988) showed that the ^{238}U and ^{226}Ra contents of the rocks are high enough to support the ^{222}Rn activity in the water. Thus, although the water itself is not in secular equilibrium, the entire water-rock system probably approaches secular equilibrium.

Unsupported concentrations of ^{222}Rn in ground water require that ^{226}Ra be located near the water-rock interface or that ^{222}Rn be efficiently transported as it is produced. Rama and Moore (1984) proposed that aqueous ^{222}Rn could be derived from ^{226}Ra residing more deeply in mineral grains if an extensive network of minuscule pores and fractures exists for the ^{222}Rn to escape. In contrast, Krishnaswami and Seidemann (1988) showed that the rate of transport of Ar in rocks was orders of magnitude less than that of ^{222}Rn, leading them to conclude that such "nanopores" do not exist. Thus, preferential concentration of ^{222}Rn parents near grain surfaces and ejection by α-recoil leads to increased emanation of ^{222}Rn compared to other gases.

Predicting areas likely to have high radionuclide concentrations in ground water remains a problem. Current information suggests that as a first approximation high ^{222}Rn concentrations are found in ground water if host rocks are enriched in U. For large-scale reconnaissance, airborne radiometric surveys may be most useful (Duval 1990), but additional measurements including ground-based sampling are required to adequately assess an area. Despite a wide range of U contents and a complex distribution of U in the rocks, Hall, Boudette, and Olszewski (1987) produced a map of New England indicating lithologies likely to produce ^{222}Rn-rich ground water. High ^{222}Rn concentrations are expected in ground water from granitic rocks and from some high-grade metamorphic rocks (Brutsaert et al. 1981; Hall, Donahue, and Eldridge 1985). In contrast, glacial deposits, low-grade metamorphics, and basic intrusive rocks are expected to have low ground-water ^{222}Rn activities.

Radium-226 concentrations in ground water generally are less than 1 pCi/L because of adsorption or precipitation. Unusually high concentrations of both ^{226}Ra and ^{228}Ra were reported by Cecil et al. (1987) for ground water in the Chickies Formation (late Proterozoic) in southeastern Pennsylvania. Here, dissolved ^{226}Ra activities are typically tens of pCi/L and in most wells ^{228}Ra concentrations were even higher. Cecil et al. (1987) attributed high Ra levels to low pH (<5.5) ground water and resultant decreased adsorption of Ra.

Uranium concentrations greater than about 10 ppb are unusual in natural waters. Highest U concentrations are found in conditions favoring desorption of U, as discussed earlier. Thus, Gascoyne (1989), Wanty et al. (1991), and Schlottmann and Breit (1992) have found high U concentrations in oxidizing ground waters rich in carbonate that are in contact with uraniferous aquifer rocks. Lower U concentrations are expected in reducing, carbonate-poor waters, especially those in contact with U-poor aquifer rocks.

Behavior of Uranium in U Deposits

Several types of U deposits will be discussed in this section to illustrate the geochemical behavior of U and its daughters in natural accumulations. A key factor in understanding natural U enrichments is to construct a reasonable geochemical model for formation of U ore deposits. This can be accomplished through an understanding of the chemical behavior of U and other elements enriched in the deposit.

Most of the U deposits in the United States are hosted by sandstone (Nash, Granger, and Adams, 1981; Adams 1991) and are located near a boundary between oxidized and reduced rock. According to a general model for sandstone-type deposits, dissolved U is carried in oxidizing ground water as uranyl ion or as a uranyl-carbonate complex, and precipitates upon reduction by hydrogen sulfide, sulfide-bearing minerals, Fe^{2+}-bearing minerals, and the like. This ore-forming process is most efficient if there is a sharp transition from oxidizing to reducing conditions that is stable for geologically significant periods of time in an aquifer (Miller, Wanty, and McHugh 1984).

A newly recognized class of U deposit is the organic-rich surficial U deposit (SUD). These deposits of Pleistocene to Holocene age form where oxidizing uraniferous ground or surface waters enter peat and organic-rich soils (International Atomic Energy Agency 1984; Culbert and Leighton 1988). Adsorption of U onto the organics, followed by reduction of U, stabilizes the deposit. These deposits are too young for secular equilibrium to have been attained in the ^{238}U-decay series. The U isotopes ^{238}U and ^{234}U are introduced together and may be near secular equilibrium, but daughters of ^{234}U are absent at the time of U accumulation. Referring to Figure 17-1 and Table 17-2, the rate-limiting step in the ingrowth of daughters is the ingrowth of ^{230}Th, with a half-life of approximately 80,000 years. Thus, SUDs would require approximately 500,000 years to approach secular equilibrium. Because SUDs are so near to the surface, environmental concerns have been raised that U may be remobilized upon perturbations to the host aquifer (Macke, Schumann, and Otton 1988; Zielinski and Meier 1988).

Natural Analog Studies of Radioactive-Waste Disposal

Since the mid-1970s considerable research has been directed toward natural migration of radionuclides as analogs to radionuclides leaking from high-level radioactive waste repositories. These studies examine more radionuclides in a greater range of concentrations than conventional ground-water studies, and most have also studied details of the hydrogeologic and mineralogic environment. Results from natural analog studies can provide the ultimate test of concepts, theories, and principles of radionuclide behavior in the environment. International studies at Alligator River, Australia (Duerden, 1991), Pocos de Caldas, Brazil (Chapman et al. 1991), Oklo, Gabon (Cowan 1978), and Cigar Lake, Saskatchewan (Cramer and Smellie, written communication, 1990) are examples of natural analog programs.

By applying concepts of radioactive equilibria for successive steps in a decay series, radionuclide redistribution by ground water can be assessed at various time scales. Once the status of radioactive equilibria has been determined in the system, specific chemical processes can be inferred to have caused or inhibited radionuclide mobility. Thus, Guthrie (1991) concluded that U redistribution by ground water has occurred within the last million years in three Australian granitoid bodies, and she proposed several physical and chemical mechanisms to explain mobilization and fixation of ^{238}U, ^{234}U, and ^{230}Th. At Alligator River, adsorption and ground-water mixing are important factors slowing migration of the ^{238}U-series elements, while ^{36}Cl and ^{129}I are more mobile (Fabryka-Martin, written communication 1991), consistent with the difference in chemical properties between easily attenuated heavy metals and easily mobile halogens. At Pocos de Caldas, redox properties play a key role. In the intensely weathered, near-surface zone, U has been leached and redistributed, but remains immobile in deeper, chemically reduced zones that contain pyrite but no Fe oxyhydroxides. Rates of weathering for U and Th are orders of magnitude slower than for the major cations, indicating U and Th are actively retarded by adsorption and precipitation. The investigations at Cigar Lake are still in progress, but results to date show that extremely high concentrations of ^{238}U-series radionuclides in the rock have been held immobile for geologic time periods by a natural mantle of Fe oxides, clays, and a silica cap (Cramer and Smellie, written communication 1990). The role of an impermeable geologic formation in impeding ground-water flow and radionuclide mobility also has been demonstrated at Oklo, the only known natural fission reactor site (Brookins 1984; Krauskopf 1988).

Perhaps the most comprehensive studies of natural radionuclide behavior in the hydrogeological environment come from the International Stripa Project (1977–1992). Exhaustive chemical and isotopic data are reported for ground waters, rocks, and gases present at an abandoned iron ore mine hosted by granites and metavolcanic sediments in central Sweden. High concentrations of ^{222}Rn and ^{226}Ra were found in Stripa ground waters at depths of 300 to 1200 m (Andrews et al. 1989b). The ultimate parent of these radionuclides, ^{238}U, is present in high concentrations in the Stripa granite. Andrews et al. (1989b) conclude that ^{222}Rn diffuses from the rock matrix to water-bearing fractures and that aqueous ^{226}Ra results from chemical dissolution rather than alpha recoil.

Radon in Ground Water as an Indoor Radon Source

In homes that rely on private or public ground-water wells for domestic supply, significant amounts of airborne ^{222}Rn may be contributed from the water (Gesell and Prichard 1980; Bruno 1983; Prichard 1987; Nazaroff et al. 1988; Lawrence, Wanty, and Nyberg 1992). In fact, the standard proposed by the USEPA for ^{222}Rn in water is based primarily on the possible contribution of dissolved ^{222}Rn to indoor air. Therefore, some homes with excessive amounts of airborne ^{222}Rn (>4 pCi/L in air, according to the U.S. Environmental Protection Agency) may be remediated by treatment of well-water supplies.

It is difficult to quantify the proportion of indoor ^{222}Rn derived from water, because many variables are involved, including ^{222}Rn concentration of the water, volume of water used indoors, air volume and exchange rate of the home, degree of mixing of indoor air, and ambient temperatures. A widely cited rule of thumb that equates 10,000 pCi/L of ^{222}Rn in water to 1 pCi/L to indoor air is so general as to be useless, yet it is the basis for many population exposure estimates and policy decisions. A more comprehensive model calculation for the contribution of waterborne ^{222}Rn (Nazaroff et al. 1988) has had limited success. The most definitive approach might be to directly measure indoor ^{222}Rn at short time intervals (<1 h) and relate variations of indoor ^{222}Rn to periods of water usage (Lawrence, Wanty, and Nyberg 1992). Application of this method appears to be quite reliable, although it is time-consuming and impractical for monitoring large numbers of homes.

Radon-222 in a well-water supply can be mitigated by one of two types of "point of entry" systems that treat the water before it enters the household plumbing. One technique employs a granular activated carbon filter, which adsorbs ^{222}Rn from solution (Lowry and Brandow 1985) but which has a severe disadvantage in that the filter may accumulate radioactivity to hazardous levels (Kinner, Malley, and Clement 1990). A preferred mitigation technique aerates the water supply and vents the extracted ^{222}Rn outdoors. This technique, although expensive at this writing, appears to be safer and more effective (Kinner, Malley, and Clement 1990; Kinner et al. 1990).

CONCLUDING REMARKS

Natural radionuclides of the U and Th decay series are ubiquitous, occurring to some extent in nearly all rocks and ground waters. In addition to the obvious enrichments near U and Th ore deposits, high concentrations of some of these radionuclides have been found in other geological environments, raising additional environmental concerns. For example, U, Ra, and Rn dissolved in ground water may pose a public health threat when the ground water is used for domestic supply. Studies of the mobility of U and its radioactive daughters in the lithosphere also will help us to better protect the environment as disposal methods are evaluated for radioactive waste and mine tailings. Current studies of waste-disposal techniques are directed toward understanding natural accumulations and mobilities of radionuclides in the lithosphere and hydrosphere.

Factors that must be considered when evaluating or predicting the mobility of U, Th, and their daughters include (1) the crystallographic residence of radionuclides in rocks and minerals; (2) the porosity, permeability, water-rock ratio, and specific surface area of rock; (3) the geochemistry of radionuclides and of coexisting waters; (4) additional processes such as alpha recoil which are specific to elements undergoing radioactive decay; and (5) the rate of in situ radionuclide production, which may lead to unexpectedly high concentrations of certain radionuclides used as ground-water tracers and geochronometers. Chemical behavior of radionuclides can be estimated by thermodynamic calculations of elemental speciation and solubility while recognizing the role of adsorption and ion exchange. In general, adsorption is limited and elemental mobilities are greater in ground water containing complexing agents. Lastly, the open or closed behavior of a system is critical as it governs whether radioactive or chemical equilibria will be established. Similarly, examination of the status of radioactive equilibria in a system may be used to indicate open or closed behavior.

References

Adams, S. S. 1991. Evolution of genetic concepts for principal types of sandstone uranium deposits in the United States. *Economic Geology Monograph* 8:225–48.

Aieta, E. M., J. E. Singley, A. R. Trussell, K. W. Thorbjarnarson, and M. J. McGuire. 1987. Radionuclides in drinking water: An overview. *Journal of the American Water Works Association* 79:144–52.

Ames, L. L., J. E. McGarrah, and B. A. Walker. 1983a. Sorption of trace constituents from aqueous solutions onto secondary minerals. I: Uranium. *Clays and Clay Minerals* 31: 321–34.

Ames, L. L., J. E. McGarrah, and B. A. Walker. 1983b. Sorption of trace constituents from aqueous solutions onto secondary minerals. II: Radium. *Clays and Clay Minerals* 31: 335–42.

Ames, L. L., J. E. McGarrah, and B. A. Walker. 1983c. Sorption of uranium and radium by biotite, muscovite, and phlogopite. *Clays and Clay Minerals* 31:343–51.

Ames, L. L., J. E. McGarrah, B. A. Walker, and P. F. Salter. 1983. Uranium and radium sorption on amorphous ferric oxyhydroxide. *Chemical Geology* 40:135–48.

Andrews J. N., S. N. Davis, J. Fabryka-Martin, J.-C. Fontes. B. E. Lehmann, H. H. Loosli, J.-L. Michelot, H. Moser, B. Smith, and M. Wolf. 1989a. The in situ production of radioisotopes in rock matrices with particular reference to the Stripa granite. *Geochimica et Cosmochimica Acta* 53(8):1803–15.

Andrews, J. N., D. J. Ford, N. Hussain, D. Trivedi, and M. J. Youngman. 1989b. Natural radioelement solution by circulating groundwaters in the Stripa granite. *Geochimica et Cosmochimica Acta* 53:1791–1802.

Asikainen, M., and H. Kahlos. 1979. Anomalously high concentrations of uranium, radium and radon in water from drilled wells in the Helsinki region. *Geochimica et Cosmochimica Acta* 43:1681–86.

Atkins, P. W. 1978. *Physical Chemistry*. San Francisco: Freeman.

Breger, I. A., M. Deul, and S. Rubinstein. 1955. Geochemistry and mineralogy of a uraniferous lignite. *Economic Geology* 50:206–26.

Brookins, D. G. 1984. *Geochemical Aspects of Radioactive Waste Disposal*. New York: Springer-Verlag.

Brownlow, A. H. 1979. *Geochemistry*. Englewood Cliffs, NJ: Prentice-Hall.

Bruno, R. C. 1983. Sources of indoor radon in houses: A review. *Journal of the Air Pollution Control Association* 33:105–9.

Brutsaert, W. F., S. A. Norton, C. T. Hess, and J. S. Williams. 1981. Geologic and hydrologic factors controlling radon-222 in ground water in Maine. *Ground Water* 19:407–17.

Cecil, L. D., R. C. I. Smith, M. A. Reilly, and A. W. Rose. 1987. Radium-228 and radium-226 in ground water of the Chickies Formation, southeastern Pennsylvania. In *Radon, Radium, and Other Radioactivity in Ground Water*, ed. B. Graves, pp. 437–47. Chelsea, MI: Lewis Publishers.

Chapman, N. A., I. G. McKinley, M. E. Shea, and J. A. T. Smellie. 1991. *The Pocos de Caldas Project: Summary and Implications for Radioactive Waste Management*. Swedish Nuclear Fuel and Waste Management Co. SKB Technical Report 90-24.

Cothern, C. R. 1987. Development of regulations for radionuclides in drinking water. In *Radon, Radium, and Other Radioactivity in Ground Water*, ed. B. Graves, pp. 1–11. Chelsea, MI: Lewis Publishers.

Coveney, R. M., P. L. Hilpman, A. V. Allen, and M. D. Glascock. 1988. Radionuclides in Pennsylvanian black shales of the midwestern United States. In *Geological Causes of Natural Radionuclide Anomalies*, eds. M. A. Marikos and R. H. Hansman, pp. 25–42. St. Louis, MO: Missouri Department of Natural Resources.

Cowan, G. A. 1978. Migration paths for Oklo reactor products and applications to the problem of geological storage of nuclear wastes. In *Natural Fission Reactors, Proceedings of the Technical Committee*, pp. 693–9. Vienna: International Atomic Energy Agency.

Crawford-Brown, D. J. 1990. Analysis of the health risk from ingested radon. In *Radon, Radium, and Uranium in Drinking Water*, eds. C. R. Cothern and P. A. Rebers, pp. 17–26. Chelsea, MI: Lewis Publishers.

Crawford-Brown, D. J., and C. R. Cothern. 1987. A Bayesian analysis or scientific judgement of uncertainties in estimating risk due to ^{222}Rn in U.S. public drinking water supplies. *Health Physics* 53:11–21.

Culbert, R. R., and D. G. Leighton. 1988. Young uranium. In *Unconventional Uranium Deposits, Ore Geology Reviews*, ed. J. W. Gabelman, pp. 313–20. Amsterdam: Elsevier.

Davina, O. A., M. Y. Yefimov, V. A. Medvedev, and I. L. Khodakovskiy. 1983. Thermochemical determination of the stability constant of $UO_2(CO_3)_3^{4-}{}_{(sol)}$ at 25–200°C. *Geokhimiya* 5:677–84.

Doerner, H. A., and W. M. Hoskins. 1925. Co-precipitation of radium and barium sulfates. *Journal of the American Chemical Society* 47:662–75.

Dongarra, G., and D. Langmuir. 1980. The stability of UO_2OH^+ and $UO_2[HPO_4]_2^{2-}$ complexes at 25°C. *Geochimica et*

Cosmochimica Acta 44:1747–51.

Duerden, P., ed. 1991. *Alligator Rivers Projects, 2nd Annual Report*, 1989–1990. Menai, Australia.

Duval, J. S. 1990. The role of National Uranium Resource Evaluation aerial gamma-ray data in assessment of radon hazard potential. In *Proceedings of a United States Geological Survey Workshop on Environmental Geochemistry*, ed. B. R. Doe, pp. 163–64. Washington, DC: U.S. Government Printing Office.

Dyck, W. 1975. Geochemistry applied to uranium exploration. *Geological Survey of Canada Paper* 75-26, pp. 33–47.

Faure, G. 1986. *Principles of Isotope Geology*, 2nd ed. New York: Wiley.

Fleischer, R. L. 1980. Isotopic disequilibrium of uranium: Alpha-recoil damage and preferential solution effects. *Science* 207:979–81.

Fleischer, R. L. 1983. Theory of alpha recoil effects on radon release and isotopic disequilibrium. *Geochimica et Cosmochimica Acta* 47(4):779–84.

Fleischer, R. L. 1988. Alpha-recoil damage: Relation to isotopic disequilibrium and leaching of radionuclides. *Geochimica et Cosmochimica Acta* 52(6):1459–66.

Fleischer, R. L., and O. G. Raabe. 1978. Recoiling alpha-emitting nuclei. Mechanisms for uranium-series disequilibrium. *Geochimica et Cosmochimica Acta* 42:973–78.

Flügge, S., and Zimens, K. E. 1939. Die bestimmung von korngrössen und diffusionskonstanten aus dem emaniervermögen (Die theorie der emaniermethode). *Zeitschrift für Physikalische Chemie (Leipzig)*, B 42:179–220.

Gascoyne, M. 1989. High levels of uranium and radium in groundwaters at Canada's Underground Research Laboratory, Lac du Bonnet, Manitoba, Canada. *Applied Geochemistry* 4(6):577–92.

Gesell, T. F., and H. M. Prichard. 1980. The contribution of radon in tap water to indoor radon concentrations. In *The Natural Radiation Environment III, Symposium Proceedings*, eds. T. F. Gesell and W. M. Lowder, pp. 5–56. Houston, TX: U.S. Department of Energy.

Goldhaber, M. B., B. S. Hemingway, A. Mohagheghi, R. L. Reynolds, and H. R. Northrop. 1987. Origin of coffinite in sedimentary rocks by a sequential adsorption-reduction mechanism. *Bulletin de Minéralogie* 110:131–44.

Gruebel, K. A., and C. S. Martens. 1984. Radon-222 tracing of sediment-water chemical transport in an estuarine sediment. *Limnology and Oceanography* 29(3):587–97.

Gundersen, L. C. S. 1989. Anomalously high radon in shear zones. In *Symposium on Radon and Radon Reduction Technology, Proceedings Volume 1*, eds. M. C. Osborne and J. Harrison, pp. V27–V44. Washington, DC: U.S. Environmental Protection Agency.

Gundersen, L. C. S., G. M. Reimer, and S. S. Agard. 1988. Correlation between geology, radon in soil gas, and indoor radon in the Reading Prong. In *Geological Causes of Natural Radionuclide Anomalies*, eds. M. A. Marikos and R. H. Hansman, pp. 91–102. St. Louis, MO: Missouri Department of Natural Resources.

Guthrie, V. 1991. Determination of recent ^{238}U, ^{234}U and ^{230}Th mobility in granitic rocks: Application of a natural analogue to the high-level waste repository environment. *Applied Geochemistry* 6(1):63–74.

Hall, F. R., E. L. Boudette, and W. J. Olszewski, Jr. 1987. Geologic controls and radon occurrence in New England. In *Radon, Radium, and Other Radioactivity in Ground Water*, ed. B. Graves, pp. 15–30. Chelsea, MI: Lewis Publishers.

Hall, F. R., P. M. Donahue, and A. L. Eldridge. 1985. Radon gas in ground water of New Hampshire. In *Proceedings of the Association of Ground Water Scientists and Engineers, Eastern Regional Ground Water Conference*, pp. 86–101. National Water Well Association.

Harvey, B. G. 1962. *Introduction to Nuclear Physics and Chemistry*. Englewood Cliffs, NJ: Prentice-Hall.

Henderson, P. 1982. *Inorganic Geochemistry*. New York: Pergamon Press.

Hostetler, P. B., and R. M. Garrels. 1962. Transportation and precipitation of uranium and vanadium at low temperatures, with special reference to sandstone-type uranium deposits. *Economic Geology* 57:137–67.

Hsi, C.-K. D. 1981. *Sorption of Uranium (VI) by*

Iron Oxides. Ph.D. thesis, Golden, CO: Colorado School of Mines.

Hsi, C.-K. D., and D. Langmuir. 1985. Adsorption of uranyl onto ferric oxyhydroxides: Application of the surface complexation site-binding model. *Geochimica et Cosmochimica Acta* 49:1931–41.

Hunter, K. A., D. J. Hawke, and L. K. Choo. 1988. Equilibrium adsorption of thorium by metal oxides in marine electrolytes. *Geochimica et Cosmochimica Acta* 52(3): 627–36.

International Atomic Energy Agency. 1984. *Surficial Uranium Deposits, IAEA Tecdoc 322.* Vienna: International Atomic Energy Agency.

James, R. O., and T. W. Healy. 1972. Adsorption of hydrolyzeable metal ions at the oxide-water interface. II: Charge reversal of SiO_2 and TiO_2 colloids by adsorbed Co(II), La(III), and Th(IV) as model systems. *Journal of Colloid and Interface Science* 40:53–64.

Janzer, V. J. 1980. Discordant gross radioactivity measurements of natural and treated waters. In *Effluent and Environmental Radiation Surveillances, ASTM STP 698*, ed. J. J. Kelly, pp. 327–41. Philadelphia, PA: American Society for Testing and Materials.

Kern, R., and A. Weisbrod. 1967. *Thermodynamics for Geologists.* San Francisco: Freeman, Cooper.

Key, R. M., N. L. Guinasso, Jr., and D. R. Schink. 1979. Emanation of radon-222 from marine sediments. *Marine Chemistry* 7: 221–250.

Kigoshi, K. 1971. Alpha-recoil thorium-234: Dissolution into water and the uranium-234/uranium-238 disequilibrium in nature. *Science* 173:47–48.

King, P. T., J. Michel, and W. S. Moore. 1982. Ground water geochemistry of ^{228}Ra, ^{226}Ra and ^{222}Rn. *Geochimica et Cosmochimica Acta* 46:1173–82.

Kinner, N. E., J. P. Malley, Jr., and J. A. Clement. 1990. *Radon Removal Using Point-of-Entry Water Treatment Techniques.* Washington, DC: U.S. Environmental Protection Agency EPA/600/S2-90/047.

Kinner, N. E., P. A. Quern, G. S. Schell, C. E. Lessard, and J. A. Clement. 1990. Treatment technology for removing radon from small community water supplies. In *Radon, Radium, and Uranium in Drinking Water*, eds. C. R. Cothern and P. A. Rebers, pp. 39–50. Chelsea, MI: Lewis Publishers.

Krauskopf, K. B. 1979. *Introduction to Geochemistry*, 2nd ed. New York: McGraw-Hill.

Krauskopf, K. B. 1988. *Radioactive Waste Disposal and Geology.* London: Chapman and Hall.

Kribek, B., and J. Podlaha. 1980. The stability constant of the UO_2^{2+}-humic acid complex. *Organic Geochemistry* 2:93–97.

Krishnaswami, S., and D. E. Seidemann. 1988. Comparative study of ^{222}Rn, ^{40}Ar, ^{39}Ar, and ^{37}Ar leakage from rocks and minerals: Implications for the role of nanopores in gas transport through natural silicates. *Geochimica et Cosmochimica Acta* 52(3): 655–58.

Lal, D. 1988. In situ produced cosmogenic isotopes in terrestrial rocks. *Annual Reviews of Earth and Planetary Sciences* 16:355–88.

Langmuir, D. 1978. Uranium solution-mineral equilibria at low temperatures with applications to sedimentary ore deposits. *Geochimica et Cosmochimica Acta* 42: 547–69.

Langmuir, D., and J. S. Herman. 1980. The mobility of thorium in natural waters at low temperatures. *Geochimica et Cosmochimica Acta* 44:1753–66.

Langmuir, D., and A. C. Riese. 1985. The thermodynamic properties of radium. *Geochimica et Cosmochimica Acta* 49: 1593–1601.

Lapp, R. E., and H. L. Andrews. 1963. *Nuclear Radiation Physics*, 3rd ed. Englewood Cliffs. NJ: Prentice-Hall.

Lawrence, E. P. 1990. *Hydrogeologic and Geochemical Processes Affecting the Distribution of ^{222}Rn and Its Parent Radionuclides in Ground Water, Conifer, Colorado.* M.S. thesis. Golden, CO: Colorado School of Mines.

Lawrence, E. P., R. B. Wanty, and P. A. Nyberg. 1992. Contribution of ^{222}Rn in domestic water supplies to ^{222}Rn in indoor air in homes in Colorado. *Health Physics* 62(2):171–77.

Lemire, R. J., and P. R. Tremaine. 1980. Uranium and plutonium equilibria in aqueous solutions to 200°C. *Journal of Chemical and Engineering Data* 25(4):361–70.

Leventhal, J. S., T. A. Daws, and J. S. Frye. 1986. Organic geochemical analysis of sedimentary organic matter associated with uranium. *Applied Geochemistry* 1:241–47.

Levinson, A. A. 1980. *Introduction to Exploration Geochemistry*, 2nd ed. Wilmette, Il: Applied Publishing.

Loomis, D. P. 1987. Radon-222 concentration and aquifer lithology in North Carolina. *Ground Water Monitoring Review*:33–39.

Lowry, J. D., and J. E. Brandow. 1985. Removal of radon from water supplies. *Journal of Environmental Engineering* 111:511–27.

Macke, D. L., R. R. Schumann, and J. K. Otton. 1988. Geology and uranium distribution in the Fish Lake surficial uranium deposit, Nevada. In *Geological Causes of Natural Radionuclide Anomalies*, eds. M. A. Marikos and R. H. Hansman, pp. 53–63. St. Louis, MO: Missouri Department of Natural Resources.

Miller, W. R., R. B. Wanty, and J. B. McHugh. 1984. Application of mineral-solution equilibria to geochemical exploration for sandstone-hosted uranium deposits in two basins in west-central Utah. *Economic Geology* 79(2):266–83.

Mohagheghi, A. 1985. *The Role of Aqueous Sulfide and Sulfate-Reducing Bacteria in the Kinetics and Mechanisms of the Reduction of Uranyl Ion*. Ph.D. thesis. Golden, CO: Colorado School of Mines.

Moore, G. W. 1954. Extraction of uranium from aqueous solution by coal and some other materials. *Economic Geology* 49:652–58.

Moreira-Nordemann, L. M. 1980. Use of $^{234}U/^{238}U$ disequilibrium in measuring chemical weathering rate of rocks. *Geochimica et Cosmochimica Acta* 44:103–8.

Nash, J. T., H. C. Granger, and S. S. Adams. 1981. Geology and concepts of genesis of important types of uranium deposits. In *Economic Geology 75th Anniversary Volume*, ed. B. J. Skinner, pp. 63–116. El Paso, TX: Economic Geology Publishing.

National Research Council. 1988. *Health Risks of Radon and Other Internally Deposited Alpha-Emitters, BEIR IV*. Washington, DC: National Academy Press.

Nazaroff, W. W., S. M. Doyle, A. V. Nero, Jr., and R. G. Sextro. 1988. Radon entry via potable water. In *Radon and Its Decay Products in Indoor Air*, eds. W. W. Nazaroff and A. V. Nero, Jr., pp. 131–60. New York: Wiley.

Nordstrom, D. K., and J. L. Munoz. 1986. *Geochemical Thermodynamics*. Menlo Park, CA: Blackwell Scientific.

Pacês, T. 1973. Steady-state kinetics and equilibrium between ground water and granitic rock. *Geochimica et Cosmochimica Acta* 37:2641–63.

Pacês, T. 1983. Rate constants of dissolution derived from the measurements of mass balance in hydrological catchments. *Geochimica et Cosmochimica Acta* 47(11):1855–63.

Petit, J. C., Y. Langevin, and J. C. Dran. 1985. $^{234}U/^{238}U$ disequilibrium in nature: Theoretical reassessment of the various proposed models. *Bulletin de Minéralogie* 108:745–53.

Plumlee, G. S., K. S. Smith, W. H. Ficklin, and P. H. Briggs. 1992. Geological and geochemical controls on the composition of mine drainages and natural drainages in mineralized areas. In *Water-Rock Interaction*, eds. Y. K. Kharaka and A. S. Maest, pp. 419–22. Rotterdam, Netherlands: A. A. Balkema.

Prichard, H. M. 1987. The transfer of radon from domestic water to indoor air. *American Water Works Association Journal* 79:159–61.

Rama, and W. S. Moore. 1984. Mechanism of transport of U-Th series radioisotopes from solids into ground water. *Geochimica et Cosmochimica Acta* 48:395–99.

Rankama, K., and T. H. Sahama. 1950. *Geochemistry*. Chicago: University of Chicago Press.

Riese, A. C. 1982. *Adsorption of Radium and Thorium onto Quartz and Kaolinite: A Comparison of Solution/Surface Equilibria Models*. Ph.D. thesis. Golden, CO: Colorado School of Mines.

Rosholt, J. N., A. P. Butler, E. L. Garner, and W. R. Shields. 1965. Isotopic fractionation of uranium in sandstone, Powder River Basin, Wyoming, and Slick Rock District, Colorado. *Economic Geology* 60(2):199–213.

Rosholt, J. N., M. Tatsumoto, and J. R. Dooley, Jr. 1965. Radioactive disequilibrium studies in sandstone, Powder River Basin, Wyoming, and Slick Rock District, Colorado. *Economic Geology* 60:477–84.

Rumbaugh, J. O., III. 1983. *Effect of Fracture Permeability on Radon-222 Concentration in Ground Water of the Reading Prong, Pennsylvania*. M.S. thesis. University Park,

PA: The Pennsylvania State University.

Schlottmann, J. L., and G. N. Breit 1992. Mobilization of As and U in the Central Oklahoma aquifer. In *Water-Rock Interaction*, eds. Y. K. Kharaka and A. S. Maest, pp. 835–38. Rotterdam: A. A. Balkema.

Schmidt-Collerus, J. J. 1967. Research in uranium geochemistry: Investigation of the relationship between organic matter and uranium deposits. Denver Research Institute. U.S. Atomic Energy Commission Contract AT(05-1)-933.

Schubert, A. J., E. R. Russell, and L. S. Myers. 1950. Dissociation constants of radium in organic acid complexes measured by ion exchange. *Journal of Biological Chemistry* 185:387–98.

Semkow, T. M. 1990. Recoil-emanation theory applied to radon release from mineral grains. *Geochimica et Cosmochimica Acta* 54:425–40.

Shanbag, P. M., and G. R. Choppin. 1981. Binding of uranyl by humic acid. *Journal of Inorganic and Nuclear Chemistry* 43:3369–72.

Tanner, A. B. 1964. Physical and chemical controls on distribution of radium-226 and radon-222 in ground water near Great Salt Lake, Utah. In *The Natural Radiation Environment*, eds. J. A. S. Adams and W. M. Lowder, pp. 253–76. Chicago: University of Chicago Press.

Thamer, B. J., K. K. Nielson, and K. Felthauser. 1981. *The Effects of Moisture on Radon Emanation*. U.S. Bureau of Mines Open-File Report 184-82.

Tsunashima, A., G. W. Brindley, and M. Bastovanov. 1981. Adsorption of uranium from solutions by montmorillonite; compositions and properties of uranyl montmorillonites. *Clays and Clay Minerals* 29(1):10–16.

U.S. Environmental Protection Agency. 1991. National primary drinking water regulations; radionuclides: Proposed rule. *Federal Register* 56:33050–127.

Waggener, W. C., and R. W. Stoughton. 1952. Chemistry of thorium in aqueous solutions. II: Chloride complexing as a function of ionic strength. *Journal of Physical Chemistry* 56:1–5.

Wanty, R. B., and L. C. S. Gundersen. 1987. Factors affecting radon concentrations in ground water: Evidence from sandstone and crystalline aquifers. *Geological Society of America Abstracts with Programs* 19(2):135.

Wanty, R. B., and L. C. S. Gundersen. 1988. Groundwater geochemistry and radon-222 distribution in two sites on the Reading Prong, eastern Pennsylvania. In *Geological Causes of Natural Radionuclide Anomalies*, eds. M. A. Marikos and R. H. Hansman, pp. 147–56. St. Louis, MO: Missouri Department of Natural Resources.

Wanty, R. B., E. P. Lawrence, and L. C. S. Gundersen. 1992. A theoretical model for the flux of radon-222 from rock to ground water. In *Geologic Controls on Radon, Geological Society of America Special Paper 271*, eds. A. Gates and L. C. S. Gundersen, pp. 73–78. Boulder, CO: Geological Society of America.

Wanty, R. B., W. R. Miller, P. H. Briggs, and J. B. McHugh. In press. Geochemical processes controlling uranium mobility in mine drainages. In *The Environmental Geochemistry of Mineral Deposits, Reviews in Economic Geology Volume 6*, eds. G. S. Plumlee and M. J. Logsdon. El Paso, TX: Economic Geology Publishing Co.

Wanty, R. B., C. A. Rice, D. Langmuir, P. Briggs, and E. P. Lawrence. 1991. Prediction of uranium adsorption by crystalline rocks: The key role of reactive surface area. In *Materials Research Society Symposium Proceedings*, pp. 695–702. Boston, MA: Materials Research Society.

Wathen, J. B. 1986. *Factors Affecting Levels of Rn-222 in Wells Drilled into Two-Mica Granite in Maine and New Hampshire*. M.S. thesis. Durham, NH: University of New Hampshire.

Weast, R. C., and M. J. Astle, eds. 1980. *CRC Handbook of Chemistry and Physics*, 61st ed. Boca Raton, FL: CRC Press.

Weigel, F. 1977. *Radium*. Gmelin Handbuch der Anorganischen Chemie: System number 31, release 2.

Weigel, F. 1978. Radon. *Chemiker Zeitung* 102:287–99.

Zebroski, E. L., H. W. Alter, and F. K. Heumann. 1951. Thorium complexes with chloride, fluoride, nitrate, phosphate, and sulfate. *Journal of the American Chemical Society* 73:5646–50.

Zielinski, R. A., and A. L. Meier. 1988. The association of uranium with organic matter in Holocene peat: An experimental leaching study. *Applied Geochemistry* 3:631–43.

18

Analysis of Ground-Water Systems in Freshwater-Saltwater Environments

Thomas E. Reilly

INTRODUCTION

Fresh ground-water systems are important sources of potable water throughout the world, and many are in contact with saline water, which, if drawn into the freshwater aquifer system, can diminish the water's potability as well as its usefulness for other purposes. In order to determine available water supplies, a quantitative understanding of the patterns of movement and mixing between fresh and saline water, and of the factors that influence these processes, is required. Additionally, an understanding of these physical mechanisms is required to establish regional monitoring systems that will adequately describe the resource and warn of any future potential water-quality problems because of the movement of the saltwater.

The purpose of this chapter is to describe the major factors that determine and influence the location and thickness of the saltwater-freshwater transition zone near or within coastal aquifers. This description will focus on the regional nature of these factors, and will use field examples to illustrate the different techniques available to analyze the processes occurring in freshwater-saltwater ground-water environments.

Although many types of saline water occur inland, this chapter deals mainly with coastal systems in contact with "seawater." However, the principles and theory set forth apply to inland saline ground water as well.

SYSTEM DEFINITION

In scientific analyses, defining the problem to be investigated and identifying the important characteristics to be considered and measured are the first steps in understanding an observed phenomenon. This section provides information on the properties that are important in freshwater-saltwater ground-water systems and briefly reviews the mathematical basis of different methods of conceptualizing, defining, and analyzing the problem.

Properties of Saltwater

Physical Properties

Density and viscosity are two physical properties of a fluid. Density is an important property in this case because it is part of the driving force that defines both the direction and rate of the fluid movement. Both density and viscosity affect the hydraulic transmitting

443

TABLE 18-1. Chlorinity and Density of Selected Natural Ground Waters of High Salinity in the United States

Location of Source	Chlorinity (parts per thousand)	Density (ρ) at 20°C (kg L^{-1})	Major Ions	Dissolved Solids (parts per thousand)
Great Salt Lake Desert, Tooele County, UT	86.6	1.11	Na, Cl	148
Trona deposit, Sweetwater County, WY	21.3	1.204	Na, Cl, CO$_3$	290
Lyons Well, Wayne County, NY	21.2	1.027	Na, Cl	38
Salado Formation brine, Eddy County, NM	16.8	1.345	Na, Mg, SO$_4$	325

Source: Adapted from White, Hem, and Waring (1963).

properties (hydraulic conductivity) which influences the rate of fluid movement.

The density of water in natural aquifer systems ranges from 0.9982 kg L^{-1} (kilograms per liter) at 20°C for pure freshwater to greater than the 1.345 kg L^{-1} reported for the Salado Formation brine of New Mexico (Table 18-1). A classification system for salty water ranging from slightly saline to brine, developed by Krieger, Hatchett, and Poole (1957), divides salty waters into four groups on the basis of total dissolved-solids concentration (Table 18-2).

Both the density and viscosity of water are dependent on the type and amount of solute. The variation in these and other properties at selected levels of salinity in seawater is given in Table 18-3. The density of "average" surface seawater ranges between 1.022 and 1.028 kg L^{-1} (Chow 1964, pp. 2–4); this value is partly dependent on the temperature as well as the solute concentration. The kinematic viscosity of average seawater is 1.826 cS (centistokes) at 0°C and 1.049 cS at 20°C.

Chemical Properties

"Average" seawater (Powell, in Chow, 1964) contains about 34.48 parts per thousand dissolved solids. The major ions that form this total are listed in Table 18-4. Although inland saline waters may have a different chemical composition than seawater, their density and viscosity are functions of their chemical compositions. A compilation of values for selected properties of several inland saline ground waters, developed by White, Hem, and Waring (1963), is given in Table 18-1.

The mixing of natural waters can result in important chemical effects, such as changes in concentration and electrical properties of the solution, shifts in the equilibria of aqueous species, and the onset of precipitation or dissolution of a solid phase (Runnells 1969). In particular, the mixing of seawater with calcium carbonate ground water has been

TABLE 18-2. Classification of Salty Waters

Description	Dissolved Solids (parts per thousand)
Slightly saline	1–3
Moderately saline	3–10
Very saline	10–35
Brine	>35

Source: Krieger, Hatchett, and Poole (1957).

TABLE 18-3. Properties of Various Concentrations of Seawater

Solute Mass Fraction (%)	Salinity (parts per thousand)	Chlorinity (parts per thousand)	Density, ρ (kg L^{-1}) at 20°C	Absolute Viscosity, μ (cP at 20°C)	Kinematic viscosity $\upsilon = \mu/\rho$ (cS at 20°C)
0.00	0.00	0.00	0.9982	1.002	1.004
0.50	4.94	2.72	1.0019	1.010	1.008
1.00	9.92	5.48	1.0057	1.018	1.012
2.00	19.89	11.00	1.0132	1.036	1.022
3.00	29.86	16.53	1.0207	1.057	1.036
3.50	34.84	19.29	1.0245	1.070	1.044
4.00	39.82	22.05	1.0283	1.080	1.050
5.00	49.79	27.57	1.0358	1.103	1.065
10.00	99.63	55.18	1.0738	—	—

Source: Adapted from Chemical Rubber Company (1982) (reprinted with permission from *Handbook of Chemistry and Physics*; copyright CRC Press, Inc.).

TABLE 18-4. Composition of Dissolved Solids in Seawater and Amount of Major Ions in "Average" Seawater

Ion	Percent Composition	Amount in "Average" Seawater (parts per million)
Chloride	55.04	18,980
Sodium	30.61	10,556
Sulfate	7.68	2,649
Magnesium	3.69	1,727
Calcium	1.16	400
Potassium	1.10	380
Bicarbonate	0.41	140
Bromide	0.19	65
Boric acid	0.07	26
Strontium	0.04	13
Others	0.01	2
Total	100.00	34,483

Source: Powell (1964) (reprinted with permission from *Handbook of Applied Hydrology*, Chow (editor); copyright McGraw-Hill Inc.).

shown to develop increased porosity and permeability in limestone aquifers and to cause dolomitization of calcium carbonate rocks (Plummer 1975; Wigley and Plummer 1976; Back et al. 1986; Sanford and Konikow 1989). These chemical effects are important. However, the focus of this chapter will be on the effects of physical properties on the interaction of saltwater-freshwater systems. Chemical effects will not be discussed further, but should be considered when evaluating any saltwater-freshwater system.

Description of the Physical System

The general class of ground-water systems addressed herein consists of a saturated porous medium containing a miscible fluid of variable density and salt concentration. This type of system occurs both inland and in coastal areas.

In relatively homogeneous porous media in a coastal area it has been observed that the denser saltwater tends to remain separated from the overlying freshwater. However, a zone of mixing, known as the zone of diffusion or dispersion (also called the transition zone) forms between the two fluids, as shown in the hypothetical cross section in Figure 18-1. In this zone of mixing, which may vary in thickness, some of the saltwater mixes with the freshwater and moves seaward, causing the saltwater to flow toward the area of mixing (Cooper et al. 1964). In coastal areas where the porous medium is heterogeneous in nature, a system of layered mixing zones can form, as shown in Figure 18-2. The three-dimensional nature of the flow system and the gradual increase in solute concentration across the zone of mixing are important factors that affect the physical aspects of the natural system.

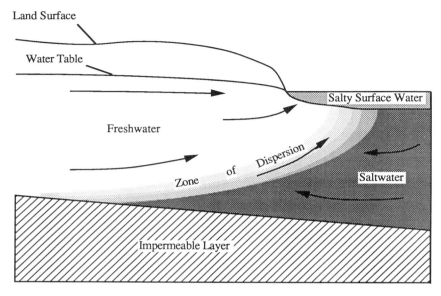

FIGURE 18-1. Hypothetical cross section showing the zone of dispersion and generalized flow patterns in a homogeneous coastal aquifer. (*Adapted from Cooper et al. 1964.*)

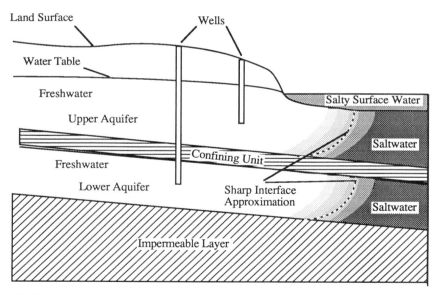

FIGURE 18-2. A ground-water system containing freshwater and saltwater in a layered coastal environment.

System Conceptualizations

In general terms, the object of quantitative analysis of these systems is to understand and describe the relation between saline and fresh ground water. The physical system just described is highly complex, and therefore is rarely treated in terms of fully three-dimensional density-dependent miscible fluid flow in a porous medium. Instead, various simplifying assumptions are usually made to facilitate a physically reasonable and tractable solution that will quantify the relationships in enough detail to increase the understanding of the phenomena under investigation. These simplifying assumptions are based on a conceptual understanding of the system and the relative importance of the many processes occurring.

Perhaps the most important assumption concerns the tendency of the freshwater and the saltwater to mix. Under certain conditions, these two miscible fluids can be considered as immiscible and separated by a sharp "interface" or boundary. This assumption of a sharp interface has been used successfully and has had an important effect on the mathematical formulation of describing the physical process.

Sharp-Interface Conceptualization

The sharp-interface conceptualization of the system assumes that the saltwater-freshwater system is composed of two completely immiscible fluids. Thus, the problem can be formulated in terms of two distinct flow fields, the freshwater flow field and the saltwater flow field. These two systems are coupled through their common boundary. This boundary is known as the saltwater interface or sharp interface, because it is assumed that fluid from one region (or system) cannot cross the interface boundary into the other region (Figure 18-3).

In his classic treatise on ground water, Hubbert (1940) proved that a potential function can be defined for a fluid where the density is constant or varies only with pressure. This fluid potential, which is the mechanical energy per unit mass of fluid, defines the direction of fluid flow and is equal to the fluid head when divided by the acceleration due to gravity. In the case of a two-fluid system, a potential or head must be defined for each fluid. Hubbert formulated mathematically the condition that must hold at the sharp interface as

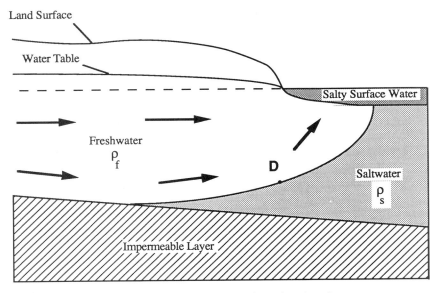

FIGURE 18-3. A hypothetical cross section showing a sharp interface.

$$Z = \frac{\rho_f}{\rho_f - \rho_s} h_f - \frac{\rho_s}{\rho_f - \rho_s} h_s \quad (18\text{-}1)$$

where Z is the vertical location of a point on the interface (L), h_f is the freshwater head (L), h_s is the saltwater head (L), ρ_f is the freshwater density (M L^{-3}), and ρ_s is the saltwater density (M L^{-3}).

The freshwater and saltwater heads are defined as

$$h_f = \frac{P}{\rho_f g} + z \quad (18\text{-}2)$$

and

$$h_s = \frac{P}{\rho_s g} + z \quad (18\text{-}3)$$

where z is the elevation of the point at which head is measured (L), g is the acceleration due to gravity (L T^{-2}), and P is the fluid pressure at the point of measurement (M L^{-1} T^{-2}). Thus, the interface condition represents the saltwater-freshwater boundary as a free-surface boundary condition where Hubbert's relation must hold true. For example, in Figure 18-3, Hubbert's relation must hold at point D, ensuring that the pressure field is continuous. In fact, Hubbert's relation is obtained by setting the pressures in Eqs. 18-2 and 18-3 equal to each other and solving for the elevation (z).

Static Saltwater

When analyzing systems under steady-state conditions, where the saltwater is assumed to be stationary and in equilibrium with the freshwater, the Hubbert relation (Eq. 18-1) can be further simplified. If the saltwater is static, there are no saltwater head gradients, so that h_s is a constant. By selecting the appropriate datum, h_s can be defined as zero, and the equation for the free surface becomes

$$Z = \frac{\rho_f}{\rho_f - \rho_s} h_f \quad (18\text{-}4)$$

This formulation then only requires the freshwater system to be analyzed and the sharp interface to move in response to freshwater heads. This analysis simplifies the two-fluid problem into a one fluid problem with a free-surface boundary condition. This method has been used by Glover (1959), Bennett and Giusti (1971), Fetter (1972), Voss (1984a), Guswa and LeBlanc (1985), Reilly et al. (1987b), and many others.

Two-Fluid Flow

When the saltwater is not static, then the problem must be formulated as the movement of two separate systems (the saltwater and the freshwater systems) with a common free-surface boundary (the saltwater interface). The interface boundary must satisfy Hubbert's (1940) relation as defined earlier (Eq. 18-1). Analysis of the problem using this conceptualization requires the solution of two simultaneous partial differential equations for the flow of freshwater and saltwater. The equation of flow in the freshwater is

$$S_f \frac{\partial h_f}{\partial t} + \nabla \cdot \mathbf{q}_f - Q_f = 0 \quad (18\text{-}5a)$$

and for saltwater it is

$$S_s \frac{\partial h_s}{\partial t} + \nabla \cdot \mathbf{q}_s - Q_s = 0 \quad (18\text{-}5b)$$

where h is the head (L), $\mathbf{q}$ is the specific discharge (L T^{-1}), S is the specific storage (L^{-1}), Q is a source-sink term (T^{-1}), $\nabla = (\partial/\partial x)\mathbf{i} + (\partial/\partial y)\mathbf{j} + (\partial/\partial z)\mathbf{k}$ (where $\mathbf{i}$, $\mathbf{j}$, and $\mathbf{k}$ are unit vectors in the x, y, and z directions, respectively), and subscripts f and s refer to freshwater and saltwater respectively. The specific discharge is determined by Darcy's law for constant density fluids as

$$\mathbf{q}_f = -\mathbf{K} \cdot \nabla h_f \quad (18\text{-}6a)$$

and

$$\mathbf{q}_s = -\mathbf{K} \cdot \nabla h_s \quad (18\text{-}6b)$$

where **K** is hydraulic conductivity (L/T).

This approach has been used by Bonnet and Sauty (1975), Pinder and Page (1977), Mercer, Larson, and Faust (1980), Wilson and Sa Da Costa (1982), and Essaid (1990a, 1990b). All but one of these applications assumed each flow domain was a two-dimensional system. Essaid (1990a, 1990b) has developed a quasi-three-dimensional approach whereby three-dimensional systems can be simulated.

Variable-Density Conceptualization

Another conceptualization of saltwater-freshwater systems is that of one miscible fluid transporting a solute that influences the density and viscosity of the fluid. This analysis of density-dependent ground-water flow and solute transport requires the solution of two simultaneous nonlinear partial differential equations, which express the conservation of mass of fluid and conservation of mass of salt. The following development of these equations follows that of Voss (1984b), with additional discussion of the equations and the conceptual and mathematical representation of dispersion.

Conservation of Mass of Fluid

The fluid mass balance is expressed as the sum of pure water and pure solute mass balances for a solid matrix in which there is negligible net movement as

$$\frac{\partial(\varepsilon\rho)}{\partial t} = -\nabla \cdot (\varepsilon\rho\mathbf{v}) + Q \qquad (18\text{-}7)$$

where ε is porosity (dimensionless), ρ is fluid density (ML^{-3}), Q is fluid mass source $[M(L^3 T)^{-1}]$, **v** is average fluid velocity $(L\ T^{-1})$ (which is the specific discharge, **q**, divided by the porosity, ε), and t is time (T). The term on the left is the total change in fluid mass contained in the void space with time. The term involving ∇ represents contributions to local fluid mass change due to excess of fluid inflows over outflows at a point. The fluid mass source term, Q, accounts for external additions or withdrawals of fluid.

This mass balance equation is the most general form expressing conservation of fluid mass. The average fluid velocity, however, depends on the fluid pressure and density as described by the variable-density form of Darcy's law:

$$\mathbf{v} = -\left(\frac{\mathbf{k}}{\varepsilon\mu}\right) \cdot (\nabla P - \rho \mathbf{g}) \qquad (18\text{-}8)$$

where **k** is the solid matrix permeability (L^2), μ is the fluid viscosity $[M(L\ T)^{-1}]$, and **g** is the gravitational acceleration vector $(L\ T^{-2})$. The gravity vector is defined in relation to the direction in which vertical elevation is measured:

$$\mathbf{g} = -|\mathbf{g}|\nabla \text{ (elevation)} \qquad (18\text{-}9)$$

where $|\mathbf{g}|$ is the magnitude of the gravitational acceleration vector.

Note that no potential function (head) is required for use of Eq. 18-8. If, however, the density is a constant, Eq. 18-8 can be simplified to take advantage of the existence of the potential. For constant-density fluids the hydraulic conductivity, **K** $(L\ T^{-1})$, is defined as

$$\mathbf{K} = \left(\frac{\mathbf{k}\rho|\mathbf{g}|}{\mu}\right) \qquad (18\text{-}10)$$

and Darcy's law can be simplified to the form given in Eqs. 18-6a and 18-6b.

Substitution of the density-dependent form of Darcy's law into the fluid mass-balance equation (Eq. 18-7) gives

$$\frac{\partial(\varepsilon\rho)}{\partial t} = +\nabla \cdot \left[\left(\frac{\rho\mathbf{k}}{\mu}\right) \cdot (\nabla P - \rho \mathbf{g})\right] + Q$$
(18-11)

The left-hand side of the equation, which represents the total change in fluid mass contained in the void space with time, can also be expressed by more basic processes dependent upon the two primary dependent variables, P (pressure) and C (concentration). Aquifer storativity under fully saturated conditions is related to $\partial(\varepsilon\rho)/\partial P$ by (Voss 1984b)

$$\frac{\partial(\varepsilon\rho)}{\partial P} = \rho S_{op} \qquad (18\text{-}12)$$

where $S_{op} = (1-\varepsilon)\alpha + \varepsilon\beta$ is specific pressure storativity $[(L\ T^2)M^{-1}]$, α is porous matrix compressibility $[(L\ T^2)M^{-1}]$, and β is fluid compressibility $[(L\ T^2)M^{-1}]$. Expanding the left-hand side, we have

$$\frac{\partial(\varepsilon\rho)}{\partial t} = \frac{\partial(\varepsilon\rho)}{\partial P}\frac{\partial P}{\partial t} + \frac{\partial(\varepsilon\rho)}{\partial C}\frac{\partial C}{\partial t} \qquad (18\text{-}13)$$

where C is fluid solute mass fraction, or solute concentration ($M_s\ M^{-1}$; mass solute per mass total fluid). Substituting in the relationship for $\partial(\varepsilon\rho)/\partial P$ gives

$$\frac{\partial(\varepsilon\rho)}{\partial t} = S_{op}\rho\frac{\partial P}{\partial t} + \varepsilon\frac{\partial\rho}{\partial C}\frac{\partial C}{\partial t} \qquad (18\text{-}14)$$

Substituting into the mass-balance equation (Eq. 18-7) gives the final equation in terms of P (pressure) and C (concentration):

$$(\rho S_{op})\frac{\partial P}{\partial t} + \left(\varepsilon\frac{\partial\rho}{\partial C}\right)\frac{\partial C}{\partial t}$$
$$- \nabla \cdot \left[\left(\frac{\rho\mathbf{k}}{\mu}\right)\cdot(\nabla P - \rho\mathbf{g})\right] = Q \qquad (18\text{-}15)$$

Conservation of Mass of Salt

The solute mass balance for a single species stored in solution can be expressed as

$$\frac{\partial(\varepsilon\rho C)}{\partial t} = -\nabla\cdot(\varepsilon\rho\mathbf{v}C)$$
$$+ \nabla\cdot[\varepsilon\rho(D_m\mathbf{I} + \mathbf{D})\cdot\nabla C] + QC^* \qquad (18\text{-}16)$$

where D_m is the apparent molecular diffusivity of solutes in solution in a porous medium including tortuosity effects ($L^2 T^{-1}$), $\mathbf{I}$ is the identity tensor, $\mathbf{D}$ is the dispersion tensor ($L^2\ T^{-1}$), C is the fluid solute mass fraction, mass solute per mass total fluid ($M_s\ M^{-1}$), and C^* is the solute concentration of fluid sources, mass fraction ($M_s\ M^{-1}$). The time derivative on the left-hand side represents the total change in solute mass with time in a volume. The term involving fluid velocity, $\mathbf{v}$, represents the average advection of solute mass into or out of the local volume. The term involving molecular diffusivity of solute, D_m, and dispersivity, $\mathbf{D}$, expresses the contribution of solute diffusion and dispersion to the local changes in solute mass. The diffusion contribution is based on an actual physical process, often negligible at field scale. The dispersion contribution is an approximation of the effect of solute advection and mixing because of variations in velocity that are not accounted for by solutes advected by the average velocity.

Dispersion

There are different ways in which dispersion itself can be conceptualized. Three of the ways are (1) constant dispersion, (2) flow-direction-independent dispersion, and (3) flow-direction-dependent dispersion. The constant-dispersion approximation assumes that the dispersion tensor, $\mathbf{D}$, is independent of velocity. A constant-dispersion approach was used by Henry (1964), Green and Cox (1966), Pinder and Cooper (1970), Lee and Cheng (1974), and Volker and Rushton (1982).

The flow-direction-independent approach defines the dispersion coefficients as proportional to the velocity in systems with isotropic permeability, which is much more physically reasonable. The dispersion tensor in this case is defined for two-dimensional flow as (Voss 1984b)

$$\mathbf{D} = \begin{bmatrix} D_{xx} & D_{xy} \\ D_{yx} & D_{yy} \end{bmatrix} \qquad (18\text{-}17)$$

where $\mathbf{D}$ is symmetric, and the diagonal elements are

$$D_{xx} = \frac{1}{v^2}(d_L v_x^2 + d_T v_y^2) \quad (18\text{-}18)$$

$$D_{yy} = \frac{1}{v^2}(d_T v_x^2 + d_L v_y^2) \quad (18\text{-}19)$$

and the off-diagonal elements are

$$D_{ij} = \frac{1}{v^2}(d_L - d_T)(v_i v_j),$$
$$i \neq j, \quad i = x, y, \quad j = x, y \quad (18\text{-}20)$$

where v is the magnitude of velocity $\mathbf{v}$ (L T^{-1}), v_x is the magnitude of the x component of $\mathbf{v}$ (L T^{-1}), v_y is the magnitude of the y component of $\mathbf{v}$ (L T^{-1}), d_L is the longitudinal dispersion coefficient (L^2 T^{-1}), and d_T is the transverse dispersion coefficient (L^2 T^{-1}). The terms d_L and d_T are called longitudinal and transverse dispersion coefficients, respectively. These are directional in nature. The term d_L causes dispersion forward and backward in the direction of flow, and the term d_T causes dispersion in the direction perpendicular to flow.

The size of the dispersion coefficients for dispersion in isotropic permeability systems depends on the absolute local magnitude of average velocity in a flowing system:

$$d_L = \alpha_L v \quad (18\text{-}21)$$

$$d_T = \alpha_T v \quad (18\text{-}22)$$

where α_L is the longitudinal dispersivity of the solid matrix (L), and α_T is the transverse dispersivity of the solid matrix (L). This representation was used in the studies of Segol, Pinder, and Gray (1975), Segol and Pinder (1976), Desai and Contractor (1977), Volker and Rushton (1982), Frind (1980, 1982), Voss (1984b), Voss and Souza (1987), and others.

The flow-direction-dependent formulation is usually appropriate in a system with anisotropic permeability or anisotropic spatial distribution of inhomogeneities. In layered aquifers, it might be appropriate for the longitudinal dispersivity to have different values parallel and perpendicular to the layers. Voss (1984b, pp. 50–54) presents an ad hoc model to represent the case of flow-direction-dependent dispersion. The ad hoc model of Voss allows the longitudinal dispersivity to vary based on direction. A longitudinal dispersivity in the direction of the maximum permeability ($\alpha_{L_{max}}$) is defined for one fluid flow direction, and another longitudinal dispersivity in the direction of the minimum permeability ($\alpha_{L_{min}}$) is defined for flow in the orthogonal direction. These two values define an ellipse such that the effective longitudinal dispersivity is dependent on the flow direction and is given as

$$\alpha_L = \frac{\alpha_{L_{max}} \alpha_{L_{min}}}{\alpha_{L_{min}} \cos^2\Theta + \alpha_{L_{max}} \sin^2\Theta} \quad (18\text{-}23)$$

where Θ is the angle from the maximum dispersivity direction. This approach has been used by Hill (1988) and Reilly (1990).

In all these formulations, the basis for dispersion is that regardless of the degree of detail included in the representation of the flow field used to calculate the ground-water velocities, local variations between actual and calculated velocities remain, which cannot be accounted for explicitly. If it were possible to generate a model or a computation that could account for all of the variations in velocity in natural aquifers, dispersive transport would not have to be considered (except for molecular diffusion); sufficiently detailed calculations of advective transport could theoretically duplicate irregular tracer advance observed in the field. In practice, however, such calculations are impossible. Field data at the macroscopic scale are never available in sufficient detail, information at the "mappable" scale is rarely complete, and descriptions of microscopic-scale variations are never possible except in a statistical sense. Even if complete data were available, an unreasonable computational effort would be required to completely define the natural velocity variations in an aquifer.

The more accurately we represent the actual permeability distribution of an aquifer, the closer the calculations of advective transport will match reality: the finer the scale of simulation, the greater the opportunity to match natural permeability variations. In most situations, however, when both data collection and computational capacity have been extended to their practical limits, calculations of advective transport will fail to match field observation. To the extent that scale variations represent random deviation from the velocity used in advective transport calculation, and to the extent that they occur on a scale which is significantly smaller than the size of the region used for advective calculation, dispersion theory may adequately describe the differences between advective calculation and field observation. However, if the velocity variations are not random, or if they occur at a scale which is large relative to the region used for advective calculation, the suitability of the dispersion approach is questionable. Moreover, even when the approach appears to be justified, determination of the coefficients needed to implement it must usually be approached empirically (for example, through model calibration). The range of validity of the quantities determined in this manner is uncertain (for a more complete discussion, see Reilly et al. 1987a).

At the present time, there is still much debate on the physical foundation for mixing in ground-water systems. The transport equation as described previously (Eq. 18-16) is generally accepted as appropriate because of the good agreement of its results with laboratory column experiments (Gillham and Cherry 1982). Difficulties do arise, however, when it is applied to field-scale problems in which undefined heterogeneities affect the flow process as previously described. Dagan

TABLE 18-5. Decisions That Determine an Appropriate Method of Quantitative Analysis

Physics of the Mixing Process	Aquifer Characteristics	Desired Scale of Resulting Analysis

(1) Properties of fluid mixing
    Sharp interface (immiscible fluids)
    Density-dependent dispersed solute (miscible fluids)
(2) Simplifications regarding dimensionality of the flow field
    Two-dimensional
    Three-dimensional
(3) Time dependence
    Steady state
    Transient state
(4) Type of saltwater flow system
    Hydrostatic saltwater
    Moving saltwater
    Moving saltwater, but density assumed a function of position only and is therefore constant with time
(5) Properties of the solute
    The solute is chemically, biologically, and physically inactive.
    The solute is chemically, biologically or physically active and may influence porosity, permeability, solute concentrations, etc.
	(1) Aquifer permeability
	    Homogeneous and isotropic
	    Homogeneous and anisotropic
	    Heterogeneous and isotropic
	    Heterogeneous and anisotropic
	(2) Scale of physical system
	    Regional flow system
	    Local flow system
		(1) Desired scale
		    Analysis applicable at regional scale
		    Analysis applicable at site-specific scale

(1982) and Gelhar and Axness (1983) have suggested a stochastic approach to dispersion, and debate on an appropriate mathematical representation of the dispersion process continues.

Variable-Density Conceptualization with a Steady-State Solute Distribution

The variable-density approach requires the simultaneous solution of a flow equation and a transport equation. An approach has been taken in large regional simulations to assume that the concentration (which defines the density) is known and will not change over the period of the analysis. Under this assumption, a solution of the transport equation is not required, and a solution of the density-dependent-flow equation (Eq. 18-15) is sufficient to define the flow system. This approach has been used by Weiss (1982), Garven and Freeze (1984), Kuiper (1985), and Kontis and Mandle (1988).

FIELD MEASUREMENTS REQUIRED FOR ANALYSIS

When the representation of the physical system is simplified into a form suitable for quantitative analysis, many decisions must be made during the problem formulation stage of analysis. These decisions will dictate or determine the field measurements required to analyze the saltwater-freshwater system and understand the regional movement. Table 18-5 lists many of these decisions and groups them into three major categories: (1) assumptions about the physics of the mixing process; (2) characteristics of the aquifer system under study; and (3) desired scale and detail of the resulting analysis. The selection of alternatives from a given category can depend on decisions in other categories. For example, Figure 18-2 shows a ground-water system containing fresh and salty water in a layered coastal environment. If regional estimates of the position of the saltwater were required, it might be appropriate to select a sharp interface (immiscible fluids) approach in an advection-dominated flow field. However, if an estimate of the chloride concentration of water pumped from the shallow well were desired, the more complex miscible-fluid approach would be required. The effects of the permeability characteristics are also depicted in Figure 18-2. If only the upper aquifer were of interest, the system might be considered homogeneous, but if the entire system were of interest, the regional heterogeneities would be needed to represent the staggered interface that can develop in natural systems.

Table 18-5 also indicates that the scale of the physical system as well as the desired scale of the results play an important role in the quantitative analysis of saltwater-freshwater systems. This is because the physical system is continuous in space and time and incorporates events occurring at all scales. In analyzing the problem the representation of the physical system must therefore be simplified, and to this end the properties are averaged over some volume of space or length of time. This averaging in conjunction with the other simplifying assumptions affects the accuracy of the mathematical description of the problem, and this effect can be an important limitation in some cases.

Data Required for All Analyses

Some data are required to formulate a working conceptualization of the system and to aid in selecting the appropriate method or methods of analysis. Table 18-6 lists in abbreviated form the steps (not necessarily in sequential order) required to perform a regional analysis of a saltwater-freshwater environment in a ground-water system. The information that is always required in a regional ground-water investigation is a definition of the hydrogeologic framework (Table 18-6, step 5) and an understanding and definition of the physical boundaries of the regional system (Table 18-6, steps 1 and 3). Information is also required on the transmitting and storage properties of the system. The exact information will depend on the

TABLE 18-6. Steps Required in the Analysis of a Ground-Water System Containing Saltwater and Freshwater

1. Identify the extent and physical boundaries of the natural (regional) ground-water flow system and develop an initial concept of its operation.
2. Identify an appropriate area around the saltwater transition zone for intensive study.
3. Define boundary conditions for the regional flow system that indicate the flow of water into and out of the system; place special emphasis on the saltwater transition zone and conceptualization of its behavior.
4. Evaluate the relationship between the regional flow system and the saltwater-freshwater transition zone by developing an estimate of the freshwater discharging at the outflow boundary defined by the transition zone.
5. Define the internal geometry (hydrogeologic framework) at the appropriate scale throughout the system.
6. Define the regional head distribution with an appreciation for the three-dimensional nature of natural flow systems. For variable-density systems, define the three-dimensional pressure, temperature, and density distribution. This is best accomplished by preparing appropriate water-level (or pressure and density) maps at different depths and cross sections.
7. Estimate the spatial distribution of hydraulic properties (hydraulic conductivity for sharp-interface approach and permeability for variable-density approach) and storage properties if transient conditions are considered.
8. Conceptualize the approximate ground-water flow pattern and estimate flow rates through both the entire regional system and the saltwater discharge area using hand calculations and compiled data.
9. Develop a numerical flow model of the system based on either a sharp-interface or a variable-density conceptualization.

system conceptualization, but spatial distributions of hydraulic conductivity (or permeability) are always required.

Required data that is unique to saltwater-freshwater systems is the distribution of fluid density (or salt concentration). This information describes the transition zone, and in conjunction with the problem definition will allow for the proper system conceptualization and method of analysis to be determined (Table 18-6, steps 2–4). In the system conceptualization stage, reconnaissance data on the thickness and location of the transition zone may be sufficient. However, more detailed information may then be required by the specific method of analysis as described next.

Data Required for Sharp-Interface Analysis

If the system is conceptualized as having a sharp interface, then the following additional information is required:

1. The location and thickness of the transition zone to confirm the conceptualization of a sharp interface and to determine an appropriate estimate of the interface location; if historical records on the movement of the interface in response to stress are available they should also be compiled.
2. The density of the saltwater; in sharp-interface approaches the saltwater system is usually assumed to have only one uniform density.
3. Freshwater heads that are sufficient to describe the freshwater flow field:
   a. When the density of the water may not be fresh, then the density or a surrogate, such as chloride concentration, should also be measured to ensure that it is a freshwater head measurement.
   b. If it is a transient problem, then available historical head data should be compiled.
4. Saltwater heads
   a. If the system is conceptualized as having a static saltwater system, then only a few head measurements are required to establish the assumed constant head in the saltwater system.
   b. If the system is conceptualized as having two moving fluids (freshwater and saltwater), then as much information on the saltwater head distribution as is available should be collected; saltwater heads measured near the transi-

tion zone but clearly in saltwater of the same density are most useful.

The information on the saltwater heads and the transition zone are usually the most scarce and difficult to obtain. This is because wells for water use are rarely completed in saltwater. In developing a plan of study for saltwater-freshwater investigations, opportunities to obtain this information should be sought and considered carefully.

Data Required for Density-Dependent Analysis

If the system is conceptualized as a single-flow field with a variable-density fluid, the data needs are focused differently. In a variable-density conceptualization the two dependent variables are pressure and density (or solute concentration). Because it is very difficult to estimate or calculate by hand the direction and rate of fluid movement in a variable-density flow field, numerical models are usually the best method of analysis. However, judicious selection of the areal and vertical location of observation points can allow for some simple estimates of direction and rates of movement.

For problems conceptualized as density-dependent flow of a single fluid, the additional data needs are as follows:

1. The three-dimensional density distribution (this is also assumed to be the three-dimensional concentration of dissolved solids).

2. Permeability distribution: any estimated hydraulic conductivities have to be converted into permeabilities based on the fluid properties that were present at the time the hydraulic conductivities were estimated.

3. Fluid pressure must be known throughout the three-dimensional flow system.

4. If any information is available or can be measured on the historical movement of the transition zone, it should be obtained. This is especially important for the simulation of flow and transport of a variable-density fluid because the only available means of estimating dispersivities is through matching historically known locations and movements of the solutes.

In field sampling, these data requirements indicate that measurements at wells should be for short well screens with known areal and vertical locations. The pressure and density of the fluid at the screened interval must also be measured. The vertical depth, pressure, and density are all very important because they enter directly into Darcy's law for variable-density fluids.

During design of sampling locations, it is very useful to have wells screened along the same horizontal plane. Along the horizontal plane, the pressure gradient indicates the direction of flow, and estimates of the rate of movement can then be made. Some previous investigators have attempted to estimate the direction and rate of flow for wells not screened along a horizontal plane by using a concept of freshwater-equivalent heads (as described here) and not accounting for the variation of vertical location; however, this can lead to erroneous results and is not recommended.

THE USE OF AND MISCONCEPTIONS REGARDING FRESHWATER-EQUIVALENT HEADS

Freshwater-equivalent heads have been used by investigators in the past to attempt to determine the direction and movement of ground-water in freshwater-saltwater environments. The appeal of the technique is that it is straightforward to calculate, the drawback is that the conditions under which it is useful are rarely found in the field, and therefore the technique is rarely applied properly. In addition, as previously stated, Hubbert (1940, p. 801) showed that a true potential that defines the direction of fluid movement can only be defined for fluids where the density is constant or varies only with pressure, which is not the case when freshwater-equivalent heads are usually used.

Freshwater-equivalent heads are calculated by adding the elevation head to the pressure head, which is calculated under the assumption that the column of water representing the pressure head is freshwater. To accomplish this, first the water pressure at a point representing the well screen, as shown in Figure 18-4, is measured or calculated as

$$P = \rho_w g l \qquad (18\text{-}24)$$

where P is the pressure (M L^{-1} T^{-2}), ρ_w is the density of fluid in the well (M L^{-3}), g is acceleration due to gravity (L T^{-2}), and l is the vertical height of fluid in the well above the point representing the well screen (L). Then a freshwater-equivalent head (h_f) is calculated as

$$h_f = z + \frac{P}{\rho_f g} \qquad (18\text{-}25)$$

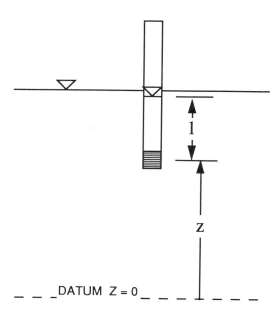

FIGURE 18-4. A well and the dimensions required to calculate freshwater-equivalent heads.

where z is the elevation above datum of the point representing the well screen, and ρ_f is the density of freshwater. Substituting the first equation into the second gives a concise formula for the calculation of freshwater head:

$$h_f = z + \frac{\rho_w}{\rho_f} l \qquad (18\text{-}26)$$

Freshwater-equivalent heads are actually a surrogate measurement for pressure. The use of freshwater-equivalent heads in the constant-density form of Darcy's law is only valid for the determination of horizontal flow, and then only if the heads are from wells screened at the same elevation.

Examination of the variable-density equations (Eqs. 18-8 and 18-9) gives some insight into why these freshwater-equivalent heads are only valid in the horizontal plane. The gravity vector, **g**, in Eqs. 18-8 and 18-9 is zero if the elevation is a constant between the points considered (Eq. 18-9), and Darcy's law (Eq. 18-8) then simplifies to only rely on the pressure gradient in the horizontal plane. Because the equivalent freshwater head is a surrogate for pressure at the well screen as has been illustrated, it can only be used to determine ground-water flow when the wells used to calculate these equivalent freshwater heads are screened at the same horizontal elevation.

The equivalent freshwater heads for the system shown in Figure 18-5 can be used to illustrate this point. The system is a closed tank of stagnant salty ground water with a uniform constant density of 1.025 kg/L. There are three piezometers open at different depths. The freshwater-equivalent heads at piezometers A, B, and C are then calculated, by using Eq. 18-26 and the density of freshwater as 1.000 kg/L, to be h_f at A = 0 m, h_f at B = 1 m, and h_f at C = 2 m.

These calculations show that the freshwater-equivalent head indicates a gradient between the wells. However, the fluid is stagnant, so there is actually no movement. Thus, any calculation of flows using the fresh-

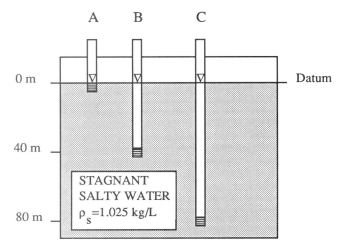

FIGURE 18-5. A diagram of a static saltwater environment with wells screened at different depths.

water-equivalent heads would provide misleading results unless additional calculations are made to correct for the effects of the vertical position of the screened interval. An example of when freshwater-equivalent heads are used appropriately is in the simulation codes of Weiss (1982), Kuiper (1985), and Kontis and Mandle (1988), which use freshwater-equivalent heads as a surrogate for pressure and "pseudosources" to account for the density component of Darcy's law. Although these codes may calculate and print freshwater-equivalent heads, the flow calculations in the simulation do account for the density component by adding additional terms in the mathematical formulation.

SUMMARIES OF SELECTED STUDIES USING THE DIFFERENT SYSTEM CONCEPTUALIZATIONS

Regional Sharp Interface

Cape Cod, Massachusetts

Ground water is the principal source of freshwater for Cape Cod, Massachusetts. A lens-shaped reservoir of fresh ground water is maintained in dynamic equilibrium by recharge from precipitation and discharge to the ocean and streams. To evaluate the eventual equilibrium impacts of regional groundwater development, quantitative analyses of the area were undertaken by Guswa and LeBlanc (1985).

The system was analyzed as a three-dimensional flow field under steady-state conditions under the assumption of a sharp interface with moving freshwater over static salt-water, as illustrated in the simplified cross section in Figure 18-6. The observed interface on Cape Cod was defined as "narrow," allowing for this approximation. The data available and used in assessing the reasonableness of the simulations were water budgets, geologic framework information, aquifer test information, water-table information, and estimates of the location of the saltwater-freshwater transition zone.

The advantage of this approach was that sufficient information was available to show and test the reasonableness of the simulations. Because only regional estimates were required, the steady-state sharp-interface approach was a reasonable approximation. The disadvantage was that the ability of individual wells to supply water with low chloride concentrations could not be evaluated. Overall, the results are useful in developing regional water-management plans.

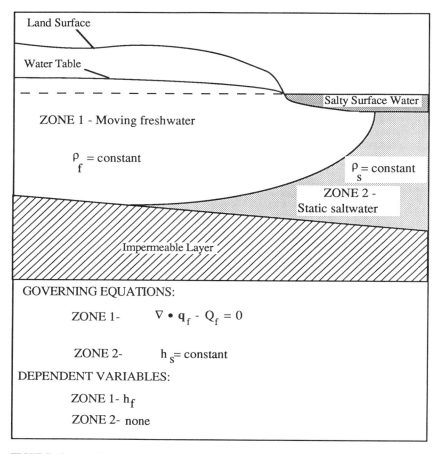

FIGURE 18-6. Sharp interface approach applied at Cape Cod, MA.

Montauk Area, Long Island, New York

The Montauk peninsula, at the extreme eastern end of Long Island, New York is a popular resort area that draws tens of thousands of vacationers annually. The sole source of fresh ground water on the peninsula is a series of Pleistocene glacial deposits that are bounded below and laterally by saltwater. The permanent population is only a few thousand; thus, the seasonal increase in population imposes a large fluctuating demand on the ground-water system. An analysis by Prince (1986) developed a simulation of the system and then estimated the response of the ground-water system to the present use and future demand.

The system was analyzed as a two-dimensional flow field under the assumption of a sharp interface with moving freshwater over moving saltwater, as illustrated in the simplified cross section in Figure 18-7. Data collected during the study indicated that the thickness of the saltwater-freshwater transition zone was usually less than 7 m, indicating that a sharp-interface approximation was reasonable. The data available and used in assessing the reasonableness of the simulations were water budgets, geologic framework information, aquifer-test information, water-table information, and estimates of the location of the saltwater-freshwater transition zone. Additional information that is useful with this approach, but rarely available, is measurements of saltwater heads.

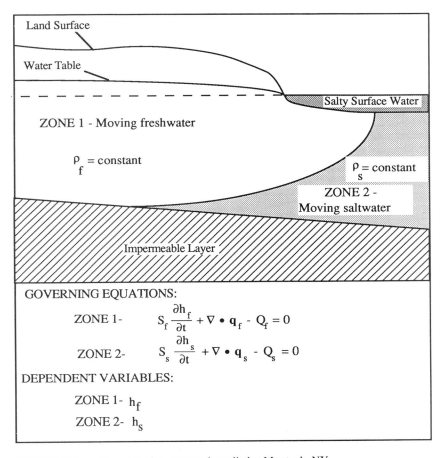

FIGURE 18-7. Sharp interface approach applied at Montauk, NY.

The advantage of this approach was that sufficient information was available to show and test the reasonableness of the simulations. Because only regional estimates were required and the transition zone was documented as very thin, the transient sharp-interface approach was a reasonable approximation. As with the Cape Cod analysis, the disadvantage was that the ability of individual wells to supply water with low chloride concentrations could not be evaluated. Overall, the results were useful in determining that the principal aquifer is capable of producing over 2.3 million L/d, and for developing regional water-management plans.

Soquel-Aptos Basin, Santa Cruz County, California

The physiography of the Soquel-Aptos Basin, Santa Cruz County, California, ranges from very steep valley slopes and angular landforms in the Santa Cruz Mountains to nearly flat marine-terraced, sea cliffs and narrow beaches along Monterey Bay. The region is mainly an urban area, and growth is projected with concurrent increases in water demand. The principal hydrologic unit of interest is the Purisima Formation, which is a layered aquifer of variable thickness. Saltwater is not yet present in the aquifer under land surface, and the position of the interface offshore is not known. An analysis was undertaken by Essaid (1990b) to quantitatively estimate the amount of freshwater flow through the system, the quantity of freshwater outflow to the sea, the undisturbed position of the saltwater interface offshore, the quantity of discharge to the sea that must be maintained to keep the interface at or near the shore, and the rate at which the interface will move due to onshore development.

The system was analyzed as a three-dimensional flow field under time-varying conditions and assuming a sharp interface with moving freshwater over moving saltwater, as illustrated in the simplified cross section in Figure 18-8. The three-dimensional approach incorporates the ability of freshwater to discharge from deeper layers into overlying layers containing saltwater. The data available and used in assessing the reasonableness of the simulations were water budgets, geologic framework information, aquifer-test information, and time-varying water-table information. An estimate of the location of the saltwater-freshwater transition zone was not available. Additional information that is also useful with this approach, but rarely available, is measurements of saltwater heads.

The advantage of this method of analysis was that the three-dimensional nature of the flow field and staggered interface location could be estimated. Also, the importance of the transient response of the system was better understood. In fact, it was shown that the present interface is probably still responding to long-term Pleistocene sea-level fluctuations and has not achieved equilibrium with present-day sea-level conditions.

Regional Variable-Density Flow with a Specified Steady-State Solute (Density) Distribution

Gulf Coastal Plain, United States

The aquifer systems of the Gulf Coastal Plain of the United States encompass 750,000 km^2 including the Mississippi embayment and offshore areas beneath the Gulf of Mexico. The gulfward thickening wedge of unconsolidated to semiconsolidated sediments is a complex interbedded sequence of sand, silt, and clay with minor beds of lignite, gravel, and limestone. The sediments crop out in bands approximately parallel to the present coastline of the Gulf of Mexico. Approximately 36 billion L/d of ground water was pumped from the system in 1980. In order to better understand this vast and complex regional flow system, a Regional Aquifer System Analysis project of the U.S. Geological Survey was undertaken. The presence of saltwater at very different concentrations in complex areal and vertical distributions necessitated the analysis of the system as a variable-density flow problem. The large area necessitated a regional approach, and a simulation was undertaken by Williamson (1987) to describe this complex system.

The system was analyzed as a three-dimensional flow field with a variable-density fluid. However, the density distribution was assumed to remain constant in space and time, as illustrated in the simplified cross section in Figure 18-9A. The three-dimensional approach incorporates the ability of freshwater to discharge from deeper layers into overlying layers containing saltwater. The variable-density approach does not treat the dispersed boundary between the freshwater and saltwater as an interface, and allows for some fluid flow to occur between and in the freshwater and saline water. The major assumption in this approach is that the density (concentration) distribution does not change in time because of the fluid flow. This assumption allows the analysis to be based on only the variable-density flow equation and does not require analysis of the salt transport. Williamson (1987) states that this approach is valid because the volume of water simulated as moving into an adjacent model block in a few decades at maximum flow rates will not significantly affect the average density in the large regional model blocks, and therefore would not affect the

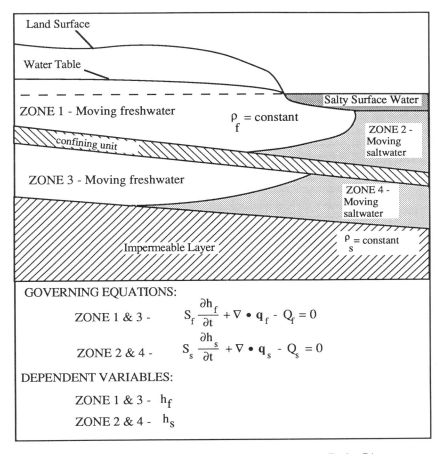

FIGURE 18-8. Sharp interface approach applied at Soquel-Aptos Basin, CA.

solution. However, if equilibrium results are required, this method of analysis may not be appropriate. The data available and used in assessing the reasonableness of the simulations were water budgets, geologic framework information, aquifer test information, and time-varying water-table information. An important data element required in this analysis, which is input as known information, is the three-dimensional density distribution for the entire system; these data are very important and are rarely known to the extent necessary. As with the previous examples, additional information that is also useful with this approach, but is rarely available, is measurements of pressures in the saltwater part of the system.

The advantage of this method of analysis is that the importance of the complex variable-density distribution on the three-dimensional flow field can be understood and quantified to some extent. This is probably the only approach that will allow this at the present time. The disadvantage is that any errors in conceptualizing the system and describing the distribution of the densities in the flow field directly affect the results and may compensate for other uncertainties. Therefore, it is difficult to assess the reasonableness or accuracy of the analysis. The analysis of this system indicated that on a regional basis, the resistance to vertical flow caused by many thin localized fine-grained beds within the permeable zones can be as important as the resistance caused by regionally mappable confining units.

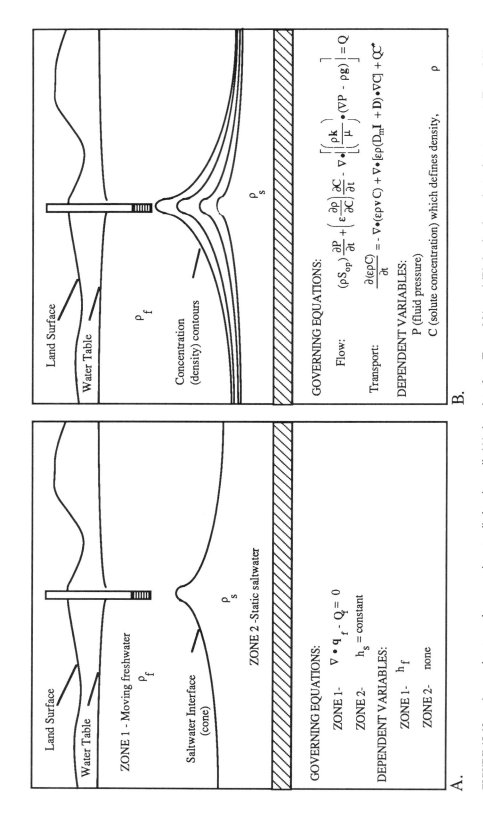

FIGURE 18-10. Approaches to analyze upconing at a discharging well: (A) sharp interface, Truro, MA, and (B) density-dependent solute transport, Truro, MA.

Regional Variable-Density Flow with Transport of Solutes

Oahu, Hawaii

The regional ground-water system of southern Oahu, Hawaii, is a thick, areally extensive freshwater lens overlying a thick saltwater zone with a transition zone between. The major aquifers on Oahu are composed of hundreds of thin lava flows that were extruded onto the land surface. The layers, generally a meter to several meters thick, form a matrix of thin overlapping tabular units commonly tens to hundreds of meters wide and up to 30 km long that dip about 5–10 degrees from the mountainous recharge areas to the ocean. The primary water supply for the island of Oahu is the southern Oahu aquifer. The transition zone varies in thickness over the area, being thin inland and thickening to about 300 m at the coast. Because of this variable-thickness transition zone, an analysis that considered the dispersed nature of this zone was undertaken by Souza and Voss (1987) and by Voss and Souza (1987).

The system was analyzed as a two-dimensional cross section with variable-density fluid flow and transport of solutes, as illustrated in the simplified cross section in Figure 18-9B. This means that both the fluid pressures and concentration of solutes (density) are treated as dependent variables and thus influence the description of the flow system. The data available and used in assessing the reasonableness of the simulations were water budgets, geologic framework information, aquifer-test information, time-varying water-table information, and estimates of the location and thickness of the saltwater-freshwater transition zone. And again, as in the previous examples, the additional information that is useful with this approach, but rarely available, is measurements of fluid pressures in the saltwater part of the system.

The advantage of this method of analysis is that the cause-and-effect relationships that influence the development of the variable-thickness transition zone can be analyzed and understood better. This density-dependent flow and transport conceptualization represents the actual physical system more completely and accurately (fewer assumptions) than previously described methods. Also, the concentration distribution of solutes is simulated, and estimates of solute concentrations in individual wells can be obtained. However, many more parameters are required to describe the system definitively, and the method is computationally more complex.

Local Analysis of Upconing beneath a Discharging Well

Although this book focuses on regional problems, it is important to note that in order for an aquifer to supply freshwater to wells, it is a necessary condition that the regional system be capable of providing the quantities of water required, but it is not a sufficient condition. Even though the regional system may be in equilibrium, local saltwater movement near discharging wells can make these wells produce unpotable water. Thus, the regional system may be capable of sustaining the rate of production, but the head declines in the vicinity of an individual well cause a local movement of saltwater to the well (called upconing). In his analysis of Montauk, Long Island, NY, Prince (1986) included some estimate of the ability of individual wells of different designs to supply water. Thus, the water management must be a combination of regional considerations and local considerations.

The methods of analysis are the same as those discussed in the regional studies; however, the scale of interest is different. A well in Truro, Cape Cod, MA, was analyzed with the sharp-interface approach (Reilly et al. 1987b) and the density-dependent flow and transport approach (Reilly and Goodman 1987), as illustrated conceptually in Figure 18-10A,B respectively. The analyses at Truro indicated that although the regional system was in equilibrium and capable of sustaining

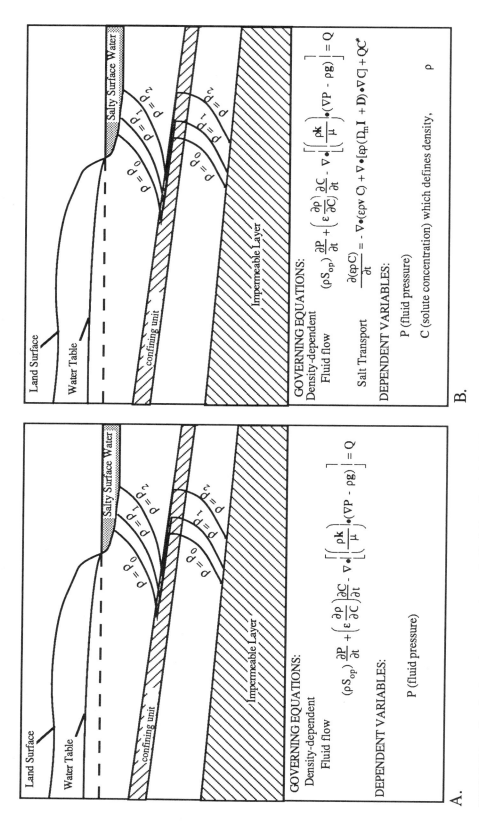

FIGURE 18-9. Density-dependent approaches applied in field studies: (A) Gulf Coast, U.S. and (B) Oahu, HI.

4.2 million L/d at a particular well field based on regional estimates (LeBlanc 1982), the actual well was not capable of this production because of local effects. A single well withdrawing the total amount can create large drawdowns that can cause the saltwater to move to the well but still be in equilibrium on a regional basis. In the Truro example, the withdrawals at the well averaged about 1 million L/d annually and were usually over 2.5 million L/d in the summer, which caused the water from the well to show some indications of saltwater contamination. Estimates from the local analysis indicated that the maximum permissible discharge from the single well is probably less than 1.8 million L/d. Thus, the withdrawal rate should be limited at the one well; and additional wells should be installed if more capacity (up to the regional capacity) is required.

MONITORING STRATEGIES IN FRESHWATER-SALTWATER SYSTEMS

In general, ground-water systems in a saltwater-freshwater environment have to be analyzed as any other ground-water system but with the additional saltwater aspect. Monitoring such systems, therefore, requires that the same information required to understand any ground-water system be collected and monitored for saltwater-freshwater systems. The unique aspect of the saltwater-freshwater environment, however, does require special considerations in developing a monitoring system.

Outpost Monitoring Wells

Outpost wells are wells located near the saltwater-freshwater transition zone. The purposes of outpost wells are to (1) serve as an early warning system of movement of the saltwater into a area of freshwater, (2) to determine natural outflow at the shoreline boundary of the flow system, (3) to detect changes in the flows at the boundary, and (4) measure rates of movement of the saltwater.

The location and screen depth of outpost monitoring wells requires careful consideration if useful information is to be obtained from them. It is best if the wells are installed with short screens because this enables pressures and concentrations to be most accurately defined in three-dimensional space. If possible, groups of wells should be established with screens set on a defined horizontal plane. For example, a group of wells could be screened at 35 m below sea level and also at 60 m below sea level. The wells should be spaced far enough apart areally, so that the heads or pressures are significantly different in order for gradients to be calculated in each horizontal plane with a minimum of noise and uncertainty. Wells arranged in triangular arrays on different horizontal planes both within the same aquifer and in different aquifers would allow for estimation of fluid flow directions and rates in the horizontal plane in both sharp-interface calculations and in variable-density fluid flow calculations using head, pressure or freshwater-equivalent head (as a surrogate for pressure). Continuous measurements of head or pressure are frequently required at these wells because these wells can be affected by the tidal cycle and fluctuate accordingly. The head, pressure, and density information from these wells would then be very useful in developing numerical simulations that would provide even more insight into the occurrence and movement of the fluid.

As indicated previously in the summaries of selected studies, head (or pressure) and density information are frequently lacking in the saltwater part of the system. This information on the saltwater is needed to conceptualize the system accurately and should be considered in the development of a monitoring system. The difficulty in obtaining this information is that the saltwater part of the system is frequently under bodies of surface water. However, occasionally islands or other surface features enable the installation of some wells. If it is possible to install wells in the saltwater part of the system, the information obtained should be used to determine the variability of density in the saltwater part of the flow system and to determine if the saltwater is stationary. These specialized wells are also part of the integrated monitoring system outlined here, and their placement is subject to the same considerations.

Upconing Monitoring Wells

Wells that are designed to monitor the occurrence and movement of saltwater in the vicinity of an individual pumping well usually provide an early warning system to prevent the water supply from becoming contaminated. These wells should have short screens and be placed some distance beneath the supply well and located as close to the supply well as feasible. If saltwater is detected in these monitoring wells, the rate of production can be decreased to "safe" rates by either trial and error or by using numerical models to estimate appropriate discharge rates.

CONCLUDING REMARKS

Saltwater-freshwater environments have been documented in almost every state in the United States. The presence of a saltwater-freshwater transition zone in equilibrium does not necessarily pose a threat to the use of the ground water. However, any development of the resource must consider the potential movement of the saltwater.

Saltwater-freshwater systems can be conceptualized in two basic ways: either sharp-interface or density-dependent flow and solute transport. The appropriateness of either of these conceptualizations and their attendant methods of analysis depends on the characteristics of the system under investigation and the questions to which answers are being sought. The case studies summarized in this chapter were used to illustrate how the various conceptualizations and methods of analysis produced useful results under different circumstances.

As with most issues related to ground water, the key to proper use of the resource is the proper conceptualization of the hydrogeologic system and boundary conditions. Numerical models can and do aid in the analysis of these systems and can provide quantitative insights as to the best management of the resource. Monitoring systems developed with knowledge of the information required to determine the direction and rate of movement are invaluable in providing both early warning of potential contamination by salty ground water and additional information that is useful in furthering the understanding of the particular system being monitored.

References

Back, W., B. B. Hanshaw, J. S. Herman, and J. N. Van Driel. 1986. Differential dissolution of a Pleistocene reef in the ground-water mixing zone of coastal Yucatan, Mexico. *Geology* 14(2):137–40.

Bennett, G. D., and E. V. Giusti. 1971. *Coastal Ground-Water Flow near Ponce, Puerto Rico*. Reston, VA: U.S. Geological Survey Professional Paper 750-D, D206–D211.

Bonnet, M., and J. P. Sauty. 1975. Un modele simplifie pour la simulation des nappes avec intrusion saline. In *Application of Mathematical Models in Hydrology and Water Resources Systems*. Proceedings of the Bratislava Symposium, International Association of Scientific Hydrology, Publication 115:45–56.

Chemical Rubber Company. 1982. *CRC Handbook of Chemistry and Physics*. 63rd ed. Boca Raton, FL: CRC Press, D-258.

Chow, V. T. 1964. *Handbook of Applied Hydrology*. New York: McGraw-Hill.

Cooper, H. H., F. A. Kohout, H. R. Henry, and R. E. Glover. 1964. *Sea Water in Coastal Aquifers*. Reston, VA: U.S. Geological Survey Water-Supply Paper 1613-C.

Dagan, G. 1982. Stochastic modeling of groundwater flow by unconditional and conditional probabilities. 2: The solute transport. *Water Resources Research* 18(4):835–48.

Desai, C. S., and D. N. Contractor. 1977. Finite element analysis of flow, diffusion, and saltwater intrusion in porous media. In *Formulation and Computational Algorithms in Finite Element Analysis*, eds. K. J. Bathe et al., pp. 958–83. Cambridge, MA: MIT Press.

Essaid, H. I. 1990a. *The Computer Model SHARP, a Quasi-Three-Dimensional Finite-Difference Model to Simulate Freshwater and Saltwater Flow in Layered Coastal Aquifer Systems*. Reston, VA: U.S. Geological Survey Water-Resources Investigations Report 90-4130.

Essaid, H. I. 1990b. A multilayered sharp interface model of coupled freshwater and saltwater flow in coastal systems: Model development and application. *Water Resources Research* 26(7):1431–54.

Fetter, C. W. 1972. Position of the saline water interface beneath oceanic islands. *Water Resources Research* 8(5):1307–15.

Frind, E. O. 1980. Seawater intrusion in continuous coastal aquifer-aquitard systems. In *Finite Elements in Water Resources*, eds. S. Y. Wang et al. Proceedings of the Third International Conference on Finite Elements in Water Resources. University, MS: University of Mississippi, 2.177–2.198.

Frind, E. O. 1982. Simulation of long-term density-dependent transport in groundwater. *Advances in Water Resources* 5:73–88.

Garven, Grant, and R. A. Freeze. 1984. Theoretical analysis of the role of groundwater flow in the genesis of stratabound ore deposits. *American Journal of Science* 284:1085–1112.

Gelhar, L. W., and C. L. Axness. 1983. Three-dimensional stochastic analysis of macrodispersion in aquifers. *Water Resources Research* 19(1):161–80.

Gillham, R. W., and J. A. Cherry. 1982. Contaminant migration in saturated unconsolidated geologic deposits, In *Recent Trends in Hydrogeology*, ed. T. N. Narasimhan. Boulder, CO: Geological Society of America, Special Paper 189:31–62.

Glover, R. E. 1959. The pattern of fresh-water flow in a coastal aquifer. *Journal of Geophysical Research* 64(4):457–59.

Green, D. W., and R. L. Cox. 1966. *Storage of Fresh Water in Underground Reservoirs Containing Saline Water—Phase 1, Project Completion Report*. Lawrence, KS: University of Kansas, Kansas Water Resources Research Institute, Contribution No. 3.

Guswa, J. H., and D. R. LeBlanc. 1985. *Digital Models of Ground-Water Flow in the Cape Cod Aquifer System, Massachusetts*. Reston, VA: U.S. Geological Survey Water-Supply Paper 2209.

Henry, H. R. 1964. Effects of dispersion on salt encroachment in coastal aquifers. In *Sea Water in Coastal Aquifers*, eds. H. H. Cooper et al. Reston, VA: U.S. Geological Survey Water-Supply Paper 1613-C.

Hill, M. C. 1988. A comparison of coupled freshwater-saltwater sharp-interface and convective-dispersive models of saltwater intrusion in a layered aquifer system, In *Proceedings of the VII International Conference on Computational Methods in Water Resources*, eds. M. A. Celia et al., pp. 211–16. New York: Elsevier.

Hubbert, M. K. 1940. The theory of ground-water motion. *Journal of Geology* 48(8):785–944.

Kontis, A.L., and R.J. Mandle. 1988. *Modification of a Three-Dimensional Ground-Water Flow Model to Account for Variable Density Water Density and Effects of Multiaquifer Wells*. Reston, VA: U.S. Geological Survey Water-Resources Investigations Report 87-4265.

Krieger, R. A., J. L. Hatchett, and J. F. Poole. 1957. *Preliminary Survey of the Saline-Water Resources of the United States*. Reston, VA: U. S. Geological Survey Water-Supply Paper 1374.

Kuiper, L. K. 1985. *Documentation of a Numerical Code for the Simulation of Variable Density Ground-Water Flow in Three Dimensions*. Reston, VA: U.S. Geological Survey Water-Resources Investigations Report 84-4302.

LeBlanc, D. R. 1982. *Potential Hydrologic Impacts of Ground-Water Withdrawal from Cape Cod National Seashore, Truro, Massachusetts*. Reston, VA: U.S. Geological Survey Open-File Report 82-438.

Lee, C., and R. T. Cheng. 1974. On seawater encroachment in coastal aquifers. *Water Resources Research* 10(5):1039–43.

Mercer, J. W., S. P. Larson, and C. R. Faust. 1980. *Finite-Difference Model to Simulate the Areal Flow of Saltwater and Freshwater*

Separated by an Interface. Reston, VA: U.S. Geological Survey Open-File Report 80-407.

Pinder, G. F., and H. H. Cooper. 1970. A numerical technique for calculating the transient position of the saltwater front. *Water Resources Research* 6(3):875–82.

Pinder, G. F., and R. H. Page. 1977. Finite element simulation of salt water intrusion on the south fork of Long Island. In *Finite Elements in Water Resources*. Proceedings of the First International Conference on Finite Elements in Water Resources. London: Pentech.

Plummer, L. N. 1975. Mixing of sea water with calcium carbonate ground water. In *Quantitative Studies in the Geological Sciences*, ed. E. H. T. Whitten. *Geological Society of America Memoir* 142:219–36.

Powell, S. T. 1964. Quality of water. In *Handbook of Applied Hydrology*, ed. V. T. Chow, pp. 19–34. New York: McGraw-Hill.

Prince, K. R. 1986. *Ground-Water Assessment of the Montauk Area, Long Island, New York*. Reston, VA: U.S. Geological Survey Water-Resources Investigations Report 85-4013.

Reilly, T. E. 1990. Simulation of dispersion in layered coastal aquifer systems. *Journal of Hydrology* 114:211–28.

Reilly, T. E., O. L. Franke, H. T. Buxton, and G. D. Bennett. 1987a. *A Conceptual Framework for Ground-Water Solute-Transport Studies with Emphasis on Physical Mechanisms of Solute Transport*. Reston, VA: U.S. Geological Survey Water-Resources Investigations Report 87-4191.

Reilly, T. E., M. H. Frimpter, D. R. LeBlanc, and A. S. Goodman. 1987b. Analysis of steady-state saltwater upconing with application at Truro Well Field, Cape Cod, Massachusetts. *Ground Water* 25(2):194–206.

Reilly, T. E., and A. S. Goodman. 1987. Analysis of saltwater upconing beneath a pumping well. *Journal of Hydrology* 89:169–204.

Runnells, D. D. 1969. Diagenesis, chemical sediments, and the mixing of natural waters. *Journal of Sedimentary Petrology* 39(3):1188–1201.

Sanford, W. E., and L. F. Konikow. 1989. Simulation of calcite dissolution and porosity changes in saltwater mixing zones in coastal aquifers. *Water Resources Research* 25(4):655–67.

Segol, Genevieve, and G. F. Pinder. 1976. Transient simulation of saltwater intrusion in southeastern Florida. *Water Resources Research* 12(1):65–70.

Segol, Genevieve, G. F. Pinder, and W. G. Gray. 1975. A Galerkin finite element technique for calculating the transient position of the saltwater front. *Water Resources Research* 11(2):343–47.

Souza, W. R., and C. I. Voss. 1987. Analysis of an anisotropic coastal aquifer system using variable-density flow and solute transport simulation. *Journal of Hydrology* 92:17–41.

Volker, R. E., and K. R. Rushton. 1982. An assessment of the importance of some parameters for seawater intrusion in aquifers and a comparison of dispersive and sharp-interface modeling approaches. *Journal of Hydrology* 56:239–50.

Voss, C.I. 1984a. *AQUIFEM-SALT: A Finite-Element Model for Aquifers Containing a Seawater Interface*. Reston, VA: U.S. Geological Survey Water-Resources Investigations Report 84-4263.

Voss, C. I. 1984b, *SUTRA—A Finite-Element Simulation Model for Saturated-Unsaturated Fluid-Density-Dependent Ground-Water Flow with Energy Transport or Chemically-Reactive Single Species Solute Transport*. Reston, VA: U.S. Geological Survey Water-Resources Investigations Report 84-4369.

Voss, C. I., and W. R. Souza. 1987. Variable density flow and solute transport simulation of regional aquifers containing a narrow freshwater-saltwater transition zone. *Water Resources Research* 23(10):1851–66.

Weiss, Emanuel. 1982. *A Model for the Simulation of Flow of Variable-Density Ground Water in Three Dimensions under Steady-State Conditions*. Reston, VA: U.S. Geological Survey Open-File Report 82-352.

White, D. E., J. D. Hem, and G. A. Waring. 1963. *Data of Geochemistry*, 6th ed. Chapter F: *Chemical Composition of Subsurface Waters*. Reston, VA: U.S. Geological Survey Professional Paper 440-F.

Wigley, T. M. L., and L. N. Plummer. 1976. Mixing of carbonate waters. *Geochimica et Cosmochimica Acta* 40:989–95.

Williamson, A. K., 1987. Preliminary simulation of ground-water flow in the Gulf Coast Aquifer Systems, South-Central United States. In *Regional Aquifer Systems of the*

United States: Aquifers of the Atlantic and Gulf Coastal Plain, eds. John Vecchioli and A. I. Johnson, pp. 119–37. Bethesda, MD: American Water Resources Association Monograph Series No. 9.

Wilson, J. L., and Antonio Sa Da Costa. 1982. Finite element simulation of a saltwater/freshwater interface with indirect toe tracking. *Water Resources Research* 18(4):1069–80.

19

Analysis of Karst Aquifers

William B. White

INTRODUCTION

Chemical reactions between circulating ground water and the soluble wall rock of carbonate and gypsum-rock aquifers enlarge joints, fractures, and bedding-plane partings, allowing increased flow velocities and localization of the flow within the conduits. Flow velocities may be sufficiently high that Darcy's law fails, there is an onset of turbulence, and coarse-grained clastic sediment may be transported through the system. Aquifers containing highly permeable conduit systems formed by dissolutional removal of the bedrock are called karst aquifers. Because of the dramatically different flow velocities between the bulk aquifer and the conduit system and because of the possibility of turbulent flow, the behavior of ground water in karst aquifers must be analyzed differently from ground-water behavior in porous media aquifers.

Localized recharge to the conduit system occurs where surface streams flow underground in sinkholes or blind valleys and where closed depressions collect overland flow and focus it into a single drain as internal runoff. The discharge of water from the conduit system often takes place at a single large spring which may form the headwaters of a substantial surface stream. There is in karst a much more intimate relationship between ground water and surface water than in other types of aquifers and the usual convention of treating ground water and surface water separately is a dangerous practice in karst systems.

This chapter discusses some of the main features of karst aquifers particularly in regard to water quality and contaminant transport. Conspicuous by their absence are much mention of the geomorphology of surface karst landforms or of caves, the accessible fragments of the conduit drainage and paleo-drainage systems. Textbooks on karst include Sweeting (1972), Jennings (1985), White (1988), and Ford and Williams (1989). Other books that deal with karst hydrology and hydrogeology include Bögli (1980), Milanovic (1981), and Bonacci (1987). Dreybrodt (1988) gives the most comprehensive treatment of the chemistry and hydraulics of the dissolution process.

PERMEABILITY IN KARST AQUIFERS

The permeability of an idealized porous media aquifer consists of interconnected pores

TABLE 19-1. Permeability in Various Aquifers

	Porous Media Aquifers	Fracture Aquifers	Karstic Aquifers
Primary	Pores and vugs	Pores and vugs	Pores and vugs
Secondary		Mechanical joints, fractures, and bedding plane partings	Solutionally enlarged joints, fractures, and bedding-plane partings
Tertiary			Integrated conduits of various sizes

and vugs in which water is stored and through which water moves by strictly laminar, Darcy flow. The permeability is considered continuous and homogeneous on a macroscopic scale so that the flow equations for continuous media can be applied. Fracture permeability is an additional secondary contribution formed by mechanical pathways due to joints, joint swarms, fractures, faults, and bedding-plane partings. Fracture permeability is usually highly anisotropic on small distance scales, and there may be an extreme contrast between permeability along the fractures and permeability through the bedrock on either side. Although fracture aquifers may also possess good primary permeability (fractured sandstones, for example), many good fracture aquifers are found in rocks with little primary permeability (fractured granites, for example). Fracture aquifers may be formed in carbonate rocks, particularly fractured dolomites. Fractures may be widened by solution, but the aperture remains sufficiently small that flow is generally laminar, although deviations from Darcy's law may be detected in field-scale measurements.

Karstic aquifers, formed in limestone, gypsum, and dolomite, may have three levels of permeability (Table 19-1). The Paleozoic limestones that make up most of the karst aquifers of the United States have very low primary permeabilities. The young limestones of Florida, Puerto Rico, the Yucatan Peninsula, and the Bahama Islands often have high primary permeabilities. In addition to the fracture permeability, karstic aquifers also have integrated and continuous systems of conduits. The crossover between a solutionally widened fracture and a conduit, for reasons to be discussed later, occurs when the aperture reaches a size of about 1 cm. Some conduits have diameters of tens of meters and they may carry flows equivalent to full-scale surface rivers. Flow in conduits is non-Darcy flow, often turbulent, and the conduit waters may be carrying a clastic load as well as a dissolved load. The residence time of water in the conduit system is very short compared with residence time in the fracture or pore systems.

KARSTIC GROUND-WATER BASINS

Delineation of Basin Boundaries

To the concepts of "aquifers" for ground-water systems and "drainage basins" for surface-water systems must be added the concept of "ground-water basin" for karst systems. Within a karst aquifer, discrete ground-water basins are often located, and each receives recharge from a specific area of land surface through sinkholes and sinking streams, and each drains to a specific spring or group of springs. Some ground-water basins may be more or less congruent with surface-water basins, although the boundaries seldom coincide exactly. In many cases, however, the ground-water basins extend under surface divides, and the flow within the ground-water basin may be in a different direction from flow in the overlying surface-water basin.

Ground-water basin boundaries are determined by highs in the water table and by divides between conduit systems. Because conduit systems evolve and change as base levels are lowered, spillover routes are common, and basin divides may shift, depending on the elevation of the water table. Ground-water basin divides usually can be determined only by an extensive campaign of water tracing. Two of the best-delineated systems are the carbonate aquifer of the Mammoth Cave region (Quinlan and Ewers 1989) and the Greenbrier Limestone aquifer of West Virginia (Jones 1973).

The Greenbrier Limestone aquifer illustrates many of the peculiarities of the karst ground-water basins (Figure 19-1). The karst basins of Figure 19-1 are formed in the Mississippian Greenbrier Limestone in Greenbrier County, West Virginia. The carbonate rocks are gently folded and are underlain by an aquiclude, the MacCready Shale, the outcrop line of which marks the eastern boundary of the ground-water basins and accounts for the odd circumstance that ground-water flow is parallel to the Greenbrier River, the main base-level surface stream. There are a series of definite drainage divides separating the karst aquifer into distinct ground-water basins. In the largest, the Davis Spring Basin, the subsurface drainage is to the south converging on Davis Spring. A series of smaller basins drain to the north and east so that there is a major basin divide between the Davis Spring Basin and the smaller basins to the north. The underground route of Culverson Creek appears to pass under the Buckeye Creek Basin, a feature that can easily occur in rocks such as the Greenbrier Limestone where carbonate units are separated by shales. Not only do karstic ground-water basins have variable divides, spillovers, and poor congruence with surface divides, but they can also be stacked one on top of another.

Flow Paths and Water Balance

A schematic profile view through a karst aquifer (Figure 19-2) forms the basis for a flow sheet for the movement of water (Figure 19-3). The flow sheet forms a conceptual model for karst aquifers and it shows the rather complicated arrangement of flow paths brought on by the interplay between surface water and ground water. Each flow path has a different residence time and water-carrying capacity, illustrating the difficulty in producing a quantitative model for ground-water flow in karst comparable to the flow net models used for porous-media aquifers.

The primary input term is the precipitation and its distribution over the land surface. A primary loss term is evapotranspiration, which will vary across the land surface and with seasonal and climatic conditions.

Most karst aquifers include inputs from surface catchments on nonkarstic rocks. These nonkarstic, but hydrologically connected, rocks are indicated schematically as

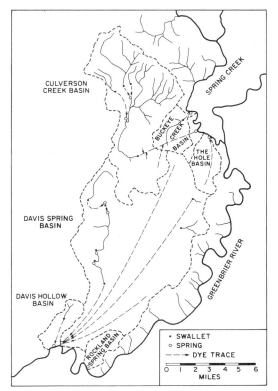

FIGURE 19-1. Ground-water basin for the Greenbrier Limestone aquifer, West Virginia. (*Adapted from a more detailed map in Jones 1973.*)

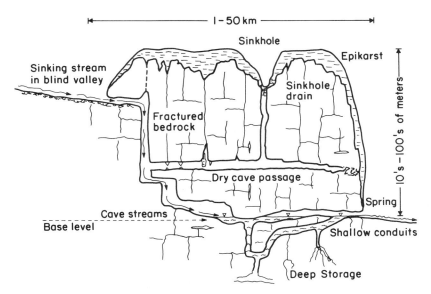

FIGURE 19-2. Profile sketch with great vertical exaggeration through a typical karst aquifer.

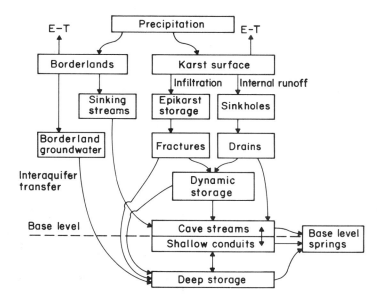

FIGURE 19-3. Flow paths through a karst aquifer. Not all of these components may be present in any given aquifer.

"borderlands" in Figure 19-3 although the reality could be quite complicated. Surface runoff from the borderlands drains through surface streams which flow onto the karst, where some sink underground at the carbonate rock contact. These are the sinking stream inputs to the karst ground-water system. Infiltration in the borderlands can produce a ground-water body in the borderland rocks from which one can allow for interaquifer transfer between the nonkarstic and the karstic aquifers. Overall, water inputs from the borderlands may be categorized as *allogenic recharge*.

A well-developed karst surface often has few or no surface streams. There is no runoff in the usual sense; overland flow during storms is diverted into closed depressions. There may be ponding in sinkholes as a form of temporary storage, but most sinkholes have efficient drains which carry the overland flow into the ground-water system as internal runoff. In the bedrock, sinkhole drains vary from solutionally widened fractures, to irregular chimneys large enough for human exploration, to circular shafts meters to tens of meters in diameter and up to hundreds of meters deep. In spite of the tremendous variety of drain morphology, they have in common a very efficient vertical transfer of water, a resultant residence time of minutes to hours, and sometimes an efficient aeration of water splashing down an air-filled shaft.

Infiltration water on the karst surface moves downward through the soil toward the bedrock contact. In most karst regions, the soil-bedrock contact is sharp, but the bedrock surface is strongly sculptured into a network of deep crevices along joints and fractures with intervening bedrock pinnacles. Although the entire rock surface may be mantled in soil with no surface expression of its complexity, the bedrock topography may have a relief from meters to tens of meters. The bedrock itself is often relatively impermeable, and the open pathways available to carry infiltration water to the ground water are limited to a few solutionally widened fractures and larger drains at fracture intersections. As a result, there is developed a temporary perched water body in what is called the *epikarst* or *subcutaneous zone* (Williams 1983). The residence time in epikarst storage is typically weeks. This is the reservoir that allows cave roofs to continue dripping even during periods of drought.

Collectively, the diffuse infiltration through the soils and the rapid internal runoff through sinkholes are categorized as *autogenic recharge*.

There is also a chemical distinction between the vertical pathways labeled "fractures" and "drains" in Figure 19-3 (Thrailkill 1968). Internal runoff often enters sinkholes with little contact with the organic rich zone of the soil and moves rapidly downward through open drains with insufficient time for chemical reaction with the carbonate wall rock. Water sampled from the walls of open shafts is usually of moderate hardness, moderate dissolved CO_2, and undersaturated with respect to calcite and dolomite. In contrast, infiltration water extracts CO_2 from the soil, where CO_2 partial pressures may be 100 times the atmospheric background and then, during the time in epikarstic storage, reacts with the carbonate minerals at the soil-bedrock interface. The infiltration water descends through the fractures with a high hardness, high dissolved CO_2, and at near saturation with respect to calcite. Such seepage waters can be sampled where they drip from cave roofs. Degassing of CO_2 into the cave atmosphere is responsible for the precipitation of secondary calcite deposits in caves.

The characteristic features of karst aquifers are the conduits that act as master drains for the ground-water system. The active conduits receive infeeders from the sinkhole drains, and there are upstream infeeders where the sinking streams connect. Conduits often lie very close to base level as determined by the elevation of the surface stream into which the spring empties. Examination of many active systems shows that the conduits undulate above and below the base-level surface as indicated schematically in Figure 19-2. Where the conduits are above base level, they appear as air-filled cave passages with streams of water on the floor. Where they lie lower, the passages are water-filled. Recent scuba exploration in many flooded conduits from Florida to Canada shows that the underwater portions have much the same morphology as the above-water portions. Water-filled conduits often lie below stream-carrying conduits. These are shown together on Figure 19-3 with a double arrow to indicate rapid exchange of conduit water.

Because of their minimal hydraulic resistance, conduit systems will not support much hydrostatic head and form a ground-water trough very close to base level. In other parts of the aquifer on both sides of the conduit, however, the ground-water levels may stand much higher. The storage volume lying between the water table and base level constitutes the dynamic storage. Much of the porosity in which the dynamic storage occurs is made up of solutionally widened fractures, small conduits, and other solutionally formed voids. The size scales of these contributions to the porosity are not known, but there is indirect evidence that there is a continuum of sizes ranging from unmodified joints 25 to 100 µm in aperture to the conduits themselves, which may range in diameter up to tens of meters. When the system receives storm input, a portion of the conduit system floods, the ground-water trough is filled, and additional water may be transferred into dynamic storage. The dynamic storage maintains spring flow during drought periods.

It is hypothesized that there is also deep storage below the shallow conduit system. The porosity for this storage consists of solutionally modified fractures, small conduits, and other solution cavities. The exact balance between stream caves, shallow water-filled conduits, and larger openings at depth below base level varies strongly with the hydrogeologic setting and is not easy to generalize. Little is known about the deep storage or about the exchange of water between deep storage and the shallow conduit system. If a karst aquifer is heavily pumped through wells that extend into the deep storage zone, the dynamic storage zone could be depleted, spring flow would cease, and the divides between ground-water basins would disappear. However, the well would continue to produce from the deep storage. The ground-water basins with their intricate flow paths and dynamic storage are shallow systems which may overlie a deeper and more regional aquifer system where the geology permits.

Aquifers with localized inputs through sinking streams and sinkholes and which have integrated conduit flow systems are termed *conduit flow aquifers*. Such aquifers show flashy storm response, discharge through springs that often become turbid or muddy, and show great variability in the chemical composition of the water. Aquifers where the conduit system is poorly developed or absent and where flow is primarily through networks of solutionally modified fractures are termed *diffuse-flow aquifers*. Springs draining such aquifers have very subdued responses to storms, remain clear, and the chemistry tends to vary little with season of the year or storm events. These are end-member types. Most real carbonate aquifers, as indicated in Figure 19-3, contain both types of permeability. Modeling the aquifer requires assigned weighting coefficients to the fraction of flow along each pathway, taking account of its characteristic response time.

Tributary and Distributary Systems

The analogy between integrated conduit drainage in the subsurface and stream channel networks on the surface must not be carried too far. Spillover routes, distributary drainage, and radial drainage are common in karst aquifers. Because of the low hydraulic gradients, master conduits often develop distributary patterns so that the discharge from the aquifer occurs at more than one spring. New springs become activated when water levels rise. The distributary pattern may be likened to channel patterns on river deltas where rivers may have many mouths.

Dye-tracing studies near the divides of karst ground-water basins show that radial drainage is extremely common. Recharge near the basin boundaries flows into all adjacent basins, showing that the concept of a ground-water basin divide, although sound on a regional scale, is somewhat fuzzy if examined in close detail.

Underground Water Tracing

Direct determination of the conduit system within a karst aquifer can be accomplished principally by two methods, cave exploration and water tracing.

A small fraction of karst springs emerge under gravity flow from open cave mouths. For these limited cases, the feeder conduit can be explored and surveyed, sometimes all the way to the infeeders. Likewise, some streams sink into open caves that can be directly surveyed. In still other instances, the active flow system can be reached from overlying dry caves with entrances somewhere on the surface. Very rarely, however, can one delineate the entire conduit system by direct survey. Streams in caves tend to sump, or passages are blocked by breakdown which act as a barrier for human explorers, although not for water. Even in highly favorable examples, one may be able to map only a few percent of the entire active conduit system. As a further limitation, most mapping has been restricted to conduits carrying free-air surface streams, although a few maps of submerged conduits have been prepared by divers. An exception is the Florida karst, where a number of excellent maps of underwater caves have been published.

Water tracing involves injecting appropriate tracers, usually organic dyes, at input points and detecting the tracers in karst windows, cave streams, and springs thus showing the input-output relationships. Multiple traces are needed to connect all input and output points and thus delineate the ground-water basin.

Water tracing with organic dyes is a highly developed technology (Smart and Laidlaw 1977; Jones 1984; Smart 1984; Mull et al. 1988; Quinlan 1989). The tracers most commonly used are sodium fluorescein (Acid Yellow 73), rhodamine WT (Acid Red 388), optical brightener (typically Fluorescent Whitening Agents 22 and 28), and, less commonly, Direct Yellow Dye 96. For most traces, dye concentrations are kept sufficiently low that there is no obvious color in the water in the discharging springs and streams.

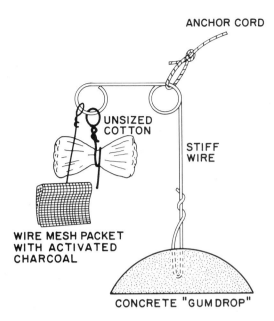

FIGURE 19-4. The "gumdrop" tracer detector developed by Quinlan (1981).

Qualitative traces use packets of activated coconut charcoal as detectors for fluorescein and rhodamine. Packets can be placed in all suspected resurgences and left there for days or weeks (Figure 19-4). The dyes are strongly adsorbed onto the charcoal and cannot be desorbed by further washing in water. The packets are collected at intervals of days to weeks and replaced by fresh packets. In the laboratory, the dye can be elutriated with alcoholic KOH, and the presence of dye determined either visually or fluorometrically. Detectors for optical brightener are wads of unsized cotton which strongly adsorb the brightener which can later be detected by its strong blue fluorescence under long-wave ultraviolet light. Optical brighteners work better than dyes in muddy water. The mud can be washed out of the cotton without stripping the adsorbed brightener. By using a combination of dyes and optical brightener, it is possible to check pathways for several inputs to all possible outputs at the same time, an important time-saving consideration if it is necessary to completely delineate a ground-water basin. A long lag time, preferably spanning several storm events, is needed to completely flush the dye from the system so that a new series of traces can be started.

Quantitative traces require placing automatic samplers at suspected resurgences so that water samples can be collected at specified time intervals. The water samples are then analyzed for dye spectrofluorometrically. With careful background corrections, and calibration of concentrations, it is possible to figure the fraction of dye recovered. Observation of the dye breakthrough curve gives a measure of the flow-through time, and the shape of the curve provides some insight into the flow path. Quantitative traces are less subjective and provide more information, but they require maintenance of automatic sampling devices.

The Karst Water Table

No single aspect of karst aquifers has received more discussion than the concept of the karst water table. There was an argument in Europe beginning in the latter years of the nineteenth century about whether the concept of a water table had any meaning in karst. One school argued that sinking streams descend swiftly through the vadose zone and terminate in an integrated ground-water body with a well-defined water table. The other school argued, mostly from alpine examples, that underground karst streams were independent with no elevational relationship between each other or with a regional-scale water table. The anti-water-table school was fond of pointing out places where air-filled caves passed beneath flowing surface streams.

In the moderate-relief terrain of the eastern United States, the karst aquifers have well-defined water-table surfaces, as measured from static water levels in wells. Much of the confusion over the karst water table arises from two sources: the highly dynamic response of the conduit system to storm events and the low permeability of the bedrock in the absence of solution openings. During low-flow conditions, the conduit may carry only a small free-surface stream marking the bottom of a ground-water trough. It is quite possible for a perched surface stream to overlie the air-filled conduit (Figure 19-5A), a circumstance that can be detected by placing piezometers in the sediments of the stream bed. If the water levels in the piezometers stand below the surface of the stream, it is clear evidence that the stream is a losing stream, perched above the conduit system by impermeable bedrock and stream channel alluvial deposits.

During a storm event, the sinking stream and internal runoff inputs rapidly carry wa-

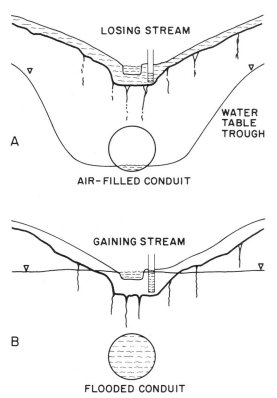

FIGURE 19-5. (A) Stream-carrying conduit with perched surface stream during low-flow conditions. (B) Same during flood flow. Piezometer shows losing or gaining stream.

ter into the base-level conduit system. It is only necessary for the floodwaters to fill the ground-water trough. As a result, cave passages flood rapidly, often in minutes, and there may be a mounding of the water table above the conduit because of the hydrostatic head provided by the upstream reaches and the relatively slow response of wall rock. Water levels in cave systems may fluctuate by tens of meters. After the storm, the conduit drains rapidly, returning to prestorm conditions on time scales from days to weeks. If the mounding of the water table reaches the surface stream, a losing stream may become a gaining stream for the duration of the storm (Figure 19-5B).

Rapid rise in the water table over conduit systems can also result in sinkhole flooding. Under low-flow conditions, sinkholes are inlets for internal runoff, which drains to the base-level conduit system below. If the conduit system is shallow, rapid filling of the ground-water trough during storm events causes a reversal of flow in the sinkhole drains (much like a backed-up sewer) with resultant ponding in the sinkholes. Sinkholes are gradually coming to be recognized as flood-prone areas, and some local authorities are beginning to zone them as such. Sinkhole flooding is a serious land-use hazard in karst and has begun to accumulate legal ramifications (Quinlan 1986).

Characterization of Karst Ground-Water Basins

The concept of the ground-water basin is extremely useful to water resources management, to water-quality protection, and to emergency responses to spills, waste releases, and other environmental hazards. The ground-water basin is also the framework on which all other characterization of the karst aquifer must be superimposed. The steps in characterization of the ground-water basin are as follows:

a. Lay out the area on topographic maps and delineate the surface drainage basin(s). Note any springs, sinking streams, and related features that may appear on the topographic maps.

b. From available geologic data determine the thickness, structure, stratigraphic position, and lithologic character of the carbonate rocks that outcrop in the basin and superimpose the karstic rock outcrop pattern onto the topographic base.

c. Walk the basin following tributary streams to look for losing reaches and swallow holes. Topographic maps are often very misleading, showing blue stream lines in channels where the surface flow has been diverted underground.

d. Walk (or wade) the base-level streams to determine the positions and discharge of springs. Large springs usually appear on topographic maps, but many small springs do not. Temperature or conductivity probes are useful for spotting obscure discharges from stream banks or stream beds.

e. Collect available information on caves, especially maps and descriptions of caves with sumps or active streams.

f. Using the accumulated knowledge of stream swallets, major sinkholes, active cave streams, and springs, undertake a program of water tracing to connect input and output locations. These should be done during both low-flow and high-flow conditions. The overall pattern of the ground-water basin can be determined with qualitative traces. When the rough pattern is known, quantitative traces can be used to resolve questionable connections and to gain further insight into the flow system.

g. Inventory water wells within the basin and determine static water levels during stable water-table conditions where possible. Use this information to prepare a water-table contour map on the topographic base.

h. With the above information in hand, it should be possible to draw the divide(s) for the ground-water basin(s). Cave streams, dye trace connections, and troughs in the water table give some indication of the layout of the conduit system. The relief of the water table above the troughs gives some information on the hydraulic conductivity of the secondary permeability that links the bulk of the aquifer with the conduit drain system.

DYNAMICAL RESPONSE OF KARST AQUIFERS

A characteristic feature of karst aquifers is their rapid response to external stimuli from storms or from seasonal variations in temperature, precipitation, and plant and soil activity. These responses can be used as a means of probing the internal structure of the aquifer.

Physical Response to Storm Events

Consider the response of a karst aquifer to a sudden intense storm that follows a long period of drought. Storm water is input to the system through sinking streams, through sinkholes, and through soil infiltration. Each of these inputs reaches the conduit drain system and through it to the spring in a different period of time. The response time of the entire aquifer will depend on the size of the ground-water basin—the distance from inputs to output—and on the degree of development of the conduit system. This leads naturally to three different storm responses as observed at the spring output (Figure 19-6):

a. Response time short with respect to the mean spacing between storms. Each storm event will produce a peaked hydrograph at the spring much like the flood hydrograph on a surface stream.

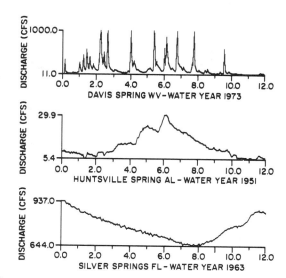

FIGURE 19-6. Hydrographs for storm response of some karst springs: (A) fast response, $t_R \ll t_{storm}$; (B) intermediate response, $t_R \cong t_{storm}$; and (C) slow response, $t_R \gg t_{storm}$.

b. Response time comparable to the mean spacing between storms. Individual storm hydrographs will tend to overlap and the sharp detail will be lost. However, there may remain broad maxima in the hydrograph corresponding to wet and dry periods.

c. Response time long compared with mean spacing between storms. The spring response becomes nearly flat with perhaps only a broad maximum reflecting seasonal variations in precipitation.

For those aquifers with sufficiently rapid response to produce storm hydrographs at the springs, a variety of parameters can be extracted from the hydrograph, as illustrated in Figure 19-7. The ratio of peak flow to baseflow, Q_{max}/Q_B, and the fitting parameters of the recession curve are most useful. The lag time is often of the same magnitude as the storm duration, and unless good continuous precipitation data are available in the recharge area it is difficult to obtain an

Dynamical Response of Karst Aquifers

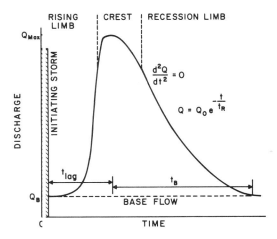

FIGURE 19-7. Schematic storm hydrograph for a karst spring defining the characteristic parameters.

accurate value. The recovery time is much larger, but is difficult to measure because of the asymptotic convergence of the recession curve into the baseflow.

The ratio of peak flow to baseflow is a measure of the "flashiness" of the aquifer. It is a function of both storm intensity and degree of conduit development within the aquifer. Fast-response springs typically drain small basins with $Q_{max}/Q_B \approx 100$, compared with about 10 for intermediate-response springs and 1–3 for slow-response groundwater basins.

The recession curves for type A and some type B aquifers can be decomposed by fitting the recession curve to an exponential function of the form

$$Q = Q_0 e^{-t/t_R}$$

or

$$Q = Q_0 e^{-\varepsilon t}$$

where t_R is the response time, ε is the exhaustion coefficient, and Q_0 is a fitting parameter, the recession curve extrapolated to zero time, but it is not equal to Q_{max} or any other measured discharge.

The storm recession hydrographs for many springs plot as several line segments (Figure 19-8). It is tempting to assign the short response segment to the conduit system and the longer response time to the diffuse-flow system that drains more gradually into the rapidly draining conduit. The numerical values of the response times depend on the storm event. Sometimes more than two straight-line segments are present. Storm recession hydrographs of karst springs cannot be said to be completely understood.

Various authors (e.g., Milanovic 1981) have shown that the exhaustion coefficient is related to the dynamic storage in the aquifer as

$$V_d = \frac{86{,}400 Q_{max}}{\varepsilon}$$

where V_d is volume of water in dynamic storage in cubic meters for peak discharge in cubic meters per second. If it can be assumed that the flow in the aquifer is laminar, the exhaustion coefficient is directly related to transmissivity, T, and storativity, S, of the aquifer:

$$\varepsilon = \frac{2T}{SL^2}$$

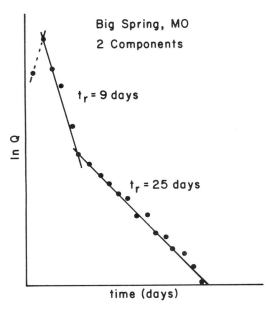

FIGURE 19-8. Storm recession curve for a typical karst spring. (*Data from U.S. Geological Survey water supply records.*)

where L is the horizontal distance from input to output.

Chemical Response to Storm and Seasonal-Scale Events

Recharge water generally enters karst aquifers undersaturated with respect to carbonate minerals. As the water moves along the various flow paths, it reacts with the carbonate rocks with a corresponding increase in hardness and saturation index. The kinetics of these reactions (see Chapters 6 and 9) are such that the time for dissolution reactions to reach equilibrium (1–10 days) is comparable to the transit time for water to move through the aquifer. Water residing in the fracture and pore system may reach equilibrium, while water moving through the conduit system is generally out of equilibrium. As a result, the chemical parameters derived from the analysis of karst spring water often fluctuate with overall discharge levels, with storm events, and with chemical changes in the recharge area brought on by land-use practices and by plant and microorganism activity changing with the seasons.

Chemical fluctuations therefore operate on several time scales. There is an annual cycle related to changing seasons and to seasonal wet and dry periods. There are fluctuations on the scale of a few weeks related to aquifer response to rainfall events and recovery to prestorm conditions. There are fluctuations on time scales of minutes to hours as storm pulses move through the system.

One of the most complete records yet published is shown in Figure 19-9 for a large spring in Missouri. Storm events initiate pulses above base flow in the spring discharge. For each of these pulses, there is a corresponding dip in the concentration of dissolved calcium, magnesium, and bicarbonate ion. The three chemical response curves (sometimes called *chemographs*) are very similar to each other. Note in particular the October storm, which was very intense and of short duration. The hydrograph for this storm returns to baseflow conditions much more quickly than the chemographs. The chemical response of the spring reflects a complex interplay of chemical reactions, mixing phenomena, and flow rates within the karst aquifer.

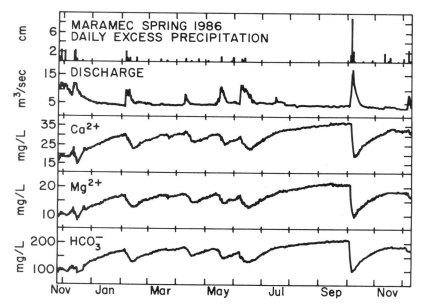

FIGURE 19-9. Precipitation, discharge, and chemical hydrographs for Maramec Spring, Missouri. (*After Dreiss 1989.*)

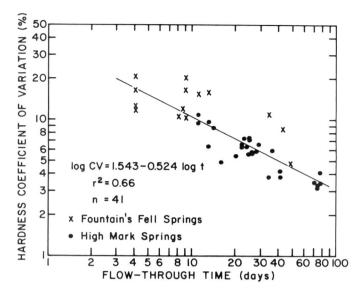

FIGURE 19-10. Relationship of coefficient of variation to flow-through time for springs in the Pennine Hills, England. (*After Ternan* 1972.)

Historically, and mainly for reasons of investigator convenience, studies of spring-water chemistry were based on weekly to monthly sampling intervals. Points taken at random from a curve such as shown in Figure 19-9 will display a considerable variability in concentrations of dissolved ions and in parameters such as the saturation index and CO_2 pressure calculated from them. The magnitude of the hardness fluctuations, expressed as the coefficient of variation, is an indicator of springs draining conduit-type karst ground-water basins with short response times (Shuster and White 1971), although the detailed interpretation depends also on the local geology (Scanlon and Thrailkill 1987). The hardness of spring waters is inversely correlated with discharge, although a specific functional relationship has not been established. The chemical response of karst springs is related to the interrelationships among the characteristic reaction time for chemical equilibrium, the flow-through or response time of the aquifer, and the spacing between storms. An inverse relationship between coefficient of variation and flow-through time has been found (Figure 19-10). Use of the coefficient of variation to distinguish conduit from diffuse-flow aquifers is limited to smaller basins, because in larger basins long flow-through times smooth out the fluctuations whether or not conduits are present.

For fast response ground-water basins, the individual dips on the chemographs often contain additional structure which can be conveniently observed by continuously recording specific conductivity (Figure 19-11). Two responses are shown. The first is an example of a storm pulse entering the system after a long period of drought. Piston flow, resulting from high hydrostatic heads in the catchment region, pushes highly saturated water from the dynamic storage in the aquifer around the conduit. As a result, the concentration of dissolved ions actually rises above the baseflow level for a short while before they drop by dilution in the flood pulse. In the second example, a single, short-duration, intense storm recharged the system and produced a sequence of small peaks on the descending limb of the chemograph. These peaks are interpreted as the arrivals of flow from different tributaries within the conduit system.

When chemical data are averaged on a

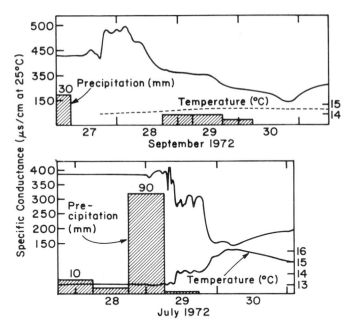

FIGURE 19-11. Fine structure in storm dilution curve for two storm events in Owl Cave, Mammoth Cave Area, Kentucky. (*Data from Hess and White 1988.*)

monthly basis, the mixing and dilution effects of individual storms tend to average out. There remains in the hardness data, and especially in the calculated P_{CO_2} data, a pronounced seasonal effect. In northern and temperate climates, P_{CO_2} is a maximum during the summer growing season and a minimum in the winter. The amplitude of the seasonal CO_2 cycle is about a factor of 10. For springs in tropical climates, where the growing season is continuous, CO_2 levels remain uniformly high except for some influence of wet and dry seasons.

SEDIMENT TRANSPORT IN KARST AQUIFERS

Transportation Mechanisms and Sediment Balance

Most caves contain thick sequences of clastic sediments. Typically, these range from clays, through silts and sands, to cobble- and boulder-size materials. There occur stream beds with a sandstone cobble armoring deep in horizontal limestone conduits far from any possible local source. Open conduit systems, like their surface counterparts, carry a sediment load as well as a dissolved load.

Clastic sediments in conduit systems move as suspended load and as bedload. The fine clays move mainly as suspended load and are found in caves in arrangements that indicate settling from standing water. Suspended-load transport, however, requires turbulent flow to keep the materials in suspension over long travel distances. The larger particle-size materials, silts, sands, and gravels, move as bedload. Consideration of the critical boundary shear for bedload transport (see White 1988) shows that flow velocities in the range of fractions of a meter per second are required to set the bed material into motion. Quantitative studies of sediment transport in conduits are sparse, but anecdotal evidence indicates that most sediment moves during flood events. Indeed, extreme flood flows may strip a conduit of

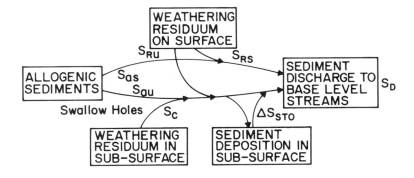

$S_D = S(\text{surface routes}) + S(\text{underground routes})$

$S(\text{surface routes}) = S_{as} + S_{RS}$

$S(\text{underground routes}) = S_{au} + S_{RU} + S_C - \Delta S_{sto}.$

FIGURE 19-12. Flow sheet for clastic sediment transport in a karstic drainage basin. S_{as} = surface sediment load from allogenic sources; S_{au} = underground sediment load from allogenic sources; S_{RS} = residual soils transported by surface routes; S_{RU} = residual soils transported by underground routes; S_c = weathering residuum generated by solution in the subsurface and transported by underground routes; S_{sto} = sediment in storage in the conduit system; S_D = total sediment discharged from the drainage basin.

its sediment filling. New material is deposited during flood recession as occurs in surface channels during floods.

Consideration of possible source areas allows writing a sediment-balance model analogous to the water-balance model (Figure 19-12). Sinking streams draining from allogenic recharge areas carry a clastic load which is discharged into swallow holes and must be carried through the conduit system. Weathering debris created from the insoluble residue of the limestone dissolved at the soil-bedrock contact is flushed into sinkholes and from there to the underlying conduit system. Enlargement of cave passages by dissolution in the subsurface also leaves behind the insoluble fraction of the bedrock. All of these materials are held in storage in the conduit system until they can be flushed out of the springs and returned to the surface drainage system. The sediment balance is important to the maintenance of the conduit drainage system. If the amount of clastic sediment injected into the karst aquifer exceeds its carrying capacity, the conduits choke up, conduit flow is blocked, the water table rises, and sinking streams return to surface routes.

Vertical Transport and Soil Instability

It is the ability of karst aquifers to transport clastic sediments that is responsible for sinkhole collapses and related land-use hazards in karst areas. Sediments accumulated in sinkholes eventually are flushed into the underlying conduit system by natural processes, but sediment transport can be enhanced over the natural background by human activities. Changing the runoff characteristics on the land surface, for example by extensive pavement, can increase the discharge into sinkholes and thus enhance the flushing of sediment into the subsurface. Alternatively, overpumping of the aquifer can lower the water table, allowing water-saturated sediments to dry, become more friable, and thus be more easily transported.

Collapses of sinkhole dumps and sinkhole collapses under landfills, waste-disposal sites, sewage lagoons, evaporation ponds, and waste-storage lagoons all have the potential of introducing both liquid and solid wastes into the conduit system. Solid wastes behave as clastic sediments. They can be transported long distances through the aquifer, where they become inaccessible sources for continuous release of contaminants as the solid waste is leached or dissolved.

EVOLUTION OF KARST AQUIFERS

The karstic character of aquifers in soluble rocks depends on the degree of development of conduit permeability, and for this reason the discussion has emphasized the behavior of the conduit system. Karst aquifers with poorly developed conduits, the diffuse-flow or fracture aquifers, behave not much differently than fracture aquifers in other rock types. The pattern of the conduit system is determined in part by the details of the geologic setting and in part by the sources of recharge and the balance between allogenic and autogenic recharge (Palmer 1991). There is in all aquifers some chemical reaction between the bedrock and the circulating ground water, but in karst aquifers the chemical reactions are responsible for the development of most of the permeability, and this takes place on timescales that are short compared with most geologic processes.

Critical Thresholds

There is a crossover from "fracture aquifers" to "conduit aquifers" when the apertures of the solutionally modified fractures and protoconduits reach a size of about 1 cm. This critical size marks three thresholds for the onset of behavior not found in nonkarstic aquifers.

a. The onset of turbulence: Application of the Darcy-Weisbach and related fluid mechanics equations to pipe systems with the hydraulic gradients typically observed in karst aquifers shows that flow velocities reach the turbulent regime when the aperture width reaches about 1 cm.

b. The kinetic threshold: The rate of carbonate-rock dissolution increases rapidly as the circulating water becomes less saturated. Once a particular pathway along joints, fractures, and bedding-plane partings becomes sufficiently enlarged to allow water with a saturation index of -0.3 or less to penetrate the aquifer, that pathway enlarges faster due to more rapid dissolution and eventually becomes a single conduit.

c. The onset of clastic sediment transport: The water flow velocities required for movement of suspended and bedload begin to occur, for typical hydraulic gradients, when the aperture reaches about 1 cm.

Timescales for Conduit Development

With the relatively good understanding of calcite and dolomite dissolution kinetics that has been developed, it is possible to calculate the time required to enlarge a fracture or bedding-plane parting to conduit size. However, the laboratory dissolution rates are usually faster than those observed in the aquifer, possibly because of the protective layers of insoluble residue that accumulate during the dissolution of rocks.

The initiation stage where mechanical fractures are enlarged to critical aperture size requires about 5,000 years. The enlargement stage, from the critical threshold to full conduit size, is of variable length, depending on the chemistry and flow behavior of the water. However, there is good evidence that fully developed conduits, meters in diameter, can form in 10,000 years. Internal runoff draining through the vadose zone is highly corrosive. Fractures acting as inlet points for sinkhole drains can enlarge to open shafts meters in diameter in a few thousand years.

Applied to the problems of the stability of foundations, impoundments, and landfills in karst areas, there would appear to be little threat from collapse due to active bedrock dissolution. Although rapid on a geologic timescale, the dissolution of carbonate rocks to form large voids is slow on a human timescale and on the scale of the usual design lifetimes of structures. The same cannot be said for gypsum karst, where active collapses due to solutional removal of bedrock can occur in tens of years.

WATER-QUALITY PROBLEMS IN KARST

Karst aquifers are threatened by all of the contamination sources that affect other aquifers. What makes karst aquifers special is the interaction between surface water and ground water due to the presence of sinking streams and the localized internal runoff through closed depressions. Localized flow in conduits keeps contaminants concentrated in a single fast-moving stream where they can migrate for long distances and sometimes reappear at unexpected places. Instead of a slowly developing plume spreading down the flow field, a contaminant slug can appear at a spring or well many kilometers away in hours or days. Because of the open conduits and high flow velocities, especially during storm events, karst aquifers can be contaminated with solid waste as well as liquid waste.

The rapid response of karst aquifers is important with regard to the cleanup of spills —jackknifed tanker trucks on the interstate, derailed railway cars, ruptured tanks, broken pipelines, and other accidents. Spills in karst areas may contaminate ground water as quickly as they contaminate surface water and require equally prompt action by cleanup personnel.

Nonpoint-Source Pollutants

Contamination of karst ground water by herbicides, insecticides, fertilizers, manure, and barnyard and feedlot runoff is not dramatically different from the contamination of other aquifers by these sources. However, agricultural chemicals washed into sinkholes by storm runoff may be carried directly into the conduit drainage system with little filtration or sorption to soils.

Water-Soluble Contaminants

Water-soluble organic compounds, such as phenols and carboxylic acids, and water-soluble inorganic compounds, such as nitrates, some cyanides, and ammonia, remain in solution in karst waters. They tend to be transmitted through the aquifer unaffected except by dilution from other inputs.

Petroleum and Other Light Hydrocarbons

Petroleum hydrocarbons—gasoline, diesel fuel, home heating oil, waste engine oil, and related materials—are less dense and only slightly soluble in water. In karst aquifers, these materials float on the water table and can be carried long distances by underground streams so long as the stream has a free-air surface. Floating organic liquids are trapped when underground streams sump and are not carried by water-filled conduits unless the flow is sufficiently turbulent to entrain the water-insoluble materials (Ewers et al. 1991).

Many petroleum hydrocarbons are volatile. Fumes can migrate upward through open fractures and solution openings in the bedrock where they can invade homes and other buildings. Some fumes, for example of chlorinated hydrocarbons, are toxic. Others may produce an explosion hazard. Flood waters invading conduits where hydrocarbon contaminants are floating on ponded water or behind sumps will float the hazardous materials upward into the overlying bedrock. The flood water-table mound may force the floating contaminants to the surface in sink-holes or into basements and other construction (Stroud et al. 1986).

Toxic Metals

Metals and semimetals from many sources are toxic to various degrees. Of concern are those of the transition elements, including chromium, nickel, copper, zinc, cadmium, and mercury as well as arsenic, antimony, and lead. These elements tend to be mobilized in acid waters and by the highly reducing conditions found in landfills and waste repositories, or else they are complexed by organic ligands in these environments. Karst aquifers tend to be well-oxygenated, near-neutral environments with much acid-neutralizing capacity. As a result, metal-bearing solutions are neutralized and the metals are oxidized to the often less-soluble higher valence states. The metals can precipitate directly as insoluble hydroxides, oxyhydroxides, carbonates, and hydroxycarbonates, which are then incorporated into the clastic sediments. The clastic sediments in many cave streams are coated with black deposits of poorly crystallized manganese oxides formed from the oxidation and precipitation of the natural manganese in the recharge water. These manganese oxides tend to be scavengers for metals and often contain percent quantities of copper, zinc, nickel, and other elements. Karst aquifers, in general, act as accumulators for metals with the exception of some organic complexes which remain in solution.

CONCLUDING REMARKS

The study of karst aquifers has undergone some odd twists and turns over the past 20 to 30 years. There was a time, up to about the 1960s, when carbonate aquifers were regarded as little different from aquifers in other kinds of rocks. Drainage through the conduit system was taken to be of little consequence and of no importance in the evaluation of water resources or their protection from pollution sources. By the 1980s, the pendulum had swung the other way, and there was, if anything, an overemphasis on the conduit system. The task at hand is to integrate these various points of view so that models may be constructed for specific aquifers that take account of all parts of the system as they are embedded into their local hydrogeologic setting. We are still some distance short of having a comprehensive theoretical model for karstic aquifers in spite of the very substantial progress that has been made.

References

Bögli, Alfred. 1980. *Karst Hydrology and Physical Speleology*. Berlin: Springer-Verlag.

Bonacci, Ognjen. 1987. *Karst Hydrology*. Berlin: Springer-Verlag.

Dreiss, S. J. 1989. Regional scale transport in a karst aquifer. 1: Component separation of spring hydrographs. 2: Linear systems and time moment analysis. *Water Resources Research* 25:117–34.

Dreybrodt, Wolfgang. 1988. *Processes in Karst Systems*. Berlin: Springer-Verlag.

Ewers, R. O., A. J. Duda, E. K. Estes, and P. J. Idstein. 1991. The transmission of light hydrocarbon contaminants in limestone (karst) aquifers. In *Proceedings of the Third Conference on Hydrogeology, Ecology, Monitoring, and Management of Ground Water in Karst Terranes*, pp. 287–305. Nashville, TN.

Ford, Derek, and Paul Williams. 1989. *Karst Geomorphology and Hydrology*. London: Unwin Hyman.

Hess, J. W., and W. B. White. 1988. Storm response of the karstic carbonate aquifer of southcentral Kentucky. *Journal of Hydrology* 99:235–52.

Jennings, J. N. 1985. *Karst Geomorphology*, 2nd ed. Oxford: Basil Blackwell.

Jones, W. K. 1973. *Hydrology of Limestone Karst*. Morgantown, WV: West Virginia Geological Survey Bulletin 36.

Jones, W. K. 1984. Dye tracer tests in karst areas. *National Speleological Society Bulletin* 46:3–9.

Milanovic, Petar T. 1981. *Karst Hydrogeology*. Littleton, CO: Water Resources Publications.

Mull, D. S., T. D. Liebermann, J. L. Smoot, and L. H. Woosley, Jr. 1988. *Application of Dye-Tracing Techniques for Determining Solute-Transport Characteristics of Ground Water in Karst Terranes*. Atlanta, GA: U.S.

Environmental Protection Agency Report EPA 904/6-88-001.

Palmer, A. N. 1991. Origin and morphology of limestone caves. *Geological Society of America Bulletin* 103:1–21.

Quinlan, J. F. 1981. Hydrologic research techniques and instrumentation used in the Mammoth Cave Region, Kentucky. In *GSA Cincinnati '81 Field Trip Guidebooks*, ed. T. G. Roberts, pp. 501–6. Falls Church, VA: American Geological Institute.

Quinlan, J. F. 1986. Legal aspects of sinkhole development and flooding in karst terranes. 1: Review and synthesis. *Environmental Geology and Water Science* 8:41–61.

Quinlan, J. F. 1989. *Ground-Water Monitoring in Karst Terranes: Recommended Protocols and Implicit Assumptions*. Las Vegas, NV: U.S. Environmental Protection Agency Report EPA/600/X-89-050.

Quinlan, J. F., and R. O. Ewers. 1989. Subsurface drainage in the Mammoth Cave area. In *Karst Hydrology: Concepts from the Mammoth Cave Area*, eds. W. B. White, and E. L. White, pp. 65–103. New York: Van Nostrand Reinhold.

Scanlon, B. R., and J. Thrailkill. 1987. Chemical similarities among physically distinct spring types in a karst terrain. *Journal of Hydrology* 89:258–79.

Shuster, E. T., and W. B. White. 1971. Seasonal fluctuations in the chemistry of limestone springs: A possible means for characterizing carbonate aquifers. *Journal of Hydrology* 14:93–128.

Smart, P. L. 1984. A review of the toxicity of twelve fluorescent dyes used for water tracing. *National Speleological Society Bulletin* 46:21–33.

Smart, P. L., and I. M. S. Laidlaw. 1977. An evaluation of some fluorescent dyes for water tracing. *Water Resources Research* 13:15–33.

Stroud, F. B., J. Gilbert, G. W. Powell, N. C. Crawford, and M. J. Rigatti. 1986. U.S. Environmental Protection Agency emergency response to toxic fumes and contaminated ground water in karst topography, Bowling Green, Kentucky. In *Proceedings of the Environmental Problems in Karst Terranes and Their Solutions Conference*, Bowling Green, KY, pp. 197–226.

Sweeting, Marjorie M. 1972. *Karst Landforms*. London: Macmillan.

Ternan, J. L. 1972. Comments on the use of a calcium hardness variability index in the study of carbonate aquifers: With reference to the Central Pennines, England. *Journal of Hydrology* 16:317–21.

Thrailkill, J. 1968. Chemical and hydrologic factors in the excavation of limestone caves. *Geological Society of America Bulletin* 79:19–46.

White, William B. 1988. *Geomorphology and Hydrology of Karst Terrains*. New York: Oxford.

Williams, P. W. 1983. The role of the subcutaneous zone in karst hydrology. *Journal of Hydrology* 61:47–67.

V
Case Studies

20

Implementation of a Statewide Survey for Agricultural Chemicals in Rural, Private Water-Supply Wells in Illinois

Dennis P. McKenna, Thomas J. Bicki, and Warren D. Goetsch

INTRODUCTION

A statewide survey to estimate the occurrence of agricultural chemicals in rural, private water-supply wells in Illinois was conducted by the Illinois Department of Agriculture (IDOA), the Cooperative Extension Service–University of Illinois at Urbana-Champaign (CES), and the Illinois State Geological Survey (ISGS). Groundwater samples were collected from approximately 340 randomly selected wells and analyzed for nitrate, nitrite, and a number of pesticides and metabolites. Sampling began in March 1991 and ended in April 1992.

The statewide survey was initiated by the IDOA in response to mandates of the Illinois Groundwater Protection Act (Ill. Rev. Stat. 1989 ch. 111 1/2, pars. 7,451 et seq.) and public concern over the potential for agricultural chemical contamination of ground water in Illinois. This concern is based on the widespread use of ground water for drinking water, especially in rural areas; the extensive use of agricultural chemicals; and reports of contamination from surrounding states with similar agricultural practices and hydrogeologic conditions.

In 1980, 97% of the rural population in Illinois relied upon ground water for drinking water (Withers, Piskin, and Student 1981). Almost half of the estimated 360,000 rural, private wells in the state are in areas where the top of the uppermost aquifer (the source of water to most private wells) lies within 15 m of land surface (McKenna et al. 1989). Aquifers within 15 m of land surface are considered most vulnerable to contamination from agricultural chemicals (McKenna and Keefer 1991; Kross et al. 1990). In addition, more than 20% of the wells in Illinois are large-diameter dug or bored wells (National Water Well Association 1986). Because of their design and generally shallow depth, these wells are especially vulnerable to contamination (Kross et al. 1990).

Shallow aquifers generally do not underlie the areas of Illinois where agricultural chemicals are used most intensively (McKenna et al. 1989). However, a significant portion of the state that is vulnerable to contamination by agricultural chemicals is cropped to corn and soybean. Only in southern Illinois are there vulnerable aquifers and minimal corn and soybean production.

Illinois ranks second in the nation in the use of herbicides (Gianessi and Puffer 1990) and first in use of nitrogen fertilizer (U.S.

Department of Agriculture 1991). Farmland comprises 11.33 million of the 15.38 million ha of land in Illinois (Illinois Department of Agriculture 1990). In more than 80% of the townships in Illinois, more than 25% of the land is planted to corn and soybean.

In 1990, an estimated 21.4 million kg of herbicides and more than 1.6 million kg of insecticides were applied to the 8.05 million ha of Illinois cropland planted to corn and soybean (Pike et al. 1991). On average, 2 of every 3 ha of land in rural areas of the state are treated with pesticides. Herbicides were applied to 98% of the corn and soybean acreage, but only a small percentage of cropland planted to small grain, hay, or pasture received pesticide applications. In 1990, more than 80% of corn acreage and nearly 30% of soybean acreage receiving preplant or preemergent weed control was treated with a herbicide that has a ground-water warning on the label due to its high potential for leaching.

Each year, Illinois farmers also apply approximately 900,000 Mg of nitrogen fertilizer (Illinois Department of Agriculture 1990). Almost all corn acreage receives applications of nitrogen; in 1989 the average application rate was 179 kg/ha.

Previous Studies

For many years, the Illinois Department of Public Health (IDPH) and county health departments have routinely analyzed well-water samples for nitrate, bacterial contamination, and various inorganic parameters. In most cases, these analyses have been conducted at the request of a well owner who suspected contamination or was concerned about exposing an infant to nitrate. As a consequence of this biased approach to selecting wells for sampling and the frequently inadequate information on well locations, data from this testing program cannot be used to assess, in a statistically reliable manner, the extent of nitrate contamination of water-supply wells.

Illinois has had no coordinated monitoring program to determine the presence of pesticides in ground water. Sampling programs have varied in scale and purpose, number of analytes and levels of detection, and degree of hydrogeologic characterization of the area(s) from which samples were collected. As a result, although more than 1,000 analyses have been performed to determine the presence of pesticides, knowledge of the extent of agricultural chemical contamination of ground water in Illinois is still limited. The following summary of sampling studies illustrates the scope of agricultural chemical monitoring programs in Illinois.

In a preliminary reconnaissance to investigate ground-water contamination by pesticides, Felsot and Mack (1984) sampled 25 private wells in five susceptible regions across the state. None of the samples contained pesticides above a 1.0-μg/L detection limit. McKenna et al. (1988) studied spatial and temporal variability in the occurrence of agricultural chemicals in a shallow unconfined aquifer underlying an intensively irrigated region in west-central Illinois. Nitrate-nitrogen levels in the ground-water samples exceeded 10 mg/L in 58% of the 281 samples from 18 monitoring wells (3–9 m deep) and 49% of the 70 samples from 19 private wells (8–12 m deep). Trace levels of pesticides were detected in shallow ground water; however, pesticide concentrations exceeded health advisory levels in only a few samples.

The U.S. Geological Survey has sampled a total of 68 public, private, and observation wells in Illinois as part of a regional reconnaissance for nitrate and selected herbicides in ground water in the midcontinental United States (Kolpin and Burkart 1991). This study was designed to determine the spatial and seasonal distribution of these compounds in near-surface aquifers (<15 m below land surface). Samples were collected in the spring and summer of 1991. Results are not yet available.

The IDPH and IDOA have sampled wells serving agricultural chemical mixing and loading facilities. In the IDPH study

(Long 1988), 77% of the 50 well samples analyzed had detectable levels of at least one pesticide; concentrations greater than 1,000 µg/L were found in some wells. The most frequently detected compounds were the commonly used corn and soybean herbicides. More than 60% of the wells sampled exceeded 10 mg/L NO_3-N; concentrations ranged from 1.2 to 1,288 mg/L with a median concentration of 25 mg/L. Initial sampling by the IDOA has detected pesticides in 12 of 38 facility wells sampled.

The Illinois Environmental Protection Agency (IEPA) has conducted a pesticide-monitoring program for public water-supply wells since 1985. Sampling was targeted initially to wells withdrawing water from sand and gravel aquifers and later to wells considered vulnerable on the basis of hydrogeologic factors and the proximity of potential point sources. The IEPA identified three public water-supply wells with trace levels of currently used herbicides and attributed these detections to agricultural chemical facilities (Clark and Sinnott 1988). The IEPA, in cooperation with the U.S. Geological Survey, is currently conducting a one-time sampling of all 3,000 public water-supply wells in the state for inorganic constituents, volatile organic compounds, and synthetic organic compounds, including 13 pesticides and two metabolites (Coupe and Warner 1991). A subset of 50 wells will be sampled on a quarterly basis to evaluate temporal trends in water quality.

Previous Monitoring Plans

Three statewide ground-water-monitoring plans have been proposed for Illinois. O'Hearn and Schock (1985) recommended sampling of public water-supply wells to characterize ambient water quality in principal aquifers. Shafer et al. (1985) proposed investigating ground-water quality in areas where contamination from hazardous materials was most likely to occur. Neither monitoring plan was designed to determine the occurrence of agricultural chemicals in ground water; nor did they target rural, private wells, which are most vulnerable to contamination from agricultural chemicals.

To provide statistically reliable estimates of the occurrence of agricultural chemicals in private wells in Illinois, McKenna et al. (1989) proposed a design for stratified random sampling of rural private wells. A primary goal of that proposal was to determine whether the extent of contamination of rural, private water-supply wells varied across the state. To maximize comparisons among strata, they recommended sampling 384 wells in each of four strata differentiated on the basis of the depth to the top of the uppermost aquifer (Figure 20-1). Their plan specifically excluded large-diameter dug or bored wells from the target population. These wells, which are very vulnerable to contamination, occur most frequently in areas where productive aquifers are not present and consequently would not be a reliable indicator of water quality in productive aquifers.

The design by McKenna et al. (1989) served as a basis for the statewide survey. The most significant differences from the previous recommendations are that, in the current study, (1) the population of wells was not stratified, (2) the target population includes all rural private wells, and (3) the selection of wells was completed in two stages.

PROJECT ORGANIZATION AND RESPONSIBILITIES

The IDOA, CES, and ISGS developed the statewide survey as a cooperative inter-agency project. Each agency assumed primary responsibility for a major component of the study. Project managers within each agency developed and supervised implementation of the study.

The ISGS was primarily responsible for development of the sampling plan for the statewide survey and procedures for sample collection. The ISGS also provided geologic

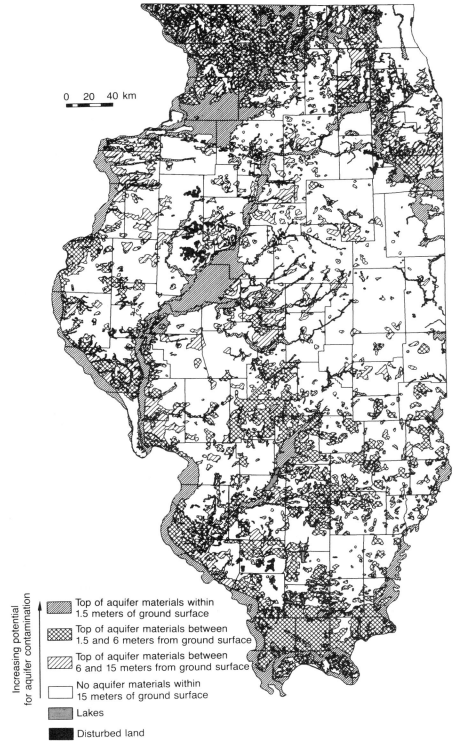

FIGURE 20-1. Potential for agricultural chemical contamination of aquifers in Illinois. (*From McKenna and Keefer 1991.*)

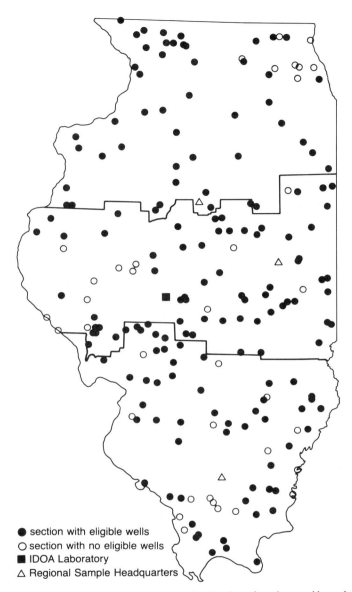

FIGURE 20-2. Location of 200 randomly selected sections and boundaries of sampling regions.

characterization of the areas in which samples were collected.

The CES was primarily responsible for conducting the initial inventory of dwellings and eligible wells, obtaining permission to sample, and administering a well-site characterization questionnaire. The CES also collected well-water samples, maintained the primary data base, and notified well users of the results of water-quality analyses. CES advisers are located in most of the 102 counties in Illinois; their knowledge and close ties to the rural population were thought to be assets in implementation of the project. Eighty of the 82 county CES offices asked to participate in the study cooperated. Project managers performed the initial dwelling inventory, administered the well-site characterization questionnaire, and requested well-user permission for sampling in the two counties unwilling to participate in the statewide survey.

Samples were collected by three regionally located sampling teams, composed of a primary sampler and an alternate sampler, assisted by county CES advisers. Illinois has an area of 149,934 km^2 with maximum dimensions of 608 km north to south and 341 km east to west. Locating samplers in northern, central, and southern Illinois (Figure 20-2) reduced each sampler's travel time to sampling sites and to the IDOA laboratory. The field sampler for the central region was appointed sampling coordinator and was responsible for scheduling and coordinating activities of the regional samplers.

The IDOA was primarily responsible for sample analyses. IDOA, the state lead agency for pesticide regulation under the Federal Insecticide, Fungicide, and Rodenticide Act, has conducted pesticide-formulation analyses since 1970 and pesticide-residue analyses since 1979. In the fall of 1989, the IDOA, in anticipation of various ground-water sampling programs planned by the state and U.S. Environmental Protection Agency proposals to require ground-water monitoring by the states, began expanding laboratory facilities for analysis of pesticides in water samples.

Each of the six analytical methods used in this study was assigned to a specific chemist. Each chemist was responsible for all components of a method, from extraction through analysis. The senior chemist ensured that all quality-control procedures internal to each analytical method were followed, reviewed reports from each analyst, and transferred analytical results to the IDOA project manager. The IDOA quality-assurance (QA) officer supervised laboratory spiking of duplicate well-water samples and ensured compliance with laboratory QA procedures. The IDOA project manager reviewed analytical results for well-water samples and external quality-control samples with the senior chemist and the QA officer, prepared final reports of analyses, and transferred results to the CES project manager.

The three project managers met periodically to review progress, resolve logistical problems, and review and sign letters notifying well users of results. A letterhead for the statewide survey was adopted to emphasize the cooperative interagency effort. Letterhead stationery was used for all correspondence to well users and for formal correspondence with members of the project team.

STUDY DESIGN

The primary goal of this study was to provide statistically reliable estimates of the occurrence of agricultural chemicals in rural, private water-supply wells in Illinois. *Occurrence* was defined as the proportion of well-water samples exceeding the method detection limit for the target pesticides, pesticide metabolites, and nitrite, or exceeding 3 mg/L for nitrate nitrogen. The survey was not designed to allow for reliable estimates to be made about the concentrations of the various analytes in wells, nor will the results be representative of wells at a regional or county level. The survey results will be an estimate of the quality of water in rural, private wells but cannot be used to estimate the quality of ground water or drinking water in Illinois.

A target confidence level of 95% with ±5% precision was chosen. The probability of occurrence of the analytes in well-water samples in Illinois was estimated to be 30% on the basis of results of private water-supply well surveys in Iowa (Kross et al. 1990), Minnesota (Klaseus, Buzicky, and Schneider 1988), and Wisconsin (LeMasters and Doyle 1989). Using these parameters, statistically reliable estimates for the state requires sampling 322 wells (see Eq. 3-11 in Chapter 3).

The target population was defined as rural, private water-supply wells in Illinois.

There are an estimated 440,000 private wells in Illinois (National Water Well Association 1986); approximately 360,000 are in rural areas of the state.

- *Rural* was defined as the area outside the boundaries of an incorporated area that has a population of 2,500 or more, as determined from 1986 Census data, and outside an urban fringe area.
- *Private* was defined as both private and semiprivate water systems. A private water system was defined as any supply that provides water for drinking, culinary, or sanitary purposes and serves an owner-occupied single-family dwelling. A semiprivate water system (1) serves other than an owner-occupied single-family dwelling, (2) has fewer than 15 service connections, or (3) serves fewer than 25 individuals (daily average) at least 60 days per year. Irrigation wells, livestock water-supply wells, public, and community water-supply wells were not inventoried nor sampled.
- *Wells* were defined as any excavation constructed for the purpose of acquiring ground water.

The target population of wells was not stratified. Stratification generally results in more precise estimates and has been used in other large-scale studies, such as the National Pesticide Survey (U.S. Environmental Protection Agency 1990) and the recent statewide survey in Iowa (Hallberg et al. 1990). The most frequently recommended stratification variables are agricultural chemical usage and aquifer vulnerability. Stratification was not used in this survey for three reasons: (1) Agricultural chemicals, especially herbicides, are used almost everywhere in rural Illinois. (2) The aquifer vulnerability map of Illinois (Figure 20-1) is not intended for use in estimating the contamination potential of areas in which the shallow large-diameter wells occur most frequently. These wells—more than 20% of the target population—withdraw water from relatively fine-grained materials that are not considered aquifers (McKenna and Keefer 1991). (3) The identification of candidate wells within each strata would have substantially increased costs. However, relations between well-water quality and both agricultural chemical usage and aquifer vulnerability will be investigated in the statistical analysis of results from the survey.

Selection of Analytes

The initial list of analytes for the statewide survey was selected on the basis of recommendations in McKenna et al. (1989). Because of the large number of pesticides used in Illinois and the high cost of analyses, chemicals selected as analytes were those with high usage in Illinois or high potential to cause ground-water contamination. However, several pesticides with suspended or canceled registration (e.g., dieldrin and endrin) were also included as analytes because of their persistence in the environment, toxicity at low concentrations, and reported detections in ground water in Illinois. The availability of a USEPA National Pesticide Survey (NPS) multiresidue method was also a criterion for analyte selection. Using NPS analytical methods will facilitate comparisons between results from the Illinois survey and results from the NPS and other studies using those methods.

The list of recommended analytes was reviewed by researchers at the University of Illinois and state agency personnel involved in regulation of pesticide use. Several compounds were excluded because of difficulties in obtaining analytical reference standards or because of very low recoveries observed during the initial demonstration of capability by the IDOA laboratory. The final list of analytes is presented in Table 20-1.

Sampling Plan

Because there is no list of rural, private water-supply wells in Illinois, a two-stage, probability sampling plan was selected. The

TABLE 20-1. Trade Names, Method Detection Limits, and Method Quantification Limits for Analytes

Analyte	Trade Name (s)[a]	MDL[b] (µg/L)	MQL[c] (µg/L)
Acifluorfen	Blazer, Tackle	0.68	3.4
Alachlor	Lasso, Bronco	1.3	2.6
Aldrin	Aldrin, Aldrex	0.004	0.016
Atrazine	Atrazine, AAtrex	0.43	0.86
Atrazine delakylated	—metabolite	1.4	1.4
Bentazon	Basagran, Laddock	0.21	1.05
Bromacil	Hyvar-X	17.0	34.0
Butylate	Sutan+, Sutazine	0.65	1.3
Carbaryl	Sevin, Savit	2.0	2.0
Carbofuran	Furadan	3.4	3.4
Carbofuran phenol	—metabolite	32.0	32.0
Carboxin	Vitavax, DCMO	2.5	5.0
Chloramben	Amiben	0.26	1.3
Cyanazine	Bladex, Extrazine	3.7	3.7
2,4-D	2,4-D	0.57	2.9
2,4-DB	Butoxone, Butyrac	4.0	20.0
Diazinon	Diazinon, Dz.n	1.7	3.4
Dicamba	Banvel, Marksman	0.16	0.80
Dieldrin	Dieldrin, aldrin metabolite	0.004	0.016
Dinoseb	Dyanap, Premerge	0.16	0.80
Endrin	Hexadrin	0.006	0.024
Endrin aldehyde	—metabolite	0.009	0.036
EPTC	Eptam, Eradicane	0.68	1.4
Ethoprop	Mocap	0.15	0.30
Heptachlor	Drinox, H-34	0.005	0.020
Heptachlor epoxide	—metabolite	0.004	0.016
Linuron	Lorox, Linex	0.65	0.65
Metolachlor	Dual, Turbo, Bicep	1.4	2.8
Metribuzin	Lexone, Sencor	0.43	0.86
Metribuzin DA	—metabolite	0.70	0.70
Metribuzin DADK	—metabolite	0.39	0.39
Metribuzin DK	—metabolite	0.28	0.28
Prometon	Pramitol	0.37	0.74
Propachlor	Ramrod, Bexton	0.11	0.44
Simazine	Princep, Simazine	0.12	0.24
Tebuthiuron	Spike	0.51	1.0
Trifluralin	Treflan	0.003	0.012
Vernolate	Reward, Vernam	0.65	1.3
Nitrate-nitrogen	—	0.10[d]	0.10[d]
Nitrite-nitrogen	—	0.10[d]	0.10[d]

[a] Includes the most commonly used products containing the analyte.
[b] Method detection limit.
[c] Method quantification limit.
[d] Reported in units of milligrams per liter.

first-stage sampling unit (primary unit) was a land section, normally 1 mi^2 (2.59 km^2). The second-stage sampling unit (population unit) was a rural, private water-supply well. In the first stage, 200 land sections were randomly selected. A relatively large number of primary units was selected in order to (1) ensure that more than 322 eligible wells (the minimum sample size required) were identified, (2) account for exclusions of land sections on the basis of land use or the absence of private wells, and (3) increase the number of counties in which primary units were located. An analysis of census data on the distribution of wells in the state indicated that, on a zip-code area basis, approximately 8% of the state has less than one well per square mile, with a median density of less than five wells per square mile.

In each section, land use (rural or urban) was determined by using defined criteria, and all dwellings and private wells in rural areas were enumerated. In stage 2, private water-supply wells were randomly selected from the rural area of each section with private wells. Because the populations of wells within each land section were not equal, the probability of selection of a well varied from section to section. The sampling plan was developed in consultation with the Illinois Statistical Office at the University of Illinois at Urbana-Champaign.

IMPLEMENTATION

The key tasks in implementation of the survey were to identify candidate wells, select wells and obtain permission to sample, characterize the well sites, collect samples, analyze samples, transmit results to the well users, and compile and interpret results (Figure 20-3). Each of these tasks and other project-management activities are briefly described in the following sections.

Development of Standard Operating Procedures and Training

Detailed standard operating procedures were developed for key field, laboratory, data-acquisition, and data-management activities. Each procedure was reviewed by the three project managers and, when appropriate, by other staff on the project or by agency

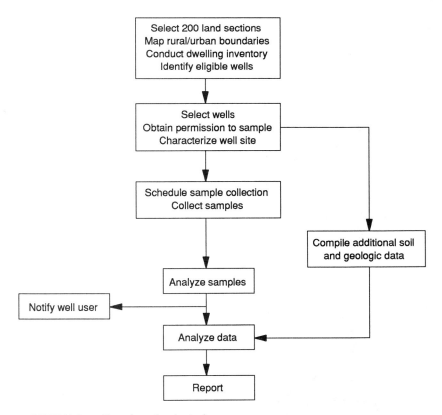

FIGURE 20-3. Overview of major tasks.

staff not involved in the survey. In some instances, implementation was hampered when formal written procedures were not completed before starting a task.

County CES staff, including agricultural, home economics, youth, and 4-H advisers conducted the initial inventory of wells, requested well-user permission to sample, and administered the well-user interview and well-site characterization form. Advisers received a packet of information describing the project, data-collection requirements, and sample copies of forms and questionnaires. Three training sessions were held using the CES Telenet system, an interactive telephone system linking the University of Illinois to county extension offices throughout the state.

A 34-page *Well-Water Sample Collection Manual* provided step-by-step instructions for all activities related to sample collection. The sample-collection manual was revised, updated, and field-tested before distribution to regional samplers. One office and two field training sessions for samplers and alternates were conducted to ensure that collection procedures were consistently and correctly followed. The reasons for using specific procedures were also explained so that project staff would be better prepared to handle unanticipated problems.

Biweekly meetings were held with all laboratory personnel associated with the survey to discuss the status of sample analyses, results of analyses of quality-control samples, and analytical problems. If necessary, corrective actions were initiated.

Well Selection

Candidate wells were identified, and written permission to sample wells was acquired before sample collection was initiated. This allowed for more efficient scheduling of sample collection and analyses.

Stage 1: Identifying Candidate Wells

The 200 land sections were randomly selected from a list of all sections in Illinois by using the ILLIMAP township and section data base (Swann et al. 1970) in the geographic information system at the Illinois State Geological Survey. A list of selected sections along with a topographic map identifying the location of the section(s) to be inventoried was sent to the CES adviser in each of the 82 counties in which a randomly selected section was located. The CES adviser was responsible for drawing the boundaries, if any, between rural and urban areas within the selected section on the topographic map. When necessary because of the relatively small scale of the topographic maps, boundaries were drawn on large-scale aerial photos.

The adviser visited each dwelling within the rural area of each randomly selected section to determine the source of water for domestic use. The adviser located the dwelling site as accurately as possible on the topographic map or aerial photo and assigned an identification number to the dwelling. The adviser then contacted each resident and recorded the following information about the dwelling: legal description; name, address, and phone number of occupant; name, address, and phone number of owner (if other than occupant); and source of water used for domestic purposes.

The purpose of this first stage was to develop a *complete* list of all dwellings and all private wells within the rural area of each of the selected sections. The completeness of the list of wells developed at this stage and the accuracy of mapping rural and urban areas was critical to obtaining accurate estimates for the entire state. Consequently, the CES advisers were provided with detailed criteria to determine if a well was eligible to be included in the sample population. Wells were not excluded because of poor construction, known water-quality problems, or location. However, a dwelling that was vacant and did not appear to be habitable was not eligible. A dwelling that was habitable but vacant was considered eligible if the dwelling was inhabited before October 1, 1990.

The CES advisers were asked to make

repeated attempts to contact the resident of each dwelling within the rural area of each selected section. This occasionally required evening and weekend telephone calls or visits. Information on the presence of an eligible well obtained from relatives, neighbors, or other informed sources was considered acceptable at this stage. Advisers were cautioned that including a well at this stage did not guarantee that it would be selected for sampling, so they should not promise a well user that a well would be sampled.

During this stage, 6 sections were determined to be entirely urban and 23 rural sections had no private wells because there were no dwellings in the section or because another water source was used, such as public water supply. One section of land was not inventoried because the field staff were concerned about their personal safety. More than 1,500 dwellings were inventoried. A total of 875 rural private wells were identified in 170 sections in 74 counties.

Stage 2: Permission to Sample and Well-User Interview

Initially, two wells were randomly selected for sampling from each selected section. If there was only one well in the section or if a well user refused permission to sample, additional wells were randomly selected from other randomly selected sections with wells. This process continued until permission to sample was acquired for approximately 340 wells. Additional wells were selected to provide alternate wells in case any of the primary well users decided not to participate in the study and to account for losses of wells or samples during the course of the study.

A list of selected wells in each section was sent to each county CES adviser, who was responsible for (1) obtaining written permission to sample the well, (2) administering the well-user interview and well-site characterization form (WUIWSCF), and (3) identifying the best available sampling point, defined as the outside faucet located closest to the well and before any water treatment or storage device.

The WUIWSCF was used to characterize conditions and management practices at well sites. Two questionnaire formats were developed: a WUIWSCF with 83 questions for rural residences (-R) and a WUIWSCF with 136 questions for farmsteads (-F). The WUIWSCF-R and WUIWSCF-F documented well construction, previous water testing, lawn and garden activities, well-setback distances from potential point sources and activities, and farming practices on adjacent land. The WUIWSCF-F documented additional information about pesticide storage, use, and disposal, fertilizer storage and use, and soil and crop management practices. To ensure the accuracy and completeness of responses, the CES adviser completed the questionnaire during an on-site interview with the well user.

Sample Collection

Sample-collection procedures were similar to those used in the National Pesticide Survey (Mason et al. 1987; U.S. Environmental Protection Agency 1990). Scheduling of sample collection to avoid potential temporal variability in water quality was patterned after the approach used in the National Pesticide Survey and the recent statewide survey in Iowa (Kross et al. 1990).

Scheduling

Samples were collected over a 13-month period from March 1991 through April 1992. A sampling schedule of 27 two-week sampling periods was used to spread sample collection across all seasons and pesticide and fertilizer application periods, meet sample holding-time requirements of the analytical methods, and not exceed laboratory capacity. During each sampling period, 13 to 15 water wells were sampled; 3 to 5 water wells were randomly selected for sampling within each of the three sampling regions (Figure 20-2). Two water wells within the same land section were not sampled in consecutive sampling periods. Also, an attempt was made to avoid sampling of water wells in adjoining counties within the same sampling period. However, in some instances, adjoining counties had such a large number of wells to be sampled that it was necessary for wells in adjoining counties to be sampled in the same sampling period.

When sample collection began, letters were sent to all well users notifying them that the statewide survey was underway. Approximately 30 days before a specific well was scheduled for sampling, the sampling coordinator sent the well user a letter stating that the well would be sampled within a specific two-day period. Copies of this letter were sent to the CES adviser and the regional sampler.

Approximately 10 days before a set of wells was scheduled to be sampled, the sampling coordinator sent the regional samplers an information packet for each well. Each packet contained the name, address, and telephone number of the well user; the legal description of the property where the well is located; a copy of the permission form signed by the well user; the topographic map identifying the location of the land section and well to be sampled; and a copy of portions of the WUIWSCF. The WUIWSCF information described the exact location of the principal sampling point identified by the CES adviser and specific comments from the well user about scheduling. In instances where a WUIWSCF was incomplete or improperly completed, a copy of the original WUIWSCF was furnished to the sampler. The questionnaire was completed by the sampler and well user at the time of sample collection.

Approximately one week before the scheduled sampling date, the regional sampler telephoned each well user to confirm that the sampling date and time were acceptable. The regional sampler also confirmed the sampling date and time with the cooperating CES adviser.

Procedures for Collecting Water Samples

Whenever possible, well-water samples were collected at an outside faucet located before any water storage or treatment device. Wells were purged prior to sample collection to ensure collection of representative samples. Teflon was used in all parts of the sampling apparatus that came into contact with water samples. All sample bottles for a well site were placed in a sample kit. Use of the sample kits simplified sample tracking and minimized bottle breakage.

A sample kit consisted of a sealed cardboard box lined with plastic and a styrofoam insert with holes for each of six sample bottles. Each sample kit was labeled with a unique four-character identifier. A plastic sleeve fitted to the top of the box contained a five-copy sample tracking form (STF) listing the sample-kit identification (ID) and the bottle ID. The STF was used to maintain chain of custody for the sample bottles in the kit. Additional sample kits were prepared for collection of field blanks, duplicate samples, and duplicate samples for spiking. Each regional sampler was also supplied with a spare kit in case a bottle was broken during sample collection.

At each well, samplers completed a well-sampling information sheet (WSIS) and well-purging parameters record (WPPR). The WSIS documented environmental conditions at the site at the time of sampling and confirmed that the principal sampling point previously identified by the county CES adviser on the WUIWSCF was the best available sampling point. The WPPR was used to record temperature, pH, and specific conductance measurements of well water before and after sample collection.

Well-purging procedures consisted of measuring the pH, temperature, and specific conductance of well water flowing continuously at approximately 2 L/min through a flow-through cell until stabilization. The flow-through cell was modified from the design of Garske and Schock (1986). To avoid overpressurizing the flow-through cell

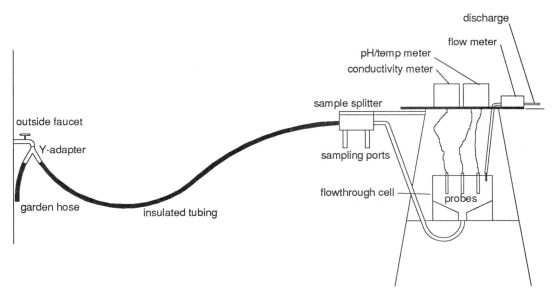

FIGURE 20-4. Apparatus for sample collection and monitoring of well-purging parameters.

while minimizing the time required to purge the well and any water-storage tank, the sampler attached a garden hose to the outside faucet with a Y-adapter. Flow through the hose was limited to approximately 40 L/min.

Stabilization was defined as the point at which three consecutive readings, taken 5 min apart, for pH, temperature, and conductance were within ± 0.1 pH units, $\pm 0.5°C$, and $\pm 5\%$, respectively. The pH meter was calibrated by the regional sampler immediately before each sampling trip. Temperature and conductivity meters were calibrated at the IDOA laboratory. After well-purging parameters had stabilized, samples were collected from a sampling port located ahead of the flow-through cell (Figure 20-4).

Occasionally, equipment malfunctions resulted in less than a full suite of well-purging parameters being measured. If the temperature probe malfunctioned, temperature was measured with a standard glass laboratory thermometer. If a hand pump was used by the resident or if the resident did not allow extended purging of low-volume wells, samplers used one of several modified well-purging procedures and noted the deviation from the standard procedures on the WPPR.

Well-water samples were collected in the types of sample bottles specified in the analytical methods. The label of each sample bottle was initialed and dated by the sampler before collecting the sample. Sample bottles were returned to the sample kit immediately after sample collection, packed in ice, and maintained at or below 4°C while being transported to the laboratory. All samples were returned to the laboratory within 36 h of collection.

If the well user was not present at the time of sampling, the sampler left a card notifying the well user that samples had been collected and that the results of water-quality analyses would be mailed to them within several months. A four-copy incident reporting form was also used (if necessary) to document any problems that might have occurred during sampling. A follow-up telephone call was made to resolve any problem to the satisfaction of the well user.

Sample Analyses

All sample analyses were performed in the IDOA Pesticide Water Laboratory in Springfield. To determine agricultural chemicals in ground-water samples, IDOA used five U.S. Environmental Protection Agency (1988) and one American Society for Testing and Materials (1986) method: USEPA 507 (nitrogen- and phosphorous-containing pesticides), 508 (chlorinated pesticides), 515.1 (chlorinated acids), 531.1 (N-methylcarbamoyloximes and N-methylcarbamates), National Pesticide Survey Method 4 (carbonates and related compounds), and ASTM D3867-85 (nitrate and nitrite). The analytes included in the state-wide survey, their common trade names, and method detection and method quantification limits are listed in Table 20-1.

Three areas of the IDOA laboratory were used for analyses of well-water samples: the General Laboratory Receiving Room, the Pesticide Residue Room, and the Pesticide Water Laboratory. Sample-kit preparation, extractions (except for method 515.1), analysis, and data reduction were performed in the Pesticide Water Laboratory. Sample kits were transferred to the samplers in the Receiving Room where supplies were replenished and sampling equipment was serviced. Spiking of duplicate well-water samples was also performed in this room, which is located away from the other project-associated laboratory areas.

The method detection limit (MDL) for an analyte was established by following the procedures used in the National Pesticide Survey (U.S. Environmental Protection Agency 1990). The minimum quantification limit (MQL) for each analyte was also estimated by using an approach similar to that used in the NPS. The MDL was multiplied by a factor determined by the analyst on the basis of the potential for interferences. The factors were 2 for method 507, 4 for method 508, 5 for method 515.1, 1 for NPS method 4, and 1 for method 531.1.

Analytical data for pesticides and metabolites were reported on the basis of the results for internal quality-control parameters (surrogate and/or internal standards) and a comparison between laboratory-determined MDLs and MQLs. A nondetection was reported as <MDL. A detection of an analyte at a concentration between the MDL and the MQL was reported as DNQ (detected, but not quantified). A detection of an analyte at a concentration above the MQL was reported as the laboratory-reported concentration.

During the initial demonstration of laboratory capabilities, low recoveries or large standard deviations in recoveries were obtained for several analytes: aciflurofen, bromacil, diazinon, metribuzin DADK, and metribuzin DK. The presence of these compounds could be reliably determined, but actual concentrations could not be reported with certainty. Results for these compounds were reported as <MDL or QUAL (qualified).

Unacceptable recoveries of internal standards were observed for some specific sets of water samples; and results for those samples were reported with an asterisk to indicate that the values reported might not be reliable.

Well-Site Characterization

In addition to data acquired from the well-user interview and well-site characterization form, soil and geologic data were compiled for each site. Published USDA–Soil Conservation Service soil survey reports and unpublished soil survey maps were used to identify the soil association present in the area around the well. The areal extent of soil mapping units within a 0.8-km radius around each well was also estimated.

The surficial geology of each land section was determined from the Stack-Unit Map of Illinois (Berg and Kempton 1988), published at a scale of 1:250,000. The mapped units depict the distribution of geologic deposits vertically from the surface to a depth of

15 m as well as horizontally over a specified area. The minimum thickness of continuous mapped units is 1.5 m, except where a unit less than 1.5 m was mapped over at least 1 km^2 (Berg and Kempton 1988).

The approach of McKenna and Keefer (1991) was used to determine the relative potential for aquifer contamination of each sequence of materials mapped in a section. Aquifer materials are defined as sand and gravel greater than 1.5 m thick, permeable sandstone greater than 3 m thick, and fractured carbonates greater than 6 m thick. They differentiated four aquifer-vulnerability groups primarily on the basis of the distance from the land surface to the top of the first continuous deposit of aquifer materials: (1) within 1.5 m of land surface, (2) between 1.5 and 6 m of land surface, (3) between 6 and 15 m of land surface, and (4) greater than 15 m from land surface.

Well-log records at the ISGS were also searched in an attempt to verify information provided by the well user on well depth, source aquifer, and well construction.

Data Management

The statewide survey generated a large volume of data, and data-management procedures were developed to manage these records accurately and efficiently. Use of a unique identifier for each site and for each sample bottle was central to data management. Each dwelling site was given a unique six-digit identifier: the first three digits corresponded to the random number assigned to the land section; the last three digits corresponded to a number sequentially assigned to each dwelling within the rural area of the section. Each sample bottle was given a unique seven-character identifier: the first four characters corresponded to a random number assigned to the sample-kit; the last three characters corresponded to the analytical method.

In the laboratory, computer-controlled, instrument-generated data (including chromatograms) were archived on a cassette-tape backup system or individual diskettes and stored in the IDOA General Chemistry Laboratory vault. The data base of final results was stored on a LAN system in a restricted-access directory.

During the study, seven separate data bases were generated: (1) dwelling inventory data (stage 1), (2) basic site data for selected wells (stage 2), (3) responses to WUIWSCF, (4) analytical results, (5) soil data, (6) geologic data, and (7) well-log data. Data bases 1, 2, 3, and 5 were maintained by the CES; data base 4 was maintained by the IDOA; and data bases 6 and 7 were maintained by the ISGS. At the conclusion of sample collection and analysis, the data bases were merged to compile summary statistics and will be used to conduct relational analyses.

Quality Assurance

A QA plan was developed, and periodic QA audits were performed on all phases of the project under the supervision of the project managers.

All information gathered during stage 1 was reviewed by a project manager, and any necessary follow-up calls were made to the CES adviser to clarify data. Five percent of all sections were also visited by a project manager to spot-check the accuracy of information submitted by the advisers. Each WUIWSCF was reviewed by the CES project manager before data were entered. Problems with missing or inconsistent responses were resolved during an interview conducted by the field sampler at the time a well was sampled.

Quality assurance for sample-collection activities required collection of field blanks at approximately 15% of the well sites, three field reviews of the procedures followed by each field sampler and alternate, laboratory measurement of the temperature of each sample kit, and review of all well-purging records to ensure that the criteria for collecting representative samples were met.

Quality assurance for sample analyses

included collection and analysis of external quality-control samples, compliance with the internal QA requirements of each analytical method, and periodic reviews of all analytical procedures and data by the senior chemist, the IDOA quality-assurance officer, and the IDOA project manager.

Two field blanks, two duplicate well-water samples, and two duplicate well-water samples for laboratory spiking were collected at randomly selected well sites during each sampling period. (Each type of QA sample was collected at approximately 15% of the well sites.) Two randomly selected analytes from method 507 and one randomly selected analyte from each of the other methods were used as spikes. Spiking concentrations were also randomly chosen at 1 to 10 times the MDL for each analyte. Only the IDOA project manager had access to external quality-control and well-sample-kit identifiers.

Analytical results were transferred from the method analyst to the senior chemist via standardized worksheets unique to each method, which included data on all pertinent internal QC parameters (surrogate standard recoveries, internal standard recoveries, laboratory-fortified blanks, and calibration mixes) as well as results for all well-water and external QC samples. The senior chemist reviewed the worksheets and other related data before they were entered into the data base.

The IDOA QA officer periodically reviewed all data generated and reported by laboratory personnel (including worksheets and chromatograms). The QA officer also supervised spiking of duplicate well-water samples, observed sample transfers between the sample custodian and samplers, and participated in reviews of final reports. These reviews included spot-checks of 5% of all data entries.

All analytical results were reviewed by the IDOA project manager, QA officer, and senior chemist. Once the data were finalized, reports of analyses of samples from each well and each set of external QC samples and summaries of analytical results for all samples collected during a sampling period were prepared. These reports were transmitted to the other project managers for review.

Questionnaires, data-transmittal forms, and field measurements were reviewed by a project manager before data entry. Verification of manual data entry into the computerized data base was accomplished by comparing 5–20% of the entries with the original hard copy.

Time Frame and Workload

The design and implementation of the survey took nearly 3 years and involved more than 100 individuals working the equivalent of 20 staff years. The project involved inventorying more than 1,500 dwellings, administering detailed questionnaires to approximately 340 well users, driving more than 70,000 miles, conducting approximately 3,000 chemical analyses, and verifying and entering thousands of data elements. Figure 20-5 presents an overview of the time frame and the staff time required for completion of the major phases of the study.

Survey design and project management required approximately 45 staff-months. Completion of stage 1 activities required approximately 18 staff-months: 1.5 days per section to inventory dwellings and determine water source and an additional 50 staff-days to prepare directions for the CES advisers, outline the selected land sections on topographic maps, conduct training, enter data into the data base, and conduct QA spot-checks. Completion of stage 2 required approximately 10 staff-months. On average, 0.5 staff-day was required to obtain permission to sample and administer the questionnaire at each selected sampling site. Completing stages 1 and 2 required driving approximately 20,000 miles.

Sample collection, including scheduling and well-user contacts, required nearly 26 staff-months. The samplers drove more than 50,000 miles. Costs for sampling

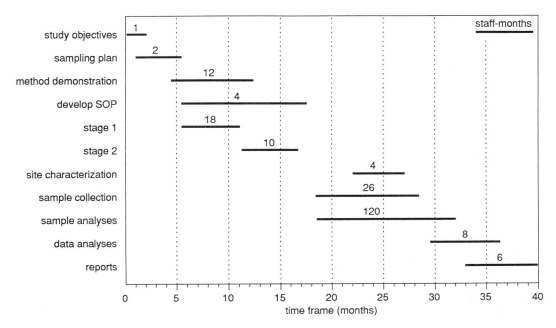

FIGURE 20-5. Time frame and estimated workload (number of staff-months required) for implementation.

equipment and commodities were about $10,000.

Ten IDOA staff members were directly involved in sample analysis. Four staff members performed administrative, quality assurance, and other part-time support functions. Four chemists and two technicians worked full-time in sample processing and analysis. The biweekly sampling schedule allowed 10 working days for the analysis of a set of approximately 20 samples. Under normal operating conditions, analyses could be completed in seven to eight working days. Data reduction, reviews of analytical results (including internal and external QC samples), and preparation of final reports of analytical results could usually be accomplished within a similar time period.

Analyses with each of the six methods of approximately 340 well-water samples and 150 QA samples required approximately 120 staff-months. Commodities, equipment maintenance, and special-purpose equipment costs totaled more than $55,000.

Compilation of soil and geologic information and well-log records required an additional four staff-months. Data entry and clerical work required approximately 18 staff-months. The estimated cost for telephone and fax charges, stationery, printing, postage, and publications was $18,000.

Problems

Implementation of the statewide survey in Illinois generally proceeded smoothly. The most significant problems were delays in completing stage 1 and stage 2 activities, a high rate of refusal by users of selected wells, and initial problems in demonstrating the analytical methods.

Significant delays were encountered in completing stage 1 because of understaffing in several counties or conflicting priorities for CES staff. In several counties, other project staff were required to complete stage 1 activities. Quality-assurance reviews also identified several cases in which the selected section was misidentified on the topographic map. These errors required additional field work by the county adviser. A significant amount of time (1 staff-month) was also spent clarifying responses to the dwelling-site inventory form. These generally involved the uses of well water or the classification of a well as private (owner-occupied) or semiprivate (rented or multiple dwellings served).

During stage 2, the rate of nonresponse was much higher than anticipated. Approximately 19% of the well users contacted refused to participate in the survey in spite of repeated assurances that confidentiality was guaranteed, no regulatory action could be taken, and no fee would be charged. This nonresponse rate is higher than the 13% nonresponse rate for rural domestic wells in the NPS (U.S. Environmental Protection Agency 1990) and the 8% rate reported for a statewide survey in Iowa (Kross et al. 1990). The most frequently cited reasons for refusal were general distrust of government agencies, unavailability due to busy work schedules, objections due to previous solicitations for testing from companies selling water-treatment devices, concern about the regulatory consequences if agricultural chemicals were detected in their wells, confidence that agricultural chemicals were not present in their water, and apathy about the condition of their drinking water. Data were compiled on all nonresponse wells to determine whether these wells are significantly different from sampled wells.

Sample Collection

The regional samplers experienced few logistical or equipment problems. The most significant problems were situations in which a well scheduled for sampling was determined to be ineligible or was unavailable for sampling at the scheduled time. The summer of 1991 was very dry through most of Illinois; in a few instances, users of shallow, large-diameter wells reported that the well had gone dry and they had recently added water from a public water supply to the well. When possible, sampling of these wells was rescheduled for the fall or winter when the addition of water would be less likely. If a well containing water from another source was sampled, the results were not used in the statistical analyses.

Occasionally, sampling of a well was rescheduled at the request of the well user. Several well users refused to allow sampling, despite previously granting written permission to sample. In two instances, one spouse had granted permission to sample without the consent of the other.

Sample Analyses

The National Pesticide Survey methods required several major modifications before they could be routinely used in the IDOA laboratory. However, subsequent analyses were performed within the method-required holding times, and only one significant equipment malfunction affected completion of analyses.

RESULTS

The following sections describe the procedure for notifying well users of the analytical results for samples from their well and plans for reporting detailed results of the statewide survey.

Well-User Notification

After the IDOA laboratory completed internal reviews and approved the release of analytical data, the three project managers reviewed all analytical results in the context of results for external quality-assurance samples. If a pesticide or metabolite for which no health advisory is available had been detected, or if more than one pesticide or metabolite had been detected, a copy of the analytical results was sent to the IDPH. The chief toxicologist at IDPH reviewed the results and prepared a letter discussing potential health effects. The confidentiality promised to well users was maintained by not identifying the well user's name and address on the copy.

Approximately three months after sample collection, each well user was sent an information packet containing a cover letter that had been signed by the three project managers and summarized the results of water-quality analyses of samples from their well. Also enclosed were the IDOA laboratory report, definitions of terms used in the laboratory report. the MDL and MQL for each analyte (Table 20-1), and a CES publication on testing and treating drinking-water supplies. If an analyte had been detected, the packet also included a copy of the USEPA Health Advisory Summary Sheet for that compound, and, if appropriate, a copy of the letter from the IDPH toxicologist. Each well user was also provided with the names, addresses, and phone numbers of individuals to contact for further information. These contacts included the three project managers, the IDPH toxicologist, representatives of the regional office of the IDPH and the county health department, and the county CES adviser. None of the contacts from outside agencies, other than the IDPH toxicologist, were provided with copies of the results sent to the well user; however, all were informed about the statewide survey and provided with samples of the materials sent to well users.

Plans for Final Reports

Overall results of the survey will be published in several formats. An initial report will document in detail the design and implementation of the statewide survey (Bicki, McKenna, and Goetsch 1993). Preliminary results will be published in several short summaries. An executive summary highlighted the overall design and implementation of the study, significant results, and preliminary interpretations of water-quality analyses (Goetsch, McKenna, and Bicki 1992). Additional reports will present detailed results and interpretations of the relationships between well-water quality and land use, well construction, and various soil and geologic factors.

CONCLUDING REMARKS

The results of the statewide survey will provide the first statistically reliable estimates of the extent of agricultural chemical contamination of rural, private water-supply wells in Illinois. The survey should also provide some information on the factors affecting contamination of private wells and the relative importance of point versus nonpoint sources of contamination.

The IDOA, as the FIFRA State Lead Agency, is responsible for the registration of all pesticides offered for sale and use in the state. The department has the authority to suspend, revoke, or cancel the registration for use of any product. The department is also responsible for developing and implementing the state's management plan to prevent pesticide contamination of ground water. A critical component of that plan will be continued monitoring of ground-water quality. By identifying areas where private wells are more vulnerable to contamination and by identifying pesticides that are more likely to cause contamination, the survey may provide a basis for more accurate targeting of future monitoring programs in the state.

References

American Society for Testing and Materials. 1986. Standard test methods for nitrite-nitrate in water. In *1986 Annual Book of ASTM Standards*, pp. 598–610. Philadelphia: American Society for Testing and Materials, Section 11 Water and Environmental Technology, Volume 11.01 Water (I), ASTM D3687-85.

Berg, R. C., and J. P. Kempton. 1988. *Stack-Unit Mapping of Geologic Materials in Illinois to a Depth of 15 Meters*. Champaign, IL: Illinois State Geological Survey Circular 542.

Bicki, T. J., D. P. McKenna, and W. D. Goetsch. 1993 (in preparation). *Design and Implementation of the Statewide Survey for Agricultural Chemicals in Rural, Private Water-Supply Wells in Illinois*. Springfield, IL: Illinois Department of Agriculture.

Clark, R. P., and C. L. Sinnott. 1988. Pesticide monitoring in Illinois community water supply wells. In *Pesticides and Pest Management—Proceedings of a Conference Held November 12–13, 1987, Chicago, IL*, pp. 125–32. Springfield, IL: Illinois Department of Energy and Natural Resources ILENR/RE-EA-88/04.

Coupe, R. H., and K. L. Warner. 1991. Groundwater quality statistics for Illinois. In *U.S. Geological Survey National Computer Technology Meeting Proceedings, Phoenix, Arizona, November 14–18, 1988*, eds. B. H. Balthrop and J. Terry. Reston, VA: U.S. Geological Survey Water Resources Investigation Report 90-1462.

Felsot, A. S., and G. O. Mack. 1984. Survey for pesticides in ground water supplies in Illinois. In *Thirty-Sixth Illinois Custom Spray Operators Training School Manual, Summaries of Presentations, January 4 and 5, 1984, Urbana, IL and 1984 Pest Control Reference Material*, pp. 135–36. Urban, IL: Cooperative Extension Service University of Illinois at Urbana-Champaign College of Agriculture in Cooperation with the Illinois Natural History Survey.

Garske, E. E., and M. R. Schock. 1986. An inexpensive flow-through cell and measurement system for monitoring selected chemical parameters in ground water. *Ground-Water Monitoring Review* 6(3):79–84.

Gianessi, L. P., and C. Puffer. 1990. *Herbicide Use in the United States*. Washington, DC: Quality of the Environment Division, Resources for the Future.

Goetsch, W. D., D. P. McKenna, and T. J. Bicki. 1992. *Statewide Survey for Agricultural Chemicals in Rural, Private Water-Supply Wells in Illinois*. Springfield, IL: Illinois Department of Agriculture.

Hallberg, G. R. et al. 1990. *The Iowa State-Wide Rural Well-Water Survey Design Report*. Iowa City, IA: Iowa Department of Natural Resources Technical Information Series 17.

Illinois Department of Agriculture. 1990. *Illinois Agricultural Statistics Annual Summary—1990*. Springfield, IL: Illinois Department of Agriculture, Division of Marketing, Bureau of Agricultural Statistics and U.S. Department of Agriculture, National Agricultural Statistics Service Bulletin 90-1.

Klaseus, T. G., G. C. Buzicky, and E. C. Schneider. 1988. *Pesticides and Groundwater: Surveys of Selected Minnesota Wells*. Minneapolis, MN: Minnesota Department of Public Health and Minnesota Department of Agriculture, Report Prepared for the Legislative Commission on Minnesota Resources.

Kolpin, D. W., and M. R. Burkart. 1991. *Work Plan for Regional Reconnaissance for Selected Herbicides and Nitrate in Ground Water of the Mid-Continent United States, 1991*. Iowa City, IA: U.S. Geological Survey Open-File Report 91-59.

Kross, B. C. et al. 1990. *The Iowa State-Wide Rural Well-Water Survey Water Quality Data: Initial Analysis*. Iowa City, IA: Iowa Department of Natural Resources Technical Information Series 19.

LeMasters, G., and D. J. Doyle. 1989. *Grade A Dairy Farm Well Water Quality Survey*. Madison, WI: Wisconsin Department of Agriculture, Trade and Consumer Protection and Wisconsin Agricultural Statistics Service.

Long, T. 1988. Groundwater contamination in the vicinity of agrichemical mixing and loading facilities. In *Pesticides and Pest Management—Proceedings of a Conference Held November 12–13, 1987, Chicago, IL*, pp. 133–50. Springfield, IL: Department of Energy and Natural Resources ILENR/RE-EA-88/04.

Mason, R. E., et al. 1987. *National Pesticide Survey, Pilot Evaluation Technical Report, Prepared for the U.S. Environmental Protection Agency*. Research Triangle Park, NC: Research Triangle Institute RTI/7801/06-02F.

McKenna, D. P., S. F. J. Chou, R. A. Griffin, J. Valkenberg, L. L. Spencer, and J. L. Gilkeson. 1988. Assessment of the occurrence of agricultural chemicals in groundwater in a part of Mason County, Illinois. In *Proceedings of the Agricultural Impacts on Groundwater—A Conference, March 21–23, 1988, Des Moines, IA*, pp. 389–406. Dublin, OH: National Water Well Association.

McKenna, D. P., and D. A. Keefer. 1991. *Potential for Agricultural Chemical Contamination of Aquifers in Illinois*. Champaign, IL: Illinois State Geological Survey Open File 1991-7R.

McKenna, D. P., S. C. Schock, E. Mehnert, S. C. Mravik, and D. A. Keefer. 1989. *Agricultural Chemicals in Rural, Private Water*

Wells in Illinois: Recommendations for a Statewide Survey. Champaign, IL: Illinois State Geological Survey and Illinois State Water Survey Cooperative Groundwater Report 11.

National Water Well Association. 1986. *Wellfax Database*. Dublin, OH: National Groundwater Information Center.

O'Hearn, M., and S. C. Schock. 1985. *Design of a Statewide Ground-Water Monitoring Network for Illinois*. Champaign, IL: Illinois State Water Survey Contract Report 354.

Pike, D. R., K. D. Glover, E. L. Knake, and D. E. Kuhlman. 1991. *Pesticide Use in Illinois: Results of a 1990 Survey of Major Crops*. Urbana, IL: University of Illinois at Urbana-Champaign College of Agriculture, Cooperative Extension Service, DP-91-1.

Shafer, J., et al. 1985. *An Assessment of Ground-Water Quality and Hazardous Substance Activities in Illinois with Recommendations for a Statewide Monitoring Strategy*. Champaign, IL: Illinois State Water Survey Contract Report 367.

Swann, D. H., P. B. DuMontelle, R. F. Mast, and L. H. Van Dyke. 1970. *ILLIMAP—A Computer-Based Mapping System for Illinois*. Urbana, IL: Illinois State Geological Survey Circular 451.

U.S. Department of Agriculture. 1991. *Agricultural Chemical Usage 1990 Field Crops Summary*. Washington, DC: U.S. Department of Agriculture, National Agricultural Statistics Service and Economic Research Service.

U.S. Environmental Protection Agency. 1988. *Methods for the Determination of Organic Compounds in Drinking Water*. Cincinnati, OH: Environmental Monitoring Systems Laboratory, Office of Research and Development, U.S. Environmental Protection Agency EPA-600/4-88/039.

U.S. Environmental Protection Agency. 1990. *National Pesticide Survey Phase I Report*. Washington, DC: U.S. Environmental Protection Agency EPA 570/9-90-015.

Withers, L. J., R. Piskin, and J. D. Student. 1981. *Ground Water Level Changes and Demographic Analysis of Ground Water in Illinois*. Springfield, IL: Illinois Environmental Protection Agency.

21

Ground-Water-Quality Monitoring in The Netherlands

Willem van Duijvenbooden

INTRODUCTION

The Netherlands is a densely populated country with about 15 million people living on an area of approximately 38,000 km^2. About 64% of the land use is for agricultural purposes, 8.5% is urbanized, and 3.6% is designated for traffic facilities. Only 12% is defined as natural regions, and even these regions are human-made and partially used as recreational areas. The country is heavily industrialized, while the agricultural use of soils is the most intense in the world. The annual use of artificial nitrogen fertilizer on agricultural soils is about 260 kg ha^{-1} yr^{-1} on average, with an annual production of animal manure of about 290 kg ha^{-1} yr^{-1} (Table 21-1). The annual use of pesticides in agriculture is about 20.6 kg ha^{-1} yr^{-1} in active ingredients.

Intensive cultivation of soils creates a serious threat for soil and ground-water quality. The ground water is polluted in broad areas, especially in sandy regions (42% of the country). In more than 70% of the agriculturally developed sandy regions, nitrate concentrations of 50 mg/L or more can be found in shallow ground water. Many pesticides are also found in ground water. As a result of acidification, nitrate concentrations in shallow ground water in about 20% of the so-called natural regions can be found in excess of the drinking-water standard of 50 mg/L, and aluminum concentrations of a few to even tens of milligrams per liter can be found in these regions. In urbanized regions, hundreds of thousands of local soil-pollution sources are indicated (Table 21-2). Many of these local sources clearly influence ground-water quality.

At present, Dutch governmental policy is directed toward the preservation and recovery of the environment, nature, and the landscape. This requires extensive monitoring of the environment.

ENVIRONMENTAL MONITORING

Several national environmental monitoring networks are used to gather environmental information necessary for implementation of Dutch environmental policy. There are national networks for monitoring the quality of air, rainwater, soil (including shallow ground water), ground water, surface water, and drinking water, as well as radiation. With the exception of surface-water quality, all of these monitoring networks are managed by the National Institute of Public Health and Environmental Protection. In addition to the national networks, several provincial, regional, and local networks

TABLE 21-1. Nitrogen Balance for Agricultural Soils (1986)

	kg/ha^{-1} yr^{-1}	%
Animal manure	291	48
Artificial fertilizer	263	43
Deposition from the air	45	7
Other sources	9	2
Total input	608	100
Removal by agricultural products	242	40
Volatilization	57	9
Surplus on soil	309	51
Total output and storage	608	100

TABLE 21-2. Some Local Soil-Pollution Sources in The Netherlands According to a Recent Inventory

Sources	Number	Number to Be Remediated
Gas generation sites	234	234
Dumping sites	3,298	~150
Car junkyards	2,100	~1,200
Former industrial sites	~400,000	~80,000
Present industrial sites	~120,000	~25,000

Source: Langeweg (1989).

exist. Independent from these networks, several biological networks managed by private organizations are directed toward birds, butterflies, plants, and so on.

All national monitoring networks have permanent monitoring locations which are sampled periodically. The frequency of sampling depends on the objectives and environmental compartment to be monitored and differs between continuous monitoring (some air and surface-water stations) and monitoring once every five years (soil sampling). General goals of the networks are description and diagnosis of environmental quality for the benefit of environmental policy, observation of changes in environmental quality, increased knowledge of environmental systems, evaluation of the effects of corrective actions performed by the Dutch government, enforcement of regulations, and facilitation of responses to emergency situations.

With respect to ground water, the networks on soil, ground-water, and drinking-water quality are especially important. In this chapter special attention is given to these networks, but first the hydrogeologic situation and soil conditions in The Netherlands are briefly discussed.

HYDROGEOLOGIC SITUATION

The Quaternary sediments in The Netherlands were mainly deposited around coastal areas, that is, in shallow seas, lower reaches of rivers, or in coastal swamps and lagoons. Sedimentary deposits were supplied partly by sea but mainly by rivers. Notably, the Rhine and Meuse rivers supplied large amounts of material. As a result, subsoils consist mainly of unconsolidated sediments, built up of fluvial sands and gravels alternating with fluvial or marine clay layers and some peaty developed layers. For this reason, The Netherlands is a relatively flat country, with the land surface generally varying between a few meters below and above sea level. In the central and eastern parts of the country, glacial thrust ridges were formed during the last glacial period (Saalien), with maximum elevations some tens of meters above sea level. The western and northern parts of the country consist mainly of diked polder areas with a land-surface elevation typically between 0 and 6 m below sea level (Figure 21-1). In 90% of the country, ground-water levels are less than 4 m below land surface (m-ls). Only in the central areas of the glacially formed hills and in the southern part of the country can deeper ground-water levels be found. Saline ground water can be found in the whole region at depths varying from a few or tens of meters in the western part to approximately 300 m below land surface in the central and eastern parts of the country (Zagwijn 1975; Engelen 1980).

Rainwater infiltration (recharge) areas are situated mainly in the sandy regions. Upward seepage occurs in low-lying regions such as the polder areas. Bank infiltration

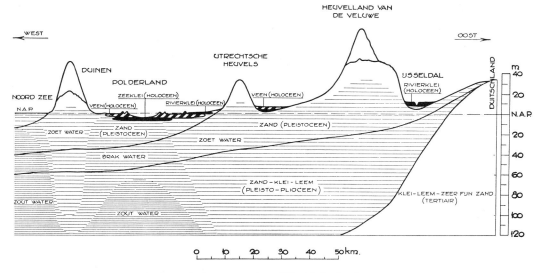

FIGURE 21-1. Schematic hydrogeologic cross section of The Netherlands.

also can be important in these regions, due to high surface-water levels relative to ground-water levels.

Ground-water velocities in The Netherlands are relatively low. Horizontal flow varies from a few to a few hundred meters per year (m/yr) with an average of approximately 30 m/yr. In recharge regions, the vertical-flow component is generally less than one m/yr.

Soil conditions in The Netherlands (see Figure 21-2) are strongly determined by the genesis of the country. Soils in the western and northern part (excluding the dune regions) are largely peaty and clayey soils (marine facies). The central and eastern parts of the country consist largely of sandy soils, while river clays are found in a zone around the branches of the Rhine and Meuse rivers (Breeuwsma and Duijvenbooden 1992).

THE DUTCH NATIONAL GROUND-WATER QUALITY MONITORING NETWORK

The Dutch National Ground-Water Quality Monitoring Network was established during the early 1980s. The objectives of the network are to (1) inventory and diagnose ground-water quality in relation to soil use, soil type, and hydrogeologic conditions, (2) indicate the extent of human influence on ground-water quality, (3) identify long-term changes in ground-water quality, and (4) provide data for quality control and ground-water management.

A monitoring strategy has been developed based on these objectives and taking into account factors determining ground-water quality (Duijvenbooden 1981, 1985, 1987). At present, the network consists of approximately 380 monitoring wells (1 per 100 km^2) with three well screens of 2-m lengths at about 10, 15, and 25 m-ls each (Figures 21-3 and 21-4). Most of the monitoring wells are located in areas with fresh ground water that can be used for drinking water. Some monitoring wells are located in regions with brackish or saline ground water.

Every monitoring well is associated with a specific combination of land use (related to contaminant loads on the land surface), soil type (related to the vulnerability of ground water to various types of pollutants), and hydrogeologic conditions (related to the possibilities for transport and geochemical processes). For the most important com-

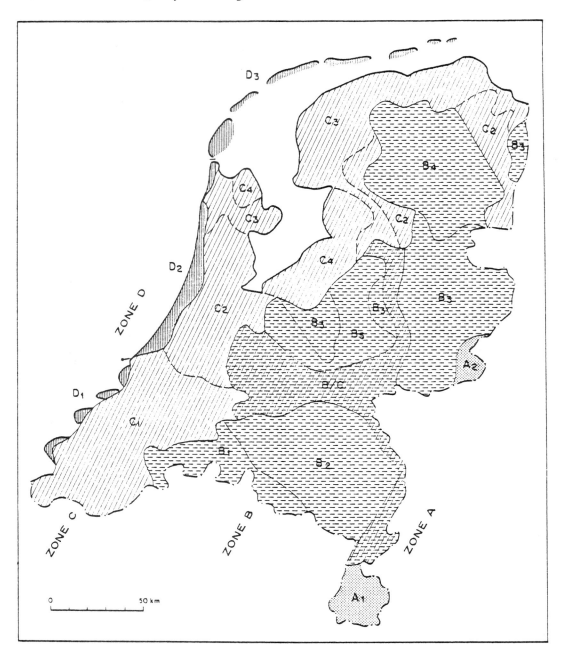

FIGURE 21-2. Major hydrologic and soil conditions in The Netherlands: (A) higher zone with bedrock close to the land surface; (B) higher freely draining areas with predominantly sandy soils at the land surface; (C) low artificially drained polder areas with mainly clayey and peaty deposits; and (D) higher freely draining sandy coastal dune belt. (*After Engelen 1980.*)

FIGURE 21-3. Location of monitoring sites of the National Ground-Water Quality Monitoring Network.

binations, sufficient monitoring wells were selected to ensure the availability of applicable mean values of most parameters and the possibility of trend analysis. The most important soil types selected were sandy soils, peaty soils, river clays, and marine clays. With regard to land use, most monitoring wells are located in intensively used grasslands, arable land, urban land, and natural regions. Regarding the hydrogeologic situation, sites were selected in recharge areas, areas with upward seepage, and zones with bank infiltration. The effects of soil type and land use are illustrated in Tables 21-3 and 21-4 with data from the monitoring network.

Table 21-3 shows that soil characteristics are of utmost importance for ground-water quality. In anaerobic areas, relatively low concentrations of oxidized elements such as NO_3^- and SO_4^{2-} are found in combination with high NH_4^+ concentrations. The genesis of marine clay is reflected in K^+, P, and NH_4^+ content. The sorption capacities of clay and peat are reflected in the low concentrations in ground water of heavy metals such as Zn. Table 21-4 indicates the effects of human activities in sandy regions.

No preexisting wells were used, partly due to the desired uniformity of monitoring sites, and partly because of the desired quality of the wells. All boreholes were

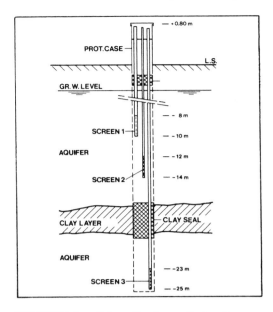

FIGURE 21-4. Schematic of a monitoring site.

drilled with a cable-tool percussion drilling system. Location of the well screen at about 10 m-ls is a compromise between the desire for early measurement of pollution and the avoidance of large numbers of monitoring wells and very frequent sampling due to large quality variations in the very shallow ground water. For these very shallow ground waters, a monitoring strategy has been developed based on nonpermanent monitoring wells (see the section "Soil-Quality Monitoring"). The well screens at 25 m-ls are always situated in the upper aquifer. The chosen length of well screens is related to the variability of ground water quality in combination with the possibility of short-circuiting flow when using long well screens.

Taking into account the very low ground-water velocities, the frequency of sampling is once per year. Special sampling instru-

TABLE 21-3. Relationship between Soil Type and Ground-Water Quality at 10 m Below the Land Surface (sites with Cl^- < 200 mg/L)

Parameter	Sand (natural)	Sand	Peat (lowlands)	River Clay	Marine Clay
K^+	1	3.8	1.8	2.6	9.6
NH_4^+	1	4.8	15.3	8	35
NO_3^-	1	4.0	0.1	0.8	0.1
SO_4^{2-}	1	1.9	0.4	1.5	1.1
Total P	1	3.0	6.0	6.0	17
Zn	1	1.6	0.4	0.4	0.4
As	1	1.7	1.0	4.0	1.0

Note: All concentrations are divided by the concentrations in natural sandy soils.

TABLE 21-4. Relationship between Land Use and Ground-Water Quality Underlying Sandy Soils at 10 m below the Land Surface (sites with Cl^- < 200 mg/L)

Parameter	Natural Areas	Pasture Land	Arable Land	Urban Areas
K^+	1	3.2	5.7	7.9
NO_3^-	1	1.1	12.7	4.9
Cl^-	1	2.5	3.1	4.3
NH_4^+	1	3.5	9.5	7
SO_4^{2-}	1	2.0	2.5	2.0
Total P	1	3	6	6
Zn	1	2.2	2.0	1.0
Ni	1	1.1	1.9	0.9
As	1	1.9	2.2	1.4

Note: All concentrations are divided by the concentrations in natural sandy soils.

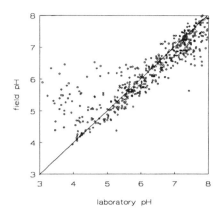

FIGURE 21-5. Relationship between in situ measurements and laboratory analyses of pH. (*1990 data from the National Ground-Water Quality Monitoring Network.*)

ments have been developed to collect samples under nitrogen gas directly before filtration. All samples are collected under standard operating procedures. The same holds for sample handling and analyses. Careful attention has been paid to quality-assurance procedures concerning all important steps in the course of monitoring, such as drilling, sampling, sample handling, analyses, data handling, and presentation.

Values of pH, electrical conductivity (EC), oxygen content, and temperature are determined in situ (in the borehole), because of differences between in situ measurements and laboratory analyses. For example, Figure 21-5 shows the relationship between in situ measurements and laboratory analyses of pH. Of special interest is the difference between the occasional very low pH values measured in the laboratory (pH 3–4) in combination with relatively high pH values (pH 5–6.5) based on in situ measurements. In all cases, these apparent differences concern anaerobic water samples with a high iron content. Acidification of these water samples occurs as a result of oxidation after sampling, resulting in a strong, unrealistic decrease of pH.

Both "basic" and "ad hoc" analytical programs are carried out. Ad hoc programs focus on specially selected groups of monitoring wells, depending on type and use of pollutants. Until now, the basic analytical program has been carried out every year on all monitoring wells. The basic analytical program for 1991 is presented in Table 21-5. Ad hoc programs have been undertaken for heavy metals (including rare earth elements), industrial organic contaminants, and several groups of pesticides. Comprehensive data are available on hundreds of contaminants.

Approximately one year after drilling the boreholes, all well screens were sampled for tritium analyses. Very low levels of tritium are present in natural ground water. Due to nuclear testing in the atmosphere carried out early in the 1960s, infiltrating rainwater contains a relatively high tritium concentration, which has caused a rise in the levels of tritium in ground water (see Figure 21-6 and Chapter 11). Due to these elevated tritium levels, it is possible to distinguish ground water which has infiltrated before and after the early 1960s.

During the establishment of the network, a specific databank program was developed. It was replaced in 1991 by a new databank manager in which the data of the other national monitoring networks of the National Institute will also be included, thus simplifying mutual use of data. Presentation of the data occurs at different levels: (1)

TABLE 21-5. Basic Analytical Program of the Ground-Water Quality Monitoring Network in 1991

Cl^-	NH_4^+	Fe	Ba	Ni	pH (field)	Benzene
NO_3^-	K^+	Mn	Sr	Cr	EC (field)	Xylenes
SO_4^{2-}	Na^+	Total P	Zn	Cu	O_2 (field)	Toluene
HCO_3^-	Ca^{2+}	DOC	Al	As	Temp. (field)	Ethylbenzene
	Mg^{2+}		Cd	Pb		

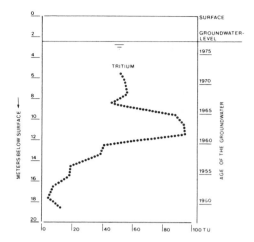

FIGURE 21-6. Vertical tritium profile in ground water at the center of a dune region.

individual data presentation, (2) annual data presentation at a provincial level, (3) annual description and diagnosis of groundwater quality in The Netherlands through statistical characterization and mapping of ground-water quality.

For instance, a clear relationship has been found between pH and trace-element concentrations in ground water. Positive correlations are generally obtained with negatively charged, oxidized trace elements, such as As, Mo, V, and W. Negative correlations with pH are found for many other trace elements, such as Cd, Cr, Cu, Ni, Pb, Zn, and Al. Aluminum, in particular, has a strong negative correlation with pH (see Figure 21-7). Because the lowest pH values in ground water are found below arable land and natural regions, the highest Al concentrations are also found in these areas, as illustrated in Figure 21-8, based on average values.

For some parameters, such as nitrate, trend analysis can be carried out with data from individual monitoring sites. As an example, Figure 21-9 gives an indication of changes in time of nitrate concentrations in ground water at monitoring site no. 16.

New techniques are being developed for extrapolating point information of the moni-

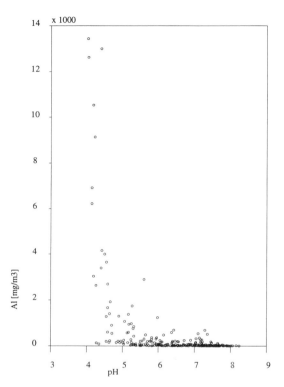

FIGURE 21-7. Relationship between pH and aluminum concentrations. (*1991 data from the National Ground-Water Quality Monitoring Network.*)

toring network in areal and depth dimensions (see Figure 21-10). Regarding these techniques, special geostatistical techniques in combination with transport modeling are important.

PROVINCIAL GROUND-WATER-QUALITY MONITORING NETWORKS

The Netherlands are divided into 12 provincial regions. At the regional level, responsibility for environmental quality is delegated to the provincial authorities, with the central government acting in a coordinating role. Although the National Ground-Water Quality Monitoring Network provides sufficient information at the national level, this is not the case for provincial policy. For this reason, since the completion of the

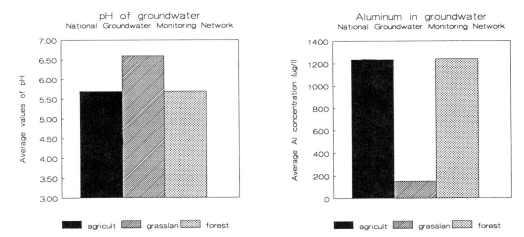

FIGURE 21-8. Average pH values and aluminum concentrations in ground water at 10 m-ls underlying arable land (agricult), intensively used grasslands (grassian), and natural (forest) regions in sandy soils. (*1991 data from the National Ground-Water Quality Monitoring Network.*)

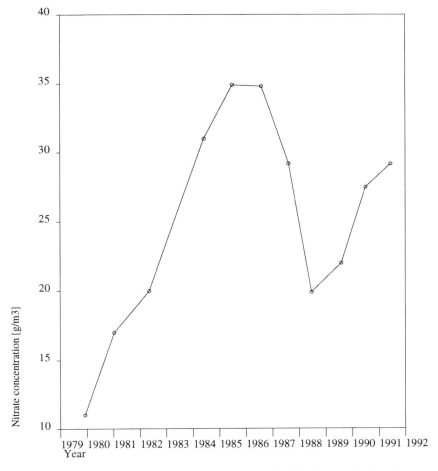

FIGURE 21-9. Nitrate concentration in ground water at 8–10 m-ls at monitoring site no. 16 (arable land, sandy soils).

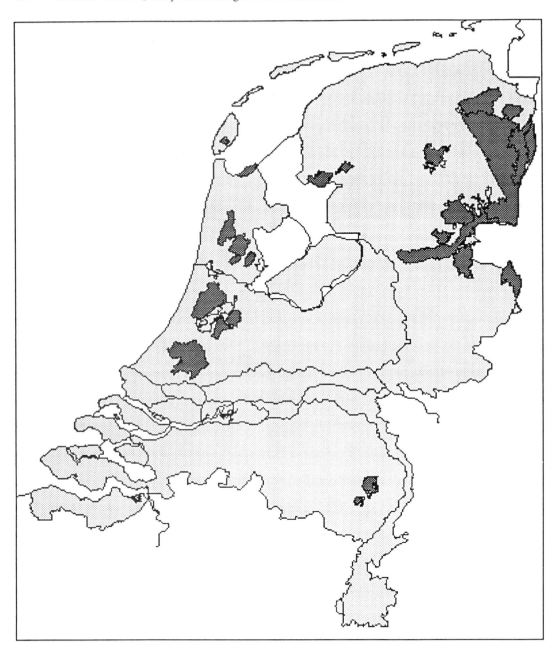

FIGURE 21-10. Spatial distribution of chromium in ground water. The area exceeding the reference value of Cr is estimated to be 37–46% of the total land surface in The Netherlands (95% confidence interval). The light gray area is not significantly different from these average national conditions; the dark gray areas have a significantly higher percentage (58–95%) of their area exceeding the reference value. (*After Boumens, in National Institute of Public Health and Environmental Protection 1992.*)

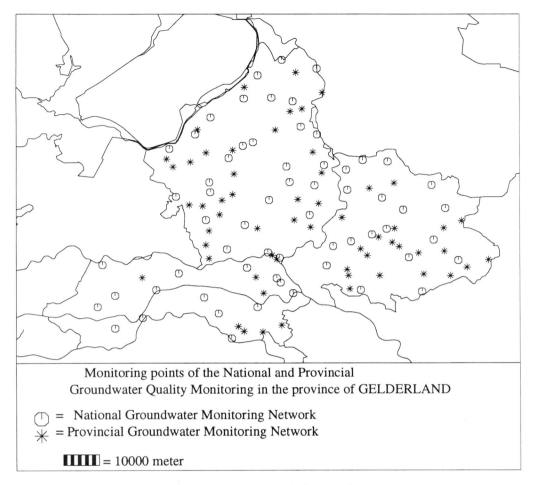

FIGURE 21-11. Monitoring sites of the national and provincial ground-water quality monitoring networks in the province of Gelderland.

national network in 1984, special attention has been paid to the development of ground-water-quality monitoring networks on a provincial level. This resulted in the establishment in 1989 of the first provincial network in the province of Gelderland (see Figure 21-11). By 1993, all 12 provincial networks will be in operation. All of these networks are developed in close cooperation with the National Institute of Public Health and Environmental Protection (RIVM), which acts as the coordinating institute for the National Ground-Water Quality Monitoring Network.

In general, the objectives of the provincial networks are identical to those of the national network. To limit the number of monitoring sites as much as possible, there has been a movement toward optimal mutual use of data among the monitoring networks. A strategy has been followed to ensure that the development and establishment of the provincial networks conform to the procedures of the national network. Much attention has been paid to the comparability and availability of data and to quality-assurance procedures. To facilitate the implementation of comparable operation procedures, the RIVM has developed standard operating procedures. Thus, it is possible to combine data from the national network with data from the provincial networks for interpretation in a specific province. Furthermore, at the national level, more reliable conclusions can be drawn on the basis of a larger amount of data.

The provincial monitoring network includes both sites from the national network and additional provincial sites (see Figure 21-11). The number of additional provincial sites varies between one and three times the number of national sites within that province, depending on the province under consideration. The provincial sites are sampled and analyzed under the responsibility of the provincial authorities and the basic sites under the responsibility of the RIVM. In most cases, sampling and analyses are coordinated by the RIVM. All data are gathered in the central data base system of the RIVM and are readily available for provincial authorities. In this central databank, other geographic data are also available, which can be used for data interpretation and presentation. In addition, all data of the national and provincial ground-water quality monitoring networks are gathered in *Monitor*, a user-friendly geographic information system (GIS) developed by order of the RIVM. The system can be installed on relatively simple personal computers and is intended as a tool for researchers and policymakers at the provincial and national levels.

SOIL-QUALITY AND SHALLOW-GROUND-WATER MONITORING

Since the second half of the 1980s, several national surveys have been carried out to obtain information on the quality of top soil and the upper few meters of (shallow) ground water. These surveys focused on the influence of several human activities on the quality of soils and shallow ground water, taking into account the vulnerability of soils and ground water to pollutants.

For "diffuse" sources of pollution, such as agricultural use of soils and atmospheric deposition, a large variability exists in the pollution load on the land surface, even within short distances. The same holds for the net recharge rate of rainwater, which infiltrates primarily on the lower terrains. Due to the large heterogeneity of soils, transport and sorption of pollutants can also differ widely within short distances. This generally results in a large variability in soil and shallow-ground-water quality. For this reason, analytical results of individual samples are not indicative of general environmental quality. This is illustrated in Figure 21-12. In this figure, individual nitrate analyses are presented from samples taken from shallow ground water at a cattle farm of approximately 10 ha in size, situated in a sandy region. Similar variations can be found even in smaller areas.

Due to the large variation in soil and shallow-ground-water quality, a description of environmental quality is only possible if a sufficient number of individual samples are taken. These samples may or may not be combined into mixed (composite) samples. The number of samples necessary depends on the variation in environmental quality and the desired reliability of its description. This can be estimated for some statistical

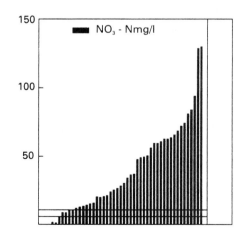

FIGURE 21-12. Cumulative frequency diagram of nitrate-N concentrations in samples of shallow ground water collected from a cattle farm in a sandy region.

characteristics. Data can be presented, for instance, as percentile values or mean values with confidence intervals.

Several national surveys have been carried out for particular diffuse sources of pollution. In agricultural regions, surveys have been done to determine the presence of heavy metals, pesticides, nitrate, phosphates, and several other major constituents in soil and shallow ground water. These constituents have been related to agricultural practices, including the load of pollutants on land surfaces. Surveys of shallow ground water in forest regions were concentrated on the presence of constituents such as heavy metals, aluminum, and nitrate.

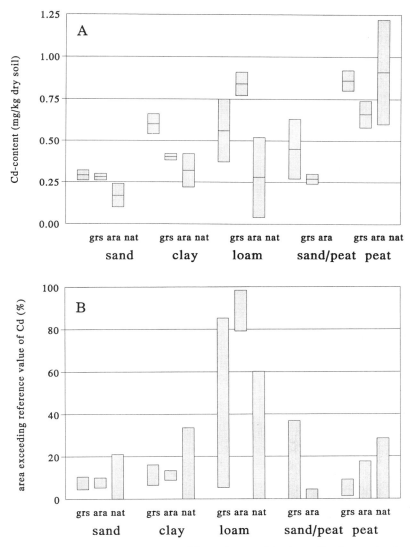

FIGURE 21-13. Cadmium in topsoil for various combinations of land use and soil type: (A) Mean values of cadmium content with plus and minus two standard deviations and (B) 95% confidence intervals for area with cadmium content exceeding the reference value (wide intervals in cases of few values or large variation). (*After van Drecht, in National Institute of Public Health and Environmental Protection 1992.*)

Another survey was carried out in the soil and ground water along highways. In addition, at the municipal level, a large amount of information is available on soil and ground-water quality in relation to numerous polluting point sources.

For illustration purposes, some results of the surveys are presented in Figures 21-13 and 21-14. In Figure 21-13A, information is presented on the 95% confidence intervals of the mean concentration of cadmium in relation to several combinations of land use and soil type. In Figure 21-13B, the precentage of Dutch soils exceeding Dutch reference levels for cadmium are presented. In Figure 21-14, areas whose percentage of soils exceeding Dutch reference levels differ significantly from the national average are indicated. The results presented are based on 2,600 samples, originating from several institutes.

In Figure 21-15, nitrate concentrations in the shallow ground water of The Netherlands are presented on the basis of calculations by a national nitrate leaching model. The model was calibrated and tested with thousands of ground-water samples and coupled with a detailed geographic information system on soil use, soil type, N load, and net recharge rate. Figure 21-16 provides information on maximum concentrations of pesticides found in shallow ground water in agricultural regions.

It must be concluded that in many cases soil and ground water are polluted to an unacceptable level due to human activities. For many other situations, calculations predict that the accumulation of several heavy metals will in time reach unacceptable levels, as can be illustrated with the help of Table 21-6. An indication is given in this table of the annual accumulation of some heavy metals in the upper 10 cm of the topsoil in sandy soils in use as pasture and arable land. This accumulation is presented in the form of the annual percentage added to the existing quantity of heavy metals in topsoil. These data are based on an extended soil survey and the annual average deposition of metals on the land surface from atmospheric deposition and agricultural use of soils.

Several regulations have been promulgated in The Netherlands to reduce human impacts on the environment. Trend monitoring is necessary to evaluate the effects of these regulations. To meet this goal in The Netherlands, special emphasis will be given to the implementation of an integrated National Soil-Quality Monitoring Network. Shallow ground water will also be sampled in this network. A strategy for establishment of a National Soil-Quality Monitoring Network has been developed on the basis of experiences with the national soil and ground-water surveys carried out in the past and some specific research. This network will consist of a few hundred monitoring fields, divided into several soil types and soil uses.

The monitoring fields are confined to areas where the pollutant loads are known. In agricultural areas, the soil-pollutant loads are generally known either for the entire farm or for a part of the farm. Several hundred soil samples and approximately 50 ground-water samples will be collected in every monitoring field to be combined into 3 or 4 composite samples. The exact number of monitoring fields necessary for a specific combination of soil use and type of soil, as well as the monitoring frequency and the number of individual samples and composite samples, can differ, depending on variability of analytical results and the objectives of the monitoring program.

Within the scope of this National Soil-Quality Monitoring Network, a long-term monitoring program was begun in 1992, concentrating on the effects on shallow ground water of national regulations designed to decrease the use of manure and artificial fertilizers on agricultural soils. The same holds true for a long-term monitoring program on soil and shallow ground water in forest regions to detect the effects of regulatory measures on the problem of acidification. Other monitoring programs, some of which are already in existence (for in-

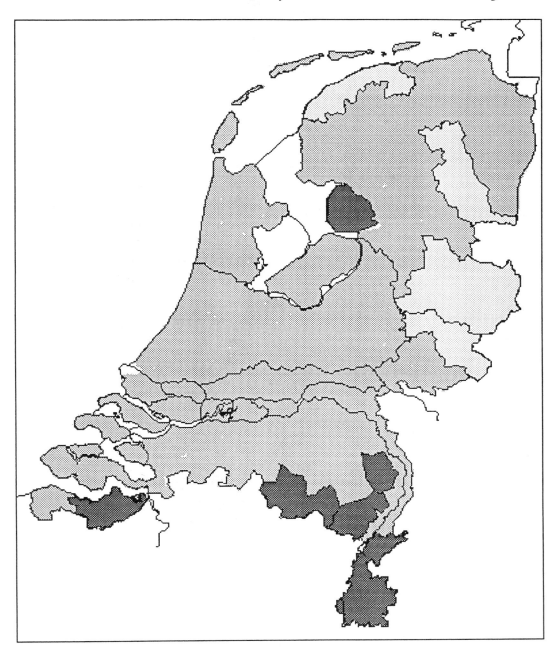

FIGURE 21-14. Spatial distribution of cadmium in topsoil. The area exceeding the reference value of Cd is estimated to be 8.5–11.1% of the total land surface in The Netherlands (95% confidence interval). The gray area is not significantly different from these average national conditions; the light gray areas have a significantly lower percentage; the dark gray areas have a significantly higher percentage. (*After van Drecht, in National Institute of Public Health and Environmental Protection 1992.*)

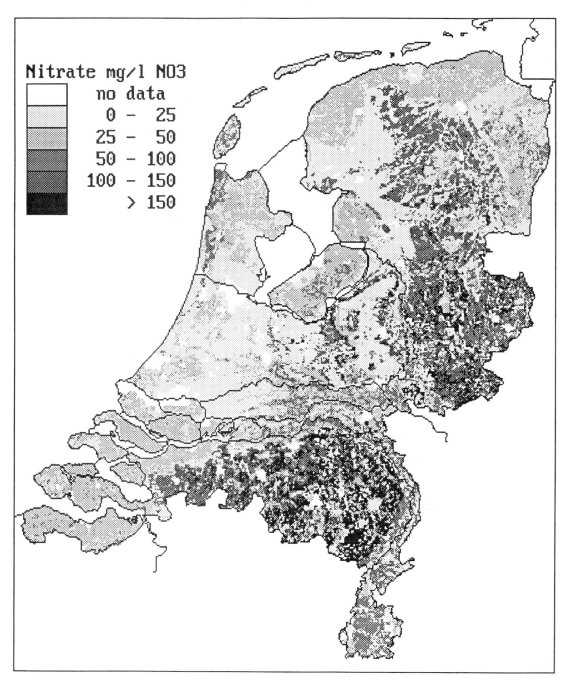

FIGURE 21-15. Nitrate concentration in shallow ground water based on 1989 data (mg/L nitrate-N). (*After van Drecht, in National Institute of Public Health and Environmental Protection 1991.*)

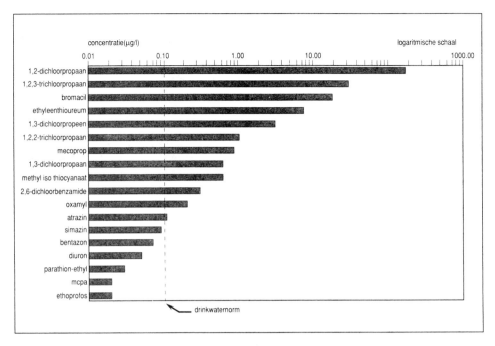

FIGURE 21-16. Maximum concentrations of pesticides found in shallow ground water during the period 1986-1990. (*After Hopman and Lagas, in National Institute of Public Health and Environmental Protection 1991.*)

TABLE 21-6. Mean Annual Load of Selected Heavy Metals on Some Sandy Soils

Parameter	Atmospheric Deposition (g ha^{-1} yr^{-1})	Agricultural Deposition (g ha^{-1} yr^{-1})		% Accumulation per Year on Topsoil (10 cm)	
		Pasture	Arable Land	Pasture	Arable Land
Cd	1.2	6.8	2.4	1.5	0
Cu	28	851	417	5	1.9
Pb	77	48	60	0.3	0.5
Zn	200	1930	780	5.5	1.0

stance, directed to pesticides), will also be added to the national network.

The national network is primarily meant to be used for monitoring, although national mapping will also be possible, based on the relationships found between quality and the factors determining that quality. Several pilot studies have been carried out to evaluate the possibilities of including biological monitoring in the national network.

In addition to the national network, it is possible to describe environmental quality of soil and ground water with the help of one-time national surveys. These surveys consist of an intensive sampling program for mapping purposes.

MONITORING OF DRINKING-WATER QUALITY

In The Netherlands, approximately 70% of the water used for public water supplies comes from ground water. The total quantity of ground water used for this purpose is about 1.0 billion m³/yr. This quantity is abstracted by approximately 240 ground-water pumping stations. Approximately one third of these pumping stations are abstracting their ground water from aquifers without overlying (clay) layers, which are vulnerable to ground-water pollution. In accordance with the Drinking Water Act, the Dutch water-supply organizations are obligated to deliver water which satisfies the Dutch drinking-water-quality standards. To check compliance, extensive monitoring of drinking water takes place before and after purification. The frequencies of sampling and analysis depend on the type of water (before or after purification) and the types of parameters to be analyzed. The sampling frequency is regulated by law and varies from daily to yearly. In total, 71 parameters are required to be analyzed, as listed in the Drinking Water Act. This list includes physical parameters, major elements, heavy metals, pesticides, trace organic compounds, and biological parameters.

Part of the compiled data (such as mean values, maxima, and minima) are included in the RIVM databank to provide the Dutch government with data for the formulation of national policy. In addition to the samples taken by the water companies, all ground-water pumping stations are sampled on a yearly basis by the RIVM. Attention also will be given to parameters not listed in the Drinking Water Act which can be of possible relevance.

Data from 1968 to the present are available in the RIVM databank. Based on these data, a gradual deterioration of ground-water quality in time has been observed for several parameters, particularly for pumping stations located in the most vulnerable ground-water areas. These parameters include major elements such as calcium (hardness), sulfate, and nitrate (see Figure 21-17), and some trace constituents such as aluminum and pesticides.

Often, for early warning purposes, information is also needed concerning the ground-water quality in the recharge area supplying the ground-water pumping station. For this reason, several ground-water pumping stations are surrounded with a monitoring network in the pumping station recharge area. In general, monitoring sites are installed on the basis of available information on the hydrogeologic and hydro-

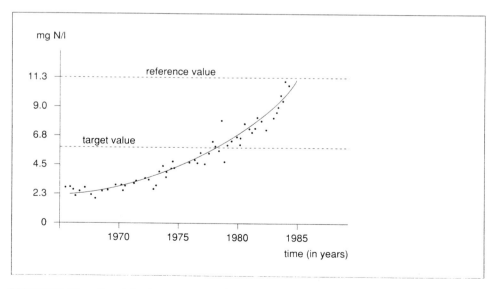

FIGURE 21-17. Trends in nitrate concentrations at the ground-water pumping station of Montferland.

chemical situation, as well as an inventory of pollution sources in the recharge area. Location of monitoring sites and well screens can be determined with the aid of ground-water models. With help of the monitoring system, in combination with the use of ground-water models, it is possible to predict water-quality changes at the ground-water pumping station.

At present, the water companies are developing a joint monitoring strategy including joint standard operating procedures for design and operation of monitoring networks around ground-water pumping stations. This strategy will incorporate the relationship of the early warning system monitoring sites around pumping stations to the standard monitoring sites of the national and provincial ground-water-quality monitoring networks.

INTERPRETATION TOOLS AND PRESENTATION TECHNIQUES

Due to the large variation in soil and ground-water quality, data from a single monitoring point can generally be considered as no information at all with regard to describing environmental quality. For a correct interpretation, sufficient point data from a sufficient number of sites should be available. These data can be interpreted and extrapolated with the help of statistical techniques and knowledge about processes in soil and ground water. Information on the pollution load, net recharge rate, and soil and ground-water conditions are also important. If a quantified relationship between environmental quality and pollution load is found, the relationship can be illustrated by means of thematic maps of soil and ground-water quality. (Geo)statistical techniques, in combination with relatively simple diagnostic or process models and detailed geographic information from a GIS, can be used to generate these thematic maps. In The Netherlands, for instance, GIS data bases are available on (1) pollution load on the land surface from deposition from the air, agricultural practices, and local pollution sources, (2) soil use, (3) vegetation, (4) net recharge rate, (5) soil characteristics, (6) geologic characteristics, (7) hydrologic characteristics, (8) topography, and so on.

INTEGRATED MONITORING

Originally, the objectives of the national environmental monitoring networks in The Netherlands were specific for their own environmental compartment. In reality, due to the mutual relationship between the environmental compartments, data from other environmental compartments are necessary for a correct interpretation. In forest regions, for instance, atmospheric deposition is relevant to soil quality. Ground-water quality will be affected by surface-water quality in regions with bank infiltration, while ground water will influence surface-water quality along draining water courses. The interrelationship of environmental compartments requires coordination and integration of the national networks of the different environmental compartments. This integration and coordination can be directed toward a fine-tuning of objectives, locations, types of pollutants to be monitored, frequency of sampling, and use of data systems and interpretative tools. At present, the national environmental monitoring networks of the RIVM are united under the umbrella of the National Integrated Environmental Monitoring Network, which has as its general objective "the integrated description of environmental quality in The Netherlands including trend development in relation to relevant factors."

At present, the monitoring sites of the national networks on radiation and on rainwater quality are integrated in the National Network on Air Quality. Furthermore, a limited number of the monitoring sites of the national networks on soil and ground-water quality are located close to monitoring sites of the national networks on air and rainwater quality. Taking into account the nature and location of the pollutant emissions and the behavior of these pollutants in the environment, evaluations on a regular basis will be made to determine which of the pollutants should be analyzed in a given monitoring network. Furthermore, increased attention will be given to the common use of data systems and interpretative tools. On a national scale, the coordination of monitoring activities takes place under the Coordination Committee for the Monitoring of Radioactivity and Xenobiotic Substances in the Environment (CCRX). This committee is a coordinating group of six Dutch ministries and 13 subordinate institutes with regard to the monitoring activities of these ministries and institutes. On a yearly basis, a National Monitoring Program will be discussed, evaluated, and, if necessary, adjusted. Within the framework of the National Monitoring Program, physical-chemical and biological monitoring will be included along with their mutual relationship.

CONCLUDING REMARKS

High population density, combined with intensive industrial and agricultural activities, clearly influence the environmental quality in The Netherlands, sometimes to unacceptable levels. On the national and provincial levels, regulations have been made to restrict the emissions of pollutants to the environment to more acceptable levels. Several national networks exist in The Netherlands for collecting information concerning environmental quality in relation to emissions, included among them are specific networks for soil and ground-water quality.

Due to the large variation in soil and ground-water quality even within short distances, statistical techniques are needed for the design of the networks and for interpretation of data from these networks. The statistical techniques can be used in combination with geographic information systems and diagnostic or process models for data interpretation and presentation. For optimal data usage, coordination and partial integration of networks for each environmental compartment on several levels is necessary (for example, between national and provincial networks). The same holds for the national networks for different environmental compartments and those for physical-chemical and biological monitoring. Coordination and integration of these networks should be stressed.

References

Breeuwsma, A., and W. van Duijvenbooden. 1992. Examples of soil environments in The Netherlands (in Dutch). In *Handbook on Soil Protection*. Alphen: Samson Tjeenk Willink.

Duijvenbooden, W. van. 1981. Groundwater quality in The Netherlands—Collection and interpretation of data. *The Science of the Total Environment* 21:221–32.

Duijvenbooden, W. van. 1985. Networks for groundwater quality. In *Design Aspects of Hydrological Networks*, ed. J. W. van der Made, pp. 112–21. The Hague, The Netherlands: TNO Committee on Hydrological Research, Proceedings and Information No. 35.

Duijvenbooden, W. van. 1987. Groundwater quality monitoring networks: Design and results. In *Vulnerability of Soil and Groundwater to Pollutants*, eds. W. van Duijvenbooden and H. G. van Waegeningh, pp. 179–91. The Hague, The Netherlands: TNO Committee on Hydrogeological Research, Proceedings and Information No. 38.

Engelen, G. B. 1980. Hydrological division of The Netherlands (in Dutch). In *Water Quality in Dutch Aquifer Systems*, ed. J. C. Hooghart. The Hague, The Netherlands: TNO Committee on Hydrological Research, Rapporten en Nota's No. 5.

Langeweg, F. 1989. *Concern for Tomorrow, National Environment Survey 1985–2010*. Bilthoven, The Netherlands. National Institute for Public Health and Environmental Protection.

National Institute of Public Health and Environmental Protection. 1991. *National Environmental Survey 2, 1990–2010* (in Dutch). Alphen: Samson Tjeenk Willink.

National Institute of Public Health and Environmental Protection. 1992. *Diagnosis of the Environment 1992, Parts I and III* (in Dutch). Reports 724801009 and 724801003.

Zagwijn, W. H. 1975. The paleographical development of The Netherlands during the past 3 million years (in Dutch). K.N.A.G. *Geografisch Tijdschrift IX 3*.

22

Multiscale Approach to Regional Ground-Water-Quality Assessment: Selenium in the San Joaquin Valley, California

Neil M. Dubrovsky, Steven J. Deverel, and Robert J. Gilliom

INTRODUCTION

Assessments of ground-water quality at regional scales must rely on inferences about large areas and volumes from relatively few direct measurements. Such inferences require a careful balance between spatially extensive measurements to describe the distribution of water quality in large areas and more intensive measurements in smaller areas to achieve an understanding of important causes and processes. Confidence in conclusions about a large regional system based on sparse measurements is gained by linking the distribution of water-quality conditions to a reasonable understanding of causes and governing processes. Conversely, the regional significance of conclusions about governing causes and processes, which are reached through detailed study at a few sites, can be understood only by evaluating the regional distribution of the conditions represented by the sites. Regional ground-water-quality assessments are best accomplished through a mixture of complementary investigations at a wide range of spatial and temporal scales and intensity that provide multiple lines of evidence to support interpretation. The purpose of this chapter is to illustrate this approach to regional ground-water-quality assessment by using studies of selenium in the San Joaquin Valley, California, as an example (Gilliom et al. 1989).

The San Joaquin Valley is a flat structural basin bounded by the Sierra Nevada on the east, the Coast Ranges to the west, the Tehachapi Mountains to the south, and the Sacramento–San Joaquin Delta to the north (Figure 22-1). Almost the entire valley floor is agricultural land, about 70% of which is irrigated. The combination of abundant water and a long growing season results in an extremely productive agricultural economy that generated about $6.8 billion in total agricultural production in 1987.

Background on Selenium

Selenium contamination of ground water in the San Joaquin Valley has attracted national attention since 1983, when selenium in water from tile-drainage systems was found to have toxic effects on waterfowl at Kesterson Reservoir. Relatively small amounts of selenium can be toxic. The Fed-

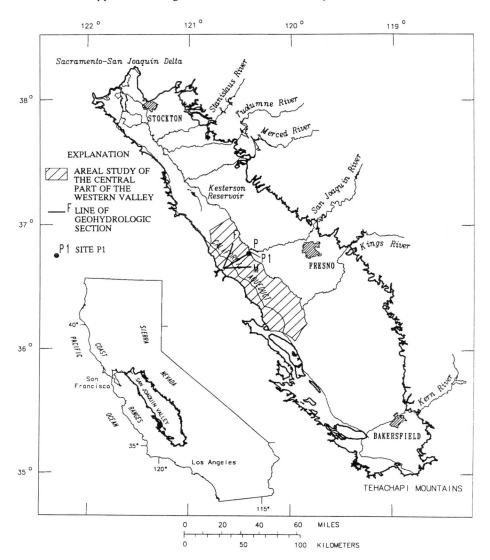

FIGURE 22-1. Location of San Joaquin Valley and locations of different scale studies.

eral standard for selenium in drinking water is 10 µg/L and the four-day average criterion for aquatic life is 5 µg/L. Concentrations of selenium in tile drainwater from parts of the central western valley were commonly greater than 100 µg/L (Gilliom et al. 1989).

Selenium can exist in four valence states: $-2, 0, +4$, and $+6$. The $+6$ and $+4$ valences occur as the oxyanions selenate (SeO_4^{2-}) and selenite (SeO_3^{2-}) under alkaline, oxidizing conditions. Under more reduced conditions, selenium can exist as elemental selenium (zero valence) and selenide (Se^{2-}). The solubility of selenate minerals generally is high (Elrashadi et al. 1987), and there are no apparent solubility constraints on selenate in shallow ground water in the western San Joaquin Valley, even in ground water saturated with respect to sulfate minerals (Deverel and Gallanthine 1989). Consequently, selenate tends to behave conservatively in oxidizing ground water. The mobility of selenite in ground water is severely constrained by adsorption onto a variety of mineral surfaces (Balistrieri and Chao 1987; Neal et al. 1987; Goldberg and Glanbig 1988). The solubilities of the reduced forms of selenium (elemental selenium and selenide) are extremely low (Elrashadi et al. 1987). Field and laboratory studies of selenium contamination at Kesterson Reservoir demonstrate selenium removal by reduction of selenate to less mobile forms (Lawrence Berkeley Laboratory 1987; White et al. 1988; Weres, Jaouni, and Tsao 1989). An understanding of these properties of selenium is a fundamental requirement for interpreting results at all scales investigated.

Overview of Study Approach

Studies were done at four primary combinations of scale and sampling strategy (Figure 22-1):

1. Valleywide sampling of existing wells that tap regional aquifers (32,000 km^2) provides a broad, but imprecise, assessment of the relation between selenium concentrations in presently used ground water and regional-scale geologic and geochemical factors.
2. Intensive sampling of shallow ground water in the central part of the western valley (crosshatched area shown in Figure 22-1) provides a detailed assessment of the areal distribution of selenium in shallow ground water where selenium contamination and agricultural drainage problems are most severe and where the influence of areally distributed factors and processes that affect shallow ground water, such as soil conditions and evapotranspiration, can be assessed.
3. Sampling of selenium distribution with depth along three transects (F, P, and M in Figure 22-1) across the central part of the western valley (each transect is about 30 km long, with wells completed at depths as great as 200 m) serves to characterize the lateral and depth distribution of selenium concentrations in the regional aquifers in relation to the areal distribution in shallow ground water, historical and present-day ground-water flow, and geochemical conditions.
4. Specialized studies at selected sites were done to address specific questions about hydrologic and geochemical processes controlling selenium concentrations and transport. The example reported in this chapter is a study of selenium mobility in a reducing environment in Sierra Nevada sediments.

Results from each scale of investigation are summarized below to illustrate their roles in providing multiple lines of supporting evidence for the assessment. Each component has distinctive strengths and weaknesses.

VALLEYWIDE SAMPLING

The distribution of selenium concentrations in regional aquifers was assessed by sampling 272 existing production wells. The goal of the sampling design was to sample two wells in each 36-mi^2 (93-km^2) township of the valley, with one well in the shallow zone of the aquifer system and one in the deep zone. The variable distribution of existing wells, both areally and with depth, resulted in an uneven distribution of sampled wells despite the uniform sampling strategy. Preliminary data analysis showed no significant differences in chemistry between upper- and lower-zone samples at the valleywide scale, with the exception of tritium. Thus, data for all well depths were combined for analyses of areal distribution of selenium concentrations.

The areal distribution of selenium was evaluated in relation to two primary surficial features: soil-selenium concentrations and depositional environment. The soil-selenium concentrations reflect the geologic sources of selenium. Evaluation of data from approximately 300 soil samples collected valleywide (Tidball, Grundy, and Sawatzky 1986) shows that almost all the soils having selenium concentrations greater than the median value of 0.13 mg/kg are in alluvial sediments derived from the Coast Ranges. In particular, the location of areas with the highest soil-selenium concentrations are associated with exposures of marine sedimentary formations in the adjacent Coast Ranges (Gilliom 1989; Presser et al. 1990). For analysis of valley-wide patterns in ground water, soil-selenium concentrations were categorized as being either greater or less than the valley median.

The two primary depositional environments in the valley, alluvial fans and basin deposits, generally have different geochemical characteristics. The alluvial-fan deposits, including areas mapped as sand dunes, are oxidized and ground water within these deposits also is oxic. Ground water in basin deposits, which include lacustrine, flood-basin, and river-channel deposits, is typically reducing because deposition occurred under inundated conditions and probably with more organic matter. This geochemical difference is important to evaluate because selenium is less mobile under reducing conditions.

Origin of Water Sampled

At all scales of investigation of ground-water quality, it is vital to understand the origin and history of water sampled. Most ground-water recharge in the San Joaquin Valley is now due to infiltration of irrigation water, much of it from surface-water sources. Prior to 1952, precipitation contained less than 5 TU (tritium units; see Chapter 11). Owing to radioactive decay, ground water from precipitation before 1952 now has less than 0.5 TU. Ground water derived from precipitation recharged since 1952, including canal water used as irrigation since 1968, commonly has a tritium concentration exceeding 10 TU. Ground water with a tritium concentration of less than 1.6 TU (twice the detection limit) either was recharged prior to 1952 or may have originated as post-1952 irrigation water from deep wells. This large contrast in tritium concentration allows the comparison of older ground water, much of which was recharged prior to agricultural development, to young water recharged since 1952 and derived from irrigation.

Most of the samples collected from the 181 production wells for which tritium was analyzed consist at least partly of water from recharge of irrigation water from surface-water sources since 1952. Median tritium concentrations were 11.3 and 9.9 TU in the alluvial-fan areas with high and low soil-selenium concentrations, respectively, and 2.4 TU in the basin area. Throughout the valley, tritium concentrations were significantly ($\alpha = 0.05$) higher in wells completed in the upper zone than in the lower zone of the regional aquifer. This vertical gradient in tritium results from the dominance of recent recharge in wells completed at the shallower depths; however, no patterns in selenium concentrations with depth are evident in the same wells.

Areal Distribution of Selenium

Selenium concentrations in ground water from the 272 wells sampled ranged from less than 1 to 120 µg/L, with a median of less than 1 µg/L (Figure 22-2). Eleven samples had a selenium concentration greater than the drinking-water standard of 10 µg/L. The distribution of selenium, dissolved solids, nitrate, and tritium concentrations among the three areas of the valley is summarized in Figure 22-3.

The highest selenium concentrations in ground-water samples were in alluvial-fan areas with soil-selenium concentrations greater than the valley median. Selenium exceeded 1 µg/L in 53% of the wells sampled in fan areas with high soil selenium; in contrast, selenium exceeded 1 µg/L in only 13% of the wells sampled in fan areas with low soil selenium. Initial data analysis showed that selenium concentrations in groundwater in the area of basin deposits are not correlated with soil-selenium concentrations; therefore, data from the basin deposits were not divided on the basis of soil-selenium

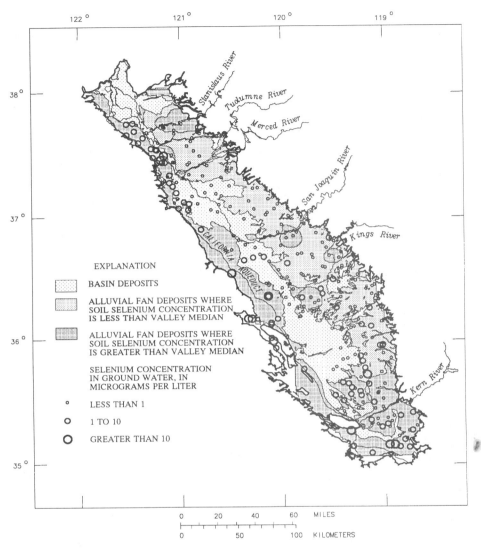

FIGURE 22-2. Selenium concentrations for water samples from existing wells in the regional-aquifer system.

concentrations. Selenium exceeded 1 µg/L in only 6% of the wells sampled in basin areas. Generally, the fan areas with high soil-selenium concentrations are composed of sediments from the Coast Ranges. The fan areas with low soil-selenium concentrations generally are composed of granitic sediments from the Sierra Nevada. Thus, there is an apparent relation between selenium in ground water and its geologic sources and concentration in soil.

Relation of Selenium Concentrations to Salinity and Redox Conditions

Selenium in ground water also is affected by processes such as leaching and evaporative concentration that affect dissolved solids and by redox conditions that affect the mobility of selenium. Soluble forms of selenium tend to be correlated with other dissolved solids, and the areal distribution of dissolved solids generally is similar to that of selenium.

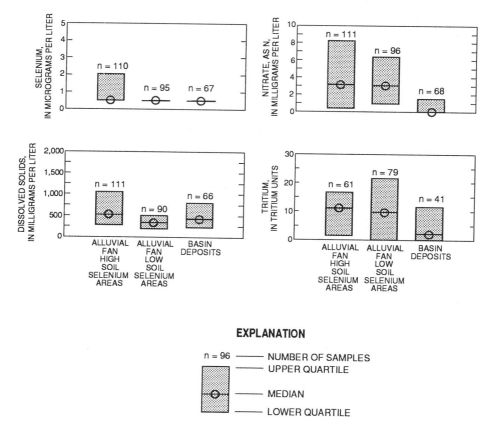

FIGURE 22-3. Selenium, dissolved solids, nitrate, and tritium data for water samples from existing wells in the regional-aquifer system.

Dissolved solids are significantly ($\alpha = 0.05$) higher in fan areas with high soil selenium (dissolved-solids median of 540 mg/L) than the other two areas and significantly higher in basin areas (dissolved-solids median of 405 mg/L) than in fan areas with lower soil-selenium concentrations (dissolved-solids median of 316 mg/L; Figure 22-3).

The uniformly low selenium concentrations in ground-water samples from the basin area, despite moderately high dissolved solids, indicate the possibility that selenium mobility may be limited by reducing conditions in the basin. The presence of reducing conditions is corroborated by the distribution of nitrate concentrations (Figure 22-3), which shows significantly ($\alpha = 0.05$) lower concentrations in the basin area (nitrate median of 0.14 mg/L as N) than in the fan areas (nitrate medians of 3.2 and 3.0 mg/L as N).

Significance

The distribution of tritium concentrations indicates a general dominance of recent recharge from irrigation in the upper zone of the regional aquifer, compared with the lower zone, but there was no similarly widespread pattern in selenium distribution with depth using existing wells at the valleywide scale. Selenium concentrations in regional aquifers generally are low and seem most affected by the regional distributions of salinity, redox conditions, and soil-selenium concentrations, the latter of which reflect the distribution of geologic sources. Generally, selenium concentrations in ground water were highest in areas of the valley where soil selenium and ground-water salinity are highest and where ground water is oxic.

The valleywide sampling of existing wells yields useful broad conclusions about the quality of water presently being used and the spatial distribution and governing factors for selenium concentrations in regional aquifers, but the scale is much too large in relation to the quantity of data available for reliable application to resource management decisions. There is no strong indication from the regional data alone of the widespread high selenium concentrations in shallow ground water of the western valley. One reason is the sparse spatial coverage in this area that results from avoidance of poor-quality ground water during selection of well locations in the past. Another is the inherent biases of existing production wells, which generally are completed at depths far below the water table and frequently withdraw water from a broad depth interval. A more detailed understanding of the distribution of selenium and the factors and processes that affect its distribution and mobility in ground water and soil requires a systematic analysis of hydrology, geochemistry, and agricultural influences within a smaller area of the valley. These more detailed studies focus on the central part of the western valley (Figure 22-1).

AREAL DISTRIBUTION OF SELENIUM IN SHALLOW GROUND WATER

The areal distribution of selenium in shallow ground water is a key limitation on alternatives for drainage management because tile-drainage systems withdraw shallow ground water to control the water table. Shallow ground water will herein be defined as water within the upper 6 m of the saturated zone in areas where the water table is less than 6 m below the land surface. This definition restricts the assessment to areas where drainage problems presently occur or may soon develop, and to the upper part of the flow system that is dominated by recent irrigation recharge. This shallow ground water in the central part of the western valley was not sampled by the existing production wells used to assess the valleywide distribution of selenium.

Reconnaissance studies of the regional distribution of selenium in shallow ground water in the central part of the western valley (Deverel et al. 1984; Deverel and Millard 1988) showed that selenium concentrations in shallow ground water are high only in Coast Ranges alluvial sediments and that samples of shallow ground water from Sierra Nevada sediments contained selenium concentrations less than 1 µg/L. These conclusions are in general agreement with findings from the valleywide assessment. The focus of this analysis is on evaluating the areal distribution of selenium in shallow ground water in Coast Ranges alluvial sediments.

Samples of shallow ground water were collected from 118 observation wells distributed throughout the central part of the western valley (Figure 22-4). The wells are 6–12 m deep and perforated over the entire depth of the well. The technique of kriging (see Chapter 4) was used to estimate the geographic distribution of selenium and dissolved solids in ground water from concentrations measured in these wells (Deverel and Gallanthine 1989).

Concentrations of selenium in shallow ground water generally are less than 20 µg/L in the middle-fan areas of the alluvial fans deposited by Cantua, Little Panoche, Los Gatos, and Panoche creeks (Figure 22-5). In the lower-fan areas, particularly at the northern and southern margins of the Panoche Creek fan, selenium concentrations in ground water are higher and are as much as several hundred micrograms per liter. Patterns of distribution are more difficult to generalize for the smaller fans of ephemeral streams (shown as shaded areas in Figures 22-4 to 22-7), largely because only a small area of

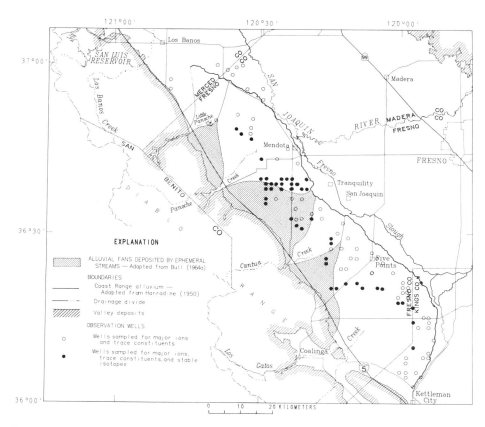

FIGURE 22-4. Areal distribution of wells sampled for chemical analyses in the central part of the western San Joaquin Valley. (*After Deverel and Gallanthine 1989.*)

these fans is underlain by shallow ground water. An exception is the lowest altitudes of the ephemeral-stream fans between the Panoche Creek and Cantua Creek fans, where some of the highest selenium concentrations are present.

Relation to Salinity

The distributions of dissolved solids and selenium in shallow ground water are similar (compare Figures 22-5 and 22-6). Dissolved solids in ground water generally are less than 3,000 mg/L in the middle-fan area of the Panoche Creek alluvial fan. Dissolved solids commonly are greater than 5,000 mg/L in the lower-fan area; concentrations are greater than 10,000 mg/L in some areas along the fan margins. Dissolved solids are similarly distributed in the Cantua Creek and Los Gatos Creek alluvial fans. Similar to selenium, concentrations of dissolved solids in areas of ephemeral-stream fans are higher than in corresponding topographic locations on adjacent fans of the major streams.

The distribution of soil salinity before most of the study area was irrigated followed the same relative distribution as dissolved solids and selenium in present-day shallow ground water. Figure 22-7 shows the distribution of subsurface soil salinity in the mid-1940s (Harradine 1950) and the area of artesian wells defined by Mendenhall, Dole, and Stabler (1916), which approximates the natural ground-water discharge area in the valley trough. Soil salinity was highest along the margins of the alluvial fans, where ground-water discharge by evapo-

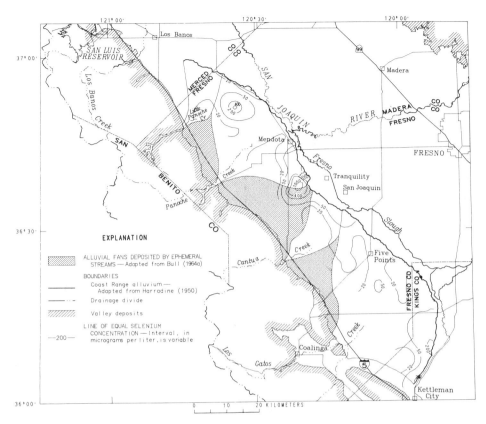

FIGURE 22-5. Selenium concentrations in ground water, 1986. (*After Deverel and Gallanthine 1989.*)

transpiration brought solutes to the surface, probably over several thousand years.

The similar distributions of historical soil salinity and shallow-ground-water salinity today are evidence that dissolved solids and selenium in shallow ground water were leached from the saline soils by irrigation water. In some places where the distribution is not similar, soils were irrigated before the 1940s soil study, and in other places shallow sampling wells tap only ground water from recent recharge through soils that have already been leached (Deverel and Gallanthine 1989). Generally, the first few decades of irrigation probably leached most of the readily soluble forms of soil selenium and other salts into the shallow ground water. Since irrigation began, there has been relatively little horizontal movement of shallow ground water on a regional scale. Thus, the present-day areal distribution of selenium in ground water generally follows the distribution of natural soil salinity in the Coast Ranges alluvial sediments, but the highest concentrations at particular places are at varying depths, depending on historical irrigation and vertical hydraulic gradients.

Relation to Soil Selenium

Total soil selenium (Figure 22-8) and selenium in shallow ground water (Figure 22-5) are poorly correlated in the areas of the Coast Ranges alluvial fans that are underlain by shallow ground water. The lack of correlation partly is because present-day total selenium in soil does not accurately represent the predevelopment distribution of soluble

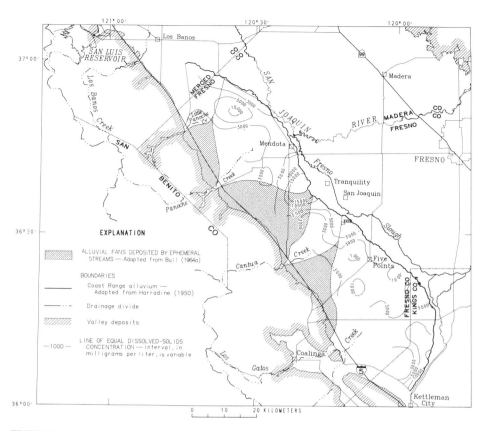

FIGURE 22-6. Dissolved-solids concentrations in shallow ground water, 1986. (*After Deverel and Gallanthine 1989.*)

forms of selenium. Soluble forms of selenium, which directly affect ground-water concentrations, are now only a small fraction of total soil selenium because the most soluble forms already have been leached from present-day soils. Amundson, Doner, and Waldron (1986) examined changes in the concentrations of readily mobile selenium in soil samples collected from the Panoche Creek alluvial fan in 1946 and 1985. Extractable selenium concentrations in soil samples collected in 1946 were higher in areas that were irrigated after 1940 than in areas that were irrigated before 1940, and soil samples collected from the same locations in 1985 showed substantial decreases in extractable selenium. Consistent with this relation, Fujii, Deverel, and Hatfield (1988) found that most selenium in presently irrigated soils is in forms that are resistant to leaching.

Evaporative Concentration

In areas where the water table is shallow, particularly at depths less than 1.5 m below land surface, evaporative concentration of dissolved solids in ground water can increase salinity and selenium concentrations far above the levels resulting from leaching of soil salts by irrigation (Deverel and Fujii 1988). Under natural conditions, when little or no recharge of ground water occurred through the soils in the lower fan, ground-water discharge by evapotranspiration resulted in accumulation of salts in the soil

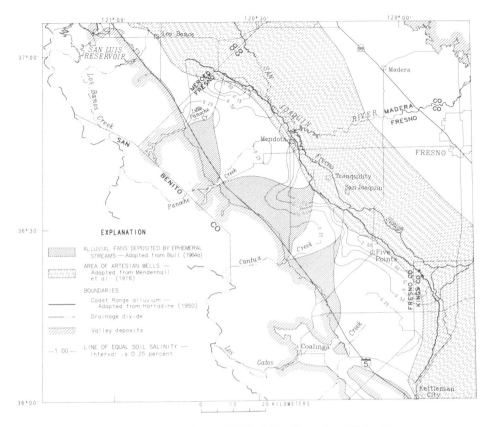

FIGURE 22-7. Subsurface soil salinity, mid-1940s. (*After Deverel and Gallanthine 1989.*)

(Figure 22-7). Under irrigated conditions, loss of water by evapotranspiration tends to concentrate salts in the ground water rather than in the soil because soils regularly are flushed by downward percolating irrigation water, and net ground-water movement generally is downward.

Figure 22-9 shows the composition of stable isotopes in 46 samples of shallow ground water collected from the central part of the western valley. Samples collected at the lowest land-surface altitudes, where the water table is shallowest, are most enriched in the heavy isotopes and follow a trend line indicating evaporative concentration. At higher land-surface altitudes where selenium concentrations are lower, the shallow ground water is less enriched in $\delta^{18}O$ than at the lower altitudes, indicating little or no evaporation.

Significance

The much greater spatial density of observations made possible by focusing on shallow ground water in a limited area of the valley greatly improves estimates of the areal distribution of selenium in shallow ground water and facilitates more detailed analysis of important ancillary information such as historical soil salinity, geomorphology of the alluvial fans, and isotope geochemistry. The combination of historical and new data for the central part of the western valley, evaluated from an interdisciplinary point of view,

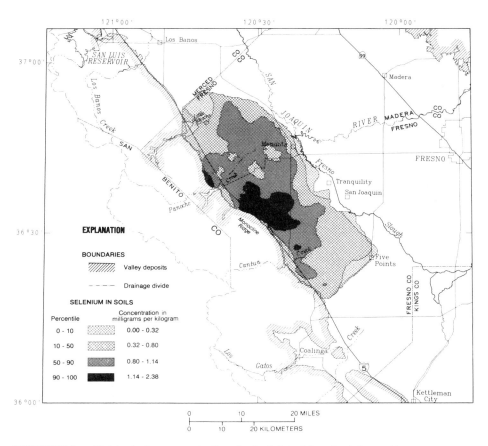

FIGURE 22-8. Total selenium concentrations for the 1.7- to 1.8-m depth interval in soil. (*After Tidball, Grundy, and Sawatzky 1986.*)

enables a detailed mapping of selenium distribution in shallow ground water and development of a conceptual model of the primary factors that govern concentrations: leaching of natural soil salts by irrigation and evaporative concentration.

The knowledge achieved by narrowing the focus of sampling to shallow ground water still is limited with respect to selenium distribution and mobility with depth or in relation to ground-water flow. These issues can only be addressed by studies that require detailed sampling of selenium, geochemical conditions, isotopic composition, and hydraulic potential at multiple depths in the aquifer system. The great cost of installing suitable sampling wells limits such efforts to selected sites that represent a range of conditions evident in shallow ground water. Findings from sampling of shallow ground water help target the sites for these more detailed studies.

DEPTH DISTRIBUTION OF SELENIUM IN GROUND WATER

Information on the distribution of selenium with depth in relation to ground-water flow and geochemical conditions is key to understanding the extent and historical movement of high-selenium shallow ground water, the distinction between natural and present-day conditions, and implications for management of drainage systems and withdrawals

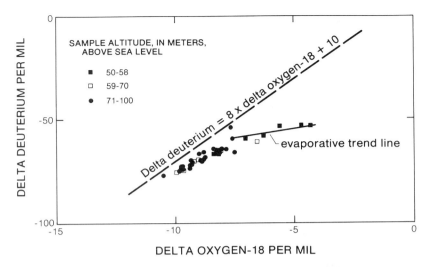

FIGURE 22-9. Comparison of delta deuterium and delta oxygen-18 for groundwater samples collected at various altitudes. (*After Deverel and Gallanthine 1989.*)

of ground water from wells. Observation wells were completed at multiple depths at several cluster sites along three geohydrologic sections through the ground-water-flow system of the Panoche Creek alluvial fan (Figure 22-1) (Dubrovsky and Deverel 1989). The cluster sites along each section are approximately aligned with the direction of natural ground-water flow. The observation wells allow high-quality measurements of hydraulic potential and geochemistry at selected depths.

Age and Origin

The history of water-table changes along the geohydrologic sections reflects the effects of increased recharge resulting from infiltration of irrigation water and changes in the rate of ground-water pumping. In 1952, the water table was within 6 m of the land surface only near the eastern limit of the P section (Figure 22-10). The water table rose substantially between 1952 and 1984 in most of the study area, resulting in extensive areas of shallow ground water between sites P1 and P3, sites M1 and M2, and sites F1 and FYR. The decline in the water table in the western areas of the sections mainly occurred between 1952 and 1967. Water levels have been increasing since surface-water supplies replaced ground water for irrigation in 1967 in the western areas, but have not yet recovered to 1952 levels.

As discussed in Chapter 2 (Figure 2-8A), the net effect of the large amounts of irrigation recharge and deep pumpage is a ground-water-flow system with high downward hydraulic gradients over much of the area. Belitz and Heimes (1990) found that the vertical downward gradient ranged from 0.003 to 1.1 in fine-grained basin deposits, from 0.08 to 0.32 in most of the midfan area, and from 0.003 to 0.07 at the fanhead. The dominance of downward gradients indicates the potential for recharge of poor-quality shallow ground water to contaminate deeper zones of the aquifer.

The distribution of tritium in ground water along the geohydrologic sections provides evidence of the depth to which recent irrigation recharge has moved. Tritium concentrations in ground water from the observation wells along the P, M, and F sections range from <0.8 to 22 TU (Figure 22-10). Tritium concentrations in ground water were greater

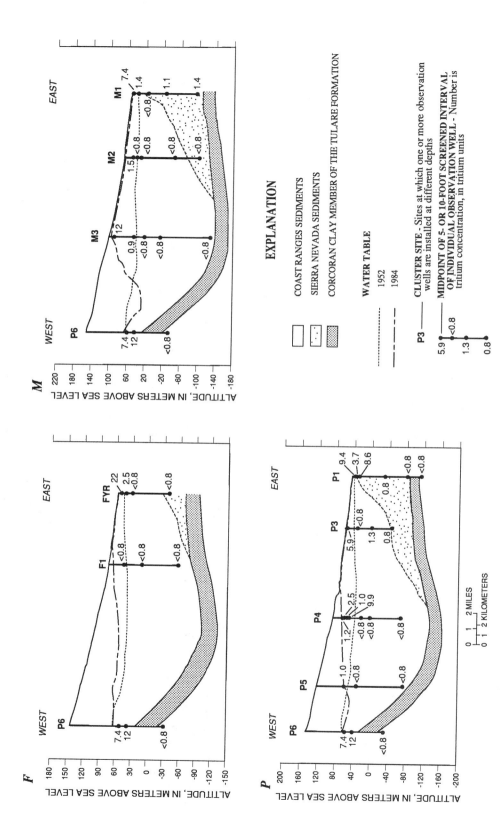

FIGURE 22-10. Tritium data for water samples from monitoring wells completed in the Panoche Creek alluvial fan in the central part of the western San Joaquin Valley. (Location of geohydrologic sections shown in Figure 22-1.)

than 1.6 TU in one or more of the shallow wells at all sites except P5, M2, and F1. The minimum depth to which post-1952 recharge has penetrated, with the exception of site P5, ranges from 3 m below the water table at site P3 to 23 m at site P6. The maximum depth of penetration of post-1952 recharge below the water table, as indicated by the shallowest sample with a tritium concentration less than 1.6 TU, ranges from 15 m below the water table at site M1 to 70 m below the water table at site P1. The data indicate that post-1952 irrigation recharge occupies a zone at the top of the aquifer typically on the order of 15 m thick.

The range of depth of the tritium front, along with the change in the position of the water table from 1952 to 1984, reflects variation in the lithology, position in the flow system, and irrigation history. Large parts of the study area also may have a considerable zone of tritium-free recharge from early irrigation with ground water that underlies the tritiated ground water. Therefore, the depth of soil salts leached by initial irrigation is likely substantially greater than the depth at which tritium is detectable.

The stable isotope composition of water is an indicator of its source and history. Samples from 15 wells completed in the middle parts of the alluvial fans in Coast Ranges deposits at depths greater than 30.5 m below the water table, referred to as the deep Coast Ranges samples, have $\delta^{18}O$ values ranging from -8.75 to -6.05, with a mean of -7.50 ± 0.70 (Figure 22-11). The low variability of this group of samples, particularly excluding four outliers, suggests a common source as the ground water recharged from streams draining the Coast Ranges under predevelopment conditions ("native" ground water).

The $\delta^{18}O$ values of ground-water samples from within 30.5 m of the water table in Coast Ranges sediments, referred to as the shallow Coast Ranges samples, have a mean of -7.63 ± 1.40 and are more variable (Figure 22-11) but not significantly different than the deep Coast Ranges ground water. The $\delta^{18}O$ values that are more negative than the native ground water probably are due to

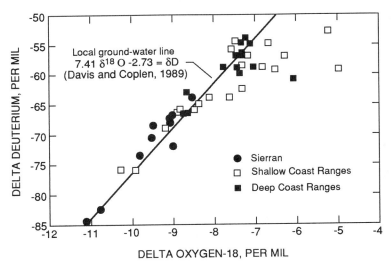

FIGURE 22-11. Delta deuterium and delta oxygen-18 data for water samples from monitoring wells completed in the Panoche Creek alluvial fan in the central part of the western San Joaquin Valley. The data are separated into three categories: Sierran—samples from wells screened in Sierra Nevada sediments; deep Coast Ranges—samples from wells screened in Coast Ranges sediments at depths greater than 30.5 m below the water table; shallow Coast Ranges—samples from wells screened in Coast Ranges sediments at depths less than 30.5 m below the water table.

irrigation recharge with depleted ground water or surface water. Data for several samples from the shallow Coast Ranges are substantially displaced to the right of the local ground-water line, evidence that the water was partly evaporated (Figure 22-11). The similarity of the $\delta^{18}O$ values in the naturally recharged ground water and shallower ground water is unexpected because of the substantial volume of recently recharged irrigation water, all of which was more depleted than the native ground water prior to application. Processes that likely contribute to this lack of contrast are mixing of irrigation and native water during recharge, irrigation with tritium-free deep ground water that is depleted in $\delta^{18}O$, and evaporation.

Salinity and Major-Ion Chemistry

Ground-water salinity is highly variable along the three sections, with the dissolved-solids concentration ranging from 983 to 35,000 mg/L. Concentrations of dissolved solids are highest near the water table at most sites and usually decrease with increased depth (Figure 22-12). Samples of ground water from the deeper ground water in Coast Ranges sediments consistently have similar proportions of the three dominant cations (sodium, magnesium, and calcium), and the proportion of sulfate in samples from these wells is consistently between 75% and 85%. Major-ion ratios vary the most at the boundaries of the study area. Samples from the wells at the eastern limit of the sections near the valley trough contain higher proportions of sodium (55–92%) than samples from wells to the west, and samples from wells within 30.5 m of the water table generally have a higher proportion of sodium and more variable sulfate than wells at greater depth.

The distribution of major ions and the relation to stable-isotope composition are similar to that described by Davis and Coplen (1989). The uniform major-ion proportions of the deeper ground water in Coast Ranges alluvial sediments is interpreted as native water that originated by recharge from Coast Ranges streams, in agreement with stable-isotope data that showed uniform and moderately enriched $\delta^{18}O$ values for these samples.

The variable nature of the major-ion chemistry in samples from within 30.5 m of the water table also is in agreement with stable-isotope data. Evaporation has contributed to the ground-water salinity in areas where the water table is shallow, but leaching is likely the dominant process that increases salinity where the depth of the water table and soil texture allow deep infiltration before significant evaporation occurs. Low dissolved-solids concentration in the shallowest water samples from sites FYR and M3 suggests that past irrigation has leached the most soluble salts from soils at these sites, leading to a reduction in the dissolved-solids concentration in the most recently infiltrated irrigation recharge. This is consistent with the relatively low concentrations of selenium and salinity in shallow ground water of the middle-fan areas (Figures 22-5 and 22-6).

Depth Distribution of Selenium

Concentrations of selenium in ground water at the cluster sites along the sections range from less than the detection limit of 1 µg/L to a maximum of 2,000 µg/L (Figure 22-13). The highest selenium concentrations are associated with the high dissolved-solids concentrations in the upper part of the aquifer. All samples from wells screened in Sierra Nevada sediments had selenium less than 1 µg/L. At sites P3, P4, M3, and FYR, the selenium concentration in the shallowest ground water sampled is less than the concentration at greater depth, illustrating the previously discussed decrease in selenium concentration in recent irrigation recharge through leached soils.

The shallow depth of the highest selenium concentrations indicates their association with irrigation recharge. Most of the wells in which selenium concentrations exceed 10 µg/L are screened near or above the 1952

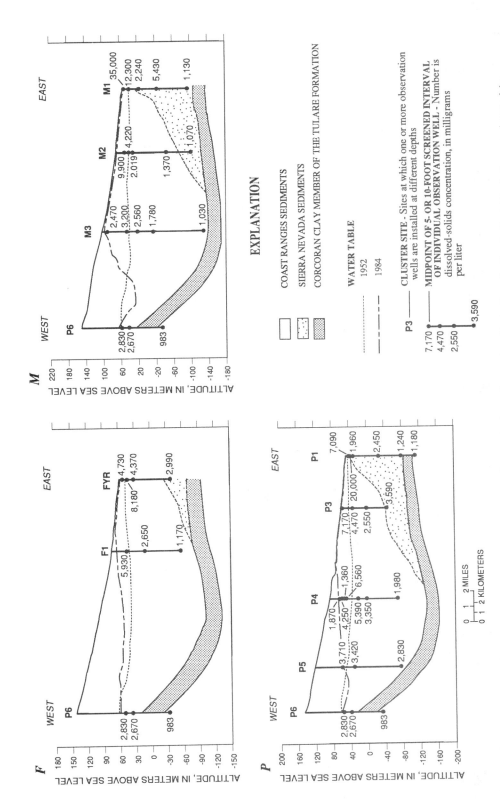

FIGURE 22-12. Dissolved-solids concentration in water samples from monitoring wells completed in the Panoche Creek alluvial fan in the central part of the western San Joaquin Valley. (Location of geohydrologic sections shown in Figure 22-1.)

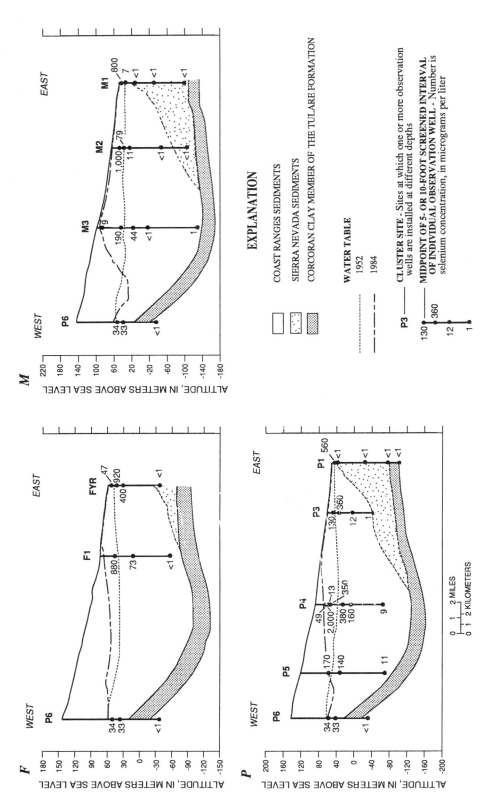

FIGURE 22-13. Selenium concentrations in water samples from monitoring wells completed in the Panoche Creek alluvial fan in the central part of the western San Joaquin Valley. (Location of geohydrologic sections shown in Figure 22-1.)

water table and generally coincide with the zone that is most variable in major-ion chemistry and stable-isotope composition. However, the difference between the selenium concentration in samples with tritium concentrations greater than 1.6 TU and samples with less than 1.6 TU is not significant, suggesting that there were high concentrations of selenium in irrigation-derived recharge prior to the introduction of imported surface water in 1968. There is also a possibility that some of the deep water with selenium concentrations of 100 µg/L or more may be derived from natural processes.

Although several of the ground-water samples show disproportionate $\delta^{18}O$ enrichment, values for $\delta^{18}O$ in samples from the geohydrologic sections explain only 24% of the variability in selenium concentrations. The weak correlation is consistent with the large depth to the water table at most of the sites prior to irrigation, as evaporation is substantial only where the water table is near the land surface. The primary cause of the high correlation between selenium and dissolved solids throughout the semiconfined zone in Coast Ranges sediments thus appears to be leaching of preexisting soil salts by irrigation recharge.

Influence of Redox Potential

The absence of detectable selenium in ground water from wells screened in Sierra Nevada sediments for a wide range of dissolved-solids concentrations indicates, as did the data for existing wells in the basin deposits, that selenium concentrations may be controlled by geochemical processes (Dubrovsky and Deverel 1989; Dubrovsky et al. 1991). As discussed earlier, the oxidized form of selenium is extremely soluble, but reduced forms of selenium are relatively immobile. Platinum electrode measurements of redox potential range from oxidizing (441 mV) in upper parts of the Coast Ranges sediments to reducing (−125 mV) in ground water in Sierra Nevada sediments. The redox potentials generally are highest in shallow wells and decrease with depth. Dissolved-oxygen and nitrate concentrations confirm the pattern. These observations are additional evidence that the absence of selenium from ground water in Sierra Nevada sediments may be due to a redox process, although the absence of selenium in ground water from Sierra Nevada sediments also could be due to a combination of the low availability of selenium in Sierra Nevada sediments and the presence of connate ground water derived from the Sierra Nevada that had little or no dissolved selenium initially.

Significance

Study of the distribution of selenium along the geohydrologic sections in relation to historical hydrology and geochemical conditions adds an understanding of vertical and long-term temporal dimensions to the areal assessment of selenium in shallow ground water. Concentrations of selenium were less than 1 µg/L in all samples from Sierra Nevada sediments, and all selenium concentrations that exceeded 10 µg/L were in the upper part of the semiconfined zone in Coast Ranges alluvial sediments.

Selenium concentrations in the upper part of the semiconfined zone are high in oxic ground water derived from irrigation recharge and are correlated with ground-water salinity. Within about 3–6 m below the water table, selenium concentrations commonly range from 10 to 50 µg/L, but are 10 to 100 times higher than this where the water table has been near the land surface for an extended period, and evaporative concentration has occurred. Water in this shallowest interval is derived principally from the most recent irrigation recharge, probably during the past 10 to 20 years. Within the range of 6–46 m below the water table, an interval of variable thickness occurs in which selenium concentrations commonly are 50 to more than 1,000 µg/L. Water in this interval is derived principally from recharge of early irrigation water but also may include some native ground water with high selenium concentrations.

Findings about the depth distribution

of selenium underscore the importance of evaluating such problems in three dimensions. Conclusions about areal distribution can be misleading without consideration of depth distribution. For example, the assessment of shallow ground water in the middle-fan areas indicated that selenium concentrations are less than 20 μg/L (Figure 22-5). However, results from the geohydrologic sections imply that, in the same areas, somewhat deeper ground water, still resulting from irrigation recharge, has substantially higher concentrations. Findings about depth distribution also reinforce the hypothesis, indicated from the valleywide data for existing wells, that selenium mobility may be limited by reducing conditions in the basin area. Testing this hypothesis requires detailed sampling at an individual site.

SITE-SPECIFIC STUDY OF SELENIUM MOBILITY IN CHEMICALLY REDUCED SIERRA NEVADA SEDIMENTS

Distribution of selenium concentrations in samples from existing wells throughout the San Joaquin Valley, and the distribution in samples from observation wells along the geohydrologic sections across the central western valley, lead to the inference that selenium is immobilized under reducing conditions of Sierra Nevada sediments in the valley trough. This is an important issue for water management in this area because of potential contamination of the heavily used aquifer in the Sierra Nevada sediments in the valley trough by recharge of high-selenium, shallow ground water from Coast Ranges sediments. The mobility of selenium in ground water that flows from oxic Coast Ranges sediments into reduced Sierra Nevada sediments was investigated in detail at site P1 on the eastern edge of the Panoche Creek alluvial fan (Figure 22-1). This site was selected because the interface of Coast Ranges and Sierra Nevada sediments is shallow, facilitating detailed sampling and because preliminary data for the site showed a high concentration of selenium in shallow ground water in Coast Ranges sediments, and low concentrations in underlying Sierra Nevada sediments. The surficial deposits at the study site consist of about 8 m of fine-grained Coast Ranges sediments deposited at the toe of the Panoche Creek alluvial fan (Lettis 1982). Medium- to coarse-grained sand zones at depths greater than 8 m are a mixture of river-channel and alluvial-fan sediments derived from the Sierra Nevada and deposited when the San Joaquin River was west of its present position.

The study site is in an area with highly saline soils in what was a ground-water-discharge zone prior to agricultural development. Following World War II, ground-water withdrawal increased to meet the needs of rapidly developing agriculture, resulting in regional declines in the water table (Belitz and Heimes 1990). By 1955, most of the area surrounding the site at a distance of 1.6 km and greater was irrigated by ground water, and interviews with local officials indicate that the site probably has never been directly irrigated. Large ground-water withdrawals, and likely additional water-table decline, continued until 1967, after which use of ground water decreased because of the availability of surface water imported by the California Aqueduct (Belitz and Heimes 1990). The water table at the site probably was drawn down 6–10 m or more from predevelopment conditions by ground-water pumping, but has entirely recovered since the decrease in ground-water pumping in 1967.

Salinity Distribution

Pore-water samples extracted from borehole cores show a zone of ground water with high but variable salinity at depths between 3.5 and 17 m. The specific conductance ranges from 7,040 to 27,100 microsiemens per centimeter (μS/cm) (Figure 22-14A). Samples from wells at a depth of 18 m and greater have a specific conductance ranging from 3,000 to 4,000 μS/cm, significantly lower than

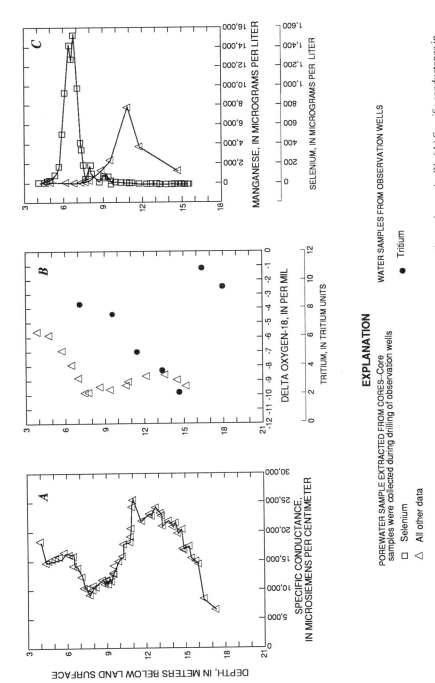

FIGURE 22-14. Concentrations of selected constituents in pore-water and ground-water samples at cluster site P1. (A) Specific conductance in pore-water samples; (B) delta oxygen-18 in pore-water samples and tritium concentrations in ground-water samples; (C) selenium and manganese concentrations in pore-water samples.

that of ground water at shallower depths. The decrease in dissolved-solids concentrations from a depth of 13–18 m is likely the result of mixing of shallow, saline ground water with lower dissolved-solids water from below, which has a different origin.

Age and Origin of Ground Water

Stable-isotope composition and tritium dating were used to evaluate the origin and evolution of ground water at different depths. The profile of the $\delta^{18}O$ data show values ranging from −9.00 to −9.95‰ at depths of 7.5 to 15 m (Figure 22-14B). The data also show a significant enrichment in $\delta^{18}O$ from −9.90‰ at 7.7 m in depth to a maximum of −5.70‰ at a depth of 4 m, slightly above the water table (Figure 22-14B). Enrichment in $\delta^{18}O$ indicates that the shallower pore water has undergone evaporation. Ground water sampled from wells at 17 and 18 m have a less enriched $\delta^{18}O$ value of −10.85‰, which is similar to the $\delta^{18}O$ of imported canal water.

Tritium concentrations in samples collected from wells at depths of 18 m or less in 1987 and 1988 range from 2.3 to 10.6 TU (Figure 22-14B). The tritium concentrations decrease from 7.5 to 15 m below land surface, but are relatively high (9.7–10.6 TU) in ground water at 16–18 m below land surface. The presence of detectable tritium indicates that water at all depths sampled contains some proportion of water that originated as precipitation after 1952. Although the exact mechanism responsible for the tritium concentration profile is not known, the data indicate increasing average age from a depth of 7.5–15 m below land surface. The abrupt change in tritium, $\delta^{18}O$, and specific conductance from a depth of 15–17 m below land surface indicate that the water at depths greater than 15 m has a different origin than the overlying shallower ground water.

Interpretation of the isotope and salinity data is summarized by a hypothesized conceptual model of hydrologic changes:

1. Under natural conditions evaporation from a shallow water table in a regional ground-water-discharge area produced saline soils.

2. Between 1943 and 1967, large withdrawals of ground water resulted in the water table declining about 6–10 m. Irrigation recharge and ground-water withdrawal probably resulted in strong downward gradients.

3. Beginning in 1967, imported canal water rapidly replaced local ground water as the primary source of irrigation water. As ground-water withdrawals decreased, the water table at the study site gradually recovered to its present level.

4. Since 1967, the lack of direct irrigation of the study site enhanced lateral flow of shallow ground water from nearby irrigated areas to the study site.

5. For the depth range investigated, the oldest ground water at the study site is now 11–15 m below land surface. The low but detectable tritium concentration and extremely high dissolved solids, despite no evidence of great evaporative enrichment, indicate that this ground water is predominantly derived from recharge from early irrigation with ground water, which leached the most soluble salts.

6. The lower salinity and higher tritium in the ground water between about 11 m below land surface and the water table probably result from recharge of imported canal water through soils that have already been irrigated for many years.

Selenium Distribution

The selenium concentration in all samples from depths of 10 m or greater is less than the detection limit of 1 µg/L. Selenium analyses for pore water extracted from undisturbed cores shows that selenium concentrations are extremely high in a narrow, sharply defined zone with maximum concentrations of about 1,500 µg/L at depths of 6.6–6.9 m below land surface (Figure 22-14C). Selenium concentrations decrease rapidly to less than 100 µg/L at depths greater than about 8.5 m and at depths less than 5.5 m. Analysis of selenium species shows that virtually all the selenium in solution is selenate (SeO_4^{2-}), with a concentration of tetravalent selenite (SeO_3^{2-}) of less than 1 µg/L in all the pore-water samples except one.

The most notable feature of the selenium distribution is that, unlike the shallow ground water in nearby areas with Coast Ranges alluvial-fan deposits, the concentration of selenium is not correlated with ground-water salinity. Selenium concentrations are highest near the water table, in sharp contrast to the very low selenium concentrations which occur in the most saline zone. The downward hydraulic gradient present at this site, which probably has existed since large ground-water withdrawals began in the 1940s, implies that the saline, low-selenium water was once near the water table. At that time, this ground water probably contained high concentrations of selenium.

Influence of Reducing Conditions

The apparent removal of selenium from the saline ground water that now occurs about 11–15 m below land surface is most likely the result of reduction of selenate to less mobile forms of selenium. The redox potential of the ground water at the study site was assessed on the basis of platinum electrode measurements and the relative abundance of several dissolved constituents that are sensitive to redox conditions. All results are discussed in detail by Dubrovsky et al. (1990). Nitrate concentrations were high in pore water above the water table, indicating that oxidizing conditions dominate in the unsaturated zone. Just below the water table, nitrate and dissolved oxygen disappear from solution, and the ground water becomes moderately reducing, as also indicated by the high concentrations of iron and manganese. This pattern is consistent with the decrease in platinum electrode redox potential with increased depth below the water table. A comparison of selenium and manganese data indicate that selenium concentrations decrease rapidly at the same depth at which manganese concentrations increase, indicating that the decrease in selenium is due to a process that occurs under reducing conditions (Figure 22-14C).

A mechanism that may be causing selenium removal under the reducing conditions is respiratory reduction of selenate to elemental selenium by anaerobic bacteria (Macy, Michel, and Kirsch 1989; Oremland et al. 1989). The presence of this type of bacterial activity in selected subsurface sediments from the study site was evaluated by measuring loss of radioactively labeled ^{75}Se-selenate from solution in samples of slurried core materials (Oremland et al. 1989). Results show that all the sediments tested have bacteria present that will reduce selenate; however, the samples from 8.5 m and deeper have a microbial population capable of reducing selenate at a much higher rate than sediments at shallower depths (Dubrovsky et al. 1990). The depth interval of high reduction capability corresponds to the zone at which the pore waters become reducing (Figure 22-14C). Preliminary investigation of solid-phase selenium concentrations for evidence of accumulation of selenium removed from solution is inconclusive because of the high degree of natural variability in solid-phase concentrations, the uncertain interval of accumulation, and the relatively small mass of selenium removed from solution compared to preexisting solid-phase levels.

The mobility of selenium in the reducing geochemical environment at the study site was more directly investigated with an injection test methodology developed by Lawrence Berkeley Laboratory (1987). One thousand liters of reduced ground water was pumped from a well at the site, spiked with a solution containing about 500 µg/L of selenate and a bromide tracer, and injected into the same zone without exposing the ground water to the atmosphere or otherwise changing the solution chemistry. The injected solution was then recovered by sampling 200-L aliquots after 2, 4, 8, 16, and 32 days. Results showed that total selenium was rapidly lost from solution, with approximately 10, 20, and 90% removal after 2, 4, and 8 days, respectively. As the selenate concentration decreased, the proportion of the total selenium present as selenite increased. Results indicate that selenium is removed from solution by reduction to a less-soluble species and support the hypothesis of bacterially mediated reduction to elemental selenium; however, an additional mechanism that needs to be evaluated is removal by adsorption of selenite.

Significance

Data from site P1 near Mendota show that selenium is not always correlated with salinity, due to mobility-limiting reactions that occur under reducing conditions. This process explains, in part, the absence of selenium in reduced geochemical environments observed at the subregional and regional scales.

The implication is that contamination of the major ground-water resources by high selenium concentrations in recharging irrigation water may be prevented by a naturally occurring bacterially mediated reduction process. This implication is pertinent to management strategies under consideration that use pumping of deep wells in Sierra Nevada sediments to control the water table. Caution must be exercised, however, in extrapolating the results of the site-specific study to the larger scales. Although the process has been demonstrated to occur, neither the rate of selenium removal in irrigation drainage nor the reductive capacity of the aquifer materials is known. Contamination could still occur if the rate of reduction is too slow relative to ground-water flow, or if the reductive capacity of the aquifer is overwhelmed by the large concentration of nitrate present in most of the shallow ground water on the west side.

CONCLUDING REMARKS

The results of four scales of investigation, ranging from regional to site-specific, provide multiple lines of supporting evidence for a comprehensive assessment of selenium in ground water in the San Joaquin Valley. Each scale of investigation has strengths and weaknesses.

Regional-scale sampling of existing wells in the valley was best suited for characterizing broad patterns in the quality of water presently used and for developing preliminary hypotheses about regionally important influences. This approach may miss important water-quality problems that have not yet affected existing wells or that have been avoided when installing wells; also, hypotheses about influences commonly are weak. In this case, the regional-scale data correctly indicated the general area where ground water has high selenium concentrations, but underestimated the concentrations in shallow ground water in the central part of the western valley by more than two orders of magnitude.

Detailed sampling of shallow ground water in the central part of the western valley allowed an affordable detailed assessment of areal patterns and variability in ground-water quality and its relation to geographically distributed causative factors such as water-table configuration and soil conditions. The focus on shallow ground water, however, leaves out consideration of historical influences on ground-water quality in relation to three-dimensional ground-water flow. The lack of data on depth distribution prevents linkage of the observed conditions in the shallow ground water to conditions observed in the regional aquifer.

Sampling of monitoring wells installed along transects provides information on the depth distribution of selenium concentrations as well as the hydrologic and historical context necessary for interpretation of the water-quality data. Data from these wells explained the discrepancies between the areal distribution of selenium in the shallow ground water of the central western valley and the regional-scale data set. The data on areal distribution are, in turn, needed to extend the specific observations made at several sites to general conclusions relevant to large areas.

Investigation of one hypothesized geochemical process at a site allows the application of extremely detailed data-collection techniques tailored to the question posed. As a result of this approach, the site-specific investigation at the eastern edge of the Panoche Creek alluvial fan demonstrated the relation between selenium transport, redox changes, and microbial activity. Conclusions based on the data from the site explain some aspects of the areal distribution of selenium at both the regional scale and the vertical distribution in the transects.

References

Amundson, R. G., H. E. Doner, and L. J. Waldron. 1986. Quantities and species of toxic elements in the soils and sediments of the Panoche Fan and their uptake by plants: Selenium immobilization, plant uptake, and volatilization. In *1985–86 Technical Progress Report, UC Salinity/Drainage Task Force*, eds. K. K. Tanji and others, pp. 152–60. Davis, CA: University of California, Division of Agriculture and Natural Resources.

Balistrieri, L. S., and T. T. Chao. 1987. Selenium adsorption by goethite. *Soil Science Society of America* 51:1145–51.

Belitz, Kenneth, and F. J. Heimes. 1990. *Character and Evolution of the Ground-Water Flow System in the Central Part of the Western San Joaquin Valley, California*. Reston, VA: U.S. Geological Survey Water-Supply Paper 2348.

Bull, W. B. 1964a. *Geomorphology of Segmented Alluvial Fans in Western Fresno County, California*. Reston, VA: U.S. Geological Survey Professional Paper 352-E.

Davis, G. H., and T. B. Coplen. 1989. *Late Cenozoic Paleohydrogeology of the Western San Joaquin Valley, California, as Related to Structural Movements in the Central Coast Ranges*. Geologic Society of America Special Paper 234.

Deverel, S. J., and Roger Fujii. 1988. Processes affecting the distribution of selenium in shallow ground water of agricultural areas, western San Joaquin Valley, California. *Water Resources Research* 24(4):516–24.

Deverel, S.J., and S. K. Gallanthine. 1989. Relation of salinity and selenium in shallow ground water to hydrologic and geochemical processes, western San Joaquin Valley, California. *Journal of Hydrology* 109:125–49.

Deverel, S. J., R. J. Gilliom, Roger Fujii, J. A. Izbicki, and J. C. Fields. 1984. *Areal Distribution of Selenium and Other Inorganic Constituents in Shallow Ground Water of the San Luis Drain Service Area, San Joaquin Valley, California: A Preliminary Study*. Sacramento, CA: U.S. Geological Survey Water-Resources Investigations Report 84-4319.

Deverel, S. J., and S. P. Millard. 1988. Distribution and mobility of selenium and other trace elements in shallow ground water of the western San Joaquin Valley, California. *Environmental Science and Technology* 22(6):697–702.

Dubrovsky, N. M., and S. J. Deverel. 1989. Selenium in ground water of the central part of the western valley. In *Preliminary Assessment of Sources, Distribution, and Mobility of Selenium in the San Joaquin Valley, California*, eds. R. J. Gilliom and others, pp. 35–66. Sacramento, CA: U.S. Geological Survey Water-Resources Investigations Report 88-4186.

Dubrovsky, N. M., J. M. Neil, Roger Fujii, R. S. Oremland, and J. T. Hollibaugh. 1990. *Influence of Redox Potential on Selenium Distribution in Ground Water, Mendota, Western San Joaquin Valley, California*. U.S. Geological Survey Open-File Report 90-138.

Dubrovsky, N. M., J. M. Neil, M. C. Welker, and K. D. Evenson. 1991. *Geochemical Relations and Distribution of Selected Trace Elements in Ground Water of the Northern Part of the Western San Joaquin Valley, California*. U.S. Geological Survey Water-Supply Paper 2380.

Elrashadi, A. M., D. C. Adriano, S. M. Workman, and W. L. Lindsay. 1987. Chemical equilibria of selenium in soils: A theoretical development. *Soil Science* 144(2):141–52.

Fujii, Roger, S. J. Deverel, and D. B. Hatfield. 1988. Distribution of selenium in soils in agricultural fields, western San Joaquin Valley, California. *Soil Science Society of America* 52(5):1274–83.

Gilliom, R. J. 1989. Geologic source of selenium and its distribution in soil. In *Preliminary Assessment of Sources, Distribution, and Mobility of Selenium in the San Joaquin Valley, California*, eds. R. J. Gilliom and others, pp. 7–11. Sacramento, CA: U.S. Geological Survey Water-Resources Investigations Report 88-4186.

Gilliom, R. J. et al. 1989. *Preliminary Assessment of Sources, Distribution, and Mobility of Selenium in the San Joaquin Valley, California*. Sacramento, CA: U.S. Geological Survey Water-Resources Investigations Report 88-4186.

Goldberg, S., and R. A. Glanbig. 1988. Anion sorption on a calcareous, montmorillonitic soil—selenium. *Soil Science Society of America* 52(4):954–58.

Harradine, F. F. 1950. *Soils of Western Fresno County*. Berkeley, CA: University of California Press.

Lawrence Berkeley Laboratory. 1987. Hydrological, geochemical, and ecological characterization of Kesterson Reservoir. In *Annual Report, October 1, 1986, through September 30, 1987*. Berkeley: University of California, Earth Sciences Division, Lawrence Berkeley Laboratory.

Lettis, W. R. 1982. *Late Cenozoic Stratigraphy and Structure of the Western Margin of the Central San Joaquin Valley, California*. Menlo Park, CA: U.S. Geological Survey Open-File Report 82-526.

Macy, J. M., T. A. Michel, and D. G. Kirsch. 1989. Selenate reduction by a *Pseudomonas* species. A new mode of anaerobic respiration. *Federation of European Microbiological Societies, Microbiological Letters* 61:195–8.

Mendenhall, W. C., R. B. Dole, and Herman Stabler. 1916. *Ground Water in the San Joaquin Valley, California*. Reston, VA: U.S. Geological Survey Water-Supply Paper 398.

Neal, R. H., G. Sposito, K. M. Holtzclaw, and S. J. Traina. 1987. Selenite adsorption on alluvial soils. I: Soil Composition and pH effect. *Soil Science Society of America* 51:1161–65.

Oremland, R. S., J. T. Hollibaugh, A. S. Maest, T. S. Presser, L. G. Miller, and C. W. Culbertson. 1989. Selenate reduction to elemental selenium by anaerobic bacteria in sediments and culture. Biogeochemical significance of a novel, sulfate-independent respiration. *Applied and Environmental Microbiology* 55:2333–43.

Presser, T. S., W. C. Swain, R. R. Tidball, and R. C. Severson. 1990. *Geologic Sources, Mobilization, and Transport of Selenium from the California Coast Ranges to the Western San Joaquin Valley: A Reconnaissance Study*. Menlo Park, CA: U.S. Geological Survey Water-Resources Investigations Report 90-4070.

Tidball, R. R., W. D. Grundy, and D. L. Sawatzky. 1986. Kriging techniques applied to element distribution in soils of the San Joaquin Valley, California: In *Proceedings, HAZTECH International Conference*, August 11–15, 1986, Denver, CO, pp. 992–1009.

Weres, O., A. R. Jaouni, and L. Tsao. 1989. The distribution, speciation, and geochemical cycling of selenium in a sedimentary environment, Kesterson Reservoir, California, U.S.A. *Applied Geochemistry* 4:543–63.

White, A. F., S. M. Benson, A. W. Yee, H. A. Wollenberg, Jr., and Steven Flexser. 1988. Groundwater contamination at the Kesterson Reservoir, California. 2: Geochemical parameters influencing selenium mobility. *Water Resources Research* 27(6):1085–98.

23

Multiscale Approach to Regional Ground-Water-Quality Assessment of the Delmarva Peninsula

Robert J. Shedlock, Pixie A. Hamilton, Judith M. Denver, and Patrick J. Phillips

INTRODUCTION

The Delmarva Peninsula (Figure 23-1) includes most of the state of Delaware and the areas of Maryland and Virginia that are east of the Chesapeake Bay. It is bordered on the east by Delaware Bay and the Atlantic Ocean. The peninsula is one of seven pilot project areas in the National Water-Quality Assessment Program (NAWQA) of the U.S. Geological Survey (Hirsch, Alley, and Wilber 1988). The project area covers about 18,000 km^2, and is essentially a coastal lowland drained by a series of short tidal streams. The peninsula is mostly in a rural setting and has a population of about 600,000. Several small cities have populations between 10,000 and 25,000, but most towns have populations of only a few thousand.

Nearly half of the peninsula is agricultural land, primarily used to grow corn and soybeans. Other crops include wheat, barley, potatoes, vegetables, and fruit. Dairy farms, nurseries, and sod farms also are found in the project area. Large poultry farms are found all across the project area, which is one of the leading production areas of broiler chickens in the United States. The agricultural lands are interspersed with woodlands, which cover about a third of the project area. The size of agricultural fields and the degree of interspersion of fields with woodlands differ across the study area and are related to differences in geomorphic and hydrologic characteristics. Other significant land uses include wetlands (about 13% of the study area) and urban or residential lands (about 7%).

The study area, which is in the Atlantic coastal-plain physiographic province, is underlain by a wedge of unconsolidated sand, silt, and clay (Figure 23-2). These sediments range from 0 to 2,500 m thick and form a complex aquifer system that consists of a series of nine confined sand aquifers overlain by an extensive surficial sand aquifer (Cushing, Kantrowitz, and Taylor 1973). Most of the water supply in the study area is derived from the ground-water system. The confined aquifers are the major sources of public water supply. The surficial aquifer is the main source for most private domestic wells, but its saturated thickness is less than 15 m in most parts of the project area. It is a major source of public water supply in the central part of the project area, however, where its saturated thickness commonly exceeds 30 m.

Ground-water-quality problems associated with nitrate in the study area have

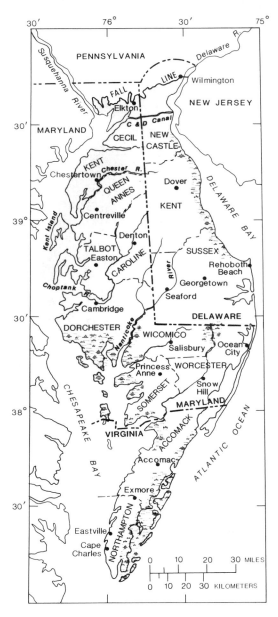

FIGURE 23-1. Delmarva Peninsula NAWQA project area.

the entire project area. This review indicated elevated nitrate concentrations to be widespread in the surficial aquifer but not in the confined aquifers. The review also indicated that nitrate concentrations in water from 18% of the wells sampled in the surficial aquifer exceeded the maximum contaminant level (MCL) for drinking water of 10 mg/L as nitrogen established by the U.S. Environmental Protection Agency (1986). Nitrogen fertilizers and poultry manure applied to farm fields are the major sources of nitrate in the project area. Effluent from septic tanks and nitrogen compounds in precipitation also contribute nitrate to the ground water.

The purpose of this chapter is to provide a brief description of the study design and selected results of the Delmarva Peninsula NAWQA project. The chapter focuses on water-quality patterns for major ions and nitrate in the surficial aquifer. Its intent is to demonstrate how data from several new sampling networks at different spatial scales were combined with data collected before this project to assess regional ground-water-quality patterns.

HYDROGEOMORPHIC REGIONS IN THE SURFICIAL AQUIFER

The project area was divided into seven subregions (Figure 23-3), referred to as *hydrogeomorphic regions* (HGMRs), which represent different hydrologic settings in the surficial aquifer (Hamilton, Shedlock, and Phillips 1991; Koterba et al. 1991; Phillips, Shedlock, and Hamilton 1992). Each HGMR has a characteristic set of geologic and geomorphic features, drainage patterns, soils, and land-use patterns. An HGMR consists of hundreds of small watersheds that, in aggregate, represent a distinct hydrologic landscape with characteristic patterns of ground-water flow and ground-water quality. Four of the HGMRs are upland regions in the interior of the peninsula, and three are coastal lowland regions. Three of the upland HGMRs encompass more than 70% of the

been documented in several previous studies (Robertson 1979; Ritter and Chirnside 1982; Bachman 1984; Denver 1986). As part of the Delmarva Peninsula NAWQA project, Hamilton, Shedlock, and Phillips (1991) reviewed available data through 1987 for

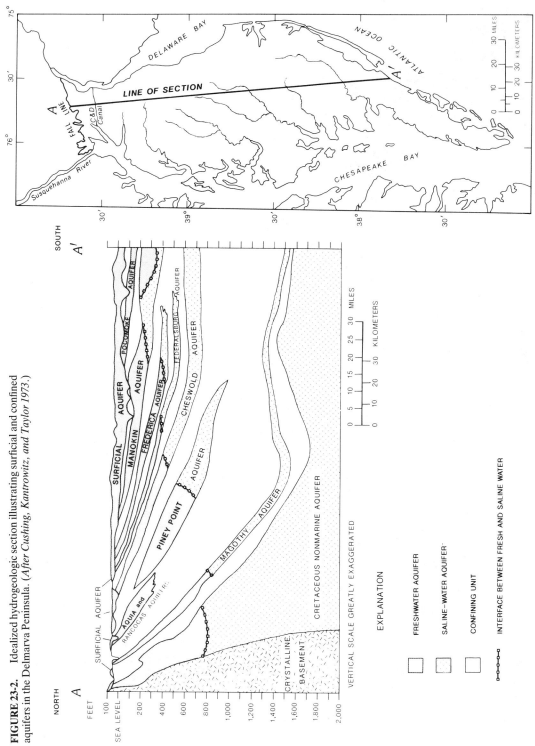

FIGURE 23-2. Idealized hydrogeologic section illustrating surficial and confined aquifers in the Delmarva Peninsula. (*After Cushing, Kantrowitz, and Taylor 1973.*)

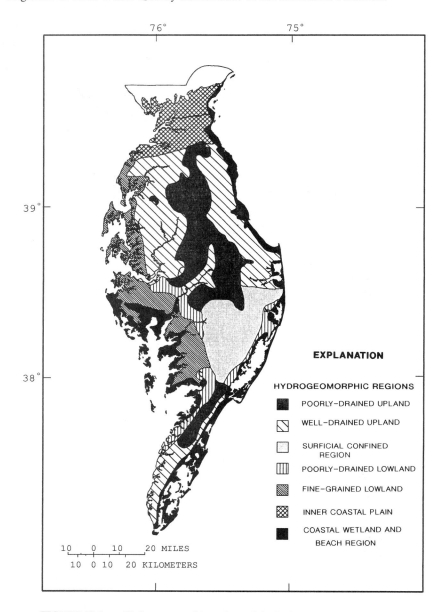

FIGURE 23-3. Hydrogeomorphic regions of the Delmarva Peninsula.

project area and contain most of the agricultural land in the project area.

The first upland HGMR, the poorly drained upland, lies along the drainage divide separating the Chesapeake Bay drainage area from the drainage areas of Delaware Bay and the Atlantic Ocean. The poorly drained upland is a headwater region characterized by shallow incisement of streams into the landscape. The area has hummocky topography, a high water table (usually within 3 m of land surface), poorly drained soil, and short ground-water-flow paths (usually less than a few hundred meters long). This region has a high percentage of woodland (43%) that is highly interspersed among agricultural fields. Many of the woodlands contain wetlands in small closed depressions that are seasonally ponded.

The second upland HGMR, the well-drained upland, flanks the eastern and western margins of the poorly drained upland and represents areas that have been more deeply incised by the tidal streams and their tributaries than the poorly drained upland. The well-drained upland is characterized by higher topographic relief, a deeper water table (usually between 3 and 10 m below land surface), more well-drained soil, and longer shallow ground-water flow paths (a few hundred to a few thousand meters) than the poorly drained upland. The region contains a lower percentage of woodland (28%) than the poorly drained upland. The woodlands in this region are primarily in riparian zones along stream valleys. Consequently, agricultural fields are larger and the percentage of agricultural land is higher in the well-drained upland than in any other HGMR in the project area.

The third upland HGMR, the surficial-confined region, is in the south-central part of the project area. This region is geomorphically similar to the poorly drained upland, with a high water table, shallow incisement of natural stream valleys, and poorly drained soils. The topography is relatively flat except for sand dunes, which form low sandy ridges that rise several meters above the surrounding landscape. This HGMR is called the surficial-confined region, because it is the only region in which confining conditions are found in the surficial aquifer. In the surficial-confined region, the surficial aquifer is more complex than in the other HGMRs and consists of three hydrogeologic units. The uppermost unit is a sheet of sand usually less than 6 m thick. In some places, peaty sand, silt, and clay are found at the base of this upper unit. The water table is usually found in the upper unit, but this unit is not thick enough to be used as a source of water supply. Ground-water-flow paths in the upper unit are generally a few hundred meters long. The middle unit consists of fine-grained deposits (interbedded marine and marginal-marine sands, silts, and clays) that form a confining layer for the underlying sands of the lower unit. In some localities of this HGMR, the middle unit is absent or entirely composed of sand. The lower unit is commonly more than 30 m thick and is the part of the surficial aquifer used for water supply in this region. Flow paths in the lower unit range in length from a few hundred to several thousand meters. The surficial-confined region has the highest percentage of woodland (55%) of the HGMRs in the project area. The woodlands are found in large tracts in uplands between streams and in riparian zones along stream valleys. Agricultural lands are interspersed among the woodlands and are heavily ditched.

The fourth upland HGMR is the inner coastal-plain region. The hydrogeology of this HGMR is complex because the sands that make up the surficial aquifer are thin and overlie sands of several of the lower confined aquifers. Some of these lower aquifers actually crop out in some of the stream valleys. The surficial deposits do not form an areally extensive aquifer in this HGMR. In this region, the confined aquifers are close to the surface, and the shallow ground-water-flow systems extend into these lower aquifers.

The three lowland HGMRs essentially cover the coastal margins of the Delmarva Peninsula and consist of the fine-grained lowland, the poorly drained lowland, and the coastal wetland and beach region. The largest of the three lowland HGMRs is the fine-grained lowland. The surficial sediments in this region are silts, sands, and organic muds deposited in Chesapeake Bay during periods when sea level was higher than at present. Because of the low permeability of the near-surface sediments, most wells in the fine-grained lowland are finished in one of the underlying confined aquifers. The poorly drained lowland consists of coarser-grained sediments than those found in the fine-grained lowland. These deposits represent marginal-marine sands or reworked sandy sediments from the margins of the upland areas. The coastal wetland and beach region covers areas underlain by a variety of marine and estuarine coastal deposits, including beaches, dunes, and tidal marshes.

Much less is known about the shallow-ground-water-flow systems in the lowland HGMRs than about those in the upland HGMRs. Throughout most areas of the lowland HGMRs, the water table is within a few meters of land surface. Because of the distribution and density of wetlands, streams, and drainage ditches in these regions, most shallow-ground-water-flow paths are probably no longer than a few hundred meters.

DESIGN OF STUDY

In the Delmarva project, the regional assessment of ground-water quality was done by assembling and analyzing a data set consisting of newly collected data and carefully screened available data. The new data consisted of analyses of ground-water samples from several different networks of wells. These networks were designed at regional and local scales so that water-quality patterns could be assessed at different spatial scales (Koterba et al. 1991).

Available Data

The available data consisted of analyses of waters from 193 wells sampled before 1987. These analyses were selected because sufficient data on site characteristics and well construction, especially depth of well screen, were associated with the wells from which the samples were collected. Other selection criteria included the availability of data for major cations and anions and a charge balance error of less than 10%.

Well Networks

Two regional-scale networks were established in the surficial aquifer. The first, referred to as the *areal network* (Figure 23-4), consisted of two wells at different depths at 35 sites geographically distributed across the peninsula. This network was designed to provide a broad areal assessment of water-quality conditions across the peninsula. The sites for this network were chosen randomly by dividing the study area into polygons of roughly equal area and by randomly choosing points within each polygon. Files of state and county well permits were then searched to locate the closest existing well suitable for sampling near the random point. At most sites, the deep well is an existing domestic supply well, and the shallow well is a new well installed within 3 m of the water table according to protocols in Hardy, Leahy, and Alley (1989).

The second regional-scale network, referred to as the *transect network*, consisted of wells along five transect lines (Figure 23-4) that cross the peninsula from east to west. The transects were spaced to provide a broad north-south coverage of the study area. Well sites in this network were specifically chosen to target areas representing the different HGMRs along each transect.

Along each transect 8–20 sites were selected. The number of sites depended on the length and number of HGMRs along each transect. A newly drilled shallow well was installed within 3 m of the water table at each site. A deeper well was installed in the surficial aquifer at one or two sites in each HGMR along each transect. Water levels were measured seasonally at most of these sites to provide data on seasonal changes in regional water-table profiles. A total of 40 wells, including nearly all the sites with 2 wells, were sampled for water-quality analysis.

Seven local-scale networks of wells, representing the three largest HGMRs, were established along the transect lines. These networks were designed to measure ground-water-quality variations on a much more detailed spatial scale than that afforded by either the areal or the transect networks. The sites in these networks were chosen to investigate ground-water-quality variations along local ground-water-flow systems in each area. These networks also were designed to provide a way to relate water-quality patterns

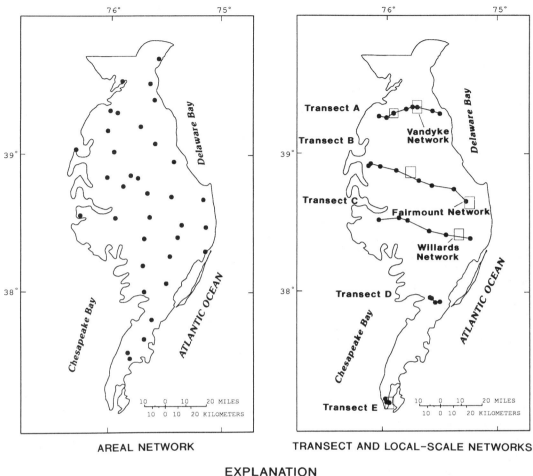

FIGURE 23-4. Locations of wells completed in the surficial aquifer in the areal, transect, and local-scale networks in the Delmarva Peninsula.

to landscape features such as land use, geomorphology, and soil type.

Each local network consists of 10–35 observation wells installed during several field seasons in areas ranging from less than 1 to 20 km². The first set of wells installed were used to map the water table and determine ground-water-flow paths. One to three additional wells were installed at some sites to provide nests of wells that were finished at

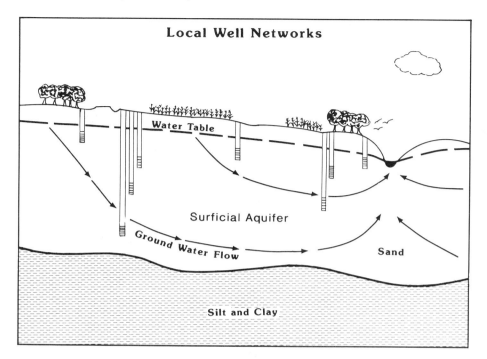

FIGURE 23-5. Local-scale well networks in the surficial aquifer in the Delmarva Peninsula.

different depths. In each case, the final network consists of several nests of wells along a ground-water-flow path from an upland recharge area to a surface-water-discharge area (Figure 23-5).

Sampling Strategy

Wells in the regional networks were sampled once each for a broad suite of constituents, including major ions, nutrients, dissolved organic carbon, selected trace elements, radon, volatile organic compounds, most commonly applied herbicides in the project area, and selected insecticides. Wells in the areal network were sampled in the summer of 1988, and wells in the transect network were sampled in the summer of 1989.

Wells in the local-scale networks were sampled under spring, summer, and fall conditions from 1988 to 1990. Water samples were collected for analysis of major ions, nutrients, dissolved organic carbon, selected trace elements, and most common herbicides.

Selected ancillary data were documented and entered into a computer data base for each site in the new well networks. Entries were made for the predominant land use within a radius of 30 m and within a radius of 400 m. Data on soil types, as categorized in state soil surveys, were entered for each site, as were measurements of water levels in the wells.

Methods of Analysis

Regional water-quality patterns were assessed by combining data from the areal and transect networks with the available data from the 193 wells sampled before 1987. This combined data set (referred to as the 1990 data set) was considered to be the regional data set and consisted of analyses from a total of 296 wells. One of the main objectives for analysis of this data set was to determine the effects of agriculture and other human activity on natural water quality. Before investigating these effects, however,

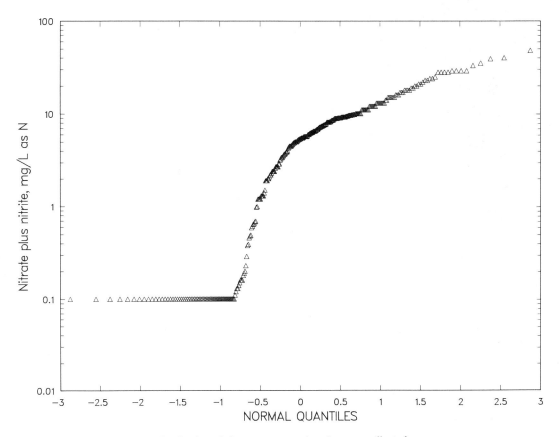

FIGURE 23-6. Probability distribution of nitrate concentrations in water collected from wells completed in the surficial aquifer in the Delmarva Peninsula.

a method was needed to distinguish water samples whose chemical composition had been significantly affected by human activity from those whose chemistry had not been affected or had been minimally affected.

These data were identified by construction of a probability plot (Gnanadesikan 1977; Kleiner and Graedel 1980) of nitrate concentrations for all of the 296 analyses (Figure 23-6). Nitrate was chosen as an indicator of human activity because it is introduced into the ground-water system in both agricultural areas (by application of fertilizer and manure) and residential areas (from septic-tank effluent and/or lawn and garden fertilizers).

The probability plot was used to define a threshold nitrate concentration that was indicative of human activity. Nitrate concentrations (as N), as shown in Figure 23-6, range from below the detection level of 0.1 to 48 mg/L. The break in slope at 0.4 mg/L is interpreted as a threshold concentration separating two populations of points on the plot. This value was selected by visual inspection. Because visual inspection can result in an approximate value, the sensitivity of the subsequent analysis to the threshold concentration was tested at a higher threshold concentration (1.0 mg/L). No statistically significant difference in the results was found. Therefore, concentrations of nitrate (as N) at or above 0.4 mg/L were considered to have been affected by human activity, and concentrations below this value were considered to have been unaffected by human activity.

About 74% of the analyses (219 of 296 samples) were considered to be affected by human activity. The remaining 77 unaffected samples were used to examine regional patterns in natural or nearly natural ground-water chemistry. This method of sorting does not consider the effects of redox conditions on the presence or absence of nitrate in ground water. Agricultural and residential activities, however, tend to change the proportions of other major cations and anions in the ground water. These changes were not detected in the 77 samples used to investigate natural ground-water chemistry.

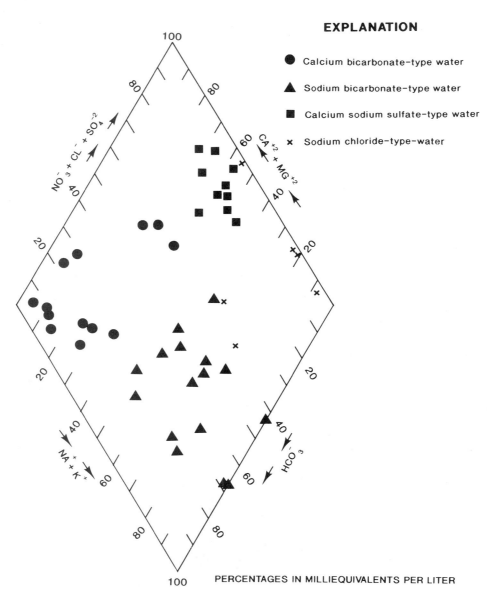

FIGURE 23-7. Hydrochemical types of natural ground water in the surficial aquifer in the Delmarva Peninsula.

GROUND-WATER-QUALITY PATTERNS

Ground water in the surficial aquifer generally is characterized by low pH, low alkalinity, and low dissolved-solids concentration (Hamilton, Shedlock, and Phillips 1991; Denver 1989; Bachman 1984). The dilute and acidic nature of the ground water is caused by the mineralogy of the aquifer materials, which are composed of silicate and aluminosilicate minerals (quartz, feldspars, and clays). In addition, the aquifer materials are nearly devoid of carbonate minerals, except in coastal areas underlain by marginal-marine sediments and in areas where the confined aquifers subcrop or outcrop in the northern part of the peninsula.

Distribution of Hydrochemical Types in the Surficial Aquifer

In spite of its dilute nature, the chemical composition of ground water in the surficial aquifer varies by region. Natural water includes calcium-bicarbonate-type water, sodium-bicarbonate-type water, calcium-sodium-sulfate-type water, and sodium-chloride-type water (Figure 23-7).

The spatial distribution of the natural-water types (Figure 23-8) is related to differences in geologic and hydrogeologic setting across the study area. The calcium-bicarbonate-type water primarily is in the northern part of the study area, where the surficial aquifer is composed of the fluvial sands of the Pensauken Formation that overlie glauconitic sands of the Aquia Formation. Calcium-bicarbonate-type water is also in coastal areas where the surficial aquifer is composed of marginal-marine deposits that contain calcium bicarbonate in the form of shell material.

Sodium-bicarbonate-type water is found primarily in the central and south-central part of the study area, where the surficial aquifer is generally thicker than in the north and is composed mainly of the Beaverdam Sand. The sodium-bicarbonate-type water is more consistently associated with the Beaverdam Sand than the Pensauken Formation. Differences in environments of deposition could explain this hydrochemical association. Some beds of the Beaverdam Sand were deposited in a shallow marine environment. In contrast, the Pensauken Formation is largely a fluvial deposit.

The calcium-sodium-sulfate-type water is found in parts of the study area where the surficial aquifer contains fine sediments deposited in riverine, estuarine, paludal, or back-barrier environments. These types of environments are more areally extensive in the southern half of the study area than in the northern half, but also can be found in the northern half of the peninsula in wooded stream valleys and forested wetlands.

The sodium-chloride-type water primarily is found in areas near tidal streams and presumably is the result of intrusion of brackish water into the surficial aquifer near the tidal reaches. None of the sodium-chloride-type water was clearly associated with road salting, which is a possible source of sodium chloride to ground water along highways.

Two water types are associated with human activity: calcium-magnesium-nitrate-type water in agricultural areas and sodium-chloride-nitrate-type water in residential areas. The calcium-magnesium-nitrate type is the most widespread and is found throughout the project area below or downgradient of agricultural fields (Denver 1989; Hamilton and Denver 1990). Applications of inorganic fertilizer, manure, and lime on farm fields produce higher concentrations of the major ions calcium, magnesium, potassium, and nitrate in ground water than in the natural-water types described (with the exception of chloride in water influenced by saltwater). The minor constituents barium and strontium also increase in concentration with agricultural applications and add to the chemical contrast between ground water affected by agriculture and natural-water types.

Therefore, ground water recharged below agricultural fields has a characteristic chemical signature, with calcium, magnesium, and

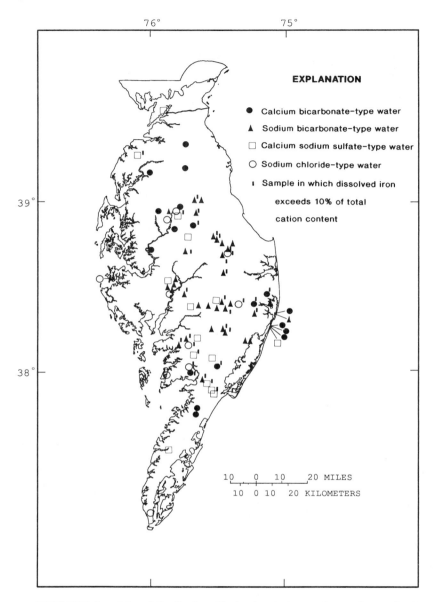

FIGURE 23-8. Map of distribution of hydrochemical types of natural ground water in the surficial aquifer in the Delmarva Peninsula.

nitrate as the dominant ions. This signature is observable in many samples because of the dilute nature of the natural ground water, which has a mean specific conductance of about 70 μS/cm. In contrast, the mean specific conductance of water with an agricultural chemical signature is 186 μS/cm. Even relatively small inputs of ions from agricultural chemicals in soil can produce water with this chemical signature. An example of geochemical mass-balance modeling of the agricultural chemical signature is given in Chapter 9.

Shallow ground water in residential areas

Ground-Water-Quality Patterns   575

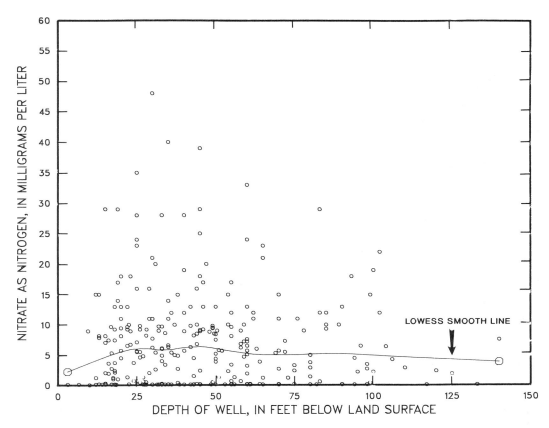

FIGURE 23-9. Relation of nitrate concentration in ground water in the surficial aquifer to depth of well.

is commonly a sodium-chloride-nitrate-type water. Each of these ions is found in septic effluent. In addition, road salt can contribute sodium and chloride.

Spatial Distribution of Nitrate in Ground Water

Variations with Depth

Hamilton, Shedlock, and Phillips (1991) found that elevated concentrations of nitrate are present at all depths in the surficial aquifer on the basis of a data set of 399 wells sampled before 1987 (referred to as the 1987 data set). They also found that nitrate concentrations generally are higher in the shallow sections of the surficial aquifer than in the deep sections. The relation between nitrate and depth in the surficial aquifer was reexamined with the 1990 data set.

The reexamination with the 1990 data set also showed that nitrates were present at all depths in the surficial aquifer, as illustrated in Figure 23-9. Although the highest concentrations are found in the shallow sections of the surficial aquifer, the analysis failed to confirm that nitrate concentrations were significantly higher in the shallow sections of the surficial aquifer than in the deep parts. A regression analysis showed no significant relation between the ranks of nitrate concentration and depths in the surficial aquifer (p value = .701). In addition, the relation between nitrate concentration and depth was graphically analyzed by generating a smooth line (referred to as a *Lowess smooth line*) through the data in Figure 23-9 with

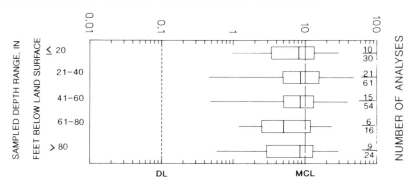

FIGURE 23-10. Nitrate concentrations in ground water affected by agriculture, grouped by depth of well in 20-ft intervals.

a method that involves the use of robust-weighted least squares (Helsel and Hirsch 1992). This smooth line is generally flat and shows no relation between nitrate concentration and depth in the surficial aquifer.

The lack of such a relation is illustrated also by the vertical distribution of nitrate concentrations in the surficial aquifer in agricultural areas (Figure 23-10). Elevated nitrate concentrations, including those that exceed the MCL for nitrate, are found at depths greater than 30 m (100 ft) in the surficial aquifer in agricultural areas. The median concentration of nitrate (as N) in water from wells screened 24 m (80 ft) or more below land surface is 8.45 mg/L. Nitrate concentrations in ground water at these depths are as high as 22 mg/L, and the MCL of 10 mg/L is exceeded in 9 of 24 samples.

Nitrate concentrations that exceed the MCL are found at these depths for several reasons. First, local ground-water-flow paths recharged at the water table seem to penetrate to the base of the surficial aquifer in most parts of the project area (Bachman 1984; Denver 1989, 1991). In addition, ground water at the base of the surficial aquifer was probably recharged after the early 1950s, as concluded from age dates inferred from concentrations of tritium (unpublished U.S. Geological Survey data) and chlorofluorocarbons (Plummer and Busenberg, written communication 1992). Therefore, nitrate that leached from the soil zone to the water table after the early 1950s has had sufficient time to migrate to deep sections of the surficial aquifer. Furthermore, the nitrate has not been chemically reduced along the flow paths because oxic conditions persist to the base of the surficial aquifer.

Concentrations of nitrate (as N) below 0.4 mg/L also are found at all depths in the surficial aquifer. These low concentrations are found in regions of the surficial aquifer along ground-water-flow paths recharged in woodlands or other nonagricultural or nonresidential land uses and in the surficial-confined region below the confining layer.

Regional Patterns

Hamilton, Shedlock, and Phillips (1991) showed that nitrate concentrations (as N) greater than 5 mg/L were found in the surficial aquifer in all of the HGMRs in the project area. With the exception of the fine-grained lowland, the MCL of 10 mg/L was exceeded in all HGMRs. However, their analysis also showed significant differences in median concentrations of nitrate among the hydrogeomorphic regions. The lowest median concentrations are in hydrogeomorphic regions with poorly drained soil or fine-grained, near-surface sediments (poorly drained lowland, 0.6 mg/L; surficial-confined region, 1.5 mg/L; and fine-grained lowland, <0.1 mg/L as N). The highest median concentrations of nitrate and greatest percentage of exceedances of the MCL were found in the well-drained upland (5.7 mg/L) and the poorly drained upland (3.1 mg/L).

The comparative analysis of nitrate concentration patterns among the HGMRs was repeated with the 1990 data set. Because of the small number of analyses in the lowland HGMRs, however, the data for these HGMRs were combined with data from the surficial confined region.

Analysis of the 1990 data set essentially confirms the general results of Hamilton, Shedlock, and Phillips (1991). Median concentrations of nitrate are statistically lower for the group containing data from the surficial-confined and lowland HGMRs than are the median concentrations of nitrate for the poorly drained uplands and the well-drained uplands. In addition, this analysis (Figure 23-11) indicates that the median concentration of nitrate is significantly higher in the well-drained upland (median = 8.9 mg/L as N) than in any other HGMR. Furthermore, the percentage of samples that exceed the MCL for nitrate is higher in the well-drained upland (about 33%) than in the poorly drained upland (about 20%).

These differences in nitrate concentrations are related, at least in part, to the ratio of woodland area to agricultural area in each HGMR (Table 23-1). For example, the well-drained upland has the lowest ratio of woodlands to agriculture, the highest median nitrate concentration, and highest frequency of high nitrate concentrations of all the HGMRs. The HGMR with the second highest median nitrate concentration, the poorly drained upland, has the third lowest ratio of woodland to agricultural land. These results also are consistent with those of Hamilton, Shedlock, and Phillips (1991), who showed that nitrate concentrations in the surficial aquifer are higher in agricultural and urban areas than in woodlands and that the percentage of samples that exceeded the MCL is highest in agricultural areas. Some of the differences in the range and median values of nitrate concentration are probably attributable to other factors such as physical and chemical differences in soils and agricultural practices. The well-drained

TABLE 23-1. Ratios of Woodland Areas to Agricultural Areas in Hydrogeomorphic Regions of the Delmarva Peninsula

Hydrogeomorphic Region	Ratio of Woodland Area to Agricultural Area
Surficial-confined region	1.23
Poorly drained lowland	1.12
Fine-grained lowland	0.82
Poorly drained upland	0.75
Inner coastal plain	0.44
Well-drained upland	0.38

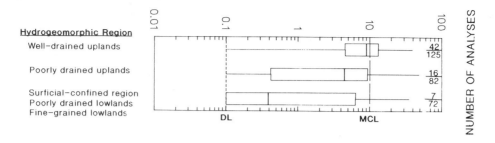

FIGURE 23-11. Nitrate concentrations in ground water in the surficial aquifer grouped by hydrogeomorphic region.

soil types in the well-drained upland are conducive to farming and could also be more favorable to nitrification than the poorly drained soil types that probably inhibit nitrification in other HGMRs.

Local Patterns

Factors affecting regional water-quality patterns and relations between ground-water flow and water quality were investigated in a series of local-scale well networks. Water-quality patterns in local networks representing the three largest HGMRs (well-drained upland, poorly drained upland, and surficial-confined region) are presented below to illustrate how local and regional water-quality patterns are related.

Well-Drained Upland

Water-quality patterns in the well-drained upland are illustrated in the Fairmount local network (Figure 23-12) investigated by Denver (1991). The network covers approximately 17 km² of interspersed agricultural and forested land in eastern Sussex County, Delaware. The land surface elevation increases from 3 m in the eastern part of the network to about 15 m in the western part. The area covering the network is typical of the well-drained upland in having relatively flat to gently rolling topography and a relatively deep water table. The soil is a well-drained sandy loam that is conducive to farming. The Fairmount area has agricultural plots that are large in comparison to watersheds in other HGMRs with smaller areas of well-drained soil. The agricultural fields are used mainly to grow corn and soybeans. Poultry manure has been applied to the fields in the past.

The surficial aquifer consists mainly of permeable quartz sand and gravel and ranges

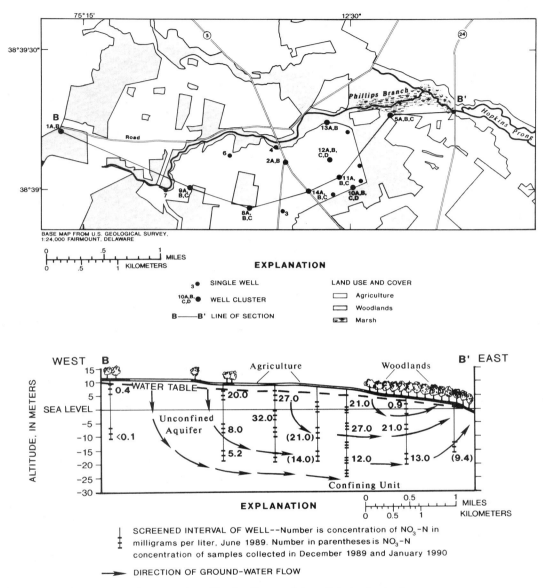

FIGURE 23-12. Map and cross section of local well network in well-drained uplands near Fairmount, Delaware, showing vertical changes of nitrate concentration.

from 24 to 48 m in thickness. Ground water generally flows from west to east, and flow paths range from less than 1 km to about 6 km long. In parts of the aquifer influenced by recharge through agricultural areas, the ground water is a calcium-magnesium-nitrate-type water. Nitrate concentrations (as N) in this ground water ranges from 1.5 to 38.0 mg/L.

The highest concentrations generally are found in near-surface ground water, directly below agricultural fields (Figure 23-12). Nitrate concentrations greater than 20 mg/L are present in the aquifer as deep as 25 m below land surface. In contrast, water that is recharged through wooded areas is a sodium-bicarbonate-type water with nitrate concen-

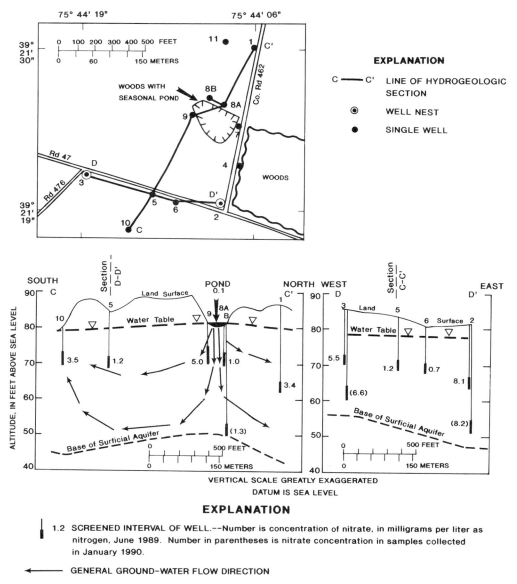

FIGURE 23-13. Map and cross sections of local well network in poorly drained uplands near Vandyke, Delaware.

trations that are less than 1 mg/L. In some places, water with low nitrate concentration overlies water with high nitrate concentration. This pattern indicates that near-surface ground water was recharged through soil covered with natural vegetation overlying ground water that was recharged through soils in upgradient agricultural areas. This pattern is seen on the eastern edge of the cross section in Figure 23-12 near the discharge area.

Poorly Drained Upland

Water-quality patterns in the poorly drained upland are illustrated in the Vandyke local network (Figure 23-13) in southwest New

Castle County, Delaware. The network covers approximately 85 ha of agricultural field with a wooded depression that is seasonally ponded. The area is characterized by hummocky topography, with small sluggish streams flowing in poorly defined shallow valleys with low gradients.

Agricultural plots in the poorly drained upland are much smaller and more highly interspersed with woodlands than in the well-drained upland. Corn and soybeans are grown in the agricultural fields. The higher elevations in the network are underlain by medium- to coarse-grained, well-drained, sandy loam, which is used for agriculture. The lower flats and depressions are underlain by poorly drained loam and are usually wooded and unsuitable for farming.

The surficial aquifer consists of fine- to medium-grained sand and is 9 to 12 m (30–40 ft) thick. The water table is flat in this area and is within 3 m (10 ft) of land surface everywhere in the network. Ground-water-flow paths are relatively short (less than 300 m long) and generally originate in the areas near the depression, and extend to streams and ditches at the margins of the sections shown in Figure 23-13.

Soil type and land use strongly affect the ground-water chemistry in this locale. Ground water recharged through well-drained agricultural soils is oxic and generally is a calcium-magnesium-nitrate-type water. Ground water recharged through poorly drained soils tends to be anoxic. In the anoxic waters, calcium and magnesium (from lime applications) also are the dominant cations. Chloride or bicarbonate are the dominant anions instead of nitrate, whose formation is inhibited by the redox conditions. In general, the concentration of dissolved iron is high in the anoxic waters, compared to oxic waters.

The concentration of nitrate in this network is <.1 to 8.2 mg/L, a range lower than that of samples from the Fairmount local-scale network in the well-drained upland. The lower range of concentrations probably is caused by several factors. First, some of the samples with high nitrate concentrations, paradoxically, contain high dissolved-iron concentrations, which indicates that these samples probably represent mixtures of oxic and anoxic water that are stratified throughout the vertical interval of the well screen. Differences in agricultural practices may also account for the differences in the range of nitrate concentration. For example, poultry manure is not applied to the fields around the Vandyke local-scale network. Soil processes also could contribute to the differences. Some of the soils in the Vandyke network have a higher organic content and a higher percentage of silt and clay than the soils in the Fairmount network. The silt, clay, and organic matter can limit or inhibit leaching of fertilizers from the soil zone. In summary, the interspersion of poorly drained soils with well-drained soils allows for the chemical evolution of both oxic and anoxic ground water in this area, producing locally high nitrate concentrations but generally lower nitrate concentrations than in the Fairmount network.

Surficial-Confined Region

Water-quality patterns in the surficial confined region are illustrated in the Willards local network (Figures 23-14 and 23-15). The Willards network covers approximately 15 km^2 of highly interspersed agricultural and forested land in eastern Wicomico County, Maryland. The area is essentially a flat, sandy plain, with low dune ridges that rise several meters above the surrounding plains. The plains contain permeable but poorly drained soils. The dune ridges contain sandy well-drained soils.

Most of the agricultural acreage rotates between corn and soybeans. Large chicken houses are located on the farms, and chicken manure is applied to the fields. Because of the low topographic relief and the poorly drained soils, the agricultural fields are covered with a network of drainage ditches that flow to the Pocomoke River on the east side of the network.

The surficial aquifer in the Willards network (Figure 23-14) is much more geologi-

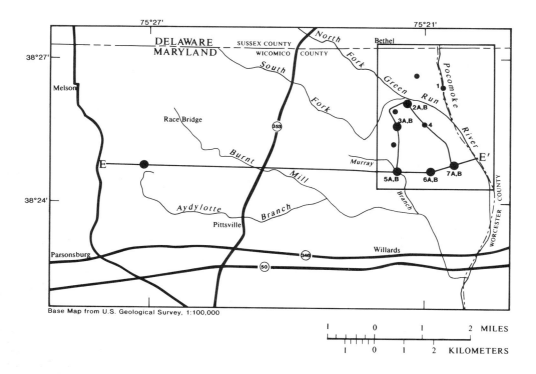
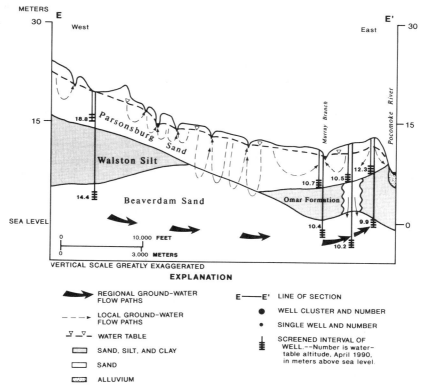

FIGURE 23-14. Map and cross section of local well network in the surficial confined region near Willards, Maryland.

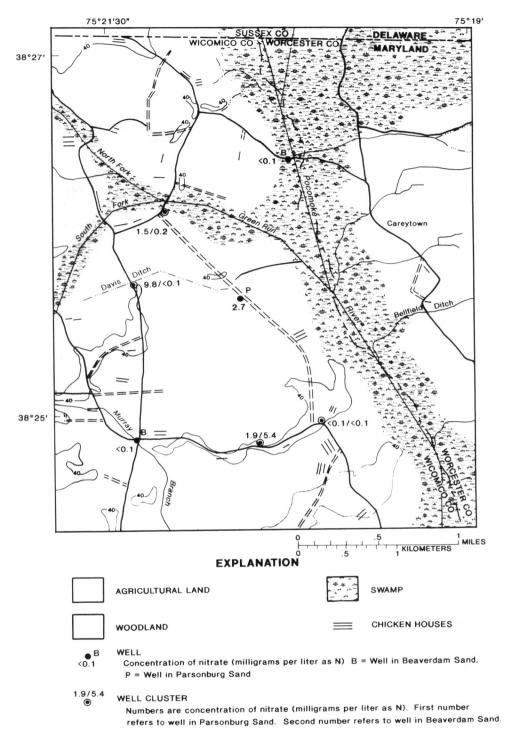

FIGURE 23-15. Map showing nitrate concentration of ground water in wells in the local network near Willards, Maryland.

cally heterogeneous than in the well-drained upland or poorly-drained upland. The major sand unit that composes the surficial aquifer, the Beaverdam Sand, is overlain by clay and silt of the Walston Silt, and by clay, silt, sand, and peat of the Omar Formation. This entire sequence is overlain by the Parsonsburg Sand, a windblown sand deposit, which forms dunes with broad sand plains between the dunes. The clay, silt, and peat deposits of the Walston Silt and the Omar Formation form a confining layer between the Parsonsburg Sand and the Beaverdam Sand. In some areas, the clay-silt deposits are absent.

Two types of ground-water-flow systems are found in the area of the Willards network. The first includes local flow systems that originate at highs in the water-table altitude in the Parsonsburg Sand. These water table highs are in interfluves between streams and ditches and in some of the dune ridges. The local flow paths are commonly less than 700 m long because of the spacing of drainage ditches in the agricultural fields. The second type of flow system is a regional system in the Beaverdam Sand. This system is recharged below topographic highs in the western part of the section shown in Figure 23-14, and is discharged to the Pocomoke River in the eastern part of the section, about 5 km from the recharge area. In most places within the surficial-confined region, the local and regional flow systems are separated by the confining layer represented by the fine-grained beds of the Walston Silt and the Omar Formation. The water-table altitude in the Parsonsburg Sand, however, is higher than hydraulic heads in the Beaverdam Sand in most places (Figure 23-14). In areas where the confining beds are absent, the local flow system probably extends into the upper sections of the Beaverdam Sand.

Shallow ground water in the Parsonsburg Sand is generally calcium-magnesium-chloride-sulfate-type water, with relatively low dissolved-iron concentrations (generally lower than 1,000 µg/l). Water in the Beaverdam Sand is a sodium-bicarbonate-type or an iron-bicarbonate-type water, with dissolved-iron concentrations generally higher than 1,000 µg/l. Concentrations of dissolved silica in ground water in the Beaverdam Sand are also higher than in ground water in the Parsonsburg Sand, which could be a function of the longer residence time of water in the Beaverdam Sand than in Parsonburg Sand.

Within the Willards local-scale network, nitrate concentrations (as N) range from less than the detection level of .1 to 9.8 mg/L in the Parsonsburg Sand. The higher concentrations are in areas of the Parsonsburg Sand where the ground water is oxic. Nitrate concentrations are generally at or near the detection level of .1 mg/L in the Beaverdam Sand, however. The one exception is at a site where the Omar Formation is sandy between the Parsonsburg Sand and the Beaverdam Sand. At this site, the nitrate concentration of water from the well in the Beaverdam Sand is 5.4 mg/L, higher than the nitrate concentration of 1.9 mg/L for the water from the adjacent well in the Parsonsburg Sand (Figure 23-15). The higher nitrate concentration of the water in the Beaverdam Sand at this site is probably caused by inflow of water from the Parsonsburg Sand under one of the adjacent agricultural fields. The low concentration of nitrate in water from the adjacent well in the Parsonsburg is probably caused by mixing of shallow ground water from an agricultural field immediately north of the well with shallow ground water from a woodland area less than 10 m south of the well.

Information Gained from Local-Scale Networks

The analysis of data from the local-scale networks suggests that the vertical distribution of nitrate in the surficial aquifer is a function of a combination of factors, including the hydrogeologic framework of the surficial aquifer, hydrochemical conditions in the soil and aquifer materials, ground-water-flow patterns, and land-use patterns. Analysis of the local networks indicates that the highest concentrations of nitrate generally are found in the upper part of the surficial aquifer below or adjacent to agricultural fields. Analysis of these networks also indicates, however, that high concentrations of nitrate in middle and lower parts of the surficial aquifer are caused by the downward migration of nitrate-rich ground water along flow paths that originate in upgradient agricultural areas.

The local networks were also used to evaluate the importance of geochemical processes in the surficial aquifer. The significance of denitrification was studied by measuring concentrations of dissolved gases (Dunkle, Plummer, and Busenberg, U.S. Geological Survey, written communication) and ratios of stable isotopes of nitrogen (Bohlke, U.S. Geological Survey, oral communication) in ground-water samples from selected wells in the local-scale networks. Dissolved-nitrogen gas concentrations and nitrogen isotope ratios consistent with denitrification were found in ground water below some, but not all, riparian zones near streams (Bohlke et al. 1992). In samples from both the local-scale and regional networks, chemical or isotopic evidence of denitrification was found in only a few cases from parts of the surficial aquifer outside of discharge zones.

Analysis of data from the local-scale networks also provides insight into the causes of water-quality patterns in each HGMR. The high nitrate concentrations and large number of samples that exceeded the MCL for ground water in the well-drained upland seem to be related to the high percentage of agricultural land and predominance of well-drained soils in this HGMR. The generally lower range of nitrate concentrations in the poorly drained upland seems to be related to the interspersion of poorly drained and well-drained soils within this HGMR and the effects of this interspersion on redox conditions in soil and ground water. Nitrate concentrations in ground water in the surficial-confined region are generally lower than in the other two upland HGMRs because of the hydrogeologic framework. The main part of the aquifer used for water supply, the Beaverdam Sand, is separated by a confining layer from the part of the aquifer that contains the water table, the Parsonsburg Sand. Locally high concentrations of nitrate, however, are found in ground water in the Beaverdam Sand in places where this confining layer is missing as in the Willards network.

This pattern in the Willards local-scale network offered an explanation for the high nitrate concentrations of many samples in the regional network from wells in the Beaverdam Sand. In all cases where geologic logs were available, occurrences of high nitrate concentrations in the Beaverdam Sand could be related to the absence of confining beds between the Beaverdam Sand and the Parsonsburg Sand.

Differences in agricultural practices also are probably responsible for some of the differences in water-quality patterns among the HGMRs. However, investigation of relations between agricultural practices and water-quality patterns was beyond the scope of this chapter.

CONCLUDING REMARKS

The design of this study provided the means to investigate ground-water-quality patterns in the Delmarva Peninsula at different spatial scales. For example, the regional networks provided data that allowed assessment of (1) regional patterns in natural ground-water chemistry and (2) regional distribution of nitrate in the ground-water system. The regional data set also provided a means to study nitrate-concentration patterns in different hydrogeologic settings and land-use categories.

The local-scale networks provided data that allowed (1) investigation of local study-area features and processes that control water-quality patterns and (2) recognition of these patterns in the regional data. Data from the local-scale networks provided an understanding of changes in nitrate concentration with depth along shallow-ground-water-flow paths in different hydrologic settings in the surficial aquifer. This information could not have been deduced only from the data from the regional-scale networks.

The hydrogeomorphic regions (HGMRs) serve as a conceptual link between regional and local patterns of water quality. Each region is an aggregate of hundreds of small watersheds with similar ground-water-flow systems and water quality. The data presented in this chapter suggest that water-quality patterns observed in the local networks repeat themselves in space in each HGMR. The repetition of the local patterns within each HGMR essentially forms the basis for the differences in regional patterns observed among the HGMRs.

In this investigation, recognition of the relations between regional-scale water-quality patterns and local-scale water-quality patterns was an iterative process. First, data from regional-scale networks were examined and analyzed for broad patterns and relations. Data from the local networks were then examined to verify patterns seen in the regional-scale data and to develop insights into the causes of water-quality patterns that were not obvious when the regional data were scanned or summarized. The regional data sets were then reanalyzed to test new hypotheses formed while evaluating the local-scale data.

This iterative analysis provided an understanding of how spatial changes in water quality are related to land-use patterns and ground-water-flow patterns. For example, the spatial distribution of nitrate in the surficial aquifer is not simply related to land-use patterns. Nitrate concentration patterns also are affected by hydrochemical conditions in the soil and aquifer materials and by shallow-ground-water-flow patterns. Elevated nitrate concentrations were commonly found in upland areas of the peninsula with well-drained soil and were only occasionally found in the coastal lowlands and areas overlain by fine-grained sediments, which tend to be dominated by poorly drained soil. Nitrates found at depth in the surficial aquifer seemed to be associated with agricultural land use upgradient in the ground-water-flow system.

For this reason, higher concentrations of nitrate are commonly found below lower concentrations of nitrate in wooded areas or in discharge areas downgradient of agricultural fields. This pattern was observed in the local-scale networks and later verified at selected sites in the regional data set.

A multiscale, multinetwork approach, as used in this investigation, can also provide an understanding of water-quality variability at different spatial scales. Knowledge of this variability is an important aspect of a regional water-quality assessment. This knowledge can be crucial to assess bias in a network, to evaluate the representativeness of data for a particular geographic region, and to design networks to monitor future trends in ground-water quality.

References
Bachman, L. J. 1984. *Nitrate in the Columbia Aquifer, Central Delmarva Peninsula, Maryland.* Towson, MD: U.S. Geological Survey Water-Resources Investigations Report 84-4322.
Bohlke, J. K., J. M. Denver, P. J. Phillips, C. J. Gwinn, L. N. Plummer, E. Busenberg, and S. A. Dunkle, 1992. Combined use of nitrogen isotopes and ground-water dating to document nitrate fluxes and transformations in small agricultural watersheds, Delmarva Peninsula, Maryland. *Eos Transactions of American Geophysical Union, Spring Meeting Supplement*, 73(14):140.
Cushing, E. M., I. H. Kantrowitz, and K. R. Taylor. 1973. *Water Resources of the Delmarva Peninsula.* Reston, VA: U.S. Geological Survey Professional Paper 822.
Denver, J. M. 1986. *Hydrogeology and Geochemistry of the Unconfined Aquifer, West-Central and Southwestern Delaware.* Newark, DE: Delaware Geological Survey Report of Investigations No. 41.
Denver, J. M. 1989. *Effects of Agricultural Practices and Septic-System Effluent on the Quality of Water in the Unconfined Aquifer in Parts of Eastern Sussex County, Delaware.* Newark, DE: Delaware Geological Survey Report of Investigations No. 45.
Denver, J. M. 1991. Groundwater-sampling network to study agrochemical effects on

water quality in the unconfined aquifer, southeastern Delaware. In *Groundwater Residue Sampling Design*, eds. R. G. Nash and A. R. Leslie, pp. 139–49. Washington, DC: American Chemical Society Symposium Series 465.

Gnanadesikan, R. 1977. *Methods for Statistical Data Analysis of Multivariate Observations*. New York: Wiley.

Hamilton, P. A., and J. M. Denver. 1990. Effects of land use and ground-water flow on shallow ground-water quality, Delmarva Peninsula, Delaware, Maryland, and Virginia. *Ground Water* 28(5):789.

Hamilton, P. A., R. J. Shedlock, and P. J. Phillips. 1991. *Water Quality Assessment of the Delmarva Peninsula, Delaware, Maryland, and Virginia—Analysis of Available Water-Quality Data through 1987*. Towson, MD: U.S. Geological Survey Water Supply Paper 2355-B.

Hardy, M. A., P. P. Leahy, and W. M. Alley. 1989. *Well Installation and Documentation, and Ground-Water Sampling Protocols for the Pilot National Water-Quality Assessment Program*. Reston, VA: U.S. Geological Survey Open-File Report 89-396.

Helsel, D. R., and R. M. Hirsch. 1992. *Statistical Methods in Water Resources*. New York: Elsevier.

Hirsch, R. M., W. M. Alley, and W. G. Wilber. 1988. *Concepts for a National Water-Quality Assessment Program*. Reston, VA: U.S. Geological Survey Circular 1021.

Kleiner, B., and T. E. Graedel. 1980. Exploratory data analysis in the geophysical sciences, Washington, DC: *Reviews of Geophysics and Space Physics* 18:699–717.

Koterba, M. T., R. J. Shedlock, L. J. Bachman, and P. J. Phillips. 1991. Regional and targeted ground-water quality networks in the Delmarva Peninsula. In *Groundwater Residue Sampling Design*, eds. R. G. Nash and A. R. Leslie, pp. 110–38. Washington, DC: American Chemical Society Symposium Series 465.

Phillips, P. J., R. J. Shedlock, and P. A. Hamilton. 1992. National water-quality assessment activities on the Delmarva Peninsula in parts of Delaware, Maryland, and Virginia. *In Proceedings of the 1988 U.S. Geological Survey Workshop on the Geology and Geohydrology of the Atlantic Coastal Plain*. Reston, VA: U.S. Geological Survey Circular 1059.

Ritter W. F., and A. E. Chirnside. 1982. *Ground-Water Quality in Selected Areas of Kent and Sussex Counties, Delaware*. Newark, DE: University of Delaware Agricultural Engineering Department.

Robertson, F. N. 1979. Evaluation of nitrate in ground water in the Delaware coastal plain. *Ground Water* 17(4):328–37.

U.S. Environmental Protection Agency. 1986. *Maximum Contaminant Levels (subpart B of part 141, National Primary Drinking-Water Regulations)*: U.S. Code of Federal Regulations, Title 40, parts 100 to 149, pp. 524–28, rev. July 1, 1986: Washington, DC: U.S. Environmental Protection Agency.

24

Ground-Water Quality in the Oklahoma City Urban Area

Scott Christenson and Alan Rea

INTRODUCTION

The Central Oklahoma aquifer underlies about 8,000 km² of central Oklahoma (Figure 24-1), where the aquifer is used extensively for municipal, industrial, commercial, and domestic water supplies. The aquifer was selected for study as part of the National Water-Quality Assessment (NAWQA) Program of the U.S. Geological Survey, and an investigation of the quality of ground water under the Oklahoma City urban area was undertaken as part of the project.

Geography of Central Oklahoma

The topography overlying the Central Oklahoma aquifer is characterized by low, rolling hills in the eastern two-thirds of the study area and by flat plains in the western third. The average annual temperature in the study area is about 16°C. The average annual precipitation is approximately 840 mm, most of which falls from April through October.

Oklahoma City ranked twenty-ninth in population among the major cities in the United States, according to data from the 1990 census. Oklahoma City (population 445,000) is surrounded by numerous smaller municipalities with populations as large as 80,000, as well as a large dispersed residential population outside the boundaries of the municipalities. Approximately 959,000 people live within the standard metropolitan statistical area. Even though Oklahoma City covers a large area and is surrounded by numerous other communities, the dominant land use in the study area is agricultural. Much of the agricultural land is used for raising livestock, although grain crops (principally wheat and sorghum) are grown in the area. Land use in the western third of the study area is agricultural and urban, and the eastern two-thirds of the study area is approximately evenly divided between agriculture and deciduous forest.

Geohydrology of the Central Oklahoma Aquifer

The Central Oklahoma aquifer consists of Permian geologic units and overlying Quaternary alluvium and terrace deposits (Figure 24-2). The Permian geologic units dip to the west at approximately 10 m/km. Most of the usable ground water is in the Garber Sandstone and the Wellington Formation. Substantial quantities of usable ground water also are in the Chase, Council Grove, and Admire Groups, which underlie the Garber

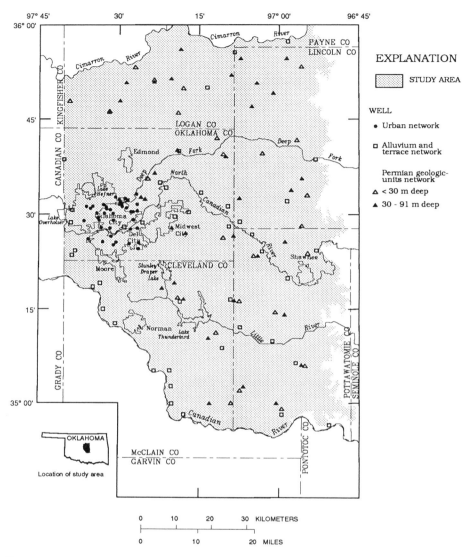

FIGURE 24-1. Geographic features of the study area and locations of wells sampled in low-density survey and urban sampling networks.

Sandstone and Wellington Formation, and in alluvium and terrace deposits, which are associated with the major streams in the study area. Most large-capacity wells completed in the Central Oklahoma aquifer are from 30 to 250 m deep and are completed in the bedrock geologic units. Domestic wells typically are from 10 to 50 m deep and are completed in either the bedrock or the alluvium and terrace deposits.

The Quaternary alluvium and terrace deposits are along streams and consist of lenticular beds of unconsolidated clay, silt, sand, and gravel. The thickness of the alluvium and terrace deposits ranges from 0 to about 30 m. The Permian geologic units consist of lenticular beds of fine-grained, cross-bedded sandstone interbedded with siltstone and mudstone. These units were deposited in a fluvial-deltaic sedimentary environment,

Design of the Sampling Program   591

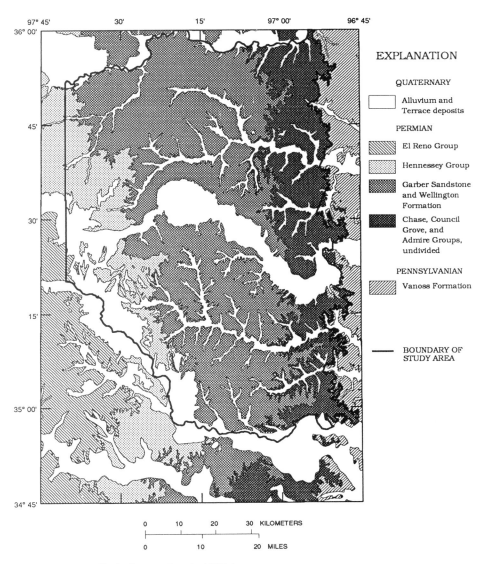

FIGURE 24-2.   Geologic map of central Oklahoma.

and the lithology is extremely variable, even over short distances.

Precipitation at a rate of about 40 mm/yr recharges the Central Oklahoma aquifer. A freshwater zone as much as 275 m thick is underlain by brines. Water in the western third of the aquifer is confined by the Hennessey Group. Most water that enters the flow system is discharged to streams in a relatively short time, but a deep, slow-moving flow system exists in the western, confined part of the aquifer.

DESIGN OF THE SAMPLING PROGRAM

Three ground-water-quality sampling networks are considered in this chapter. The sampling networks were designed to compare ground-water quality in the entire aquifer to ground-water quality in the urban area. Two networks were established to assess water quality in the entire aquifer: a network of wells completed in Permian geologic units, and a network of wells completed in the Quaternary alluvium and terrace deposits. These two networks are referred to as low-density survey sampling networks. The third network, an urban sampling network, was established within the central part of Oklahoma City and included the downtown business district, industrial areas, and older residential areas. The density of wells (number of wells per unit area) in the urban network was higher than the density of wells in the low-density survey sampling networks. All wells sampled in all three networks were existing water-supply wells equipped with operational pumps, and were generally, but not exclusively, domestic wells. Wells in these networks were sampled between April 5, 1988, and July 28, 1989, using methods described by Hardy, Leahy, and Alley (1989).

Low-Density Survey Sampling Networks

Random-selection techniques were employed for the two low-density survey sampling networks. These random-selection techniques were used to obtain water samples that were distributed throughout the study area, both areally and vertically, so that a representative sample of the water resource could be collected.

Wells for sampling were randomly selected using a computer program (Scott 1990). The study area was subdivided into areas where either the Permian geologic units or the alluvium and terrace deposits are present at the land surface. The computer program was designed to select points (one primary and two alternate points) such that all points in an area had an equal probability of being selected. The well nearest the primary point was located by field personnel and sampled if it met certain suitability requirements. A well was considered suitable for sampling if (1) it was completed in the appropriate geologic unit, (2) it was of the appropriate depth for the network, (3) it was equipped with a submersible pump, and (4) a water sample could be taken near the wellhead before the water passed through water-treatment systems or pressure tanks. A few wells were sampled that did not meet these criteria; extra purging time was allowed for those wells. If a suitable well could not be found near the computer-selected primary point, alternative points were used.

The freshwater zone in the Permian geologic units is as much as 275 m thick. To obtain water samples distributed vertically as well as areally, we divided the low-density survey sampling of Permian geologic units into three categories based on well depth: (1) shallow wells, less than 30 m (100 ft) deep, (2) intermediate-depth wells, 30 m to 91 m (300 ft) deep, and (3) deep wells, greater than 91 m deep. The results of the sampling of deep wells are not discussed in this chapter, because water from deep wells probably is unaffected by urbanization and because none of the wells in the urban sampling network were deeper than 91 m. Forty-two wells completed in alluvium and terrace deposits were sampled, and 25 shallow and 35 intermediate-depth wells completed in Permian geologic units were sampled (Figure 24-1).

Urban Sampling Network

A different random selection strategy was used to select wells for the urban network. Locating wells suitable for sampling is difficult in the Oklahoma City urban area because most of the water use in the urban area is from public water supplies, and wells frequently are not visible from the street. An announcement was placed in a local newspaper in an effort to locate wells. The announcement briefly explained the project and requested that well owners in the area contact the U.S. Geological Survey if they were interested in having their wells sampled. A local television station also had a news story on the project and asked well owners to telephone the U.S. Geological

Survey if they wished to have their wells sampled. Candidate wells were subject to the same sampling suitability requirements as the wells in the low-density survey sampling networks and were all less than 91 m (300 ft) deep. All wells located in this manner were plotted on a map of the urban area. A grid was placed over the map of the urban area, and wells were selected randomly from each grid cell. Forty-one wells were sampled for the urban network (Figure 24-1).

Sampling Biases

Although efforts were made to obtain a random, unbiased sampling of wells in each sampling network, some biases exist. One known bias is that the sampling was limited to wells that were operational and had pumps installed. If a known or obvious water-quality problem existed in an area, it is likely that wells would not be in use in that area and therefore such an area would not have been sampled. Other biases might be related to the willingness of the well owners to allow their wells to be sampled or to the similarity of wells constructed in a particular area, many of which may have been drilled by the same driller.

The manner in which the sampled wells were constructed was thought to bias the observed water quality. Domestic supply wells completed in the Central Oklahoma aquifer generally are drilled by rotary methods, cased to the full depth of the well, screened or perforated at depths opposite the most permeable strata, gravel packed from the bottom of the well to 3 m (10 ft) below land surface, and sealed with cement to the land surface. This method of construction increases the yield of a well, particularly wells completed in the Permian geologic units, where individual sandstone layers may produce only small amounts of water. This type of construction, however, also increases the likelihood that a well can be contaminated from substances introduced at or near the land surface. Even if most of the well's water is coming from deep strata, a contaminant introduced at or near the land surface could migrate down the gravel pack and be present in the produced water.

Oklahoma City is located at the eastern edge of the outcrop of the Hennessey Group. Thus, much of the Oklahoma City urban area is situated where a thin layer of the Hennessey Group overlies the Garber Sandstone, and wells are drilled through the Hennessey Group and completed in the Garber Sandstone. Outside the urban area, few wells were sampled where the Garber Sandstone is overlain by the Hennessey Group, so there could be a geologic bias in the urban sampling network. However, the concentrations of the inorganic constituents within the urban sampling network did not appear to be spatially correlated with the Hennessey Group.

Quality-Assurance Sampling

Quality-assurance sampling was used to evaluate the precision and accuracy of the well sampling. Several types of quality-assurance sampling were used:

1. *Blanks* Blanks are solutions that initially are free of the constituents of interest and subsequently processed in a manner designed to determine if the sampling or analytical processes introduce contaminants into the samples. Two different types of blanks were used in the field sampling program. *Trip blanks* consist of blank solutions that are put in the same type of bottle as those used for environmental samples. The trip blanks are kept with the environmental sample bottles both before and after sample collection. Trip blanks identify contaminants that might be introduced by the sample bottles or by diffusion into the sample bottle while the bottle is being transported. Trip blanks were used only for volatile organic compounds (VOCs), because VOCs were thought to be the only constituents likely to enter samples by these processes. *Field blanks* are blank solutions that are subjected to the same processes of sample collection, field processing, preservation, transportation, and laboratory handling as an environmental sample. Field blanks identify contaminants introduced by the process of collecting an environmental sample. Field blanks generally are used in conjunction with other types of quality-assurance samples to identify sources of contamination.

2. *Spikes* Spikes are samples to which known concentrations of specific constituents are added. *Field spikes* are used to determine the effects of preservation and transportation to the laboratory on the constituents of interest. The field spikes were added to samples immediately after the samples were collected and always in duplicate to assess the reproducibility of the spiking process. *Laboratory spikes* were added to samples immediately before the samples were analyzed and were used in conjunction with field spikes to evaluate the degradation of constituents from the time the samples were taken until laboratory analysis. Laboratory spikes also measure the effect of the chemical composition of the sample on the spiked compounds.

3. *Blind samples* Blind samples are samples of known composition that are submitted for analysis. Blind samples test the accuracy of the analytical procedures.

4. *Duplicate samples* Duplicate samples consisted of two or more samples collected from the same source, one immediately after the other, and analyzed in exactly the same manner. No sample splitting device was used. Duplicate samples measure the variability caused by the sampling and analytical procedures.

5. *Repeated samples* At some wells where the analytical determinations were equivocal, a second sample was collected and analyzed. This second sample was collected weeks or months after the first sample, and the quality of water in the well could have changed during that time.

Problems with sampling for inorganic constituents and VOCs were identified by the use of quality-assurance samples. Field blanks indicated that part of the sampling apparatus introduced dissolved zinc into samples. The filter used to remove suspended sediment from samples was suspected to have introduced the zinc, but because filtration is required by the analytical procedure, filtration of samples continued. Analytical determinations of dissolved zinc were not used in any analysis of data. A few field blanks were found to contain very small concentrations (at or just above the minimum reporting level) of dissolved copper and lead. About 90% of the environmental samples had concentrations of lead and copper below the minimum reporting level, so samples were not contaminated substantially by copper and lead.

Contaminated field blanks indicated that Teflon[1] tubing used as part of the sampling equipment during the alluvium and terrace network sampling was contaminated with trichlorofluoromethane. Additional rinsing of the Teflon tubing prior to sampling eliminated trichlorofluoromethane in subsequent field blanks. Trichlorofluoromethane reported in water samples collected with contaminated tubing was not included in the analyses used in subsequent discussions in this chapter. The term *reported* is used instead of *detected* because sometimes the presence of a constituent can be detected but the concentration is below the laboratory minimum reporting level.

Early in the sampling program methylene chloride was reported in water used for rinsing sampling equipment. Environmental samples collected during this period could have been contaminated by this water. A different source of water was used for subsequent sampling. Methylene chloride is a

---

[1] Any use of trade, product, or firm names is for descriptive purposes only and does not imply endorsement by the U.S. Government.

common laboratory solvent and is thought to have contaminated several environmental samples. The presence of methylene chloride in an environmental sample could have been caused by contamination of the sample and was not considered to indicate that methylene chloride was present in the aquifer.

Recovery rates for spiked pesticide samples ranged from 20% to 160%, with an average of 86%. Recovery rates for spiked VOC samples ranged from 36% to 102%, with an average of 70%. At one well, some spiking compounds in the VOC field-spiking solution were completely lost from the solution, and new compounds were reported that were not present in either the environmental sample or the spiking solution. The reasons for this are unknown, and the results of this spiked sample are not included in the recovery statistics. Blind samples produced data that compared favorably with the most-probable values of reference materials having known compositions.

In summary, the quality-assurance sampling indicated that results for pesticides and inorganic constituents (except for zinc) were reproducible and reasonably accurate. Problems with sampling for VOCs were identified by the use of quality-assurance sampling and corrected.

METHODS USED FOR CHARACTERIZING URBAN LAND USE

Several methods were used in this investigation for characterizing the land use near the sampled wells. Because there is no clear best method, the results of the different methods are compared. If several of the methods indicate the same relations between urban land use and ground-water quality, then confidence in the validity of those relations is increased. Similarly, if the results of the different methods do not agree, the results could be an artifact of the methods used rather than cause-and-effect relations in the hydrologic system. The methods of categorizing urban land use that are discussed are based on the sampling network, field sheets, point-in-polygon overlay, buffer-overlay analysis, and particle-tracking analysis.

Sampling Network

As discussed previously, the sampling networks were designed to compare groundwater quality in the urban area to groundwater quality in the entire aquifer. The urban sampling network targeted the Oklahoma City urban area, and the low-density survey sampling networks were designed to assess overall water quality in different geologic units of the Central Oklahoma aquifer. Therefore, the first method of characterizing urban land use is based on comparing ground water from the urban network to ground water from the low-density survey networks. Because the concentrations of inorganic constituents were found to be dependent on geology (Parkhurst, Christenson, and Schlottmann 1989), comparisons for inorganic constituents were made only between wells completed in the Garber Sandstone or Wellington Formation.

Field Sheets

Field sheets were filled out by project personnel describing predominant land use within a 400-m (quarter-mile) radius of each well. Land uses from the field sheets were reclassified into urban and nonurban categories.

Point-in-Polygon Overlay

Another method used to characterize land use near the sampled wells was to overlay the well locations on a map of land use. This process was automated by using a computerized geographic information system (GIS) to store and manipulate map information, known as *themes*, in digital form. The land-use map used was the Geographic Information Retrieval and Analysis System (GIRAS) Land-Use and Land-Cover Map (Fegeas et al. 1983). This map was compiled and digitized from high-altitude aerial photography (taken in 1975 and 1981) at a base scale of 1:250,000.

The simplest method of overlaying well locations on land-use maps is known as a point-in-polygon overlay. With this method, the land use assigned to a well is that of the polygon which the well location overlays on the land-use map. If there are errors in the well location or errors in the boundaries on the land-use map, there is a chance the land use at the well will be misclassified.

Misclassification also can occur as a result of the generalization that is inherent in any map. The amount of generalization is mainly a function of map scale. For the GIRAS land-use map, polygons of some land-use categories having areas smaller than 4 ha or widths less than 200 m were not digitized. The minimum delineation area for other land-use categories was 16 ha, and minimum width was 400 m (Fegeas et al. 1983). This generalization eliminated polygons too small to distinguish easily at the base map scale of 1:250,000. Therefore, any well located in a land-use area smaller than the minimum delineation area would be misclassified when overlaid on the GIRAS land-use map.

Buffer-Overlay Analysis

One common approach used to reduce errors associated with misclassification is to draw a circular buffer around the well location and overlay the buffer on the other map themes. The buffer area also serves as a simplified representation of the recharge area for a well. By use of the GIS, buffers with a radius of 400 m (one-quarter mile) were generated around each well. The area of every land-use category within each buffer was calculated, and from this the predominant (largest area) land-use category within each buffer, as well as the presence or absence of urban land use in each buffer, was determined. In effect, the buffer-overlay analysis is used to summarize the land use in the vicinity of each well and in the simplified representation of the well's recharge area. This summarization reduces the effects of misclassification

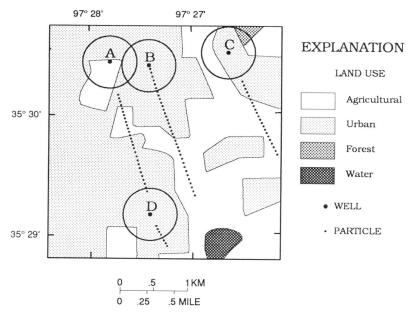

FIGURE 24-3. Four urban network wells with 400-m-radius buffers and particles overlaid on land use.

errors. As an example, four wells in the urban network and their buffers overlaid on the GIRAS land-use map are shown in Figure 24-3.

A buffer-overlay analysis also was used with a digital map of sewered areas, which areas were considered to be a measure of urban land use. Urban areas generally are served by municipal sewage-disposal systems. In contrast, septic systems generally are used in less populated areas. The areas served by municipal sewer systems also generally coincide with the areas served by municipal water-supply systems. The map of sewered areas was based on information from county and municipal health departments. The information included generalized boundaries of the areas served by municipal sewer systems. The base map used was at the 1:250,000 scale, but the boundaries were generalized, so the effective scale of the map is considered to be smaller than 1:250,000.

A few problems exist with the buffer-overlay approach. Hay and Battaglin (1990) determined that buffer size affected the statistical significance of correlations between land use and ground-water quality. Barringer et al. (1990) suggested eliminating wells with overlapping buffers from statistical analyses because of spatial autocorrelation. Of the wells in the sampling networks presented in this chapter, only 10 pairs of buffers overlapped; the maximum amount of overlap was about 22% of the area of a buffer. Because the buffer areas were used only to categorize the land use for a well as urban or nonurban, and the number of overlaps was small, spatial autocorrelation was not considered to be a problem.

Other problems associated with the buffer-overlay analysis are that equal weight is given to land uses in all directions from a well, equal weight is given to land uses at all distances within the buffer radius from the well, and the circular buffer is a poor representation of a well's recharge area. To ameliorate the effects of these problems, a particle-tracking analysis was developed and is presented here.

Particle-Tracking Analysis

A ground-water-flow model for the study area was developed so that the recharge area for each well could be estimated. The ground-water-flow model (McDonald and Harbaugh 1988) uses a finite-difference approach; the Central Oklahoma aquifer was discretized into three-dimensional cells 2 km on each side and 30 m (100 ft) thick. A particle-tracking program (Pollock 1989) was used in conjunction with the ground-water-flow model to trace hypothetical particles of water backward from each well to its recharge area. Particles were placed along each simulated well between the base of the well and the water table, as interpolated from the water-table map in Christenson, Morton, and Mesander (1990).

Particles shown on Figure 24-3 represent the recharge locations for water flowing into each well calculated by the ground-water-flow model and particle-tracking program. The finite-difference model is based on the assumption of steady-state conditions, with no pumping of the individual wells. Under these assumptions, the model best shows regional ground-water flow, not the flow conditions near a pumped well, but improving the model to better simulate conditions around each well is beyond the scope of this investigation. In a manner similar to the buffer-overlay analysis, the use of a large number of particles leads to a summarization process in the categorization of the land use associated with wells. This summarization reduces the effects of misclassification errors. The recharge area for each well was categorized as having an urban or nonurban land use, based either on predominance (the category associated with the largest number of particles) or on the presence or absence of any particle recharging from an area of urban land use.

Figure 24-3 illustrates how different methods of classifying urban and nonurban land use may yield different results for some wells. The land use at well A is classified as agricultural by a simple point-in-polygon overlay, but is classified as predominantly urban by a buffer-overlay or particle-tracking analysis. The land use at well C is classified as urban by either point-in-polygon or buffer overlay, but the majority of particles associated with the well recharge through agricultural land. All four wells are classified as urban using a presence-absence analysis with either a buffer-overlay or particle-tracking analysis.

Land use at wells was classified as either urban or nonurban using nine different combinations of characterization methods and map themes. The sewered-areas map was not used in a point-in-polygon overlay or presence-absence overlay with either well buffers or particles, because of its low level of accuracy relative to the other map themes used. The different methods of categorizing urban land use resulted in somewhat different sets of wells being categorized as urban and nonurban.

GROUND-WATER QUALITY

The results of the chemical analyses for the sampling networks are summarized in Tables 24-1, 24-2, and 24-3. All samples were analyzed by the U.S. Geological Survey's National Water-Quality Laboratory using standard methods.

TABLE 24-1. Median Values of Selected Common Constituents, Nutrients, and Trace Elements

	Sampling Network			
		Permian Geologic Units		
Constituents and Properties	Urban	Shallow	Intermediate Depth	Alluvium and Terrace Deposits
pH (standard units)	7.2	6.9	7.3	7.2
Oxygen, dissolved (mg/L)	5.2	4.9	6.9	2.3
Alkalinity, total (mg/L)	310.	259.	280.	258.
Calcium, dissolved (mg/L)	88.	42.	54.	55.
Magnesium, dissolved (mg/L)	42.	21.	26.	20.
Sodium, dissolved (mg/L)	72.	30.	30.	39.
Potassium, dissolved (mg/L)	1.9	1.1	1.0	1.3
Bicarbonate, total (mg/L)	378.	315.	342.	315.
Carbonate, total (mg/L)	0.	0.	0.	0.
Sulfate (mg/L)	37.	18.	22.	34.
Chloride (mg/L)	76.	19.	20.	22.
Fluoride, dissolved (mg/L)	0.2	0.3	0.3	0.3
Nitrite plus nitrate (mg/L)	2.0	0.75	0.93	0.87
Arsenic, dissolved (μg/L)	1.	<1.	1.	<1.
Barium, dissolved (μg/L)	150.	180.	240.	230.
Cadmium, dissolved (μg/L)	<1.	<1.	<1.	<1.
Chromium, dissolved (μg/L)	<5.	<5.	<5.	<5.
Copper, dissolved (μg/L)	<10.	<10.	<10.	<10.
Iron, dissolved (μg/L)	16.	5.	3.	11.
Lead, dissolved (μg/L)	<10.	<10.	<10.	<10.
Manganese, dissolved (μg/L)	1.	<1.	<1.	6.
Mercury, dissolved (μg/L)	<0.1	<0.1	<0.1	<0.1
Selenium, dissolved (μg/L)	1.	<1.	<1.	<1.
Silver, dissolved (μg/L)	<1.	<1.	<1.	<1.
Gross-alpha particle activity (pCi/L)	9.1	3.4	5.	3.1
Radon, total (pCi/L)	200.	150.	120.	200.
Uranium, dissolved (pCi/L)	3.05	1.09	0.96	1.51
Number of wells in network	41	25	35	42

TABLE 24-2. Number of Wells Where Organic Compounds Were Reported in Water Samples

	Sampling Network				Laboratory Reporting Level (μg/L)	Maximum Reported Concentration (μg/L)
	Urban	Permian Geologic Units		Alluvium and Terrace Deposits		
Compound		Shallow Depth	Intermediate Depth			
Organochlorine Pesticides						
Aldrin	1	0	0	0	0.01	0.01
Chlordane	5	1	0	0	0.1	0.8
Dieldrin	6	1	1	0	0.01	0.58
DDE	0	1	0	0	0.01	0.01
Phenoxy-Acid Herbicides						
2,4-D	2	2	1	0	0.01	0.02
2,4-DP	0	0	0	1	0.01	0.01
2,4,5-T	0	0	0	1	0.01	0.01
Picloram	0	1	0	1	0.01	0.01
Dicamba	0	0	1	1	0.01	0.01
Triazine and Other Nitrogen-Containing Herbicides						
Atrazine	0	0	1	2	0.1	0.3
Prometone	4	0	0	1	0.1	3.5
Volatile Organic Compounds						
Bromoform	2	0	1	0	0.2	2.1
Carbon tetrachloride	3	0	0	0	0.2	1.6
Chlorodibromomethane	2	0	1	0	0.2	11.
Chloroform	19	2	3	0	0.2	25.
Chloroethane	0	0	0	1	0.2	0.2
Chloromethane	0	0	1	0	0.2	1.0
1,2-Dibromoethane	1	0	0	0	0.2	0.2
Dichlorobromomethane	5	0	1	0	0.2	25.
Dichlorodifluoromethane	2	0	0	1	0.2	0.7
1,1-Dichloroethane	0	0	0	2	0.2	2.5
1,2-Dichloroethane	2	1	0	1	0.2	0.4
1,2-Dichloroethene	1	0	0	2	0.2	1.0
1,1-Dichloroethylene	0	0	0	2	0.2	45.
Methylene chloride	2	1	0	2	0.2	0.7
Styrene	1	0	0	0	0.2	0.2
1,1,2,2,-Tetrachloroethane	0	0	1	0	0.2	0.3
Tetrachloroethylene	6	0	0	0	0.2	2.1
Toluene	0	1	0	0	0.2	0.2
1,1,1-Trichloroethane	0	0	0	1	0.2	200.
Trichloroethylene	2	0	0	3	0.2	90.
Trichlorofluoromethane	0	0	2	1	0.2	0.3
Number of wells in network	41	25	35	42		

TABLE 24-3. **Organic Compounds Not Reported in Any Ground-Water Samples**

Volatile Organic Compounds[a]
 Benzene
 Bromobenzene
 Chlorobenzene
 1,2-Dichlorobenzene
 1,3-Dichlorobenzene
 1,4-Dichlorobenzene
 Ethylbenzene
 1,1,1,2-Tetrachloroethane
 1,1,2-Trichloroethane
 Dibromomethane
 Methyl bromide
 1,2-Dichloropropane
 1,3-Dichloropropane
 2,2-Dichloropropane
 1,2,3-Trichloropropane
 1,1-Dichloropropene
 cis-1,3-Dichloropropene
 trans-1,3-Dichloropropene
 ortho-Chlorotoluene
 para-Chlorotoluene
 Vinyl chloride
 Xylene

Carbamate Insecticides[b]
 Aldicarb
 Aldicarb sulfone
 Aldicarb sulfoxide
 Carbofuran
 3-Hydroxycarbofuran
 Methiocarb
 Methomyl
 1-Napththol
 Oxamyl
 Propham
 Propoxur
 Sevin

Organochlorine Pesticides[c]
 DDD
 DDT
 Endosulfan
 Endrin
 Heptachlor
 Lindane

Organochlorine Pesticides (continued)
 Methoxychlor
 Mirex
 Naphthalenes, polychlorinated
 Perthane
 Toxaphene

Phenoxy-Acid Herbicides[c]
 Silvex

Organophosphorus Insecticides[c]
 DEF
 Diazinon
 Disulfoton
 Ethion
 Trithion
 Malathion
 Methyl parathion
 Methyl trithion
 Parathion
 Phorate

Triazine and Other Nitrogen-Containing Herbicides[d]
 Alachlor
 Ametryne
 Bromacil
 Butachlor
 Butylate
 Carboxin
 Cyanazine
 Cycloate
 Diphenamid
 Hexazinine
 Metolachlor
 Metribuzine
 Prometryne
 Propachlor
 Propazine
 Simazine
 Simetryne
 Terbacil
 Trifluralin
 Vernolate

[a] Minimum reporting level = 0.2 micrograms per liter (µg/L).
[b] Minimum reporting level = 0.5 µg/L.
[c] Minimum reporting level = 0.01 µg/L, except polychlorinated naphthalenes = 0.1 µg/L, and toxaphene = 1.0 µg/L.
[d] Minimum reporting level = 0.1 µg/L.

Inorganic Constituents

A summary of the results of the chemical analyses for inorganic constituents is shown in Table 24-1. Table 24-1 lists median values for major cations and anions and constituents for which the U.S. Environmental Protection Agency has established maximum contaminant levels (MCLs) or secondary maximum contaminant levels (except for zinc; see the "Quality-Assurance Sampling" section).

Parkhurst, Christenson, and Schlottmann (1989) found that in the Central Oklahoma aquifer, the concentrations of certain ions in ground water were related to geologic unit. For example, concentrations of arsenic, chromium, and selenium (trace elements that are a substantial problem in central Oklahoma) commonly exceeded the MCLs in wells completed in deeper zones of the Garber Sandstone and Wellington Formation, but rarely exceeded the MCLs in other geologic units. Because the effect of geology on inorganic constituents is well documented by Parkhurst, Christenson, and Schlottmann (1989), no further analysis of this factor was done for this chapter. When comparing inorganic constituents in subsequent parts of this chapter, only analyses from wells completed in the Garber Sandstone and Wellington Formation were compared.

Organic Constituents

The number of wells in each network where organic compounds were reported and the maximum concentrations reported are listed in Table 24-2. Eleven different pesticides and 21 different VOCs were reported. Most organic compounds were reported at relatively small concentrations, with the exception of several VOCs. The large concentrations of 1,1,1-trichloroethane, trichloroethylene, and 1,1-dichloroethylene were all from a single well, and the next-largest concentration of each of these three compounds was considerably lower than the maximum concentration. This particular well was a domestic well that apparently was contaminated by a solvent spill approximately 50 m from the well. The well owners were unaware of the problem at the time the well was sampled.

Table 24-3 shows organic constituents for which samples were analyzed but not reported by the laboratory in any water samples. Because most organic constituents that were reported were at concentrations just above the laboratory reporting level (Table 24-2), it is possible that some wells produce water containing some of the organic constituents in Table 24-3 at concentrations too small to be reported with the analytical procedures used.

Other studies have reported some of the same organic constituents in urban ground water as were reported in this study. Lawrence and Whitney (1990) reported prometone, tetrachloroethylene, and trichloroethylene in ground water in the Carson City, Nevada, urban area. Eckhardt, Siwiec, and Cauller (1989) reported dieldrin, chlordane, 1,1,1-trichloroethane, tetrachloroethylene, and trichloroethylene in urban and suburban areas on Long Island, New York.

FACTORS RELATED TO GROUND-WATER QUALITY

One of the primary objectives of the project was to determine the natural and anthropogenic factors related to ground-water quality. The factors discussed in this chapter are tritium concentration, geology, and urban land use.

Tritium Concentration

The tritium concentration in a water sample is not a causative factor (a factor that may affect the quality of water produced by a well) but was found to be strongly related to water quality, especially the presence or absence of pesticides. Water samples from wells with tritium concentrations above the laboratory minimum reporting level of 0.3 picocurie per liter (pCi/L) are considered to contain some water that entered the ground-water-flow system after 1952 (see Chapter 11 for a discussion of tritium in ground water).

Pesticides were reported only in samples from wells that had reportable tritium. This was expected because organic pesticides were not commonly used in the United States until about the time tritium concentrations became elevated. Thus, tritium concentration is an indicator of a well's vulnerability to pesticide contamination. The presence of reportable tritium does not indicate that a well will be contaminated with pesticides; rather, reportable tritium indicates that there are flow paths to the well by which substantial quantities of post-1952 recharge water enter the well. The absence of reportable tritium indicates that there are no flow paths to the well by which substantial quantities of post-1952 recharge water enter the well.

In contrast, VOCs were reported in water samples from five wells that had no measurable tritium. The VOCs reported in these five wells were 1,2-dichloroethane, 1,2-dichloroethene, trichloroethylene, bromoform, chlorodibromomethane, chloroform, dichlorobromomethane, chloromethane, trichlorofluoromethane, and 1,1,2,2-tetrachloroethane. These VOCs are trihalomethanes, organic solvents, or chlorofluorocarbons, all of which were in use prior to 1952. Thus, the presence or absence of measurable tritium in a well cannot be taken to indicate a well's vulnerability to VOC contamination.

Geology

The relation between the geologic unit in which the sampled wells were completed and the presence or absence of organic constituents was tested using a contingency-table analysis with a chi-squared test statistic (P-STAT, Inc. 1989). Contingency tables are used with categorical variables. Geologic units were divided into three categories, depending on the primary aquifer contributing water to the well: alluvium and terrace deposits, Garber Sandstone and Wellington Formation, or Chase, Council Grove, and Admire Groups. Wells were categorized either as having or not having organic compounds. Wells were categorized as having organic compounds if any compound in the group of organic compounds was reported in the sample. Organic constituents were grouped as pesticides and VOCs because, with the exception of chloroform, no single constituent was detected in more than a few samples. If the expected value for any cell of a contingency table is too small, the assumptions of the contingency-table analysis are not valid.

The null hypothesis for the contingency-table analysis was that the proportion of wells in which organic constituents were reported was the same in each geologic unit, and the alternative hypothesis was that the proportion of wells in which organic constituents were reported was different among geologic units. The null hypothesis was rejected if the p-value of the test was below a significance level of .05. For pesticides, the calculated p-value was .695, and for VOCs the p-value was .979, both much larger than .05. Thus the null hypothesis that the proportion of wells in which organic constituents were reported was the same in each geologic unit was accepted. The lack of a statistically significant relation between geologic units and the detection of organic constituents was unexpected because the Permian geologic units are very different from the alluvium and terrace deposits. A possible cause of this unexpected result is that the method of construction of the sampled wells may allow contaminants to enter the well around the shallow cement seal and through the gravel pack, regardless of the geologic unit in which the well is completed (see the discussion in the "Sampling Biases" section).

Urban Land Use

Relations between urban and nonurban land use and water quality were assessed using the Mann-Whitney test (P-STAT, Inc. 1989) for inorganic constituents and contingency tables with a chi-squared statistic for organic constituents. The null hypothesis, in general, was that the populations, in this case the concentrations or presence of chemical constituents in ground-water samples, were the same in the urban and nonurban wells. The alternative hypothesis was that the populations were different in the urban and nonurban wells. The null hypothesis was rejected if the p-value of the test was less than .05.

Inorganic Constituents

The Mann-Whitney test was used for comparing inorganic constituents because the concentrations of inorganic constituents are continuous and because this nonparametric test does not require assumptions about the population distributions. The p-values from the Mann-Whitney tests for the nine different methods of categorizing urban land use are shown in Table 24-4. The following constituents always had a p-value greater than .05: dissolved oxygen, carbonate, sulfate, nitrite plus nitrate, arsenic, barium, copper, lead, mercury, selenium, and silver. For these constituents, the null hypothesis was accepted; that is, the populations of these constituent concentrations were the same in urban and nonurban wells.

The following constituents always had a p-value less than .05: alkalinity, calcium, magnesium, sodium, potassium, bicarbonate, chloride, iron, gross-alpha radioactivity, and uranium. For these constituents, the null hypothesis was rejected and the alternative hypothesis was accepted; that is, the concentrations of these constituents were significantly different in urban wells than those in nonurban wells regardless of the method used to categorize urban land use. Also, the median concentrations of these constituents were always higher in the urban wells than in the nonurban wells regardless of the method used to categorize urban land use. The remaining constituents, pH, fluoride, cadmium, chromium, manganese, and radon, had p-values both greater than and less than .05, depending on the method of land-use categorization.

Organic Constituents

Almost all the organic compounds in water samples were reported at concentrations at or slightly above the laboratory reporting level. The presence or absence of organic compounds in water samples and the land use at wells are categorical variables and may be compared by using contingency tables with a chi-squared test statistic. The null hypothesis was that the proportion of wells with reportable organic compounds was the same in urban areas as in nonurban areas. The alternative hypothesis was that the proportion of wells with reportable organic constituents was different in urban and nonurban areas.

Organic constituents were grouped together because almost all individual organic constituents were reported in only a few water samples. Pesticides were treated as a group, as were VOCs. In addition, chloroform was reported in 19 wells in the urban network and 5 wells in the Permian geologic unit networks and could be used in the contingency tables by itself. Thus, contingency tables were prepared for the following groupings: (1) pesticides, (2) VOCs, (3) VOCs except chloroform, and (4) chloroform.

The p-values calculated from the contingency tables are listed in Table 24-5. The land-use categorizations were the same as for inorganic constituents. For chloroform by itself and for VOCs, the p-values were less than .05 for all land-use categorizations. Thus, the null hypothesis was rejected and the alternative hypothesis accepted; that is, the proportion of wells with reportable chloroform or reportable VOCs was different in the urban and nonurban areas. For the pesticide and the VOCs except chloroform groupings, some p-values were greater than and some less than .05, depending on which land-use categorization method was used. Thus the null hypothesis was either accepted or rejected for these two groupings, depending on the method of land-use categorization.

Discussion

The comparisons by sampling network (which represent the basis of the sampling design) consistently indicated that the presence or absence of organic constituents, as well as the concentrations of many of the inorganic constituents, were significantly different in urban and nonurban wells. Additionally, the median concentrations of all inorganic constituents that were significantly different for any method of land-use categorization

TABLE 24-4. *P*-Values from Mann-Whitney Tests Comparing the Concentrations of Inorganic Constituents in Urban and Nonurban Land Uses

Constituents and Properties	Networks	Field Sheets	Point-in-Polygon GIRAS	400-m Buffers			Particle-Tracking Analysis		
				GIRAS		Sewered Areas	GIRAS		Sewered Areas
				Predominant	Presence/Absence		Predominant	Presence/Absence	
pH	.0167	.2567	.1108	.1140	.0029	.0234	.1103	.2100	.0309
Oxygen, dissolved	.6740	.5943	.8479	.4467	.4614	.7107	.4758	.3557	.7057
Alkalinity	.0001	.0030	.0021	.0002	.0007	.0039	.0000	.0004	.0053
Calcium	.0000	.0226	.0018	.0003	.0020	.0017	.0000	.0001	.0023
Magnesium	.0000	.0120	.0006	.0003	.0007	.0007	.0000	.0001	.0009
Sodium	.0002	.0002	.0001	.0000	.0001	.0046	.0000	.0002	.0056
Potassium	.0001	.0010	.0001	.0000	.0002	.0004	.0000	.0000	.0004
Bicarbonate	.0001	.0028	.0019	.0002	.0006	.0037	.0000	.0003	.0050
Carbonate	.4334	.6685	.5600	.5312	.3787	.4749	.4842	.4142	.4701
Sulfate	.6611	.1741	.2114	.1539	.3242	.6527	.1485	.1734	.6205
Chloride	.0002	.0025	.0001	.0001	.0004	.0025	.0000	.0000	.0030
Fluoride	.0126	.2564	.2713	.2904	.0454	.0450	.0347	.0285	.0372
Nitrite plus nitrate	.8823	.4041	.5344	.8781	.7498	.4189	.7262	.7191	.4176
Arsenic	.5490	.6355	.6422	.9349	.3262	.7807	.7197	.9593	.7127
Barium	.5071	.9493	.8222	.8730	.7942	.4365	.6661	.8526	.4883
Cadmium	.0037	.7586	.0686	.1060	.0078	.0014	.1878	.0685	.0016
Chromium	.2461	.6571	.6357	.7971	.0290	.1379	.5140	.3582	.1480

Constituent	Networks	Field Sheets	Point-in-Polygon GIRAS	Predominant	Presence/Absence	Sewered Areas	Predominant	Presence/Absence	Sewered Areas
Copper	.0555	.1443	.2550	.1885	.1255	.3237	.1508	.2461	.4191
Iron	**.0000**	**.0012**	**.0000**	**.0001**	**.0000**	**.0001**	**.0000**	**.0003**	**.0001**
Lead	.3023	.3698	.7304	.8241	.3885	.2238	.9713	.3416	.2323
Manganese	**.0085**	.2203	**.0241**	**.0137**	**.0113**	**.0025**	**.0068**	**.0099**	**.0031**
Mercury	.4698	.3773	.3994	.5295	.2541	.7735	.2823	.5506	.7967
Selenium	.7564	.1241	.3280	.5903	.5703	.3684	.5142	.6734	.4037
Silver	.1729	.7307	.8819	.9345	.3024	.0854	.2803	.2266	.0926
Gross-alpha	**.0001**	**.0015**	**.0013**	**.0008**	**.0001**	**.0003**	**.0002**	**.0013**	**.0005**
Radon	**.0018**	.6555	.2921	.0919	**.0129**	**.0089**	.0278	.0696	**.0118**
Uranium	**.0023**	**.0085**	**.0142**	**.0124**	**.0008**	**.0021**	**.0060**	.0246	**.0029**

Notes: GIRAS = Geographic Information Retrieval and Analysis System.
Underlined numbers indicate concentrations are significantly different (*p*-value less than .05) in wells in urban and nonurban land uses.

TABLE 24-5. *P*-Values from Contingency-Table Analyses Comparing the Presence or Absence of Organic Constituents in Urban and Nonurban Land-Use Categories

| | | | | 400-m Buffers | | | Particle-Tracking Analysis | | |
| | | | | GIRAS | | | GIRAS | | |
Constituents	Networks	Field Sheets	Point-in-Polygon GIRAS	Predominant	Presence/Absence	Sewered Areas	Predominant	Presence/Absence	Sewered Areas
Pesticides	**.042**	.069	**.024**	**.041**	.128	**.006**	.530	.927	.238
All VOCs	**.000**	**.000**	**.000**	**.000**	**.000**	**.000**	**.001**	**.012**	**.010**
VOCs except chloroform	**.000**	**.000**	**.000**	**.000**	**.000**	**.000**	.165	.422	.238
Chloroform	**.000**	**.000**	**.000**	**.000**	**.000**	**.000**	**.000**	**.003**	**.010**

Note: Underlined numbers indicate that the proportion of wells with reportable organic constituents is significantly different in urban and nonurban land uses.

were larger in urban wells than in nonurban wells, except fluoride, cadmium, and chromium. For the land-use categorizations that had significant differences, median concentrations of fluoride were smaller in the urban wells, and the median concentrations of cadmium and chromium were the same in urban and nonurban wells. The median concentrations of cadmium and chromium were less than the minimum reporting level, but the upper ends of the distributions showed significant differences. The proportion of wells with reportable organic compounds was higher among the urban wells than the nonurban wells regardless of the method of land-use categorization.

Initially, the particle-tracking analysis was expected to provide the most accurate description of urban land use in the recharge area of wells, and thus was expected to show the highest statistical significance in the comparisons of urban and nonurban ground-water quality. The organic constituents shown in Table 24-5, however, seem to show a stronger relation (smaller p-values) between urban land use and water quality if buffer or point-in-polygon overlays are used rather than the particle-tracking analysis. In contrast, from Table 24-4 it can be seen that for inorganic constituents, 22 of 27 p-values are smaller for the particle-tracking analysis than for the buffer-overlay analysis if predominant land use is considered.

These results might mean that the particle-tracking analysis better represents the portion of water flowing from the predominant land use in the recharge area, as evidenced by the results for the major inorganic ions. Because organic constituents are reported at very small concentrations, their presence in wells may be the result of direct contamination by small quantities of the substance either at the wellhead or around a shallow surface seal into the gravel pack. Another interpretation is that sorption or degradation of organic compounds in the aquifer may be important processes that were not accounted for in the particle-tracking analysis.

POSSIBLE SOURCES OF CONTAMINANTS

The term *contaminant* is defined as any substance in ground water that was introduced as a result of human activity. Many different substances can contaminate ground water, and the use of many of these substances tends to be more concentrated in urban areas. Some of the possible sources of contaminants and how these sources are related to urban land use are described here.

Inorganic Constituents

Paved surfaces in urban areas increase runoff and decrease the amount of precipitation that recharges the aquifer. Decreasing the amount of recharge from precipitation, which is low in dissolved constituents, may result in larger concentrations of dissolved constituents in ground water. In contrast, lawn watering, a common practice in residential areas in the Oklahoma City urban area, could increase recharge. Because lawns usually are watered during the summer months when evapotranspiration is greatest, it is unknown how much water from lawns recharges the aquifer and how much is lost to evapotranspiration.

Treated water from municipal treatment plants can enter the aquifer from lawn watering or from leaking waterlines. The concentrations of sodium, potassium, chloride, and sulfate from two of Oklahoma City's three surface-water reservoirs (Betty Fox, City of Oklahoma City, written communication, 1991) tend to be higher than the background concentrations in the aquifer.

Leakage from sewer lines and effluent from septic tanks could contaminate ground water. Oklahoma City does not measure concentrations of major cations and anions in water entering its sewage-treatment plants, but chemists at the city's wastewater treatment plants believe the major-ion concentrations would resemble treated public supply water because municipal water is the major component of residential sewage (Willard

Keith, City of Oklahoma City, oral communication, 1991).

Road salt is another possible source of contamination in the urban area. Oklahoma City uses a mixture of salt and sand on roads during the winter. In recent years, the road salt has been purchased from a mine in Hutchinson, Kansas. The chemical composition of the salt is 95–97% sodium chloride, 2–4% calcium sulfate, 1–2% "water insolubles," and less than 1% magnesium chloride, iron oxide, calcium chloride, magnesium sulfate, and moisture (Max Liby, Hutchinson Salt Company, written communication, 1991). Road salt could account for the elevated concentrations of sodium and chloride in samples from urban wells.

Fertilizer is a possible source of the elevated concentrations of inorganic constituents in the ground water under the urban area. Lawns and gardens in the Oklahoma City area commonly are fertilized with chemical fertilizers. Several inorganic constituents reported at elevated concentrations in urban wells, including calcium, magnesium, sodium, potassium, and chloride, are ingredients in common fertilizer and fertilizer carrier compounds (Brady 1974). There is evidence that phosphate in some fertilizers contains substantial quantities of uranium (Spalding and Sackett 1972), which could account for some of the elevated uranium and gross-alpha particle activity in urban wells.

One common fertilizer for which the composition of the carrier compound is known is potash, or potassium chloride. A total of 1,792 tons of potash was sold in Oklahoma County between July 1, 1989, and July 1, 1990 (Oklahoma Department of Agriculture 1990). Assuming that one quarter of that total was applied evenly over the Oklahoma City urban area and mixed with an estimated recharge of 40 mm, a solution containing 14 mg/L potassium and 13 mg/L chloride would result. Although it is not possible to verify the assumptions made in this calculation, it is clear that fertilizer could provide enough of many inorganic constituents to raise the concentrations of those constituents in the ground water under the Oklahoma City urban area.

Brines from oil and gas wells could contaminate ground water. Oil and gas have been produced in the study area since the beginning of the century, and approximately 13,000 oil and gas wells penetrate the Central Oklahoma aquifer. The Oklahoma City oil field, located in the southeastern part of the urban area, is one of the 10 largest oil fields in the conterminous United States (Landes 1970). Most oil wells produce brines as well as hydrocarbons. At present (1992), these brines are reinjected into deep strata, generally through old production wells that have been converted into injection wells. In the past, these brines have been spread on the land surface or dumped in holding pits and may have leaked into the ground water. Although oil and gas wells are common throughout the study area, wells in the Oklahoma City urban area tend to have been drilled before stricter regulations regarding casing requirements and brine disposal were put in place. Median concentrations of major ions in brines produced from some oil and gas wells in Oklahoma County (Parkhurst and Christenson 1987) are as follows: pH, 6.3; alkalinity, 46.4 mg/L as $CaCO_3$; calcium, 13,000 mg/L; magnesium, 2,510 mg/L; sodium, 75,500 mg/L; chloride, 146,000 mg/L; sulfate, 232 mg/L; and dissolved solids, 237,000 mg/L. Snavely (1989) reported that brines produced during hydrocarbon production are enriched in radium isotopes. Radium-228 concentrations in 113 wells in Oklahoma, Texas, and along the Gulf Coast ranged from 19 to 1,507 pCi/L, and radium-226 concentrations ranged from 0.1 to 1,620 pCi/L. Radium-226 is an alpha-particle emitter and is part of the uranium-238 decay series, which also includes radon-222. Contamination from brines could account for elevated concentrations of major ions as well as radionuclides.

In summary, the elevated concentrations of inorganic constituents in urban wells are difficult to attribute to a single source. Although only samples from wells completed in the Garber Sandstone and Wellington Formation were compared, the possibility cannot be completely discounted that differences within those geologic units could account for the differences in concentrations. Sufficient data have not been collected during the current investigation to assess the effects of changes in recharge volume in the urban area on the chemical quality of ground water. Leakage from water supply and sewage lines could account for elevated sodium, potassium, and chloride concentrations, but these sources cannot account for elevated gross-alpha particle activity or the elevated calcium and magnesium concentrations. Road salt could account for elevated sodium and chloride concentrations but not the elevated concentrations of other ions. Fertilizers could account for the elevated inorganic constituents and the radionuclides, although nitrate concentrations are not also elevated. Nitrogen species, however, constantly are being consumed by vegetation and biological activity in the soil, so the absence of differences in nitrate concentrations does not preclude fertilizer as a source of inorganic contaminants. Contamination from oil-field brines could account for all the elevated major-ion concentrations and the radionuclides.

Organic Constituents

Pesticides reported in water samples from the Central Oklahoma aquifer clearly represent a form of contamination because there are no natural sources of these compounds. All the pesticides reported in water samples from the Central Oklahoma aquifer are known to have been used in the study area. Although much attention has been directed at the agricultural use of pesticides, many pesticides are used in urban areas on residential lawns and gardens, along roads and utility corridors, and in houses and other buildings. A contributing factor to pesticide use in urban areas could be that the cost of pesticides is a much less important economic constraint on a homeowner than it is on a farmer. No data are available concerning the amount of pesticides used in central Oklahoma, but it is possible that larger amounts of pesticides are applied per acre in the urban area than in the outlying agricultural areas.

The organochlorine insecticides reported in water samples collected as part of this investigation were in common use in central Oklahoma, although their use is declining at present (1992). Chlordane and dieldrin were used as termaticides, and DDE is a breakdown product of DDT, which was used extensively as an insecticide. The phenoxyacid herbicides commonly are used for weed control. Of the triazine herbicides, atrazine is a common agricultural herbicide and prometone is used for weed control on lawns and gardens, as well as for the complete elimination of vegetation along roads and around utility poles.

The volatile organic compounds reported in water from the aquifer also are contaminants. Most of the VOCs reported in water samples are in use or have been used as solvents, propellants, fumigants, or in manufacturing processes. Styrene is a compound used in the manufacturing of many different types of plastics (Verschueren 1983). Trihalomethanes, such as bromoform, chlorodibromomethane, dichlorobromomethane, and in particular, chloroform, are the organic constituents most commonly reported in urban wells. One common source of trihalomethane compounds is the chlorination of drinking water for city water supplies, which may be a source of recharge water. The trihalomethanes form when chlorine or bromine react with dissolved organic carbon (Thurman 1985). If the bromide ion is present, chlorine oxidizes bromide to bromine, which reacts with dissolved organic carbon (Thurman 1985). Water in the Oklahoma City distribution system has had total concentrations of trihalomethane compounds as large as 262 µg/L (Judith Duncan, Oklahoma State Department of Health, written communication, 1990). Current regulations require the total trihalomethane concentration in public water supplies to be less than 100 µg/L. Trihalomethanes also are associated with manufacturing, particularly the manufacturing of chlorofluorocarbon refrigerants and fire extinguishers (Verschueren 1983). Dichlorodifluoromethane and trichlorofluoromethane are refrigerants and are present in many sources, including air conditioners, refrigerators, sewage, and the atmosphere.

Gasoline from leaking underground storage tanks was expected to be a common contaminant in the urban area. The composition of gasoline varies greatly among refiners and has changed over time. Some of the VOCs reported in water sampled for this study are present in some gasolines, including 1,2-dibromoethane, 1,1-dichloroethane, 1,2-dichloroethane, and toluene, but these compounds have many other uses as solvents, fumigants, and manufacturing agents (Lucius et al. 1989). Toluene is a major component in gasoline, accounting for as much as 30% by weight of gasoline (Kreamer and Stetzenbach 1990). Benzene, ethylbenzene, and xylene also are major components of gasoline, but none of these were reported in any samples. Thus, it is not possible to determine if the compounds reported in samples represent gasoline contamination.

During the first year of sampling (1988), all wells where pesticides or VOCs were reported were revisited by field personnel in an effort to identify sources of the contaminants. At about half of these wells, a likely source of contamination was identified in the immediate vicinity of the wellhead. In many cases the well owner had used the contaminant near the well or was storing chemicals near the wellhead. Because some wells apparently were contaminated from sources close to the wellhead, it is thought that the contaminants could be entering the well directly or around the shallow cement seal and through the gravel pack.

CONCLUDING REMARKS

The Central Oklahoma aquifer underlies about 8,000 km^2 of central Oklahoma, where the aquifer is used extensively for municipal, industrial, commercial, and domestic water supplies. The effects of urban land use on the quality of water in the Central Oklahoma aquifer were investigated.

Sampling networks were established to investigate the quality of the ground water under the Oklahoma City urban area and the overall water quality of the Central Oklahoma aquifer. An urban network was established in the central part of the urban area. Low-density survey sampling networks were established to assess the overall water quality of Permian geologic units and alluvium and terrace deposits. Wells included in these networks were randomly selected. All wells sampled were existing water-supply wells, mostly, but not exclusively, domestic supply wells. Water samples were collected and analyzed for inorganic constituents, volatile organic compounds, and pesticides.

The presence of pesticides in wells was strongly related to the presence of measurable concentrations of tritium. Pesticides were reported only in wells that had reportable concentrations of tritium. Tritium is thus a directly measurable indicator of a well's vulnerability to pesticide contamination. The presence of reportable tritium does not indicate that a well will be contaminated with pesticides, but simply that there are flow paths to the well by which substantial quantities of post-1952 recharge water enter the well. The absence of reportable tritium indicates that there are no flow paths to the well by which substantial quantities of post-1952 recharge water enter the well.

A contingency-table analysis indicated that the geologic unit in which the wells were completed did not affect the proportion of wells with reportable concentrations of volatile organic compounds or pesticides. This result was unexpected because the Permian geologic units are very different from the alluvium and terrace deposits.

Mann-Whitney tests were used to compare selected inorganic constituent concentrations in urban and nonurban wells. The tests indicated that there were statistically significant differences between urban and nonurban wells in the concentrations of alkalinity, calcium, magnesium, sodium, potassium, bicarbonate, chloride, iron, gross-alpha radioactivity, and uranium. Median concentrations of these constituents were larger in urban than in nonurban wells. Concentrations of dissolved oxygen, carbonate, sulfate, nitrite plus nitrate, arsenic, barium, copper, lead, mercury, selenium, and silver had no significant differences in urban and nonurban wells. Whether the concentrations of pH, fluoride, cadmium, chromium, manganese, and radon, were significantly different or not in urban and nonurban wells depended on the method of land-use categorization.

A contingency-table analysis indicated that the proportion of wells with reportable pesticides or volatile organic compounds might or might not be different in urban and nonurban land uses, depending on the method used to categorize urban land use and how the volatile organic compounds were grouped. On the basis of the results of statistical tests and field work, there is evidence that the sources of organic contaminants commonly were near the wellhead, which might indicate that the shallow seals and gravel packs provided pathways for contaminants to enter the wells.

The comparisons by sampling network consistently indicated that the presence or absence of organic constituents, as well as the concentrations of many inorganic constituents, were significantly different in the samples from urban and nonurban wells. Additionally, the median concentrations of all inorganic constituents that were significantly different for any method of land-use categorization, except fluoride, cadmium, and chromium, were larger in samples from urban wells than in those from nonurban wells. The proportion of wells with reportable organic compounds was higher among the urban wells than the nonurban wells regardless of the method of land-use categorization.

Potential sources of inorganic and organic contamination include changes in recharge, lawn watering, leakage from municipal water and sewer lines, road salt, fertilizer, oil-field brines, pesticide usage, manufacturing, and leaking underground storage tanks. The observed contamination probably results from multiple sources; the dominant sources could not be determined.

References

Barringer, Thomas, Dennis Dunn, William Battaglin, and Eric Vowinkel. 1990. Problems and methods involved in relating land use to ground-water quality. *Water Resources Bulletin* 26(1):1–9.

Brady, N.C. 1974. *The Nature and Properties of Soils*. New York: Macmillan.

Christenson, S. C., R. B. Morton, and B. A. Mesander. 1990. *Hydrogeologic Maps of the Central Oklahoma Aquifer, Oklahoma*. Oklahoma City, OK: U.S. Geological Survey Open-File Report 90-579.

Eckhardt, D. A., S. F. Siwiec, and S. J. Cauller. 1989. Regional appraisal of ground-water quality in five different land-use areas, Long Island, New York. In *U.S. Geological Survey Toxic Substances Hydrology Program—Proceedings of the Technical Meeting, Phoenix, Arizona, September 26–30, 1988*, pp. 397–403. Reston, VA: U.S. Geological Survey Water-Resources Investigations Report 88-4220.

Fegeas, R. G., R. W. Claire, S. C. Guptill, K. E. Anderson, and C. A. Hallam. 1983. *Land Use and Land Cover Digital Data*. Alexandria, VA: U.S. Geological Survey Circular 895-E.

Hardy, M. A., P. P. Leahy, and W. M. Alley. 1989. *Well Installation and Documentation, and Ground-Water Sampling Protocols for the Pilot National Water-Quality Assessment Program*. Reston, VA: U.S. Geological Survey Open-File Report 89-396.

Hay, L. E., and W. A. Battaglin. 1990. *Effects of Land-Use Buffer Size on Spearman's Partial Correlations of Land Use and Shallow Ground-Water Quality*. Trenton, NJ: U.S. Geological Survey Water-Resources Investigations Report 89-4163.

Kreamer, D. K., and K. J. Stetzenbach. 1990. Development of a standard, pure-compound base gasoline mixture for use as a reference in field and laboratory experiments. *Ground Water Monitoring Review* 10(2):135–45.

Landes, K. K. 1970. *Petroleum Geology of the United States*. New York: Wiley.

Lawrence, S. J., and Rita Whitney. 1990. *Shallow Ground-Water Quality in the Vicinity of a Small Urban Area in West-Central*

Nevada. Paper read at Nevada Water Resources Annual Conference, February 1990 in Las Vegas, Nevada.

Lucius, J. E., G. R. Olhoeft, P. L. Hill, and S. K. Duke. 1989. *Properties and Hazards of 108 Selected Substances*. Denver, CO: U.S. Geological Survey Open-File Report 89-491

McDonald, M. G., and A. W. Harbaugh. 1988. *A Modular Three-Dimensional Finite-Difference Ground-Water Flow Model*. Reston, VA: Techniques of Water-Resources Investigations of the U.S. Geological Survey, Book 6, Chapter A1.

Oklahoma Department of Agriculture. 1990. *Tonnage Distribution of Fertilizer in Oklahoma Counties by Grade and Material, Annual, for the Period July 1, 1989 to July 1, 1990*. Oklahoma City, OK: Oklahoma Department of Agriculture.

Parkhurst, D. L., S. C. Christenson, and J. L. Schlottmann. 1989. *Ground-Water Quality Assessment of the Central Oklahoma Aquifer, Oklahoma—Analysis of Available Water-Quality Data Through 1987*. Oklahoma City, OK: U.S. Geological Survey Open-File Report 88-728.

Parkhurst, R. S., and S. C. Christenson. 1987. *Selected Chemical Analyses of Water from Formations of Mesozoic and Paleozoic Age in Parts of Oklahoma, Northern Texas, and Union County, New Mexico*. Oklahoma City, OK: U.S. Geological Survey Water-Resources Investigations Report 86-4355.

Pollock, D. W. 1989. *Documentation of Computer Programs to Compute and Display Pathlines Using Results from the U.S. Geological Survey Modular Three-Dimensional Finite-Difference Ground-Water Flow Model*. Reston, VA: U.S. Geological Survey Open-File Report 89-381.

P-STAT, Inc. 1989. *P-STAT User's Manual*. Princeton, NJ: P-STAT, Inc.

Scott, J. C. 1990. *Computerized Stratified Random Site-Selection Approaches for Design of a Ground-Water-Quality Sampling Network*. Oklahoma City, OK: U.S. Geological Survey Water-Resources Investigations Report 90-4101.

Snavely, E. S. 1989. *Radionuclides in Produced Water, a Literature Review*. Dallas, TX: American Petroleum Institute.

Spalding, R. F., and W. M. Sackett. 1972. Uranium in runoff from the Gulf of Mexico distributive province: Anomalous concentrations. *Science* 175:629–631.

Thurman, E. M. 1985. *Organic Chemistry of Natural Waters*. Dordrecht, The Netherlands: Martinus Nijhoff/Dr. W. Junk.

Verschueren, Karel. 1983. *Handbook of Environmental Data on Organic Chemicals*. New York: Van Nostrand Reinhold.

25

Uses and Limitations of Existing Ground-Water-Quality Data

Pixie A. Hamilton, Alan H. Welch, Scott C. Christenson, and William M. Alley

INTRODUCTION

One of the initial activities undertaken in the U.S. Geological Survey's National Water-Quality Assessment (NAWQA) Program was to compile, screen, and interpret the large amount of water-quality data available for the regional hydrologic systems of the Carson River Basin in Nevada and California, the central Oklahoma aquifer in Oklahoma, and the Delmarva Peninsula in Delaware, Maryland, and Virginia. These data, referred to as "existing data," include analyses of ground-water samples collected by local, state, and federal agencies prior to the initiation of the NAWQA Program in 1986. Assessment of the existing data helped to provide an initial description of ground-water quality in these areas, to develop hypotheses about major factors that affect ground-water quality, and to define data needs for additional sampling. Results of these assessments are described in reports by Welch et al. (1989) for the Carson River Basin, by Parkhurst, Christenson, and Schlottmann (1989) for the central Oklahoma aquifer, and by Hamilton, Shedlock, and Phillips (1991) for the Delmarva Peninsula.

The three ground-water study areas represent diverse hydrologic environments. Two of the areas, the Delmarva Peninsula and central Oklahoma, are presented as case studies in Chapters 23 and 24. The third area, the Carson River Basin, encompasses about 10,300 km^2 in the western Great Basin and eastern Sierra Nevada. The Carson River Basin consists of mountainous headwaters and five major hydrographic areas. Each hydrographic area generally includes an alluvial valley, bordered by hills or mountains, through which the Carson River flows. The ground-water system in each alluvial valley generally consists of a shallow water-table aquifer and one or more deeper confined aquifers as typified in Figure 13-1.

This chapter focuses on the uses and limitations of existing data for regional ground-water-quality assessment. No attempt is made to summarize water-quality conditions in the three study areas. The chapter discusses the sources of existing data, the screening procedures used, the suitability of existing data for regional ground-water-quality assessment, and selected applications of the data. The discussions should be useful to investigators who plan to use existing data to assess ground-water quality in large regions.

TABLE 25-1. Characteristics of Major Water-Quality Data Bases Used by the Three Projects

Data Base	Number of Sites	Number of Samples	Data Availability	General Purpose for Data Collection	Site Location Information	Spatial Distribution of Sites
				Carson River Basin		
Desert Research Institute	101	132	Published reports	General ground-water research	Generally township/range/section	Primarily in one hydrographic area
Nevada State Health Laboratory	557	1,307	Paper files	Primarily a public service	Optional user supplied township/range/section or lat./long.	Distributed throughout the basin, both areally and with depth
U.S. Geological Survey	243	489	Computer files	General ground-water research	Latitude and longitude	Highly variable among the hydrographic areas
				Central Oklahoma Aquifer		
Association of Central Oklahoma Governments	188	582	Computer files	Water resources planning	Municipal well identification	Predominantly from municipal wells in western part of study area
Oklahoma State Department of Health	401	1,658	Computer files	Regulatory purposes	Distribution system location	Predominantly from water distribution systems
Oklahoma Water Resources Board	104	156	Computer files	Water resources planning	Legal description of site locations	Distributed throughout aquifer, both areally and with depth
National Uranium Resource Evaluation (NURE)	510	510	Computer files	Assessment of national uranium resources	Decimal latitude and longitude (0.001°)	Shallow wells throughout study area
U.S. Department of Defense	192	289	Computer files	Identification and elimination of contamination of ground water at an Air Force base	Latitude and longitude estimated from maps	Tinker Air Force Base and vicinity
U.S. Geological Survey	588	650	Computer files	General ground-water research	Latitude and longitude	Predominantly from alluvial and terrace deposits along major rivers

Delmarva Peninsula

Delaware Water Resources Center	681	681	Paper files	General ground-water research	State plane coordinates	Shallow wells distributed thoughout Eastern Sussex County, Delaware
Virginia Water Control Board	203	203	Computer files	General ground-water research	Latitude and longitude	Wells distributed throughout Virginia part of the peninsula
Virginia Water Project, Inc.	48	48	Paper files	General ground-water research	Latitude and longitude	Shallow dug wells in rural agricultural areas, all about 30 ft deep
Wicomico County Health Department, Maryland	Unknown	2,584	Computer files	Regulatory purposes	State-plane grid coordinates accurate to nearest 1,000 ft	Shallow wells distributed throughout Wicomico County
National Uranium Resource Evaluation (NURE)	710	710	Computer files	Assessment of national uranium resources	Latitude and longitude	Shallow wells located throughout Delaware and Maryland, about 1 well per 5 mi^2
U.S. Geological Survey	976	976	Computer files	General ground-water research	Latitude and longitude	Wells distributed throughout the peninsula and among different depths

SOURCES OF GROUND-WATER-QUALITY DATA

Large amounts of ground-water-quality data were available in each study area. The data were collected by various organizations to meet diverse objectives, ranging from monitoring for compliance with local, state, and federal drinking-water standards or criteria to conducting research on specific ground-water issues. These data included analyses of 2,300 water samples from more than 1,000 wells and a few springs in the Carson River Basin, 4,439 water samples from 1,604 wells and 409 distribution systems in the central Oklahoma aquifer, and more than 5,000 water samples from wells in the Delmarva Peninsula.

Major sources of water-quality data are shown in Table 25-1, along with a brief description of the purpose for data collection and the availability of data for water-quality assessment. Although the U.S. Geological Survey and the U.S. Environmental Protection Agency maintain large national data bases, the largest sources of data in each study area were from state and county health agencies. For example, 70–90% of the analyses for all but the most downstream part of the Carson River Basin were from the Nevada State Health Laboratory. About one half of the analyses for the Delmarva Peninsula were obtained from the Wicomico County Health Department of Maryland.

Most of the data were obtained from computer files. Special software commonly was needed to access, transfer, and restructure the computerized data bases. Data in paper files that were organized in a central filing system with a manageable number of analyses also were included in the evaluation. These data required manual entry into computer files. It was not feasible to use data from all agencies because some of the data were housed in different offices within the same agency or were in filing systems that were not dedicated exclusively to water-quality data.

DATA SCREENING

Screening procedures were used to verify the integrity of the data. A primary screening procedure used in all three projects involved charge-balance checks between major cations and anions (see Chapter 6). Cation-anion balance checks were computed for all analyses for which a complete suite of major ions was available. Those analytical results that exceeded a balance error of 10% were excluded from the interpretations of major-ion chemistry. Other screening procedures involved cross-checking data on magnetic tape with original laboratory sheets and identifying samples from sources affected by water-treatment processes or water softeners. Interaction with the agencies providing the data bases proved to be invaluable in screening the data and in placing appropriate caveats on their use.

SUITABILITY OF DATA FOR REGIONAL WATER-QUALITY ASSESSMENT

Certain characteristics of the data bases limit their usefulness for regional assessment of ground-water quality. These characteristics include sampling procedures and laboratory analytical methods, sampling design and objectives, spatial coverage of the sampled sites, water-quality constituent coverage, and availability of ancillary site information. Limitations of existing data bases described in this chapter refer only to the use of the data for regional water-quality assessment. Data from particular programs might fully meet the specific objectives for which they were collected, yet these data might have limited relevance to water-quality questions of regional scope.

Sampling Procedures and Laboratory Analytical Methods

Sampling methods and laboratory analytical methods commonly differed among agencies in each of the three study areas. Many of the differences involved the analysis and reporting of "total" as opposed to "dissolved" concentrations. Assessment of those analyses that included both dissolved and total concentrations in samples collected from the same well (but commonly not collected at the same time) indicated that the difference between dissolved and total concentrations generally was minimal; most of the values were within 10%. Differences in dissolved and total iron concentrations, however, were much greater. For example, the median concentration of total iron was an order of magnitude larger than the median concentration of dissolved iron in data from the Carson River Basin. Total iron might include iron hydroxide precipitates and rust particles from pipes.

Differences in sampling and analytical procedures among agencies in the three study areas also reflected use of laboratory as opposed to field measurements of physical properties such as pH. Preservation techniques also differed among agencies. For example, preservation of samples collected for determination of nitrate concentrations by some agencies involved only chilling of the sample at the time of collection. Preservation techniques used by other agencies included chilling and adding mercuric chloride, which improves the stability of nitrate concentrations. Differences in detection limits, not only among agencies but within agencies over time, also complicated the interpretations.

Sampling and analytical procedures commonly were poorly documented. In addition, descriptions of quality-assurance procedures and documentation of quality-assurance results commonly were lacking.

Sampling Design and Objectives

Some of the data were collected to meet specific objectives of agencies and to address known or suspected water-quality concerns. Such data, which might overrepresent water-quality problems, are not suitable for drawing statistical inferences about the overall quality of ground water in a region. Many times, the purpose for collecting the data was not documented, particularly in some large multipurpose data bases that included data from numerous projects. Knowledge of the sampling design is needed to distinguish data with site-selection bias from those without such bias.

Spatial Coverage of Sampled Sites

The spatial distribution of sampling sites, both areally and with depth, differed widely among agencies (see Table 25-1). Differences in the distribution of sampling sites could cause possible problems with aggregation and comparison of results from different data bases. For example, comparison of data bases from two different areas might indicate that nitrate concentrations in one area are higher than those in another simply because the bulk of the data are from shallow wells in one area and from deep wells in the other.

The suitability of available data for interpretation of water quality at different spatial scales was limited. Summary of water-quality data for the entire ground-water system within each study area was of limited utility given the distinct differences in physiography and geology in each area. Data were available to characterize specific subregions that represent different hydrogeologic or land-use settings; however, sampling programs were not designed specifically to characterize these subregions and, therefore, few data existed for some subregions of interest. Data also were lacking at a local scale, such as along specific ground-water-flow paths. Local-scale studies had not been done in any of the three study areas with the explicit purpose of representing typical ground-water quality. Local-scale studies had been done only to study known or suspected water-quality problems.

Water-Quality Constituent Coverage

Although samples had been collected from many sites in each of the three study areas, the chemical analyses commonly included a

TABLE 25-2. Number of Sites for Which Analyses Were Done for Selected Properties and Water-Quality Constituents

Water-Quality Property or Constituent	Number of Sites		
	Carson River Basin	Central Oklahoma Aquifer	Delmarva Peninsula
Properties and Major Constituents			
pH (field)	1,127	961	2,189
Specific conductance	337	1,232	1,716
Calcium	1,083	954	909
Chloride	1,110	835	1,749
Complete suite of major ions	811 (786)[a]	559 (539)[a]	718 (468)[a,b]
Nitrate (or nitrate plus nitrite)	872	860	4,325
Major Metals and Trace Elements			
Arsenic	859	810	54
Boron	297	620	159
Iron (dissolved)	129	546	527
Iron (total)	761	290	2,177
Molybdenum	57	507	120
Selenium	101	918	8
Vanadium	40	507	823
Radionuclides			
Gross-alpha radioactivity	0	216	0
Radon-222	8	0	0
Uranium	8	559	710

[a] Number shown in parentheses is number of sites for which the major-ion balance was within 10%.
[b] If nitrate is included in the ionic balance, approximately 576 meet a 10% ionic-balance criterion.

limited set of constituents. Analyses for pesticides and other organic compounds were rare. Lack of data for pesticides is a major limitation in the analysis of existing data for the central Oklahoma aquifer and in the Delmarva Peninsula; selected pesticides were found at environmentally significant levels in the sampling programs that followed in each of these areas. Table 25-2 lists the number of sites for which analyses were made for selected properties and inorganic water-quality constituents. Several features to note are that (1) the numbers of sites differed considerably among constituents and among study areas, (2) a complete suite of major ions for water-type classification was available at less than 50% of the sites, (3) measurements of radon-222 were made for only eight sites, and (4) measurements of gross-alpha radioactivity were available only for the central Oklahoma aquifer. Radon and gross-alpha radioactivity were found at environmentally significant levels in the sampling program that followed in the Carson River Basin.

Availability of Ancillary Site Information

Ancillary site information is needed for regional interpretation of available water-quality data. For example, information on well construction is needed to evaluate the suitability of a well for sampling, to identify the hydrogeologic unit(s) from which the well is producing, and to relate the observed water quality to local hydrogeology. Accurate well locations are needed to relate observed water quality to natural and human factors such as soils and land use.

Information on site characteristics commonly was limited, and the reliability of the information unknown. Some efforts were made to supplement site information on selected wells. For example, site locations in the Carson River Basin and in central Oklahoma were verified in the field, where possible. Results of these efforts were mixed. Site locations commonly were associated with an address of a well owner that differed from the well location, and site locations commonly were recorded incorrectly or imprecisely. Site identification commonly was specified by legal descriptions or other systems (see Table 25-1) and required conversion into latitude and longitude coordinates.

Table 25-3 illustrates the availability of information on three basic site characteristics: well depth, depth to water, and aquifer(s) or depth interval(s) to which the well is open. Information on well depth was the most commonly available site characteristic. Many of the data bases included information on the other two characteristics at less than 50% of the wells. For example, about 60% of the data bases (8/13) had measurements of depth to water for less than 10% of the sampled wells.

TABLE 25-3. Availability of Selected Site Characteristics in Data Bases

	Number of Data Bases Having Indicated Percentage of Sites with Information on		
Percentage of Sites	Well Depth	Depth to Water	Aquifer(s) or Depth Interval(s) to Which Well Is Open
<10%	0	8	4
10–49%	0	2	2
50–90%	7	3	1
>90%	6	0	6
Total no. of sites	13	13	13

APPLICATIONS

Despite their limitations, the existing data were useful for some simple assessments of regional ground-water quality in each area. The interpretations were mostly descriptive rather than explanatory because of uncertainties associated with the water-quality data and limitations in the ancillary information. The main applications of the data were development of (1) maps that showed sites where measured concentrations of selected properties and constituents exceeded drinking-water standards or criteria, (2) maps and geochemical diagrams (such as trilinear diagrams) that depicted general chemical characteristics of ground water in different areas, and (3) graphs, boxplots, and statistics that displayed general differences in ground-water quality with depth and among different hydrogeologic settings and land uses. Temporal trends in concentrations were not investigated for any water-quality constituent in the study areas because of a lack of time-series data and uncertainties about the effects of changes in sample collection, preservation, and analytical techniques.

Data were used to estimate summary statistics for selected settings in each of the three study areas. Because the data were not obtained from a single sampling program specifically designed for this purpose, the results are subject to many possible biases. The extent of these biases in each study area was investigated by comparing the existing data and data representing a grid-based regional survey with respect to the percentage of samples exceeding drinking-water standards and criteria established by the U.S. Environmental Protection Agency (1986a, 1986b). The grid-based regional-survey data were collected by each of the projects in the sampling program that followed the analysis of existing data and provide consistently collected data specifically designed for regional water-quality assessment. Selected constituents of primary water-quality concern in each study area (such as nitrate, arsenic, selenium, and iron) were included in the com-

TABLE 25-4. Percentage of Wells Where U.S. Environmental Protection Agency Maximum Contaminant Level (MCL) or Secondary Maximum Contaminant Level (SMCL) Was Exceeded for Selected Constituents Using Existing Data and Regional-Survey Data

Water-Quality Constituent and Corresponding MCL or SMCL[a]	Hydrogeologic Setting	Percentage of Wells Where MCL or SMCL Was Exceeded	
		Existing Data	Regional Survey
Carson River Basin			
Nitrate (MCL, 10 mg/L as N)	Carson and Eagle valleys	<1	1
	Carson Desert	4	6
Arsenic[b] (MCL, 0.05 mg/L)	Carson and Eagle valleys	2	3
	Carson Desert	56	47
Central Oklahoma Aquifer			
Nitrate (MCL, 10 mg/L as N)	Bedrock wells		
	<100 ft deep	26	12
	100–300 ft	13	5
	>300 ft	0	0
	Alluvium and terrace deposits	10	10
Selenium[b] (MCL, 0.01 mg/L)	Bedrock wells		
	<100 ft deep	0	9
	100–300 ft	9	14
	>300 ft	41	37
	Alluvium and terrace deposits	3	4
Delmarva Peninsula			
Nitrate (MCL, 10 mg/L as N)	Surficial aquifer	18	12
Dissolved iron (SMCL, 0.03 mg/L)	Surficial aquifer	22	36

[a] U.S. Environmental Protection Agency 1986a and 1986b
[b] Existing data include filtered and unfiltered analyses; regional survey data are filtered analyses.

parisons. The results, grouped by constituent and hydrogeologic setting, are summarized in Table 25-4. Some broad similarities between the existing and regional-survey data are noted here, as are indications of biases that might be present in the existing data. Some of the differences between the two data bases also might result from spatial and temporal biases in the regional-survey sampling.

Carson River Basin

Two hydrogeologic settings are considered in the Carson River Basin: Carson and Eagle valleys in the upper part of the basin and the Carson Desert, a hydrologically closed basin, in the lower part (Table 25-4). Nitrate and arsenic concentrations in the existing data base are similar to those in the regional-survey data base. Specifically, both data bases show that the percentage of analyses in which arsenic concentration exceeds the MCL (maximum contaminant level) are notably different between the two hydrogeologic settings, and the percentage of analyses in which nitrate concentration exceeds the MCL is low in both parts of the basin. The similarity of results between existing and regional-survey data probably reflects the areally distributed nature of the sampled sites in both data bases.

Central Oklahoma Aquifer

Two hydrogeologic settings are considered in the central Oklahoma aquifer: a bedrock aquifer (with three ranges of well depths) and alluvial and terrace deposits along major streams (Table 25-4). Both data bases show that concentrations of selenium increase with depth and concentrations of nitrate decrease with depth in the bedrock aquifer. The percentage of analyses in which selenium concentration exceeds the MCL is somewhat similar in both data bases. Concentrations of nitrate in the bedrock aquifer exceed the MCL in a higher percentage of samples in the existing data base than in the regional-survey data base. This difference probably results from spatial bias in the sampling design for the collection of the existing data. Concentrations of selenium and nitrate in the alluvial and terrace deposits are similar in existing and regional-survey data bases. This similarity probably results from the relatively uniform distribution of sites throughout the alluvial and terrace deposits.

Delmarva Peninsula

The surficial aquifer covers much of the peninsula and is treated as one hydrogeologic setting (Table 25-4). Concentrations of nitrate exceed the MCL more frequently in the existing data base (18%) than in the regional-survey data base (12%). The percentage is higher in the existing data base despite a greater median depth of the wells in the existing data base (17 m) than in the regional-survey data base (10 m). The difference in the percentage of samples in which nitrate concentration exceeds the MCL probably results from biases in the sampling design of the existing data base. Nitrate in water has been a primary water-quality concern on the peninsula for many years, and many of the existing data probably were collected from targeted areas of known or suspected contamination of nitrate. The data would, therefore, be expected to yield overestimates of the extent of peninsula-wide contamination of ground water. In contrast to concentrations of nitrate, concentrations of dissolved iron are higher and exceed the SMCL (secondary maximum contaminant level) more frequently in the regional-survey data base (36%) than in the existing data base (22%). Elevated concentrations of dissolved iron are associated with reducing environments where nitrate is not stable and where nitrate concentrations are expected to be minimal.

CONCLUDING REMARKS

A large amount of ground-water-quality data has been collected by a diverse group of organizations for a wide range of purposes. Many of these data are useful for regional ground-water-quality assessments, provided that care is taken to assess the manner in which the data were collected and analyzed and to identify the individual hydrologic settings and sampling objectives the data represent.

Because of diverse network designs and differences in sampling and laboratory methods, existing ground-water-quality data are difficult to aggregate in an automatic fashion to provide meaningful results. This is particularly true for large regional or national summaries. It is usually necessary to analyze the data bases separately and to look for consistencies in conclusions drawn from each. Close interaction with the agencies that collected the data is required to understand limitations of the data.

Significant limitations in the spatial and

temporal coverage of the existing data bases and in the water-quality-constituent coverages are common to all three study areas. The data commonly are geographically clustered, and few data on many pesticides and other organic compounds of recent concern are available.

Analyses of existing information may prove valuable in constructing an initial description of ground-water quality, in developing hypotheses about major factors that affect ground-water quality, and in defining additional data needs. The usefulness of an analysis of existing data is enhanced if the hypotheses and conclusions derived from the analysis can be tested by comparison to consistently collected data specifically designed for regional water-quality assessment.

References

Hamilton, P. A., R. J. Shedlock, and P. J. Phillips. 1991. *Water-Quality Assessment of the Delmarva Peninsula, Delaware, Maryland, and Virginia—Analysis of Available Ground-Water-Quality Data through 1987*. Towson, MD: U.S. Geological Survey Water-Supply Paper 2355-B.

Parkhurst, D. L., S. C. Christenson, and J. L. Schlottmann. 1989. *Ground-Water-Quality Assessment of the Central Oklahoma Aquifer, Oklahoma—Analysis of Available Water-Quality Data through 1987*. Oklahoma City, OK: U.S. Geological Survey Open-File Report 88-728.

U.S. Environmental Protection Agency. 1986a. *Maximum Contaminant Levels* (subpart B of part 141, National Interim Primary Drinking-Water Regulations): U.S. Code of Federal Regulations, Title 40, Parts 100 to 149, revised as of July 1, 1986, pp. 524–28.

U.S. Environmental Protection Agency. 1986b. *Secondary Maximum Contaminant Levels* (section 143.3 of part 143, National Secondary Drinking-Water Regulations): U.S. Code of Federal Regulations, Title 40, Parts 100 to 149, revised as of July 1, 1986, pp. 587–90.

Welch, A. H., R. W. Plume, E. A. Frick, and J. L. Hughes. 1989. *Ground-Water-Quality Assessment of the Carson River Basin, Nevada and California—Analysis of Available Water-Quality Data through 1987*. Carson City, NV: U.S. Geological Survey Open-File Report 89-382.

Index

ABS surfactants, as event markers, 280–281
Acid-base equilibrium expression, 164
Acid-base reactions, 144
Acidic ground water
   countermeasures, 417
   health effects, 413–414
Acidic soil, 408–409
Acidification
   ground water, 405
   relation between ground water and surface water, 410–411
   soil, 406–408
      and ground-water modeling, 414–416
   by sulfur deposition, 411
Acid precipitation, 405–418
   ground-water composition trends, 411–413
   plant uptake of nutrients, 410
   weathering, 408–410
Adsorption, pathogens, 393–395
Advection-dispersion models, tritium movement, 265
Age. *See also* Dating
   CFCs, 269–270
      interpretation, 270–271
   ground-water
      definition, 256
      nitrate concentration, 314
Agriculture
   nitrate, 304–307
      irrigation, 306–307
   pesticide use, 359–360
Aldicarb, Central Sand Plain, WI, 366–367
Aliphatic hydrocarbons, 157
Alkalinity, 411–413
Alluvial aquifer. *See* Aquifers, alluvial
Altitude, effect on isotopic content of precipitation, 233–234
Aluminum, 413
   relation with pH, 522–523
Alzheimer's disease, 413
Analysis of uncertainty, kriging, 95–98
Analytical flow models, 36–37
Animal wastes, nitrate, 303–304
Anoxic zone, 125
Aqueous complex, 137
Aqueous-phase transport, inorganic chemicals, 175
Aquia aquifer, mass-balance modeling, 211–212
Aquifers
   alluvial
      atrazine biodegradation, 194–196
      contaminant plume of nitrate, 308
      Fountain Creek, CO, 32–33
   aquia, mass-balance modeling, 211–212
   carbonate, 132
      sulfate reduction and methanogenesis distribution, 191–194
   Central Oklahoma
      geography and geohydrology, 589–591
      ground-water-quality data application, 621
   coastal-plain
      mass-balance modeling, 207–210
      sources and sinks, 117–119
      TEAPs distribution and high-iron ground water zonation, 187–191
   conduit flow, 476
   contaminated
      reaction sequence, 125
      reclamation, 126
   diffuse-flow, 476
   Floridan, 38–39
      sulfate reduction and methanogenesis distribution, 191–194
   Greenbrier Limestone, WV, 473
   karst. *See* Karst aquifers
   leakage between, 245–246
   Madison, mass-balance modeling, 213
   Middendorf, SC, 188–190
   potential for agricultural chemical contamination, 495–496
   property description and measurement, 114–115
   recharging, tritium input, 258, 260
   sensitivity to acidification, 414–415
   surficial
      hydrochemical types distribution, 573–575
      hydrogeomorphic regions, 564–568
      Willards network, 581–582, 584
   temporal variations, 10–11
   tritium movement, 260–265
Areal network, 568–569
Arid zones, recharge and flow rate, 241
Arizona, organic contaminants surveys, 337
Aromatic compounds, 157
Atrazine, 160
   biodegradation, 194–196
   as event markers, 283–286
   river-aquifer interactions, 367–370
Attenuation factor index, 351–352

Bacteria
   factors influencing fate in subsurface, 389
   filtration, 392–393
Bicarbonate, sources and sinks in coastal-plain aquifers, 117–119
Biochemical zonation, 125–126

624   Index

Biodegradation
  atrazine, alluvial-aquifer sediments, 194–196
  inorganic chemicals, 168–169
Blanks, 593–594
Blind samples, 594
Block kriging, 99
Boxplots, 44, 47–48
Buffer-overlay analysis, 596–597

^{13}C, 117–119, 237–238
^{14}C, 238
  dating, mass-balance modeling 212–214
Cadmium, in topsoil, 527–529
Calcium, sources and sinks in coastal-plain aquifers, 117–118
California
  organic contaminants surveys, 335–337
  San Joaquin Valley, 371–372, 537–538. *See also* Selenium
  Soquel-Aptos Basin, freshwater-saltwater environments, 460–462
Candidate population, 4
Capture zone, 35–36
Carbonate aquifer. *See* Aquifers, carbonate
Carbon-14 dating, mass-balance modeling, 212–214
Carson River basin, 620–621
Central Oklahoma aquifer. *See* Aquifers, Central Oklahoma
Central Sand Plain, WI, aldicarb, 366–367
Central Valley Aquifer System, CA, 37–38
CFCs. *See* Chlorofluorocarbons
C_4H_8, isomers, 161
Chelate, 137, 139
Chemical elements, in ground water, 141–142
Chemical hydrogeology, scales, 111–128
  aquifer property description and measurement, 114–115
  change from oxic to anoxic conditions, 125
  chemical modeling, 115
  decreases of scale, 115
  evaluation of controls by physical and chemical heterogeneity, 124, 127
  historic evolution, 113–116
  local scale, 114, 121–123
    radium concentrations, 123
    sampling density, 121
  principles, 126–127
  reactions in coastal plain, 127
  regional scale, 114, 116–121
    bicarbonate sources and sinks, 117–119
    calcium sources and sinks, 117–118
    hydrochemical facies, 119, 121
    local flow cells, 117
    magnesium sources and sinks, 119–120
    mass balance approach, 117
    sodium sources and sinks, 119–120
    sulfate sources and sinks, 119–120
  simplifying assumptions, 116
  site scale, 114, 123–126
    biochemical zonation, 125–126
    chemical heterogeneity, 128
    organic contaminant role, 124
    subsurface sources of organic compounds, 124
    thin zone sampling, 128
Chemical units, ground water, 140–141
Chemographs, 482–483
Chlorofluorocarbons, 268–277, 288
  additional sources and sinks, 272–273
  age dating, 269–270
  applications, 269
  environments and conditions best suited for age dating, 277
  interpretation of CFC age, 270–271
  limitations of usefulness, 278–279
  physical processes, 273–277
  recharge temperature, 270–272
Chromium, spatial distribution in ground water, 522, 524
^{36}Cl, 240
Cluster analysis, 50–51
Cluster and staged selection, 80–81
CO_2, dissolution and exsolution, 146
Coastal-plain aquifers. *See* Aquifers, coastal-plain
Cokriging, 105–106
Cometabolism, 169
Compartment models, tritium movement, 263–265
Complexation reactions, 144
Conduit flow aquifers, 476
Contaminants. *See also* Organic contaminants
  water-soluble, 487
Contamination, 3
  anthropogenic sources, 3–4
  entry routes, 3
  extent of, 4
  transport, role of wells, 15
Contributing areas, pumping wells, 35–37
Copper, 413–414
Covariance, geostatistical models, 89
Critical loads, sulfur and nitrogen, 416–417
Cross validation, kriging, 98–99
Cumulative distribution function, 88

Darcy's law, 449
Dating
  carbon-14, mass-balance modeling, 212–214
  ground water, 247–248. *See also* Environmental tracers
  uses, 255
DBCP. *See* Dibromochloropropane
DCB, 282, 284
DCPA metabolites, 373
Debye–Hückel theory, 200–201
Deforestation, 113

Degradation, pesticides, 349–351
Dehydrohalogenation, 167–168
Delmarva Peninsula, regional water quality assessment, 563–586
 available data, 568
 distribution of hydrochemical types in surficial aquifer, 573–575
 ground-water-quality data application, 621
 hydrogeomorphic regions, 564–568
 local-scale networks, information from, 584–585
 methods of analysis, 570–572
 nitrate, spatial distribution, 575–578
 poorly drained upland, 580–581
 sampling strategy, 570
 surficial-confined region, 581–585
 well-drained upland, 578–580
 well networks, 568–570
Denitrification, 299–300, 313–315
 ground water, 300
Density, inorganic chemicals, 162
Density-dependent analysis, data requirements, 455
Desethylatrazine, 367
Desorption, 147–148
Diagenetic alterations, 142
Dibromochloropropane, 335
 contaminated water and soils, 25
 San Joaquin Valley, CA, 371–372
Diffuse-flow aquifers, 476
Disjunctive kriging, 101–102, 397–399
Dispersion, freshwater-saltwater environments, 450–453
Dissolution, NAPL-to-water, 174
Dissolution gases, 146–147
Dissolved constituents, 135, 137
 computer modeling, 137–140
 sources, 246–247
Distribution functions, geostatistical models, 88
Documentation, wells and well sites, 17–18
Double selection, 81
DRASTIC index, 360–361, 363, 373
Drift, kriging in presence of, 99–100
Drinking water
 quality monitoring, Netherlands, 531–533
 radionuclides, 432
Duplicate samples, 594
Durov diagrams, 44–46

Effluent, nitrate, 304
Electrical neutrality, 140
Elements, in ground water, 141–142
Elimination, inorganic chemicals, 167–168
Environmental isotopes, 227–249
 applications, 241–249
  dissolved constituent sources, 246–247
  evaluation of ground-water flow and storage characteristics, 246
  ground-water dating, 247–249
  ground-water flow in fractured rocks, 246
  ground-water mass characterization, 244
  identification of recharge area or source of water, 244–245
  leakage between aquifers, 245–246
  recharge and flow rate, 241, 244
  stream discharge separation in components, 246
 carbon, 237–238
 chlorine, 240
 δ values, 229–230
 hydrogen isotopes, 232–236
 isotopic content of precipitation, 233–234
 isotopic fractionation, 227–228
  factor, 229
 nitrogen, 238–239
 oxygen isotopes, 232–236
 radioactive decay law, 231
 radioactive isotopes, 231
 Rayleigh distillation equations, 230
 relationship between δD and $\delta^{18}O$, 234–236
 strontium, 240–241
 sulfur, 239–240
 terrestrial abundance, 228
Environmental monitoring, Netherlands, 515–517
Environmental tracers, 255–288. See also Chlorofluorocarbons; Organic compounds; Tritium
 classes, 255–288
 comparison, 277–279
 event markers, 256–257
 krypton-85, 266–268
 nuclear event markers, 279
 range in years, 256–257
 tritium–helium-3, 265–266
Ethylene dibromide, 335
Evaporation, from surface-water bodies, 235
Evaporative concentration, selenium and, 546–547
Evapotranspiration, 473
Event markers. See also Organic compounds
 environmental tracers, 256–257
 herbicides, 282–287
  atrazine, 283–286
 nuclear, 279
 organic tracers, future studies, 287
 surfactants, 280–282
Exsolution, gases, 146–147

Facilitated transport, 175
Fe(III)-reducing bacteria, 190
Fertilizers, nitrate, 302–303
"Fictitious point" method, 98
Finite population correction, 69–70
Floridan Aquifer System. See Aquifers, Floridan
Fluid mass balance, freshwater-saltwater environments, 449–450
Forage, nitrate, 302–303

Forest land, nitrate, 301–302
Forward modeling, 214–222
 combined with inverse modeling, 219–220
 reaction-path modeling, 215–220
 reaction-transport modeling, 220–222
Fountain Creek alluvial aquifer, CO, nitrite and nitrate concentration, 32–33
Fracture, depth of water flow, 27
Fractured-rock systems, 39–41
 ground-water flow, 246
Fracture zones, 41
Free ions, 138–139
 in water, 135
Freshwater, saltwater dispersive flow into, 221–222
Freshwater-equivalent heads, 455–457
Freshwater-saltwater environments, 443–466
 conceptualizations, 447
 conservation of mass
  of fluid, 449–450
  of salt, 450
 dispersion, 450–453
 field measurements required for analysis, 453–455
 local analysis of upcoming beneath discharging well, 463–465
 outpost monitoring wells, 465–466
 physical system description, 445–446
 regional sharp interface, 457–459
 regional variable-density flow
  with specified steady-state solute distribution, 459–462
  with transport of solutes, 463
 saltwater properties, 443–445
 sharp-interface
  analysis, data requirements, 454–455
  conceptualization, 447–449
 static saltwater, 448
 two-fluid flow, 448–449
 upcoming monitoring wells, 466
 use of and misconceptions regarding freshwater-equivalent heads, 455–457
 variable-density conceptualization, 449–453
  with steady-state solute distribution, 453
Freundlich isotherms, 394
Functional groups, 157–160
 water solubility and, 162–163

Gases, dissolution and exsolution, 146–147
Geochemical data, 42–55
 boxplots, 44, 47–48
 cluster analysis, 50–51
 Durov diagrams, 44–46
 multivariate statistical analyses, 47, 49–52
 principal components and factor analysis, 49–51
 relating ground-water quality to land use, 52–55
 simple map displays, 42
 sources of land-use data, 54–55
 Stiff patterns, 42–43
 trilinear diagrams, 43–44
Geochemical methods, subsurface microbiology, 184–186
Geochemical models, 199–223
 aqueous species in water, 137–140
 forward modeling, 214–222
 inverse modeling, 199–214
Geology, 113
Geostatistical models, 87–106
 covariance and correlation, 89
 distribution functions, 88
 kriging, 94–106
 mathematical expectation, 88–89
 microbial transport, 396–400
 regionalized variables and intrinsic hypothesis, 89–90
 stationarity, 89
 variograms, 90–94
Geostatistical theory, 87
Geostatistics, 87
Geothermal exchange, 235
Grasslands, nitrate, 302–303
Gravel pack, hydrogeologic unit effect, 16
Greenbrier Limestone aquifer, WV, 473
Grid-based search sampling, 81–84
Gross-α and gross-β activity, radionuclides, 432
Ground water
 chemical elements in, 141–142
 chemical units, 140–141
 contamination studies, 127–128
 dissolved species, 135, 137
  computer modeling, 137–140
 environmental conditions, 155–156
 forms of chemical elements, 135–136
 mass, characterization, 244
Ground-water basin. See Karstic ground-water basins
Ground-water-flow systems, 27–42
 broad-scale heterogeneity effects, 29–30
 Central Valley Aquifer System, CA, 37–38
 contributing areas to pumping wells, 35–37
 Floridan Aquifer System, 38–39
 fractured-rock systems, 39–41
 geologic controls on flow patterns, 28–30
 karst aquifers, 41–42
 large-scale ground-water development effects, 37–39
 local-scale, 31
 scales, 27–28
 surface-water–ground-water relations, 31–35
 transient flow effects, 30–31
Ground-water-quality surveys, 63–84
 Delmarva Peninsula, 563–586
 error, types and sources, 65–66

future network modifications, 69
grid-based search sampling, 81–84
Illinois, 439–511
Netherlands, 515–534
organic contaminants, 335–337
probability sampling approaches, 69–81
  cluster and staged selection, 80
  grid-based approaches, 76–80
  random selection with auxiliary variables, 80–81
  simple random selection, 69–73
  stratified random selection, 73–76
random selection role, 64–65
sampling frame, 67–68
spatial correlation effects, 66
target populations, 64
temporal variability, 68–69
terminology, 64
well suitability for sample collection, 67
volatile organic compounds, 334–335
Gulf coastal plain, freshwater-saltwater environments, 459–460

Hawaii, Oahu, freshwater-saltwater environments, 463
Health effects, acidic ground water, 413–414
Heavy metals, load on sandy soils, 531
Henry's law constant, 146, 166, 172, 174, 357
Herbicides, as event markers, 282–287
Hermite polynomials, 101–102
High-iron ground water, zonation in coastal-plain aquifers, 187–191
H_2S, dissolution and exsolution, 146–147
Hydraulic conductivity, spatial trends, 39
Hydrocarbons, in karst aquifers, 487
Hydrochemical facies, 119, 121
Hydrogen isotopes, 232–236
Hydrogeochemical cycle, 111–112
Hydrogeology, 113–114. *See also* Chemical hydrogeology, scales
  Netherlands, 516–518
  organic contaminants, 324
  settings, subdivision by, 8
  systems, definition, 111
Hydrogeomorphic regions, surficial aquifer, 564–568
Hydrologic cycle, 111, 300
Hydrolysis, 144–145
  inorganic chemicals, 167
Hydrophobicity, inorganic chemicals, 164, 170
Hyporheic zone, 34–35

Illinois, survey for agricultural chemicals in rural wells, 493–511
  analyte selection, 499–500
  final reports, 511
  implementation, 501–510
    data management, 507

  problems, 509–510
  quality assurance, 507–508
  sample analysis, 506
  sample collection, 503–505
  standard operating procedures and training, 501–502
  time frame and workload, 508–509
  well selection, 502–503
  well-site characterization, 506–507
previous monitoring plans, 495–496
previous studies, 494–495
project organization and responsibilities, 495, 497–498
results, 510–511
sampling plan, 499–501
study design, 498–501
well-user notification, 510–511
Inactivation, pathogens, 389–392
Index method, regional assessment, 360–363
Indicator kriging, 102–105
Inorganic chemical processes and reactions, 131–149
  acid-base reactions, 144
  complexation reactions, 144
  diagenetic alterations, 142
  dissolution and exsolution of gases, 146–147
  history of idea development, 131–133
  ion filtration and osmosis, 145–146
  kinetic versus equilibrium considerations, 148–149
  oxidation-reduction reactions, 145
  precipitation-dissolution reactions, 142–144
  saturation index, 142–143
  sorption and desorption, 147–148
  substitution-hydrolysis reactions, 144–145
Inorganic chemicals, 155–176
  aqueous-phase transport, 175
  biodegradation, 168–169
  carbon bonding, 155, 157
  density, 162
  dissociation constant, 164–165
  elimination, 167–168
  equilibrium distribution, 170–171
  functional groups, 157–160
  Henry's law constant, 166
  hydrolysis, 167
  hydrophobicity, 170
  isomerism, 160–162
  NAPL-to-air vaporization, 174
  NAPL-to-water dissolution, 174
  nonvolatile, 165
  nucleophilic substitution reactions, 167
  Oklahoma City urban area, 600–601, 603
    sources, 606–608
  oxidation-reduction, 168
  phase-transfer processes, 169–172
  physical/chemical properties, 158
  semivolatile, 165

628   Index

Inorganic chemicals (cont.)
    solubility and chemical hydrophobicity, 162–164
    sorption, 169–172
    structure and nomenclature, 157, 159
    transformation processes, 166–169
    transport processes, 174–175
    vapor-phase transport, 175–176
    vapor pressure, 165
    vapor sorption, 172
    volatile, 165
    volatilization from ground water, 172–174
Insecticides, Long Island, 370, 372
Intrinsic hypothesis, 90
Inverse modeling, 199–214
    combined with inverse modeling, 219–220
    mass-balance modeling, 205–214
    speciation modeling, 200–205
Ion exchange, 147–148
Ion filtration, 145–146
Ion pair, in water, 137
Iowa, river-aquifer interactions, 367–370
Isomerism, 160–162
Isotopes. See also Environmental isotopes
    content of precipitation, 233–234
    fractionation, 227–228
        factor, 229
    radioactive, 231
    stable and radioactive, 133

Japan, volatile organic compounds surveys, 334–335
Joints, 41

Karst aquifers, 41–42, 471–488
    chemical response to storm and seasonal-scale events, 482–484
    conduit development timescales, 486–487
    conduit flow aquifers, 476
    conduits, 475–476
    critical thresholds, 486
    diffuse-flow aquifers, 476
    evolution, 486–487
    karstic ground-water basins, 472–480
    nonpoint-source pollutants, 487
    permeability, 471–472
    petroleum and other light hydrocarbons, 487
    physical response to storm events, 480–482
    sediment transport, 484–486
    toxic metals, 488
    water-quality problems, 487–488
    water-soluble contaminants, 487
Karstic ground-water basins, 472–480
    allogenic recharge, 473–475
    autogenic recharge, 475
    boundary delineation, 473–474
    characterization, 479–480
    conduits, 475–476
    flow paths and water balance, 473–476

    karst water table, 478–479
    tributary and distributary streams, 476
    underground water tracing, 477–478
Karst water table, 478–479
Kinetic adsorption, pathogens, 395
Kriging, 94–106
    analysis of uncertainty, 95–98
    block versus point, 99
    cokriging, 105–106
    cross validation, 98–99
    disjunctive, 101–102, 397–399
    error map, 95–97
    "fictitious point" method, 98
    indicator, 102–105
    lognormal, 101
    "order relations problems," 103
    in presence of drift, 99–100
    probability, 105
    universal technique, 100
    variance, 95–97
        disjunctive kriging, 102
Krypton-85, 266–268, 288

Lake water, exchange with ground water, 241
Land use, relating ground-water quality to, 52–55
Langmuir equilibrium isotherm, 393–394
Latitude, isotopic content of precipitation, 234
Ligand, 137
Livestock production, nitrate, 307
Local scale, 114
Lognormal kriging, 101
Long Island, NY
    freshwater-saltwater environments, 458–459
    insecticides, 370, 372

Madison aquifer, mass-balance modeling, 213
Magnesium, sources and sinks in coastal-plain aquifers, 119–120
Major-ion chemistry, selenium depth distribution, 552
Manganese, high levels, 121–123
Mann–Whitney tests, 603–605
Map displays, 42
Massachusetts, Cape Cod, freshwater-saltwater environments, 457–458
Mass-balance equations, 116
Mass-balance modeling, 205–214
    carbon-14 dating, 212–214
    equation, 205–206
    example, 207–210
    redox mass-balance equation, 206
    stable isotope data, 210–212
Measurement bias, 66
Metathetical reactions, 144–145
Methanogenesis, fractured carbonate aquifer, 191–194
Microbiology. See also Subsurface microbiology
    ecology, ground-water systems, 183–186

Microorganisms. *See also* Pathogens
  characteristics, 383–384
  sources, 384–385
  sustaining life functions, 184–186
Middendorf aquifer, SC, 188–190
Mineralization-immobilization, nitrogen cycling, 297–298
Minnesota, pesticides in wells, 372
Mobile particles, 175
Molality, 141
Molecular bonding, 157
Monitoring. *See* Netherlands
Monitoring wells
  organic contaminants, 329–330
  outpost, 465–466
  upconing, 466
Montauk area, Long Island, NY, freshwater-saltwater environments, 458–459
Multivariate statistical analyses, 47, 49–52
  cluster analysis, 50–52
  principal components and factor analysis, 49–51
Municipal sludge, microorganism source, 385

NAPL-to-air vaporization, 174
NAPL-to-water dissolution, 174
Near-stream environment, features, 35
Nebraska
  organic contaminants surveys, 338
  river-aquifer interactions, 367–370
Netherlands, ground-water-quality monitoring, 515–534
  ad hoc programs, 521
  basic analytical program, 521
  drinking-water quality, 531–533
  Dutch National Ground-Water Quality Monitoring Network, 517, 519–524
  environmental monitoring, 515–517
  fields, 528
  hydrogeologic situation, 516–518
  integrated monitoring, 533–534
  interpretation tools and presentation techniques, 533
  local soil-pollution sources, 516
  long-term monitoring program, 528
  provincial network, 522, 525–526
  relationship between land use and ground-water quality, 519–520
  relationship between soil type and ground-water quality, 519–520
  site location, 519
  soil-quality and shallow-ground-water monitoring, 526–531
NETPATH, 207, 211, 213–214
Neutral molecules, in water, 135, 137
New York, Long Island
  freshwater-saltwater environments, 458–459
  insecticides, 370, 372

Nitrate, 297–317
  agriculture, 304–307
  animal wastes, 303–304
  concentrations, 532
    ground-water age, 314
    long-term trends, 12–13
    probability distribution, 571
  crop and livestock production, 307
  distribution and variability, 307–316
    denitrification and time, 313–315
    depth and time, 313
    depth distribution, 308–309
    real variability and depth distribution, 309–310
    temporal variations, 310–313
    variability among wells, 315–316
  forest land, 301–302
  geological nitrogen, 301
  grasslands, forage, and pastoral agriculture, 302–303
  leaching, 305
  management, 307
  nitrogen cycling, 297–300
  organic wastes, 303
  row-crop production, 305–306
  septic tanks, 304
  sewage sludge and effluent, 304
  sources, 300–307
  spatial distribution, 575–578
    local patterns, 578
    regional patterns, 577–578
    variations with depth, 575–577
  vegetable crops, 306
  waste materials, 303–304
Nitrification, 298–299
Nitrogen
  critical loads for ground water, 416–417
  in Dutch soils, 515–516
  geological, 301
Nitrogen cycling, 297–300
  denitrification, 299–300
  hydrologic cycle, 300
  mineralization-immobilization, 297–298
  nitrification, 298–299
  spatial complexity, 299–300
Nitrogen isotopes, 238–239
Nonaqueous phase liquids, 124–126
  zonation and characteristic reactions and processes, 125
Nucleophilic substitution reactions, 167
Numerical flow models, 37
Nutrients, plant uptake, 410

Oahu, Hawaii, freshwater-saltwater environments, 463
Oklahoma City urban area
  buffer-overlay analysis, 596–597
  characterizing urban land use, 595–598
  field sheets, 595

630 Index

Oklahoma City urban area (*cont.*)
    ground-water quality, 598–601
        geology, 602
        organic constituents, 600–601, 603
        tritium concentration, 601–602
        urban land use, 602–606
    inorganic contaminant sources, 606–608
    organic constituent sources, 608–609
    particle-tracking analysis, 597–598
    point-in-polygon overlay, 595–596
    sampling
        biases, 593
        low-density survey networks, 592
        network, 595
        program design, 591–595
        quality-assurance, 593–595
        urban network, 592–593
    water quality, 589–610
Organic compounds
    agricultural application, 124
    as event markers
        future studies of, 287
        herbicides, 282–287
        surfactants, 280–282
    Oklahoma City urban area, 601, 603
    sources, 608–609
    subsurface sources, 124
Organic contaminants, 323–341. *See also* Volatile organic compounds
    hydrogeology, 324
    limitations of existing quality data, 338–340
    nonaqueous-phase liquids, 325, 327–328
    phases, 325
    pollutant sources, 325–328
    role, 124
    state surveys, 335
        California, 335–337
        other states, 337–338
    subsurface monitoring, 328–332
    types, 325
    vadose-zone gas monitoring, 331
Organic wastes, nitrate, 303
Osmosis, 145–146
Otis Air Force Base, contaminant plume, 339–340
Outpost monitoring wells, 465–466
Overlay method
    point-in-polygon, 595–596
    regional assessment, 360–363
Overlying low-permeability units, 26–27
Oxidation, atrazine, 195
Oxidation-reduction reactions, 145
    inorganic chemicals, 168
Oxygen isotopes, 232–236, 551–552

Particle-tracking analysis, 37, 597–598
Pathogens, 383–400. *See also* Microorganisms
    detection in water, 387–388
    factors affecting microbial fate and transport, 388–395
        adsorption, 393–395
        die-off rate constant, 392
        inactivation, 389–392
        kinetic adsorption, 395
        physical filtration, 392–393
    modeling transport, 395–400
        geostatistical approaches, 396–400
        regional transport applications, 395–396
    setback distance, 397, 399–400
PCE, 282, 284
Permeability, karst aquifers, 471–472
Pesticide Root-Zone Model, 353
Pesticides, 160, 345–376
    analyte detection limits, 374–375
    analytes, 374
    association with nitrate contamination, 376
    attenuation factor index, 351–352
    comparison of regional studies, 375
    degradation, 349–351
    episodic leaching, 354–355
    episodic water flow and solute transport effects, 353–355
    factors influencing contamination potential, 345–360
        agricultural practices and pesticide use, 359–360
        behavior and fate in vadose zone, 346–357
        point-source contamination, 357–359
    field monitoring techniques, 375
    implications for regional-scale studies, 373–376
    local-scale studies, 375–376
    metabolites, 374
    nonagricultural uses, 359–360
    occurrence, 374
    persistence, 348–349
    preferential flow effects, 355–356
    pseudo-first-order kinetic model, 350
    regional assessment of water vulnerability, 360–366
        overlay and index methods, 360–363
        simulation models, 363–364
        statistical methods, 364
        uncertainty, 364–366
    retardation factor, 351
    seasonal patterns, 358
    in shallow ground water, 531
    simple indices of residence time and degradation rate, 351–353
    sorption, 346–349
    studies
        aldicarb in Central Sand Plain, WI, 366–367
        DBCP in San Joaquin Valley, CA, 371–372
        insecticides on Long Island, 370, 372
        Minnesota wells, 372

river-aquifer interactions in Iowa and
    Nebraska, 367–370
  USEPA National Pesticide Survey, 372–373
  temporal variability, 375
  transformation processes, 346
  transport processes, 346
  uptake by plants, 357
  volatilization, 357
  well type, 376
Petroleum hydrocarbons, 336
  in karst aquifers, 487
pH, relation with aluminum concentration,
    522–523
Phase-transfer processes, inorganic chemicals,
    169–172
Physical filtration, pathogens, 392–393
Picher mining area, reaction-path modeling,
    215–218
Piston-flow approaches, tritium movement, 260,
    262
Plant, uptake of nutrients, 410
Point-in-polygon overlay, 595–596
Point kriging, 99
Point-source contamination, pesticides, 357–359
Pollutants, nonpoint-source, karst aquifers, 487
Precipitation. *See also* Acid precipitation
  environmental isotopes studies, 234–236
  isotopic content, 233–234
Precipitation-dissolution reactions, 142–144
Preferential flow, pesticides and, 355–356
Principal components and factor analysis, 49–51
Probability density function, 88
Probability kriging, 105
Protozoan parasites, 384
Pseudo-first-order kinetic model, pesticides, 350
Pumping, quality variations during, 11–12
Pumping wells, contributing areas, 35–37
Pure culture methods, subsurface microbiology,
    183–184

Q-mode factor analysis, 50–51
Quality-assurance programs, 18–19
  Illinois survey, 507–508
Quality studies. *See also specific locations*
  biases from well-screen placement, 16
  existing versus specially constructed wells,
    14–17
  ground-water-resource assessments, 6
  limitations of existing data, 338–340
  near water table, 16
  point of use studies, 5
  regional investigations, 4–5, 19
  relating to land use, 52–55
  soil characteristics and, 519–520
  spatial scales, 6–9
    sampling density, 9
    subunits, 8–9
    survey sampling, 6–7

  targeted sampling, 7–8
  temporal scales, 9–14
    long-term trends, 12–14
    seasonal variations, 10–12
    short-term variability, 9–10
  uses and limitations of existing data, 613–622
    ancillary site information, 618–619
    applications, 619–621
    data base characteristics, 614–615
    data screening, 616
    data sources, 616
    sampling design and objectives, 617
    sampling procedures and laboratory
      analytical methods, 616–617
    spatial coverage of sampled sites, 617
    suitability of data for regional water-quality
      assessment, 616–619
    water-quality constituent coverage, 617–618
  vadose zone relation, 23–24
  well-water surveys, 5–6

Radioactive decay, 231, 425–427
Radioactive equilibria, 427–429
Radioactive-waste disposal, natural analog
    studies, 434–435
Radionuclides, 231, 423–436
  abundances in rocks and ground water,
    423–425
  drinking-water supplies, 432
  environmental screening for gross-α and
    gross-β, 432
  ground water, 433–434
  natural analog studies of radioactive-waste
    disposal, 434–435
  nuclear versus chemical properties, 429–430
  properties, 426
  radioactive decay, 425–427
  radioactive equilibria, 427–429
  radium, 431
  radon, 431–432
    in ground water, 435–436
  secular equilibrium, 428–429
  thorium, 431
  uranium, 430–431
    behavior in U deposits, 434
Radium, 431
  concentrations, 123
Radon, 431–432
  in ground water, 435–436
Random measurement errors, 66
Random sampling error, 65–66
Random selection
  with auxiliary variables, 80–81
  ground-water-quality surveys, 69–73
    mean estimation, 69–72
    median estimation, 72
    proportion estimation, 72–73
    role in, 64–65

Random selection (*cont.*)
  stratified, 73–76
  from irregular polygons, 77–80
  within blocks, 76
Random variables, types, 88
Ratio and regression estimators, 80–81
Rayleigh distillation equations, 230
Reaction-path modeling, 215–220
  combining forward and inverse modeling, 219–220
  example, 215–219
  mass transfers, 216
Reaction-transport modeling, 220–222
Recharge
  allogenic, 473–475
  areal and temporal distribution, 30–31
  autogenic, 475
  environmental isotope use, 241, 244
  ground-water-flow systems effects, 28–29
  seasonal variations and, 10
  temperature, chlorofluorocarbons, 270–272
Redox potential, selenium depth distribution and, 555
Redox state, solution, 206
Regional Aquifer System Analysis Program, 132
Regional hydrochemical facies, introduction of concept, 132
Regional microbial transport, applying transport models, 395–396
Regional scale, 114
Regional water-quality assessment, data suitability, 616–619
Reservoir models, tritium movement, 262–263
Rhode Island, high-manganese water, 121–123
River-aquifer interactions, Iowa and Nebraska, 367–370
River water, exchange with ground water, 241
Rocks, radionuclide abundances, 423–425

Salinity, selenium
  depth distribution, 552–553
  distribution and, 544–545
  mobility and, 556–558
Saltwater
  chemical properties, 444–445
  chlorinity and density, 444
  classification, 444
  composition of dissolved solids, 445
  dispersive flow into freshwater, 221–222
  physical properties, 443–444
  static, 448
Saltwater–fresh water interface, 113. *See also* Freshwater-saltwater environments
Sample collection, Illinois survey, 503–505
Sampling density, 9
Sampling frame, 67–68
San Joaquin Valley, 371–372, 537–538. *See also* Selenium

Saturation index, 201
Seasonal-scale events, chemical response of karst aquifers, 482–484
Seasonal variations, 10–12
  isotopic content of precipitation, 234
  masking long-term trends, 12
Sediment
  Sierra Nevada, chemically reduced, selenium mobility, 556–560
  transport, karst aquifers, 484–486
    mechanisms, 484–485
    vertical, 485–486
Selection bias, 66
Selenium, 537–539
  areal distribution in shallow ground water, 543–548
    evaporative concentration, 546–547
    salinity and, 544–545
    significance, 547–548
    soil selenium and, 545–546
  depth distribution in ground water, 548–556
    age and origin, 549–552
    redox potential and, 555
    salinity and major-ion chemistry, 552–553
    significance, 555–556
  mobility in chemically reduced Sierra Nevada sediments, 556–560
  San Joaquin Valley, CA, 537–561
  valleywide sampling, 539–543
    areal distribution, 540–541
    concentration and salinity redox conditions, 541–542
    significance, 542–543
    water origin, 540
Septic-tank effluent
  bacterial and virus source, 384–385
  nitrate, 304
  placement, 397–399
Sewage sludge, nitrate, 304
Shallow ground water
  quality
    temporal variations, 10
    variation, 526–527
  selenium areal distribution, 543–548
Shear zones, 41
Sierra Nevada sediments, chemically reduced, selenium mobility, 556–560
Simulation models, pesticides, 363–364
Site scale, 114
Sodium, sources and sinks in coastal-plain aquifers, 119–120
Soil
  acidification, 406–408
  acidity, 408–409
  characteristics and ground-water quality, 519–520
  critical loads, 417
  geographic data bases, 24–25

instability, sediment transport in karst
    aquifers, 485–486
mapping units, 24–26
mineralogic and petrologic classification, 416
quality, variation, 526–527
regional characterization, 24–26
selenium and ground water distribution,
    545–546
Solubility, inorganic chemicals, 162–164
Solute mass balance, freshwater-saltwater
    environments, 450
Solute transport, pesticides and, 353–355
Soquel-Aptos Basin, freshwater-saltwater
    environments, 460–462
Sorption, 147–148
    inorganic chemicals, 169–172
    pesticides, 346–349
Spatial correlation, ground-water-quality surveys
    effects, 66
Speciation modeling, 200–205
    example, 201–203
    mass-action equation, 200
    saturation-index equation, 204
    uncertainties, 203–205
Spikes, 594
Stable isotope data, mass-balance modeling,
    210–212
Stationarity, geostatistical models, 89
Statistical methods, pesticides, 364
Statistical strata, selection, 75–76
Stiff patterns, 42–43
Storm events
    chemical response of karst aquifers, 482–484
    karst aquifer response, 480–482
Streamflow, components, 32
Streams
    discharge, ground-water and surface-water
        components, 246
    hyporheic zone, 34–35
    quality, trends, 14
    seepage to and from, 31
Streptomycetes conidia, 393
Stringfellow Acid Pits, 340
Strontium isotopes, 240–241
Suboxic zone, 126
Substitution-hydrolysis reactions, 144–145
Subsurface microbiology, 181–196
    atrazine biodegradation in alluvial-aquifer
        sediments, 194–196
    ecology, 183–186
    geochemical methods, 184–186
    historical overview, 182–183
    pure culture methods, 183–184
    terminal electron-accepting process
        distribution, 186–194
Subsurface monitoring, organic contaminants,
    328–332
Subunits, 8–9

Sulfate
    reduction, distribution, 191–194
    sources and sinks in coastal-plain aquifers,
        119–120
Sulfate-reducing bacteria, 190
Sulfur
    critical loads for ground water, 416–417
    isotopes, 239–240
Surface water, acidification, relation with ground
    water, 410–411
Surface-water–ground-water relations, 31–35
    analysis methods, 33
    contaminant distribution, 32
    downgradient release, 32–33
    during dry periods, 33
    ground-water movement near gaining stream,
        33–34
    spatial and temporal variability, 35
Surfactants, as event markers, 280–282
Surficial aquifer. *See* Aquifers, surficial
Survey sampling, 6–7
Systematic grid selection, 76–80

Targeted sampling, 7–8
Target population, 4
    ground-water-quality surveys, 64
TCE, 282, 284
Temporal variability, ground-water-quality
    surveys, 68–69
Terminal electron-accepting processes,
    distribution, 186–194
    coastal-plain aquifers, 187–191
    fractured carbonate aquifer, 191–194
Thorium, 431
    abundances in rocks and minerals, 424
Toxic metals, karst aquifers, 488
Tracers. *See* Environmental tracers
Transect network, 568–569
Transport processes, inorganic chemicals,
    174–175
Trichloroethylene, 181
Trihalomethanes, 335–336
Trilinear diagrams, 43–44
Tritium, 256–265, 287–288
    concentration, Oklahoma City urban area,
        601–602
    distribution in ground water, 549–551
    history of input, 256–260
    movement through aquifers, 260–265
        advection-dispersion models, 265
        compartment models, 263–265
        piston-flow approaches, 260, 262
        reservoir models, 262–263
    seasonal effect, 258
Tritium–helium-3, 265–266

Uncertainty, vulnerability assessment, 364–366

Underground water tracing, karstic ground-water basins, 477–478
United States
  volatile organic compounds surveys, 332–334
  waterborne disease, 385–387
Universal kriging technique, 100
Unsaturated hydrocarbons, 157
Unsaturated zone, recharge and flow rate, 241
Upconing, 463–465
  monitoring wells, 466
Upland, water-quality patterns
  poorly drained, 580–581
  well-drained, 578–580
Uranium, 430–431
  abundances in rocks and minerals, 424
  behavior in U deposits, 434
USEPA National Pesticide Survey, 372–373

Vadose zone, 23–27
  intermediate, 26
  gas monitoring, 331
  overlying low-permeability units, 26–27
  pesticide behavior and fate, 346–357
  regional characterization, 24–27
  relation to ground-water quality, 23–24
  relation of properties to water quality, 27
Vaporization, NAPL-to-air, 174
Vapor-phase transport, inorganic chemicals, 175–176
Vapor pressure, inorganic chemicals, 165
Vapor sorption, inorganic chemicals, 172
Variograms, 90–94
  experimental, 91
  models, 91–94
Vegetable crops, nitrate, 306
Viruses, 383–384
  factors influencing fate in subsurface, 390
  inactivation rates, 390–392
Volatile organic compounds, 323, 325
  California surveys, 336–337
  Japanese surveys, 334–335
  national surveys, 332–335
  United States surveys, 332–334

Volatilization
  inorganic chemicals from ground water, 172–174
  pesticides, 357

Waste
  disposal, in landfills and lagoons, 124
  nitrate, 303–304
Water
  flow, episodic, pesticides and, 353–355
  physical properties, 133–135
  physiochemical properties, 134
  quality problems, karst aquifers, 487–488
  structure, 133
Water balance, karstic ground-water basins, 473–476
Waterborne-disease outbreaks, 384–387
  causative agents, 386–387
Water-rock exchange, low-temperature, 235–236
Water table
  karst, 478–479
  sampling near, 16
Water tracing, karstic ground-water basins, 477–478
Weathering, 408–410
Wells
  contaminant transport, 15
  discharging, upconing beneath, 463–465
  documentation, 17–18
  existing versus specially constructed wells, 14–17
  Illinois survey. *See* Illinois
  nitrate variability, 315–316
  open interval, 15–17
  purging, 9–10
  selection for Illinois survey, 502–503
  sites, documentation, 17–18
  suitability for water-quality sample collection, 67
  water surveys, 5–6
Wisconsin, organic contaminants surveys, 337–338